Table		III	IV	V	VI	VII	0
							2 He (2)
		5 B	6 C	7 N	8 O	9 F	10 Ne (2,8)
		13 Al	14 Si	15 P	16 S	17 Cl	18 Ar (2,8,8)
29 Cu	30 Zn	31 Ga	32 Ge	33 As	34 Se	35 Br	36 Kr (2,8,18,8)
47 Ag	48 Cd	49 In	50 Sn	51 Sb	52 Te	53 I	54 (2,8,18, Xe 18,8)
79 Au	80 Hg	81 Tl	82 Pb	83 Bi	84 Po	85 At	86 (2,8,18, Rn 32,18,8)

66 Dy	67 Ho	68 Er	69 Tm	70 Yb	71 Lu

98 Cf	99 E	100 Fm	101 Mv	102 No	103 Lw

Chemistry
Molecules That Matter

Chemistry
Molecules That Matter

Eugene G. Rochow
Harvard University

George Fleck
Smith College

Thomas R. Blackburn
Hobart and William Smith Colleges

HOLT, RINEHART AND WINSTON, INC.
New York Chicago San Francisco Atlanta
Dallas Montreal Toronto London Sydney

Copyright © 1974 by Holt, Rinehart and Winston, Inc.
All Rights Reserved
Library of Congress Catalog Number: 73-9222
ISBN: 0-03-084395-2
Printed in the United States of America
9 8 7 6 5 4 3 2 1   071   7 6 5 4

Typography: Fototronic Laurel by New England Typographic Service, Inc.
Editor: Holly Massey
Designer: Jerry Tillett
Production Manager: Marion Palen
Printer and Binder: Kingsport Press
Drawings: George Kelvin
Cover: Edmée Froment

Preface

Ever since a strange-looking hominid picked up the first smooth flint pebble and carried it elsewhere to serve his purposes, man has been disturbing the natural order of his surroundings. When he learned to use fire, he started to pollute the atmosphere with smoke, and when he began to cultivate the soil, he disturbed the natural vegetation. Trash and garbage were left on the ground outside his cave where they accumulated to impressive depths (and, incidentally, provided us with a record of his age and his activities). Man's struggle to survive and prosper had begun. That struggle continues to this day with the same purposes and the same consequences. Now, however, there are more people, and the consequences are more noticeable and more serious. Man has not learned to live in any greater harmony with his environment, nor does he use its materials any more sensibly.

It is the *interaction* of man with his environment that shaped his evolution, and it is with that interaction that we are concerned here. Stone Age man found his prey becoming more elusive, so he devised more effective hunting weapons. He learned to put a cutting edge on flint, and then on copper and bronze. Later came mining, which evolved from the rather simple process of searching for lumps of natural copper to a more elaborate search for the ores of metals. But the reduction of ore to metal is a chemical process that requires fuel, so trees were cut down to make charcoal, and eventually coal had to be mined for the purpose. Even more fuel was required to make bricks and mortar. In order to perform these and all the other tasks of a complex society, large numbers of people had to be transported to their places of work, which required still more metals and fuel for the transportation system. And, above all, man expanded his technology and his use of natural resources to make war. All these activities disturbed nature more and more, inevitably leading to the present polluted state of the air, the water, and the land, and to the depletion of natural resources.

Man now stands at a crossroads. He must learn to live in some kind of equilibrium with nature. Thinking people see that man cannot continue to consume and despoil. We face a painful reorientation of our everyday life. Many would put it more strongly: either we learn to manage our environment, or it will kill us off. The frightening growth of world population adds further to the urgency of this conclusion. The population of the world has nearly quadrupled in the last century. Since not all of the earth's population is adequately fed even now, it becomes obvious that there will be great difficulty in feeding the six to seven billion expected to be living by the year 2000.

How can we learn to manage things better? There are many economic, social, and political questions to be considered, of course, but in this book we shall be concerned with the *materials* used for food, clothing, housing, and transportation. This is the province of chemistry. To learn about materials and their conversion we need first to know something about matter: its structure, its inherent tendencies, and the ways by which it is changed into the materials we need and use. From this book, the student can learn the basic principles that underlie the composition of matter — principles that are surprisingly simple and orderly. Then he or she can go on to consider the processes by which materials are isolated and converted to other materials. Various ways of improving the production, the use, and the reuse of materials are suggested, and the student is encouraged to consider improvements or radically new concepts. Without new ideas and constructive action, we are all in for a very bad time.

A further word about this course of study is in order. It is not the intent of this book to embrace all of chemistry, for that is a huge subject. Neither can this book present all the fundamental principles essential to the training of a professional chemist. This is a book with a slant. It presents enough fundamental chemistry to enable the student to understand some of the major problems of our times, and it presents chemical information, models, and concepts relevant to those problems. Its four basic goals are:

1. To point to some of the problems — intellectual, social, and technological — whose solutions require understanding of chemical systems and the manipulation and control of such systems.

2. To describe and define the conceptual tools that chemists use as they work to construct and use realistic chemical models of the world as they attack its problems.

3. To present a sampling of the results of chemical investigation which permit insights into the nature of matter, and glimpses of the potential for creating new materials and processes.

4. To supply problems and exercises by which the student (and the instructor) can measure his or her progress in understanding concepts.*

---

* However, we do not believe that the only ideas worth learning are those that can be broken down to the dimensions of a homework problem or a weekly quiz. Consequently, there are sections in this book that are accompanied by a few or no such exercises. It is certainly not our intention that this material be assigned for memorization and later disgorgement. Rather, in these sections we have felt that the best purposes of chemical education will be served by a straightforward narrative of our present concepts and models. Our objective in presenting this material is not, in the current jargon, "behavioral," but intellectual: we aim for an appreciation (possibly nondemonstrable) and an understanding of nature *as well as* for the development of manipulative skills.

Chemistry is a changing science, partly because of the delightful unexpected discoveries which open new research opportunities, and partly because each new generation changes the direction in both research and applications. This book is directed primarily to those who do not plan to enter the professional study of chemistry, with the hope that some of its ideas will enable them to live in their chemical world according to a reasonably reliable chemical model. If the student wants to go further in the study of chemistry, he or she will find help in the Suggestions for Further Reading, and further help in more advanced courses in chemistry, to which this book is a partial, but truthful, introduction.

E. G. R.

G. F.

T. R. B.

Contents

Preface   v

1   About Nitrogen Fixation, the Food Problem, and the Rest of this Book   1

1-1 Introduction, 2 · **1-2** The Great Nitrogen Shortage, 1898–1914, 3 · **1-3** Room-Temperature Fixation of Nitrogen, 9 · **1-4** Other Essential Elements, 11 · **1-5** What Else Can Be Done?, 11

2   Atoms and the Chemical Elements   15

2-1 Abstract, 16 · **2-2** Introduction, 16 · **2-3** Atomic Weights, 18 · **2-4** The Gram-Atomic Weight, 22 · **2-5** The Sizes of Atoms, 24 · **2-6** Atomic Structure I: Electrons, 25 · **2-7** Atomic Structure II: The Nucleus, 27 · **2-8** Atomic Structure III: The Arrangement of Electrons, 31 · **2-9** Atomic Structure IV: The Periodic System, 41 · **2-10** Atomic Structure V: Energy Levels and Atomic Spectra, 45 · **2-11** Summary, 50 Glossary, 51 · Problems, 53 · Suggestions for Further Reading, 55

3   Molecules: The Formation of Chemical Compounds   57

3-1 Abstract, 58 · **3-2** Unwinding Watchsprings, Falling Bricks, and the Formation of Chemical Compounds, 58 · **3-3** Ionic Bonding: The Consequences of Outright Transfer of Electrons, 60 · **3-4** The Names of Chemical Compounds, 64 · **3-5** Covalent Bonding, 66 · **3-6** Coordinate Covalent Bonds, 77 · **3-7** Molecular Weights and the Mole, 79 · **3-8** Formulas and Their Quantitative Significance. Chemical Arithmetic Concerning Formulas, 80 · **3-9** Chemical Equations, 83 · **3-10** Weight Relations. Chemical Arithmetic Concerning Equations, 85 · **3-11** Reactions in Water

Solution. Molarity, 86 · **3-12** Reactions Involving Gases, 88
3-13 Summary, 91 · Glossary, 93 · Problems, 94
Suggestions for Further Reading, 96

4 Molecular Structure and Symmetry 97

4-1 Abstract, 98 · **4-2** On the Importance of the Right
Arrangement, 98 · **4-3** The Importance of Symmetry, 101
4-4 Molecules, Models, and Chemical Bonds, 103 · **4-5** The
Elements of Molecular Symmetry, 106 · **4-6** The Pauli
Principle and the Shapes of Molecules, 112 · **4-7** Internal
Motions of Molecules, 118 · **4-8** Summary, 123
Glossary, 124 · Problems, 125 · Suggestions for Further
Reading, 128

5 Properties of Tangible Matter 129

5-1 Abstract, 130 · **5-2** Solids, Liquids, and Gases, 130
5-3 Ionic Substances, 132 · **5-4** Covalent Substances, 133
5-5 Molecular Substances, 134 · **5-6** Metals, 138
5-7 Recapitulation: Chemical Categories of Matter, 141
5-8 Orderliness and Structure in Solids, 142 · **5-9** Structures
Arising from Nondirectional Bonding, 144 · **5-10** Structures
Derived from Directed Bonding, 151 · **5-11** Gases: The
Consequences of Chaos, 153 · **5-12** Changes of Phase, 157
5-13 Summary, 164 · Glossary, 165 · Problems, 166
Suggestions for Further Reading, 167

6 Periodicity and Chemical Properties 169

6-1 Abstract, 170 · **6-2** Periodicity of the Ionization
Potential, 170 · **6-3** Electron Affinity, 176 · **6-4** Atomic
Sizes, 177 · **6-5** Polarizability, 183 · **6-6** Scales of
Electronegativity, 184 · **6-7** Periodicity and Chemical
Reactivity, 188 · **6-8** Summary, 196 · Glossary, 197
Problems, 198 · Suggestions for Further Reading, 199

7 Chemical Equilibrium 201

7-1 Abstract, 202 · **7-2** The Phenomenon of Chemical
Equilibrium, 202 · **7-3** The Consequences of Equilibrium, 204
7-4 Gas-Phase Reactions and the Effect of Pressure, 208
7-5 Some Solution Equilibria, 209 · **7-6** Equilibria Within
Solids, 225 · **7-7** The Drift Toward Equilibrium, 225
7-8 Summary, 226 · Glossary, 227 · Problems, 228
Suggestions for Further Reading, 230

8 Oxidation and Reduction: Electrons on the Move 233

8-1 Abstract, 234 · **8-2** Electron-Transfer Reactions in General, 234 · **8-3** The Meaning of Oxidation and Reduction, 235 · **8-4** Half-Reactions, 237 · **8-5** Oxidation Number, 237 · **8-6** Balancing Oxidation-Reduction Equations, 239 · **8-7** Chemical Energy from Oxidation-Reduction Reactions, 242 · **8-8** The Concept of Entropy, 262 **8-9** Practical Considerations About Electric Power from Chemical Reactions, 267 · **8-10** Some Practical Batteries, 268 **8-11** Fuel Cells, 272 · **8-12** Chemicals from Electricity: Electrolysis, 275 · **8-13** Summary, 281 · Glossary, 281 Problems, 283 · Suggestions for Further Reading, 285

9 Energy and Chaos: Why Chemical Reactions Happen 287

9-1 Abstract, 288 · **9-2** Heat, Work, and Energy in Chemical Change, 288 · **9-3** Conservation of Energy: The First Law of Thermodynamics, 290 · **9-4** The Energy of Formation of Chemical Compounds, 291 · **9-5** The Enthalpy Function, 292 · **9-6** The Ongoing Nature of Things, 297 **9-7** Chaos, Probability, and Equilibrium States of Matter, 299 **9-8** The Second Law of Thermodynamics: The Entropy Function, 303 · **9-9** The Entropy of the World and Chemical Equilibrium, 306 · **9-10** The Second Law and Living Systems, 312 · **9-11** Summary, 315 · Glossary, 316 Problems, 316 · Suggestions for Further Reading, 318

10 The Ultimate Source of Energy 321

10-1 Abstract, 322 · **10-2** Energy Sources and Energy Flows, 323 · **10-3** Stable Versus Unstable Nuclei, 324 **10-4** The Isotope Chart, 334 · **10-5** Theories of Nuclear Structure, 336 · **10-6** Nuclear Reactions, 339 · **10-7** Fission, 341 · **10-8** Fusion, 348 · **10-9** Effects of Radiation, 349 **10-10** Summary, 352 · Glossary, 355 · Problems, 356 Suggestions for Further Reading, 357

11 Mechanisms of Reactions 359

11-1 Abstract, 360 · **11-2** A Model to Simulate Global Population Change, 360 · **11-3** The Production and Decay of Radioactive Carbon: A Key to Dating the Past, 366 **11-4** A Model for a Chemical Reaction, 371 · **11-5** Computer Simulation of Our Chemical-Reaction Model, 375

11-6 Evaluation of Rate Constants from Experimental Data, 380 · **11-7** Testing a Reaction Mechanism, 382 **11-8** Little Rate Constants, Big Rate Constants, and Activation Energies, 382 · **11-9** Summary, 384 · Glossary, 385 Problems, 386 · Suggestions for Further Reading, 389

12 Molecular Orbitals — 391

12-1 Abstract, 392 · **12-2** Orbitals, 392 · **12-3** The Three-Body Dilemma and the Necessity for Relying on Hydrogen-Atom Calculations, 394 · **12-4** Pencil-and-Paper Mapping of the Interior of an Orbital, 396 · **12-5** Linear Combinations, 402 · **12-6** Electron Distribution in the Water Molecule, 403 · **12-7** Antibonding Orbitals and the Photodissociation of Water, 410 · **12-8** Summary, 412 Glossary, 412 · Problems, 413 · Suggestions for Further Reading, 414

13 Around the Coordination Sphere — 415

13-1 Abstract, 416 · **13-2** Kekulé, His Models, and Carbon Compounds, 416 · **13-3** Werner: Expansion of the Notion of Combining Capacity, 418 · **13-4** *cis* and *trans* Isomerism in Octahedral Coordination Complexes, 419 · **13-5** Optical Rotation and Isomerism, 421 · **13-6** G. N. Lewis: Bonding by Sharing Electrons, 423 · **13-7** Isomerization of an Octahedral Complex, 425 · **13-8** Reaction of Cobalt Ions with EDTA: Complexation with a Polyfunctional Ligand, 430 · **13-9** Vitamin B_{12}: Another Cobalt Complex, 432 · **13-10** Summary, 434 · Glossary, 435 Problems, 436 · Suggestions for Further Reading, 437

14 Organic Chemistry — 439

14-1 Abstract, 440 · **14-2** The Myth of Vital Force, 440 **14-3** Is Carbon Different?, 442 · **14-4** Some Simple Compounds of Carbon, 445 · **14-5** The Naming of Organic Compounds, 448 · **14-6** Isomerism in Organic Compounds, 451 · **14-7** The Functional Groups of Organic Chemistry, 455 · **14-8** Reaction Mechanisms in Organic Chemistry, 466 · **14-9** Some Practical Examples, 471 **14-10** Summary, 479 · Glossary, 480 · Problems, 481 Suggestions for Further Reading, 483

15 The Design of Materials   485

15-1 Abstract, 486 · **15-2** Metals, 486 · **15-3** Ceramics, 493
15-4 Polymers and Plastics, 496 · **15-5** Elastomers, 510
15-6 Dyes, 514 · **15-7** Flavorings, Perfumes, and Such: Challenges for the Future, 517 · **15-8** Summary, 518
Glossary, 519 · Problems, 520 · Suggestions for Further Reading, 521

16 Measurements on Chemical Systems: The Chemical Study of What is in Sea Water, and How Much   523

16-1 Abstract, 524 · **16-2** The Significance of Chemical Measurements, 525 · **16-3** Methods Based on Reaction Stoichiometry, 526 · **16-4** Ion-Selective Electrodes, 529
16-5 Spectrophotometry: Analysis Via the Absorption of Light, 535 · **16-6** Chromatography and the Measurement of Traces of Volatile Substances, 539 · **16-7** The Reliability of Chemical Measurements, 545 · **16-8** Summary, 547
Glossary, 549 · Problems, 550 · Suggestions for Further Reading, 551

17 Chemical Control in a Living Cell   553

17-1 Abstract, 554 · **17-2** General Aspects of Biochemistry, 554 · **17-3** Biological Catalysts, 555
17-4 Proteins: Molecules of Immense Variety, 561
17-5 Synthesis of Proteins, 566 · **17-6** Summary, 569
Glossary, 570 · Problems, 571 · Suggestions for Further Reading, 571

18 Case Studies in Environmental Chemistry   573

18-1 Abstract, 574 · **18-2** Equilibrium and Cyclic Models of the Ecosphere, 574 · **18-3** Limestone, Carbonic Acid, and Carbon Dioxide, 576 · **18-4** Cycles of Nutrients, 578
18-5 Air Pollution and Photochemical Smogs, 583
18-6 Thermodynamics and Ecology, 587 · **18-7** Science, Technology, Politics, and Information, 592 · **18-8** Summary, 592 · Glossary, 593 · Problems, 594 · Suggestions for Further Reading, 595

Epilogue   601

Appendix I Chemical Compound Handbook   **603**

Appendix II Mathematics in Chemistry   **621**

Appendix III Some Physics Background for Chemistry   **637**

Index   **655**

Chemistry
Molecules That Matter

1
About Nitrogen Fixation, The Food Problem, and The Rest of This Book

1-1 Introduction

As set forth in the Preface,[1] this is not just a book about the principles of chemistry, but rather a book concerned with *your view* of chemistry, the chemistry in your life, and in your world in your time. There are plenty of good, sound textbooks that teach the classical fundamentals of chemistry (some at great length and remarkable depth), but this one seeks to do more than that: It seeks to involve you in a productive study of chemistry as it relates to the world's real problems. This is a cooperative venture, and it needs your help from the beginning. You are asked to apply what you already know about science at the very start. If that should turn out to be little, you can find instruction all along the way as you proceed through the rest of the book, and you can find help in all that you need to learn. Should you find that you want to go far beyond the confines of this book in some relevant aspects that begin to interest you, we shall guide you also in that exploration; you can go as far and as fast as you like. All that is necessary at the outset is a willingness or eagerness to learn.

We begin with a literary experiment. This chapter deals with the most fundamental of the world's problems, and it tells of some of the triumphs and lapses of chemistry in that area. It tells something of what has been done, what needs to be done, and what is being done right now in the long attempt to match the world's food supply to the world's population. We suggest you read it straight through, pausing only to mark or make note of what you do not understand in it. Everyone knows some chemistry, if only that which he absorbs from newspapers and novels. Every reader also has had at least some formal instruction in science, and the level of that instruction has increased tremendously in the last two decades. But even if you insist that you know nothing of chemistry or remember nothing, the "plot line" of the chapter will be apparent. Mark the terms and concepts that are new and strange, and then check them off as you master them later on. The chapters that follow this one take up all the fundamental concepts, one by one, with ample chance to test your grasp of them. As for the new terms, you will find them in glossaries at the end of the pertinent chapters—there is a glossary after every chapter (except this one, of course). The properties of all the chemical compounds mentioned in the book are listed in Appendix I.

Mark the things you don't understand. Use the margins.

You may be pleasantly surprised to find that you understand many of the technical details in this chapter, rather than just a few of them, and that nearly all of the words seem at least vaguely familiar. Good! You will be able to make faster progress through the rest of the course than your less fortunate neighbor, and will have more time to explore related areas. Be honest with your self-testing, though, as you read this chapter. Let your instructor and the resources of this book help you get the whole story. You are going to need it, and not just to pass this course.

And now, read on.

---

[1] Please read it at once, if you have not already.

1-2 The Great Nitrogen Shortage, 1898–1914

On a chill September evening in 1898, Sir William Crookes arose before the British Association for the Advancement of Science at its annual meeting at Bristol to deliver his presidential address. Previous presidents had contented themselves with a survey of the past year's achievements in science and engineering, but Sir William deliberately chose this particular occasion to raise in a very dramatic form a question he thought should be brought before the entire world, a question which he called very simply "The Wheat Problem." As the talk unfolded, he showed by an impressive array of facts and figures that the world reserve of wheat was declining, and that almost all the available wheat lands were already under cultivation, so that within a rather short time, the entire wheat-eating population of the world faced a shortage of bread. In short, there would soon be widespread hunger and famine. The growing population, which was enjoying the fruits of rapid advancement in medicine and public health and the economic benefits of the industrial age, would of necessity have to level off its numbers by processes of hardship and even tragedy.

Sir William was not a demagogue nor an irresponsible person. A photograph of him, with his carefully trimmed beard and his penetrating gaze through steel-rimmed spectacles, shows him to be the very picture of the respected Victorian scientist. At the time he delivered the presidential address which was to rock the entire thinking world, he was perhaps the most honored scientist of his country, and he spoke out of a sense of public duty. Much criticism was leveled at him at the time, especially by people who labored under the illusion that practically all of the vast North American continent could be planted to wheat if necessary. A year later, however, he collected all of his facts and references, together with the statements of agricultural experts both in Europe and the United States, and brought out a little book about the matter still called *The Wheat Problem*, published by John Murray in London, 1899. In it he again states his conclusion in very clear terms:

This question of food supply is of urgent importance today, and it is a life-and-death question for generations to come. Many of my statements you may think are of the alarmist order; certainly they are depressing, but they are founded on stubborn facts. They show that England and all civilized nations stand in deadly peril of not having enough to eat. As mouths multiply, food resources dwindle. Land is a limited quantity, and the land that will grow wheat is absolutely dependent on difficult and capricious natural phenomena. I am constrained to show that our wheat-producing soil is totally unequal to the strain put upon it. After wearying you with a survey of the universal dearth to be expected, I hope to point a way out of the colossal dilemma. It is the chemist who must come to the rescue of the threatened communities. It is through the laboratory that starvation may ultimately be turned into plenty.

By that last remark Sir William meant that the problem of the food supply was readily seen to be a chemical one. Wheat needs many different nutrients and growth factors in order to grow, but the most limiting one is *nitrogen* in some suitable form of chemical combination. This comes about because

Bread and chemistry

4

ABOUT NITROGEN FIXATION, THE FOOD PROBLEM, AND THE REST OF THIS BOOK

all living organisms are based on proteins, and all proteins are composed of α-amino acids (see Chapters 14 and 17), which contain the NH group and are based on nitrogen. Hence nitrogen is an absolutely essential element in all plants and animals. The nitrogen content of the various plant and animal protein tissues varies from 14 to 17 percent, so it is obvious at once that a very substantial quantity of nitrogen in some suitably available form is necessary to grow wheat at a yield of 100 bushels per acre. Although some species of bacteria can absorb molecular nitrogen from the air and convert it to nitrate salts suitable for plant nutrients, the higher plants by themselves cannot do this; they must have their nutrient nitrogen in the form of water-soluble compounds. Animals, in turn, are parasitic upon the plant life of the world; they get their protein material from the plants or other animals they eat. Eventually, the nitrogenous waste products of the animals decay, if they are not used by plants, and release N_2 to the atmosphere. The entire nitrogen cycle is shown in Figure 1-1.

FIGURE 1-1 The nitrogen cycle. Lightning provides energy for the combination of some nitrogen and oxygen in the atmosphere, producing NO_2 which is washed from the air by rain which falls to the earth, entering the soil. Soil contains nitrates from the NO_2, from nitrogen-fixing bacteria, and from decayed vegetation. Soil also contains $NaNH_4HPO_4$ and urea, $(H_2N)_2C=O$, from animal wastes. Cultivated fields also have nitrogen, phosphorus, and potassium added as fertilizers. Plants obtain nitrogen from the soil, incorporating that nitrogen into plant proteins. Animals eat the plants, absorb amino acids from the plant proteins, and incorporate the amino acids into animal proteins. People eat the plants and eat meat. Nitrogenous wastes from plants, animals, and people return in part to the soil, and in part to the atmosphere.

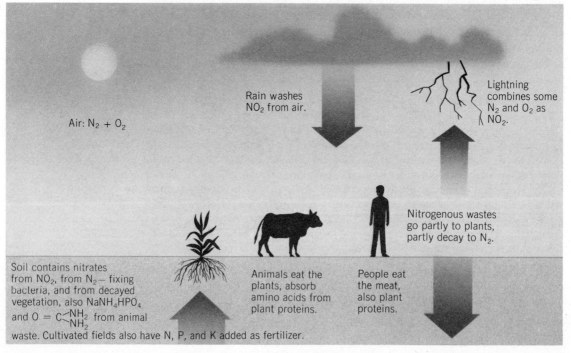

The land receives only so much fixed nitrogen annually from rainfall and from nitrogenous wastes. The bacteriological fixation of atmospheric nitrogen on the roots of leguminous plants constitutes an admirable step in the right direction, as had been pointed out by agricultural experts a full century and a half before, but there is a limit to the amount of nitrogen that crop rotation with clover or alfalfa can supply each year, and rotation, of course, takes the land out of wheat cultivation for that time. Any additional nitrogen, therefore, must be furnished by the farmer himself in the form of nitrates or salts of ammonia. Such addition has indeed been proved very effective, as Crookes cited from experiments in England. Laws and Gilbert had planted an experimental field at Rothemstead with wheat for 13 consecutive years without any fertilizer and had obtained an average yield of 11.9 bushels per acre. For the next 13 years, it was again sown with wheat each year, but in addition 500 lb of sodium nitrate per acre was spread on the ground. The average yield for those 13 years was 36.4 bushels per acre. That is, the output was more than tripled. To put it in another way, 20 lb of sodium nitrate produced an increase of one bushel of wheat. At this rate, about 150 lb of sodium nitrate or its equivalent would have to be added to each acre of wheat land annually in order to raise the world yield of wheat from the contemporary long-term average of 13 bushels per acre to 20 bushels per acre, which would be sufficient, Crookes said, to maintain the population for at least three generations to come. Simple computation showed that 12 million tons of sodium nitrate would be required annually to put off starvation to this extent.

The question then became how to get 12 million tons of sodium nitrate per year. It was clear that the supply of natural sodium nitrate from Chile was inadequate. It was being mined at the rate of 1,200,000 tons per year, only one-tenth of what was required, and even if it could be produced at a sufficient rate, the supply would last only three to four years. It was obvious that the only alternative was to develop some chemical process for taking inert gaseous nitrogen from the air and making from it a nitrogenous nutrient for plants. The resources of elementary nitrogen in the atmosphere are actually limitless; not only is it available to the extent of 4 million billion tons at any one time, but all of the combined nitrogen eventually finds its way back to the atmosphere once more. The way out, therefore, could be stated very clearly by Sir William in the following terms, which are quoted from page 45 of his book:

The fixation of atmospheric nitrogen therefore is one of the great discoveries awaiting the ingenuity of chemists. It is certainly deeply important in its practical bearings on the future welfare and happiness of the civilized races of mankind. This unsolved problem, which so far has eluded the strenuous attempts of those who have tried to wrest the secret from nature, differs materially from other chemical discoveries which are in the air, so to speak, but are not yet matured. The fixation of nitrogen is vital to the progress of civilized humanity. Other discoveries minister to our increased intellectual comfort, our luxury, or our convenience; they serve to make life easier, to hasten the acquisition of wealth, or to save time, health, or worry. The fixation of nitrogen is a question of the not far-distant future. Unless we can class it among certainties to come, we shall be squeezed out of existence.

CHAPTER 1

"Fixed" means chemically combined, not free in the air.

Any organic or inorganic source will do.

Note that a bushel of wheat weighs about 80 lbs!

6
ABOUT NITROGEN FIXATION, THE FOOD PROBLEM, AND THE REST OF THIS BOOK

It was not necessary that some startling new development of science be brought into the picture, although some were soon to come. By the already-known method for bringing about atmospheric combination of nitrogen and oxygen with an electric arc, sodium nitrate could be made (although at rather high cost):

A calorie is a unit of energy. A "kcal" (kilocalorie) is 1,000 cal.

$$43 \text{ kcal of heat} + N_2 + O_2 \xrightarrow[\text{rapid chilling}]{2000°C} 2NO$$

$$2NO + O_2 \longrightarrow 2NO_2$$

This is just chemical shorthand.

$$3NO_2 + H_2O \longrightarrow 2HNO_3 + NO \text{ (returns to process)}$$

$$HNO_3 + NaOH \longrightarrow NaNO_3 + H_2O$$

It was calculated that it would cost 26 English pounds to produce one ton of sodium nitrate in this way. Given a much cheaper source of electric power, as at Niagara Falls, this might be brought down to 5 pounds sterling per ton (and incidentally, one plant at Niagara Falls, it was calculated, would have sufficed to supply the entire world with the combined nitrogen needed to avert famine).

This process, which certainly offered a way out, was soon superseded by a newer chemical process invented by Adolph Frank and Nikoden Caro, in which very much less electric power was required. In the new process, lime was reduced with coke in an electric furnace to make calcium carbide, a material already in production for the acetylene lamps of the day. The same calcium carbide heated to red heat in an atmosphere of pure nitrogen formed a compound called calcium cyanamid, which is still being made in enormous quantities today. When treated with steam, this substance yielded ammonia, which, in the form of any of its common salts, could be used directly by growing plants as a source of nitrogen:

$$CaO + 3C \xrightarrow{1000°C} CaC_2 + CO$$

$$CaC_2 + N_2 \longrightarrow CaCN_2 + C$$

$$CaCN_2 + 3H_2O \longrightarrow CaCO_3 + 2NH_3$$

Chemical equations are explained in Chapter 3.

Early in the twentieth century, factories were established in Italy, Germany, France, Norway, Japan, and Canada, but none was built in the United States because it was considered that our potential wheat supply was adequate for our own needs. Fifteen years later, however, a large plant was built at Muscle Shoals on the Tennessee River in Alabama, because by that time nitric acid for explosives had become essential to the waging of World War I. Half a century later, cyanamid plants here and in Canada are still used for the production of chemicals for mining, for the manufacture of acrylonitrile fibers, and for the production of melamine and related plastics. But as a method for producing cheap nitrogenous fertilizer, the cyanamid process soon yielded to another method developed in Germany.

It should be remembered that during the period 1900–1914, there were high political and military stakes involved, for Germany was already thinking in terms of war and had to develop some reliable domestic supply of nitric acid for explosives that would make it independent of Chile saltpeter in case of an English naval blockade. It was Fritz Haber who came up with a procedure which involved no electric power; it required only coal, air, and water to produce ammonia and nitric acid. In the first step, red-hot coal was blown with steam to produce carbon monoxide and hydrogen, and by further reaction with water vapor, this yielded carbon dioxide and hydrogen. The hydrogen was separated and then was brought into combination with nitrogen (distilled from liquid air) at high pressure and elevated temperature in the presence of a catalyst[2] to form ammonia directly:

$$C + H_2O \xrightarrow{800°C} CO + H_2$$

$$CO + H_2O \xrightarrow[\text{catalyst}]{500°C} CO_2 + H_2$$

$$3H_2 + N_2 \xrightarrow[450°C, 500 \text{ atm}]{Fe + 1.0\% (K_2O + Al_2O_3)} 2NH_3$$

The equations summarize the prose above.

Part of the ammonia could then be combined with atmospheric oxygen over a catalyst to produce (eventually) nitric acid, and combination of ammonia and nitric acid gave ammonium nitrate, which could be used directly as a nitrogen-rich inorganic agricultural fertilizer:

$$4NH_3 + 5O_2 \xrightarrow{Pt} 4NO + 6H_2O$$

$$2NO + O_2 \longrightarrow 2NO_2$$

$$3NO_2 + H_2O \text{ (hot)} \longrightarrow 2HNO_3 + NO$$

$$HNO_3 + NH_3 \longrightarrow NH_4NO_3$$

NH_4NO_3 is also useful for making gunpowder.

The immediate use of the process, however, was military; as soon as the first commercial plant was in steady operation in Germany, war was declared, and the tragic years of 1914–1918 ran their course. The process lived on to supply postwar Europe (and now the rest of the world) with ample amounts of fixed nitrogen at a cost so low that almost any farm can be made more productive with it. Ten new ammonia plants were built in the United States alone during the decade 1940–1950, and many have been erected in other countries since that time. The total world capacity for combined nitrogen now exceeds 4 million tons per year, equivalent to 24 million tons of sodium nitrate, which is twice Sir William Crookes' estimate of what was necessary in 1898 for the next three generations. Some 85 percent of all this ammonia and ammonium nitrate are made by the Haber process. So within 16 years of Sir William's talk, not one but three processes were able to meet the situation, each in competition with the other to lower the cost.

---

[2] Mostly metallic iron, made by reducing Fe_3O_4 (containing 0.5 percent Al_2O_3 and 0.5 percent K_2O) with hydrogen.

8
ABOUT NITROGEN FIXATION, THE FOOD PROBLEM, AND THE REST OF THIS BOOK

What happened to the great nitrogen shortage of the early 1900s then? In four simple words, the problem was solved. What is equally important, the shortage is still being met in the United States, despite the pressures of greatly increased population. So quietly and efficiently do the factories for the chemical combination of nitrogen operate that many people today are not even aware that they are completely dependent upon these plants. If something should happen today to put them all out of commission, a month's shutdown would markedly affect the agricultural markets of the following year, and if the plants were to be shut down for a whole year, there would be widespread hunger. With the plants operating steadily, this has not happened. Instead, for 40 years the United States had a problem with its agricultural *surplus*, a problem which had a happy ending when the surplus grain could be shipped to the Soviet Union and to India in time of great need.

It is interesting, before we go on, to take further note of some other results of the aftermath of Sir William's talk. The effect on the size of the world's population and the distribution of peoples over the surface of the earth has been very great indeed; it could very well be said that the principle of population control laid down so firmly by Thomas Malthus 150 years ago has been held in abeyance all this while. That principle, which expected starvation to control the population at a size fixed by the maximum agricultural production Malthus could foresee, failed to take effect because Malthus did not take into account future scientific developments which came to light as a result of social pressure. Today there are many other philosophers, historians, and economists whose conclusions might turn out to be similarly inaccurate for the same lack of some elementary knowledge of the workings of science, and this provides one more very good reason for strengthening the teaching of science throughout the world.

Note: not *refuted!*

Turning now to the present, it is evident that there is need for still greater agricultural production in order to keep up with the growing population, and many more factories for the chemical fixation of nitrogen will have to be built within the coming two decades. This is especially true of the underdeveloped countries of Asia and South America. Within the more industrialized countries, there is now a sufficient surplus of chemical compounds of nitrogen to permit the production of nitrogenous plastics, fibers, dishes, dyes, and weed killers on a very large scale; all of these should have been impossible luxuries, according to Malthus. They may still become impossible luxuries several decades hence, of course. Meanwhile, the starker problems of survival face the underdeveloped countries, and we should take a look at the possibilities for obtaining sufficient combined nitrogen to raise the crops to feed their people.

In Asia (exclusive of China and Japan), the estimated[3] grain requirement will be 245 million metric tons annually by 1980, and 20 million tons of fertilizer (calculated as NH_4NO_3) will be required to produce this much grain. This is about *10 times* as much fertilizer as was consumed annually (much of it imported) during the decade 1960–1970. How can that much fixed nitro-

---

[3] *Chemical and Engineering News*, vol. 46 (1968): Feb. 19, p. 28; Nov. 25, p. 28; Oct. 7, p. 118; Oct. 14, p. 90.

gen be produced? If the fertilizer is to be made from ammonia synthesized by the Haber process, two serious problems arise at once:

1. The Haber process requires hydrogen as a reactant to produce ammonia. There has to be a source of gaseous hydrogen for all that ammonia. In the United States, hydrogen is a by-product from the refining of petroleum, and in Germany, it is obtained from water and coal. But in a land that has neither oil nor coal, where will the H_2 come from? Water could be electrolysed, but electrolysis requires lots of electric power, and in these countries power is very scarce. The only way out would seem to be the generation of electric power from nuclear energy, and the use of that power to produce hydrogen from water. Nuclear power generation involves advanced technology and will require substantial technical and economic assistance.

2. The Haber synthesis also requires very high pressures and rather high temperatures. Expensive plant equipment, including reaction vessels that operate reliably, efficiently, and safely at high pressure and temperature, is needed. This equipment requires large capital outlay, representing a severe drain on scarce money and materials.

The situation in terms of the total cost and availability of raw materials, energy, and manufacturing facilities is not much different for the cyanamid method, or for the arc method. The outlook is indeed not favorable for a massive extension of present methods for combining atmospheric nitrogen and producing fertilizers. Our hope for the future lies in exploring other chemical possibilities.

1-3 Room-Temperature Fixation of Nitrogen

The bacteria that live in symbiotic relationship on the roots of alfalfa plants and clover plants are able to absorb molecules of nitrogen, N_2, and convert them to soluble, nutritive compounds of nitrogen by some chemical process which requires no heat and no high pressures. We do not know how they do this, but acting on the premise that "if bacteria can do it, we can do it too," much searching has gone on for the bacterial secret. Some cyclical "model reactions" have been uncovered,[4] which use complicated coordination compounds of platinum or ruthenium, and these have led to future speculations on how the bacteria do it. However, duplication of the actual feat has not been accomplished at the time of this writing. Coordination compounds of molecular N_2 are known (and this should not be surprising, because N_2 is isoelectronic with CO, CN^-, and NO^+), but almost always it comes off as N_2 again when the complex is decomposed. The key to the matter is that some kind of reducing agent is needed to transform the N_2 to $-NH_2$ groups within the complex, and so an extensive search for an economical reduction reaction has begun.

One promising lead in the search has been the use of titanium compounds. In 1964 it was found that the reduction of $TiCl_4$, $MoCl_5$, WCl_6, $FeCl_3$,

More on this in Ch's 3 and 13.

i.e., its electron arrangement is the same. See Ch. 3.

"Reducing": Ch. 8.

---

[4] *Chemical and Engineering News*, vol. 45 (1967), Jan. 30, p. 32.

10
ABOUT NITROGEN FIXATION, THE FOOD PROBLEM, AND THE REST OF THIS BOOK

or $CrCl_3$ with lithium aluminum hydride gives complexes which absorb N_2 and give ammonia upon hydrolysis. But lithium aluminum hydride is a very expensive reducing agent, so it was good news when several years later, organomagnesium compounds were found to work as well. For example,

"Organomagnesium": Ch. 14.

$$2(C_5H_5)_2TiCl_2 + C_2H_5MgCl \longrightarrow \begin{array}{c} C_5H_5 \quad H \quad C_5H_5 \\ \diagdown \diagup \diagdown \diagup \\ Ti \quad Ti \\ \diagup \diagdown \diagup \diagdown \\ C_5H_5 \quad H \quad C_5H_5 \end{array}$$

Then absorption of N_2 by the titanium hydride complex gives the substance

$$\begin{array}{c} C_5H_5 \quad NH \quad C_5H_5 \\ \diagdown \diagup \diagdown \diagup \\ Ti \quad Ti \\ \diagup \diagdown \diagup \diagdown \\ C_5H_5 \quad NH \quad C_5H_5 \end{array}$$

Further reduction with ethyl magnesium chloride gives two molecules of $(C_5H_5)_2TiNH_2$, and hydrolysis of this gives ammonia.

A further advance came in 1968, when it was found that metallic sodium can be used as the reducing agent.[5] Titanium isopropoxide, $Ti(OPr)_4$, is reduced first with sodium dissolved in naphthalene:

Pr = "propyl," C_3H_7

$$2Na + Ti(OPr)_4 \longrightarrow Ti(OPr)_2 + 2PrONa$$

The Ti(II) compound has empty orbitals available for coordinating with N_2 molecules as electron donors:

$$N \equiv N + Ti(OPr)_2 \xrightarrow{25°C} Ti(OPr)_2N_2$$

The nitrogen compound is then reduced further by four more equivalents of sodium dissolved in naphthalene, producing a titanium nitride. Treatment of this with isopropyl alcohol causes alcoholysis (analogous to hydrolysis), forming ammonia:

You have arrived at the level of Ch. 15.

$$Ti(OPr)_2N_2 + 4Na \longrightarrow Na_2TiN_2 + 2NaOPr$$

$$Na_2TiN_2 + 6PrOH \longrightarrow 2PrONa + Ti(OPr)_4 + 2NH_3$$

$Ti(OPr)_4$

where PrOH is isopropyl alcohol, and OPr is isopropoxide ion. The titanium tetraisopropoxide goes back into the process, and hydrolysis of the sodium isopropoxide gives isopropanol for the last step (plus sodium hydroxide, which can be dried and electrolysed to get back the sodium):

$$PrONa + H_2O \longrightarrow PrOH + NaOH$$

Van Tamelen found that with the indicated proportions of 6 moles of sodium to 1 mole of titanium tetraisopropoxide, the yield of ammonia was 65 percent of theoretical. This is very encouraging. Enough electric power is needed to regenerate the sodium; the rest of the nitrogen fixation can be accomplished

---

[5] E. E. van Tamelen, *Journal of the American Chemical Society*, vol. 91, p. 1551 (1969).

at room temperature and without high pressures. This would indeed be a welcome process to relieve the next nitrogen shortage, which will be upon the world very soon.

1-4 Other Essential Elements

Plants do not grow on nitrogen alone. They need carbon, of course, which they get as carbon dioxide and reduce by the wonderful pathways of photosynthesis.[6] The hydrogen they get from water. Phosphorus in the form of organic or inorganic phosphates is essential for the photosynthetic process and as a structural element for many of the catalytic substances (enzymes and coenzymes) necessary to the functioning of the life cycle of the plant. Potassium salts must also be provided, because potassium is so readily washed out of the soils that usually not enough is left to maintain the right ionic balance in the organism. Magnesium, iron, and calcium are required, as are a few "trace elements" needed in low concentrations. Sulfur also is required, but usually there is plenty of sulfate ion available in the soil or water.

Trace: a small amount

Of the various elements (besides nitrogen) that must be available to growing plants, phosphorus has received the most attention. Aldous Huxley, in his popular book *Brave New World,* described a fictional phosphorus shortage which made it necessary to collect corpses and recover the phosphorus from them as P_2O_5. Actually, there has been no phosphorus shortage within the time Huxley predicted, because we now have methods for the recovery of phosphorus in useful agricultural form from vast deposits of phosphate rock, which are not likely to run out for a very long time.

It is difficult to think of any other shortage of a particular element that could not be met by the application of present chemical knowledge through economic and practicable channels, *if* the social pressure to do so is strong enough. Much has been said about the impact of science on society, but we see from the events related in this chapter that society also has a tremendous impact on science. Left to themselves, scientists will work principally to satisfy their own curiosity, but economic factors and social pressures can and do influence them, strongly. Faced with a national or worldwide problem, they very often have been able to apply the results of past research to the solution of that problem. It seems important to keep science healthy, for that reason alone.

1-5 What Else Can Be Done?

If the threat of mass destruction of human life through war can somehow be averted, and if mankind learns to live in greater harmony and equilibrium with his environment, we may expect that the world's population will continue to grow at the rate of about 2 to 3 percent per year, doubling every

---

[6] See J. A. Bassham and M. Calvin, *The Path of Carbon in Photosynthesis,* Englewood Cliffs, N.J.: Prentice-Hall (1957).

12
ABOUT NITROGEN FIXATION, THE FOOD PROBLEM, AND THE REST OF THIS BOOK

30 to 40 years. No one likes to think of limiting the population by any enforced means; it is much more attractive to us to find some way of feeding, clothing, and housing all these new people until such a time as their numbers may be stabilized by *voluntary* methods, by colonization of other planets, or by whatever else the future may have in store for us. Let us now look to what might be done in this direction, using well-established and matter-of-fact chemical principles of today which do not require any "breakthrough" for their application. What really could be done if we had twice as many people on the earth some years hence, or indeed 100 times as many people?

It has been pointed out often[7] that agricultural output has increased enormously in the past 40 years, through mechanization, development of better seed strains, and adequate fertilization. However, there must inevitably come a time when present agricultural procedures will fail us, and it will not be possible to produce the daily necessities by raising plants in cultivated soil. What might we do at such a time?

Basically, it is no harder to make *food* from coal, oil, air, and water than it is to make dyes or rubber or plastics. In fact, the synthesis of good nourishing substances is a great deal simpler than the synthesis of fibers like nylon. Edible fats, for example, which have always been the first items to run short in countries where the food supply is interrupted, could be made (and indeed have been made) from coal and water. The combination of acetylene and formaldehyde yields glycerol. This is an essential component of every edible fat. Further, the combination of carbon monoxide and hydrogen (from the action of steam on coal) produces alcohols, which are then oxidized to form a mixture of organic acids. The indigestible branched-chain carboxylic acids are removed. The remaining straight-chain organic acids are finally combined with glycerol, producing perfectly good edible fats.

Branched chains, etc.: see Ch. 14.

Similarly, the nutritionally useful parts of proteins, the α-amino acids, could also be synthesized from coal or petroleum. The coal can be converted by way of acetylene to acetaldehyde, which by combination with ammonia from the fixation of atmospheric nitrogen would lead to amino acids. These are the small molecules used by the human body to build up its own protein structures; about 14 of them are essential to human nutrition. The actual proteins that we consume in the form of milk or meat or vegetable protein are converted by the digestive processes into these separate amino acid building blocks. Indeed, that whole process sometimes is bypassed in severe cases of digestive disturbance by the injection of the necessary amino acids directly into the bloodstream. Most recipes for appetizing food depend on macromolecules such as proteins and starches for texture and structure. So amino acids will not serve very generally in present recipes; either amino acids must be combined into macromolecules, or else some innovative recipes must be developed. It will be very interesting to see what concoctions actually will be produced by future cooks from the materials they will buy at the market, and indeed what materials will be available there.

Macromolecules: Big molecules. See Ch's. 5, 14, and 15.

---

[7] See, for instance, J. B. Billard and J. P. Blair, "The Revolution in American Agriculture," *The National Geographic Magazine*, vol. 137, p. 147 (Feb. 1970).

An alternative to synthesizing amino acids is to let microorganisms do the entire job of converting nitrate salts to protein, on an around-the-clock basis and without light. This is possible and practical. Some yeasts will grow at the interface of an oil-water mixture, deriving their carbon and hydrogen from hydrocarbons in the oil, and their nitrogen and oxygen from the water solution. The necessary nitrates, phosphates, and other nutrients are dissolved in the water. To compete with fish meal and soybean meal, the British Petroleum Company has developed a process to make chicken feed and cattle feed supplements from petroleum. They use *Candida* yeast, feeding it on pure C_{10} to C_{18} hydrocarbons from an oil-refinery stream. The product is 65 percent protein, considerably higher than the protein content of soybean meal. Since one ton of yeast product is obtained from one ton of hydrocarbon, the exchange is highly favorable and promises to help the worldwide protein shortage a great deal. The first operative plant, producing 4000 tons per year, started production in Grangemouth, England, in 1970; a 16,000-ton plant in Lavéra, France, followed in 1971.[8]

Carbohydrates present more of a problem, but still they are only twice as complicated as glycerol to make from our starting materials. It might well turn out, however, that the world's supply of carbohydrates will be met for a very long time to come by conversion of cellulose, which is the chief constituent of wood, straw, and agricultural waste. The cellulose can be separated from the resinous lignin by the processes now used in the manufacture of paper. The purified cellulose could then be hydrolyzed by cellulase enzymes or by boiling with slightly acidulated water, yielding glucose, fructose, and the other simple sugars. Considerable refinement of the process would be necessary to adapt it for making sweets for humans, but we would then be able to use all of the enormous amounts of cellulose that are discarded or burned as fuel. We could also grow much more. Trees and bushes will grow on mountains, in deserts, and among rocks, where cereal grains and other vegetable crops cannot be grown.

There is a great deal more to healthful human nutrition than a supply of carbohydrates, fats, and proteins. There are vitamins, for example, but some of these are even now being synthesized for our daily use. There are many other growth factors, but chemical exploration of these continues to tell us more about their structures, and we may reasonably hope that eventually all of them can be synthesized. Even if they cannot be synthesized, they are needed in such small amounts that they could be produced by conventional agriculture. That way we could devote *all* our lands to this purpose, and make our "bread-and-butter" items in large food factories.

What, then, is really required? For one thing, carbon is needed in a reduced and usable form—carbon in the form of coal and petroleum, for instance. Therefore, any continued use of coal and petroleum for heating, transportation, and power deprives future generations of potential foodstuffs. Coal and oil are even now very inviting as chemical raw materials for synthesis of plastics, fibers, and many other everyday products. For another thing, we

VERY important!

---

[8] *Chemical and Engineering News*, vol. 48 (1970), Feb. 2, p. 21.

14

ABOUT NITROGEN FIXATION, THE FOOD PROBLEM, AND THE REST OF THIS BOOK

need to replace the present practice of burning wood for fuel, since we may want to use the wood as a source of food. It behooves us, then, to hurry along with the development of solar heat, and of safe methods for obtaining power from nuclear energy, so that we no longer need to use up our reserves of fossil fuels and the products of our forests for heat and power.

As for clothing, many textiles are already made from synthetic fibers, and we could just as well make all of them, if necessary. A man eats his own weight of food every 20 or 21 days. His requirements for clothing are very much smaller, so the problems of supply of raw materials for clothing are that much simpler. As a matter of fact, there is no reason why the chemical raw materials, which were put to use in the manufacture of clothing, could not be recovered and used for the subsequent manufacture of food, if that were to become necessary. The further development of cheap sources of heat and power may also change drastically our habits of living, making exposure to cold weather much less necessary and thereby simplifying the clothing problem.

In a similar way, housing for all these new people offers much less difficulty if we have a ready source of cheap energy available. With such energy we could make metals more cheaply; we could fuse mineral substances into brick and building blocks the way we now make glass, and indeed turn out houses plentifully and economically. They should, of course, be inorganic houses; we have seen that we shall want to use wood only as a raw material for food. Since we now build our schools and offices and public buildings out of metal, glass, and ceramics, all of which are inorganic materials, it would be no great change to use similar inorganic, fireproof construction for dwellings as well.

Read it again; mark it up for emphasis.

What really is the outlook then? Even so brief a survey as this one tells us that just as the great nitrogen shortage of 1898 was met, and continues to be met every day of our lives by a mixture of chemistry and ingenuity, so could the material needs of tomorrow's populations be met. Actually, we could feed, clothe, and house many, many times the present world's population if we wanted to set our minds to it. What is required is not a whole new collection of scientific facts and principles, but rather the will to direct our own thoughts and practices toward the conservation of our irreplaceable resources and the gradual development of new habits of living and governing, so that many future generations of people also may live.

2

Atoms and the Chemical Elements

The edifice of science requires not only material but also a plan.

DMITRI IVANOVITCH MENDELEEF, 1870

2-1 Abstract

The basic unit of matter under ordinary conditions is the atom. After some general discussion of the properties of atoms, we shall describe the three most important particles of which all atoms are composed, and then consider the structure of the atom from five points of view: (1) the separation of electrical charge within the atom, (2) the atomic nucleus and the particles of which it is composed, (3) the rules governing the arrangement of electrons in the atom, (4) the remarkably simple way in which the fundamental particles combine according to five quantum rules to produce the 100-odd kinds of atoms of which the entire known universe is made, and (5) the summary of atomic structure that is contained in the **Periodic Table** of the elements.

2-2 Introduction

Now we start at the beginning.

All matter is composed of ultimate chemical building blocks called atoms, the existence of which is generally conceded.[1] The evidence for the existence of atoms is indirect but overwhelming. For example, it would be hard to devise a reasonable alternative explanation for the regular array of objects in Figure 2-1, which shows a metal crystal highly magnified by an instrument called a field ion emission electron microscope.

It might be supposed beforehand that this complex planet, with its mountains and seas, its mantle and core, would be composed of an infinite variety of atoms, but this is not so. Only 89 kinds of atom are found in weighable amounts in all the stuff of the earth. About 15 more kinds of atom can be derived from more common materials by involved processes of nuclear physics and nuclear chemistry. Since almost everything is possible (at least in theory), some of these processes may also be duplicated in nature without benefit of cyclotrons or nuclear reactors, resulting in minute traces of some of these more exotic kinds of atom being present along with the more common and more "natural" 89. Even taking into account these traces, we can say that the entire earth is made up of only about 100 kinds of atoms distinguished by unique and readily recognizable chemical behavior.

---

[1] This statement is made only in the interest of saving time and words. The reader is free to differ, and need not accept the concept of atoms on faith. If he wants to read about the early struggles which led to the atomic theory and to its adoption, he is referred to H. M. Leicester, *The Historical Background of Chemistry*, New York: Dover (1971). An excellent account of the quantitative investigations which convinced chemists about atoms is given by L. K. Nash in *Stoichiometry*, Reading, Mass.: Addison-Wesley (1966).

FIGURE 2-1 Photograph of the actual array of atoms in a crystal of indium, taken by Prof. Erwin W. Mueller of the Pennsylvania State University by means of an instrument which he invented, the field-emission electron (or ion) microscope. Within a large evacuated glass tube, a thin rod of metal, sharpened and etched to an extremely fine point, concentrates on its tip a high-voltage electric charge of such intensity that electrons are emitted from all of the atoms in the single crystal of metal that constitutes the tip. These electrons (or ions produced by them) fly in straight lines to a large screen some distance away, reproducing a picture (with magnification of 1,000,000 or more) of the original placement of atoms on the surface of the metal.

Even rocks from the moon contain the same kinds of atom as those on earth, although in somewhat different proportions. Indeed, so far as we can tell by the most exhaustive study of the characteristics of light from the sun and the stars, and from the most detailed quantitative analysis of those visiting bits of matter from outer space which we call meteorites, the whole universe

Elegant cosmic unity!

is constructed of the same kinds of atom as we know here, with nothing new or strange thrown in. This is encouraging evidence of a basic simplicity in natural matter.

All atoms of one kind (that is, all atoms that have the same chemical behavior) constitute one chemical **element**. Each element is given a name (for identification) and a symbol (for convenience). Hence we may say that the substances that make up the earth are composed of only about 100 chemical elements, and that these are the same elements as those found in all other matter in the universe. It has taken hundreds of years to separate and identify all of these elements, with most of the progress being made since 1850. Indeed, only in the nuclear age, since about 1930, has it been possible to synthesize new chemical elements (and also the old ones) in the laboratory. In Chapter 10, we shall consider how this is done. Right now we are concerned only with the tendencies of the chemical elements to combine with one another and to form the compound substances which we find in nature, or which we either want to make (for a particular benefit or necessity) or want to prevent being made (because of poisonous or undesirable characteristics).

It was recognized a long time ago that the atoms of a chemical element have a particular average **atomic weight**,[2] relative to the atoms of another element, and that these weights govern in a precise way the proportions by which one element combined with another. It was also recognized (by Mendeleef in 1869) that the atomic weight of an element appears to have some kind of fundamental influence on the chemical behavior of that element: *how* the element combines with other elements, and with *which* ones it usually combines. Such combining preferences proved difficult to explain in detail solely on the basis of weights, however, and now we realize that we must turn to details of *structure*, rather than weight, in order to gain a clear understanding of the chemical behavior of atoms. It also turns out that other details of structure explain why most atomic weights are averages of a set of values, rather than being a single figure corresponding to a single chemical behavior. We shall come to that when we turn to the subject of atomic structure later in this chapter. Before that point, we should consider the more simple aspects of atoms, such as their characteristic sizes and weights.

2-3 Atomic Weights

Table 2-1 lists all the chemical elements in the alphabetical order of their names and gives their symbols (usually abbreviations of the names) and their accepted atomic weights. In a very real way, this table may be said to list the names and numbers of all the players in the game of matter with which this book is concerned.

---

[2] It would be more strictly correct to speak of the *mass* of an atom, for the property remains the same throughout the universe. However, since *weight* is a measure of the earth's gravitational attraction for that mass, we can compare masses at the earth's surface by comparing weights, and so it is all right to speak of relative atomic weights.

TABLE 2-1
The Chemical Elements, With Their Symbols, Their Atomic Numbers, and Their Atomic Weights

Name	Symbol	Atomic Number	Atomic Weight
Actinium	Ac	89	...
Aluminum	Al	13	26.9815
Americium	Am	95	...
Antimony	Sb	51	121.75
Argon	Ar	18	39.948
Arsenic	As	33	74.9216
Astatine	At	85	...
Barium	Ba	56	137.34
Berkelium	Bk	97	...
Beryllium	Be	4	9.0122
Bismuth	Bi	83	208.980
Boron	B	5	10.811 ± 0.003
Bromine	Br	35	79.909
Cadmium	Cd	48	112.40
Calcium	Ca	20	40.08
Californium	Cf	98	...
Carbon	C	6	12.01115 ± 0.00005
Cerium	Ce	58	140.12
Cesium	Cs	55	132.905
Chlorine	Cl	17	35.453
Chromium	Cr	24	51.996
Cobalt	Co	27	58.9332
Copper	Cu	29	63.54
Curium	Cm	96	...
Dysprosium	Dy	66	162.50
Einsteinium	Es	99	...
Erbium	Er	68	167.26
Europium	Eu	63	151.96
Fermium	Fm	100	...
Fluorine	F	9	18.9984
Francium	Fr	87	...
Gadolinium	Gd	64	157.25
Gallium	Ga	31	69.72
Germanium	Ge	32	72.59
Gold	Au	79	196.967
Hafnium	Hf	72	178.49
Helium	He	2	4.0026
Holmium	Ho	67	164.930
Hydrogen	H	1	1.00797 ± 0.00001
Indium	In	49	114.82
Iodine	I	53	126.9044
Iridium	Ir	77	192.2
Iron	Fe	26	55.847
Krypton	Kr	36	83.80
Lanthanum	La	57	138.91

TABLE 2-1 (Continued)
The Chemical Elements, With Their Symbols, Their Atomic Numbers, and Their Atomic Weights

Name	Symbol	Atomic Number	Atomic Weight
Lead	Pb	82	207.19
Lithium	Li	3	6.939
Lutetium	Lu	71	174.97
Magnesium	Mg	12	24.312
Manganese	Mn	25	54.9380
Mendelevium	Md	101	...
Mercury	Hg	80	200.59
Molybdenum	Mo	42	95.94
Neodymium	Nd	60	144.24
Neon	Ne	10	20.183
Neptunium	Np	93	...
Nickel	Ni	28	58.71
Niobium	Nb	41	92.906
Nitrogen	N	7	14.0067
Nobelium	No	102	...
Osmium	Os	76	190.2
Oxygen	O	8	15.9994 ± 0.0001
Palladium	Pd	46	106.4
Phosphorus	P	15	30.9738
Platinum	Pt	78	195.09
Plutonium	Pu	94	...
Polonium	Po	84	...
Potassium	K	19	39.102
Praseodymium	Pr	59	140.907
Promethium	Pm	61	...
Protactinium	Pa	91	...
Radium	Ra	88	...
Radon	Rn	86	...
Rhenium	Re	75	186.2
Rhodium	Rh	45	102.905
Rubidium	Rb	37	85.47
Ruthenium	Ru	44	101.07
Samarium	Sm	62	150.35
Scandium	Sc	21	44.956
Selenium	Se	34	78.96
Silicon	Si	14	28.086 ± 0.001
Silver	Ag	47	107.870
Sodium	Na	11	22.9898
Strontium	Sr	38	87.62
Sulfur	S	16	32.064 ± 0.003
Tantalum	Ta	73	180.948
Technetium	Tc	43	...
Tellurium	Te	52	127.60
Terbium	Tb	65	158.924
Thallium	Tl	81	204.37

TABLE 2-1 (Continued)
The Chemical Elements, With Their Symbols, Their Atomic Numbers, and Their Atomic Weights

Name	Symbol	Atomic Number	Atomic Weight
Thorium	Th	90	232.038
Thulium	Tm	69	168.934
Tin	Sn	50	118.69
Titanium	Ti	22	47.90
Tungsten	W	74	183.85
Uranium	U	92	238.03
Vanadium	V	23	50.942
Xenon	Xe	54	131.30
Ytterbium	Yb	70	173.04
Yttrium	Y	39	88.905
Zinc	Zn	30	65.37
Zirconium	Zr	40	91.22

Atomic weights are no longer determined by comparing equivalent (but large) numbers of atoms of two different elements by means of a laboratory balance, because no balance can be made so precise as to compete in sensitivity and precision with an instrument known as the mass spectrograph. Hence the values given in Table 2-1 were all determined by mass spectroscopy. The mass spectrograph is a complicated electromagnetic instrument, which operates on two completely reliable physical principles: the attraction of electrostatic charges of opposite sign and the deflection of a moving electric charge by a magnetic field.[3] The actual construction of the instrument is far too complex to be detailed here, but the schematic drawing shown in Figure 2-2 will give you the essential idea.[4] Although a fair estimate of the actual mass[5] of an atom can be made by careful measurement of the intensities of the electric and magnetic fields and by noting the angle of deflection of the atom within the instrument, the most precise determinations are made by *comparison* of the deflection of an atom with that of known atoms taken as standard. The values given in Table 2-1 are based on carbon-12 taken as standard, in accordance with a decision made by the International Union of Pure and Applied Chemistry.

Some weights are given with six figures.

---

[3] See Appendix III for a summary and explanation of physical principles essential to this text.

[4] See also *McGraw-Hill Encyclopedia of Science and Technology*, vol. 8, pp. 164–166. (This encyclopedia, published in 1960 but with annual *Year Book* supplements, is available in almost all libraries and is a good secondary reference on all technical matters.)

[5] Since the mass spectrograph does not involve the earth's gravitation as a force, it is atomic *mass* rather than atomic *weight* that is actually measured in the instrument. However, a ratio of such masses is exactly equal to the ratio of the corresponding weights.

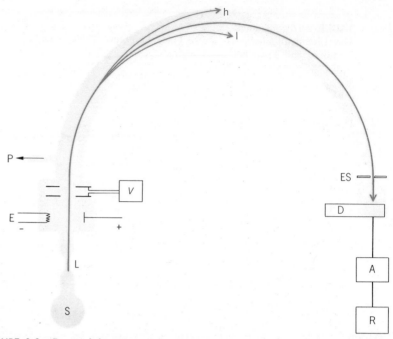

FIGURE 2-2 Essential features of a mass spectrograph. A source (S) of the atoms to be examined is connected to the instrument through a narrow leak (L), which allows a small, steady stream of atoms to pass. The atoms are given a positive charge by bombardment with electrons (see later discussion) at E. An electrical field supplied by voltage V fires the charged atoms into the curved chamber (which is kept evacuated by a pump at P). A magnetic field is applied perpendicular to the book page by magnetic poles above and below the curved chamber; this field causes the path of the charged atoms to curve to an extent that depends on the strength of the magnetic field, the accelerating voltage V, the charge on the atoms, and the mass of the atom. For given values of the first three variables, only those atoms with a certain mass follow the proper curvature to hit the exit slit ES. Light atoms strike the inner wall (I); heavier ones strike the outer wall (h). The value of V is adjusted continuously while a detector D, amplifier A, and recorder R register the passage of charged atoms through the exit slit.

2-4 The Gram-Atomic Weight

Since the values given in Table 2-1 are *relative* values (relative to the standard), they are ratios, and hence are pure numbers without a label or dimension of grams or ounces. Nevertheless, the student may properly inquire what is the actual weight (in familiar units) of one hydrogen atom, for example, if its atomic weight is given in the table as 1.008. The equations that govern the operation of the mass spectrograph provide an answer: A common hydrogen atom weighs only 1.673×10^{-24} grams (g). One single atom of fluorine weighs 3.154×10^{-24} g, and so on. In fact, each "unit" in the scale of relative atomic weights is actually equivalent to 1.660×10^{-24} g, so the weight in common units[6] of any kind of atom may readily be calculated.

---

[6] Since 1 oz equals 28.35 g, the weight in ounces or pounds or tons can be figured out accordingly, if desired. See Appendix II about exponents as a way of writing very large and very small numbers.

It follows that single atoms are very light, and so there must be enormous numbers of them packed in a familiar weight or volume of matter. For example, 1 cm³ of gold weighs 19.32 g, and each gold atom is seen to weigh 326.97×10^{-24} g. Hence there must be $19.32 \div (326.97 \times 10^{-24})$ atoms of gold in that cubic centimeter, or 59,000,000,000,000,000,000,000 atoms. All in a space $\frac{1}{16}$th of a cubic inch! Similar calculations show that all atoms are *very* tiny, far smaller than can be seen with the naked eye, or even with the most powerful optical microscope. They are also so light individually that we cannot feel them, or come anywhere near weighing them even with our most sensitive laboratory balance.[7] Hence we must deal with *enormous numbers of atoms* any time we want to examine a speck of dust, or make a gold ring, or build a steel bridge.

How enormous these numbers of atoms must be is difficult to comprehend, because the use of exponential numbers (or strings of zeros) hides their magnitude from us; we do not imagine them in terms of daily experience. We need to do a few operations in order to understand such numbers. For example, if we could take all of the atoms in that 1 cm³ of gold and lay them side by side in a single line, the line would extend 6 trillion miles. Or, if we took an ordinary light bulb, pumped out all the gas in it, and then drilled a hole in it large enough to admit 7 million atoms of the gases in the air *each* second, the bulb would take 100 million years to fill to atmospheric pressure. Or we might take an ordinary tumbler of drinking water, and, supposing we could mark all of the atoms of oxygen in the water it contains, pour the glass of water into the ocean (any ocean, since they are all connected). Let it mix for a few thousand years. Then withdraw a glass of ocean water from *anywhere*, and there would be 2000 marked atoms in that glass!

Some useful number games

With such huge numbers of atoms in all visible and tractable samples of matter which we might choose for laboratory study, how can we keep track of those numbers while we study the combining ratios of the atoms? How can we make a quantitative study of chemical change, if we cannot see how many atoms take part? There is a way, and it is very simple in principle: We fix upon a conveniently large number of atoms, sufficiently large to see and to handle and measure, and we always use that number of atoms (or a simple multiple or fraction of it) in our think-problems or our experiments. That is, instead of taking 1 g, each of elements A, B, and C (which quantity would contain different numbers of atoms A, B, and C because their atomic weights are different), we take an agreed-upon large number of atoms of A, B, and C, and then see how they interact. The agreed-upon large number is not a thousand, or a million, or a million billion, but 6.023×10^{23}. The reason for this seemingly strange number is that it is the reciprocal of 1.660×10^{-24}, the conversion factor we used for getting grams from atomic weight units. Hence, if we take 6.023×10^{23} atoms (which is 602,300,000,000,000,000,000,000 atoms) of *any* element, the quantity will have a weight in grams exactly equal to the atomic weight of that element. This makes it easy to remember, as well as being a practical unit of convenient size: A quantity

Keeping the numbers straight

The most important large number

---

[7] Very special balances will weigh to the nearest 0.000000001 g, but that much gold would contain about 3,000,000,000,000 atoms.

of *any* element, having a weight in grams numerically equal to its atomic weight, always contains 6.023×10^{23} atoms of that element.[8] This quantity is called the **gram-atomic weight**.

For example, the atomic weight of gold is 196.697, so one gram-atomic weight of gold is 196.697 g of gold. And this quantity of gold contains as many atoms as does 126.9044 g of iodine (whose atoms weigh 126.9044 units of atomic mass).

Obviously, we could think about, and could use, a *pound*-atomic weight or a *ton*-atomic weight of an element, if we had a balance (or "scale") that measured pounds or tons instead of grams. For laboratory use, however, the metric system of measurement is universal. Moreover, the gram-atomic weight is convenient. A gram-atomic weight of powdered sulfur is about a cupful, and a gram-atomic weight of iron is about $\frac{1}{2}$ in.^{3} of the solid metal.

2-5 The Sizes of Atoms

We know less about the exact sizes of atoms than we do about their exact weights. This comes about because we do not know how tightly packed the atoms are in liquid elements, nor even in solid elements (where the atoms do not move about as they do in liquids). The best we usually can do is to assume that the atoms of metal in a solid metallic element like gold or iron are actually touching,[9] and to take half of the center-to-center distance as the radius of one atom. (See Chapter 6 for a longer discussion of atomic radii.) In liquids, the situation is more fuzzy, for we cannot be sure that the atoms touch all the time; in gases, we are sure that they do *not* touch. Hence we have to use less direct methods for estimating the sizes of atoms in a gas such as helium, or in a vapor such as that of mercury.

For solids, the best information about sizes of the constituent atoms comes from examination of the exact structure of the solid by X-ray diffraction.[10] In this work, the positions of atoms are determined precisely, and the spacing of their centers in the solid-geometry framework is known quite well. From this spacing we get an atomic radius. (Some elements, which ordinarily are gaseous or liquid, have been frozen and then examined by X-ray diffraction, yielding information as accurate as that for metals.) The conclusions are:

---

[8] The number 6.023×10^{-23} is known as Avogadro's number because the Italian physicist Amadeo Avogadro deduced it in 1811 from the combining volumes of gases. The number may be determined precisely by, for example, calculating the theoretical density of a metal from its interatomic spacing (determined by X-ray diffraction), and from this the number of atoms in one gram-atomic weight of that metal.

[9] Whatever "touching" means! Since atoms have no skin and are not hard solid spheres (as we soon shall see), it is difficult to specify just where an atom begins. We may know where its center lies, but we can only approximate the radius.

[10] For an explanation of the phenomenon of diffraction, see Appendix III. The diffraction of X-rays by atoms arranged in a three-dimensional solid lattice is more complicated than diffraction of light in one plane by a grating, and hence the interpretation of the diffraction pattern is much more difficult.

1. The radii of those atoms that are readily available to us vary from 0.74×10^{-8} cm (for hydrogen) to 4.5×10^{-8} cm (for cesium)—none greater nor less than these.
2. The sizes are not proportional to the atomic weights of the elements, but vary in a cyclic fashion with increasing atomic weight.
3. The range of sizes is so small that the distance 10^{-8} cm (commonly called one **angstrom**, or 1 Å) may be taken as a convenient measure of atomic dimensions.

The small picture

2-6 Atomic Structure I: Electrons

If all atoms were hard spheres, like marbles, we should be hard put to explain their combining with one another in definite proportions (and in definite geometric patterns), as they are known to do. Fortunately, atoms are found to have structure, and all chemical and physical properties may be related decisively to that structure. The result is a coherent theory of the constitution of matter so comprehensive and so successful that it represents one of the greatest triumphs of the human intellect. We shall put portions of that theory to use at once.

In the first place, all atoms are known to contain **electrons**. These electrons constitute one of the three principal kinds of elementary particle[11] of which all atoms, no matter how complex, are composed. Electrons are bits of negative charge, about 10^{-13} cm in radius and weighing 9.10×10^{-28} g. The fact that all atoms do contain electrons is readily evident from several lines of everyday experience:

Evidence for electrons

1. Electrons are emitted by all kinds of atom when heated to a sufficiently high temperature. The emission of electrons from white-hot tungsten filament in a TV picture tube, and from a red-hot thimble coated with barium oxide in a radio vacuum tube, provides but two examples out of many.
2. Electrons are emitted from atoms under investigation in a mass spectrometer, leaving positively charged residues called **ions**. Similarly, electrons are emitted from atoms of neon under electrical stress, and the recombination of such electrons with the positive ions of neon gives forth light in every neon sign.
3. Electrons are emitted by many metal surfaces (such as those of sodium, zinc, and cesium) when light of proper wavelength impinges on the surface.

---

[11] The term "particle" is an unavoidable half-truth. Matter of all kinds behaves either like a collection of particles or like a collection of waves, depending on the observations that are made on it. The philosophical problems raised by this remarkable situation are still unresolved, but either the particle or the wave language may be properly used as long as it is appropriate to the experimental observation to be described. In the words of the Danish physicist Niels Bohr, the wave and particle models of matter are complementary. We shall use particle language for the present, and switch to the wave picture when it is more useful (in describing the domains within which electrons appear to move within the atom). For more information on the wave-particle duality, see "Suggestions for Further Reading."

This is the *photoelectric effect,* the basis for photoelectric cells which count cars on the highway and turn on lights at dusk.

4. Metals conduct electricity by passage of a stream of electrons, as through a copper wire. If the introduction of a few electrons at one end of a wire results in the emission of the same number of electrons at the other end, as is found, and if the atoms of copper are stacked solidly one up against the next, the only way electrons can be conceived to get through is to be passed from one atom to the next, or exchanged as regular inhabitants of the atoms.

5. Beams of electrons are the means of taking electron micrographs in the electron microscope, producing pictures which most students have seen.

The simplicity of nature again!

All electrons are alike, no matter what atoms they come from. All carry unit electrical charge, 1.602×10^{-19} coulombs (c); this constitutes a sort of atom of electricity and shows that electrical charge, like matter, may not be subdivided indefinitely. Hence electrons from any source will fit into and belong to any atoms where there is a place for them. The exchange of such electrons between atoms becomes the basis for the bonding together of atoms to form the aggregates called **molecules**, which we shall consider in Chapter 3.

Having agreed that all atoms contain electrons, the next question is: *how many?* We might suspect, intuitively, that the lightest atoms (those of the element hydrogen) contain only one electron each, and that the heavier elements are more complex, containing perhaps scores of electrons. This is so, but the task of settling upon exactly how many electrons are present characteristically in each kind of atom fell to a British physicist, Henry G. J. Moseley, who determined in 1910–1914 what he called the *atomic number* of each element he could study. His method was to use a solid sample of the element as the target for a stream of fast-moving electrons in an X-ray tube and to measure the frequency of the X-rays that were emitted.

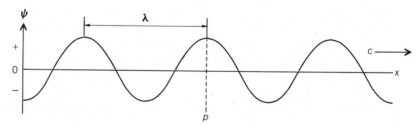

FIGURE 2-3   A wave, of wavelength λ, moving to the right with velocity c. A crest, or any other chosen part of the wave form, will pass point p with a frequency of ν times per second. The displacement from zero at any point is ψ. This drawing is a graph of ψ as a function of distance x.

X-rays are a form of electromagnetic radiation (see Appendix III). They are a wave phenomenon, as shown in Figure 2-3. The distance between repeating parts of a wave (for example, from crest to crest) is called the *wavelength.* The time it takes the wave to travel a distance equal to its wavelength is called the *period* of the wave, and if one watched a train of waves passing a given point in space, a number ν (ν is the Greek letter *nu*) of wavelengths would pass that point every second. The number ν is called the *frequency* of the wave. X-rays have wavelengths of about 10^{-8} cm and frequencies in the neighborhood of 10^{18} sec^{-1}. The notation sec^{-1} may be read "per second"; that is, 10^{18} wave crests of an X-ray pass a given point in space (such as a cavity in your teeth) per second. Other forms of electromagnetic radiation are, in order of increasing wavelength; ultraviolet light, ordinary (visible) light, infrared radiation, microwaves, and radio waves.

Moseley found that each element emitted a characteristic frequency, and that the many frequencies fitted into a pattern such that

$$\nu = A(Z - B)^2 \qquad (2\text{-}1)$$

where ν is the frequency of the X-rays emitted by a given element, and Z is the number of electrons which atoms of that element contain in the neutral state, A and B being constants which are the same for all elements. The atomic number Z is 1 for hydrogen, 2 for helium, 3 for lithium, and so on, up to 92 for uranium and higher numbers for the transuranium elements. Such atomic numbers are listed in Table 2-1;[12] in effect, they are serial numbers for the elements, in the order of increasing complexity.

2-7 Atomic Structure II: The Nucleus

If electrically neutral atoms contain one or more electrons in numbers corresponding to Z, then there must be a corresponding quantity of positive charge somewhere within the atom, because in the absence of any electrical or chemical stress all atoms are neutral. At first, it was believed that the positive charge resided in an all-pervasive microcosmic "jelly," in which the electrons were imbedded. That view was changed drastically in 1910, when Geiger and Marsden did experiments on the scattering of alpha particles (emitted by radium and by other naturally radioactive elements) by thin foils of heavy metals. These experiments showed that the fast-moving, positively charged and massive alpha particles did not always go right through a layer of gold, suffering only small deflections such as would be expected from passing through (or near) tenuous patches of positively charged "jelly," but instead sometimes suffered very large deflections, and in a few cases actually were turned back toward the source (see Figure 2-4). Ernest Rutherford, who had proposed the experiment, saw that this result could be interpreted in only one way: There must be a very massive and highly concentrated region within the atom, embodying most of the mass of the atom and all of the positive charge required to hold the electrons. Only collision with such a dense, massive core (which Rutherford called the **nucleus**) could send an alpha particle

What was the jelly made of?

---

[12] An arrangement of the elements in order of atomic numbers was published in 1912 when Moseley was only 26 years old. An enjoyable biography of this scientist is B. Jaffe, *Moseley and the Numbering of the Elements*, New York: Doubleday (1971), available in paperback.

bouncing backward or off at a large angle, as had been observed. In this way, the idea of the nuclear atom was born, and with it the concept of an *atomic number* as the number of unit positive charges held by the nucleus. We have seen how Moseley was able to produce unambiguous integral values for these atomic numbers, for a large number of elements, only a few years later.

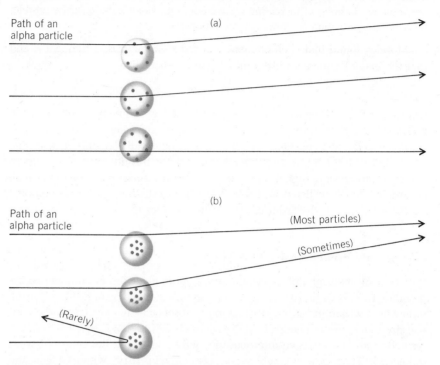

FIGURE 2-4 Predicted scattering patterns of alpha particles. Only small deflections would be expected if matter were composed of positive charges distributed rather uniformly throughout each atom, as pictured in (a). But if all the positive charge of an atom were concentrated in a minute nucleus (b), then sometimes a near miss or a direct hit with the highly charged nucleus would result in a very large deflection.

The factor A in Equation (2-1) is, among other things, what is required to convert the number $(Z - B)^2$ to the dimensions appropriate for frequency ($\sec^{-1}$). That is, A is a proportionality factor. On the other hand, B has a different origin and significance. Subtracted from the atomic number Z, it has the effect of making the X-ray frequencies smaller than they might otherwise be. As we shall see later in this chapter, radiation (including X-rays) is emitted by an atom when electrons move closer to the nucleus under the influence of its positive charge. The fact that the frequencies of the X-rays emitted are proportional to $(Z - B)^2$ and not to Z^2 alone implies that the electrons are not "feeling" the whole nuclear charge Z, but only part of it, $(Z - B)$. The quantity B is known as a *screening constant;* it reflects the fact that the negative charges of other electrons in the atom partly offset, or screen, the nuclear charge of $+Z$. We shall see some important chemical consequences of screening in Chapter 6.

Further scattering experiments soon revealed the approximate dimensions of atomic nuclei: about 10^{-13} cm, or 10^{-5} Å. Knowing the general size of electrons, it became evident that the inside of the atom is mostly empty space, since atoms are 1 or 2 Å in diameter. To scale, if the nucleus of a hydrogen atom be represented by a golf ball, then the single electron would be revolving about the golf ball at a distance of 20 miles! We cannot be sure that the electron is revolving about the nucleus, either; we only know its average distance away. The situation of the electrons within an oxygen atom has been likened to a few flies buzzing about in a cathedral—such is the picture to scale.

The nuclear matter itself is equally astonishing: From the known masses and dimensions, the density of nuclear matter is calculated to be about 1.6 billion tons per cubic inch. The nucleus of the simplest atom, an atom of hydrogen, bears one positive charge (equal but opposite to the charge on an electron) and is by far the smallest nucleus. It does not seem to be made up of anything else; like the electron, it appears to be an ultimate elementary particle. It is called the **proton**. Protons have a mass of 1.007825 on the scale of atomic weights, and a diameter somewhat less than that of the electron.

Protons are believed to be present in all nuclei, because they can be expelled from nuclei heavier than that of hydrogen during nuclear reactions, and because the nuclear charge is always an integral multiple of the charge on one proton. However, protons do not account for all of the mass of a nucleus of an element above hydrogen. From helium ($Z = 2$) to calcium ($Z = 20$), most nuclei have masses equal to approximately twice the collective masses of the protons they contain, and above calcium the nuclei all have masses which are more than twice the sum of the proton masses. So there must be other massive constituents in the nuclei of all atoms except those of ordinary hydrogen. At first it was thought that there might be a number of protons equal to approximately twice the atomic number, and that a number of electrons about equal to Z resided right within the nucleus too; this situation would bring the net positive charge down to Z, but would allow a mass which agreed with the facts. However, the proposal is untenable in view of the fact that electrons are too big to fit inside nuclei. Several lines of reasoning[13] point to the impossibility of electrons residing within the nucleus. There had to be some other explanation.

A puzzlement

The perplexity ceased when Chadwick discovered the **neutron** in 1932. Neutrons are elementary particles (of the third and last kind that chemists ordinarily need to be concerned about) with a mass of 1.008665 on the atomic weight scale, and with no electric charge. They ordinarily do not exist alone, but rather are tightly bound to protons within atomic nuclei. The bonding is not the same as that which bonds together atoms to form molecules of ordinary matter; nuclear bonding forces are much shorter in range and are over a thousand times as forceful within the nucleus itself. The special condi-

The puzzle solved

---

[13] In the first place, it is possible to calculate the radius an electron should have if it is assumed to be a particle obeying Coulomb's law of electrical forces (see Appendix III). This radius, 1.4×10^{-13} cm, is about the same size as the entire nucleus. Second, we shall shortly see that particles also behave like waves; in the problems at the end of this chapter, you will find what happens when you assume that the wavelength of an electron is short enough to fit into a nucleus.

tions that allow us to rearrange nuclear bonds and to exploit nuclear bonding energy will be taken up in Chapter 10; it is a fascinating and very important area of the science of matter and energy, but it had best wait until ordinary chemical bonding has been considered.

For our present purposes then, we may think of atoms being made up of electrons and nuclei, with the nuclei composed of Z protons plus Z (or more than Z) neutrons tightly bound together. The numbers of elementary particles are definite. All atoms of fluorine, for example, contain 9 electrons arranged around a nucleus composed of 9 protons and 10 neutrons. The atomic weight of the entire aggregate is 18.9984.

This total mass is less than the sum of the component parts, as we can see by adding them up:

$$
\begin{array}{lll}
\text{Nine protons:} & 9 \times 1.007825 = & 9.070425 \\
\text{Ten neutrons:} & 10 \times 1.008665 = & 10.08665 \\
\text{Nine electrons:} & 9 \times 0.000549 = & \underline{0.00494} \\
& & 19.16201 \\
\text{Mass of F atom:} & & \underline{18.9984} \\
\text{"Missing" mass:} & & 0.1636 \\
\end{array}
$$

For every atom of fluorine, this represents a *mass defect* of 0.1636 relative to its elementary-particle makeup, and a mass defect is observed for all other kinds of atoms as well. The energy, which, through Einstein's famous equation,

$$E = mc^2 \qquad (2\text{-}2)$$

(where m is mass, and c is the velocity of light) is equivalent to the missing mass, is the energy that would have to be supplied to dismantle a fluorine atom into its component parts. Therefore, this energy is called the *binding energy*. Since most of the mass of the atom is in its nucleus, most of the binding energy arises from the assembly of the nucleus. Changes in nuclear composition thus involve the uptake or release of large quantities of energy. The practical consequences of nuclear energy are well known to everyone; the situation is discussed further in Chapters 10 and 18.

A discussion of isotopes, at last

So far, the discussion in this chapter has carefully skirted around the subject of **isotopes**, but now the reader is in a position to understand them. Isotopes are atoms of different weight that have the same atomic number Z (and hence the same number of electrons and the same chemical behavior), but have different masses because of different numbers of neutrons in their nuclei. Many elements have isotopes, but not all do. The element fluorine was appropriate in the preceding paragraph because it is anisotopic; all its naturally occuring atoms have masses of 18.9984. Similarly, gold is anisotopic, and so are the common elements iodine and arsenic. Hydrogen and oxygen have stable isotopes, however, and so do most of the other elements. Natural hydrogen is comprised of two stable isotopes, of masses 1.007825 and 2.01410, and a third isotope of mass 3.016049, which is unstable and which decomposes by radioactive disintegration. These isotopes of hydrogen differ only in the number of neutrons in their nuclei: The lightest isotope, of mass number 1,[14] has a nucleus composed of a single proton and no neutrons; the second isotope, of mass number 2, has one proton and one neutron in its nucleus; and the third isotope has one proton and two neutrons in its nucleus. The lightest isotope makes up 99.985 percent of natural hydrogen, and the second isotope 0.015 percent, making for an average atomic weight of 1.00797. (The third isotope can be made only in a nuclear reactor, and possibly by action of cosmic rays; it does not constitute enough of natural hydrogen to measure.)

In much the same way, natural oxygen consists of three stable isotopes, of mass numbers 16, 17, and 18. It is composed mostly of oxygen-16, which constitutes 99.76 percent of the oxygen in air. In contrast, the element bromine is composed of almost equal proportions of two isotopes of mass numbers 79 and 81, so its atomic weight averages out to 79.91. For elements with isotopes, the atomic weights in Table 2-1 are averages for the natural distribution ratio of their isotopes. Usually this ratio is exceedingly constant, but in a few cases (which are indicated in the table), some variation of the ratio produces uncertainty in the atomic weight. For convenience and the saving of time, isotopes usually are designated by symbols in which the mass number properly appears at the *upper left* of the element's symbol, and the atomic number (if needed) at the lower left: note $^{16}_{8}O$, $^{17}_{8}O$, and $^{18}_{8}O$. This is by international agreement. The space to the upper right is left for the charge on an ion, as in $^{2}_{1}H^+$, and the lower right corner is reserved for a composition coefficient indicating atomic ratios, as in H_2O or H_2SO_4.

2-8 Atomic Structure III: The Arrangement of Electrons

Having learned about the general features of atomic structure, and having determined how many electrons are present in atoms of each chemical element, the only remaining question is, how are the electrons arranged or disposed about the nucleus? Are they in a random or haphazard configuration? Are all of equal potential energy, and hence all equally available for exchange or removal? Are the electrons moving, or do they occupy precisely fixed places? What really *is* the detailed structure of a given atom?

Where the chemistry originates from

All these questions were pondered during the period 1915 to 1955, and reliable answers have been obtained to most of them. The situation is not completely clear, because chemistry is not a finished science. Enough of the recent developments and the currently held views is summarized here to enable the student to appreciate the implications of atomic structure, and to use these in order to make predictions about chemical reactions.

The first comprehensive theory concerning the distribution of electrons in atoms was that of Niels Bohr, proposed in 1913. Bohr's fundamental postulates were:

---

[14] "Mass number" is a convenient way of indicating the total number of neutrons and protons in a nucleus. It is actually just the atomic weight or atomic mass rounded off to the nearest whole number. Mass number is frequently given the symbol A, just as atomic number is Z.

1. The Rutherford concept of a nuclear atom is valid (it was very new and controversial then).
2. Planck's theory about the quantization of energy applies. (In 1900, in order to explain radiation as a mode of transferring energy, Max Planck proposed that energy is atomistic and is radiated in the form of discrete chunks which he called quanta, the energy content of each quantum being such that $E = h\nu$, where E is the amount of energy; ν is the frequency of the associated radiation; and h (Planck's constant) is a universal constant of nature, having a value of 6.623×10^{-27} erg-sec.)
3. Parts of the classical theory of the motions of electrically charged objects are rejected as not applicable to objects of the tiny dimensions of atoms.
4. The electrons are assumed to revolve in circular orbits around the nucleus, in a sort of miniature solar system. (The orbits were later changed to elliptical ones, to account for more detailed observation of the light emitted by atoms; see Postulate 6.)
5. Only certain orbits are permitted and not others.

Bohr's criterion for an allowed orbit involved an arbitrarily introduced *quantization* condition: The circumferences of the circular orbits should be only integral multiples of the quantity h/mv, where h is Planck's constant; m is the electron's mass; and v is its velocity. In 1923, 10 years after Bohr's theory of the atom was evolved, Louis de Broglie showed that a moving particle could also be considered a wave, with wavelength h/mv. Thus, Bohr's rule for allowed orbits amounts to the condition that an integral number of **de Broglie wavelengths** must fit into the circumference of the orbit (Figure 2-5). Later we shall show that when wave motion is confined by some such boundary condition, quantized behavior is displayed even by ordinary-sized objects.

The wave must fit.

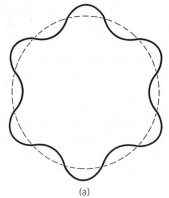 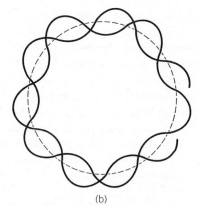

(a)   (b)

FIGURE 2-5 De Broglie waves for the circular orbits of a stationary state (a), and for a quantum mechanically excluded state where the waves are destroyed by interference (b). [Source: J. J. Lagowski, *The Chemical Bond*, Boston: Houghton-Mifflin (1966). Used by permission.]

6. Radiant energy (the light of characteristic frequency emitted by excited atoms in their emission spectra) is emitted or absorbed only when an electron jumps from one permitted orbit to another. (Putting excitation energy into the atom, by heat or by electrical means, or by absorption of radiation,

causes an electron to jump to an orbit farther away from the nucleus. Return of a displaced electron to an orbit closer to the nucleus causes emission of radiant energy such that the change in energy, ΔE, of the system is

$$\Delta E = h\nu$$

where ν again is the frequency of emitted radiation.)

7. In all other respects but these, classical physics is assumed to hold (that is, in other respects the laws of motion hold, and the law of electrostatic attraction remains valid).

Bohr's theory is still popular—

The electronic energies predicted by Bohr's model are in agreement with the energy differences (ΔE) implied by the radiation emitted by hydrogen. Such *quantitative verification* is an important clue that Bohr was on the right track. But though his model of the atom was reasonably helpful in describing a few other simple atoms, it failed irreparably when even moderately complex atoms were considered. The discovery of wave properties for atoms and electrons afterward made the theory obsolete, but for many years it still remained popular with students of general science and even in chemical circles. This is surprising, because the Bohr theory, useful as it was to spectroscopists, said nothing about molecules and did not lead to a theory of chemical bonding.

—but it does nothing for chemists.

The experiments of Davisson and Germer in 1926 showed that streams of electrons impinging on crystals of nickel oxide produced diffraction effects like those from beams of X-rays; this phenomenon could be explained only by attributing wavelike properties to electrons. Thus, the wave-particle duality that had been proposed three years earlier by de Broglie was *experimentally demonstrated*. In fact, the demonstration made it clear that all particles could be expected to have wavelike properties, in accord with de Broglie's general theory. Soon protons were found to act like waves too, and then neutrons were found to produce diffraction patterns when they struck crystals.[15] Today we make use of the wave properties of electrons in the electron microscope, where such waves replace the light waves of the ordinary optical microscope.

To understand the significance of this development, suppose we calculate the wavelength of an electron accelerated to a velocity of 5.9×10^9 cm/sec in an actual electron microscope, and compare this with the wavelength of a baseball pitched at maximum speed in a World Series game. For the electron

$$\lambda = \frac{h}{mv} = \frac{6.623 \times 10^{-27} \text{ erg-sec}}{9.1 \times 10^{-28} \text{ g} \times 5.9 \times 10^9 \text{ cm/sec}}$$

$$= \frac{6.623 \times 10^{-27}}{53.5 \times 10^{-19}}$$

$$= 0.12 \times 10^{-8} \text{ cm, or } 0.12 \text{ Å}$$

---

[15] A good account of these (and many other) aspects of the elaboration of atomic structure, with illuminating quotations from the original reports, is given in a paperback by J. J. Lagowski, *The Chemical Bond*, Boston: Houghton-Mifflin (1966), No. G-2 in the series *Classical Researches in General Chemistry*.

Here the wavelength of 0.12 Å is much shorter than the wavelength of visible light (the wavelength of yellow light is about 5900 Å), and so use of electrons enables us to see much smaller objects than with an optical microscope. An electron has a wavelength of appropriate size to interact with the objects we want to investigate. By contrast, the baseball having a regulation mass of about 140 g and traveling at about 25 m/sec has a corresponding wavelength

$$\lambda = \frac{h}{mv} = \frac{6.623 \times 10^{-27} \text{ erg-sec}}{140 \text{ g} \times 2.5 \times 10^3 \text{ cm/sec}}$$

$$= 1.9 \times 10^{-32} \text{ cm, or } 1.9 \times 10^{-24} \text{ Å}$$

This wavelength is so tiny as to be undetectable and utterly without effect as the batter tries to hit the ball.

So electrons may have wavelengths of about atomic dimensions, and therefore we must consider their wave properties rather than just think about them as hard spheres. Hence wave mechanics becomes important in considering atomic structure, and we shall take a look at its conclusions. Before doing so, however, let us rearrange the equation for wavelength slightly, and then look at its implications:

$$h = (mv)\lambda$$

Since h is a constant, the wavelength and the product mv (which is the particle's momentum—see Appendix III) are rigidly locked in a reciprocal relationship. Bohr and Heisenberg showed that it is impossible to measure the position of a tiny subatomic particle without jostling it so severely that a large uncertainty in its momentum results. The momentum may be measured accurately in another experiment, but only at the cost of a complementary uncertainty in the particle's position. It is impossible to know position *and* velocity with high accuracy at the same time. This is the Heisenberg **uncertainty principle**, and it means that we have to abandon all hope of delineating accurately the positions and velocities of electrons in atoms. The best we can do is to think in terms of very large numbers of atoms, and then to express only *probability* of finding an electron at a given place at a given time. The principles of statistics take over, just as they do in other areas where we must deal with probabilities (in life insurance, for example).

Maybe a hope you never had, but it is important

THE WAVE-MECHANICAL MODEL OF ATOMS We can now turn to the contribution of wave mechanics to the structure of atoms. We must consider wave motion in three dimensions. It is assumed that the amplitude of the electron wave, ψ, can be expressed as a function of the spatial coordinates x, y, and z around the nucleus. Because of the successes of the Bohr model in describing emission spectra in terms of stable states of atoms, we shall seek *stationary states* of the atom in which the functional relationship between ψ and the coordinates x, y, and z is unchanging. These stationary states are completely analogous to those of sound waves in an organ pipe or other wind instrument: They have a form that is described mathematically by a *wave equation*, a form that depends on certain physical constraints called boundary conditions. (We shall return shortly to a familiar example of boundary conditions on a wave.)

For those readers who are familiar with the notations of differential calculus,[16] we present the wave equation applicable to the de Broglie waves:

$$\frac{\partial^2 \psi}{\partial x^2} + \frac{\partial^2 \psi}{\partial y^2} + \frac{\partial^2 \psi}{\partial z^2} + \frac{8\pi^2 m}{h^2}(E - V)\psi = 0 \qquad (2\text{-}3)$$

where E is the energy of the electron; V is its potential energy (see Appendix III) in the electrical field of the nucleus; m is the mass of the electron; and h is Planck's constant, as before. This equation was proposed by Erwin Schrödinger.

The goal of the wave-mechanical approach is to find a mathematical function relating ψ to x, y, and z that obeys Equation (2-3). This is relatively easy for very simple situations, such as a single particle moving without changing forces on it (roughly speaking, like a cue ball rolling across a billiard table), and it is, fortunately, possible (though not easy) for the case of a single electron moving in the nuclear electrical field, that is, for the hydrogen atom. No one has been able to find a function for any system containing more than three particles (for example, a helium atom, a rock, or a daisy) relating ψ to x, y, and z.

The reason for this failure is easy to understand. The term V (the potential energy) in Equation (2-3) is made up, for atomic systems, of all of the electrical forces between particles. These in turn (Appendix III) depend on the distance between the particles. We may locate one of the particles (the nucleus) at the center of the coordinate system; the location of one electron with respect to the nucleus is thus built into the problem in terms of its x, y, and z coordinates. But if there are two or more electrons in the atom, we cannot know where they are with respect to each other (and thus, how much their like charges repel each other) unless we know the spatial distribution of their wave functions ψ—and *that* was the object of the exercise to begin with!

From this failure, it would seem that we are no better off than when we had only the Bohr model for atoms. However, the wave-mechanical model of one-electron atoms has been extended to describe complex atoms by treating each electron in the latter as though it were alone. Many of the properties of interest to chemists do not depend strongly on the repulsion of the other electrons in the atoms. The only justifications for this simplification of complex atoms are: (1) There is no satisfactory alternative at present; (2) the properties of many-electron atoms are in remarkable agreement with this treatment. The one-electron model for complex atoms, and its usefulness for describing their chemical properties, will be the subject of the rest of this chapter, and will recur throughout the book.

Let us now examine the wave-mechanical hydrogen atom that serves as such a fruitful model for all of chemistry. For much-needed convenience in solving it, Equation (2-3) is transformed into polar coordinates (see Appendix II), and boundary conditions are imposed in the form of integer constants derived from earlier considerations of quantum mechanics (which, in turn,

It's the best we have, and it works.

---

[16] Unfamiliarity with the language need not detract from appreciation of the result. Neither does it lessen the reliability of the conclusions, for the operations have been checked by thousands, and have been found correct.

ATOMS AND THE CHEMICAL ELEMENTS

The four quantum numbers are essential.

inherited them from earlier studies in spectroscopy). These integers are called **quantum numbers**. They are *extremely important* to the student who wants to understand atomic structure, and are used constantly to designate the configuration of electrons in atoms. Although quantum numbers have a highly theoretical sound, they are in fact an everyday phenomenon. Whenever wave motion is subject to boundary conditions, quantum numbers arise automatically.

For example, a string (such as a guitar string) will sound a single fundamental tone and a series of overtones when it is plucked. A reasonably adept player can isolate these overtones ("harmonics") by plucking the string, while a finger of his other hand is lightly placed at points halfway, one-third way, one-fourth way, and so on, down the string. The vibratory patterns of the string are:

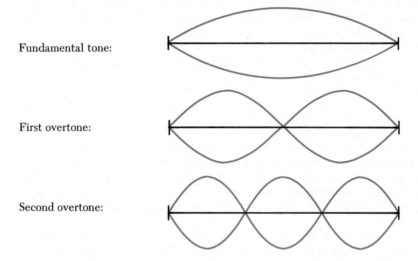

and so on.

Note that each of these patterns requires points of zero vibration ("nodes" at the ends of the string, and for the overtones, one or more additional nodes between the ends). The nodes are, of course, required by the fact that the string is firmly clamped at each end. "Clamping at the ends" is the "boundary condition." A string not so clamped will not sound. The string's motion is wavelike, and the quantum condition on a clamped vibrating string is that an integral number of half-wavelengths must fit between the clamped ends. That is, $\frac{1}{2}$ wavelength, 1 ($\frac{2}{2}$) wavelength, $\frac{3}{2}$, ..., $\frac{n}{2}$ wavelengths must fit exactly into the string length for a stable vibration. The number n in this series is a quantum number, which can take any positive integral value. For no other value of n is a stable wave possible. (Why?)

A guitar string is a one-dimensional object; for our purposes, at least, only its length matters. Similar considerations apply to vibrating two-dimensional objects (like drumheads), but since a two-dimensional system may support stable wave patterns in two directions (north-south and east-west), two quantum numbers are required.

An atom (or any other "real" object) is, in a full physical sense, four-dimensional; it has three spatial dimensions and persists through time. As a four-dimensional system of bounded waves (electrons), it exhibits four quantum numbers. These are:

n (the *principal* quantum number) can have only positive integers for values, starting with one. So $n = 1, 2, 3,$ or 4, and so on, with no set maximum (although $n = 7$ sets a practical limit to our interest at present). The principal quantum number may be thought of as a serial number applied to the successive *shells* or sets of electron patterns, starting nearest the nucleus and progressing outward (that is, toward higher potential energy).

l (the *orbital* quantum number) gives the orbital angular momentum (in a more figurative sense than in Bohr's theory) of a single electron in any designated shell. The permissible values of l are limited by the value of n. Specifically, l may have values of 0, 1, 2, 3, and so on, up to $(n-1)$, but not higher. So if $n = 3$, l can have values of 0, 1, and 2, but *only* these values.

m_l (the *magnetic* quantum number) measures the component of the total angular momentum of an electron, projected in a fixed direction. As such, it represents certain subdivisions of the orbitals and derives its limits from l. Specifically, m_l has values which range from $-l$ through zero to $+l$, so that if $l = 2$, $m_l = -2, -1, 0, +1,$ and $+2$.

m_s (the *spin* quantum number) also has to do with magnetic properties, but represents only the *spin* angular momentum as the electron spins (we think) on its axis. For an electron, this quantized spin momentum is limited to a single magnitude, although it may have different values for other elementary particles. Since the spin may be in either the left or right direction, but only one of these, m_s has a value of $+\frac{1}{2}$ or $-\frac{1}{2}$, and nothing else. Its value is not dependent on values of n, l, or m.

With the use of these quantum numbers as constants, solutions to the Schrödinger equation are obtained which are called *wave-function orbitals*, or just **orbitals** for short. The results can be shown graphically by plotting the radial distribution of probability of finding an electron in a given small region of space; that probability is proportional to ψ^2. Alternatively, the solutions can be transformed back into the familiar Cartesian (rectilinear *xyz*) coordinates, and the results can be shown as figures of revolution about the *x*, *y*, and *z* axes. Such figures denote the regions of highest probability for finding an electron resident there.[17] Figure 2-6 shows the most common of these figures of revolution, and in popular usage these geometric shapes themselves are spoken of as the atomic orbitals, rather than the mathematical solutions from which the shapes were derived.

It is seen from Figure 2-6 that the shapes of atomic orbitals differ markedly, depending on the values for the quantum number l. When l is zero, there is uniform angular probability for containing an electron, so the orbital is pictured always as a sphere. This is true no matter in what shell we find the sphere; n may be 1, or 2, or any higher integer, and still the $l = 0$ orbital

---

[17] The orbital is there whether any electrons are in it or not. That is, wave mechanics gives us the regions where electrons are likely to be found, if there are any, but these regions are the same whether empty or filled.

is a sphere. The probability varies radially, as we move outward from the nucleus in any direction, but no particular direction is preferred by the electron over any other. Such spherical orbitals, corresponding to $l = 0$, are always called s orbitals. The $1s$ orbital is smaller[18] than the $2s$ orbital, and the $2s$ is smaller than the $3s$, and so on.

Remember s is spherical.

When l has a value of 1 (again, no matter whether n is 2, 3, 4, or more), the atomic orbitals are *three* in number, for when $l = 1$ we note three values for m_l, namely, -1, 0, and $+1$. These $l = 1$ orbitals are called p orbitals and are readily recognized and remembered because each one has two lobes extending in opposite directions from the nucleus. The three p orbitals are at right angles to one another, and so are designated conveniently as p_x, p_y, and p_z orbitals, as is shown in Figure 2-6. When l has a value of 2, the orbitals are *five* in number, corresponding to all the possible values of m_l, namely, -2, -1, 0, $+1$, and $+2$. These are called d orbitals[19] and are shown at the bottom of Figure 2-6. Notice that each d orbital has *four* lobes, except the d_{z^2}. A sense of symmetry would lead us to expect six possible arrangements; a set of three interaxial orbitals (d_{xy}, d_{xz}, and d_{yz}) and a set of three axial orbitals (directed along each of the three pairs of axes). However, there are only five permissible values of m_l, as given above, and hence only five d orbitals are possible. Hence the d_{z^2} orbital is really a combination of two others (in a mathematical sense) and is distinctive in shape (the "doughnut and dumbbell" orbital).

p orbitals point perpendicularly.

d orbitals are doubly directed.

We could go on and depict the seven f orbitals that correspond to $l = 3$, with $m_l = -3, -2, -1, 0, +1, +2,$ and $+3$. However, these are still more complicated; some of them have eight lobes, and all are more difficult to envision than what is depicted in Figure 2-6. Furthermore, f orbitals are seldom used in chemical bonding, and so are of lesser importance to our primary task of inquiring how matter is put together. For these reasons, we shall concentrate on s, p, and d orbitals for the rest of this chapter and the next.[20]

Each of the orbitals, of any type or designation, may contain two electrons, but never more than two. This comes about from the limitation on the values of m_s given above and from the requirement (see below) that no two electrons in the same atom may have the same set of values for all four quantum numbers. For any orbital, then, after we have fixed n, l, and m_l, the orbital may contain a maximum of two electrons, which may be thought of (in particle imagery) as circulating through the orbital and being found most often in regions of maximum probability within that orbital. The orbital might be empty, or contain only one electron, but never more than two electrons. When

---

[18] An orbital does not really have any surface, and hence has no absolute limit on size; the probability simply approaches zero. In order to put an arbitrary limit on size, for purely practical reasons, we depict a volume which encloses 95 percent of the probability that an electron will be there.

[19] The designations s, p, d, and f, come from old-time classification of spectral lines as being in the *s*harp, *p*rincipal, *d*iffuse, and *f*undamental series.

[20] For a representation of the shapes of f orbitals, see the *Journal of Chemical Education*, vol. **41**, pp. 354, 358 (July 1964).

there *are* two electrons, they must have opposite spins (corresponding to values of m_s of $-\frac{1}{2}$ and $+\frac{1}{2}$) and are said to be paired. An unpaired electron exerts a magnetic force to be expected of a moving electric charge, but paired electrons exert equal and opposite magnetic forces, which cancel out. Hence an atom or a molecule containing one or more unpaired electrons can be recognized by the associated magnetic field.

With the numbers and kinds of atomic orbitals in mind, and with the interrelation of the quantum numbers also firmly in mind, we might begin the intellectual exercise of *designing* chemical elements, as though none were already at hand. That is, we could start with the simplest atom, of atomic number 1, and proceed to build up atoms that are more and more complicated, designating the distribution of electrons in them as we go. Only two points need to be settled as we proceed with this exercise: In what order are the *s*, *p*, and *d* orbitals filled? And if we have a set of equivalent orbitals, how shall successive electrons be placed in them?

The first point probably could not have been settled in advance by mere speculation, but examination of the properties of the actual natural elements makes it very clear. The distribution takes place not in the straightforward *s*, *p*, *d* sequence in any given shell, but in a somewhat overlapping sequence which goes 1*s*; 2*s*, 2*p*; 3*s*, 3*p*; 4*s*, 3*d*, 4*p*; 5*s*, 4*d*, 5*p*; 6*s*, 4*f*, 5*d*, 6*p*; 7*s*, 5*f*, and so on. While the student should be aware of the fact that the *f* orbitals are buried down in the interior of the atom, so to speak, there is no need to memorize the sequence at this time because it is obvious from the Periodic Table, soon to be considered.

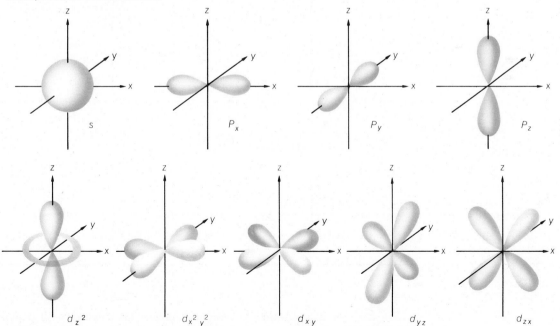

FIGURE 2-6 Representations of the distribution of electronic density in *s*, *p*, and *d* atomic orbitals. [*Source:* J. J. Lagowski, *The Chemical Bond*, Boston: Houghton-Mifflin, p. 175 (1966). Used by permission.]

The second point is covered by the rule of maximum multiplicity, which states that electrons being placed in a set of equivalent orbitals will go into successive orbitals *singly* until each orbital of the set has one electron in it. Then (and only then) additional electrons are paired with the previously unpaired ones. Hence a set of three 2p orbitals, for example, will receive single electrons in all three before any pairing of electrons occurs.

TABLE 2-2
Application of Principles of Atomic Structure to a Table of Electron Configurations

n	l	m_l	name of orbital	m_s	number of electrons		total electrons in shell
1	0	0	s	$-\frac{1}{2},+\frac{1}{2}$	2		2
2	0	0	s	"	2		
	1	-1		"	2		
		0	p	"	2	6	
		$+1$		"	2		8
3	0	0	s	"	2		
	1	-1		"	2		
		0	p	"	2	6	
		$+1$		"	2		
	2	-2		"	2		
		-1		"	2		
		0	d	"	2	10	
		$+1$		"	2		
		$+2$		"	2		18
4	0	0	s	"	2		
	1	$-1,0,+1$	p	"	6		
	2	$-2,-1,0,+1,+2$	d	"	10		
	3	-3		"	2		
		-2		"	2		
		-1		"	2		
		0	f	"	2	14	
		$+1$		"	2		
		$+2$		"	2		
		$+3$		"	2		32

We are now in a position to design a universe, as it were, with pencil and paper. We remember that no two bodies can occupy the same space at the same time, and expect that no two electrons can occupy exactly the same place within an atom at the same time. That is, no two electrons can have all four quantum numbers exactly the same;[21] each electron within an atom is unique in this respect. So when we start with $n = 1$ (that is, the first shell), we are limited by the rules governing quantum numbers to $l = 0$, $m_l = 0$, and $m_s = -\frac{1}{2}$ and $+\frac{1}{2}$. Hence there can be only two electrons in the $1s$ orbital and, therefore, in the entire shell. That ends the first sequence of elements: $Z = 1$ (hydrogen, 1 electron) and $Z = 2$ (helium, 2 electrons). We then begin the second shell, with $n = 2$ and hence $l = 1$ and $l = 0$. The actual placement of electrons in this **aufbau** (building-up) process is best shown in a chart (Table 2-2). The process could go on further than is shown in Table 2-2, and indeed does, in nature, but the general pattern is revealed in the development of the four shells shown there. Notice that the shells get progressively larger and more complicated, and that there is a repetition of structural elements in the successive shells. The total number of electrons accommodated in the shells increases in the order 2, 8, 18, and 32, a sequence which may sound odd to ears accustomed to a decimal system, but whose mechanics are given by the simple formula $2n^2$ (where n again is the principal quantum number).

Simplicity again!

2-9 Atomic Structure IV: The Periodic System

In the preceding section, we found considerable fundamental simplicity, harmony, and symmetry in the ultimate structure of matter, and now we turn to further evidence of it. If we take the successive lines of Table 2-2, and match the progressive filling of orbitals (with 1, 2, 3, 4 electrons) with the actual chemical elements having corresponding progressively increasing atomic numbers (see Table 2-1), with due regard for the order of filling orbitals noted in the previous section, we arrive at a chart or table of the elements in the order of increasing complexity of atomic structure. While it is enlightening for the student to do this, we recognize that it already has been done for us, in the form of the Periodic System which Dmitri Ivanovich Mendeleef devised more than a hundred years ago.[22] Nothing was known about the electronic structure of atoms a century ago, of course, and Mendeleef's Periodic System made no assumptions about such structure. (Therein lies its strength, actually, for it stands firmly on the rock of universal and timeless observation, and has outlasted many drastic changes in our views of the structure of matter.) Nevertheless, it must agree with any independent but internally consistent theory of atomic structure, or else there is something wrong with our picture. Chemistry is now at the stage where there are no unresolved differences, and so we may look to the Periodic System as an embodiment of the principles of electronic configuration covered in this chapter.

Understanding wave functions

---

[21] This was first stated by Wolfgang Pauli in 1925 and is generally known as the Pauli exclusion principle.

[22] See Mendeleef, *Annalen der Chemie und Pharmacie*, suppl. vol. 8, p. 151 (1872) for his Table II, which embraced 66 elements. Mendeleef's earliest announcement of the Periodic Law came in 1869, and a world-wide centennial was observed in 1969.

Figure 2-7 shows a current form of Mendeleef's table, embracing all of the known chemical elements, both naturally occuring and synthetic. The elements are arranged in the order of increasing atomic number. This is almost exactly the order of increasing atomic weight. Mendeleef's belief concerning the chemical properties of an element being dependent upon its *weight* is almost exactly correct today, although we now recognize that nuclear and electronic structure more closely govern the behavior of the elements; atomic weight is now seen to be a *consequence* of structure, rather than a *governing principle*.

The relation of Figure 2-7 to Table 2-2 is apparent at once: The Periodic Table consists of sequences of 2, 8, 18, and 32 elements, all arranged in the order of increasing complexity of electron configuration. In order to avoid a table of awkward length, the longest sequences shown are those of 18 elements; in the last two sequences, there are inserts, which are shown in detail at the bottom of the chart. The horizontal sequences Mendeleef named **periods**, and we note that seven periods embrace all the known elements. The vertical columns were named **groups**, and these are indicated by Roman numerals at the tops of the columns.

Periodic Table

	I	II										III	IV	V	VI	VII	0
1	1 H																2 He (2)
2	3 Li	4 Be										5 B	6 C	7 N	8 O	9 F	10 Ne (2,8)
3	11 Na	12 Mg										13 Al	14 Si	15 P	16 S	17 Cl	18 Ar (2,8,8)
4	19 K	20 Ca	21 Sc	22 Ti	23 V	24 Cr	25 Mn	26 Fe	27 Co	28 Ni	29 Cu	30 Zn	31 Ga	32 Ge	33 As	34 Se	35 Br	36 Kr (2,8,18,8)
5	37 Rb	38 Sr	39 Y	40 Zr	41 Cb	42 Mo	43 Tc	44 Ru	45 Rh	46 Pd	47 Ag	48 Cd	49 In	50 Sn	51 Sb	52 Te	53 I	54 Xe (2,8,18, 18,8)
6	55 Cs	56 Ba	57 La *	72 Hf	73 Ta	74 W	75 Re	76 Os	77 Ir	78 Pt	79 Au	80 Hg	81 Tl	82 Pb	83 Bi	84 Po	85 At	86 Rn (2,8,18, 32,18,8)
7	87 Fr	88 Ra	89 Ac †															

* Rare earth metals	58 Ce	59 Pr	60 Nd	61 Pm	62 Sm	63 Eu	64 Gd	65 Tb	66 Dy	67 Ho	68 Er	69 Tm	70 Yb	71 Lu

† Uranium metals	90 Th	91 Pa	92 U	93 Np	94 Pu	95 Am	96 Cm	97 Bk	98 Cf	99 E	100 Fm	101 Mv	102 No	103 Lw

FIGURE 2-7 Current form of Mendeleef's Periodic Table, including all known chemical elements. The Roman numerals across the top label the *groups* (the vertical columns of elements), and the numbers at the far left label the *periods* (the horizontal rows of elements).

	s						Periodic Table						p					
1	1 H																	2 He (2)
2	3 Li	4 Be											5 B	6 C	7 N	8 O	9 F	10 Ne (2,8)
3	11 Na	12 Mg				d							13 Al	14 Si	15 P	16 S	17 Cl	18 A (2,8,8)
4	19 K	20 Ca	21 Sc	22 Ti	23 V	24 Cr	25 Mn	26 Fe	27 Co	28 Ni	29 Cu	30 Zn	31 Ga	32 Ge	33 As	34 Se	35 Br	36 Kr (2,8,18,8)
5	37 Rb	38 Sr	39 Y	40 Zr	41 Cb	42 Mo	43 Tc	44 Ru	45 Rh	46 Pd	47 Ag	48 Cd	49 In	50 Sn	51 Sb	52 Te	53 I	54 Xe (2,8,18,18,8)
6	55 Cs	56 Ba	57 La *	72 Hf	73 Ta	74 W	75 Re	76 Os	77 Ir	78 Pt	79 Au	80 Hg	81 Tl	82 Pb	83 Bi	84 Po	85 At	86 Rn (2,8,18,32,18,8)
7	87 Fr	88 Ra	89 Ac * *															

* Rare earth metals	58 Ce	59 Pr	60 Nd	61 Pm	62 Sm	63 Eu	64 Gd	65 Tb	66 Dy	67 Ho	68 Er	69 Tm	70 Yb	71 Lu	**4f**

* * Uranium metals	90 Th	91 Pa	92 U	93 Np	94 Pu	95 Am	96 Cm	97 Bk	98 Cf	99 E	100 Fm	101 Mv	102 No	103 Lw	**5f**

FIGURE 2-8 The Periodic Table, showing blocks corresponding to the filling of s, p, d, or f orbitals.

After Mendeleef's ideas were accepted, the lightest elements (hydrogen through chlorine) received the lion's share of attention when people began to ponder about the basis of the Periodic System, and these became known as the *representative elements*. They occur in the two short periods of eight elements each. The three sequences of 10 elements each which live in the middle of the longer periods, between Groups II and III, became known as *transition elements,* and, since they are all metals, are still called **transition metals**. In the two very long periods, of 32 elements each, the two inserts of 14 elements each are printed at the bottom of the table. These are given their most common names in Figure 2-7, but are sometimes known as the **lanthanide** and the *actinide* series, from the names of their immediate predecessors, lanthanum (57) and actinium (89).

In order to make the correspondence between the table and modern features of atomic structure completely clear, we may mark off parts of the Periodic Table as shown in Figure 2-8. We see that Groups I and II correspond to placement of one and two electrons, respectively, in the 1s, 2s, 3s, 4s, 5s, 6s, and 7s orbitals, while Groups III, IV, V, VI, VII, and 0 correspond to progressive filling of the three p orbitals in each of the periods. The transition elements are those in which the heretofore empty d orbitals are filled in progression, and hence we have a 3d series of transition metals, a 4d series,

No conflict between modern theory and the century-old table

A handy Periodic Table is inside the front cover.

and a 5d series, which occur in the 4th, 5th, and 6th periods. Lastly, the 14 rare earth metals (lanthanides) correspond to progressive filling of the seven 4f orbitals, and the uranium metals (actinides) correspond to the filling of the 5f orbitals. It is seen that in Period 4, the 4s orbitals fill first, then the 3d orbitals, and last the 4p orbitals. Further down, in Period 6 the 6s orbitals fill first; the 5d series begins with La; and then the 4f series (Ce-Lu) intervenes. Only after the 4f orbitals are full do the 5d elements resume, and then the 6p elements last. Hence the order of filling orbitals, in the ascending order of their potential energy in terms of their distance from the nucleus, is apparent from the Periodic System and need not be memorized in the abstract. As for the Periodic Table of elements itself, undoubtedly it is a help to be able to visualize or reproduce the table at will, but it is not essential at this stage to commit it to memory. Constant usage of the table will gradually make it familiar, and to facilitate this you should have a copy before you whenever you study.

We have seen the close interrelationship between Mendeleef's historic Periodic System and the modern theory of atomic structure. The greatest contribution of the Periodic System is still to come, however; it summarizes the chemical behavior of all the elements by organizing them in groups and periods, allowing us to predict the general combining characteristics of an element as it bonds itself to others to make the compound substances that make up almost all matter. The modes of such combination are the subject of Chapter 3, but while we are considering the utility of the Periodic Table, it is just as well to look forward a bit and to classify the elements into general categories in accordance with their positions in the table. To do this, we can make use of an abbreviated form of the Periodic Table shown in Figure 2-9.

Periodic Table

1	1 H																	2 He
2	3 Li	4 Be											5 B	6 C	7 N	8 O	9 F	10 Ne
3	11 Na	12 Mg											13 Al	14 Si	15 P	16 S	17 Cl	18 Ar
4	19 K	20 Ca	21 Sc	22 Ti	23 V	24 Cr	25 Mn	26 Fe	27 Co	28 Ni	29 Cu	30 Zn	31 Ga	32 Ge	33 As	34 Se	35 Br	36 Kr
5	37 Rb	38 Sr	39 Y	40 Zr	41 Nb	42 Mo	43 Tc	44 Ru	45 Rh	46 Pd	47 Ag	48 Cd	49 In	50 Sn	51 Sb	52 Te	53 I	54 Xe
6	55 Cs	56 Ba	57 La*	72 Hf	73 Ta	74 W	75 Re	76 Os	77 Ir	78 Pt	79 Au	80 Hg	81 Tl	82 Pb	83 Bi	84 Po	85 At	86 Rn
7	87 Fr	88 Ra	89 Ac															

*Lanthanide series

FIGURE 2-9 Abbreviated form of the Periodic Table, showing placement of metals, metalloids, and nonmetals in the Periodic System. Metalloids are identified by shaded squares.

The main points to be noted are:

1. Elements at the left and in the center of the Periodic Table are *metallic* in their elementary form; examples are the well-known elements iron, copper, and mercury. Each has its own distinctive physical properties such as melting point, hardness, and so on. In terms of atomic structure, the metals all have plenty of vacant orbitals, especially p orbitals. They also can lose electrons from their orbitals under sufficiently potent electrical stress, or by action of an habitual electron grabber among the other elements. Loss of one or more electrons necessarily leaves a positive ion,[23] so the metals are often referred to as *electropositive* elements. The larger a metal atom is, the more remote its outer electrons are from the nucleus, and the more screened they are from the positive charge of the nucleus by completed shells of electrons. (For a discussion of screening, see Section 2-7). Hence the electropositive characteristic generally increases *to the left and downward* in the Periodic Table.[24]

Metals

2. The elements in the upper right-hand corner of the table are of the opposite kind. For reasons that we shall develop in Chapter 6, they acquire electrons readily to fill up their almost-complete orbitals, and so form negative ions. These are the *electronegative* elements; examples are the familiar elements oxygen and chlorine. Their electronegativity falls off among the heavier elements of the right-hand group, as shown, for the reasons given in Point 1. It is seen that only one-fifth of the elements are nonmetals, but these few are very important.

Nonmetals

3. In between the two major classifications, there is a diagonal region in the table where the elements are neither nonmetallic nor decidedly metallic. This is the region of the **metalloids**, elements that look something like metals, but seldom form positive ions. The well-known semiconductors silicon and germanium fall in this category. The metalloids behave chemically both like metals and like nonmetals.

Metalloids

We shall return for a much more detailed look at chemical periodicity in Chapters 5 and 6 after we have considered the formation of chemical compounds in Chapter 3.

2-10 Atomic Structure V: Energy Levels and Atomic Spectra

There is still another way of depicting the electronic structure of atoms, and that is by drawing diagrams of potential energy. The average distance of an electron from the nucleus is considered, and according to the inverse square law of electrostatic attraction,[25] this gives a value for the potential energy (just as a book held some distance away from the earth's surface has a measurable potential—gravitational—energy, and will move toward the earth unless

---

[23] An ion is an atom that has acquired an electric charge, either positive or negative, by loss or gain of electrons. See earlier discussion of electrons in atoms, at the beginning of this chapter.

[24] We shall meet some exceptions later, all of them in the very center of the table.

[25] See Appendix III.

restrained). Electrons in various kinds of orbitals will then be seen to have different potential energies, and we may indicate these relative energies on a vertical scale, as in Figure 2-10. It is but a short step from this to a display of all the *s, p, d,* and *f* orbital energy levels, as shown in Figure 2-11. Having such an energy-level diagram, we can then indicate the electron population of the various levels by putting on the lines the required number of dots or crosses to represent the total number of electrons in that atom (or ion). Better still, we can use little arrows to indicate the two kinds of spin, and so make clear the pairing of electrons in the various orbitals. Using this convention, we may designate the electronic configuration of a lithium atom (Z = 3) by

$$
\begin{array}{ll}
3p & \underline{}\ \underline{}\ \underline{} \\
3s & \underline{} \\
2p & \underline{}\ \underline{}\ \underline{} \\
2s & \underline{\uparrow} \\
1s & \underline{\uparrow\downarrow}
\end{array}
$$

whereas a nitrogen atom would have the arrangement

$$
\begin{array}{ll}
3p & \underline{}\ \underline{}\ \underline{} \\
3s & \underline{} \\
2p & \underline{\uparrow}\ \underline{\uparrow}\ \underline{\uparrow} \\
2s & \underline{\uparrow\downarrow} \\
1s & \underline{\uparrow\downarrow}
\end{array}
$$

and the configuration of a chlorine atom would be indicated by

$$
\begin{array}{ll}
3p & \underline{\uparrow\downarrow}\ \underline{\uparrow\downarrow}\ \underline{\uparrow} \\
3s & \underline{\uparrow\downarrow} \\
2p & \underline{\uparrow\downarrow}\ \underline{\uparrow\downarrow}\ \underline{\uparrow\downarrow} \\
2s & \underline{\uparrow\downarrow} \\
1s & \underline{\uparrow\downarrow}
\end{array}
$$

Two shorthand schemes

This scheme is more informative than the designation $1s^2\,2s^2\,2p^6\,3s^2\,3p^5$ for chlorine, although either may be used. Obviously, the configuration of **ions** may also be shown in the same way: A chloride ion, Cl^-, would have one more electron in the last $3p$ level.

The order of energy levels in an atom depends on the value of the nuclear charge. For example, the electronic configuration of a calcium atom (Z = 20) is $1s^2\,2s^2\,2p^6\,3s^3\,3p^6\,4s^2$, indicating that for this species the $4s$ orbitals are lower in energy than the $3d$, as is indicated by Figure 2-10. On the other hand, the removal of two electrons from a titanium atom produces an ion Ti^{2+} with the same number of electrons (20) as a calcium atom, but with Z = 22. Spectroscopic observations on Ti^{2+} indicate that its electronic structure is $1s^2\,2s^2\,2p^6\,3s^2\,3p^6\,3d^2$, showing that for Z = 22, the $3d$ orbitals are lower in energy than the $4s$.

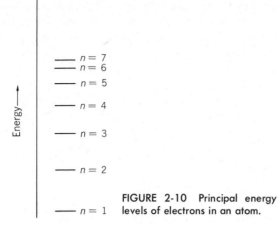

FIGURE 2-10 Principal energy levels of electrons in an atom.

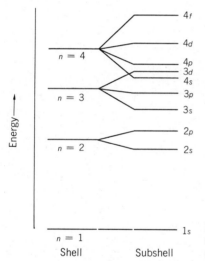

FIGURE 2-11 Energy levels of electrons in the principal shells and subshells in an atom.

Energy-level diagrams are also used to show the internal electron transitions involved in the emission of line spectra. Here an electron in its customary energy level is raised temporarily to a higher energy level by an input of energy (but not taken away entirely, as it would be in the formation of a positive ion), and then returns to a lower energy level by emitting an appropriate quantum of energy. The atom in its normal condition is said to be in its *ground state,* and it is *excited* by input energy (as for instance, in a high-temperature electric arc), resulting in *radiation* of energy at a frequency corresponding to the difference in energy $\triangle E$ between the excited-state level and the ground level. As an example, consider Figure 2-12, where the various principal energy levels for a hydrogen atom are shown, together with the actual energy values in ergs. An electron raised to any of the higher energy levels may return to the ground state by radiating a quantum of energy at the indicated wavelength. Depending on the degree of excitation, hydrogen atoms are seen to be capable of emitting light at many wavelengths. Furthermore, an electron might not return all the way to the ground state in one jump; if it returned only to the $n = 2$ level, light of other wavelengths would be emitted as shown. In each case, the frequency is precisely defined by $\triangle E = h\nu$, and so the emission spectrum of an isolated hydrogen atom consists of a pattern of bright sharp lines whose wavelengths are indicated in Figure 2-12. (If the atom were subjected to strong external electric or magnetic fields, or surrounded closely by other atoms which exert their fields, the lines would be broadened and shifted.)

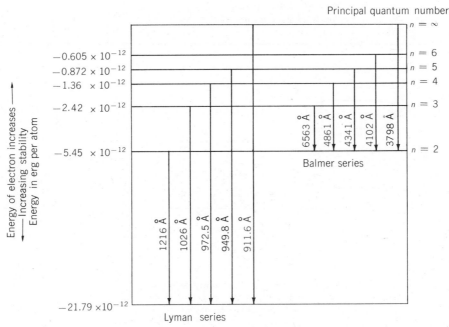

FIGURE 2-12 The energy-level diagram of the hydrogen atom. Wavelengths are in angstrom units. The energy is expressed as the work necessary to separate the electron from the hydrogen atom against the attractive force of the nucleus. Zero energy then means that no work is required to remove the electron when the distance between the proton and the electron is infinite; that is, when $n = \infty$. [Source: A. Turk, H. Meislich, F. Brescia, and A. Arents, *Introduction to Chemistry*, New York: Academic Press (1968). Used by permission.]

What has been described for a hydrogen atom may be extended to an atom of any other element by using an energy-level diagram with spaces and levels appropriate to that atom, and the result would be a diagram of electron transitions, which would describe and explain the arc emission spectrum of that element. Since the number of electrons and their respective spacings on an energy-level diagram are both uniquely characteristic of that element, it follows that each element emits a characteristic bright-line spectrum when heated very hot. We may use these characteristic patterns of varicolored lines to recognize the various elements in any sample of matter, even though the sample may be far away in the sun or stars. Hence the recording and interpretation of emission spectra are a valuable method of analysis, capable of revealing the elementary composition of any substance. The technique is relatively simple: A tiny sample of the substance is vaporized in an electric arc, and the emitted light is focused on the slit of a spectroscope or spectrograph[26] (see Figure 2-13). The light, which passes through the slit, is directed as a parallel beam into a prism or a diffraction grating, and there is split into its component colors. A subsequent optical system then allows an image of the slit to be seen or photographed in each of the component colors (that is, a bright-line spectrum is observed or recorded). People accustomed to doing spectrographic analyses can recognize the spectra of various elements at a glance, but even a novice can measure the wavelength of a selected line and look it up in the wavelength tables to see what element emitted it. For the relationship between color and wavelength, see Figure 2-14. Furthermore, relative *quantities* of elements present in the sample may be estimated from intensities of the lines relative to those observed at known concentrations of the element. So atomic spectra not only have helped us to understand and to map out the structures of atoms, but also have provided us with a method of qualitative and quantitative analysis. Among the applications of this technique is the detection of heavy metal pollutants such as lead and mercury in soil, water, and biological samples.

How to analyze the entire universe

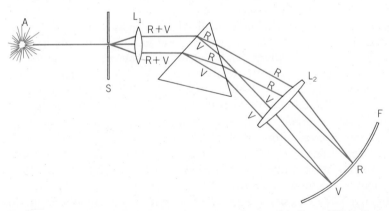

FIGURE 2-13 Essential features of a spectrograph. An electric arc (A) excites atoms of the sample; these atoms emit their characteristic spectrum in returning to the ground state. For purposes of this illustration, it is assumed that two lines are emitted, one red (R) and the other violet (V). The mixture of R + V is passed through a slit (S) and formed into a parallel beam by lens L_1. Prism P separates R and V into two diverging beams, and lens L_2 focuses these onto the photographic film F. R and V fall on the film in the form of colored line images of the slit; this is the origin of the term "line" for light of a single wavelength.

---

[26] If we observe the pattern of colored lines visually, the instrument is a spectroscope; if the spectrum is photographed, it is a spectrograph; if the spectrum is detected with a photoelectric detector such as a photomultiplier tube, the instrument is a spectrophotometer.

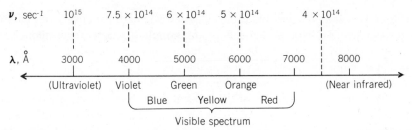

FIGURE 2-14 A portion of the electromagnetic spectrum, showing the relationship of wavelength λ and frequency ν to color in the visible range.

2-11 Summary

In this chapter, we have begun our study of matter by describing the atoms of which all matter is composed. The various kinds of atoms are identifiable by their common physical and chemical properties. All atoms that have the same nuclear charge (see below) and thus the same chemical properties constitute a chemical element. There are at present 89 elements occurring naturally on the earth; another 15 have been synthesized; and there are more appearing as the years go by.

Each kind of atom has a similar structure, with a nucleus containing nearly all of the mass of the atom, surrounded by a diffuse region of rapidly moving, negatively charged electrons. The nucleus is composed of positively charged elementary particles (protons) and neutral ones (neutrons). The number of protons in the nucleus is called the atomic number (Z) and defines the chemical identity (and name) of the element.

Because atoms are exceedingly small, ordinary mass units such as grams or pounds are too large for convenient use, and so a relative scale has been established on which the proton and neutron each have masses of about one unit. The mass of an atom on this relative scale is known as its *atomic weight*. A *gram-atomic weight* of an element consists of an agreed-on large number (6.023×10^{23}) of atoms of that element; the number is that required to make up a total mass, in grams, numerically equal to the average atomic weight of the atoms. Because atoms of the same element may contain different numbers of neutrons, and thus have different atomic weights, the observed atomic weights are averages of the various isotopic weights represented in the element as it is found in nature.

Although X-ray emission frequencies and (with a few exceptions) atomic weights increase in the order of the atomic number of the elements, most other properties vary in a cyclic, or periodic, fashion with increasing Z. The explanation for this fact lies in the recurring shell structure of the electrons. The first successful theory of the electronic shell structure was devised by Bohr, who envisaged the electrons traveling in fixed orbits, with energies and paths around the nucleus that are only a small "allowed" number of the continuous range of imaginable possibilities.

This purely particle-oriented model did not take into account the wave nature of matter. A mathematical theory devised by Schrödinger, based on this wave character (which is very pronounced for small, light particles like electrons), was as successful as the Bohr model for simple atoms, and more easily extended to complex atoms containing many electrons. The Schrödinger wave model of the atom depicts the wavelike electrons occupying regions in space (orbitals) around the nucleus. These electrons have fixed energies and are characterized by the values of four quantum numbers (n, l, m_l, and m_s). These numbers determine, respectively, the shell number, the shape and orientation of the orbital, and the spin of the electron.

To preserve a particulate view of matter, the intensity of the wave function at a point in space may be thought of as related to the probability of finding the electron at that point, or, equivalently, to the fraction of its time that the electron spends near that point.

The electronic structure of atoms may (in imagination) be built up by placing electrons, one of each spin, in these orbitals in order of increasing energy. The resulting electronic structures account very nicely for the observed periodicity of chemical properties of the elements with increasing Z. In detail, the elements may be organized into a Periodic Table, in which the horizontal rows (periods) correspond to the partial or complete exhaustion of the possible values of l, m_l, and m_s for a given value of n (shell number), and the vertical columns (groups) consist of elements with analogous electronic structures in successively higher shell numbers.

This general theory of electronic structure is verified by atomic spectra. When an electron moves from a higher-energy to a lower-energy orbital, light is emitted whose energy is equal to the decrease in energy of the atom. When the electrons in large numbers of atoms undergo transitions among the various energy levels allowed by the quantum numbers, a series of light emissions occurs, which can be sorted in order of frequency or wavelength by a spectrograph, forming an emission spectrum. Since emission spectra are completely characteristic of the allowed orbitals (determined by the Schrödinger equation) and their energies (determined by the nuclear charge Z and the quantum numbers), they are useful for the identification of elements when they occur in substances of unknown composition.

GLOSSARY

angstrom: 10^{-8} cm, a unit of length that is convenient for describing distances on the atomic scale; symbolized by Å.

atomic number: the number of protons in the nucleus of an atom, and the number of electrons in the electrically neutral atom; symbolized by the letter Z.

atomic weight: the mass of an atom on a relative scale based on the mass of $_{12}^{6}C$ = 12.00 ... .

aufbau: the imaginary process of adding electrons to atomic orbitals one at a time, in order of increasing orbital energy, to arrive at the electronic configuration of the complete atom. The name is German for "buildup."

de Broglie wavelength: according to a theory first formulated by Louis de Broglie, any particle should behave like a wave, with wavelength h/mv. This prediction has been verified for a wide variety of particles.

electron: an elementary particle with negative charge and mass 5.49×10^{-4} on the atomic mass scale.

element: a class of atoms, characterized by its atomic number Z.

gram-atomic weight: a number of atoms (6.023×10^{23}) sufficient to make up a mass, in grams, numerically equal to the average atomic weight of the atoms. It is defined formally as the number of carbon atoms in 12.00000 g of $^{12}_{6}C$.

group (in the Periodic Table): all of the elements lying in a vertical column of the Periodic Table.

ion: an atom containing a number of electrons different from its atomic number, so that it carries a net electrical charge.

isotopes: atoms with the same atomic number, but differing atomic weights, due to different numbers of neutrons in the nuclei.

lanthanide: an element with atomic number in the range 58 to 71. In this range, electrons are successively entering the $4f$ orbitals as atomic number increases, so that lanthanide elements are characterized by a partly filled $4f$ subshell.

metalloid: an element whose properties are intermediate between those of metals and nonmetals. See Figure 2-9.

molecule: a stable association of atoms, held together by electrostatic forces (called chemical bonds) between electrons and the nuclei of the bonded atoms.

neutron: an elementary particle with zero charge and mass 1.008665 on the atomic-mass scale.

nucleus: the positively charged center of an atom, composed (except for the hydrogen nucleus, which is a single proton) of protons and neutrons.

orbital: a solution to the Schrödinger equation for electrons in an atom, consisting of a domain in space and characterized by a definite fixed energy, both of which are determined by the values of the quantum numbers for the particular solution.

period: a horizontal row in the Periodic Table, containing elements whose outermost (s and p) electrons occupy orbitals with the same principal quantum number.

Periodic Table: an arrangement of elements in order of increasing atomic number, with repeating rows designed to show the chemical groups. See Figure 2-7.

proton: an elementary particle with positive electrical charge (equal and opposite to that of the electron) and mass 1.007825 on the atomic-mass scale.

quantum numbers: integers (or half-integers) that arise as part of the solution to the Schrödinger equation. For a list of atomic quantum numbers, see Section 2-8.

transition metal: an element characterized by a partly filled d subshell.

uncertainty principle: the quantum-mechanical theorem that expresses the impossibility of simultaneous precise measurement of the position and momentum of a particle. This principle arises from the fact that measurements cannot be made infinitely delicately (with disturbing effects infinitely small), since matter and energy are quantized. For discussion, see Section 2-8.

PROBLEMS

2-1 Although the Bohr model of the atom has been superseded, both for quantitative calculations and for informal discussion, by the wave-mechanical model, its predictions of many properties of one-electron atoms remain accurate, and calculations based on it will help you to get a feel for the magnitude and behavior of atomic properties. This group of calculations and those in Problem 2-4 are included for that reason.

The Bohr model yields the following expression for the energies of stable electron orbits in a hydrogen atom:

$$E = -\frac{109{,}680\, hc}{n^2}$$

where h (Planck's constant) = 6.623×10^{-27} erg-sec, c (the velocity of light) = 2.998×10^{10} cm/sec, and n is the quantum number of the orbit.

a. Calculate the energy (in ergs) of an electron in the orbit with $n = 1$, and in an "orbit" with $n = \infty$ (the latter would represent an electron completely removed from the atom).
b. The difference between the energies calculated in Part a is called the ionization energy. Explain why this name makes sense. (The experimental value for the H atom is 2.18×10^{-11} erg. Does this agree with your calculation?)
c. Remembering the equation relating the energy and frequency of light, show that the frequency of light emitted by a hydrogen atom in an electron transition from quantum level n_1 to level n_2 is

$$\nu = 109{,}680 c \left(\frac{1}{n_2^2} - \frac{1}{n_1^2} \right)$$

d. Calculate the frequency of the light emitted during a transition from $n = 3$ to $n = 2$, and from $n = 4$ to $n = 2$. What colors are these emission lines? (The observed spectrum shows, among many others, lines at $\nu = 4.567 \times 10^{14}$ sec^{-1} and 6.165×10^{14} sec^{-1}.)

2-2 From the sketches in Figure 2-6, try to find nodes in the electron wave functions depicted. Relate the number of nodes you find to the value of quantum number. (If you cannot find any nodes after a reasonable try, see Appendix II.)

2-3 Show that the velocity with which any wave moves through space is given by the product $\lambda \cdot \nu$.

2-4 According to the Bohr model of the hydrogen atom, the electron revolves around the nucleus in orbits of radius $(0.528 \times 10^{-8})n^2$ cm, where n is the quantum number of the orbit and can have only positive integral values.

a. What are the radii of the first three orbits predicted by this model? Sketch them to scale.
b. According to the same model, the electron performs $(3.3 \times 10^{15})/n^3$ revolutions per second (rev/sec) in its orbit. Calculate the *speed* of the electron (in cm/sec) in the orbit whose n is 1.
c. From the information in this problem, derive a general equation for the speed of an electron in a Bohr orbit. For approximately what value of n is the orbital speed 1 cm/sec? How large is the orbit for that value of n?

2-5 Assuming that an electron could have a de Broglie wavelength small enough to fit inside an atomic nucleus, calculate the mass it would have if its velocity were limited (by relativistic effects about which you need not worry) to the velocity of light, 3×10^{10} cm/sec. Does this result make sense?

2-6 The element helium is named for the sun (Greek *helios*) because its atomic spectrum was first observed in sunlight. Explain why the observation of new spectral lines is good evidence for the existence of a previously unknown element.

2-7 The emission spectrum of sodium includes two very strong lines whose wavelengths are 5890 and 5896 Å.

  a. What color does this light have?
  b. For either line, calculate
    i. the frequency (see Problem 2-3);
    ii. the energy change (in ergs) of a sodium atom when light of this wavelength is emitted. (Is this change an increase or decrease in energy?)
    iii. the total change in energy that would occur if one gram-atomic weight of sodium emitted light of this wavelength.
  c. (Laboratory exercise) The next time you have a burner going, drop some powdered sodium chloride into it. The flashes of color that you see are the 5890 and 5896 Å emission lines of sodium. (You can elicit the same lines by rapping the burner base gently on the bench top. Why?)

2-8 Explain the mechanism by which an emission spectrum is produced when a collection of atoms is heated to a high temperature.

2-9 When the light from the sun (and other stars) is examined in a spectrograph, a number of "dark lines" are observed corresponding to wavelengths at which relatively little light is coming from the sun. This phenomenon is attributed by astronomers to light being *absorbed* by cool, gaseous atoms in the sun's outer atmosphere. By means of an energy-level diagram, show what happens to produce an absorption, or dark-line, spectrum.

2-10 Given below are the atomic numbers of some elements. For each, use the *aufbau* process to write down the electronic structure ($1s^2\ 2s^2$ ...); decide in which block of the Periodic Table (see Figure 2-8) the element lies, and whether it is a metal, a nonmetal, or a metalloid. Atomic numbers: 5, 10, 15, 20, 30, 40, 50, 60, 80, 100.

2-11 Here are some relatively unfamiliar elements (identified by their symbols). Find them on the Periodic Table and (without going through the *aufbau* process) write down the electronic structure of their outermost electrons. In a few words, characterize their main properties: Kr, Sb, Os, Ho, Lw.

2-12 What is the mass of 5 gram-atomic weights of Mg? At 28.35 g/oz, how many ounces is this? How many Mg atoms are contained in this quantity of Mg?

2-13 Suppose we choose the mass of one slice of bread as an arbitrary mass standard, and call it one "slice." Suppose, further, that we establish (by weighing bologna sandwiches) that a slice of bologna has a mass, on this arbitrary scale, of 1.5 "slice." We shall now define a number (analogous to Avogadro's number) that is the number of slices of bread in one ton of bread. This number turns out to be about 4×10^4. Let us call it one ton-slice weight. Now suppose that you are responsible for ordering enough bread and bologna to feed 40,000 people (at the university annual picnic) two bologna sandwiches each.

a. How many tons of bread will this require?
b. How many ton-slice weights of bread is that?
c. How many slices of bologna (at one slice per sandwich) will be required?
d. How many ton-slice weights of bologna is that?
e. How many tons of bologna should you order?

(We shall return to weight relationships in Chapter 3, but in the meantime, bear in mind the famous theorem that the properties of bologna are invariant under the slicing operation.)

SUGGESTIONS FOR FURTHER READING

Short books about the structure of atoms and molecules: Quite a number of these, intended for freshmen, have appeared in the past decade. Try:

SEBERA, D. K., *Electronic Structure and Chemical Bonding*, New York: Blaisdell (1964);

GRAY, H., *Electrons and Chemical Bonding*, New York: W. A. Benjamin (1964); and

HOCHSTRASSER, R. M., *Behavior of Electrons in Atoms*, New York: W. A. Benjamin (1964).

A longer and more detailed book, for a slightly more ambitious reader: ANDERSON, J. M., *Introduction to Quantum Chemistry*, New York: W. A. Benjamin (1969). There is really no way to clarify the mysteries of wave mechanics but to plunge into the details, and Anderson's book, written for undergraduates, will help you as humanely as possible.

At the opposite pole from Anderson's book is: HOFFMANN, B., *The Strange Story of the Quantum*, 2nd revised ed., New York: Dover (1959). Hoffmann attempts to give the flavor of the quantum revolution in a totally nonmathematical way by means of witty and suggestive analogy. The drawback is that one emerges with only an analogy of understanding, but the book is fun to read.

For a survey of atomic orbitals with a complete bibliography, see: BERRY, R. S., "Atomic Orbitals," *Journal of Chemical Education*, vol. 43, pp. 283–299 (June 1966).

The Heisenberg uncertainty principle and the idea that knowledge of one sort may exclude other kinds of knowledge (as in the position and velocity of an atomic particle) have caused wide philosophical repercussions. If your interests lie in this direction, try:

SCHRÖDINGER, E., *What is Life?* Cambridge: Cambridge University Press (1945). The author of Equation (2-3) is unconvinced of the validity of Niels Bohr's idea of complementary knowledge and gives his reasons in the chapter "Do Quantum Jumps Occur?" (Schrödinger thinks not.)

A procomplementarity discussion is given by: HEISENBERG, WERNER, "Quantum Theory and Its Interpretation," in *Niels Bohr: His Life and Works as Seen by His Friends and Colleagues*, S. Rozental, ed., New York: Wiley (1967). This volume also includes a very funny description of Schrödinger's visit to Bohr's laboratory; Bohr's persistence in arguing the complementarity philosophy drove Schrödinger to his bed, but didn't stop even in the sickroom. Schrödinger's parting cry was "If we must go on with these damned quantum jumps, then I am sorry I ever started to work on atomic theory!"

For arguments extending the complementarity philosophy beyond physics, see:

Bohr, N., *Atomic Physics and Human Knowledge* (1958), and *Essays, 1958–1962 On Atomic Physics and Human Knowledge,* New York: Wiley (1962);

Holton, G., "The Roots of Complementarity," *Daedalus* (Fall 1970); and

Blackburn, T. R., "Sensuous-Intellectual Complementarity in Science," *Science,* vol. **172**, p. 1003 (June 4, 1971).

3
Molecules: The Formation of Chemical Compounds

3-1 Abstract

The purpose of the rather detailed review of atomic structure in the previous chapter was to provide a background for understanding the interaction of atoms as they form **compounds**. Rarely are isolated atoms of any element useful or important; only when they are bonded to other atoms by rearrangement or exchange of their electrons do they become the stuff of which planets and people are made. The present chapter takes up the two principal questions about chemical combination: *Why* atoms combine, and *how* they do so. The question of *why* involves primarily the subject of energy transfer, and this can be considered only briefly here; after more background, a more satisfying treatment will be given in Chapters 8, 9, and 10. The other aspect, *how* atoms combine, can be presented at this time in a simplified, nonmathematical manner. We shall consider the formation of compounds by the transfer of electrons from the orbital(s) of one atom to the orbital(s) of another, followed by overlap of atomic orbitals in such a way that the sharing of *pairs* of electrons is made possible. The resulting compounds need names, so the essentials of a systematic scheme of nomenclature are outlined. The useful shorthand device called a **chemical equation** then is introduced. Lastly, the quantitative concepts of **molecular weight** and the **mole** are introduced, and their utility in making simple calculations of chemical quantities is explained.

3-2 Unwinding Watchsprings, Falling Bricks, and the Formation of Chemical Compounds

Wound-up watchsprings unwind if permitted by the watch's escapement mechanism. Objects that are elevated fall down unless held up. Hot things always get colder, and batteries always run down in use. Chemists have generalized from these sorts of experiences and observations and now take it as axiomatic that all systems, whether they be mechanical or electrical or chemical, when changing by themselves, change to conditions of lower potential energy. Systems in states of high potential energy undergo such changes spontaneously unless considerable trouble is taken to restrain them. That is, when we cement bricks into place high up on a building, they may stay there for a long time, holding onto the extra potential energy given them when they were hoisted into place by the brick mason. But eventually they will return to earth, and in so doing give up that energy.

Atoms often combine spontaneously.

And so it is with atoms, which follow the same universal laws. An atom of just about any element is in a state of high potential energy if isolated, and in states of lower potential energy in many compounds. Most atoms combine with other atoms to form aggregates which we call **molecules**, and, in combining, release some of their potential energy. Since the resulting molecule has less energy than did its constituent parts when they were separated, the molecule is closer to ground level or the zero-energy level, and so is in a more stable condition.

Sometimes we want to detach atoms from their previous states of combination, even going to considerable trouble to accomplish this, in order to guide them into new and more desired combinations. Occasionally, we may even want to exercise our ingenuity further and, by deliberately putting in considerable amounts of expensive energy, make atoms combine in a less likely way to give desired but unstable substances.[1] In any case, the combination of atoms is called a **compound**. If the process of compound formation is accompanied by evolution of heat (as the chemical system moves toward lower potential energy), then we describe that chemical reaction as **exothermic** (from the Greek: *exo*, out; *therm*, heat), meaning that heat is transferred *out* of the system [see Figure 3-1(a)]. On the other hand, if we need to put extra heat *into* the chemical system in order to achieve the desired compound, as in Figure 3-1(b), we speak of the formation reaction as **endothermic** (and expect that eventually the compound will come tumbling down the energy hill to ground level, giving up the extra potential energy that we put into it in order to build the structure). Sometimes we get out or put in electrical energy (or radiant energy, or mechanical energy) instead of transferring only heat, and then the corresponding terms are **exoergic** and **endoergic**, meaning that energy, in a general sense, comes out or goes into the chemical system during the particular reaction being described.

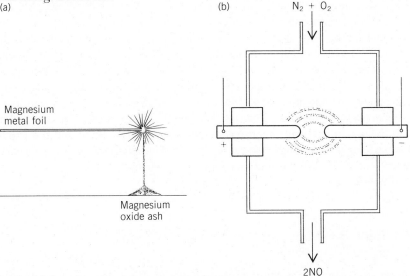

FIGURE 3-1 (a) A strip of magnesium ribbon, after being ignited with a match, burns in air. The flame is brilliant, too bright to observe directly with a naked eye, and it is very hot. The white ash which is produced is a powder, a compound of oxygen (from the air) and magnesium, called magnesium oxide. The process of formation of magnesium oxide from its constituent atoms is an *exothermic reaction*, because heat is released during the reaction. (b) Oxygen and nitrogen abound in the atmosphere, yet they do not combine of themselves. Only when the mixed gases are heated to 3000°C in an electric arc do the elements combine to form a compound, nitrogen oxide, which issues (along with unchanged N_2 and O_2) from the arc chamber. When the electric arc is shut off, formation of nitrogen oxide ceases immediately. Since the formation of the compound requires continual input of energy in order to proceed, the process is an *endothermic reaction*.

---

[1] Systems can get more potential energy, but they have to be given that energy. Watch-springs do get wound, but not by themselves. Bricks do rise from the street to the top of a building under construction, but not unaided. The welcome hot cup of coffee was made with cold water which got hot, but it did not get hot by itself.

Atoms combine with one another *by rearranging their electron structures in concert*. This usually results in a reduction of the potential energy of the electronic distribution. The rearrangement of electrons may take place in three different ways: Electrons may be given up entirely (forming positive ions); electrons may be acquired outright from other atoms and kept (forming negative ions); or electrons may be *shared* by two or more atoms. When all the electron rearrangement occurs by electron sharing, the resulting aggregate of nuclei and electrons is a **molecule** and has no net electrical charge. There may also be a combination of electron sharing with outright loss or gain of charge, yielding an aggregation of atoms with electric charge, a polyatomic ion. All three modes of electron transfer are important, and each imparts particular physical, chemical, and structural properties to the resulting compound. We shall take up the types of bonding in order, and then consider how the mechanism of electron rearrangement may be represented on paper.

It should be emphasized at the outset that not all atoms seek to change or rearrange their complement of electrons. In some cases, the readily available orbitals (those accessible on the energy scale) are already filled, as in helium or neon, so no electron transfer is feasible; the elements simply persist as isolated atoms. All of the elements in Group 0 (He, Ne, Ar, Kr, Xe, and Rn) exist as monoatomic gases, but the last three can be forced into chemical combination by extremely electron-seeking reagents, especially fluorine; the first three resist even that drastic treatment. Moreover, there are some elements which have electrons and orbitals available for bonding, but which are "disinterested" in the sense that there is little stability to be gained by electron rearrangement; the amount of potential energy that can be lost by such rearrangement is so small as to be not worthwhile, energetically speaking. Mercury is such an element; its atoms remain single in the liquid and the vapor. Even when chemical combination is induced in mercury, the compounds come apart readily. So we must expect that the tendency of elements to combine will vary widely.

3-3 Ionic Bonding: The Consequences of Outright Transfer of Electrons

If one atom can reduce the potential energy of its electron configuration by gaining additional electrons for its unfilled orbitals, it must get those electrons from some other atom, one that gives up electrons readily. The electrons cannot be manufactured out of nothingness. Even if electrons are provided by a battery or a dynamo, something had to give up the electrons in the first place. So we expect (and find) that electron gain and loss occur together, and that the number of electrons gained by acceptor atoms is exactly equal to the number of electrons lost by donor atoms.

Often the process of electron gain and loss is one of direct transfer of electrons, yielding positive and negative ions simultaneously. For example, a lithium atom has the electron configuration ($1s^2\ 2s^1$), and thus a single electron in its $2s$ orbital. Lithium readily loses this electron, becoming the positive

Li$^+$ ion, and reverting to the helium electronic structure ($1s^2\,2s^0$). No chemical compound of helium has yet been made, and this fact is strong evidence for the stability of the ($1s^2\,2s^0$) configuration. No chemical force can extract an electron from the depths of the filled $1s$ orbital of either helium or lithium. On the other hand, fluorine with the electron configuration ($1s^2\,2s^2\,2p^5$) lacks only a single electron to complete the filling of its $2p$ orbitals, and so attain the stable neon structure ($1s^2\,2s^2\,2p^6$). You will recall that, like helium, neon has eluded all attempts of chemists to form compounds. Fluorine gives up a great deal of energy when it gets that extra electron, becomes F$^-$, and achieves a stable energy state of lowered potential energy.

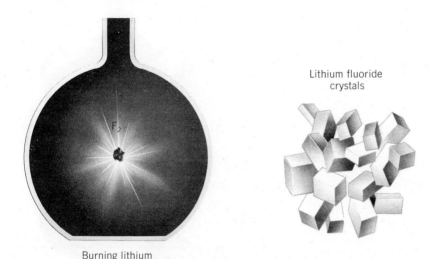

FIGURE 3-2 A small piece of the soft metal lithium, blazing brilliantly as it burns in the presence of the gas fluorine. The resulting ash is the white powder lithium fluoride, a salt which dissolves in water to give the ions Li$^+$ and F$^-$. If a saturated solution of LiF is prepared, and then water slowly allowed to evaporate, **crystals** of LiF form. Each crystal is an orderly arrangement, in three dimensions, of alternating positive and negative ions. The ions arrange themselves into this repeating pattern as they crystallize from solution, creating a natural chemical object of great regularity and often of striking beauty. The shape of the crystal is a direct consequence of the arrangement of the myriad of constituent ions.

Hence it is no surprise to find that the strongly **electronegative** element fluorine combines vigorously with the markedly **electropositive** element lithium to transfer one electron per atom and form the compound LiF, lithium fluoride (see Figure 3-2). The compound consists entirely of Li$^+$ ions and F$^-$ ions, arranged in equal numbers in a cubic **lattice** (see Figure 3-3) to form a **crystalline** white solid. The vast array of ions is strikingly orderly. Each Li$^+$ ion is surrounded by six F$^-$ ions in the lattice, and each F$^-$ ion is, in turn, surrounded by six Li$^+$ ions, so that each positive charge is shielded from other positive charges (thus repulsion is minimized) and surrounded by negative charges (thus attraction is maximized). The entire solid is held firmly

A simple example

together by the electrostatic forces of attraction between ions of opposite charge. The charges are in the positions which make the potential energy of the whole crystal as small as possible. The arrangement is so stable that a temperature of 870°C is required to overcome the strong attractive forces and cause the solid to melt. (The melting points, boiling points, and physical states of all the pure compounds mentioned in this book are given in Appendix I, so that it will not be necessary to include them in the text.) It is also significant to note that the combination of one gram-atomic weight of lithium with one gram-atomic weight of fluorine to give lithium fluoride evolves 145,540 **cal**—a decidedly exothermic reaction, and hence one that forms a very stable compound.

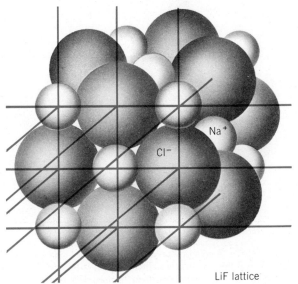

FIGURE 3-3 The crystal lattice of LiF, compared to Na$^+$ and Cl$^-$ ions drawn to the same scale.

Chlorine, element 17, is seen to have an electron configuration very similar to that of fluorine, so we expect lithium to combine with chlorine to form lithium chloride, LiCl. This it does, with liberation of 97,420 cal for each gram-atomic weight of lithium and chlorine. Similarly, all of the Group I elements (called the "alkali metals" because they react with water to give strongly alkaline compounds) react vigorously with all of the Group VII elements (the "halogens," meaning "the salt-formers") to form stable white solids made up of unipositive and uninegative ions. These are the archetypal ionic compounds, and all have similar properties: high melting points, cubic ionic structures[2] soluble in water, and high electrical conductance in the melted condition.

---

[2] All the alkali halides except CsCl, CsBr, and CsI have the same packing arrangement of ions (the "rocksalt structure" of Figure 3-3). In the three exceptions, the large Cs$^+$ ions are surrounded by eight negative ions, not six.

In Group II, the occupied orbital with highest energy is an *s* orbital, and in each case the orbital is filled with a *pair* of electrons. They are not lost from the atom as readily as from the corresponding Group I metals of comparable atomic weight. When these elements do become ions, *both* electrons are lost, and dipositive ions are formed. For instance, Mg becomes Mg^{2+}, thereby attaining the neon electronic configuration. Hence we find many Group II halides, such as MgF_2, $MgCl_2$, $MgBr_2$, MgI_2, CaF_2, $CaCl_2$, $CaBr_2$, and so on, all excellent examples of ionic compounds. (Beryllium atoms are so small that they act somewhat differently, as we shall see later.)

The elements of Group VI all have places for two additional electrons in their *p* orbitals, and so can acquire two electrons and form dinegative ions. Oxide ion, O^{2-}, and sulfide ion, S^{2-}, are the most common examples. We now contemplate the white crystalline solid MgO, made of Mg^{2+} and O^{2-} ions, and ask: Should it have a higher melting point than the comparable NaF? Is the solid packed any more densely? That is, does it make any difference if all the ions have *two* charges apiece instead of one charge? A study of the properties of MgO gives the answers very quickly:

Compound	Melting Point	Density
MgO	2800°C	3.58 g/cm^3
NaF	988°C	2.56 g/cm^3

These experimental data show that the Mg^{2+} and O^{2-} ions in magnesium oxide are indeed much more tightly bound together and much more closely packed than are the Na^+ and F^- ions in sodium fluoride. The electrostatic attraction is four times as great at any particular separation when the ions have two charges apiece, and so the ions pack more closely and resist the disruptive effect of heat to a much greater extent. The geometry of the crystal structure, by the way, is exactly the same in both compounds, but in MgO the distance between ion centers is less.

In general, oxides and fluorides of most of the metals are ionic, although the chlorides, bromides, and iodides of many transition metals are not. Most oxygen-containing ternary[3] compounds of *all* the metals, such as their nitrates, sulfates, and phosphates, are also ionic. There are literally thousands of ionic compounds, and from their properties we can draw some conclusions about ionic bonding:

Some cautious generalizations

1. **Ionic bonds** are not specific individual links between pairs of atoms, but are generalized electrostatic attractions which hold the positive and negative ions in place.
2. Ionic compounds are almost always crystalline solids at room temperature, and most have high melting points. The melting points vary a great deal, however, being exceedingly high for some oxides and nitrides (2800°C for MgO, 2715°C for ZrO_2, 2530°C for BeO, 2200°C for AlN)

---

[3] Binary compounds contain only two elements; ternary compounds contain three elements; quaternary compounds contain four elements, and so on.

and rather low for Group I nitrates and hydroxides (272°C for CsOH, 255°C for $LiNO_3$).

3. Ionic compounds conduct electricity when melted, because their ions migrate in an electric field and transfer electric charge. Only compounds composed of ions can do this.

4. Many ionic compounds are soluble in water, but not all; those that have very high lattice-bonding energy (such as MgO, Al_2O_3, and Fe_2O_3) are not soluble in water, but those of low melting point often are soluble. Those ionic compounds that do dissolve in water provide electrically conducting solutions.

A more detailed treatment of ionic compounds will appear in Chapter 5.

3-4 The Names of Chemical Compounds

Some remarks about the names that have been used in this chapter (and which will be used in the rest of this book) are now in order. There is one system for naming "inorganic" compounds (compounds of the metals with electronegative elements, compounds of the nonmetallic elements with one another, and so on, and in fact all compounds except certain ones containing carbon), and there is another system for "organic" compounds (those particular compounds of carbon which are treated in Chapters 14 to 19). We shall consider the system for naming inorganic compounds here; the scheme for organic compounds will be taken up in Chapter 14.

A helpful, systematic language

There need be no mystery in the naming of ordinary inorganic compounds if a few general rules are understood. The rules outlined below are those adopted by the International Union of Pure and Applied Chemistry and approved by a committee of the American Chemical Society. This nomenclature system attempts to provide rules so that the name of a compound can always be derived from its formula, and a formula can always be written when the name is known.

I. Elements

The symbols for the chemical elements are given in Table 1-1, on the Periodic Table of Chapter 1, and also inside the back cover of this book.

II. Compounds of Two Elements

A. *Formulas* are written with the symbol of the more positive (metallic) element first, and suitable subscripts to show the atomic proportions: CuCl, $CuCl_2$, CF_4, H_2O.

B. *Names* are written in the same order, with Greek prefixes to indicate the proportions and with the ending *-ide:*

NO	nitrogen oxide
N_2O	dinitrogen monoxide
NO_2	nitrogen dioxide
NCl_3	nitrogen trichloride
N_2O_4	dinitrogen tetroxide
PCl_5	phosphorus pentachloride
SF_6	sulfur hexafluoride
FeS_2	iron disulfide

Other prefixes: 7 hepta, 8 octa, 9 ennea, 10 deca, 11 undeca, 12 dodeca. Mono is usually omitted unless it should be emphasized.

C. In speech, either the formula or name may be used. When there is little possibility of error, or when the elements have one fixed combining capacity (a single **oxidation state**), the prefixes may be omitted:

NaCl	sodium chloride
CaO	calcium oxide
$CaCl_2$	calcium chloride
H_2S	hydrogen sulfide

D. For elements with variable combining capacity (with more than one oxidation state), the oxidation state is shown by the Greek prefixes in the names, or by Roman numerals placed after the name:

$FeCl_2$, FeO, and Fe^{2+} X^{2-} are iron(II) compounds.
$FeCl_3$, Fe_2O_3, and Fe^{3+} X^{3-} are iron(III) compounds,

where X is any appropriate element.

E. Common or trivial names are retained for some common substances like water and ammonia, but all other compounds are named by the general method.

III. Compounds of Three or More Elements

A. *Formulas* and *names* follow the same order as above with the symbol of the most positive (most metallic) element placed first. Complex ions having special names are treated like elements:

NH_4Cl	ammonium chloride
$CaCO_3$	calcium carbonate
$NaHCO_3$	sodium hydrogen carbonate (not sodium bicarbonate)
$LiAlSi_2O_6$	lithium aluminum silicate
$Fe(OCN)_3$	iron(III) cyanate
$Fe(SCN)_3$	iron(III) thiocyanate

B. Acids containing oxygen are given particular names:

H_2SO_3	sulfurous acid
H_2SO_4	sulfuric acid
$H_2S_2O_3$	thiosulfuric acid (*thio* comes from the Greek θεῖow, meaning sulfur)
H_2SO_5	peroxy(mono)sulfuric acid
$H_2S_2O_8$	peroxydisulfuric acid
HNO_2	nitrous acid
HNO_3	nitric acid

IV. Coordination compounds with complex ions as the negative ions use the names of the complex ions as modifiers, and sometimes use the Latin names of metals for euphony:

$K_2[PtCl_6]$	potassium hexachloroplatinate
$K_4[Fe(CN)_6]$	potassium hexacyanoferrate(II)
$K_3[Fe(CN)_6]$	potassium hexacyanoferrate(III) or tripotassium hexacyanoferrate

For hydrates and ammoniates:

$CaCl_2 \cdot 6H_2O$	calcium chloride hexahydrate
$AlCl_3 \cdot NH_3$	aluminum chloride ammoniate
$[Cr(NH_3)_6]Cl_3$	hexamminechromium(III) chloride

66

MOLECULES: THE FORMATION OF CHEMICAL COMPOUNDS

V. For naming other classes of compounds not mentioned in this outline, see "Rules for Naming Inorganic Compounds," *Journal of the American Chemical Society*, vol. 63, p. 889 (1941); or *Nomenclature of Inorganic Chemistry,* a report of the International Union of Pure and Applied Chemistry, London: Butterworths (1959). Rules for naming carbon compounds appear in Chapter 14.

The purpose of all the rules given above is not to give the student something more to memorize, but to guide him so that he can recognize chemical compounds from their names as well as from their formulas. Often the names are easier to pronounce than the formulas, and certainly they are in more general usage; names of compounds appear far more often in newspapers and magazines than formulas. There is but one criterion for a name: The reader ought to be able to understand what is meant. The names given here meet that criterion, and are acceptable in all social and scientific circles. Many archaic and slang terms are not.

3-5 Covalent Bonding

The other type of bonding

Rearrangement of the electron configurations of atoms to form compounds (again, usually with decrease of potential energy within the system and with evolution of equivalent observable energy) takes place more often by *sharing* or *exchange* of electrons than by outright loss or gain of electrons to form ions. The shared electrons then constitute a bond to hold together the sharing partners, by attracting and holding their nuclei. Such a process of linking together atoms to form molecules is called **covalent bonding** and is exceedingly prevalent. Every element that engages in ionic bonding also is capable of forming covalent bonds, especially of a type where *pairs* of electrons are contributed to empty orbitals of a positive ion, so we find covalent bonding widespread among the metals. It also is a common form of bonding between the electronegative elements themselves, and even between halogen elements and metals, if the metal has a small size and its orbitals are already about half populated with electrons. The compounds of beryllium are almost all covalent, because of its small size. Among the metalloids, covalent bonding predominates, and for carbon covalent bonding is almost the only form (to the tune of a million organic compounds). The formation of shared-pair bonds is a very important subject to study.

Covalent bonding is not the exclusive province of some elements.

(Neither is ionic bonding)

We can conveniently approach an understanding of covalent bonding by means of a thought experiment. Let us visualize the orbitals of two atoms and imagine what might happen when the atoms approach each other very closely. Suppose we choose hydrogen and oxygen as the interacting elements and visualize the events that might occur during the stepwise formation of their very familiar compound, H_2O, from three isolated atoms. Let us imagine the two hydrogen atoms approaching the oxygen atom along perpendicular paths, which we shall label the y and z axes. Thus, we set the stage:

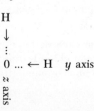

Our scenario calls for the hydrogen atom moving along the y axis to arrive first at the oxygen, there to interact with an unfilled p orbital centered on the oxygen nucleus. Pairing of electrons will occur, and a covalent bond will be formed.

First, we recall that oxygen has two unpaired electrons in its $2p$ orbitals:

$2p$ ⥮ ↑ ↑
$2s$ ⥮
$1s$ ⥮

Electron configuration of unbonded oxygen atom

Let us suppose that this particular oxygen atom, being confronted by hydrogen atoms moving in along the y and z directions, has two electrons in the $2p_x$ orbital (a p orbital with lobes extending in front and behind the textbook page), and one electron in each of the other p orbitals ($2p_y$ and $2p_z$ orbitals) directed along the y and z encounter directions. An approaching hydrogen atom can interact with only one of these half-filled orbitals; it requires a second hydrogen atom to interact with the other p orbital.

A hydrogen atom has only one electron, and that in a $1s$ orbital with spherical symmetry. The hydrogen $1s$ orbital is half filled. As the encounter along the y axis proceeds to the point where the two nuclei get within 1 Å of each other, the spherical hydrogen orbital will overlap the protruding p_y orbital to a considerable extent, as depicted in Figure 3-4. Since the hydrogen orbital contained one electron, and the p_y orbital of oxygen contained one electron, the overlap region provides space of high probability for accommodating both electrons. The two atomic orbitals can now be considered to be filled, at least on a sharing basis. The pair of electrons will reside for a substantial fraction of the time between the two nuclei, and this pair constitutes a bond. The nuclei cannot approach more closely because of their mutual repulsion; they both have positive charge. The nuclei are kept from flying apart by their mutual attraction toward the pair of electrons. The result is that both the electrons and the nuclei settle into an equilibrium position. Meanwhile, much potential energy has been converted to heat energy given up to the surroundings, and a stable covalent bond has been formed.

But a second encounter is also underway. Since the p_z orbital of oxygen also contains one electron, the same process can be repeated along the z axis, as shown in the last diagram of Figure 3-4. This gives a stable molecule of the composition H_2O and a structure approximately like that of Figure 3-4 with the two O–H bonds at about right angles to each other[4] and the atoms 0.958 Å apart. As for stability, the combination of one gram-atomic weight

Two nuclei and an electron pair bind themselves together.

---

[4] This description is too simple, for the

$$O \begin{matrix} \diagup H \\ \diagdown H \end{matrix}$$

angle is found by actual measurement to be 104.45°, not 90°. The reason for the greater separation of the hydrogen atoms will appear later in this chapter.

of oxygen with sufficient hydrogen gives off 68,387 cal. The molecule bears no net positive or negative charge; it is electrically neutral and is a separate entity. Single molecules of H_2O wander around independently in water vapor and are only partly restrained in liquid water. At low temperatures, the H_2O molecules line up in rows and frameworks, because there is more regional negative electric charge diametrically opposite the two hydrogen atoms than in the vicinity of those H atoms, so each H_2O molecule is a little electric **dipole**, and these dipoles line up:

$$\begin{array}{c}H\\\delta+\quad\diagdown\\\quad\quad O\delta-\\\diagup\\H\end{array}\quad\begin{array}{c}H\\\delta+\quad\diagdown\\\quad\quad O\delta-\\\diagup\\H\end{array}\quad\begin{array}{c}H\\\delta+\quad\diagdown\\\quad\quad O\delta-\\\diagup\\H\end{array}\quad\begin{array}{c}H\\\delta+\quad\diagdown\\\quad\quad O\delta-\\\diagup\\H\end{array}\quad\begin{array}{c}H\\\delta+\quad\diagdown\\\quad\quad O\delta-\\\diagup\\H\end{array}$$

The lowercase Greek delta, δ, indicates a small fraction of one electric charge. Such attraction is readily overcome by thermal agitation, and the solid melts at 0°C. These water molecules are said to be **polar**, because each molecule is electrically polarized, with a slight separation of positive and negative charge.

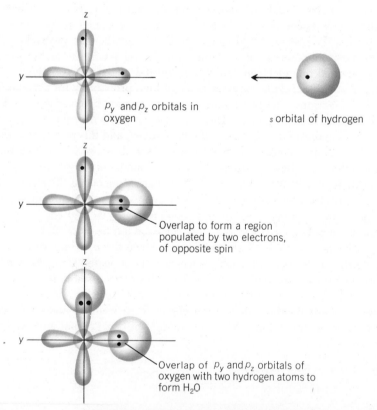

FIGURE 3-4 Formation of covalent bonds in H_2O.

Just as a matter of electron bookkeeping, we could depict the combination of an oxygen atom with two hydrogen atoms in a purely static way by drawing an energy-level diagram to show where the hydrogen electrons go:

A bookkeeper's accounting of the electrons

Oxygen atom H_2O molecule

Here we use dots to show the electrons that came from oxygen and crosses to show the electrons that came from hydrogen. The result should not be taken as a mechanistic picture of the true situation with the H_2O molecule, however; it ignores the hydrogen atoms after combination, and it treats individual electrons as though their ancestry could be traced back to a particular parent atom. A much better way of keeping track of the electrons is to concentrate only on these electrons that are in the highest energy, outermost shell, and to group these around the symbols of the participating elements in what is called a *Lewis dot diagram:*[5]

$$\begin{matrix} & H & \\ :\ddot{O}: & H & \end{matrix}$$

Actually, all of these representations are too crude, even though the overlap of atomic orbitals is a much better representation than the latter two diagrams. For a more exact explanation of what really takes place during the formation of a covalent bond, it is necessary to use wave mechanics again and to consider in detail (by mathematical methods) the interaction of the atomic wave functions as they are combined mathematically to give *molecular orbitals.* We do not need this now, for we can rely upon the simple overlap concept; later on, in Chapter 12, molecular-orbital theory will be taken up.

Covalent bonding is not limited to dissimilar atoms. All the aspects of atomic orbitals can pertain equally well to combination of two identical atoms, provided there are electrons available and suitable orbitals in which to put them. Consider as an example two fluorine atoms. We know by now that two of the $2p$ orbitals of fluorine are filled, but the third is only half filled. Let us call the half-filled orbital $2p_z$. Then the approach of two fluorine atoms can result in p_z–p_z overlap, as depicted in Figure 3-5. If the two electrons involved in this overlap have opposite spins, then they can each share the entire domains of both orbitals. All the conditions for covalent bonding are met once more, and a pair of electrons is shared between the two fluorine atoms. In this case, the internuclear distance is found experimentally to be 1.42 Å (42 percent greater than in H_2O), and the bond is not so strong as an O–H bond, but it suffices to keep all fluorine atoms tied up

Homo-bonds, as it were

---

[5] After G. N. Lewis of Berkeley, who pioneered in the shared-electron-pair interpretation of the covalent bond.

in the form of F_2 molecules in elementary fluorine. Indeed, all chlorine exists in the elementary state as Cl_2, all bromine as Br_2, all iodine as I_2, all nitrogen as N_2, and all oxygen as O_2. Even hydrogen atoms bond with each other to form H_2 molecules, if there is nothing else to combine with, and the H–H bond is very strong—104,200 cal are required to separate 2.02 g of hydrogen into 2 gram-atomic weights of hydrogen atoms. So there is nothing about covalent bonding that requires the two partners of the bond to be different, or of different electron-attracting ability, the way they must be for ionic bonding.

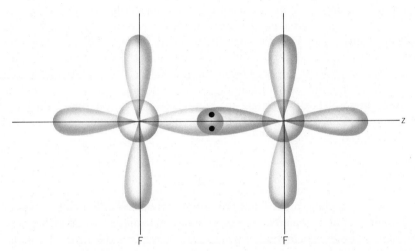

FIGURE 3-5 Formation of a covalent bond between fluorine atoms to give an F_2 molecule.

Hetero-bonds—

At the same time, if the atoms that form a covalent bond are *not* atoms of the same element, there must necessarily be unequal attraction for the shared pair of electrons. This is just to say that in a molecule of Cl_2, the two chlorine atoms have equal electron-attracting ability, so we picture the shared pair of electrons as residing halfway between the two atoms, on the average. But in a molecule of HF, the fluorine atom attracts electrons much more strongly than the hydrogen atom does, so we picture the shared pair as being much closer to the fluorine, on the average, than to the hydrogen. Hence there will be an unequal distribution of charge within the molecule; one end has a higher concentration of negative electricity than the other end does. We say that the molecule of HF has an **electric dipole moment**, and expect that it will orient itself in an electric field with the negative end pointing toward a positive charge elsewhere, and the positive end pointing toward a negative charge elsewhere. It follows that molecules of HF will align themselves in the fashion

—and their consequences

$$\overset{\delta+}{\ldots\,H} - \overset{\delta-}{F} \ldots \overset{\delta+}{H} - \overset{\delta-}{F} \ldots \overset{\delta+}{H} - \overset{\delta-}{F} \ldots$$

They do this so much, in fact, that the molecules in liquid HF are held there, despite thermal agitation, up to +19°C, where they finally boil away. This boiling point, the temperature of a cool room, is high for the hydride of a light element; compare CH_4, which boils at −162°, but has a dipole moment of zero. The polar compound HF obviously resembles water, where a large dipole moment causes the molecules to attract each other and stay in liquid form. Further effects of electric dipole moments on the physical properties of matter are taken up in Chapter 5.

The degree of charge separation between dissimilar covalently bonded atoms can be expressed not only in terms of the measurable dipole moment, but can also be calculated by a method devised by Pauling[6] and expressed as "percent ionic character," which means percent of one full electron charge separation, as in Li^+F^-. The percent ionic character of a covalent bond can vary from zero in Cl–Cl or H–H to 14 percent in an H–Cl bond and 46 percent in an H–F bond.

MULTIPLE COVALENT BONDS. So far, we have considered only the overlap of the single atomic orbitals, and especially the *axial overlap* (overlap along the internuclear axis) of s and p orbitals. We have depicted each orbital by sketching a line which encloses a region in which the probability of finding an electron (or the pair of electrons) is always greater than some particular value. If we draw p orbitals so as to enclose a wider range of probability, however, it becomes obvious from a pencil sketch that there can be *sideways overlap* of adjacent p orbitals which are involved in the axial-overlap bonding. As a hypothetical example, consider p–p overlap such as that depicted in Figure 3-5, but now draw the p orbitals which are perpendicular to the internuclear axis much fatter (that is, draw them to enclose a wider range of probability for finding the electrons). To simplify our discussion, let us leave out the p orbitals that overlap end to end along the internuclear axis, and just draw a solid line to indicate that covalent bond (see Figure 3-6). We can now concentrate our attention on the degree of sideways overlap of the perpendicular p orbitals. Considered in this way, Figure 3-6 makes it clear that there can be supplemental covalent bonding, along with the primary covalent bond, *if* the newly overlapping orbitals have one electron in each, or if one orbital is vacant and the other has two electrons to contribute. Such supplemental bonding, where it is possible, strengthens the direct bond. When we remember that there is also a *third* set of p orbitals, directed out in front of the paper and behind it, and that these can overlap too and form yet another bond if suitable vacancies and electrons are available, we see that there can be *two* such supplemental bonds if all conditions are met. Moreover, we are not limited to p orbitals for this kind of overlap; obviously, d orbitals can overlap with one another, or with p orbitals.

Several electron pairs can help bond two nuclei together.

---

[6] L. Pauling, *The Nature of the Chemical Bond*, 3rd ed., Ithaca, NY.: Cornell University Press, p. 98 (1960). A more recent (though abridged) version is L. Pauling, *The Chemical Bond*, Ithaca, N.Y.: Cornell University Press (1967).

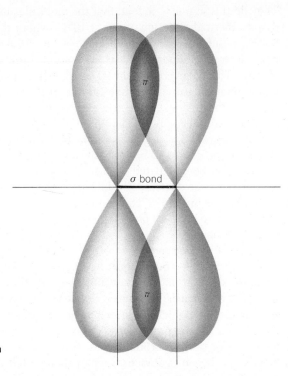

FIGURE 3-6 Formation of a π bond.

π bonds come into the picture.

The general term for such a supplemental covalent bond is π bond.[7] Note that the electron density that constitutes a π bond is not cylindrically symmetrical around a line drawn between the two nuclei, but instead consists of two overlap regions, which are mirror images of each other (see Figure 3-7). The "mirror" is a plane in which the electron density from the two p orbitals is zero. The original bond formed by direct axial overlap (Figure 3-5) is called a σ bond. The electron density in a σ bond is cylindrically symmetrical about the internuclear axis. It is customary in chemical shorthand to indicate a σ bond simply by a dash, as in H–Br, and to indicate the presence of one σ bond plus one π bond by two dashes, C=C (a "double bond"). When two π bonds are formed along with the σ bond, they are shown as a triple bond, as in C≡C. Multiple bonding is prevalent in organic chemistry, where double and triple bonds have long been recognized and shown as such. Multiple bonding also is very common in the inorganic chemistry of the transition metals, and also in the compounds of silicon, of phosphorus, and of sulfur, but unfortunately it seldom is indicated in conventional formulas. A chemist should get into the habit of visualizing all the orbitals involved or available when covalent bonds are formed, and learn when to expect multiple bonds as a means of strengthening molecules.

---

[7] Students can keep π bonds and σ bonds straight in their minds by associating π (Greek lowercase **pi**) with the sideways overlap of **p** orbitals, and σ (Greek lowercase sigma) with s orbitals, which cannot overlap sideways.

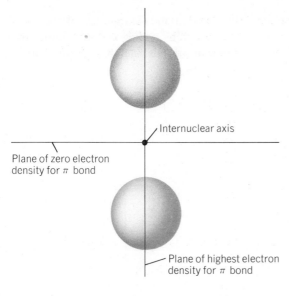

Plane of zero electron density for π bond

Internuclear axis

Plane of highest electron density for π bond

FIGURE 3-7 Another view of the π bond introduced in Figure 3-6. This time the internuclear axis is perpendicular to the page, protruding above and below the paper. The regions of overlap have mirror reflection symmetry across a plane of zero electron density. There is no cylindrical symmetry around the internuclear axis.

HYBRIDIZATION. We come now to the question of *how many* covalent bonds an atom can form. The answer is not always immediately clear from a simple energy-level diagram. Consider the classical case of carbon, a favorite example: The electron distribution in the ground state is

$2p$ —↿— —↿— ——
$2s$ —↿⇂—
$1s$ —↿⇂—

From this, we would expect that carbon could unite with two atoms of hydrogen to form CH_2, pairing off the $2p$ electrons. But we have already encountered the compound CH_4, and the alert student must already have asked, "How is it possible for carbon to combine with four hydrogen atoms?" The answer lies in **hybridization**, or mixing, of the outermost orbitals.

A slight complication

Our task is to explain the experimental observations that CH_4 is a highly symmetric molecule, with all four hydrogens in equivalent positions. All carbon-hydrogen internuclear distances are the same. All hydrogen-carbon-hydrogen internuclear angles are the same. All hydrogen-hydrogen internuclear distances are the same. All this must surely mean that all four carbon-hydrogen bonds are identical. Hybridization of the four $2s$ and $2p$ orbitals yields four orbitals differing only in their direction in space; we shall make immediate use of these four equivalent hybrid orbitals in our attempts to describe four equivalent chemical bonds in CH_4.

The significant mathematical fact underlying hybridization is that the set of orbitals whose properties are displayed in Figure 2-6 *is just one of the many sets* of solutions of the Schrödinger equation for an isolated atom. Other sets of valid solutions can be obtained by addition and subtraction[8] of members of the set of Figure 2-6. One such set, formed by adding and subtracting the $2s$ and the three $2p$ orbitals, consists of four orbitals, each having the shape

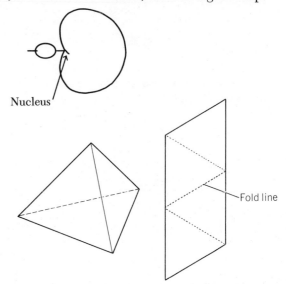

Nucleus

FIGURE 3-8 A model of a regular tetrahedron. If you have trouble conjuring up an image of a tetrahedron in your mind, try making a model which you can then hold in your hand. A piece of heavy paper will do.

Fold line

If the carbon atom is imagined at the center of a regular tetrahedron, then the proper relative orientations of the four sp^3 orbitals can be visualized by having each orbital pointing out from the carbon nucleus toward one of the vertices of the tetrahedron (see Figure 3-8). In terms of an energy-level diagram, again with dots to show the original electrons and crosses to show the electrons contributed by hydrogen in forming a compound, the steps would be

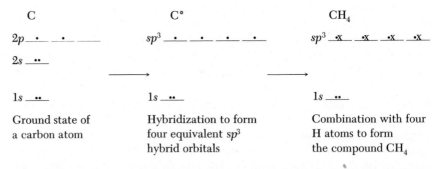

C	C°	CH$_4$
Ground state of a carbon atom	Hybridization to form four equivalent sp^3 hybrid orbitals	Combination with four H atoms to form the compound CH$_4$

So we associate sp^3 hybridization with tetrahedral structure. The sp^3 hybridization, which describes the covalent bonding in CH$_4$, serves as a prototype

---

[8] It sounds more sophisticated to call these operations of addition and subtraction *the formation of linear combinations*. Thus, we might (and shall) speak of "the formation of four appropriate linear combinations of s and p orbitals to yield four sp^3 hybrid orbitals." For more of this, see Chapter 12.

for most of the compounds of carbon. The usual geometry in carbon chemistry, as it develops into organic chemistry, is based on a tetrahedral configuration around each carbon atom.[9]

Four bonds to carbon are *possible*. But we still have not spoken to the question of why CH_4 is the preferred hydride, not CH, CH_2, or CH_3. As we have already found in other cases, energy considerations provide the reasons. Energy is required to make the $2s$ electrons available for covalent bonding, but then a great deal more energy is released by the formation of four carbon-hydrogen bonds rather than one, or two, or three. A collection of four hydrogen atoms and one carbon atom is in its lowest potential energy state when *four* carbon-hydrogen bonds have been formed. The process stops at four, because then the possibilities for electron sharing have been exhausted. All carbon orbitals are filled, and the carbon nucleus is surrounded by a neon configuration of electrons.

There are four electron pairs, and there are four bonds.

Hybridization of s and p orbitals on other atoms also is possible, and for the same considerations of energy. We find sp^3 hybridization of nitrogen in ammonia (NH_3), with one of the four equivalent hybrid orbitals occupied simply by a pair of electrons which were present in the original nitrogen atom. Indeed, there probably is sp^3 hybridization of the atomic orbitals of oxygen when water is formed, for the H–O–H bond angle is found to be 104.5°, which is much closer to the theoretical tetrahedral angle of 109° (center to two corners) than to the 90° expected from the diagram of Figure 3-4. That is why the simple overlap of Figure 3-4 is insufficient, as pointed out earlier on page 67. Hybridization of the s and p orbitals of oxygen is logical, even though it does not lead to the formation of more covalent bonds, because it allows the hydrogen atoms of H_2O to take up positions further from each other than would be possible from simple overlap of p orbitals, and the mutual repulsion of hydrogen nuclei provides a strong reason for maximum separation. Thus, a state of lower potential energy is achieved at the same time; the H_2O molecule cannot be linear (H–O–H) because there are two pairs of electrons in the already-filled sp^3 orbitals, and these repel each other *and* the bonding electrons.

Hybridization can also be used fruitfully to provide alternative descriptions of multiple bonding. For instance, our σ-π description of a double bond was in terms of two different bonds between the same nuclei, bonds different in energy and different in shape. Hybridization can yield two equivalent orbitals to describe the same double bond. The wave functions for the σ bond and the π bond can be combined to produce two new hybrid bonding orbitals that are identical except for their orientation in space. These hybrid bonding orbitals look like those shown by contour lines in Figure 3-9. The electrons in such bonds would then tend to stay as far apart as possible while remaining in the internuclear bonding region, thus satisfying both the requirement that like electrical charges repel each other, and the requirement of the Pauli

Another description of the same situation

---

[9] When carbon forms four σ bonds, the hybridization is always sp^3 and the structure tetrahedral. But when double bonds are formed, the hybridization is necessarily limited to sp^2 (triangular), and when triple bonds are formed, the hybridization is sp (linear).

exclusion principle that electrons tend to arrange themselves in spin pairs separated in space from other pairs. In the same manner, hybridization of one σ orbital and two π orbitals yields three identical-in-shape bonding orbitals, differing only by being rotated a third of the way around the internuclear axis.

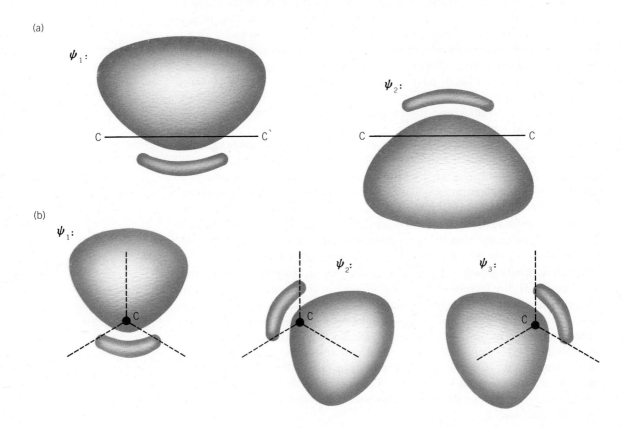

FIGURE 3-9 (a) Side views of a carbon-carbon double bond: linear combination of a σ orbital and a π orbital provides two equivalent σ-π hybrid bonding orbitals, labeled ψ_1 and ψ_2. (b) End views of a carbon-carbon triple bond: linear combination of a σ orbital and two π orbitals produces three equivalent σ-$π^2$ hybrid bonding orbitals, labeled ψ_1, ψ_2, and ψ_3.

Bonds like those in Figure 3-9 are called, from their shapes, "bent" bonds or, less formally, "banana" bonds. Chemists used to thinking of double bonds in terms of the σ-π picture are generally skeptical of the banana-bond picture, but the beginner is in a position to realize that both are only mathematical and pictorial models, each having its particular usefulness and limitations as a description of what we guess to be reality.

Other forms of hybridization are recognized in chemistry and are incorporated into the operations of wave mechanics to produce a comprehensive theory of molecular orbitals. Some aspects of this theory will be considered in Chapter 13, where they are relevant to the structure and function of coordination compounds, substances that are essential to living organisms and very useful in chemical synthesis. The *formation* of coordination compounds, or rather the formation of coordinate covalent bonds, is pertinent right now, however, and need not be delayed until Chapter 13.

3-6 Coordinate Covalent Bonds

Covalency is not limited to situations in which two atoms can each furnish one electron to form a shared pair that becomes a bond. Far from it. One molecule, or ion, or atom, may have a pair of unused electrons which it can donate to an empty orbital of another molecule or ion or atom, and so a shared-pair bond can be set up in this way. For example, carbon monoxide has a pair of electrons which are readily donated to metal atoms or ions, and this is exactly why carbon monoxide is such a dangerous poison as an air pollutant. It gets into the air as a result of incomplete combustion,[10] especially incomplete combustion of the gasoline in automobile engines—typical exhaust gas contains 3.5 percent CO. The gas is colorless, odorless, and tasteless; its presence can be discerned only by its effects. It causes dizziness and headache at first, then unconsciousness, and finally death due to lack of oxygen. This comes about because the hemoglobin of human blood has 200 times as much affinity for carbon monoxide as for oxygen, and so takes up CO rapidly and holds on to it. The oxygen-transport mechanism thereby is impaired, and although the victim still looks ruddy and healthy (the hemoglobin-CO compound is bright pink), he will die of anoxemia if the concentration of CO in the air he breathes is 0.2 percent or more. The strong attachment of CO to hemoglobin is a result of donation of a pair of electrons on the carbon atom to an empty d orbital of iron in the hemoglobin. Perhaps the structural formula for hemoglobin would appear too formidable to the reader at this stage, but the principle can be illustrated just as well by considering an isolated iron atom and following its mode of combination with CO molecules. The structure of CO can be written as $:C\equiv O$, in which carbon is bound to oxygen by a σ bond and two π bonds, and in which a pair of unused electrons occupies one of the carbon orbitals and is indicated by two dots. Elemental iron has a d^6 configuration, with room for four more electrons in the $3d$ orbitals and for six electrons in the $4p$ orbitals. The filled but unused orbital of carbon may overlap one of the unfilled d or p orbitals of iron, setting up a covalent bond by a donor-acceptor mechanism:

$$Fe + :C\equiv O \longrightarrow Fe-C\equiv O$$

Since carbon monoxide contributes two electrons per molecule of CO, and since iron has room for exactly 10 more electrons before its $3d$ and $4p$ orbitals

Variation on the theme

Compare 3.5 percent in exhaust gas!

---

[10] Even the smoke from a cigarette contains carbon monoxide, and it is estimated that steady smokers inactivate some 10 percent of their hemoglobin by chronic CO poisoning.

are full (verify this by counting off the spaces in the periodic table), the process can be repeated until we arrive at the compound

$$\begin{array}{c} O \\ \| \\ C \\ | \\ O \equiv C - Fe \\ | \\ C \\ \| \\ O \end{array} \begin{array}{c} C \equiv O \\ \\ C \equiv O \end{array}$$

Pentacarbonyliron(0)

(yellow liquid, melting point $-21°C$, boiling point $103°C$)

in which the orbitals of iron are saturated. The five carbon monoxide groups are distributed essentially uniformly around the central iron atom.[11] This is a good example of a **coordination compound**, and the bonds are spoken of as **coordinate covalent bonds**, or **donor-acceptor bonds**.

A simpler example of coordinate covalency is provided when ammonia is dissolved in water to form an alkaline solution ("ammonium hydroxide"). Here the $:NH_3$ molecule, with a pair of unused electrons in the fourth sp^3 hybrid orbital of nitrogen, donates a pair of electrons to the empty $1s$ orbital of a hydrogen ion, H^+:

$$H^+ + :NH_3 \longrightarrow NH_4^+, \text{ ammonium ion}$$

The ammonium ion is tetrahedral.

Once the coordinate covalent bond is formed, it is the same as the other three covalent bonds, and the four hydrogen atoms are equidistant from the nitrogen atom and from one another. Notice that the NH_3 molecule is neutral, so its coordination to H^+ neither diminishes nor increases that charge. In the same way, neutral NH_3 molecules can (and do) coordinate to Cu^{2+} ions, Ni^{2+} ions, Fe^{3+} ions, Co^{2+} ions, and indeed almost all transition-metal ions to form coordination compounds of distinctive colors. Similarly, H_2O molecules can donate pairs of electrons to positive metal ions to form hydrated salts, and to neutral acceptor molecules to form hydration products. Coordination is often a first step toward further reaction, for a coordinated ion or molecule is in a good position to undergo a more profound rearrangement of electrons with the other parts of the molecule.

There can be still other types of covalent bonds, such as delocalized-electron bonds and three-center bonds, but these are comparative rarities, and discussion of them is best left to specific examples. Considering just ordinary one-electron-per-atom covalent bonds and coordinate covalent bonds, we see that there are some general characteristics which can be listed for such bonds, no matter which way they are formed:

---

[11] There is no way to distribute five groups uniformly around a central point! The actual arrangement with pentacarbonyliron (0) has the symmetry of a trigonal bipyramid, a geometrical solid which can be constructed by gluing together a face from each of two regular tetrahedrons. This geometry comes from hybridization of the iron orbitals.

1. Covalent bonds are *specific links* between individual atoms, and not general electrostatic attractions such as hold together ionic compounds.
2. As a consequence of the orbital overlap, covalent bonds are directional in space.
3. There is often a possibility of π bonding as well as σ bonding, leading to multiple bonds.
4. Covalent bonds between dissimilar atoms always involve some internal distribution of charge, leading to dipole moments in unsymmetrical structures.
5. Many covalently bonded compounds are gaseous, but many others are liquids or solids at room temperature. Some are even very high-melting solids (SiC, mp 2600°C). The range of physical properties is greater in covalent compounds than in ionic compounds.
6. Covalent compounds do not conduct an electric current readily, the way liquified or dissolved ionic compounds do, but they may *react* with water or other liquids to produce ionic products that do conduct.

Again, some generalizations

3-7 Molecular Weights and the Mole

When an atom of hydrogen (atomic weight 1.008) combines with an atom of chlorine (atomic weight 35.453) to form a molecule of HCl, that molecule obviously has a **molecular weight** equal to the sum of atomic weights, or 36.461.[12] Similarly, H_2O has a molecular weight of 15.999 + 2(1.008), or 18.015, and F_2 has a molecular weight of 37.997. Each molecule, of any composition, has a molecular weight equal to the sum of the atomic weights of the atoms it contains.

Since the molecular weights of covalent compounds are so useful (indeed, so necessary) in any quantitative consideration of chemical reations, or in figuring how much of one substance may be made from a given amount of another substance, we shall need a similar concept which can apply to *ionic* compounds. The easiest step to take is to consider *all* compounds as though they were indeed covalent, as least for the purposes of following the weight relations during chemical reactions. It makes no difference to the weights of atoms whether they are bonded by ionic or covalent or any other kind of bonds, so it is no distortion of the truth to adopt the convenient view that the concept of molecular weights shall embrace all kinds of compounds, and that *molecular weights shall include formula weights of ionic compounds*. So we say that sodium chloride, NaCl, has a molecular weight of 58.443 (which is the sum of the atomic weights of Na and Cl), even though it contains no molecules.

A simplifying assumption

Just as we needed a convenient weighable, seeable, touchable quantity of atoms to use in laboratory experiments and chose 6.023×10^{23} atoms (Avogadro's number) as the most logical number of atoms to use, so we may

---

[12] The energy lost, 22 kcal (22,000 cal) per gram-atomic weight of Cl reacting, is equivalent (by $E = mc^2$) to only 0.000000001 g, and so is not significant here.

apply the same arguments of Chapter 2 to our present problems and choose 6.023×10^{23} *molecules* of a compound as a convenient quantity for people to work with. We call this quantity of compound a **mole** (after the Latin *moles:* a mass or pile of material). It follows that a mole is a **gram-molecular weight,** a mass of substance having a weight in grams numerically equal to the molecular weight (and hence the formula weight, if the substance is an ionic compound). It also follows that the weight of one mole of a compound is equal to the sum of the gram-atomic weights of all the elements it contains, in the proportions indicated by the formula. So a mole of water weighs 18.015 g, and a mole of fluorine weighs 37.997 g. A mole of UF_6 weighs 352.02 g, or $\frac{7}{9}$ths of a pound, and this seems so much that it points to the necessity of using *millimoles,* or one-thousandth of a mole, at times, just for convenience.

It may seem unnecessary to talk about moles and gram-atomic weights by separate names, when both involve 6.023×10^{23} units. True, in modern chemical usage the word *mole* covers both, and the term is even extended further. So the reader will probably hear chemists speak of a mole of helium (even though it consists of only atoms), and also of "a mole of electrons," when they mean 6.023×10^{23} electrons (equivalent to one faraday). No matter how widely the word "mole" is used in chemical parlance, however, it always has a precise quantitative meaning, for it refers to Avogadro's number of the items being considered.

3-8 Formulas and Their Quantitative Significance. Chemical Arithmetic Concerning Formulas

The preceding sections have considered the ways in which atoms combine to form compounds. The discussion was quite theoretical in places, but now the theory of chemical combination leads to a very practical area: the quantitative aspects of chemical change. We shall take up the methods for recording and summarizing the interactions between substances, and the simple chemical arithmetic that governs the quantities of materials involved.

Before considering the writing of **chemical equations** and the quantitative interpretation of them, it would be well to notice that a chemical formula gives us *quantitative information* along with the elementary composition. The formula H_2O tells us not only that the compound we call water contains in each of its molecules one atom of oxygen and two atoms of hydrogen, but also tells us *the proportions by weight of these elements.* If we were to take one mole of water (18.0 g), the hydrogen in that much water (2.02 g of H) would constitute

$$\frac{2.02 \text{ g}}{18.0 \text{ g}} \times 100 = 11.2 \text{ percent of the weight}$$

Similarly, the oxygen would constitute the balance, or

$$\frac{16.0 \text{ g}}{18.0 \text{ g}} \times 100 = 88.8 \text{ percent}$$

So we can obtain the proportions by weight of the constituent elements, if we know the formula.

When new compounds are made for the first time, we do not know their formulas. We may *hope* that we know their compositions, but these must immediately be *proved* in order to have any acceptance. The usual (and universally applicable) method for determining the formula is to analyze the compound in two ways. First, it is analyzed to determine what elements are present, and second, quantitative analytical procedures are used to determine the percentage of each element by weight. The methods for accomplishing these two aims comprise the subject matter of *analytical chemistry*, a large branch of the science of chemistry. We are not concerned here with the methods, but only with the results. Suppose, for example, that carbon tetrafluoride had never been made before, and we knew only that a gas could be isolated from the action of a large excess of fluorine on organic materials. After purification, the gas was subjected to analysis for the likely elements and was found to contain only carbon and fluorine. Quantitative analysis then showed that the gas contained 13.6 percent carbon, by weight, and 86.4 percent fluorine. What is its formula?

The analyst

If we took exactly 100 g of the compound, that 100 g would contain 13.6 g of carbon and 86.4 g of fluorine. Those are the proportions by weight. The *atomic* proportions can be determined by finding how many gram-atomic weights of carbon and fluorine are present in that 100 g of compound. These will be

$$\frac{13.6 \text{ g of carbon}}{12.0 \text{ g/g--at.wt. of C}} = 1.13 \text{ g--at.wt. of C}$$

and

$$\frac{86.4 \text{ g of fluorine}}{19.0 \text{ g/g--at.wt. of F}} = 4.54 \text{ g--at.wt. of F}$$

Here are the atomic proportions, but they look very strange because we always write formulas with the elements in integral proportions. We can convert our result to integers by dividing both numbers by the lower one, thereby making that one unity:

$$\frac{1.13}{1.13} = 1.0 \text{ atoms of carbon}$$

$$\frac{4.54}{1.13} = 4.0 \text{ atoms of fluorine}$$

Now we can write the formula CF_4 for the compound.

It might be, of course, that a compound contains atoms in the proportion of small whole numbers *other than one*, as in Cr_2O_3 or Fe_3O_4. In such a case, the situation usually will make itself clear, as in this example:

We analyze a white sapphire, and find that it contains only aluminum (52.9 percent) and oxygen (47.1 percent). What is the chemical formula? Proceeding as before,

$$\frac{52.9 \text{ g of Al}}{27.0 \text{ g/g--at.wt. of Al}} = 1.96 \text{ g--at.wt. of Al}$$

and

$$\frac{47.1 \text{ g of O}}{16.0 \text{ g/g–at.wt. of O}} = 2.94 \text{ g–at.wt. of O}$$

We divide these numbers by the smaller, and we have

$$\frac{1.96}{1.96} = 1.00 \text{ atoms of Al}$$

and

$$\frac{2.94}{1.96} = 1.50 \text{ atoms of O}$$

But we cannot write the formula as $AlO_{1.50}$, because we cannot have half of an atom of oxygen involved in the chemical bonding. There must be an integral number of whole atoms. Yet the ratio of atoms is clear enough: It is 1.0 to 1.5, or 2 or 3. So we may write Al_2O_3 as the chemical composition of sapphire. A blue sapphire, by the way, would show much the same composition on chemical analysis, but careful spectrographic study (see Chapter 2) would show a small amount (probably about 0.1%) of titanium as impurity, and possibly a trace of iron as well. If the impurity were chromium (0.2 to 0.3 percent), the aluminum oxide crystal would be colored red, and we would call it a ruby.[13]

A few "impurities" make all the difference.

Any such calculation based on composition by weight can give only the simplest possible formula, called the **empirical formula,** which tells us only the names and proportions of the elements in the compound, and nothing of its structure or complexity. If we have further information, gained by experiments that can reveal molecular weight and actual placement of atoms, we may be able to go beyond the empirical formula and write a **structural formula.** In any case, however, the study of a chemical compound always begins with elementary analysis and the determination of the empirical formula. A final example will make this clear. In an actual research project, the reaction of methyl alcohol with heated silicon produced a volatile liquid. The liquid was analyzed and was found to contain 7.80 percent hydrogen, 31.8 percent carbon, 18.4 percent silicon, and 42.0 percent oxygen. What is it?

This is a true story.

The empirical formula of the mysterious substance was first calculated. Taking 100 g of the substance, we would have

$$\frac{7.80 \text{ g of hydrogen}}{1.008 \text{ g/g–at.wt. of H}} = 7.74 \text{ g–at.wt. of H}$$

$$\frac{31.8 \text{ g of carbon}}{12.01 \text{ g/g–at.wt. of C}} = 2.65 \text{ g–at.wt. of C}$$

$$\frac{18.4 \text{ g of silicon}}{28.09 \text{ g/g–at.wt. of Si}} = 0.65 \text{ g–at.wt. of Si}$$

$$\frac{42.0 \text{ g of oxygen}}{16.00 \text{ g/g–at.wt. of O}} = 2.62 \text{ g–at.wt. of O}$$

---

[13] The study of gems and semiprecious stones can provide a fascinating introduction to chemical composition and the structure of crystalline compounds. See any textbook of mineralogy, or the *McGraw-Hill Encyclopedia of Science and Technology.*

These four numbers must then be divided by the smallest one, 0.65, in order to get the atomic ratios as small whole numbers. When this is done, the result is found to be 11.9 atoms of hydrogen, 4.08 atoms of carbon, 1.00 atoms of silicon, and 4.03 atoms of oxygen. Allowing for small errors in the analysis, it is evident that the empirical formula is $SiC_4H_{12}O_4$. But we cannot understand so complicated a formula without some inkling of how the compound is put together. How is all that hydrogen bonded, for example? Is it attached to carbon or to oxygen? Or to both? To answer these questions, we must do further experiments. We soon find that the compound reacts with water, especially if some acid is present, and a product of the reaction is the familiar, rather simple, easily recognized substance *methyl alcohol*, CH_4O, our starting material. This alcohol is known (from hundreds of experiments during the past 100 years) to have the structure

$$\begin{array}{c} H \\ | \\ H - C - OH \\ | \\ H \end{array}$$

and when it reacts with metals and metalloids, it does so by losing the hydrogen atom from its OH groups and attaching the remaining CH_3O- group to the metal or metalloid. So we know that the carbon, hydrogen, and oxygen in our new compound are present in the form of four CH_3O- groups. Since silicon is directly below carbon in the Periodic Table and may be expected to participate in the same sort of hybridization described in Section 3-5 for carbon, we expect a tetrahedral structure for our new compound, with four CH_3O- groups arranged at the corners of a tetrahedron surrounding the silicon atom. To write it on a flat surface according to the usual convention, we have the *structural formula*

$$\begin{array}{c} OCH_3 \\ | \\ H_3CO - Si - OCH_3 \\ | \\ OCH_3 \end{array}$$

A few numbers led to a structure.

3-9 Chemical Equations

When elements combine to form compounds, or when compounds undergo chemical change to form new substances, the event is called a chemical **reaction.** We distinguish between the substances (elements or compounds) we started with (called **reactants**) and the new substances that come out of the reaction (called **products**). At first we seek only a clear and concise way of indicating what happened. For an example of a reaction, we might take the production of aluminum by a process invented by Charles M. Hall in 1886 while he was a student at Oberlin College. In this process, the source of aluminum is its oxide, a high-melting solid which is insoluble in water but is soluble in red-hot liquid cryolite (sodium aluminum fluoride). **Electrolysis** of the hot solution gives molten aluminum and liberates oxygen. We could describe the successive events in regular prose at first, in words such as these:

Dialuminum trioxide dissolves in melted trisodium aluminum hexafluoride to provide independently mobile tripositive aluminum ions and dinegative oxide ions. If carbon electrodes[14] are placed in the red-hot solution, and a strong electric current is passed through the solution at low voltage, the tripositive aluminum ions are attracted to the negative electrode, where they receive electrons which neutralize their charges and convert them to aluminum atoms. Hence globules of molten metallic aluminum form on the negative electrode and drop to the bottom of the container. At the same time, the negative oxide ions are attracted toward the positively charged electrode and migrate toward it through the solution. Upon reaching the surface of the positive electrode, two electrons are withdrawn from each oxide ion by the force of the electric field, leaving an oxygen atom. If the temperature were lower, two atoms of oxygen could combine to form a diatomic molecule of oxygen, but at red heat the oxygen atoms combine with the carbon of the electrode and form carbon monoxide, which soon burns to carbon dioxide. Molten aluminum is withdrawn from the bottom of the vessel periodically, and fresh dialuminum trioxide is added to the solution from above, so the process is continuous.

Much of the information contained in the preceding paragraph could be summarized by using formulas and equations, with a large saving of time and patience and with no loss in exactness. The prose description is not incorrect, it is just needlessly long drawn out, and conveys almost no quantitative information. By using formulas instead of names for the compounds, and writing **chemical equations** to show the rearrangement of atoms, we can describe the reaction in terms of reactants, conditions of reaction, and products. Such a description would go this way:

$$Al_2O_3(s) \xrightarrow[1000°C]{Na_3AlF_6} 2Al^{3+}(l) + 3O^{2-}(l)$$

$$Al^{3+}(l) + 3e^- \xrightarrow{electrolysis} \underline{Al}(l)$$

$$O^{2-}(l) \xrightarrow{electrolysis} [O] + 2e^-$$

$$C(s) + [O] \xrightarrow{1000°C} CO\uparrow(g)$$

$$2CO(g) + O_2(g) \longrightarrow 2CO_2(g)$$

The state of the substance is indicated by the notation (s) for solid, (l) for liquid, and (g) for gas. (The line under the symbol Al means that it has separated from the solvent and will settle to the bottom, whereas the upward arrow written after CO means that the CO goes off as a gas.) To condense further, the essentials of the process could also be summarized in just one line, which shows the overall net change:

$$Al_2O_3(s) + 3C(s) \xrightarrow[electrolysis]{Na_3AlF_6 \; 1000°C} 2\underline{Al}(l) + 3CO\uparrow(g)$$

---

[14] *Electrode* (literally, a door by which electrons enter or leave), a conducting solid introduced into a chemical system for the purpose of passing an electric current through the system. Some of the other words in this paragraph may not be familiar to the student at this point, but they do not need to be. The main reason for the paragraph is to show how long and involved a prose description of a reaction can be, and how much is saved by writing an equation.

This concise summary, for all its brevity, tells us this much *more* than the prose account:

1. It tells us that three atoms of carbon are consumed for every "molecule" of Al_2O_3, giving three molecules of CO as by-product for every two atoms of aluminum produced.
2. If we start with one *mole* of Al_2O_3, we shall require 3 gram-atomic weights (or 3 moles) of carbon, and we shall obtain 2 gram-atomic weights (or 2 moles) of aluminum, along with 3 moles of by-product CO.

This second conclusion follows directly from the first and is unavoidable. When we say "one mole of Al_2O_3," we mean exactly Avogadro's number of Al_2O_3 units, and so we necessarily produce twice Avogadro's number of Al atoms and thrice Avogadro's number of CO molecules, because no atoms can be destroyed or created along the way. The chemical equation is not a mathematical equation, to be sure, but it is exact in this respect: The same number of atoms of each particular element must appear on the right side of the arrow as appear on the left side. Only when all atoms are accounted for in this way is the equation said to be *balanced*.

3-10 Weight Relations. Chemical Arithmetic Concerning Equations

We can go one step further with our one-line summary known as a chemical equation; we may indicate the *actual weights* in grams which correspond to the number of moles involved according to our conclusion 2 in the preceding section:

$$Al_2O_3 + 3C \xrightarrow[\text{electrolysis}]{Na_3AlF_6 \; 1000°C} 2Al + 3CO\uparrow$$

$$101.96 \text{ g} \quad 36.03 \text{ g} \qquad\qquad 53.96 \text{ g} \quad 84.03 \text{ g}$$

This statement makes the significance of the equation vividly clear, for it tells us exactly how much aluminum we can get, under the best possible conditions,[15] from 101.96 g of aluminum oxide. But suppose we do not start with 101.96 g of Al_2O_3. Suppose we have on hand, and want to use, twice that much, or ten times that much, or a thousand times that much? Then the amount of aluminum that can be obtained will obviously be in the same proportion, because the equation applies just as exactly to every mole or millimole or 1000 moles of Al_2O_3 used. Doubling the amount of starting material doubles the amount of product to be expected, and so on, by simple arithmetical proportion. To put it generally, every balanced chemical equation represents a quantitative statement about the weight relations of reactants and products. This is the basis of **stoichiometry**, the science that deals with

This is all there is to a problem in chemistry.

---

[15] The equation assumes 100 percent efficiency in the process and makes no allowance for factors that might lower efficiency. Hency the yield indicated by the equation represents a theoretical maximum, never quite attained in practice.

the measurement of relative proportions of elements and compounds in chemical reactions.

The practical importance of stoichiometry can be appreciated readily from a simple example. Suppose we ask this question: What weight of carbon electrodes will be consumed in the making of a metric ton (1000 kg) of metallic aluminum? A thousand kilograms is 1,000,000/53.96 times larger than the 53.96 g of aluminum specified in our equation, so that operation will require 1,000,000/53.96 times as much carbon. That is,

Where do we get all the carbon electrodes?

$$\frac{1,000,000}{53.96} \times 36.03 \text{ g C} = 668 \text{ kg of carbon required}$$

So even though the method produces aluminum electrolytically, and not by reduction of the ore with coke in a blast furnace, the way iron is produced, the manager of the plant still has to count on two-thirds of a ton of carbon being consumed for each ton of aluminum made, in addition to the cost of the electricity. This is one big reason why aluminum costs more per pound than iron.

3-11 Reactions in Water Solution. Molarity

Most of the reactions in the human body take place in water solution, and most of the changes that take place upon the surface of the earth take place through the agency of ground water or within the oceans themselves. Furthermore, many of the laboratory experiments that are carried on in chemistry and medicine use solutions of reagents. So we need to extend stoichiometry in a way that will allow us to determine quantities by measuring *volumes of solutions*, rather than by weighing everything.

A **solution** consists of a *solvent* and one or more substances (called *solutes*) dissolved in it. In order to carry out quantitative work, we need to know *how much* of a particular solute is present in a given volume of solution. Such an expression could be in percent by weight, or in pounds per gallon, or in grams per liter, or in any other system of units. However, we have seen particular advantage in expressing quantities of chemical substances in *moles*, so it would be convenient to adopt a system of expressing concentration in terms of *moles of solute per liter of solution*. This is called the *molarity* of that substance and is expressed by a number followed by M. Thus, a 1.00 M solution of sulfuric acid contains exactly one mole (98.08 g) of H_2SO_4 per liter of the solution, and a 0.10 M solution of H_2SO_4 contains 9.81 g of H_2SO_4 per liter of solution, and so on. If we always express the volume in liters, the product of volume times molarity will give us the actual number of moles; from there on, we can proceed just as in any problem involving weight relations.

Molarity is a convenient unit of concentration.

EXAMPLE: Some industrial hydrochloric acid must be neutralized before disposing of it. What weight of sodium hydroxide will be needed to neutralize 1000 gal (3785 liters) of 5 M hydrochloric acid?

Solution: First we need a balanced equation, which is

$$HCl + NaOH \to NaCl + H_2O$$

From this we see that sodium hydroxide (NaOH) reacts with hydrochloric acid mole for mole. The number of moles of HCl is

$$3785 \text{ liters} \times 5 \text{ moles/liter} = 18{,}925 \text{ moles}$$

Hence we shall need 18,925 moles of NaOH, and that means

$$18{,}925 \text{ moles} \times 40.0 \text{ g/mole} = 757{,}000 \text{ g of NaOH} = 757 \text{ kg of NaOH}$$

or

$$757 \text{ kg} \times 2.20 \text{ lb/kg} = 1665 \text{ lb NaOH}$$

If we were to ask *what volume* of 5 *M* solution of NaOH will neutralize 1000 gal of 5 *M* HCl, the answer would be 1000 gal. Or, if we asked what volume of 10 *M* NaOH will neutralize 1000 gal of 5 *M* HCl, the answer would be 500 gal (and so on, in proportion, for any such problem). All we need to know is the proper molar ratio of reactants for the desired reaction, and we can figure out the weight of substance or the volume of solution (of known concentration) that is required.

Would it be all right to dump the neutralized solution in a river?

Sometimes the operation illustrated above is turned around, especially in chemical analysis; we might find out experimentally what volume of Solution A reacts exactly with a known volume of Solution B (of known concentration), and we want to know the *concentration* in Solution A.

Example: It is found that 50 ml of a solution of NaOH will neutralize exactly 100 ml of a 0.5 *M* solution of sulfuric acid, H_2SO_4. What is the concentration of the solution of NaOH?

Solution: The equation is

$$2NaOH + H_2SO_4 \longrightarrow Na_2SO_4 + 2H_2O$$

from which we see that 2 moles of NaOH react with one mole of H_2SO_4. The experiment has used

$$0.1 \text{ liter} \times 0.5 \text{ moles/liter} = 0.05 \text{ moles of } H_2SO_4$$

so twice this number of moles of NaOH will be required, or 0.10 mole.

But this weight of NaOH must be contained in the 50 ml of NaOH solution which sufficed for the neutralization. With 0.10 mole of sodium hydroxide in 50 ml of solution, there must be 1.0 mole in 500 ml, and hence 2.0 moles in 1000 ml. Therefore, the concentration of NaOH in the solution is 2.0 moles/liter, or, in shorter terminology, 2.0 *M*.

The method of reasoning used in the last example was that of simple proportion, which comes logically and naturally to most students. For those who like everything reduced to a mathematical equation, the operation may be expressed this way:

$$\frac{0.10 \text{ mole NaOH}}{50 \text{ ml}} \times \frac{1000 \text{ ml}}{\text{liter}} = 2.0 \text{ moles/liter, or } 2.0 \text{ } M$$

It follows from these examples that the volume of a solution times its molarity gives us the number of moles, and, knowing the number of moles and having the balanced equation, we are on safe ground for doing all the chemical arithmetic of solutions.

3-12 Reactions Involving Gases

There is one more aspect of stoichiometry which should be considered at this time, and that is how to deal with *gases* when they react with other gases, with liquids, or with solids. It is difficult to weigh a gas; it is much easier to measure its volume. But the volume of a gas changes with pressure, so we need to record the pressure every time we measure the volume of a gas. Knowing the existing pressure, we can then calculate what volume the gas would have under *standard* pressure, which is one standard atmosphere (the average reading of a mercury barometer at sea level, and hence the pressure of atmospheric air that will support a column of mercury 760 mm high). Most students were taught how to do this in a general science course; for others, it need only be pointed out that the volume of a given weight of gas varies inversely with the pressure. Hence the observed volume is "corrected" to the volume it would have at standard pressure by multiplying it by the appropriate ratio of pressures.

Example: 100 ml of oxygen is collected in the laboratory at a barometric pressure of 740 mm. To find what volume would be occupied by this sample of oxygen at 760 mm, we multiply by the ratio 740/760:

$$100 \text{ ml} \times \frac{740 \text{ mm}}{760 \text{ mm}} = 97.4 \text{ ml}$$

We know the ratio must be 740/760, and not the reverse, because 100 ml of gas squeezed to a higher pressure (the standard 760 mm) would necessarily occupy a smaller volume. If the gas had been collected at 800 mm pressure, the ratio would have to be above unity, because at 760 mm the gas would occupy more volume.

Gases also expand when heated, and so the volume of a particular weight of gas varies with the temperature. The reason for this, together with the meaning of temperature and the reasons for different temperature scales, are all best left to the general discussion of the physical properties of matter in Chapter 5; if you already know or will believe (easily observed and verified by you) that the common gases expand or contract in direct proportion to the change in temperature in degrees Kelvin (°K), you will see (1) the need to correct all measured volumes of gases to volume at a standard temperature, and (2) how to do it. The agreed-upon standard temperature is the freezing point of water at one atmosphere pressure, which is 0°C or 273°K (273° on the absolute or Kelvin scale). Hence we record the existing temperature when a gas is collected and multiply its observed volume by the appropriate ratio of temperatures in degrees Kelvin to get the volume it would occupy at 0°C. (It always is necessary to translate to °K to achieve the direct proportionality; °K = °C + 273.)

EXAMPLE: In the previous example, the 100 ml of oxygen was collected at a room temperature of 30°C, which is 303°K. At 0°C (273°K), obviously it would have a smaller volume, being cooler, so we multiply the volume by the ratio 273/303. We may as well use the volume already corrected to standard pressure, which we found to be 97.4 ml:

$$97.4 \text{ ml} \times \frac{273°K}{303°K} = 87.7 \text{ ml at } 0°C$$

We could have just as well carried out both operations at once, namely,

$$100 \text{ ml} \times \frac{740 \text{ mm}}{760 \text{ mm}} \times \frac{273°K}{303°K} = 87.7 \text{ ml at STP}$$

The volume of gas is then said to be "corrected to standard temperature and pressure," written STP for short.

The reason for going to all this trouble is that once we know the volume of a particular quantity of gas at STP, *we know the number of moles.* This happy state of affairs comes about because, as Amadeo Avogadro discovered long ago, equal volumes of gases (of any sort) under the same conditions of temperature and pressure contain the same number of molecules. Specifically, if we adhere to *standard* temperature and pressure, one mole of gas occupies exactly 22.4 liters, whether the gas be hydrogen, oxygen, nitrogen, carbon monoxide, or whatever.[16] Hence 2.24 liters of any gas at STP will contain one-tenth of a mole, and 1.12 liters at STP will contain one-twentieth of a mole, and so on. Knowing this, we are in a position to deal with the stoichiometry of reactions that involve gases. Instead of knowing only that the reaction of one mole of sulfuric acid (98.1 g) with one gram-atomic weight of zinc (65.4 g)

$$H_2SO_4 + Zn \longrightarrow ZnSO_4 + H_2$$

gives one mole of hydrogen (2.02 g), we know also that it gives *22.4 liters of hydrogen* at STP. Hence if we need to prepare some hydrogen in the laboratory, we know how much zinc and sulfuric acid we shall need in order to get the desired quantity of hydrogen.

The figure 22.4 liters at STP is known as the *molar volume,* and its utility extends far beyond the petty considerations of laboratory experimentation no matter how important those may seem to the experimenter at the time. The stoichiometry of gases governs such enormous operations as burning coal in power stations, producing unwanted sulfur dioxide in the metallurgy of copper, and consuming air in the burning of gasoline in automobile engines. One everyday example may make this clear.

Another important number

EXAMPLE: How much air is required to burn one gallon of gasoline? (We shall assume this to mean complete combustion to carbon dioxide and water, but the same

---

[16] The "exactly 22.4 liters" is true for the gases mentioned, but the "whatever" is an oversimplification. Under conditions near the point of liquifaction, gases do not behave in "ideal" fashion, and deviations occur. Our statement holds for most common gases, though.

To burn gas in a car, how much air will you use?

Automobiles or people? We have to decide on priorities.

principles of stoichiometry of gases could be applied to finding out how much sulfur dioxide and carbon monoxide are produced, and so on.) Gasoline is a mixture of hydrocarbons, but its composition may be represented fairly well by the formula C_7H_{16}, with a molecular weight of $(7 \times 12) + (16 \times 1)$, or 100. We write the balanced equation for complete oxidation,

$$C_7H_{16} + 11\,O_2 \longrightarrow 7CO_2 + 8H_2O$$

From this we see that one mole of C_7H_{16} (100 g) requires 11 moles of oxygen, or $11 \times 32 = 352$ g of oxygen. But we could just as well write 11×22.4 *liters* of oxygen, or 246.4 liters of O_2 at STP. Since a gallon of gasoline weighs about 2800 g, a gallon will require 28 times as much oxygen as 100 g, so we write

$$\frac{2800\text{ g}}{100\text{ g}} \times 11 \text{ moles} \times 22.4 \text{ liters/mole} = 6900 \text{ liters of } O_2 \text{ at STP}$$

Since only a fifth of air is oxygen, we write

$$0.20 \times \text{total volume of air} = 6900 \text{ liters at STP (the oxygen)}$$

and the total volume of air is

$$\text{volume} = \frac{6900 \text{ liters}}{0.20} = 34{,}500 \text{ liters at STP}$$

This 34,500 liters of air is equal to about *9000 gal* of air, still at STP. At the more usual 30°C and prevailing average pressures, the volume of air is about 10,000 gal, just to burn one gallon of gasoline. The example shows us what enormous quantities of nitrogen (which is four-fifths of air) must be pushed through the automobile cylinders, heated up, partially oxidized to objectionable oxides of nitrogen, cooled again, and finally discharged (loaded with impurities) into the atmosphere. And this is the air we breathe!

You may find it interesting to figure out from the same chemical equation the volume of carbon dioxide produced by burning that one gallon of gasoline—you will be surprised at the quantity! Moreover, consideration of the huge volume of air that must be sent through the engine leads to an appreciation of how hard it is to get enough oxygen to the right place at the right temperature to get truly complete combustion. The usual result is that the exhaust gas contains 3.0 to 3.4 percent carbon monoxide, a most unpleasant substance to have around, for reasons we have already seen. Coping with the carbon monoxide, the residual hydrocarbons, the oxides of nitrogen and sulfur,[17] and the dispersed oil droplets in automobile exhaust is one of the major engineering problems of our time. Multiplying these considerations by 50 (because there are about that many more chemical problems of equivalent magnitude con-

---

[17] Crude oil contains anywhere from 0.2 to 7 percent sulfur, and although most of this is removed during refining, some still is present in gasoline. That which is removed must still be disposed of, of course.

nected with urban life)[18] will give you many good reasons for studying stoichiometry, for believing in it, and for using what you know about it, every day.

3-13 Summary

Atoms by themselves are but the means to an end: combination with other atoms to form the myriad chemical compounds which make up the animal, vegetable, and mineral worlds. In this chapter, we have had a short look at why atoms combine with other atoms, in terms of the release of energy (a view soon to be explored in detail in Chapter 9). The combination of atoms to form compounds takes place by one or a combination of several mechanisms, each of which involves the rearrangement of electrons from their atomic distributions to a new molecular or ionic distribution. Early in the chapter, the actual changes in arrangement of the electrons within several representative atoms were considered in detail as these atoms become the positive and negative ions of typical ionic compounds, and from these examples (and many more) some general characteristics of *ionic* compounds were deduced: Such compounds almost always are crystalline solids of moderate-to-high melting point, and those of moderate melting point usually are soluble in water (the ones of high melting point have such high lattice energies that their ions may not separate under the influence of water). Either in water solution or when melted, ionic compounds were seen to conduct electricity through actual migration of their ions in the electric field.

Next, the direct coupling of atoms by the process of *covalent* bonding was considered, and it was seen that the close approach of two atoms (with proper orientation) could lead to the overlap of singly-occupied atomic orbitals, resulting in a covalent bond between the two atoms. Although the bond itself consists in the attraction of both nuclei for the pair of shared electrons, and hence is electrostatic in origin, the covalent bond is distinguished over ionic bonds in several important respects: It is an individual bond between a particular pair of atoms; it is oriented in space with respect to any other covalent bonds of the same atoms, because the original atomic orbitals were so oriented; and the bond always has some degree of polarization of electric charge if the atoms are dissimilar, just because the two partners exert different attrac-

---

[18] Smog-producing hydrocarbons in the air, sulfur oxides and sulfuric acid from the burning of coal, the disposal of solid rubbish, the disposal of sewage, the removal of mercury and other poisonous metals from industrial waste, the elimination of smoke from industrial processes, the disposal of glass bottles and "tin" cans (and their separation from waste paper, plastics, and garbage if the combustible rubbish is to be burned), the recovery of useful sulfur from the flue gas of power stations, the recovery of steel, copper, and aluminum from worn-out automobiles, the elimination of phosphates from waste water and water supplies, the runoff of salt and other reagents used to melt ice on streets, the disposal of solid waste from mining operations, the recovery of phosphorus and potassium from minerals for use in fertilizer, obtaining nitrogenous plant food from atmospheric nitrogen, and so on, and so on.

tions for the shared pair of electrons. This latter consideration explains the dipolar effects of many covalent compounds.

When an empty atomic orbital overlaps a doubly occupied orbital of another atom, a *coordinate* covalent bond can result. Such a bond has all the characteristics of any other covalent bond, but in addition, it always is polar because of the outright donation of an electron pair. Coordinate covalent bonds are the basis of coordination compounds, a large class of substances (often involving transition metals) which will be treated in Chapter 13.

Single covalent bonds (σ bonds) sometimes are reinforced by subsidiary π bonds, which come about by the sidewise overlap of adjacent p orbitals on the atomic partners of the bond. A necessary condition for π bonds is just like that for σ bonds: Both of the overlapping atomic orbitals must contain single electrons, or else one of the orbitals must be empty and the other doubly occupied. There can never be more than two π bonds in addition to a σ bond because of the geometry of p orbitals.

Some atoms, such as carbon, silicon, and most transition metals, form more covalent bonds than would be possible by simple overlap of appropriately occupied atomic orbitals. This phenomenon is explained by a preliminary equal distribution of single electrons in a combination of available orbitals, a process called *hybridization* of the orbitals. Such a redistribution requires energy because it goes against the normal order of filling orbitals, but the necessary energy is furnished by the formation of a larger number of covalent bonds.

Lastly, the theory of atomic structure and chemical bonding provides not only an explanation of the material world around us, but also the basis for a *quantitative* description of chemical compounds and their interactions. This chapter describes some of the more elementary arithmetical operations made possible by that basis. If we have the chemical formula of a compound, we can easily calculate the percentage of each element in that compound; if we have the percentage composition by analysis, we can derive the empirical formula. Further, when substances react to form new substances, a balanced chemical *equation* can be written for the change; and when we have such an equation, we can calculate the *weights* of reactants and products that correspond to that change. When gases are involved, the task is simplified by making use of the universal *molar volume* of a gas, which is 22.4 liters under standard conditions. If we know the volume of a gas only at some other temperature and pressure, we can readily compute its volume at STP and then use that volume to calculate the quantities involved in a chemical reaction.

Not all the various aspects of chemical arithmetic have been presented in this chapter. Other kinds of calculation, important in the areas of electrochemistry and nuclear chemistry, will be considered later at the appropriate places. Too much chemical arithmetic at this point might lead to the conclusion that chemistry is all figures. This is not so; even professional chemists do not spend all their lives doing chemical calculations. The creative and utilitarian aspects of chemistry are too important for that. At the same time, every student should realize that the quantitative aspects are always there and cannot be avoided. Everything we do or make or hope to accomplish in a chemical way is governed by the weight relations which are inherent in chemical reactions.

Just because this chapter is purposely brief, the student would do well to choose a chemical workbook, or a book of programmed instruction in chemical arithmetic, and to work out some useful problems on his own. Any college bookstore or library has such books; a few are suggested in the reading list. You should pick one that corresponds to your own interests and your own study habits. The few problems that follow may be considered the barest start along the road to proficiency.

GLOSSARY

calorie: international unit of heat or energy; the amount of heat that will raise the temperature of one gram of water one degree centigrade. Abbreviation: cal.

chemical equation: a brief shorthand statement of the overall changes that occur in a particular chemical reaction, including the formulas of reactants and products, the molar proportions of each, and (usually) the conditions for the reactions.

compound (chemical compound): any combination of two or more atoms of different elements held together by chemical bonds.

coordinate covalent bond: a covalent bond in which *both* electrons of the bond come from one partner of the bond, rather than one electron from each partner.

coordination compound: a compound in which at least some of the atoms of the compound are held together by coordinate covalent bonds.

covalent bond: a specific and directional chemical bond between two atoms, arising from the sharing of a pair of electrons. A covalent bond becomes possible when atomic orbitals overlap, and two electrons are available to occupy the resulting molecular orbital (see Chapter 12).

crystal: a solid body composed of a regular, repetitive, three-dimensional array of atoms, molecules, or ions.

crystalline: made up of, or pertaining to, crystals.

dipole: a structure of molecular dimensions which have two regions of opposite electric charge.

donor-acceptor bond: a coordinate covalent bond.

electrolysis: a reaction which takes place upon the passage of an electric current through the reaction mixture.

electronegative: electron seeking; capable of attracting electrons to itself.

electropositive: electron releasing; characterized by a tendency to lose electrons and become positively charged.

empirical formula: the simplest formula that can be written for a compound, giving only the symbols of the constituent elements and their atomic proportions.

endoergic: characterized by an absorption of energy.

endothermic: characterized by an absorption of heat.

exoergic: characterized by an evolution of energy.

exothermic: characterized by an evolution of heat.

gram-molecular weight: a number (6.023×10^{23}) of molecules (or ions, if an ionic compound) sufficient to make up a mass, in grams, numerically equal to the molecular weight.

hybridization: a mixing of atomic orbitals to produce the same number of hybrid orbitals, all equivalent, of higher energy. The necessary energy usually comes from formation of a larger number of covalent bonds, by an exoergic process.

ionic bond: not a specific bond between a pair of atoms at all, but a summation of the electrostatic attraction which holds together compounds made up of positive and negative ions in an ionic compound.

lattice: an orderly, repetitive, three-dimensional array of atoms or molecules or ions, as in a crystal.

mole: short for gram-molecular weight; or, more generally, 6.023×10^{23} molecules of a particular compound. The word is also used loosely for 6.023×10^{23} atoms, electrons, or anything.

molecular weight: the sum of the atomic weights of the atoms that make up a given compound.

molecule: a stable association of atoms, held together by electostatic forces (called chemical bonds) between electrons and the nuclei of the bonded atoms.

oxidation state: the number of electrons lost or gained by an element in forming a compound (see Chapter 8 for details).

polar molecule: a molecule having two regions of opposite electric charge, due to its distribution of electrons.

product: substance formed during a chemical reaction.

reactant: one of the substances taking part in a reaction.

reaction: any process of chemical change, in which the atoms of the reactant(s) rearrange to form the product(s).

solution: a homogeneous mixture (usually liquid, but sometimes solid) of two or more substances, each dispersed molecularly in the other(s). The component present in largest proportion is usually called the *solvent*, and the other components are called *solutes*, but sometimes large amounts of solid may be dissolved in a small amount of liquid solvent.

stoichiometry: the portion of chemistry that deals with the quantitative proportions of elements in compounds, or of elements and compounds involved in a reaction.

structural formula: a graphic arrangement of atomic symbols designed to show the relative positions of the atoms in a molecule or an ion.

PROBLEMS

3-1 Ammonium nitrate (which is NH_4NO_3), ammonium chloride (which is NH_4Cl), and urea (which is $(NH_2)_2CO$ or N_2H_4CO) are three compounds which are produced in large quantities for various purposes. Each can serve also as a source of combined nitrogen for plant nutrition. Which of the three contains the largest proportion of nitrogen?

3-2 Dangerous explosives such as dynamite are no longer used in the mining of coal and salt; they have been displaced by a much safer and cheaper blasting material, which is a simple mixture of ammonium nitrate and fuel oil. Although it smells of fuel oil, the mixture looks like dry ammonium nitrate because it contains only 5 percent of the oil, soaked up by the dry powder. Five percent of oil may seem too little for maximum effect. On the other hand, it is essential to oxidize all of the carbon in the oil to CO_2 rather than CO, lest the miners be subjected to carbon monoxide poisoning. Is a mixture of 95 percent NH_4NO_3 and 5 percent $C_{12}H_{26}$ in strict stoichiometric proportion, or does it contain a surplus (even a deficiency) of ammonium nitrate? (*Hints:* First write a balanced equation for the reaction of NH_4NO_3 with $C_{12}H_{26}$ to give CO_2, H_2O, and N_2. Then find out how much NH_4NO_3 is required to react with precisely 5 g of $C_{12}H_{26}$, and see whether this is more or less than 95 g.)

3-3 Two colorful semiprecious stones are the bright green *malachite* $(Cu_2(OH)_2CO_3)$ and the deep blue *azurite* $(Cu_3(OH)_2(CO_3)_2)$. Which contains more copper, by weight? How does its percentage of copper compare with that of the common ore mineral of copper called *bornite* which is Cu_5FeS_4?

3-4 A common type of plywood adhesive is made from synthetic urea (see Problem 3-1) and formaldehyde (which has the empirical formula CH_2O). The adhesive is made by mixing 1200 kg of a 30-percent solution of formaldehyde in water with 720 kg of a 50-percent solution of urea in water. From these proportions by weight, calculate the molar proportions (that is, find out how many moles of formaldehyde combine with one mole of urea to make the adhesive).

3-5 A valuable copper-bearing ore contains the mineral *chalcopyrite*, which has 34.62 percent Cu, 30.45 percent Fe, and 34.93 percent S. What is the empirical formula of the mineral?

3-6 If coal were pure carbon, what volume of oxygen at STP would be required to burn one metric ton (1000 kg) of the coal? What volume of air (which is 20 percent O_2) at STP? What volume of air at 20°C and 735 mm pressure? How much CO_2 is produced, also at 20°C and 735 mm?

3-7 If the coal of Problem 3-6 were not pure carbon, but contained various sulfide minerals to the extent of 5 percent total sulfur, how much sulfur dioxide would be discharged into the air during the burning of 1000 metric tons of such coal in a power station? Give the answer both in grams and in liters at stack temperature (200°C) and pressure (780 mm).

3-8 A current plan to absorb sulfur dioxide from the stack gases of power stations depends upon the oxidation of SO_2 to SO_3, and the capture of the SO_3 by $Ca(OH)_2$ derived from limestone to form solid $CaSO_4$:

$$2SO_2 + O_2 = 2SO_3(g)$$

$$SO_3 + Ca(OH)_2 = H_2O + CaSO_4(s)$$

How much solid $CaSO_4$ must be disposed of each day, somehow, if 1000 metric tons of coal (5 percent sulfur) are burned at the station daily?

3-9 Trace (that is, very small) concentrations of substances in solutions or mixtures are often expressed using a simple weight fraction. This practice is especially prevalent in the context of environmental chemistry. For example, the United States Public Health Service maximum tolerance for lead (Pb) in drinking water is 0.05 part per million (ppm). That is, drinking water should contain no more than 0.05 g of Pb per million grams of water.

a. To what molar concentration of Pb^{2+} would this tolerance correspond? (You may assume that one liter of drinking water weighs 1000 g.)

b. What mass of lead would be contained in a municipal 10-million-gal water tank if the water in it contained 10 parts per *billion* lead?

3-10 Recent studies of Seneca Lake, New York, indicate that the chloride concentration of its water may have increased by 50 ppm (see Problem 3-9) during the past five years. Seneca Lake, one of the Finger Lakes, has a volume of 1.5×10^{13} liters. What mass of sodium chloride would have had to dissolve in the lake to produce this increase in chloride content? Use Appendix I to calculate the size of a cube that would contain this much salt.

SUGGESTIONS FOR FURTHER READING

Much has been written about chemical bonds, and just a sample is presented here as an introduction to an important section of the chemical library:

GILLESPIE, R. J., "Bond Angles and the Spatial Correlation of Electrons," *Journal of the American Chemical Society*, vol. **82**, p. 5978 (1960); "The Valence-Shell Electron-Pair Repulsion Theory of Directed Valency," *Journal of Chemical Education*, vol. **40**, p. 295 (1963).

LAGOWSKI, J. J., *The Chemical Bond*, Boston: Houghton-Mifflin (1966).

PAULING, L., *The Nature of the Chemical Bond*, 3rd ed., Ithaca, N.Y.: Cornell University Press (1960). There is a later, abridged paperback edition (Cornell University Press, 1967).

WALTERS, E. A., "Models for the Double Bond," *Journal of Chemical Education*, vol. **43**, p. 134 (1966).

There are available several excellent books written to help a student understand and master the concepts introduced in Sections 3-7 to 3-12. Among these are:

BARROW, G. M., et al., *Understanding Chemistry*, New York: W. A. Benjamin (1969). A sequence of programmed instruction in concepts and in chemical arithmetic (2 vols., paperback).

KIEFFER, W. F., *The Mole Concept in Chemistry*, New York: Van Nostrand Reinhold (1962, paperback).

NASH, L., *Stoichiometry*, Reading, Mass.: Addison-Wesley (1966, paperback).

QUICK, F. J., *Workbook for Introductory College Chemistry*, New York: Macmillan (1966).

4
Molecular Structure and Symmetry

*God, thou great symmetry
Who put a biting lust in me
From whence my sorrows spring,
For all the frittered days
That I have spent in shapeless ways
Give me one perfect thing.*

ANNA WICKHAM

The Man with a Hammer. From *The Contemplative Quarry* by Anna Wickham. Reprinted by permission of Harcourt Brace Jovanovich, Inc.

4-1 Abstract

We have now seen how atoms are constituted, and how they combine to form molecules. We have noted some consequences of the reshuffling of atomic combinations that we call a chemical reaction. We shall want to know why the properties of atoms and molecules result in the observable physical and chemical properties of matter as we find it all around us in the world. Before we can take that step, we need to look more deeply at the structure of individual molecules. We shall find that profound simplicity and order result from the apparently chaotic dance of nuclei and electrons in molecules. The sizes and shapes of molecules are governed by principles that, although still not fully understood by anyone, are surprisingly simple in their main points.

As a framework for our discussion of molecular geometry, we shall use the unifying theme of *symmetry*. Although some of the formal language of symmetry will appear strange at first, you will find that it is as useful and flexible as the decimal system of counting, and not much harder to understand. After discussing the shapes and symmetries of molecules in repose, we shall conclude with a brief introduction to internal molecular motions: vibrations and rotations.

4-2 On the Importance of the Right Arrangement

Composing a musical tune is not just a matter of choosing what notes of the scale are to be used. It matters a great deal how the notes are arranged. Random tapping of keys on a piano seldom produces satisfying music, because most ways of arranging sequences of notes yield noise, not music. Yet even when most of the conceivable combinations are judged to be noise and discarded, there are still vast possibilities and rich variety left to be available for the creative use of composers. Whether it be a simple tune picked out one note after the next, or a highly patterned fugue, piano music results from sequences, combinations, and structuring of the rather small number of units that are the notes of the keyboard.

Only about a hundred building units—the atoms of the chemical elements—suffice to yield all the substances of the material world. In fact, the atoms of just hydrogen, oxygen, carbon, sulfur, nitrogen, and phosphorus can be structured together into patterns in space that yield *almost* all of the molecules needed to constitute a living organism. Yet with all the variety that can be found in the molecules of nature, the combinations of atoms that actually result are not just any arrangements that might be imagined. There are, in fact, severe restrictions placed on the structures of molecules. Of all the ways that one might imagine atoms jumbled together, only a very small fraction constitutes structures that could exist or that do exist.

So it is not just a question of what atoms are collected together to form a molecule. It matters a great deal how the atoms are arranged. In overall composition, there is not much difference between vinegar and sugar syrup, although vinegar and syrup are unmistakably different. A molecule of cane sugar (sucrose) has the composition $C_{12}(H_2O)_{11}$; a molecule of grape sugar (glucose) is $C_6(H_2O)_6$; and a molecule of acetic acid (vinegar is acetic acid diluted with water, with traces of other compounds that affect the flavor and sometimes produce a colored solution) is $C_2(H_2O)_2$. Glucose and acetic acid have the same relative numbers of carbon, hydrogen, and oxygen atoms, but the arrangement in space of the atoms is different. As is so often the case in chemistry, the properties of each of these substances are consequences of the structure of its constituent molecules.

Consider, as another example, a collection of six hydrogen atoms, two carbon atoms, and an oxygen atom. A stable molecule can be assembled in two different ways from these atoms. The two possibilities are to be described by drawing structural formulas in which lines connect atoms that are close to one another. Thus, we can write the two **isomers**

$$\begin{array}{cc} \text{H} & \text{H} \\ | & | \\ \text{H}-\text{C}-\text{O}-\text{C}-\text{H} \\ | & | \\ \text{H} & \text{H} \end{array} \qquad \begin{array}{cc} \text{H} & \text{H} \\ | & | \\ \text{H}-\text{C}-\text{C}-\text{O}-\text{H} \\ | & | \\ \text{H} & \text{H} \end{array}$$

Dimethyl ether   Ethyl alcohol (ethanol)

Dimethyl ether is a highly flammable gas, bp −23.6°C, whereas ethyl alcohol is a liquid at room temperature, bp 78.3°C. Both compounds are narcotics. Dimethyl ether has the same effects on humans as the closely related diethyl ether, once a widely used anesthetic. Ethyl alcohol, more properly called ethanol, is the alcohol in alcoholic beverages.

Seemingly subtle differences in geometry can have profound effects on the reactions and properties of molecules. Many biologically active molecules have mirror-image counterparts which fail to react in biologically acceptable ways. (As Alice remarked, "Perhaps looking-glass milk isn't good to drink.") Similarly, two hydrocarbon molecules, differing only in geometry, may have quite different combustion behavior in an automobile engine. Two ions containing chromium, differing only in geometry, can have different colors. More on these subjects lies ahead in this book. See Chapter 17 for biochemistry, Chapter 14 for isomerism in organic compounds, and Chapter 13 for isomerism in metal complexes.

The shape determines the difference.

Let us begin our discussion of geometry and symmetry by examining the shape and structure of a small molecule, an isolated water molecule. A water molecule, as we saw, is composed of two atoms of hydrogen and an atom of oxygen. Actually, it can be misleading to speak of the structure of a molecule solely in terms of the positions of *atoms*. Individual atoms lose much of their individuality when embodied and embedded in a molecule. The nucleus and the inner kernel of electrons stay together, but there is often substantial sharing of the outer valence electrons. There are experimental methods[1] that can give precise and reliable information about positions of the nuclei, and we shall use some of this information as we focus our attention on the geometry and symmetry of the collection of nuclei in the water molecule. As we saw in Chapter 3, the three nuclei are in a V-shaped arrangement.[2] We add the fact that the average distance between a hydrogen nucleus and an oxygen nucleus is 0.958 Å, and, as we know, the angle formed by the intersection of lines drawn between hydrogen and oxygen nuclei is 104.45°.

How are we to depict this collection of nuclei? One possibility is to draw the picture

$$\circ$$
$$\bullet \quad \bullet$$

where ○ denotes an oxygen nucleus and ● stands for a hydrogen nucleus. For some other molecule that contains many different elements, we would not have enough symbols as simple as ○ and ●, so instead the letter symbol for the appropriate element is used to mark the location of each nucleus. Experience also has shown that a reader's eyes need visual clues about the geometry of such a structure, and so conventionally we write

We need (and we have) a simple way to write structures.

$$\text{H} \diagup \overset{\text{O}}{\diagdown} \text{H}$$

Lines are drawn connecting certain of the nearest-neighbor nuclei to convey something of the geometry of the arrangement of nuclei and thus of the molecule.

Actually, even more than nuclear geometry is customarily conveyed by these lines. If two nuclei are to be very close to each other in a molecule, and yet not be repelled by each other so strongly as to destroy the molecule, then there must be electrons between these positively charged particles. These electrons, engaged in mutual attraction with both nuclei, bind the two nuclei together. So a bond does exist between these two nuclei, and the line joining the nuclei locates the position of this directional covalent chemical bond. Indeed, the line itself is loosely referred to as a "bond."

Recall Ch. 3.

---

[1] The symmetry of a molecule and its internuclear dimensions can often be determined reliably by a combination of some rather sophisticated experiments involving diffraction and spectroscopy. The principles are discussed briefly in Appendix III. For additional information, see W. S. Brey, Jr., *Physical Methods for Determining Molecular Geometry*, New York: Reinhold (1965).

[2] Structural information presented in this chapter has been taken from *Tables of Interatomic Distances and Configuration in Molecules and Ions*, London: The Chemical Society, Special Publications No. 11 (1958) and No. 18 (1965). These two volumes give references to the original research papers in which the experimental data and interpretations are reported.

4-3 The Importance of Symmetry

A striking feature of the water molecule is its symmetrical structure. From side to side, and from front to back, one-half of the molecule is the mirror image of the other half. We shall find that a high degree of symmetry is characteristic of the geometry of most simple molecules (and of small parts, at least, of complex molecules).

Symmetry has fascinated men since ancient times. For some, symmetry has been closely associated with meaningful form, with ideal beauty, and with perfection. Many things in nature are beautifully symmetric. Animals, flowers, snowflakes, and mineral crystals are familiar examples of this recurrent theme. Doubtless, you have been introduced to the biological terms *bilateral* and *radial* to describe the symmetries of human beings and starfish, respectively. Molecules also adopt bilateral and radial symmetry, but we shall use a more precise language to do justice to the rich variety of symmetries that can be found in molecules.

Electrons and nuclei, pulled and pushed by their own attractions and repulsions, might be expected to swarm in random motion and shapeless ways. Yet molecules have quite definite shape, form, and symmetry. The highly symmetric dynamic forms that actually exist are seen by some people to be perfect things, reflections of a transcendent Great Symmetry. Others see the use of the vocabulary of symmetry to be a masterful stroke by which man has imposed simple and orderly interpretations on chaotic electron motion.

One fundamental distinction may be helpful in focusing our thoughts. A traveling wave, a trail of footprints, and a line of telegraph poles have *translational*, or *glide symmetry;* a particular pattern is repeated "endlessly" with a fixed distance (for waves, the wavelength) between repeating parts of the pattern. Molecules, on the other hand, like animals and flowers, generally exhibit *point symmetry*. Rather than an endlessly repeating pattern, point symmetries involve reflection, inversion, or rotation around a central point. This kind of symmetry is self-contained and finite, like the molecules themselves. (Glide symmetry will become important in our discussion of crystals found in Chapter 5.)

This is just a different and helpful way of looking at things.

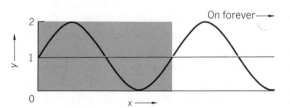

FIGURE 4-1 Graph of an endless sine function, illustrating glide symmetry. This function can be imagined as formed from a unit section (shaded in this figure), which is repeated again and again to produce an unending sine wave. (See also Figure 2-3.)

Symmetries can be found in mathematical expressions also. The equation

$$y = 1 + \sin x$$

may be graphed to produce a sine wave which has glide symmetry (Figure 4-1). On the other hand, the graph of

$$y = x^2$$

is a parabola (Figure 4-2), which shows an element of point symmetry; the left side is the mirror image of the right side. We have seen (Chapters 2 and 3) that the distribution of electrons in atoms and molecules is described mathematically by the Schrödinger wave equation. It is an important rule of wave mechanics that solutions of this equation must display the same symmetry characteristics as the molecule itself does. We shall return to this rule later in this chapter and again in Chapter 12.

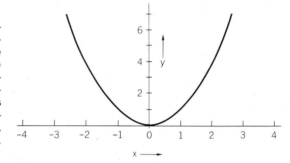

FIGURE 4-2 Graph of a parabola, illustrating point symmetry. The graph for negative values of x is the mirror image of the graph for positive values of x. A mirror, placed perpendicular to the page with its edge along the y axis, will visually complete the full graph, using just half of the graph. Try it!

It is a fascinating, inexpensive, and, for chemists, useful hobby to classify together objects having the same symmetry. For example, a seagull, a human being, and a spider, whatever their differences, all possess a plane of symmetry, with respect to which the left and right halves of their bodies are as mirror images of each other.[3] The same is true of a pair of eyeglasses, most articles of clothing, and a bicycle. Similarly, a five-armed starfish and a five-bulb chandelier have the same kind of symmetry.

The chemical significance of symmetry is considerable. We have already remarked that the biological activity of many compounds depends on their symmetry or lack of it. On a simpler level, the fact that the arrangement of nuclei in a molecule is symmetrical implies that the electrons (including both bonding and nonbonding electrons) must be distributed with identical symmetry. For example, we noted that one-half of a water molecule is the mirror image of the other half; this means that one of the O—H bonds must be identical to the other one, both in arrangement of electrons and in strength. To put it another way, the two hydrogen atoms in a water molecule must be chemically identical, because they are symmetrically interchangeable. The same kind of statement may be made of atoms, or groups of atoms, in any molecule when they occupy symmetrically identical positions. This realization makes the task of understanding the structure and chemical behavior of molecules much easier, both for the beginner and for the professional chemist. We shall return to this topic after we have had a look at some tangible ways of modeling molecules.

What it means to be the same

---

[3] This statement ignores, for simplicity, the asymmetric arrangement of internal organs, and the unique facial asymmetry humans develop in the course of living.

4-4 Molecules, Models, and Chemical Bonds

The wonder is that any molecule stays together. Each molecule is nothing more than a collection of atomic nuclei and lots[4] of electrons, but the fact is that molecules do stay together. The nuclei and electrons stay in particular orderly configurations, sometimes for very long periods of time. The relative positions of the nuclei in a molecule often remain the same, within a few percent fluctuation due to molecular vibrations, for days, years, and for some compounds for millenia. One theme of this textbook is the development of theories which explain, or at least make plausible, the observed fact that the molecules of many chemical compounds are stable systems of particles in which the forces of attraction dominate over the forces of repulsion, in which the tendencies toward order and stability dominate over the tendencies toward chaos and instability.

Speculative chemists have long been searching for explanations for this stability of compounds. A century ago, even the best explanations presented in terms of atoms and molecules were unconvincing to most physical chemists, and their response was to search for alternatives to chemical atomism.[5] It is possible to talk about chemical compounds and chemical reactions without talking about atoms or molecules. It is possible to write about molecules without speculating about the forces that hold molecules together. It is possible to write formulas for compounds without indicating any commitment to ideas about chemical bonds.

Yet from the 1850s to the present, chemists have been drawing formulas of compounds in such a way as to imply the structure of individual molecules. And with only a few exceptions, most representations of molecules have appeared to indicate just where the chemical bonds are located in the molecules. The pictures that chemists drew of molecules strongly influenced the way chemists thought about chemical bonding. The chemists already had a self-consistent, if not well-supported, picture of a great many molecules and their bonds when the revolutionary quantum physicists discovered sound theoretical foundations for chemical bonding between 1900 and 1925, and instruments for measuring bond angles and distances were invented.

Thought experiments of a century ago

In those early days of structural chemistry, chemists were guided by geometry. Many learned about the geometry of molecules by building three-dimensional models of molecules. As early as 1810, John Dalton was using wooden models made of balls and sticks to illustrate the combination of atoms to form compounds. He found these models to be effective teaching tools. A. W. Hofmann gave a lecture to the weekly evening meeting of the Royal Institution of Great Britain in April 1865, and brought along a collection of

---

[4] There is only a single electron in the hydrogen molecule ion, H_2^+, and only two electrons in the hydrogen molecule H_2. For all the rest of the molecules in the world, however, it is certainly fair to say that there are lots of electrons. How many electrons are there in a single molecule of CH_4?

[5] For an examination of atomism in late nineteenth-century physical chemistry, see G. M. Fleck, *Journal of the History of Ideas*, vol. **24**, p. 106 (1963).

croquet-ball models to illustrate some chemical points (Figure 4-3). Several of Hofmann's bond angles were incorrect (the geometry around each carbon atom is tetrahedral, not planar), but the main feature of his models was clear and correct: Nuclei in a molecule are arrayed in space, and his models directly communicated this fact. The idea of using wooden balls with holes drilled in them and sticks to fit into the holes caught on. The balls represent nuclei. The sticks are bonds. Almost every chemist now keeps a set of similar molecular models close to his desk, knowing that often the fastest way to learn about the geometry of a molecule is to make a scale model of it.

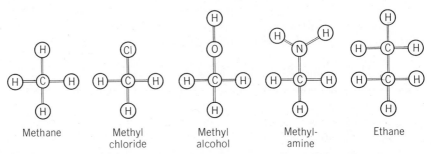

FIGURE 4-3 Models of molecules, constructed from croquet balls, used to illustrate a lecture in 1865. A twentieth-century chemist would construct the models so as to achieve a tetrahedral arrangement of balls around each carbon atom. [Source: A. W. Hofmann, *Proceedings of the Royal Institution of Great Britain*, vol. 4, p. 416 (1865).]

Proof that you must not believe everything you see

Ball-and-stick models can tell us many things about the possible ways of arranging nuclei in space, although the models may imply more than the model builder intends. If the balls are to represent nuclei, they grossly exaggerate the size of the nuclei relative to the overall size of the molecule. If the balls are to represent a portion of the electron cloud around the nuclei, then the viewer is asked to imagine a diffuse cloud when he sees a hard ball. Nobody believes that bonds in a real molecule are much like the sticks that hold the wooden balls together. Yet the sticks represent—sometimes accurately and sometimes inaccurately—the places in space where bonds might be.

Ball-and-stick models can be faithful to some important properties of actual molecules, and these are the properties on which we shall now focus. Properly constructed, a ball-and-stick model will be a physical object whose symmetry properties are identical to the symmetry properties of the actual molecule. If attention is paid to the lengths of the connecting sticks, these models give accurate information about the connectedness of the structure, about the relative location of the nearest neighbors of any particular nucleus, and about *some* of the possible internal twistings and turnings that might transform one structure into another. Thus, construction and manipulation of ball-and-stick models, in addition to being an amusing distraction, is an informative way to learn about the symmetry and geometry of molecules.

Some sets of molecular models permit construction of models of molecules in which the bonds are explicitly shown. Illustrated in Figure 4-4 is a model of ethene, C_2H_4, constructed with plastic polyhedra, plastic balls, and plastic "bonds" from the Benjamin/Maruzen HGS® models (W. A. Benjamin, Inc.).

FIGURE 4-4 A plastic model of a molecule of ethene, constructed from Benjamin/Maruzen HGS® models. Photographed with permission of W. A. Benjamin, Inc.

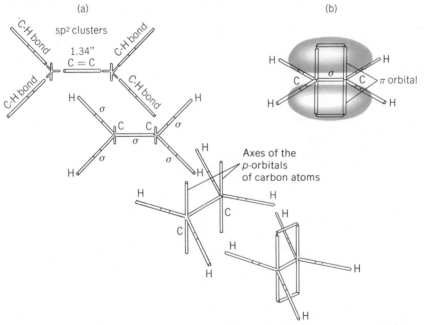

FIGURE 4-5 (a) Construction of a frame model of ethene using metal connector valence clusters and pieces of plastic tubing from Framework Molecular Models®. Used with permission of Prentice-Hall, Inc. (b) The relationship between the plastic frame and a representation of a π orbital.

The relative positions of the atomic nuclei are indicated by the positions of the polyhedra and balls. Two bent plastic rods joining the two carbon polyhedra are meant to describe or to represent a double bond. Certainly, a double bond is not accurately and completely described by two bent plastic rods. Yet this model is hopefully telling us something about the nature of a double bond in ethene. In Figure 4-5, a step-by-step construction scheme is presented for making a molecular model of ethene with Framework Molecular Models® (Prentice-Hall, Inc.), in which pieces of plastic tubing are connected with metal connectors. These plastic rods are supposed to be a representation of the chemical bonding in ethene. The feature common to the two models and to an actual molecule of ethene is a set of identical symmetry properties.

All the commonly used molecular models represent faithfully the symmetry of the molecules. Each has its own special advantages. The wary model builder will be aware of the limitations and will use models with an informed and lively imagination. Focusing on the symmetry properties, and deemphasizing the material features of the model, we can use models to gain insights into the nature of these unseeable and untouchable molecules.

4-5 The Elements of Molecular Symmetry

The statement that a water molecule is V shaped is a concise way of saying that the water molecule has two mutually perpendicular **reflection planes** and a twofold **axis of rotational symmetry**. These three **symmetry elements** are possessed by the molecule as a whole, and also separately by the arrangement of nuclei and by the electronic distribution. All molecules (in fact, all objects) that possess this particular set of symmetry elements are classified together as having identical symmetry. Actually this book has exactly those symmetry elements, if the printing is ignored. It and all other normally shaped books have the same symmetry as the H_2O molecule.

A **symmetry operation** is an operation which transforms the orientation (position) of a molecule into a new orientation which is indistinguishable from the original. Every symmetry operation is associated with a symmetry element.[6] There are four symmetry elements associated with actual changes in the orientation of the whole molecule or of parts of the molecule—**rotation, reflection, inversion,** and **improper rotation**—and a fifth symmetry element, the **identity** element, which is associated with the pseudooperation of not doing anything. These fundamental elements and operations are listed in Table 4-1. The operation of rotation can actually be performed on a real, solid object by spinning it around its rotational axis. Reflection, inversion, and improper rotation must ordinarily be treated as thought operations that cannot be illustrated by manipulating a real object.

ROTATION. If rotation of a molecule about some axis passing through the molecule results in a new orientation (a different position) indistinguishable from the original, then the axis (this line) is a symmetry element of the molecule. If the molecule can assume n different superimposed orientations about this axis during a full rotation, then the axis is said to be an n-fold rotational axis, and it is symbolized by C_n. Thus, the two-handled sugar bowl in Figure

---
[6] We distinguish between "operation" and "element" for clarity in speaking about symmetry. If an object contains a symmetry element, we can perform the corresponding symmetry operation on that object. We cannot tell the difference between the object before and after the symmetry operation has been performed.

4-6 has a twofold axis of rotation, because the bowl passes through two positions that are indistinguishable during a full revolution about an imaginary line drawn through the bowl. We call the element a C_2 axis, and the operation a C_2 rotation. The equilateral triangle in Figure 4-7 has a C_3 axis perpendicular to the plane of the triangle, and also three C_2 axes lying in the plane of the triangle. Every object has C_1 axes (in fact, an unlimited number of C_1 axes), because a full rotation about any axis returns any object to its original orientation. A circle has a C_∞ axis and an an unlimited number of C_2 axes. A sphere has an unlimited number of C_∞ axes.

C_2 symmetry axis
(a symmetry element)

C_2 rotation
(a symmetry operation)

FIGURE 4-6 A two-handled sugar bowl with a twofold axis of symmetry. This C_2 axis is indicated by the line which we can imagine piercing the lid and the bowl through their centers. If the bowl is rotated a half turn around this axis, the bowl will be in new orientation, indistinguishable from its original orientation. In a full turn, the bowl passes through new and equivalent orientations twice; hence the axis is a twofold axis, or an axis of order two.

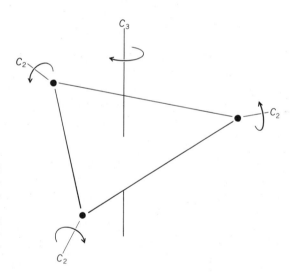

FIGURE 4-7 An equilateral triangle, showing the threefold C_3 rotational axis and the three twofold C_2 rotational axes. Each of the lines is a symmetry element. Each of the rotations is a symmetry operation.

REFLECTION. A plane that divides a molecule into two halves that are mirror images of each other is a symmetry element of the molecule, and is given the symbol σ, the Greek letter sigma. The pattern of letters

        MIRROR   ЯOЯЯIM

Not the same as the σ bond of Ch. 3

has a reflection plane—a plane of symmetry. This plane acts exactly like an ordinary looking-glass mirror. Putting the symmetry element in place, we have

                   σ
        MIRROR | ЯOЯЯIM

A coffee cup has a plane of symmetry. It is a vertical plane, slicing the cup into two mirror-image halves, each of which has a slice of the handle. If you glued one of these halves to a mirror, with the cut edges in contact with the mirror, the real half and the reflected half (in the mirror) would generate the image of a complete coffee cup. The sugar bowl in Figure 4-6 has two planes of symmetry. One slices each of the handles in half. The other (perpendicular to the first) reflects one handle into the other. *The two-handled sugar bowl has the same symmetry as a water molecule.*

INVERSION. If an object has a **center of symmetry**, then each point when reflected through this central point is transformed into an equivalent point. This operation of *inversion* yields an object indistinguishable from the original object. The letters I, O, S, H, X, and Z all have a center of symmetry. Each letter can be inverted and remains the same figure as before inversion. Note, for example, that each point on the letter Z can be connected to an equivalent point by a line whose midpoint is at the center of symmetry:

The operation of inversion consists of interchanging all of these points; that is what is meant by reflection through the center of symmetry. If a collection of nuclei has the symmetry element i (an inversion point, or center of symmetry), then each nucleus is transformed into an equivalent nucleus when the operation of inversion is performed.

Look around for objects that contain the symmetry element i. A simple periscope, an unopened "tin" can, a bar of soap, a soda straw, a football, and a pencil sharpened on both ends—dissimilar as these objects appear to be, they share in common the presence of the symmetry element i.

IMPROPER ROTATIONS. Imagine a triangular sandwich of the type often served as an appetizer. Such a sandwich has a C_3 axis if the slices of bread are equilateral triangles [Figure 4-8(a)]. Now twist the top piece of bread through one-sixth of a revolution with respect to the bottom slice, so that the corners of the top slice overhang the edges of the bottom slice. The resulting skewed canapé still has a C_3 axis, and it also has an S_6 improper axis [Figure 4-8(b)].

The existence of an improper rotation axis (a rotation-reflection or alternating axis) means that rotation about that axis, followed by reflection in a mirror plane perpendicular to the axis, produces a transformed object that is indistinguishable from the original. Not being just a pure rotation, this symmetry operation is called an improper rotation. If the rotation is a C_6 rotation, then the combined rotation-reflection operation is called an S_6 rotation. It is not necessary for either the axis of rotation or the mirror plane *alone* to be a symmetry element of the object; the two transformations are considered together.

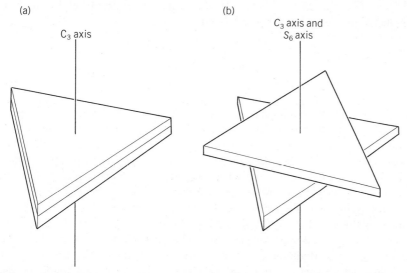

FIGURE 4-8 (a) A proper triangular sandwich, with a C_3 rotational axis perpendicular to the bread. (b) An improper sandwich, with the top slice given a one-sixth turn. The C_3 axis remains, but in addition, the same line has become an S_6 axis.

To verify that the skewed sandwich has an S_6 axis, let us look at one of the corners of the upper slice of bread. Imagine a C_6 rotation of the whole sandwich, bringing that corner *over* a position where there was originally a corner of the lower triangle of bread. Reflection of the whole sandwich through the plane of the filling brings the rotated sandwich into a position indistinguishable from the original. You can make yourself an object with an S axis in a similar manner from any sandwich made from slices of bread cut in the shape of a regular polygon.

TABLE 4-1
Symmetry Operations and Symmetry Elements

Symmetry Operation	Symmetry Element	Symbol
Rotation	Axis of symmetry (axis of rotation)	C
Reflection	Plane of symmetry (mirror plane) (reflection plane)	σ
Inversion	Center of symmetry (center of inversion)	i
Improper rotation	Improper axis (alternating axis) (rotation-reflection axis)	S
Identity	Identity	I

SYMMETRY ELEMENTS OF A MOLECULE.

The structural formula of a molecule of the hydrocarbon ethane is

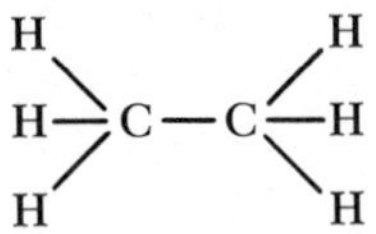

$$\begin{array}{c} H \quad\quad H \\ H-C-C-H \\ H \quad\quad H \end{array}$$

In its most stable conformation (see Section 4-7), the two CH_3 groups are staggered with respect to each other. That is, the molecule possesses a center of inversion and an S_6 improper axis of rotation. In Figures 4-9 to 4-13, various symmetry elements are located and identified. In addition to the symmetry elements depicted, ethane also has two additional mirror planes and three different C_2 axes. Can you locate them? If you have difficulty in finding all of these symmetry elements, you should construct a model which faithfully portrays the positions of the nuclei and the symmetry of the structure. Such a model may be a perspective drawing on paper, but be forewarned that such drawings can lead you astray. A set of molecular models[7] is extremely useful in constructing such representations. Toothpicks, nondrying modeling clay, and patience will also produce satisfactory results. Styrofoam balls (available in some chemistry departments and most five-and-tens, at least during the months just before Christmas) and pipe cleaners make excellent models.

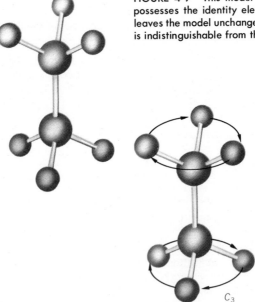

FIGURE 4-9 This model of ethane, like all other objects, possesses the identity element, *I*. The act of doing nothing leaves the model unchanged. The object before the operation is indistinguishable from the object after the operation.

FIGURE 4-10 Rotation through an angle of 120° around the C_3 axis yields a model that is indistinguishable from the model before rotation. Therefore, this model of ethane possesses a C_3 symmetry element.

---

[7] Commercially available atomic and molecular models are surveyed by A. J. Gordon, *Journal of Chemical Education*, vol. **47**, p. 30 (1970).

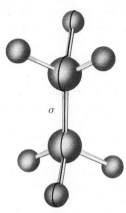

CHAPTER 4

FIGURE 4-11 Reflection through the mirror plane σ interchanges some of the hydrogen nuclei. For the nuclei that lie in the σ plane, reflection interchanges the two halves of the electron cloud around each nucleus. After reflection, the ethane molecule is indistinguishable from the molecule before reflection. Thus, the molecule possesses a reflection plane as a symmetry element. There are actually some other reflection planes in ethane; can you find them?

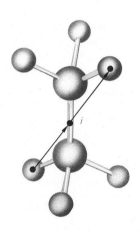

FIGURE 4-12 Inversion is reflection through the point i. Example: Start with some point and draw a straight line from that point through i. Continue the line an equal distance beyond i. You have just located a point which is indistinguishable from the starting point.

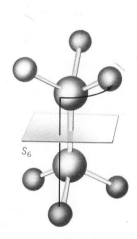

FIGURE 4-13 This improper rotation is a combination of rotation and reflection. First, we rotate the model through an angle of 60° (one-sixth of a full circle). Then we reflect through a mirror plane perpendicular to the rotational axis. The result is a model indistinguishable from the starting model. Because the rotation part of the operation was a C_6 rotation, we call the total operation an S_6 improper rotation.

For another molecular example, we can return to the water molecule. An isolated water molecule is depicted in Figure 4-14 with its three nuclei lying in the σ plane. The 10 electrons together constitute a swarming cloud of negative charge in front of and in back of the σ plane. The electron distribution in front of the σ plane is the mirror image of the electron distribution behind the plane.

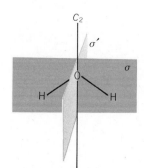

FIGURE 4-14 Symmetry elements of a water molecule. All three nuclei lie in the σ plane. The σ' plane is perpendicular to the plane of the nuclei. The C_2 axis is the line of intersection of the two mirror planes.

A second mirror plane (which we shall call the σ′ plane, the sigma-prime plane), perpendicular to the σ plane, reflects one of the hydrogen nuclei to give a mirror image which coincides with the other hydrogen nucleus. The σ′ plane divides the nuclear configuration, the electronic distribution, and thus the whole molecule into mirror-image halves. The two hydrogen nuclei are said to be **symmetry equivalent**, because they can be transformed into each other by the symmetry operations that transform the water molecule into indistinguishable configurations.

The C_2 axis lies at the intersection of the two mirror planes, and the axis passes through the oxygen nucleus. A rotation of the molecule (or a rotation of the collection of nuclei, or a rotation of the distribution of electrons) 180° around the C_2 axis yields a new molecule superimposed upon, and indistinguishable from the original. The two symmetry-equivalent hydrogen nuclei are interchanged by the C_2 operation.

4-6 The Pauli Principle and the Shapes of Molecules

We now extend the description of bonding presented in Chapter 3, this time with greater emphasis on how the Pauli exclusion principle operates to influence the geometry of particular molecules. For illustrations, we shall return to three small molecules—methane, ammonia, and water—the simplest hydrides of carbon, nitrogen, and oxygen.

Several different shapes can be imagined for methane.

Methane is the chief constituent of natural gas.

Hypothetical structures of methane

Square pyramidal   Square planar

The experimentally verified structure of methane is neither square pyramidal nor square planar, but instead has the four hydrogen nuclei surrounding the central carbon nucleus in a highly symmetric manner. This cage of four hydrogen nuclei can be thought of as consisting of a hydrogen atom at each of the alternate corners of a cube.

Actual cage around the carbon atom

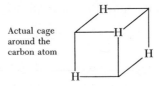

Exactly the same arrangement of the four hydrogen nuclei can be thought of in terms of a nucleus at each apex of a regular tetrahedron.

Actual cage around the carbon atom

Evidence for a tetrahedral shape of four bonds around carbon was found in the nineteenth century, long before the era of sophisticated electronic instrumentation and complex theory. This chemical evidence is based on the fact that only one substance with the formula CH_2Cl_2 has ever been found (that is, there are no isomers). If this compound had square pyramidal or square planar shape, then it should be possible to find (or to make) two different CH_2Cl_2's: one with the structure

Hypothetical trans-structures of dichloromethane

with the chlorine nuclei across *(trans)* from each other, and one with the structure

Hypothetical cis-structures of dichloromethane

with the chlorine nuclei adjacent *(cis)* to each other.

In the actual structure, since each of the four corners of a tetrahedron is equidistant from the other three, any arrangement of two hydrogen nuclei and two chlorine nuclei at the apices of a tetrahedron is equivalent to any other. You should verify this fact of geometry for yourself with the aid of models. This tetrahedral arrangement places the four nuclei around the carbon as far apart as possible, consistent with a C—H internuclear distance of 1.068 Å and a C—Cl internuclear distance of 1.772 Å. Nineteenth-century chemists reasoned that if dichloromethane had to have a tetrahedral structure in order to conform to the facts, then other compounds in which carbon is bonded to four other atoms, including methane, should be tetrahedral, too. Later experiments, using diffraction and spectroscopic methods, have substantiated this expectation.

This cannot be a regular tetrahedron. Why?

We have seen that a molecule of methane can be described in terms of a cage of hydrogen nuclei arranged on alternate corners of a cube, or equivalently, arranged on all the corners of a tetrahedron. A third way to think about this cage of hydrogen nuclei is to visualize an imaginary sphere[8] of radius 1.091 Å, with the carbon nucleus at its center, and with the four hydrogen nuclei lying on its surface. The tetrahedral shape places each hydrogen nucleus as far from the other hydrogens as possible on the surface of the sphere. As you can verify, each of these hydrogen nuclei is symmetry equivalent to each of the others.

How can we explain the observed tetrahedral symmetry of methane? Why does the methane molecule not have a square pyramidal structure, or a square planar structure? We shall find an answer by focusing on the electrons in the molecule, and by employing the Pauli exclusion principle.

---

[8] A slated sphere, like the ones used in geography classes for drawing maps of the earth with chalk, is helpful in drawing a picture of this hydrogen-nucleus cage.

There are 10 electrons in the methane molecule, and all except the two $1s$ electrons localized near the carbon nucleus are involved in bonding. The Pauli exclusion principle requires each of these eight bonding electrons to have a different set of quantum numbers. Recall (Chapter 2) that the first three quantum numbers govern the spatial orientation of an orbital, while the fourth determines the spin of the electron in the orbital. One way to satisfy the Pauli principle is to have four sets of spin-paired electrons in which each set occupies an orbital that is in shape identical to, but in orientation different from, the other three orbitals. The special feature of the Pauli principle is that it allows spin-paired electrons to occupy the same region of space, but tends to exclude all other electrons from that region. The result is that electrons of the same spin tend to keep as far apart as possible. Because electrons repel one another, because the Pauli principle restricts just two electrons to an orbital, and because each of the four orbitals has its greatest concentration of electron density in its own special region of space, the four orbitals have the same shape but different orientations. These orientations are determined by maximization of the average distances between the orbitals. It is as though the individual orbitals repel one another.[9]

An isolated carbon atom has the full symmetry of a sphere; there is nothing to prevent the six electrons from spreading around the single nucleus with spherical symmetry. For the same reason, we would expect the extremely reactive (and therefore extremely unstable) C^{4-} ion to be spherically symmetrical, with electrons evenly distributed around the nucleus. Since there is no way to achieve full spherical symmetry when the cage of four protons is placed around the C^{4-} ion to produce a methane molecule, it is clear that in the process of forming methane, some electron localization must occur. We shall say that directed bonding orbitals are formed.

What shapes a methane molecule?

Let us imagine a hypothetical process in which a spherically symmetrical C^{4-} ion (a carbon nucleus with 10 electrons) reacts with 4 protons to form a methane molecule. Schematically, we can represent the changes in the orientation of the eight bonding electrons as

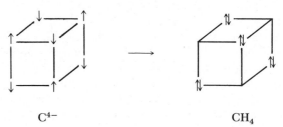

C^{4-} CH_4

Without the four protons to influence the symmetry, the eight electrons in C^{4-} move as far apart as possible, and this fact can be stated by pinpointing each electron at the corner of a cube. (These points are the places where

---

[9] The geometrical model being developed in this section has been called the *electron-pair repulsion* theory. For further information, see R. J. Gillespie, *Journal of Chemical Education*, vol. **40**, p. 295 (1963); *ibid.*, vol. **47**, p. 18 (1970).

the particular electron is supposed to be most often; however, the average locations of all the eight electrons form a total electron distribution that is spread out into complete spherical symmetry.) Not only do the electrons stay as far apart as possible, but *electrons with the same spin* are as far apart as possible; four electrons of the same spin have a maximum probability of being at the corners of a regular tetrahedron. We have indicated these positions by arrows pointed in the same direction. The other four electrons have a maximum probability of being at the corners of a second tetrahedron. The composite picture of the C^{4-} is of electrons at the corners of a cube, with successive corners occupied by electrons of opposite spin.

Diagonally opposite corners of the cube thus are occupied by electrons of opposite spin. We can, therefore, imagine spin-paired electrons arranged in orbitals of essentially the shape

$$\big(\downarrow\big) \; C \; \big(\uparrow\big)$$

When the four protons combine with the C^{4-} ion, the two electrons of opposite spin in each orbital are attracted toward a positive proton. The same thing happens with each orbital and each proton, and the result is that the two tetrahedra of electrons and the tetrahedron of protons are brought into coincidence. Each bonding orbital assumes essentially the shape

$$\big(C \updownarrow H\big)$$

The substantial concentration of electron density between the two nuclei stabilizes the structure of the molecule, because both positively charged nuclei attract the negatively charged electrons. The electron pair acts as electrostatic glue, overcoming the force of repulsion between the two nuclei, and actually adding some additional stability. We describe this process by saying that *four chemical bonds have been formed*.

Now review pages 73–75 for comparison.

A structure that is precisely tetrahedral has a

angle of 109.5°. Experimental findings for methane are in accord with an internuclear angle of 109.5°. Let us compare this "tetrahedral angle" with the corresponding angles found in two other hydrides, NH_3 and H_2O. The

To be almost, but not quite

angles in ammonia are 107.3°, and the

angle in water is 104.5°. Both angles are close to the tetrahedral angle. We shall look at an explanation of why the angles are almost tetrahedral, *and* an explanation of why they are not exactly tetrahedral.

AMMONIA. We shall describe the bonding in NH_3 by the hypothetical reaction of an N^{3-} ion with three protons. There is a full set of eight bonding electrons around the nitrogen nucleus (together with the two $1s$ electrons of the nitrogen kernel), but there are only three protons to be bound. The bonding can be described in terms of three N—H orbitals and a single nonbonding (lone-pair) orbital. In each of the N—H orbitals, the pair of electrons is under the influence of two nuclei, and this tug of war between two electron attractors restricts the electron pair largely to the region between nuclei. Because the lone-pair electrons are under the influence of just one nucleus, they occupy a less-confined orbital. This orbital takes up more than one-fourth of the space around the nitrogen nucleus, and thus effectively repels the other three orbitals more strongly.

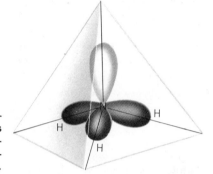

FIGURE 4-15 Repulsion between the nonbonding lone-pair electrons and the electrons in each of the three bonding orbitals in an ammonia molecule, resulting in distortion of internuclear angle from exact tetrahedral geometry.

This is the lone pair donated in forming a coordinate covalent bond—see p. 78.

The lone-pair orbital influences the geometry of an ammonia molecule by squashing the other three orbitals together, reducing the

angles to about two degrees less than the tetrahedral angle (see Figure 4-15). The three protons adopt *almost* the positions of three corners of a tetrahedron surrounding the nitrogen nucleus, because there are four sets of spin-paired electrons that *almost* exactly divide the space around the nitrogen nucleus into equal shares. The deviation from exact tetrahedral geometry comes about because the nonbonding pair of electrons is able to occupy more than its fair share of the space, since it is not constrained by interactions with two nuclei as are the bonding pairs.

WATER. We shall describe the bonding in H_2O by the hypothetical reaction of an O^{2-} ion with two protons. Again, there are eight bonding electrons, but a water molecule has only two protons. We shall describe this bonding in terms of two O—H orbitals and two lone-pair orbitals. The lone-pair orbitals occupy more space than the bonding orbitals, and thus the two lone-pair orbitals take up more than half of the total space around the oxygen nucleus. The two O—H orbitals are squashed together even more than was observed with the N—H orbitals in ammonia. The angle in a water molecule is found to be about four degrees less than the tetrahedral angle.

In water, as in ammonia, the presence of lone pairs of electrons in nonbonding orbitals gives a structure that is approximately tetrahedral, because the total number of orbitals to be fitted into the space is four. The nonequivalence of the two types of orbitals results in distortion of the tetrahedral angles (see Figure 4-16).

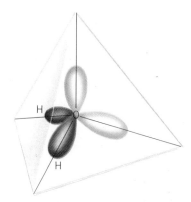

FIGURE 4-16 Electrons in two nonbonding orbitals of a water molecule repel electrons in two bonding orbitals, thereby reducing the O$<^H_H$ internuclear angle to less than the "tetrahedral angle."

FIGURE 4-17 (a) Pictorial representations of some representative molecules selected from Table 4-2. Compare the geometry of the molecules ICl_3, BF_3, and NH_3, noting the effect of lone pairs of nonbonding electrons. The nonbonding pairs act as place holders, forcing the remaining bonding orbitals into the remaining places around the bonding sphere. The nonbonding electrons, under the attractive influence of just one nucleus, occupy more than their fair share of the space, thus crowding the bonding electrons and therefore the bound nuclei. (b) Compare the geometry of PBr_5 with that of IF_5. Why the difference?

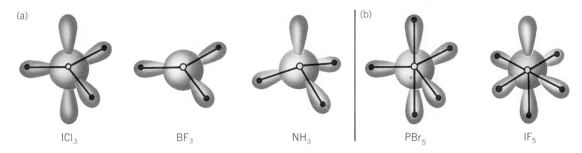

Although very many compounds, including most compounds of biological significance, feature tetrahedral geometry around many of their atoms, other fundamental geometries are found where there are more or fewer than four electron-pair domains around an atom. The tendency for electron pairs to adopt an arrangement that maximizes their average distance apart results in the geometric arrangements listed in Table 4-2 for various numbers of pairs of bonding electrons. When the existence of lone-pair orbitals, or of different nuclei bonded with the various orbitals, makes the orbitals nonequivalent, it is anticipated that the actual geometry of the molecule will resemble the geometry given in the table, but that there will be some distortion. Further use of the electron-repulsion theory will be made in Chapter 13 when we examine molecules in which the central atom is surrounded by a large cluster of bonded atoms.

TABLE 4-2
PREDICTED GEOMETRIES FOR SIMPLE MOLECULES

Number of Electron Pairs	Internuclear Angle (when all bonds are equivalent)	Shape	Associated Geometric Figure or Solid	Examples († denotes molecules with lone pairs)
2	180°	linear	line	$Hg(CH_3)_2$
3	120°	 triangular	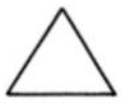 equilateral triangle	BF_3
4	109.5°	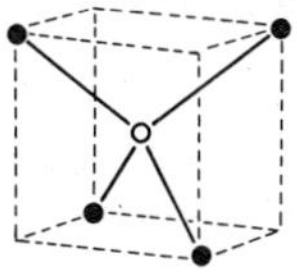 tetrahedral	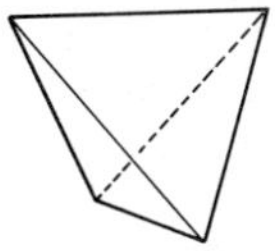 tetrahedron	CH_4, NH_3,† H_2O†
5	90°, 120°	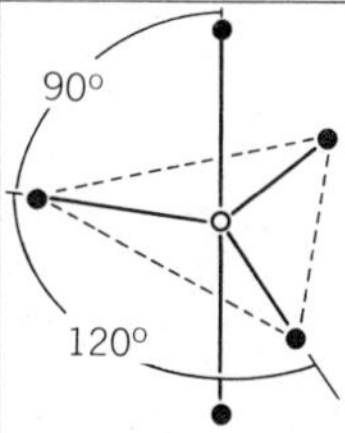 trigonal bipyramidal	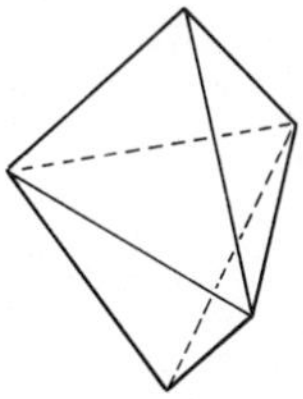 trigonal bipyramid	PBr_5, ICl_3†
6	90°	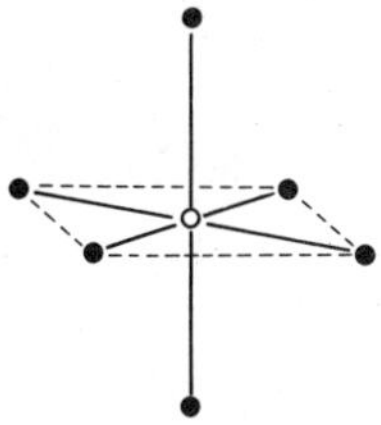 octahedral	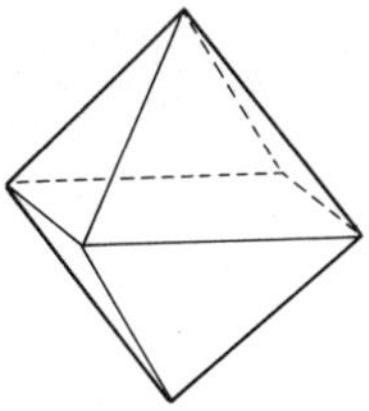 octahedron	SF_6, IF_5,† XeF_4†

4-7 Internal Motions of Molecules

Build yourself a molecular model of ethane,

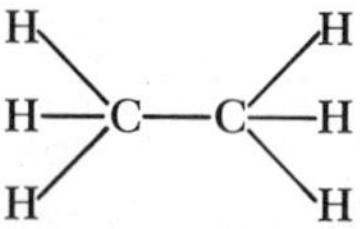

$$H_3C-CH_3$$

Using any of the type of models mentioned in this chapter, you will have built a model that can be considered to be two CH_3 groups (two methyl groups) attached together with a rod, a stick, or a piece of plastic tubing. In such a model, the two methyl groups can be rotated with respect to each other around the C—C axis. In the model, there is the possibility of *internal rotation*.

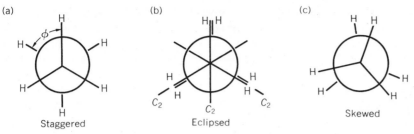

FIGURE 4-18 Newman projections of ethane conformations. Each is a schematic representation of an end view along the C—C axis. In the eclipsed conformation, there are three C_2 elements which are shown. In the staggered conformation, there are three corresponding S_2 axes. The staggered conformation does have three C_2 axes. You may have to use a model to find them. There is a continuous range of skewed forms, only one of which is shown here.

To what extent does this internal rotation feature of the model correspond to a property of the ethane molecule? The pictures drawn in Figure 4-18 are schematic views along this axis. These end-on views are called Newman projections of the molecule. An infinite number of different rotational conformations are possible: three equivalent staggered conformations [Figure 4-18(a)] in which the C—C axis is both an S_6 axis and a C_3 axis, three equivalent eclipsed conformations [Figure 4-18(b)] in which the C—C axis is only a C_3 axis, and a continuous range of skewed conformations such as the one shown in Figure 4-18(c) and the ones implied in Figure 4-19 between "staggered" and "eclipsed" minima and maxima.

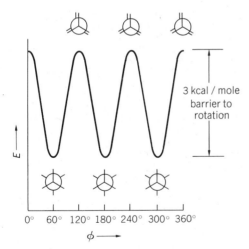

FIGURE 4-19 Energy barriers to free rotation in ethane. Energy is plotted versus ϕ, the skewing angle. Angles of internal rotation that represent the staggered and the eclipsed conformations are indicated by the appropriate Newman projections.

It has been found experimentally that the staggered conformation has the lowest energy. Being most stable, it is the conformation most likely to occur. The eclipsed conformation is least stable and hence is least likely to occur. However, the energy differences are only a little greater than the energy available from a typical molecular collision at room temperature. Thus, a particular molecule at room temperature will spend most of its time in staggered conformations, passing to slightly skewed conformations very often, and occasionally will get enough energy to pass to an eclipsed conformation and then quickly over (or back) to a staggered conformation. These energy relationships are diagrammed in Figure 4-19, where energy of the molecule is plotted versus ϕ (Greek letter phi), the angle of rotation around the C—C axis. At very low temperatures, molecular collisions seldom have enough energy to allow an ethane molecule to pass over the eclipsed-conformation energy barrier. At such low temperatures, one of these particular staggered conformations is preserved—frozen in—and there is essentially no rotation about the C—C bond.

This is an important concept in organic chemistry.

When the temperature is high enough so that the available energy is large compared to the energy barrier for rotation, we say that there is *free rotation* about the C—C bond. At lower temperatures, the dominant motion is not complete rotation, but instead a back-and-forth rotational oscillation. In the terms of Figure 4-19, the molecule is constrained to stay near the valleys of the energy curve. The molecule stays near a staggered conformation, but is periodically slightly skewed, alternately counterclockwise and clockwise twisted.

Simple periodic motion is a phrase that describes the continuing periodic oscillations of the two methyl groups, twisting back and forth around their lowest-energy (their equilibrium) position. Simple periodic motion is a phrase that also describes the motion of a clock pendulum, the motion of the ratchet of a watch escapement, the swinging of a swing, and the jiggling of a weight suspended by a spring. In a word, these motions are all **vibrations**.

There are other vibrations occurring in the ethane molecule, but these vibrations are not revealed by a ball-and-stick model. For purposes of illustrating the many vibrational motions of a molecule, a molecular model can be constructed by replacing the sticks with stiff springs. All nuclei are in constant motion. Every internuclear angle and every internuclear distance is constantly changing. Most of these changes are small but never ceasing.

How to think about vibrations

The whole molecule shakes like a bowlful of jelly. The jiggling of each nucleus appears to be random and disorderly, without form or meaning, and the electron cloud quivers along with the nuclei. The clue to finding form and order in the shaking and quivering is to think of the nuclear motions as combinations of simple vibrations. And the way to make the motions seem most orderly and simple is to take advantage of the symmetry properties of the molecule and look for modes of vibration that have some of these same symmetry properties.

It is an amazing fact[10] that *any* small displacements of nuclei from their equilibrium positions can be described as combinations of **symmetry-correct modes of vibration**, vibrational motions of the whole molecule that individ-

---

[10] Reasons for this important result are developed in an excellent paperback book, H. H. Jaffé and M. Orchin, *Symmetry in Chemistry*, New York: Wiley (1965). For further details, see F. A. Cotton, *Chemical Applications of Group Theory*, 2nd ed., New York: Wiley (1970); L. H. Hall, *Group Theory and Symmetry in Chemistry*, New York: McGraw-Hill (1969); and J. R. Ferraro and J. S. Ziomek, *Introductory Group Theory and Its Application to Molecular Structure*, New York: Plenum Press (1969).

ually have symmetry properties related to the symmetry properties of the molecule. We shall introduce two technical terms—**symmetric** and **antisymmetric**—to describe the symmetry of vibrational motion. And then to see how symmetry principles help to simplify the description of molecular vibrations, we shall return to the water molecule to see how the procedure works out in practice. (The same principles apply to ethane and to all other molecules; however, it is easier to see what is happening with a small molecule like water.)

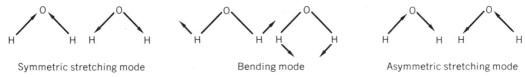

Symmetric stretching mode   Bending mode   Asymmetric stretching mode

FIGURE 4-20 Symmetry-correct modes of vibration of a water molecule. The arrows indicate the direction of motion of the hydrogen nuclei.

The three symmetry-correct modes of vibration for a water molecule are pictured in Figure 4-20. To be symmetry correct, the vibrational mode must be either symmetric or antisymmetric with respect to each of the symmetry operations of the molecule. A mode depicted with arrows as in Figure 4-20 is *symmetric* with respect to a symmetry operation if that operation transforms the two arrows into a new set that is indistinguishable from the original set of arrows. A mode is *antisymmetric* with respect to an operation if that operation transforms the arrows into a new set of arrows, each indistinguishable in orientation but opposite in direction; the transformed arrows will be superimposed on the original arrows, but the arrowheads will be on opposite ends. Thus, the **symmetric stretching mode** and the **bending mode** are each symmetric with respect to every one of the symmetry operations of the molecule: identity, C_2 rotation, and reflection in either the σ plane or the σ' plane (Figure 4-14). The **asymmetric stretching mode** is symmetric only with respect to the identity operation and to reflection in the σ plane, and is antisymmetric with respect to a C_2 rotation and to reflection in the σ' plane. (For more on the difference between the terms "symmetric" and "antisymmetric," see Problem 4-8.) Any small displacements of the nuclei from their equilibrium positions, no matter how chaotic and random these motions may be, can be described as a superposition of these three symmetry-correct modes of vibration.

Back to H_2O—how does it wiggle?

Chemists often think about these modes of vibration in terms of mechanical analogues (see Figure 4-21). Mechanical models, constructed with springs, can simulate accurately the motions involved in molecular vibrations. The energies of vibrations of molecules, however, are regulated by the laws of quantum mechanics. The Schrödinger equation (Chapter 2) predicts that, for each of the vibrational modes, the energy is quantized, and that the energy difference $\triangle E$ between two successive energy levels is given by

$$\triangle E = \frac{h\kappa}{4\pi}$$

where h is Planck's constant, and the constant κ has a value characteristic of the particular molecule and mode concerned and is related to the masses of the vibrating nuclei and to the strengths of the chemical bonds that hold these nuclei together.

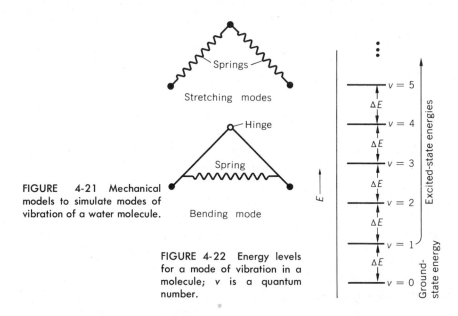

FIGURE 4-21 Mechanical models to simulate modes of vibration of a water molecule.

FIGURE 4-22 Energy levels for a mode of vibration in a molecule; v is a quantum number.

Light having energy equal to $\triangle E$ can be absorbed by a molecule, exciting the molecule to its next higher vibrational energy state (up one rung on the energy ladder depicted in Figure 4-22) for the particular mode. For many molecular vibration modes, infrared radiation has the right energy to be absorbed. Experimental determination of the energies that can be absorbed by a molecule in becoming vibrationally excited is conveniently made with an infrared absorption spectrophotometer.[11] By careful interpretation of these infrared data, a great deal of information about molecular symmetry and about bond strengths has been obtained. In addition, because each compound has different and characteristic vibrations, the frequencies of infrared light will, in general, be different. In fact, these differences are distinctive enough so that infrared absorption spectroscopy can serve to detect and to identify many substances in mixtures.[12]

---

[11] For a general discussion of absorption spectrophotometry, see Appendix III.

[12] In a similar manner, distinctive transitions between *rotational* energy levels can also be used to identify molecules. The energy of rotation (the energy of the turning or spinning of the whole molecule as it moves through space) is also quantized, and the permitted energies can be depicted by a diagram much like Figure 4-22. Radiation having energy equal to the difference between two energy levels can be absorbed or emitted; such energy differences for rotation correspond to microwave radiation with wavelengths of a few millimeters. Microwave signals received by radiotelescopes have revealed 23 different molecules in interstellar clouds, astronomers interpreting these signals as arising from molecules emitting radiation as they change rotational states. See P. M. Solomon, "Interstellar Molecules," *Physics Today*, vol. 26, no. 3, pp. 32–40 (March 1973).

The forces that hold a vibrating molecule together have the same symmetry as the nuclear arrangement and as the electron distribution. The internal motions of a molecule, as we have just seen, also have this same symmetry. The mathematical equations that are the solutions of the Schrödinger equation for this molecule also have this same symmetry. In every sense, the molecule itself has this symmetry. Since the hand-size molecular models commonly used to represent this molecule also have the same symmetry, we see that the language of symmetry is a natural language for talking about molecules. It is helpful for the beginning student of chemistry, and it is invaluable for the research chemist in the field of molecular structure and molecular spectroscopy, to make full use of this language of symmetry.

The shapes of models and motions and molecules

4-8 Summary

"In the beginning was symmetry," said the distinguished atomic physicist Werner Heisenberg.[13] Using the water molecule and the ethane molecules as examples, we have explored the static geometry of molecules and the dynamic motion of the same molecules, using the language of symmetry to focus our attention and to stimulate our imagination. Tangible models are often used to represent molecules, and these are very useful in understanding structural features of molecules that may be difficult to visualize abstractly. Whatever other degree of realism models may have, they must represent faithfully the symmetry elements of the actual molecule.

The origin of molecular symmetry lies in the arrangement of neighboring atoms around each particular atom, and in the angles defined by the neighboring bonded nuclei. We found that a rather simple scheme explains the main features of the pattern of internuclear angles found in molecules, namely, that the geometry adopted around a particular atom depends mainly on the number of pairs of electrons (bonding and nonbonding) in that atom's outer shell. The geometry of an actual molecule is that which maximizes the angles (that is, which minimizes the repulsions) between adjacent electron pairs.

We examined the use of models in depicting the three-dimensional structure of molecules, and then returned to the consideration of symmetry in a rather formal way. We saw that there are five fundamental symmetry elements that a molecule may possess: the identity element (possessed by all objects), rotations, reflections, inversion, and improper rotations. The particular combination of symmetry elements an object has governs its overall symmetry and classifies it together with other objects of the same symmetry.

The atoms in every molecule are in constant vibration around their average positions. In considering molecular vibrations, we found an immediate use for the symmetry properties of molecules: The apparently random jiggles

---

[13] W. Heisenberg, *Physics and Beyond: Encounters and Conversations,* New York: Harper & Row, p. 240 (1971). Heisenberg was commenting on the significance of the high symmetry displayed by subatomic particles, as well as critical asymmetry, and the asymmetry of the vast majority of biological molecules. In this emphasis on the supreme importance of form and an ordering principle in nature, his words seem to be in the same spirit as "In the beginning was the Word," *Revised Standard Version of the Bible,* John 1:1.

of the nuclei in a molecule can be analyzed into a combination of a few simultaneously occurring modes of vibration. These vibrational modes must each have the symmetry of the molecule itself, the motion being either symmetric or antisymmetric with respect to each of the symmetry operations of the molecule. Since the energy of vibration, like all atomic and molecular energies, is quantized, molecules may be excited from one vibrational state to another by a quantum of light, just as electrons in single atoms may. The energy differences between allowed vibrational states correspond to the energies of light quanta in the infrared spectral region. Thus, molecular vibrations may be investigated by determining the absorption of specific frequencies of infrared light.

Many molecules also have structures that allow rotation around single bonds, and an important facet of molecular geometry is the presence of internal rotations. Sometimes there is free rotation, allowing a complicated molecule to assume many different conformations. Sometimes there is restricted rotation, and then there is often an internal vibrational mode that corresponds to rotational oscillations around a bond.

GLOSSARY

antisymmetric: having the property of reversing direction or sign as a result of a symmetry operation; said of a vibration, a mathematical function, or (in Chapter 12) an orbital.

asymmetric stretching mode: a molecular vibration in which two symmetry-equivalent nuclei move together in synchronization, with one internuclear distance stretching, while the symmetry-equivalent internuclear distance is shrinking.

axis of rotational symmetry: an imaginary line passing through an object. An object possesses an element of rotational symmetry of order n if the rotated object becomes indistinguishable from the original object n times while turning through 360°. Symbolized by C_n, where n is the order of the rotation. All objects possess C_1 axes.

bending mode: a molecular vibration in which internuclear distances between bonded pairs of nuclei remain constant, but internuclear angles periodically increase and decrease in size.

center of symmetry: an imaginary point in an object, through which any real point in the object can be reflected to coincide with an identical, symmetry-equivalent point. The center of symmetry is the midpoint of a line joining a point and its reflected (inverted) point. An object possesses a center of symmetry (a center of inversion) if that object is indistinguishable from the object produced by inverting (reflecting through the center) every point. Symbolized by the letter i.

identity: a symmetry operation consisting of doing nothing to an object; also, the associated symmetry element. Every object has the identity element. Symbolized by the letter I.

improper rotation: a composite symmetry operation, consisting of a rotation followed by a reflection in a plane perpendicular to the rotation axis; also, the associated symmetry element. Symbolized by S_n, where n is the order of the rotation.

inversion: a symmetry operation in which each point in an object is reflected through a center of symmetry *(q.v.)*. Symbolized by the letter *i*.

isomers: compounds which have different structures and different properties, but the same empirical (simple) formula.

reflection: a symmetry operation in which each point in an object is reflected across a plane. Symbolized by the letter σ.

reflection plane: a symmetry element, an imaginary mirror. An object possesses a reflection element if a plane can bisect the object, producing two mirror-image halves. Symbolized by the letter σ.

rotation: a symmetry operation in which an object is turned around an axis of rotational symmetry *(q.v.)*. Symbolized by C_n, where *n* is the order of the rotation.

symmetric: (1) possessing symmetry elements which include at least one of the following: reflection plane, rotation axis of order two or greater, inversion point, improper rotation axis; (2) having the property of being indistinguishable from the result of a symmetry operation; said of a vibration, a mathematical function, or (Chapter 12) an orbital.

symmetric stretching mode: a molecular vibration in which the full symmetry of the molecule is maintained at all times. When one of the changing internuclear distances increases, all other changing internuclear distances also increase.

symmetry-correct mode of vibration: an oscillating motion of nuclei in which the motion is either symmetric or antisymmetric with respect to each of the symmetry operations of the molecule. (The motion may be symmetric with respect to some of the operations, antisymmetric with respect to others.)

symmetry element: an imaginary geometrical entity (such as a point, a line, or a plane) with respect to which a symmetry operation can be performed. An object contains a symmetry element if the associated symmetry operation transforms the object into a new object indistinguishable from the original.

symmetry equivalent: said of two or more nuclei that can be interchanged by a symmetry operation of the molecule; also said of any other feature of a molecule, such as a bond or an orbital. Any two symmetry-equivalent features are identical in all respects.

symmetry operation: a process (identity, rotation, reflection, inversion, or improper rotation) which converts an object (or a mathematical function, or an orbital) into a transformed object equivalent to, indistinguishable from, but not necessarily identical to, the original.

vibration: a periodic motion in which displacements of particles (or distortions of an object) occur alternately in opposite directions from the equilibrium positions (or equilibrium state).

PROBLEMS

4-1 The letter A is its own mirror image, indistinguishable from its own reflection in a mirror. List as many as you can of the other letters that share this symmetry property with A.

4-2 The word TO|OT has a reflection plane. Another word with a reflection plane is TϕT. How many more such words can you find? Make a list.

4-3 List and identify the orientation of all symmetry elements of each of the following molecules:

a. H–O–H (bent)   Both H—O internuclear distances are equal.

b. NH₃ (pyramidal)   All N⟨H,H internuclear angles are equal. All H—N internuclear distances are equal. The nitrogen nucleus is not in the plane of the hydrogen nuclei.

c. benzene (C₆H₆)   All internuclear angles involving hydrogen and its nearest carbons are 120°. All nuclei are coplanar. All C—C internuclear distances involving nearest neighbors are equal. All C—H internuclear distances involving nearest neighbors are equal. All C⟨C,C internuclear angles involving nearest neighbors are 120°.

d. O—N—Cl   The three nuclei are not colinear.

e. H₂C=O style (formaldehyde-like)   Both C⟨O,H internuclear angles are 121°. The C⟨H,H internuclear angle is 118°. All nuclei are coplanar. Both C—H internuclear distances are equal.

f. CH₄   All C⟨H,H internuclear angles are equal. All C—H internuclear distances are equal. The nuclei do not all lie in the same plane.

g. CH₂F₂   The C⟨F,F internuclear angle is 108.3°. The C⟨H,H internuclear angle is 111.9°. All C⟨F,H angles are equal.

h. H—C≡C—H   All nuclei are colinear. Both H—C internuclear distances are equal.

4-4 Select any three objects (pencil, lamp, and so on) from your desk and locate all of the symmetry elements possessed by each. Now find, for each of those objects, some other thing that has an identical set of symmetry elements.

4-5 We characterized the symmetry of an isolated water molecule on the basis of the arrangement of the nuclei alone. However, there are two nonbonding electron pairs on the oxygen (Figure 4-16). If these lone pairs are taken into account, does H₂O retain the same set of symmetry elements? *Suggestion:* make a model of H₂O, using empty pegs to represent the positions of the nonbonding pairs of electrons, and try each of the symmetry operations on the model.

4-6 The structures of NH$_3$ and ICl$_3$ are very closely related; both molecules are depicted in Figure 4-17. Yet the two molecules do not have identical symmetry. Explain the similarities and the differences by finding all the symmetry elements of each molecule, pointing out certain symmetry elements that are contained in one molecule but missing in the other.

4-7 Phosphorus pentachloride, PCl$_5$, has a molecular structure in which five chlorine atoms cluster around a single phosphorus atom. In the cage of chlorine nuclei, there are three symmetry-equivalent nuclei, and two other symmetry-equivalent nuclei. Three of the P—Cl internuclear distances are 2.04 ± 0.06 Å, and the other two P—Cl internuclear distances are 2.19 ± 0.02 Å. There are several symmetry planes in the molecule. One symmetry plane contains the phosphorus nucleus and three symmetry-equivalent chlorine nuclei,

$$\text{Cl} - \text{P}(\text{Cl})(\text{Cl})$$

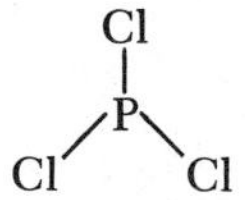

There are three other symmetry planes that contain one of those chlorine nuclei, the phosphorus nucleus, and the remaining two symmetry-equivalent chlorine nuclei,

$$\text{Cl} - \text{P}(\text{Cl}) - \text{Cl}$$

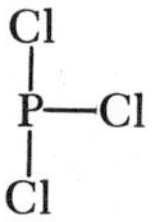

a. Explain on the basis of a reasonable bond structure and in terms of valence shell electron-pair repulsions why PCl$_5$ has the shape it does.
b. Locate all the symmetry elements of the molecule. What are the symmetry elements of a molecule formed by replacing one of the three equivalent (equatorial) chlorine atoms by another atom (for example, bromine)? What if one of the two equivalent (polar) chlorine nuclei is replaced by another atom?

4-8 Consider the following mathematical functions:
a. $y = x$
b. $y = x^2$
c. $y = x^3$
d. $y = \sin x$
e. $y = \cos x$

Imagine each function plotted on a piece of graph paper with a mirror plane, perpendicular to the paper, containing the y axis and separating the positive x and the negative x regions. Classify each function as symmetric or antisymmetric with respect to reflection in the mirror plane. (**symmetric:** $y_{\text{original}} = y_{\text{reflected}}$; **antisymmetric:** $y_{\text{original}} = -y_{\text{reflected}}$.)

4-9 Assume that a set of capital letters is made of some material with thickness, so that every letter has a mirror plane splitting the thickness of the material. Under this assumption, how many capital letters have the same symmetry as a water molecule? List them.

4-10 Every object contains C_1 axes. To what other symmetry operation is a C_1 operation equivalent? To what other symmetry operation is an S_1 operation equivalent? To what other symmetry operation is an S_2 operation equivalent?

4-11 Dichlorosilane has the molecular structure

$$\begin{array}{c} Cl \\ | \\ H-Si-H \\ | \\ Cl \end{array}$$

The Si$<^{Cl}_{Cl}$ angle is 110 ± 1°. Both Si—Cl internuclear distances are 2.02 ± 0.03 Å. The symmetry-equivalent hydrogens are as far as they can get from the silicon atoms and as far as they can get from each other, consistent with a Si—H internuclear distance of about 1.48 Å. Make a model, or a perspective drawing, of a molecule of dichlorosilane, and locate all symmetry elements. Show that whenever you perform two successive symmetry operations on the molecule, the result is the same as if you had performed some single symmetry operation of the molecule. Do not forget the identity operation.

4-12 A molecule of diborane has the structure

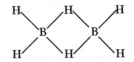

This molecule is not planar. It has the identity element, three mutually perpendicular reflection planes, a center of symmetry, and three mutually perpendicular C_2 axes. Make a model, or a perspective drawing, of the molecule, verifying all of the symmetry elements.

4-13 An isolated molecule of hydrogen peroxide, H_2O_2, has only two symmetry elements, I and C_2. Make a model, or draw a picture, of a structure with this symmetry.

4-14 A molecule of formaldehyde has the structure

$$\begin{array}{c} O \\ \| \\ C \\ / \backslash \\ H H \end{array}$$

All nuclei lie in the same plane, and the two hydrogen nuclei are symmetry equivalent. Sketch some symmetry-correct modes of vibration for this molecule. There is a total of six vibrational modes for formaldehyde.

SUGGESTIONS FOR FURTHER READING

Each scientist is entitled to his own ideas about how to draw pictures of molecules, but it is particularly interesting to see how Linus Pauling thinks such pictures should look. Professor Pauling has been one of the most influential physical chemists in the mid-twentieth century in formulating and fostering notions about chemical quantum theory and the shapes and features of molecules. He and artist Roger Hayward collaborated to produce *The Architecture of Molecules,* San Francisco: Freeman, paperback edition (1970), a collection of 57 full-color representations of molecules, crystals, and such.

Three paperbacks that will provide further reading at a level about one notch above this chapter are: GRAY, H. B., *Chemical Bonds,* Menlo Park, Calif.: W. A. Benjamin (1973); JAFFÉ, H. H. and ORCHIN, M., *Symmetry in Chemistry,* New York: Wiley (1965); and MISLOW, K., *Introduction to Stereochemistry,* New York: W. A. Benjamin (1966).

5
Properties of Tangible Matter

5-1 Abstract

We claim to be engaged in the study of matter. Yet it is hard to recognize in our abstract and idealized notions of atomic and molecular structure the vast and various world about us. Even if we come in from the rain, slush, or sunshine, from the grass and cement world outdoors into the relative simplicity of a laboratory, we do not find atoms and molecules dancing before us. We find bulk matter in a variety of states, textures, densities, colors, shapes, and reactivities. In this chapter, we shall look at some of the ways in which the variousness of "real" tangible matter can be understood on the basis of what we surmise about the structure of its constituent atoms or molecules. We shall discuss the three states of matter, noting that gases are constituted of atoms and molecules that stay apart, but that liquids and solids are made of particles that somehow stay together. We shall investigate some of the forces of intermolecular attraction that hold matter together. Solids are often very regularly ordered internally, and the molecular reasons for this natural ordering are examined in detail. Finally, we shall explore the role of energy in changes of state.

5-2 Solids, Liquids, and Gases

Most of the physical properties, and many of the chemical properties, of a substance are primarily determined by its *state of aggregation,* that is, whether it is a *solid,* a *liquid,* or a *gas.* This state, in turn, must depend primarily on the strength and nature of the forces that exist among the particles (atoms, ions, or molecules) of the substance. Therefore, we shall find ourselves spending most of this chapter thinking about the forces among particles and the arrangements of particles that result from the forces. Since these forces depend, as we have seen, on the structures of the particles, there is hope that we can make sense out of the observed properties of bulk matter on the basis of the properties of its constituent particles.

Suppose we begin by thinking about an assemblage of atoms or molecules in a flask. Almost all of the particles will be moving; the higher the temperature, the higher the velocity of an average particle. In the absence of forces of attraction or repulsion, each moving particle continues to move in a straight line until it collides with another particle or with the wall of the flask. These motions in all directions will quickly disperse the particles evenly throughout the flask. Thus, in the absence of interparticle forces, matter would be expected to disperse to fill all available space. Matter would act like, and therefore be, a gas.

But what if there are forces of attraction among the particles? The answer depends on whether the attractive forces are sufficiently strong to hold together two particles during a typical collision with a moving, unattached particle. The energy imparted by an average collision increases with increasing temperature, so forces sufficient to hold together partners at low temperature may not be strong enough to allow the partnership to survive a typical collision at a higher temperature. Thus, we have the familiar notion that a given substance may be found as a *solid* at low temperatures (interparticle forces in a solid are much stronger than those experienced during an average collision at that temperature, a "thermal collision"; thus, a tight, regular structure is allowed to persist, and seldom will a collision dislodge a particle of this solid), a *liquid* at temperatures above the **melting point**[1] (thermal collisions are violent enough in a liquid to disrupt solid structures, but not to separate the particles, which cling together in an amorphous, free-flowing mass), and a *gas* at still higher temperatures above the **boiling point** (thermal collisions are violent enough in a gas to knock apart the particle partnerships and to prevent those partners from sticking together during subsequent collisions). These three stages—these three states of matter—are depicted in Figure 5-1.

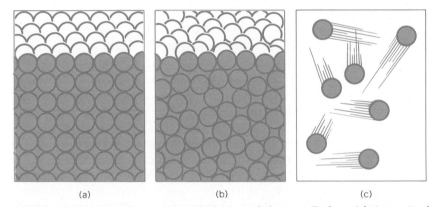

FIGURE 5-1 (a) Particles in a *solid*, packed in an orderly array. Each particle is permitted only very limited freedom of movement. The solid is constantly being bombarded by external particles, but at this temperature the energy of a bombarding particle is too small to disrupt the structure of the solid. (b) Particles in a *liquid*, packed together very compactly. Individual particles are held as part of the whole mass, but a particular particle's neighbors keep changing. This mass flows to fill the bottom of the container. (c) Particles in a *gas*, almost entirely separated from one another. Occasional pairs persist for only short periods of time during a collision.

We see that the question whether a particular substance is to be found as a solid, a liquid, or a gas at some temperature is really a question of whether the prevailing temperature is below the melting point (the substance is a solid), between the melting point and the boiling point (the substance is a liquid), or above the boiling point (the substance is a gas). In ordinary conversation, statements such as "mercury is a liquid," "gold is a solid," "oxygen is a gas"

---

[1] The temperature at which a substance melts and boils depends on the pressure it is under; one can freeze a liquid or condense a gas at a higher temperature by squeezing on them, provided the solid is denser than the liquid (true of almost all substances) and that the liquid is denser than the gas (true of all substances). We shall explore this notion later in this chapter. For the present, we shall assume a constant pressure of one atmosphere.

imply that the prevailing temperature is near room temperature, about 20 to 23°C.

There are several ways in which electrical forces can hold together particles. Let us consider the kinds of bulk properties each of these confers on matter.

5-3 Ionic Substances

The particles that are the structural units of **ionic crystals** are positive and negative ions. These ions are held together by the attraction of opposite charges. The ions may be simple ones, as for instance Na^+ and Cl^- in NaCl. They may be polyatomic ions, such as NH_4^+ and SO_4^{2-} in the salt $(NH_4)_2SO_4$, with each having its own stable internal structure maintained by strong covalent bonds. In either case, but especially for simple monatomic ions, the forces between the ions in an ionic solid are nearly nondirectional—the forces are essentially the same in all directions around an ion. Thus, a given positive ion, attracting negative charge in all directions, can be surrounded on all sides by negative ions, and each of these negative ions may likewise be surrounded by positive ions. This pattern is continued for great distances (great on the atomic scale), as shown in Figure 5-2, with all units attracting their nearest neighbors. The result of all this attraction is that assemblages of ions are difficult to disrupt. Ionic solids often are high melting; all ionic substances are solid at room temperature, though mixtures of certain salts melt only slightly above it. Melting an ionic substance produces an amorphous assemblage of ions in which individual ions can move past one another while still remaining in contact. Thus, an ionic liquid conducts electric current readily (Figure 5-3), while an ionic solid with ions constrained to fixed positions has a much lower electrical conductance.

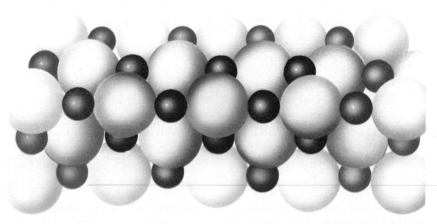

FIGURE 5-2 A schematic representation of part of one possible form of an ionic crystal. Note that each positive ion (darker shading) is completely surrounded by negative ions, and vice versa.

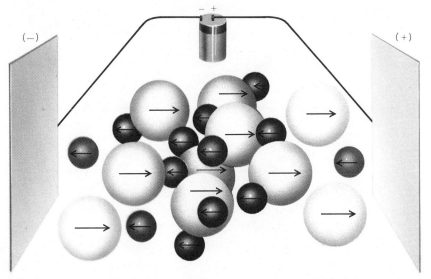

FIGURE 5-3 Electrical conduction in a portion of a molten ionic substance. Direction of drift is marked on the ions: Positive ions (darker shading) migrate toward the negative electrode, negative ions toward the positive electrode. Charge is transferred to the indicated external circuit by reactions at the surfaces of the electrodes.

5-4 Covalent Substances

Some substances are composed of atoms each of which has the ability to form two or more covalent bonds with its neighbors. Examples are carbon, tungsten carbide (WC), and silicon dioxide (SiO_2). With each atom in the substance bonded to more than one neighbor, the possibility arises of "infinite" chains, sheets, or three-dimensional structures of atoms (see Figure 5-4). As in the case of **ionic substances,** this welding of the assemblage of atoms into a single gigantic structure (gigantic on the atomic scale) confers high melting point and hardness to the solid. Again, all such covalent substances are solids at room temperature. When they do melt (quartz, one form of SiO_2, melts at 1610°C), the bonding between atoms, being directional and requiring electron sharing, cannot remain functional as it does in the case of molten salts. Thus, molten silicon dioxide consists of temporary clumping of Si and O atoms, in which covalent bonds persist for short times, and regular structure exists over short distances. These bonds are continually broken by thermal collisions and vibrations, and then reformed with new bonding partners. Molten silicon dioxide reflects this situation in being very viscous (nonrunny) just above the melting point, with viscosity decreasing with increasing temperature as the clumps become smaller and have shorter lifetimes.

What can you use for a container for molten silicon dioxide? See Appendix I.

Many familiar rocks and minerals, including the clay minerals of the soil, contain "infinite" covalent assemblages of Si, O, Al, Mg, and others, forming a rigid, negatively charged framework with holes into which positive ions

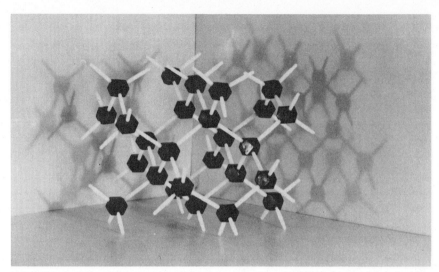

FIGURE 5-4 Diamond, a covalent substance, is represented here by a model showing each carbon atom covalently bonded to four others. The resulting structure is extremely hard. (Photograph: John H. M. Hill. Model from Benjamin/Maruzen HGS Molecular Structure Models, copyright © 1969, W. A. Benjamin, Inc., Menlo Park, Calif.)

fit to form the electrically neutral whole. These covalent-ionic hybrid structures share the properties of purely covalent or ionic structures: They are hard as rock, and high melting.

5-5 Molecular Substances

Suppose that a substance is composed of electrically neutral molecules, molecules whose exposed atoms are incapable of further bonding, either ionic or covalent (Figure 5-5). Then neither kind of force mentioned above is available for formation of a solid or liquid. We might expect that all such substances would also be found as gases, but this is not so, because of relatively weak forces (called **van der Waals forces**) which serve to hold the molecules of these **molecular substances** gathered together as long as the temperature is not too high. In contrast to ionic or covalent substances, molecular solids are generally soft and low melting because of the weakness of these forces.

These are secondary forces, real enough, but not on a par with chemical bonds.

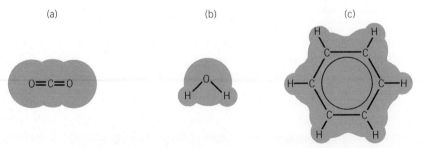

FIGURE 5-5 Representations of some molecular substances. (a) Carbon dioxide, CO_2. (b) Water, H_2O. (c) Benzene, C_6H_6.

The general term "van der Waals forces" covers a range of weak forces between molecules, all of which are electrical, and all of which are much weaker than covalent or ionic bonds. They were first postulated to exist by J. D. van der Waals to account for observations on the pressure and volume of gases. Gases with very weak interparticle forces, such as He and Ne, obey quite closely the ideal gas equation (see Section 5-11).

$$P = \frac{RT}{V}$$

where P is pressure; T is temperature on the Kelvin scale; V is volume per mole; and R is a proportionality constant. Because some other gases exert lower pressure than (RT/V), van der Waals proposed that weak attractive forces exist among these gas molecules. Such forces would cause each molecule to be attracted toward the center of the mass of gas, resulting in weaker collisions with the container walls and thus lower pressure at a given volume and temperature. Modern structural investigations and quantum-mechanical arguments have suggested several explanations for such forces; we shall discuss *dipole interaction, hydrogen bonds, and London dispersion forces.*

Why is the attraction away from the walls?

DIPOLE INTERACTIONS. If the molecules we are considering are polar (Chapter 3), as for example $CHCl_3$, nitrobenzene, or $(CH_3)_3N$ [Figure 5-6(a)], then two of them may align in a crystal or during a collision in the gas phase in such a way as to optimize the forces of attraction between the charged ends of the molecules. We can imagine either head-to-tail orientation or alignment with opposite charges adjacent [Figure 5-6(b)]. In both configurations, the favorable (attractive) interactions of opposite charges outweigh the repulsions of like charges because the opposite charges are closer together than the like ones, and electrical forces decrease with the square of the distance between charges (Appendix III).

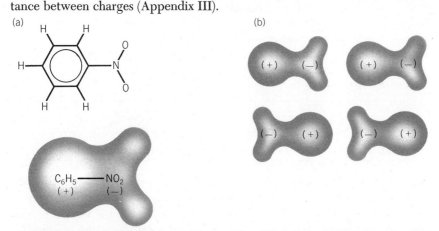

FIGURE 5-6 Charge separation in the polar molecule nitrobenzene, $C_6H_5NO_2$. (a) Two representations of a nitrobenzene molecule. The structural formula of nitrobenzene reveals a concentration of electronegative atoms (N and O) at one end of the molecule. The relatively poorly shielded nuclear charge in these atoms causes a concentration of electronic charge at the "nitro" end of the molecule, leaving the "benzene" end slightly positive. (b) Favorable alignments of polar molecules. Both head-to-tail and adjacent orientations place opposite charges close to each other, lowering the potential energy of a collection of molecules and stabilizing the grouping.

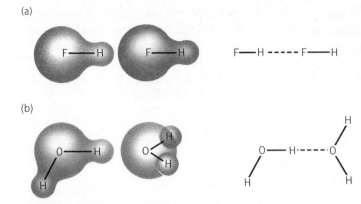

FIGURE 5-7 Hydrogen bonding in (a) HF and (b) H₂O. Both the structural alignment of the dipoles and the conventional representation using a dashed line for a hydrogen bond are shown. Approximate contour lines connecting regions of equal electron density are shown on the left, illustrating the low electron density near the hydrogen nucleus.

HYDROGEN BONDS. A special case of dipole interaction is found in molecules containing hydrogen covalently bonded to oxygen, nitrogen, fluorine, or chlorine, as for example in the compounds H_2O, NH_3, HF, and HCl. This bond is, of course, polar, with hydrogen having less than its "fair share" of electron density. When this loss of electron density happens to any atom, it becomes the center of a positive electrical field. In the case of hydrogen, this field is stronger than for the other elements, because of its small size and (what amounts to the same thing) absence of a shell of electrons, close to the nucleus and uninvolved in bonding, to act as a negative electrostatic shield around the positive nucleus. The nearly unshielded hydrogen nucleus, a virtually naked proton, can then be approached closely by negative sites on another molecule to create an unusually strong dipole interaction. Figure 5-7 shows what happens in two simple cases. Hydrogen bonding plays a crucial role in the chemistry of water solutions (Chapter 7) and of living organisms (Chapter 17). The three-dimensional structures of proteins and nucleic acids are largely determined by hydrogen bonds, and the chemical mechanisms of cellular regulation and reproduction depend critically on hydrogen bonding.

Hydrogen is unique, this way.

LONDON DISPERSION FORCES. Neither of the forces we have considered so far can account for the fact that symmetrical, nonpolar molecules and atoms also exhibit van der Waals forces. For example, even the monatomic gases (helium, neon, argon, krypton, xenon, and radon) can be liquified if the temperature is lowered sufficiently. Thus, even with these noble gases, there must exist weak forces among their atoms. A clue is provided if we notice that their boiling points (remember that the boiling point is also the liquification point) lie in the same order as the number of electrons in the atoms[2] (see

---

[2] This is, of course, also the order of their atomic weights. Resist the temptation to conclude that heavier particles naturally "fall" into liquid or solid states. We are dealing here with tiny, easily scattered atoms and molecules, not with rocks and balloons!

Table 5-1). The explanation for this phenomenon was provided by Fritz London in 1930 on the basis of electronic motions around the nucleus. Although we have generally been describing the atoms as having tiny *spherical* electron clouds, this symmetric shape is, after all, only the time-average distribution of electrical charge. At any instant, there may exist an unbalance of charge to one side of the nucleus, and at another time, to the other side [Figure 5-8(a),(b)]. The result of this oscillation of charge from one side of the symmetrical structure to another is an oscillating electrical field. If there is another atom nearby, this oscillating electrical field will encourage it to match its oscillations with those of the first atom [Figure 5-8(c)], just as permanent dipoles tend to align themselves. Two such atoms oscillating in harmony will always see each other as presenting a favorable dipole; they will experience a *mutually induced dipolar attraction*. The energy of such an interaction will, of course, be greater the closer together the atoms come, since the strength of the electrical field that they generate increases with decreasing distance, and the stronger their dipolar fields, the more they encourage each other to become more polar yet. The correlation of boiling point of nonpolar molecules with number of electrons in the molecule reveals that these forces are stronger, the larger and more complex the electron clouds of the atoms or molecules of the substance. Further reasons for this will become evident in the next chapter.

Literally a dance of the atoms

TABLE 5-1
Correlation of Boiling Point With Complexity of Electronic Structure For Some Nonpolar Substances

	Substance	Number of Electrons	Boiling Point (°C)*
Monatomic Noble Gases	He	2	−268.6
	Ne	10	−245.9
	Ar	18	−185.7
	Kr	36	−152.3
	Xe	54	−107.1
	Rn	86	−61.8
Diatomic Halogens	F_2	18	−188.1
	Cl_2	34	−34.6
	Br_2	70	+58.8
	I_2	106	+184.3
Hydrides of Group IV	CH_4	10	−161.5
	SiH_4	18	−111.8
	GeH_4	36	−88.5
	SnH_4	54	−52

* One atmosphere pressure.

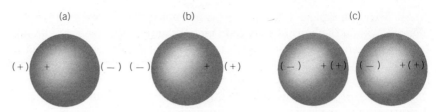

FIGURE 5-8 London forces between nonpolar atoms. (a) Instantaneous dipole created by a momentary displacement of the center of negative charge (determined by the instantaneous distribution of electrons in the atomic orbitals) from the center of positive charge (the nucleus). (b) The situation approximately 10^{-15} sec after the distribution shown in (a). (c) Two adjacent atoms, in a condensed phase or during a collision in the gas phase, mutually inducing aligned dipoles (see text).

5-6 Metals

Try to define a metal.

We still have not considered a large class of substances found nearly everywhere: the metals. Most of the properties of metals are familiar to everyone—their ability to reflect light, rather than to transmit or absorb it, their variable (but often high) densities, their excellent electrical and thermal conductance, and their wide range of hardness and melting points. Each of us easily recognizes a metal. Can we also find an explanation of the properties that we recognize as common to metals and at the same time account for the variability of their properties? (An example of this variability: Tungsten and mercury are both metals, but we use mercury, which melts at −38.87°C, as the liquid in a thermometer that we may read by the light of an incandescent solid tungsten filament operating at a temperature well over 2500°C.)

Each molecular orbital extends over the whole crystal.

The bonding that holds together an aggregation of metal atoms presents us with still another new phenomenon, the metallic bond. A metallic bond is a covalent bond (that is, a sharing of electrons) that extends in three dimensions over an entire "infinite" array of metal atoms. Note that this situation is different from that described for a **covalent substance,** in which an "infinite" three-dimensional net of localized two-atom bonds connects the assemblage of atoms. We shall explore further the reasons why a given substance is a metal or a nonmetal in the next chapter; suffice it to say for now that the bonding electrons in a metallic substance are held so weakly by individual metal atoms that they are quite free to move through the entire assemblage. A metallic crystal, which you can hold in your hand and look at, is in fact a single macroscopic, three-dimensional molecule, in which the bonding between the atoms is by three-dimensional **molecular orbitals** (Chapter 12) large enough to see (Figure 5-9).

The electrons used to form these metallic bonds are those held most loosely by an individual metal atom. They are the atom's highest-energy electrons, the electrons in its outer electron shell. What is left when the outer electrons are used for the metallic bond is, of course, a positive ion. Thus, a metallic crystal can be thought of as an array of positive centers held together by a kind of matrix of negative charge composed of the electrons in the all-encompassing molecular orbitals surrounding the positive centers. If the quantity of charge lost from each atom in forming this sea of delocalized electrons is large, then the force between the atoms and the delocalized electrons will be large, and the result will be a stable assemblage. Such a metal will be relatively hard and have a high melting point. Metals that have only a few electrons in their outer shells (sodium, calcium, and aluminum, to name a few) should be, and are, relatively soft and low melting. Intermediate cases will, in general, be metals intermediate in hardness and melting point. We shall come back to this question in more detail in the next chapter when we consider the influence of electronic structure on chemical and physical properties in a bit more systematic way.

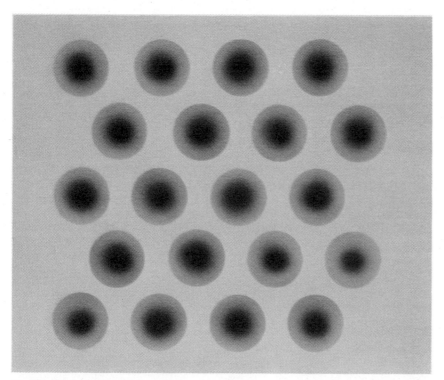

FIGURE 5-9 Metallic bonding. Spheres represent the nuclei and the completed inner shells of electrons; outer-shell electrons (lightest color) occupy delocalized orbitals filling the entire crystal. These delocalized electrons are shared by all of the nuclei in the crystal. The attraction between the positive nuclei and the delocalized electrons holds the crystal together.

Metals conduct electrical current. They do so because of the mobility of the electrons in their macroscopic molecular orbitals. That is, the electrons in a metallic substance can move freely from one end of such a substance to the other. This is true also of liquid metals, so that the switch that turns on the tungsten light bulb may, in fact, contain a mercury contact.

All metals are shiny, at least when they are clean. Light which falls on a chunk of metal does not pass through. The light is either absorbed or re-

Without metals, how could we distribute electricity?

flected, and we know that with a flat piece of polished metal—a mirror—most of the visible light is reflected.[3] (Many metal surfaces do not stay bright and shiny very long, quickly reacting with oxygen in the air and becoming coated with films of metal oxide. Silver is a scavenger of H_2S from the atmosphere, and shiny silverware tarnishes, becoming coated with a thin film of Ag_2S.) Light does pass through many substances. A crystal of sodium chloride (an ionic substance), a crystal of quartz (a covalent substance), and liquid water (a molecular substance) are three examples of substances that are transparent. (Finely divided salt or quartz, or a thick mist of minute water droplets, seems white because of surface reflections, as does a scratched crystal.) Metals as a class appear to be a special case, and we shall search for an explanation for their inability to transmit light by contrasting their structure with the structures of other substances already considered in this chapter.

Table salt is transparent?

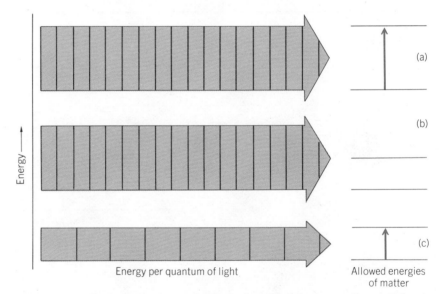

FIGURE 5-10 Energy relationships in the absorption of light by matter. The vertical scale is proportional to energy. On the left, the vertical thickness of the arrow represents the energy per photon of light whose wavelength is represented by the vertical lines across the arrow. On the right, the horizontal lines represent the energies of quantum-mechanical stationary states (See Chapter 2) of the atoms of which the matter is composed. The vertical distance between these lines represents the difference in energy between stationary states. (a) Light, whose energy per photon matches the difference in energy between two quantum states of an atom, is absorbed, with light energy raising the atom from the lower to the upper energy state (colored arrow). (b) Light of the same wavelength and energy cannot be absorbed by matter when the difference in allowed energy states of the matter does not match the energy of the light. This light is transmitted through the matter. (c) Light of a longer wavelength (and therefore lower energy per photon) *does* match the energy structure of the matter shown in (b). This light is absorbed, with the simultaneous promotion of the atom from a lower to a higher energy level (colored arrow).

---

[3] As anyone who can tell a penny from a dime knows, some metals are colored; this fact implies some absorption of light over part of the visible spectrum. This effect is weak, however, compared to the general ability of metals to reflect light of all wavelengths.

Associated with any light wave is an oscillating electrical field. That field will interact with the electrical charges in matter under certain circumstances. If interaction occurs, the light will be either absorbed by the matter or reflected. Light can be absorbed by the matter only if the energy of the light (per photon) matches the difference in energy between the existing state of an electron in the matter and some possible higher energy state (Figure 5-10). In matter in which the electrons are all closely confined either in bonds or in lone pairs, the gaps between allowed energy levels are relatively large. That is why most covalent molecular and ionic substances transmit visible light.

When the oscillating electrical field of a light wave falls on a metallic crystal, electrons in the delocalized molecular orbitals are free to oscillate in step with the light wave. This interaction prevents the electrical forces from penetrating into the crystal; they are, in effect, shorted out by the motion of the delocalized electrons. And, for most metals, the energy of visible light does not match the energy difference between allowed quantum states of the atoms in the crystal. Unable either to pass through the crystal or to be absorbed, the light must do the only other possible thing—it bounces off the surface of the crystal.

5-7 Recapitulation: Chemical Categories of Matter

We have described four types of substances, basing our classification on the types of chemical forces that exert attraction between gas particles when the substances are at temperatures above their boiling points and the forces that hold the particles close together in the liquid state and the solid state. We have investigated relationships between these forces and the observable properties of substances. Table 5-2 summarizes some physical properties of these four types of substances in broad generalities; there are plentiful exceptions.

Think of examples for each category.

Naturally, there remain many materials which do not fit neatly into any one of these classes. As we have seen, many minerals are both covalent and ionic. The *metalloid* silicon is a covalent substance, which is an electrical semiconductor and is opaque and shiny (metallike) to visible light; its optical behavior with infrared light is more typical of covalent substances. We shall return to the correlation of observable properties with atomic and molecular structure in more detail in the next chapter.

TABLE 5-2
Properties of Typical Substances

Type of Substance	Hardness	Melting point	Electrical Conductance	Optical Properties
Ionic	Variable but high	Variable but high	Low (except molten)	Transparent
Covalent	Very high	High	Low	Transparent
Molecular	Low	Low	Low	Transparent
Metallic	Variable; Flexible and malleable	Variable	High	Opaque and shiny

5-8 Orderliness and Structure in Solids

Many familiar solids show striking and obvious regularity of form (Figure 5-11). Thus, when we see a solid object that occurs naturally with flat faces meeting in straight edges *at angles that are uniform from sample to sample*, we may suspect that we are seeing the outward and visible manifestation of an inward regularity. Such regularity, highly symmetric and extending over arrays of vast numbers of atoms, is found in many substances.

FIGURE 5-11 Crystals in nature. (a) Snow crystals. Top pictures photographed by U. Nakaya, reproduced by permission of Mrs. Ukichiro Nakaya, copyright 1954 by the President and Fellows of Harvard College. Lower photographs courtesy of Carl Zeiss, Inc. (b) Geode. Photograph: Gertrude Catlin.

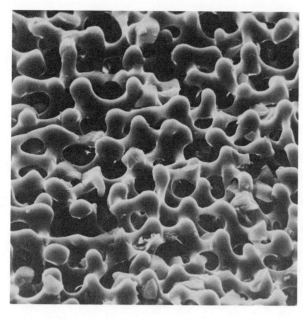

FIGURE 5-12 Photomicrograph (taken by a scanning electron microscope) of a portion of the skeleton of the sea urchin *sphaerechinus granularis*. Although the region shown constitutes a single crystal of calcite ($CaCO_3$), no plane surfaces or reproducible angles are shown. In such a case, only the diffraction of waves such as X-rays reveals the crystalline character of the solid. [*Source:* Hans-Ude Nissen, *Science,* vol. **166,** p. 1151 (Nov. 28, 1969). Used by permission.]

Solids whose internal arrangement of atoms or molecules shows a regular repetition in any direction through a large distance on the atomic scale are called **crystals.** This definition, based on a long-range order at the atomic level, includes many objects that do not appear crystalline to the eye, such as a worn and rounded grain of sea sand. Indeed, it has recently been shown that some parts of the skeleton of shellfish such as the sea urchin are highly crystalline, even though no large-scale evidence of angularity or planarity can be detected (Figure 5-12). In this section, we shall consider the common crystal structures and imagine how they may be created through simple and regular assemblies of spherical particles.

The crystal and its lattice

THE UNIT CELL. Because of the regularity with which the microscopic structural elements are repeated within a crystal, we can describe the whole structure simply by taking advantage of the *translational symmetry* of the arrangement. We need to know the structure of a small part of the crystal, and we need a rule for repeating this unit of structure throughout three dimensions. Let us take a familiar example from the macroscopic world, and confine our attention for the time being to just two dimensions—let us consider a brick wall.

or glide symmetry

A brick wall presents a structure that is made by placing small, repeating units into a regular array. However, a description of the shape of a single brick is not sufficient to tell us about the construction of the wall, even though we know that the wall is an assemblage of bricks only (let us leave the mortar out of this). For example, there is nothing in the nature of our unit brick to tell us whether the wall is laid with bricks overlapping, or merely stacked up (Figure 5-13). For a complete description, we need a **unit cell,** which is the smallest structural unit of the wall (or crystal) which, when propagated in specified directions, will generate the whole structure. Unit cells of two methods of constructing a brick wall are shown in Figure 5-14. Note that, in either case, the required information, in addition to any internal structure of the unit cell, is an angle at which the unit cell is to be propagated and a repeat distance along the axes defining the angle. (The assumption in Figure 5-14 is that the wall is a two-dimensional structure containing only one course of bricks. How could the unit cell be modified if a wall several bricks deep were to be described? Make a sketch.) Any crystal has a repeating array of unit cells, and we call the whole structure a *lattice*.

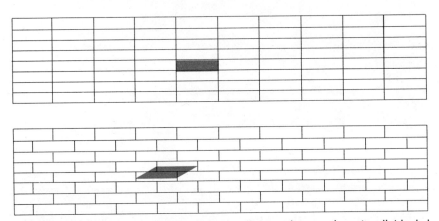

FIGURE 5-13 Two ways of building a brick wall. In each case, the unit cell (shaded in color) is required in order to describe unambiguously how the wall is to be constructed. If the wall is treated as a two-dimensional structure, how many bricks are contained in each unit cell?

5-9 Structures Arising from Nondirectional Bonding

Many of the common crystal structures arise because the bonding forces in the crystal pull the particles (atoms or molecules) of the substance into the closest possible contact. If you play with the packing of spheres, such as ping pong balls or tennis balls, you will soon find that some strikingly straight-edged structures are produced by such forces. (Indeed, it is recommended that the following be read with some uniform and packable—wood or styrofoam—balls at hand; much of what we shall say is believed much more easily when it is seen.) We shall now examine some close-packing structures.

Look for what isn't there.

TYPES OF HOLES. Of course, whenever we pack round objects together, there is bound to be some empty space. In understanding the structure of matter, as in many other spheres of action, what is missing is as important as what is there. If we place three spherical objects as close together as we can, we generate a *trigonal hole* whose center lies in the plane of the centers of the spheres (Figure 5-14). If we add a fourth sphere to this group above (or below) that plane, we have created a *tetrahedral hole* (Figure 5-15), so called because it lies at the center of a regular tetrahedron. An alternative way of covering our original trigonal hole generates an *octahedral hole* (Figure 5-16). Note that of these three, the octahedral hole is the largest, and the trigonal the smallest. All close-packed structures made of spheres must contain these three types of holes, and only these.

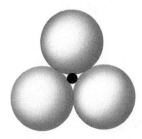

FIGURE 5-14 A trigonal hole. This is the hole generated by the close packing of three spheres. The center of the hole lies in the same plane as the centers of the spheres. In this illustration, a small sphere is shown fitting into the hole, with its center coinciding with the center of the hole.

(a)

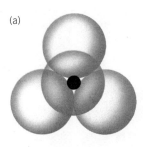

(b)

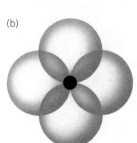

FIGURE 5-15 A tetrahedral hole. Again, a small sphere is shown inserted in the hole. This tetrahedral hole is generated by the close packing of four spheres. Note that the center of the hole does not lie in the plane of any three of the four close-packed spheres. For this reason, a tetrahedral hole is actually larger than a trigonal one, and a larger sphere can be fit into it. (a) The tetrahedral hole viewed along one of its four C_3 rotational symmetry axes, that is, from directly above the plane of three of the spheres. (b) a view oblique to the plane of three of the spheres. What sort of symmetry axis is revealed in this view? How many such axes do you think there are?

Recall definitions of symmetry elements in Ch. 4.

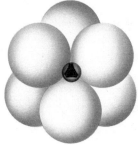

FIGURE 5-16 An octahedral hole. The center of this hole lies midway between the planes defined by the centers of groups of three spheres. The view in this figure is down a C_3 rotational axis (which is also an S_6 improper rotational axis) perpendicular to the planes of two trigonal groups of spheres. The axis passes through the center of the octahedral hole, while it is also the center of the small sphere shown occupying the octahedral hole.

CUBIC CLOSE PACKING (CCP). Lest you think that all this attention to holes is much ado about nothing, let us proceed to create a solid structure from them. An octahedron of six spheres has, like all octahedra, eight faces (hence the name); each face is a triangle of three spheres, centering on a trigonal hole. If each of these eight trigonal holes is covered with a sphere

How to make a cube from spheres

to make a tetrahedron, the resulting array of spheres is cubical. We could readily imagine that such a cube, propagated in all directions, might generate a cubical object (Figure 5-17). Note that to create the whole structure, we would need to propagate not the entire round-cornered cube shown in Figure 5-18, but a true cube with the round corners squared off.[4] Two representations of such an object are shown in Figure 5-19; they represent the unit cell of cubic close packing. This unit cell, for reasons that should be evident, is also called a *face-centered cube*. The second drawing in Figure 5-19, which shows only the locations of the centers of the spheres, consists of what are called *lattice points*. A lattice is an imaginary array of geometric points regularly arranged in space; a crystal is created when atoms or molecules occupy lattice points.

FIGURE 5-17 Cubic close packing. A unit cell of cubic close packing. (Photograph: Gertrude Catlin.) →

← FIGURE 5-18 Cubic close packing. Eight unit cells of cubic close packing, showing the sharing of spheres by adjacent unit cells. (Photograph: Gertrude Catlin. Model by LeMont Scientific, Inc.)

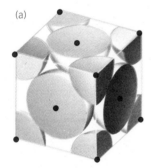

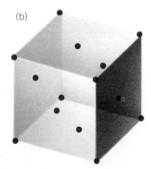

FIGURE 5-19 Two representations of a unit cell of cubic close packing (compare with Figures 5-17 and 5-18). (a) In this picture, portions of spheres lying outside the conventional boundaries of the unit cell have been shaved off. (b) The same unit cell, with only the centers of the spheres (the so-called lattice points) shown.

---

[4] Those parts of the spheres that are cut off by the boundaries of the unit cell belong to neighboring unit cells. For example, since two neighboring cubes share a face, the sphere centered in that face is exactly halved by the unit-cell boundary. Spheres centered on an edge are quartered, and those centered on a corner are cut into eighths (why?). Unit cells need not be built up from sphere centers; by a diagonal shift of the cell boundaries, the same unit cell can be made with its corners in holes, so that complete spheres are included within the cell boundaries. The unit cells shown here are the ones usually chosen, but choice of the location of the unit cell is purely conventional.

FIGURE 5-20 Cubic close packing. The unit cell of Figure 5-19(a) has been tipped up so that the close-packed planes are horizontal, to aid in visualizing them. (Photograph: Gertrude Catlin.)

We can perform one further imaginary (or real) experiment with our cube of spheres. If we tilt it up on one corner, we find that we can think of it as a 14-sphere sample out of an array of a least four planes of close-packed spheres (Figure 5-20). If we now carefully examine these planes, we find that (1) no sphere of the second plane lies above any sphere of the first (this is always true of close-packed spheres); (2) no sphere of the third plane lies above any sphere of either of the first two (such an arrangement is not necessarily the case with close-packed spheres, but is always true of cubic close packing); (3) every sphere of the fourth layer (in our cube there is only one) does lie exactly above a sphere of the first layer. Thus, the fourth layer is an exact replica of the first, and of course, the fifth would be a replica of the 2nd, the 6th of the third, and so forth, in a perfect crystal. If we labeled corresponding layers, as literary people are fond of doing, with rhyming lines of poetry, we would say that a cubic close-packed structure could be characterized by the scheme ... ABCABCA ..., with every third layer repeating. Such a scheme is the fundamental definition of cubic close packing.

Packing atoms in the least possible space

HEXAGONAL CLOSE PACKING (HCP). If we look down on two nested close-packed layers, we find (Figure 5-21) that there are two sets of hollows available for a third layer. Placing one sphere in one hollow of a given set forces all of the other spheres in that layer also to occupy hollows in the same set if close packing is to be maintained. In cubic close packing, we found that the set of hollows that lie over trigonal holes in the first layer (marked c in Figure 5-21) is used by the third layer. Notice, though, that the alternate set of hollows (marked h) lies directly above spheres of the first layer, and it is this set that is used in *hexagonal close packing*. The packing order is then ... ABABA ... in contrast to cubic's ... ABCABCA .... The unit cell in the hexagonal case has two faces lying in planes of close-packed layers. It is the diamond-shaped solid shown in Figure 5-22. The natural hexagonal symmetry of close-packed circles emerges in this arrangement, and the unit cell is one-third of a hexagon (Figure 5-23).

Another closest way

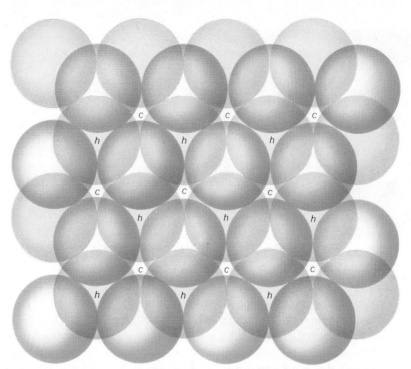

FIGURE 5-21 Top view of two close-packed layers of spheres. Use of the hollows marked c in the top layer (in color) by a third layer of spheres leads to cubic close packing. Use of the hollows marked h by a third layer leads to hexagonal close packing (see text).

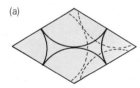

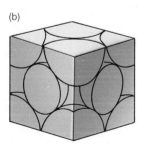

FIGURE 5-22 The conventional unit cell of hexagonal close packing. (a) Oblique view. Although in a two-dimensional drawing the unit cell looks cubical, it is actually lozenge shaped, as can be seen in the top view (b). As in Figure 5-19, portions of spheres lying outside the boundaries of the unit cell are shown "shaved off."

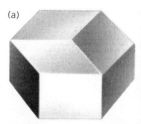

FIGURE 5-23 Hexangonal close packing. (a) Three of the lozenge-shaped unit cells in Figure 5-22 packed together to make a hexagonal prism. (b) Many hexagonal prisms pack together to retain the hexagonal outline, a phenomenon familiar from the contemplation of floor tiles in, for example, bathrooms. (Photograph: Gertrude Catlin. Model by LeMont Scientific, Inc.)

SOME EXAMPLES OF CLOSE-PACKED SOLIDS. Many substances whose fundamental unit is monatomic, such as metals and the noble gases, adopt close-packed structures in the solid. Helium, neon, argon, and krypton all form CCP crystals, as do copper, silver, platinum, and other metals. HCP is adopted by magnesium, beryllium, and many of the rare earth metals, among others. Nickel has two forms: cubic and hexagonal.

Besides these relatively simple cases of close-packed spheres, many substances form close-packed crystals in which the unit sphere is not a single atom, but a nonspherical molecule which, by rapid tumbling rotation in its lattice site, shoulders out for itself a spherical domain. Molecular hydrogen forms HCP crystals in this way, and CH_4 forms cubic ones.

IONIC STRUCTURES DERIVED FROM CLOSE PACKING. Ionic substances, of course, must contain at least two different kinds of ion. Very often, the negative ions are considerably larger than the positive ions. The electrons in any atom repel one another and tend to spread out; this effect is exaggerated in negative ions relative to positive ions, in which the overbalance of positive charge in the nucleus draws the electrons more closely in to the center. Consequently, many ionic crystal structures can be thought of as the result of squeezing positive ions into the holes in a lattice of negative ions. The binding energy of an ionic crystal comes from the mutual attraction of opposite charges and is weakened by the repulsion of like charges. Thus, a crystal in which the negative ions are actually in contact in a close-packed arrangement will be relatively unstable (though some do exist); a better arrangement is for the positive ions to be somewhat too large for the holes they occupy, so that the negative lattice is forced apart. This insures close contact of opposite charges and separation of like charges. Nevertheless, the geometry of close-packed structures is retained. Although the number of substances having such structures is legion, perhaps three examples will suffice.

Anions are often much larger than comparable cations.

1. The sodium chloride (rock salt) structure. Sodium chloride and most of the other alkali halides, as well as the oxides of the alkaline earth metals (for example, MgO), crystallize with the metal ions occupying octahedral holes in a lattice of negative ions. Since a close-packed lattice contains just as many octahedral holes as it does spheres, every octahedral hole[5] is occupied by a metal ion. The resulting structure is shown in Figure 5-24. Note that both the positive ion and the negative ion are surrounded octahedrally by ions of opposite charge.

Can you convince yourself that there are just as many holes as spheres?

2. The fluorite and antifluorite structures. There are twice as many tetrahedral holes in a close-packed structure as there are spheres. Many substances with 2:1 stoichiometry take advantage of this fact to form crystals derived from close packing like NaCl, but using the tetrahedral holes. The cubic version of this structure (Figure 5-25) is called *antifluorite* if positive ions occupy the tetrahedral holes in a negative lattice (for example, as in Na_2O), and *fluorite* if negative ions occupy tetrahedral holes in a positive lattice, as

---

[5] At least in an ideal crystal. Real crystals exhibit defects of various kinds, such as the inclusion of foreign atoms and the presence of vacant holes where there should be an ion.

in CaF_2 (the mineral fluorite). Since Ca^{2+} ions are smaller than F^- ions, it is stretching matters a bit to call the positive lattice of Ca^{2+} "close packed." Still, the geometry of Ca^{2+} in fluorite is the same as if it were close packed.

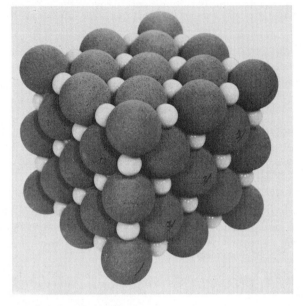

FIGURE 5-24 A model of the sodium chloride structure. Positive ions (light-colored spheres) occupy all of the octahedral holes in a face-centered cubic array of negative ions (dark spheres), and vice versa. Compare Figure 5-19(b). (Model by LeMont Scientific, Inc. Photograph: Gertrude Catlin.)

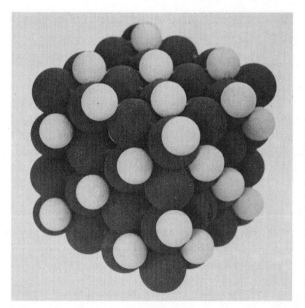

FIGURE 5-25 The fluorite structure. Negative ions (dark spheres) occupy the tetrahedral holes in a cubic array of positive ions (light spheres). Note that, although greatly expanded by the insertion of negative ions, the positive ion lattice is identical to that of cubic close packing (Figure 5-18). (Model by LeMont Scientific, Inc. Photograph: Gertrude Catlin.)

3. Zinc sulfide: wurtzite and zinc blende. Zinc sulfide crystallizes in two forms, in both of which *half* of the available tetrahedral holes are utilized, consonant with the one-to-one stoichiometry of ZnS. The hexagonal form is

called wurtzite and is based on HCP of the sulfide ions (Figure 5-26); the cubic form is called zinc blende, or sphalerite.

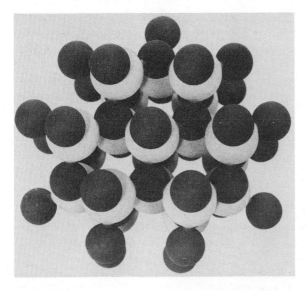

FIGURE 5-26 The wurtzite structure. Each ion is surrounded tetrahedrally by ions of the opposite charge. The overall geometry is provided by a hexagonal array of negative ions (light spheres). (Model by LeMont Scientific, Inc. Photograph: Gertrude Catlin.)

5-10 Structures Derived from Directed Bonding

When the forces holding the crystal together are directed to specific angles, the geometry dictated by close packing may not correspond to the bond angle dictated by valence-shell repulsions (Chapter 4). In this case, somewhat more open structures, as various as the substances forming them, are observed. A few examples may serve to illustrate the point.

DIAMOND: A CUBIC STRUCTURE WITH TETRAHEDRAL COVALENT BONDING.

The diamond crystal structure, including a unit cell, is shown in Figure 5-27. The fact that each carbon atom has four bonding regions in its valence shell makes a tetrahedral arrangement of these regions lowest in energy. One way to arrange such bonding (but not the only, as we shall see) is in the cubic pattern shown here.

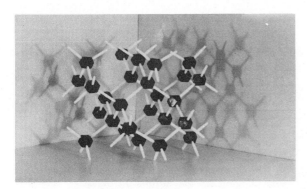

FIGURE 5-27 The crystal structure of diamond. Each carbon atom is bonded covalently to four others at the corners of a tetrahedron. The overall structure is cubical. (Photograph: John H. M. Hill. Model from Benjamin/Maruzen HGS Molecular Structure Models, copyright © 1969, W. A. Benjamin, Inc., Menlo Park, Calif.

ICE: A HEXAGONAL STRUCTURE WITH TETRAHEDRAL HYDROGEN BONDING.

In Ice I (the familiar form stable below 0°C at one atmosphere pressure), every oxygen atom is hydrogen bonded to four others, two through its own hydrogen atoms and two through its nonbonding electron pairs to the hydrogen atoms of a neighboring water molecule. Again, the octet outer shell structure of the oxygen favors tetrahedral geometry, though in this case, the requirement is somewhat relaxed, since the hydrogen bonds need not be straight. (In fact, they are, for lowest energy.) In Ice I, a hexagonal form (Figure 5-28) is adopted and is reflected in the familiar and beautiful hexagonal symmetry of the snowflake (Figure 5-11). Note the resemblance of this structure to that of wurtzite (Figure 5-26). That it is not close packed, though, is shown by the well-known lightness of ice; liquid water is more dense than ice because this open hydrogen-bonded structure of ice is partly destroyed on melting. With the geometric restrictions relaxed in the liquid state, water molecules can pack closer together.

FIGURE 5-28 In Ice I, oxygen atoms (light-colored spheres) are connected by hydrogen bonds (represented in this model by the glue holding the spheres together). Note that the adoption of this structure results in much empty space in the form of hexagonal channels that run right through the crystal. (Model by LeMont Scientific, Inc. Photograph: Gertrude Catlin.)

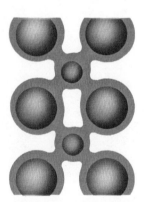

FIGURE 5-29 The crystal structure of $PdCl_2$. Palladium atoms (small spheres) are covalently bonded to four bridging chlorine atoms (large spheres) at the corners of a square, and the whole "infinite" chain lies in a single plane. The chains are packed parallel to each other to make up the whole crystal.

$PdCl_2$: SQUARE PLANAR COVALENT BONDING IN "INFINITE" CHAINS.

In palladium dichloride, the d-electron structure of the metal gives it a preference for bonding that is directed toward the corners of a square. These squares are fused in one dimension to form chains whose length is as great as the dimension of the crystal (Figure 5-29).

More complex chains, sheets, and three-dimensional structures are formed by the aluminosilicates, minerals containing aluminum and silicon covalently bonded to oxygen, often with other metals ionically bonded to these covalent groups. Chain aluminosilicates account for the fibrous properties of asbestos; sheet structures evidence themselves in the flaky crystals of mica.

5-11 Gases: The Consequences of Chaos

The ultimate in disorder

After the superb order and symmetry of crystalline solids, there would seem to be little that can be said about gases beyond the qualitative picture of molecular chaos presented in Figure 5-1(c). However, just the fact that the physical properties of gases are dominated by the random flight of their molecules, and not by the internal structure or packing of those molecules, means that there is a remarkable parallelism in the behavior of all gases. For example, we described in Chapter 3 the dependence of the volume of a mole of gas on its temperature and pressure and claimed there that almost all gases behaved the same in this respect. Let us now return to the subject.

"THE SPRING OF THE AIR."   Although it is (and doubtless always has been) a common observation that gases are compressible and always fill completely any container, it was the seventeenth-century chemist Robert Boyle who first showed that the volume occupied by a given sample of a gas is (approximately) inversely proportional to the pressure under which it is held, as long as the temperature is kept constant. This relationship, which becomes exactly true only at very low pressures, may be expressed mathematically by the equation

$$PV = \text{constant (at constant temperature)} \quad (5\text{-}1)$$

A hard look at Equation (5-1) reveals that any change in the pressure or the volume of a gas requires a reciprocal change in the other. Thus, doubling the pressure can be achieved by halving the volume, a 10-percent increase in the volume will result in a 10-precent decrease in the pressure, and so forth. Boyle expressed this behavior by referring to air [whose mixture of nitrogen and oxygen obeys Equation (5-1) quite closely at pressures near one atmosphere and at ordinary temperatures] as like a spring.

TEMPERATURE: THE KELVIN SCALE.   We have said that a constant temperature is required for Equation (5-1) to be obeyed by a gas. The natural next question is, what if the temperature changes? What happens to the PV product then? Some data for two gases are shown in Figure 5-30. Note that, although the two gases behave quite differently, the PV product of H_2 appears to change more simply than that of CO_2, and that their PV products are converging to the same behavior (dashed line) at high temperature. In fact, at lower *pressures*, both sets of data merge with the dashed line. Taking the dashed line as an approximation or limiting rule at low-enough pressure and high-enough temperature,[6] we note that, plotted against centigrade temperature, the PV product describes a straight line with a positive intercept on the vertical axis. The equation describing such behavior for one mole of gas is

$$PV = Rt + b \quad (5\text{-}2)$$

t is temperature in °C.

where R and b are numerical constants—the slope and intercept,[7] respectively,

---

[6] What is meant by "low enough" and "high enough" clearly, from Figure 5-30, depends on which gas we are discussing.

[7] See Appendix II for a discussion of these terms if they are unfamiliar to you.

of the dashed line. Note that this relationship has the property observed by Boyle—the *PV* product is a constant at constant temperature.

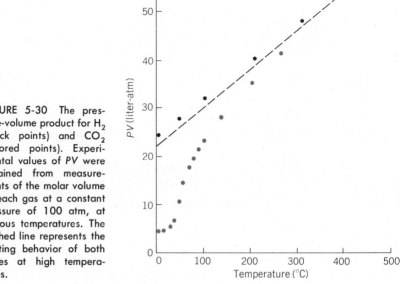

FIGURE 5-30 The pressure-volume product for H_2 (black points) and CO_2 (colored points). Experimental values of *PV* were obtained from measurements of the molar volume of each gas at a constant pressure of 100 atm, at various temperatures. The dashed line represents the limiting behavior of both gases at high temperatures.

The Kelvin scale is fundamental and essential. Why have any other?

Contemplation of this kind of behavior led physical chemists to the realization that Equation (5-2) could be made more simple, and thus more elegant and expressive, by redefining the temperature scale. If the dashed line in Figure 5-30 were extended to the left, it would reach a value of zero at some negative temperature on the centigrade scale [Figure 5-31(a)]. Such a zero might indicate the existence of a natural or "absolute" zero of temperature, as opposed to the arbitrarily chosen zero of the centigrade scale, the freezing point of water. Careful measurement of the *PV* product of gases at very low pressures produces a value of −273.2° for the centigrade temperature at which the *PV* product would reach zero, if it continued to follow the dashed line of Figure 5-31(a). Of course, any real gas condenses to a liquid long before the temperature decreases to −273°C. You can see this happening in Figure 5-30 to CO_2, which has relatively large intermolecular London forces; the simpler molecule H_2 does not form a liquid until much lower temperatures. And since no real gas follows the dashed line exactly, but only approaches it at high temperatures and low pressures, the behavior it describes is called "**ideal gas** behavior."

A simpler form of Equation (5-2) now results if we simply shift the origin of the temperature scale from the freezing point of water (0°C) to the zero point of the ideal gas *PV* product [Figure 5-31(b)]. The new scale of temperature so defined is called the Kelvin scale (or sometimes the absolute scale).

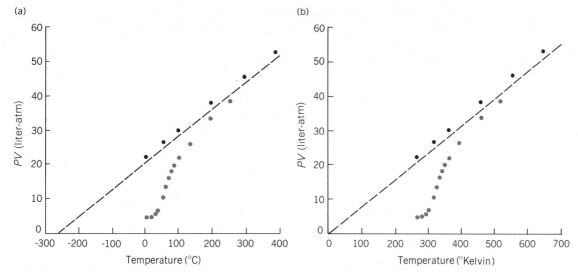

FIGURE 5-31 The PV product for H_2 and CO_2 plotted versus temperature. (a) The dashed line of Figure 5-30 has been extended to its intercept with the temperature axis at $-273°C$. (b) The same data as in (a), but the temperature scale has been shifted so that $0°$ coincides with the intercept of the dashed line on the temperature scale. This intercept defines the origin of the Kelvin scale of temperature.

The size of a degree remains unchanged from the centigrade scale, so that $0°C$ is $273.2°K$ (or more simply, 273.2 kelvins); $100°C$ is $373.2°K$ (or 373.2 kelvins), and so forth. With a vertical intercept of zero at $0°$, the ideal gas line is now described by the equation

$$PV = RT \qquad (5\text{-}3)$$

where capital T is reserved for temperatures on the Kelvin scale.

THE COMPLETE IDEAL GAS LAW. The PV products we have been discussing refer to a fixed quantity (mass) of gas—one mole, in Equations (5-2) and (5-3). Yet it is only common sense that 2 moles of a gas under given conditions of temperature and pressure occupy twice as much volume as one mole, and so forth. Rewriting Equation (5-3) in the form

$$V = \frac{RT}{P}$$

for one mole, we can take account of the proportionality between V and the number of moles of gas specifically:

$$V = n\frac{RT}{P}$$

or

$$PV = nRT \qquad (5\text{-}4)$$

for n moles of an ideal gas.

Remember that ideal gases are as rare as ideal people.

MIXTURES OF GASES: PARTIAL PRESSURES. It is remarkable but true that Equation (5-4) applies equally well to mixtures of gases as to chemically pure ones, as long as n represents the total number of moles of gases present.

Suppose, for simplicity, that we consider air to be a mixture of 79 mole percent N_2 and 21 mole percent O_2. Since at low enough-pressures and high-enough temperatures, air obeys the ideal gas law, we may write Equation (5-4) as

$$P_{air} = (n_{N_2} + n_{O_2})\frac{RT}{V} \quad (5\text{-}5)$$

But for a given sample of air (that in your lungs, for example), the values of T and V are the same for all of the gases in the mixture. Also, the constant R is the same for all gases. Thus, we may break the right-hand side of Equation (5-5) into two parts:

$$P_{air} = n_{N_2}\frac{RT}{V} + n_{O_2}\frac{RT}{V} \quad (5\text{-}6)$$

The two terms on the right-hand side of Equation (5-6) are just the ideal gas law expressions for the pressures that the two gases would exert if present by themselves in the same volume at the same temperature:

$$p_{N_2} = n_{N_2}\frac{RT}{V}$$

We can consider the individual gases independently.

and

$$p_{O_2} = n_{O_2}\frac{RT}{V}$$

These two pressures are referred to as the *partial pressures* of the gases N_2 and O_2 in the mixture we call air. The reason for this terminology is more apparent if we plug them back into Equation (5-6):

$$P_{air} = p_{N_2} + p_{O_2}$$

That is, the total pressure in a mixture of gases may be thought of as the sum of the partial pressures of the component gases, each partial pressure being the pressure that the gas would exert if it were alone in the same container.

We can see, qualitatively, the reason for this if we return to Figure 5-1(c) (page 131) and imagine that a certain fraction of the atoms shown there is a different chemical species from the rest. The total pressure exerted by the gas results from the billions upon billions of collisions of these atoms with the wall of the container during each second. The average force exerted by each kind of atom during a collision with the wall is the same (see Problem 5-4), so that the total pressure is just the sum of the pressures produced by each kind of atom. Removal of some of the atoms would decrease the pressure in proportion to the numbers of atoms removed, regardless of their chemical identity.

While we are on the subject of partial pressures, we shall develop one more relationship which will be useful in Chapter 7. Let us compare the equations for the total pressure p and any one of the partial pressures p_1 in a mixture of two (or, in principle, any number) of gases:

$$P = (n_1 + n_2)\frac{RT}{V}$$

and

$$p_1 = n_1 \frac{RT}{V}$$

Dividing the second of these equations by the first, we find

$$\frac{p_1}{P} = \frac{n_1 RT/V}{(n_1 + n_2)RT/V}$$

$$= \frac{n_1}{(n_1 + n_2)}$$

But the ratio $n_1/(n_1 + n_2)$ is just the *mole fraction* of component 1 in the gas mixture. Using the symbol X_1 for this mole fraction, we may then say

$$\frac{p_1}{P} = X_1$$

or

$$p_1 = X_1 P$$

And, of course, the analogous relationship applies for component 2 and indeed for all of the components in any mixture of ideally behaving gases.

5-12 Changes of Phase

Suppose you put a glass of water outdoors in the wintertime. After a cold night, you may find that the water has frozen; later, you may come back and find that the ice has melted; and much, much later, that the water has disappeared, escaped from the confines of the glass into the air. The familiar processes of freezing and evaporation both, in fact, can be thought of as the result of water molecules' escaping from the liquid phase. It is perfectly reasonable, therefore, to postulate that one property of any substance is its *escaping tendency*, that is, the tendency of molecules to leave that substance in a given form and to enter a different form. When the new form is the same chemical substance as the one escaped from, as in freezing, melting, evaporation, and condensation, the escaping process is called a change of **phase**; if the new form is a different chemical substance, a chemical reaction has occurred. We shall consider the relation of escaping tendency to chemical reactions in Chapters 7 and 9; for now, let us stick to the simpler changes of phase to see some of the factors that influence these "escapes."

Clearly, one such influence is temperature, as we saw with our glass of water. This makes sense, because temperature, as we recall, is a measure of the average molecular energy of vibration, rotation, and other motion within a phase. Escape from a phase may involve doing work against intermolecular forces, and the necessary energy is supplied by these molecular motions.

Another slightly less familiar factor that influences escaping tendency is pressure. A gas under high pressure certainly has a greater escaping tendency than one under lower pressure, thus the well-known behavior of inflated balloons when their nozzles are untied. An only slightly subtler phenomenon is the effect of pressure on changes of phase. Pressure cookers work because more thermal energy (a higher temperature) is required to boil water under higher pressure than at one atmosphere. Apparently, the escaping tendency of steam is increased more by high pressure than is that of water. Shortly, we shall see why.

Finally, escaping tendencies depend on composition. For example, water containing some dissolved salt or alcohol will not escape to the solid phase (freeze) at 0°C. Its escaping tendency is reduced by the presence of the solute, thus the success of "road salt" and antifreeze in keeping cars running in the winter.

Now suppose we examine a substance at a given temperature, pressure, and composition. We should find that its escaping tendency is the same as that of another sample of the same substance under the same conditions. For example, suppose we get ourselves a fresh glass of pure water and observe it at 25°C. If we close the top of the glass with an air-tight seal, pump out the air in the space above the water, and attach a manometer to measure pressure (Figure 5-32), we shall find that, as soon as we shut off our pump, the pressure in the "empty" space begins to rise. Evidently, some water molecules are escaping from the liquid, and water vapor is accumulating in this space. After a time, the pressure ceases rising and remains steady at 0.0312 atm. A second closed container, similarly treated, gives exactly the same result, even if the size and shape of the whole system is different from our first example. We would also get the same result with water from another source, so long as it is pure water at 25°C. *At 25°C and 0.0312 atm pressure, the escaping tendency of water vapor is exactly the same as that of liquid water at the same temperature and pressure.* There is no tendency for any more liquid water to disappear or to appear. The two phases have reached **equilibrium**.

What are the two phases in this experiment?

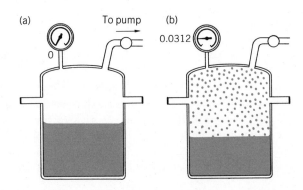

FIGURE 5-32 (a) With the space above the water evacuated, the manometer reads 0 atm. (b) When the pump is stopped and the system closed, the pressure gradually rises to a steady value of 0.0312 atm.

To say that two portions of matter have equal escaping tendencies says nothing about *how* such a state is achieved. Water molecules escape from the liquid phase when they acquire, through thermal collisions, enough energy to break the bonds holding them to the other molecules in the liquid; they escape from the gas phase when they collide with the liquid surface and come under the influence of those same forces again. There is no reason for believing that these processes stop when equilibrium is reached; thermal agitations continue within the liquid phase, and gas-phase molecules continue to rain onto the liquid surface. The logical conclusion is that evaporation and condensation continue in the state of equilibrium, but that these two opposing processes occur at equal rates, so that no net change in the system is observed with time. We may write a chemical equation for the process

$$H_2O(l) = H_2O(g) \qquad (5\text{-}7)$$

The equality sign in such an equation means not that the two sides of the equation are the same thing (which is not true), but that they have reached *equilibrium* with each other. Reaction (5-7) goes to the right and to the left at equal rates at equilibrium.

Let us perform another experiment and change the temperature of one of our liquid-gas equilibrium systems. If we raise the temperature, more of the molecules in the liquid phase will have enough energy to escape from the intermolecular forces in the liquid phase, so that the rate at which molecules leave the liquid will increase. The greater rate of evaporation will not be equaled by the rate of condensation until more molecules have accumulated in the gas phase. This increased population in the gas phase exerts a higher pressure, and the equilibrium vapor pressure is thus larger at higher temperatures. Lowering the temperature below 25°C will, of course, result in a lower equilibrium vapor pressure, by similar reasoning.

The experiments described above have been carried out many times, and the results are plotted in Figure 5-33(a). Notice that the equilibrium vapor pressure continues to increase up to 374.2°C, where it has reached 217.7 atm. Above this point (called the *critical point*), no phase boundary can be observed between the liquid gas states at any pressure. The thermal agitations above the *critical temperature* are so great that a separate liquid phase cannot be maintained; the liquid expands to fill all the space available to it; that is, it behaves just like a gas, and any distinction between the two is lost.

Solid substances also have an equilibrium vapor pressure; this is the phenomenon behind freeze-drying. The vapor pressures of ice and liquid water are shown together in Figure 5-33(b). Note that the curves cross at 0°C; at that point, both ice and liquid water are in equilibrium with the same pressure of water vapor, 0.006 atm. Thus, it would be possible to create the three-phase assembly shown in Figure 5-34. If both liquid water and ice have the same escaping tendency as water vapor under these conditions, they must have the same escaping tendency as one another, and thus must be in equilibrium with one another. *Two substances in equilibrium with a third substance are in equilibrium with one another;* in the language of mathematicians, equilibrium is transitive. If we remove the barrier in Figure 5-34, no change will occur

Opposing processes are balanced at equilibrium.

Fig. 5-33(b) is called a phase diagram.

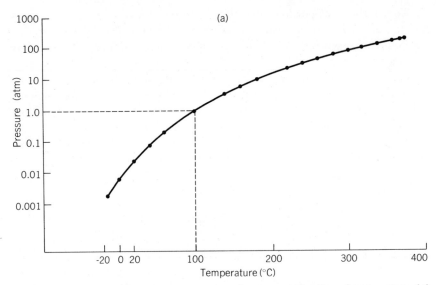

FIGURE 5-33 The equilibrium vapor pressure of water as a function of temperature. (a) The data over a very wide range of temperatures. In order to fit all points clearly on a single graph, the pressure has been plotted exponentially, that is, with evenly increasing powers of 10. (b) The vapor pressures of water and of ice plotted together in the region near 0°C. Since it is possible under some conditions to "supercool" water below its freezing point without causing the formation of ice, some data are plotted for liquid water below 0°C. Data for ice above 0°C are extrapolated from real data at and below 0°C. To show the region immediately around 0°C more clearly, it has been magnified in the large circle. Since only a small range of pressure is represented in this graph, these data are plotted linearly, rather than exponentially.

in the ice-water system (except that the ice will bob to the surface); equilibrium is a physical as well as a chemical state.

Before we go on to a final imaginary experiment, let us note a final feature of Figure 5-33(a): The equilibrium pressure reaches one atmosphere at 100°C, which means that it is possible for bubbles of vapor to form under the liquid surface without being collapsed by an external pressure of one atmosphere. This bubble formation we call boiling, and the "normal" boiling point (that is, the boiling point under one atmosphere pressure) is thus 100°C.

To examine the effect of composition on the vapor pressure, suppose we dissolve some salt in the liquid water. It is always true that the formation of a solution decreases the escaping tendency of the components of the solution. Thus, the escaping tendency of the water is decreased by the addition of salt (or any other solute, for that matter). In terms of the rate of evaporation from the liquid phase, we can understand this if we reflect that, of all particles in the liquid solution with the thermal energy required for a water molecule to leave the phase, only a fraction *is* water molecules, and the rest is Na^+ and Cl^- ions. Thus, the rate at which water molecules leave the liquid is decreased compared to the rate at which pure water at the same temperature evaporates. On the other hand, the Na^+ and Cl^- ions are held in the

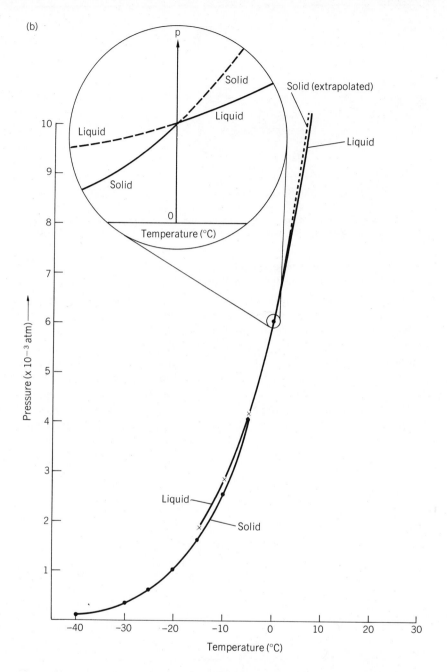

liquid phase by strong electrical forces. Thus, the salt has an extremely small tendency to evaporate, so the properties of the gas phase are unchanged by the addition of salt to the water. The rate of evaporation of water will thus be less than that of condensation for a time, until the number of molecules in the gas phase has decreased. Thus, when equilibrium is reestablished, the equilibrium vapor pressure will be lower.

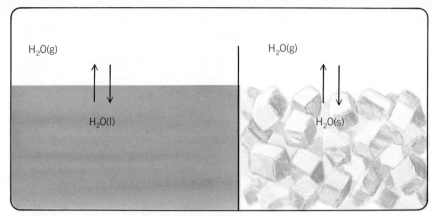

FIGURE 5-34 Three phases of water at equilibrium in a closed system. This very special situation is realizable only at a temperature of 0.01 °C and a water vapor pressure of 0.006 atm.

This phenomenon is observed at all temperatures, so that the vapor-pressure curve is displaced as shown in Figure 5-35. Two new phenomena now manifest themselves. First, the intersection of the liquid vapor-pressure curve with the line representing a pressure of one atmosphere must occur at a higher temperature; that is, the boiling point is elevated by the addition of a solute to the liquid. Second, when ice crystals form from such a solution, they are still nearly pure ice; there is very little possibility of inclusion of sodium and chloride ions in the hydrogen-bonded ice structure (Figure 5-28). The properties of the ice, including its vapor pressure, are then practically unaffected by the presence of salt in the water. Thus, the intersection of the ice and water vapor-pressure curves that marks the freezing point is shifted to a lower

This is a way of getting pure water from sea water.

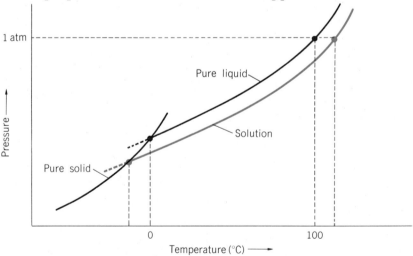

FIGURE 5-35 The effect of a solute on the vapor pressure of water. The equilibrium vapor pressure of pure liquid water and pure ice (solid black lines) and of water above an aqueous solution (solid colored line) are shown schematically. As before, the intersection of liquid water's vapor-pressure line with the line representing one atmosphere pressure is the *boiling point*, and its intersection with the vapor-pressure line for solid water is the *freezing point*. Note the effect of added solute on the vapor pressure, the boiling point, and the freezing point of the solution [compare Figure 5-31(b)].

temperature, which accounts for the use of salt in lowering the freezing point of water in ice-cream freezers and on highways.

HEAT OF FUSION AND VAPORIZATION. Reaching a certain temperature is not a sufficient condition for melting a solid or vaporizing a liquid. When a solid crystal collapses to a liquid, or the molecules of a liquid break contact and fly apart into a gas, work is done against the bonding forces that have held them together. The energy to do this work must come from outside the melting or evaporating substance, in the form of heat. It is this heat, withdrawn from the skin, that gives evaporating perspiration its cooling effect. Some **heats of fusion and vaporization** are listed in Table 5-3; note that their magnitudes reflect the forces holding the structure together.

To freeze is to fuse.

TABLE 5-3
Heats (Enthalpies*) of Fusion and Vaporization

	Substance	Heat of Fusion (kcal/mole)	Heat of Vaporization (kcal/mole)
Nonpolar molecular (or atomic) substances	He	0.005	0.02
	Ne	0.08	0.43
	Xe	0.55	3.02
	CH_4	0.23	2.21
	CCl_4	0.64	7.14
	SO_3	1.90	9.51
Hydrogen-bonding substances	H_2O	1.44	9.70
	NH_3	1.43	5.59
	CH_3OH	0.53	8.42
Ionic substances	NaCl	6.21	40.8
	MgO	18.51	—
	$MgCl_2$	10.3	32.6
Metals	Na	0.63	25.1
	Mg	1.74	41.2
	Fe	2.74	90
	Ni	4.27	90

* The enthalpy function is introduced in Chapter 9.

CHANGES OF CRYSTALLINE FORM. Many substances have more than one crystalline form, each stable within a particular range of pressures or temperatures. For example, the metals cobalt and nickel each adopt cubic and hexagonal close-packed structures; and ice displays at least 11 crystalline forms, labeled Ice I, Ice II, ..., Ice XI.

More compact, tightly bonded structures are generally found at low temperatures and high pressures. When rhombic sulfur, the form stable at low temperatures, converts to monoclinic sulfur, the high-temperature form, an input of 96 cal of energy per mole is required, reflecting a loosening of the crystalline structure.

LE CHATELIER'S PRINCIPLE. Ice (that is, ordinary Ice I) is less dense than water. When Ice I at $-1°C$ is put under sufficient pressure, it melts, with a decrease in volume. (That is, pressure depresses the freezing point of water.) Most solids are *more* dense than their liquids, and the liquid can be made to *freeze* above the normal melting point by the application of sufficient pressure. As we have seen, the boiling point of water is elevated by increasing the pressure on it (as in a pressure cooker). We might summarize all these observations by saying that phase equilibria shift in such a way that the substance occupies the smallest possible volume at high pressures. Putting it another way, when we squeeze on substances, they tend to take up a minimum of room.

A related observation is that, if we place a mixture of ice and water (say, a tall, cool drink) in a warm environment, the phase equilibrium shifts toward the liquid phase; the ice melts. If we put it in a cold environment, the water freezes. Both of these processes have taken place in such a way as to accommodate the stress placed on the equilibrium system; in a warm environment, the solid melts, *absorbing* the heat of fusion; in a cold one (in which heat would tend to flow out of the mixture of phases), it freezes, *releasing* this same heat of fusion. Of course, the analogous observations can be made about a liquid-gas equilibrium system.

See Ch. 11 for more on feedback.

The accommodation of phase equilibria to stresses placed on them in the form of altered pressure or temperature is an example of *Le Chatelier's principle*, which in essence states that systems in equilibrium will shift in the direction that they are pushed. Le Chatelier's principle has very broad applications in the chemistry of equilibrium systems, as we shall see in Chapter 7. It is, in fact, a manifestation of *negative feedback*, which is a feature of all stable systems, not just chemical ones. Place a glass of ice water in an oven, and it absorbs heat by melting some of the ice; place it in a freezer, and it releases heat by freezing some of the water. Place it under pressure; very well, it takes up less room. You should give a few minutes to considering what would happen if the opposite responses occurred (positive feedback) in the case of phase equilibria under stress or, for that matter, in any similar situation (suppose, for example, that the prices of commodities increased as the demand for them decreased).

5-13 Summary

We began and ended this chapter with some answers to a critical question in our study of matter: Why is some matter gaseous, some liquid, and other solid? We noted that all matter would be a gas except for the existence of attractive intermolecular forces. The nature of these forces—directional or nondirectional, strong or weak—determines many properties of a particular substance. We compared and contrasted and began to explain the properties of ionic, covalent, molecular, and metallic substances.

But not all of the story is told in terms of forces. Many of the properties of matter derive from the internal ordering of constituent particles, and it turns out that three-dimensional space permits only a few different orderly arrangements of anything. We are interested in atoms and molecules, but we found it convenient to investigate the geometry of the filling of space with spheres which we can draw and which we can pick up and manipulate. A sampling of structures of common crystals was presented to give a glimpse of the immense richness of detail about structures that is known. At the opposite extreme of no molecular order, we found some useful regularities in the behavior of gases.

Finally, we asked some questions about freezing and melting, boiling and condensing. The answers this time were less in terms of individual molecular behavior, and more in terms of the collective behavior of tangible portions of matter. We shall encounter more of this latter approach in Chapter 9.

GLOSSARY

boiling point: the temperature at which liquid and gaseous forms of a substance are in equilibrium, at a specified pressure, usually one atmosphere.

covalent substance: a substance each of whose atoms is bonded to two or more neighbors, so that networks of covalent bonding extend through the entire crystal.

crystal: a solid whose internal arrangement of atoms or molecules shows a regular repetition in any direction through a large distance on the atomic scale.

equilibrium: a state in which there is no tendency or possibility for any spontaneous change to occur.

heat of fusion (vaporization): the energy input required to melt (vaporize) a substance.

ideal gas: a (hypothetical) gas whose pressure, volume, and temperature obey exactly the equation $PV = nRT$, where P is pressure; V is volume; T is temperature; R is a constant; and n the number of moles of a gas present. All real gases approach ideality at low pressure and high temperature.

ionic substance: a substance whose constituent particles are ions (electrically charged molecules or atoms).

melting point: the temperature at which solid and liquid forms of a substance are in equilibrium, at a specified pressure, usually one atmosphere.

molecular orbital: an orbital whose domain includes, rather than just nuclei, all or a significant portion of an entire molecule. These orbitals are discussed fully in Chapter 12.

molecular substance: a substance whose constituent particles are molecules.

phase: a mechanically separate, homogeneous part of a heterogeneous system.

unit cell: the smallest portion of a crystal which, when repeated in specified directions, reproduces the structure of the entire crystal.

van der Waals forces: weak forces between atoms and molecules arising from permanent or induced electrical polarity.

PROBLEMS

5-1 Tile floors are often composed of square tiles (tiles with C_4 symmetry), laid so as to fill two-dimensional space (that is, the whole floor). Tiles with other shapes can also be laid without leaving gaps. Sketch tile floors laid with tiles having C_2, C_3, C_4, C_5, C_6, and C_7 symmetry elements. (If you succeed with all six cases, you will have succeeded where all previous attempts have failed. Which tile symmetry(ies) cannot be used for covering the whole floor?)

5-2 Ordinarily, masons build walls with blocks or brick that are either cubes or rectangular solids. Sketch some other shapes that could be used for constructing a wall several bricks thick. These bricks must fit together to fill all space, leaving no voids. Will a tetrahedron do? A trigonal bipyramid? An octahedron?

5-3 Portions of 10 different brickwork patterns are displayed in Figure 5-36. For each pattern, sketch a unit cell which could serve to generate the entire two-dimensional repeating structure of a very large brick wall.

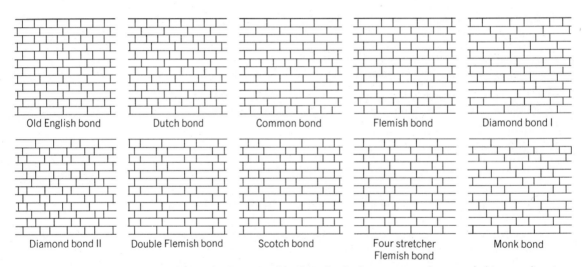

FIGURE 5-36 Sections of brick walls. Each pattern can be extended in two directions as far as a mason can lay bricks. The resulting wall is a structure with translational symmetry.

5-4 If, as is asserted on page 156, each molecule in any ideal gas mixture contributes equally to the pressure, and if the pressure results from collisions of molecules with the container wall, why do heavy molecules not cause a higher pressure than light ones? (*Hint:* what must be true of the average speed of heavy molecules compared to light ones? Does this seem "reasonable"?)

5-5 From the dashed line in Figure 5-30, calculate approximate values for R and b in Equation (5-2).

5-6 Using the value of R obtained in Problem 5-5, calculate the volume of one mole of water vapor in equilibrium with the liquid at 25°C. (See Figure 5-32).

5-7 a. By what factor does H_2O expand when it evaporates at 25°C? (See Problem 5-6; you will also need to figure out the volume of one mole of water, which you are equipped to do without hints.)
b. Assuming that water molecules are in contact in the liquid, about how far apart are they, on the average, in the vapor under the conditions of Problem 5-6?

5-8 On a day when the atmospheric pressure is 0.96 atm, what are the partial pressures of O_2 and N_2, assuming mole percentages of 21 and 79 percent respectively?

5-9 A steel tank containing 5 moles of helium is connected by accident to a supply of neon gas, and an additional 5 moles of neon flow into the tank, without allowing any helium to escape. What happens to

a. the total pressure in the steel tank?
b. the mole fraction of helium?
c. the partial pressure of helium?

5-10 About the largest concentration of the atmospheric pollutant SO_2 most people can tolerate is 5 ppm; that is, 5×10^{-6} g SO_2 per gram of air.

a. Assuming that the average molecular weight of molecules in the air is 29, how much volume does a gram of air occupy?
b. To what mole fraction of SO_2 in the air does 5 ppm correspond?
c. Assuming an active lung capacity of 0.2 liter, how many moles of SO_2 would you inhale during each breath during a 5 ppm SO_2 smog alert? ($T = 300°K$, $P = 1$ atm.)
d. At 6.02×10^{23} molecules per mole, how many molecules of SO_2 would you inhale?

5-11 The next time you are driving during a blizzard, compare and contrast the bonding, hardness, and melting points of

a. steel
b. sand
c. salt
d. H_2O.

Which of the forms of bonding used in these substances is used predominantly in your body? (If you do not know, see Chapter 17, and drive carefully!)

SUGGESTIONS FOR FURTHER READING

"The mysteries of matter have stimulated the great intellectual exploration of our time. There are two reasons why we should share in its excitements. One is for the sheer fun, the esthetic pleasure, call it what you like, of reaching deeper into the unknown. The other is for the understanding to be gained as a result," says C. P. Snow, in the introduction to: Lapp, R. E., *Matter*, New York: Time-Life Books (1969), a magnificently illustrated introduction to the properties of tangible matter. There are also many beautiful photographs in: Nakaya, U., *Snow Crystals: Natural and*

Artificial, Cambridge, Mass.: Harvard University Press (1954), a book which explores one of the most common and yet most fascinating crystalline materials.

Symmetry as a universal theme is developed by: WEYL, H., *Symmetry,* Princeton: Princeton University Press (1952), starting with art, architecture, plants, and animals, and concluding with crystals and mathematics.

Two introductions to the atomic structure of metals are: MARTIN, J. W., *Elementary Science of Metals,* London: Wykeham (1969); and HUME-ROTHERY, W., *Electrons, Atoms, Metals and Alloys,* 3rd ed., New York: Dover (1963).

Properties and structures of crystals are discussed in two very readable books: BUNN, C., *Crystals: Their Role in Nature and in Science,* New York: Academic (1964); and BENNETT, A., HAMILTON, D., MARADUDIN, A., MILLER, R., and MURPHY, J., *Crystals: Perfect and Imperfect,* New York: Walker and Company (1965). Many structures are drawn, described, and discussed in KREBS, H., *Fundamentals of Inorganic Crystal Chemistry,* London: McGraw-Hill (1968).

6
Periodicity and Chemical Properties

6-1 Abstract

Even with the generalizations we have used so far, the task of understanding the properties of matter would be impossible even for professional chemists if it were not for periodicity and the Periodic Table of the elements. In Chapter 2, we saw that the fundamental basis of the Periodic Table lies in some simply stated quantum rules governing the configuration of the highest-energy electrons of atoms. We also found that the overall pattern of the table—with electron-releasing metals on the left and electron-attracting nonmetals on the right—is a direct consequence of the successive filling of energy levels in a shell structure of the electrons in atoms. Now that we have seen some ways in which atoms combine to form molecules (Chapter 3), and these in turn form liquids and solids (Chapter 5), we are ready to explore some finer details of chemical periodicity that account for the individualities of the hundred or so elements and the millions of compounds that constitute our material world. We shall examine, in turn, periodicity of ionization potential, electron affinity, atomic sizes, polarizability, electronegativity, and chemical reactivity.

6-2 Periodicity of the Ionization Potential

We have seen that the process of forming a chemical bond, whether covalent or ionic, involves rearranging electronic structures of the atoms that form the bond. All isolated atoms are arrangements of positive and negative charge, so we need to know how easily these arrangements can be disrupted in the process of forming a bond. One measurable property that gives us a handle on this atomic information is the **ionization potential**. The ionization potential expresses the electrical force[1] (in volts) required to remove one electron from an atom in the process

$$X \longrightarrow X^+ + e^- \qquad (6\text{-}1)$$

where X is an atom of any element, and e^- is an electron.

Figure 6-1 is a graph of ionization potentials of the elements in order of their atomic number. Periodicity is clearly evident. We shall find that even the minor wiggles in this graph faithfully reflect the filling of the electronic energy levels according to the quantum rules. Most striking, however, is the overall pattern represented in Figure 6-2. Ionization potential is so intimately connected with chemical bonding (and therefore with chemical stability and chemical reactivity) that this periodicity of ionization potential is manifested in periodicity of many observable chemical properties.

---

[1] More precisely, the ionization potential is the electrical voltage that is just able to pull an electron off of the atom X. The *energy* expended (the work done) when the electron is actually removed is the ionization *energy*. This ionization energy is usually expressed in *electron volts*, and in these units the ionization potential (volts) and the ionization energy (electron volts) have values that are numerically equal. For a discussion of potential, work, force, and energy, see Appendix III.

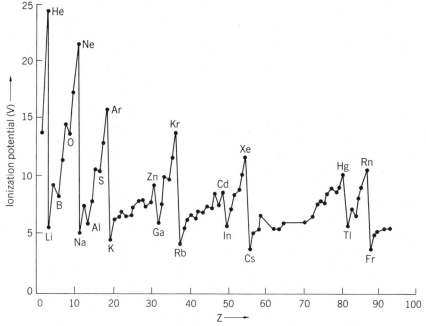

FIGURE 6-1 Ionization potential of the chemical elements, plotted versus atomic number Z. The observed periodicity is a consequence of the orderly filling of the electronic energy levels according to the rules that come from quantum theory. The values plotted here are taken from Figure 6-2.

As explained in Chapter 2, progression across each period of the Periodic Table involves the filling of orbitals, with one electron added for each unit increase in atomic number. We shall pay particular attention to the filling of the s and p orbitals in the highest-energy outer **valence shell** of the atom. *The occupancy of the valence-shell orbitals determines the chemical nature of an atom.* Metals have nearly empty s and p orbitals in their valence shell; nonmetals have these orbitals nearly filled.

Elements with nearly empty valence shells have, of course, smaller nuclear charges compared to the elements farther to the right in the same period. The ionization potential depends primarily on the relationship between the nuclear charge and the principal quantum number of the outermost electron in the atom. For example, lithium and beryllium have their outermost electrons in the second principal quantum level, have nuclear charges of 3 and 4, respectively, and are metals with low ionization potentials. In the same period, oxygen and fluorine have their outermost electrons in the same second level, but have nuclear charges of 8 and 9, and hold those electrons so tightly that they are nonmetals with high ionization potentials. Similar comparisons of general chemical behavior hold true throughout the Periodic Table, even on a more local basis. Thus, in the fifth period, indium and tin, with outermost electrons in the $4p$ level, are metals, but iodine, with a nuclear charge three units higher than tin's, is a nonmetal.

"Valence" is a word often used loosely. See the glossary at the end of the chapter.

1																	2
H 13.60																	He 24.58
3 Li 5.39	4 Be 9.32											5 B 8.30	6 C 11.26	7 N 14.54	8 O 13.61	9 F 17.42	10 Ne 21.56
11 Na 5.14	12 Mg 7.64											13 Al 5.98	14 Si 8.15	15 P 10.55	16 S 10.36	17 Cl 13.01	18 Ar 15.76
19 K 4.34	20 Ca 6.11	21 Sc 6.54	22 Ti 6.82	23 V 6.74	24 Cr 6.76	25 Mn 7.43	26 Fe 7.90	27 Co 7.86	28 Ni 7.63	29 Cu 7.72	30 Zn 9.39	31 Ga 6.00	32 Ge 7.88	33 As 9.81	34 Se 9.75	35 Br 11.84	36 Kr 14.00
37 Rb 4.18	38 Sr 5.69	39 Y 6.38	40 Zr 6.84	41 Nb 6.88	42 Mo 7.10	43 Tc 7.28	44 Ru 7.36	45 Rh 7.46	46 Pd 8.33	47 Ag 7.57	48 Cd 8.99	49 In 5.79	50 Sn 7.34	51 Sb 8.64	52 Te 9.01	53 I 10.45	54 Xe 12.13
55 Cs 3.89	56 Ba 5.21	57 La* 5.61	72 Hf 5.5	73 Ta 7.88	74 W 7.98	75 Re 7.87	76 Os 8.7	77 Ir 9	78 Pt 9.0	79 Au 9.22	80 Hg 10.43	81 Tl 6.11	82 Pb 7.42	83 Bi 7.29	84 Po 8.43	85 At 9.5	86 Rn 10.75
87 Fr 3.83	88 Ra 5.28	89 Ac† 6.9															

*Rare earth metals

58 Ce 6.5	59 Pr 5.7	60 Nd 5.7	61 Pm	62 Sm 5.64	63 Eu 5.67	64 Gd 6.16	65 Tb 6.7	66 Dy 6.8	67 Ho	68 Er 6.08	69 Tm 5.81	70 Yb 6.22	71 Lu 6.15

†Uranium metals

90 Th 6.95	91 Pa	92 U 6.08	93 Np	94 Pu 5.8	95 Am 6.0	96 Cm	97 Bk	98 Cf	99 E	100 Fm	101 Mv	102 No	103 Lw

FIGURE 6-2 Correlation of the ionization potential of an element with its position in the Periodic Table. Each square contains an element's symbol, its atomic number, and its ionization potential (in volts) if known. [*Source of data:* M. C. Day and J. Selbin, *Theoretical Inorganic Chemistry*, 2nd ed., New York: Reinhold, pp. 122–123 (1969). By permission.]

Looking again at Figure 6-2, we notice that our arguments do not say all that needs to be said. Thus, in Group V, nitrogen and phosphorus are nonmetals, but bismuth is a metal (arsenic and antimony are intermediate in character, showing increasing metallic character with increasing atomic number). All of the elements in Group V have analogous electronic structures, with two s electrons and three p electrons in the valence shell, yet the ionization potentials in this group show a steady decrease from 14.54 V for nitrogen to 7.29 V for bismuth. In concentrating on the outermost electrons of atoms, where the chemical actions of ionization and bond formation take place, we ignored the screening effect of the inner shells of electrons (Chapter 2). Abetting this screening effect is the fact that the outermost electrons are farther from the nucleus, and so are subject to smaller nuclear attraction, just because electrical forces decrease with the square of the increasing distance (see Appendix III). We can compare the effect of screening and the effect of distance from the nucleus very easily in the relatively simple case of lithium. A neutral atom of lithium (atomic number 3) has two electrons in the 1s subshell and one in the 2s; in our shorthand (Section 2-10), the configuration is $1s^2 2s$. When lithium is ionized, either directly or by reaction with an electron-attracting element like fluorine (Section 3-3), it is the 2s electron that is removed. From Figure 6-2 we find that removal of this first electron requires an energy of 5.39 electron volts. We may compare this number with two related numbers: (1) the energy that would be required to remove an electron from the 2s subshell of lithium *if* there were no other electrons in the atom, and (2) the energy that would be required to remove an electron from the 1s subshell *if* there were no other electrons in the atom. These two different quantities are indicated schematically in Figure 6-3. The ionization potential of a lithium atom containing only a 2s electron turns out to be 30.6 V; the ionization potential of a lithium atom with just a 1s electron is 122 V.

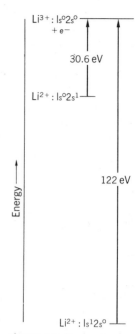

FIGURE 6-3 Energy changes for removing a single electron from a Li^{2+} ion. This ion has only one electron, and the result of the removal of this electron is a naked nucleus, a Li^{3+} ion. The energy required for removal of the electron depends on the original energy state of the Li^{2+} ion.

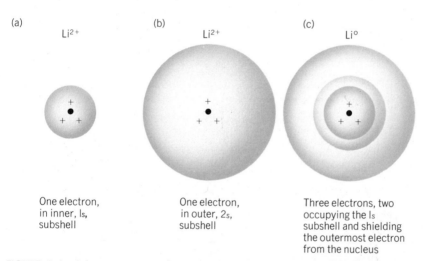

FIGURE 6-4 Schematic representation of (a) a Li^{2+} ion with its single electron closely held in the 1s orbital. (b) a Li^{2+} ion in a higher energy state, with the single electron farther from the nucleus in a 2s orbital; this electron, being less tightly held in the positive field of the nucleus, can be removed more easily. (c) a Li atom, showing the two 1s electrons shielding the 2s electron from the positive charge of the nucleus.

If these data are confusing, the interpretation is indicated in Figure 6-4. If an electron were alone in a 1s subshell around a lithium nucleus with its charge of +3, 122 eV of energy would be required to dislodge it. If it were alone in a 2s subshell, the energy required would be only 30.6 eV. The difference between these two quantities represents the factor of distance from the nucleus, amounting to a fourfold decrease in stability on going from the 1s to the 2s subshell. Still, an atom with ionization energy of 30.6 eV is more stable than a helium atom (ionization energy 24.58 eV). The rest of the difference, between the calculated properties of the hypothetical $Li^{2+}(2s)$ and the observed properties of Li(1s²2s) with its experimentally measured ionization energy of 5.39 eV, is clearly attributable to the screening effect of the pair of electrons in the 1s subshell on the single electron in the 2s subshell. This difference in ionization potential between 30.6 eV and 5.39 eV makes the difference between a hypothetical electron-attracting ion and the actual easily ionized metal.

We have identified two major factors that influence the value of the ionization energy: the number of positive charges on the nucleus and the degree of screening by the electrons close to the nucleus. Increasing nuclear charge makes an electron more difficult to remove from atoms as one considers elements farther to the right in any period. Screening makes removal of an electron easier as one considers the progression of elements down most[2] groups of the Periodic Table, moving down toward elements whose outer electrons are in high principal quantum levels and are screened from the nucleus by increasingly thick layers of electrons.

"Screening" is shielding or insulating.

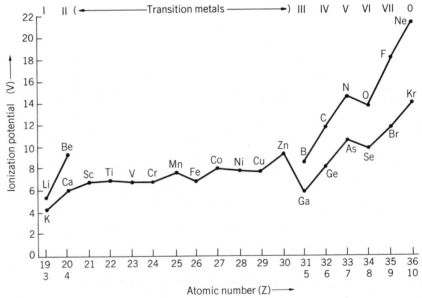

FIGURE 6-5   Ionization potential plotted versus atomic number Z for two periods of elements. This is an expanded portion of Figure 6-1.

On closer examination, you will note some still finer details in Figure 6-1. The portions from Li to Ne (Z from 3 to 10) and from K to Kr (Z from 19 to 36) will repay close scrutiny, and they are shown expanded in Figure 6-5. In the first series, note that B has a slightly lower ionization energy than Be,

---

[2] The exceptions to this rule, which occur in the center of the Periodic Table (and notably in the series Cu–Ag–Au, Ni–Pd–Pt, and Zn–Cd–Hg), are due to the lanthanide contraction, a phenomenon that involves filling of the f orbitals. It is discussed later in this chapter.

despite having a higher nuclear charge attracting the electrons in the same ($n = 2$) quantum shell. In a boron atom, the electron removed on ionization is a $2p$ electron, while in beryllium it is a $2s$. Because of the lobed shape of p orbitals (see Figure 2-6), electrons in them spend less time near the nucleus than do those in s orbitals. The $2p$ electron is rather effectively screened by the $2s$ electrons in the same shell, and thus the $2p$ subshell in boron is higher in energy than the $2s$ in beryllium despite the larger nuclear charge of boron.

A slight hitch in the smoothly rising trend after boron comes between N and O, the point where double occupancy of orbitals in the $2p$ shell first occurs. That is, a nitrogen atom in its most stable state has one electron in each of the three $2p$ orbitals; its electronic configuration is, in the language of Chapter 2, $1s^2 2s^2 2p^3$. When a fourth $2p$ electron is added in oxygen, it must share an orbital with another electron. The repulsion between these two paired electrons makes either of them unexpectedly easy to remove; that is, the ionization energy decreases. The sharp-eyed reader should be able to spot this effect between the third and fourth members of a "p block" of elements each time it occurs in Figure 6-1.

It is this that leads to strong bonding in N_2, and makes it so hard to "fix" for our nutrition (see Ch. 1).

Turning to the second series in Figure 6-5, from K to Kr, we find illustrated the effect of filling the d subshell between Sc ($Z = 21$) and Zn ($Z = 30$). We find a peculiar situation in the d block of elements, the transition metals. Proceeding toward the right in a new period, each element has one more d electron than the one before; between Sc and Zn, it is the $3d$ subshell that is being filled. But the outermost electron in each element is an s electron; in our series, it is a $4s$ electron. Thus, ionization potentials in this group of elements measure the energy required to remove $4s$ electrons from atoms with increasing nuclear charge and an increasingly dense screen of $3d$ electrons inside the $4s$ subshell. One might expect almost no change in ionization potential across such a period, since the nuclear charge and the number of screening electrons both are increasing in step. However, the $4s$ electrons do spend some fraction of their time very near the nucleus, as do all s electrons, and this so-called *penetration* of the $4s$ orbitals makes the effect of the nuclear charge predominate. The ionization energy slowly increases as the nuclear charge increases.

All these distinctions and exceptions make chemical individuals out of the elements.

Finally, note that the d block is completed at zinc ($Z = 30$). The next element (Ga) shows the decrease in ionization energy we expect when a p electron, rather than an s electron, is being removed. In an atom of gallium, it is a $4p$ electron that has the highest energy and that is removed most easily.

Ionization potentials of the Group 0 elements Ne and Kr are in no way out of line with the others, though they are, of course, the highest in their respective periods. At one time, it was thought that a filled quantum level conferred "special" stability on an atom, leading to the observed reluctance of the Group 0 elements to form compounds, and to the stability of ions like K^+, F^-, and O^{2-} with "inert-gas" structures. But we can see from Figure 6-5 that the ionization energies of Ne and Kr are just what one would predict if asked to extrapolate from the preceding elements. Group 0 elements, in fact, are not known to form negative ions, and from the time of their discovery (Sir William Ramsey and his collaborators found helium, neon, argon, krypton, and xenon between 1894 and 1898) until the 1960s, it was thought that these elements could not react to form chemical compounds. Then in 1962 the discovery of the fluorides and oxides of xenon laid to rest the myth of immutable aloofness of this group, and set chemists looking for alternative names to the no-longer-appropriate "inert gases."

6-3 Electron Affinity

For elements far to the right in any period, the ionization potentials are high, because the nuclear charge is at a maximum for that period. One might surmise that even a neutral atom of such an element would tend to attract electrons, so long as there were room in the valence shell to accomodate them. This is, in fact, the case, and these elements are well known for their tendency to form *negative* ions in the process

$$X(g) + e^- \rightarrow X^-(g) \tag{6-2}$$

Electron affinity is the energy *released* when an electron is added to an atom. Electron affinity is rather more difficult to measure experimentally than the ionization energy (often it is easier to arrange to knock something apart than to put two things together), and so experimental values of electron affinity are relatively scarce. Some values are listed in Table 6-1. Note that, by rather confusing chemical convention, a positive value of the electron affinity signifies the *release* of energy, while a positive value of the ionization energy represents the *input* of energy required to form a positive ion by removal of an electron.

Another convention in need of reform

TABLE 6-1
Some Electron Affinities of Atoms

H	0.747	eV [°]
F	3.45	eV [†]
Cl	3.61	eV [†]
Br	3.36	eV [†]
I	3.06	eV [†]
O	1.47	eV [§]
N	~ -0.1	eV [¶]
S	2.07	eV [‖]
Li	0.54	eV [°]
Na	0.74	eV [°]
Be	~ -0.6	eV [°]
Mg	~ -0.3	eV [°]

[°] H. A. Skinner and H. O. Pritchard, *Transactions of the Faraday Society*, vol. 49, p. 1254 (1953).
[†] R. S. Berry and C. W. Riemann, *Journal of Chemical Physics*, vol. 38, p. 1540 (1963).
[§] L. M. Branscomb, *Nature*, vol. 182, p. 248 (1958).
[¶] A. P. Ginsberg and J. M. Miller, *Journal of Inorganic and Nuclear Chemistry*, vol. 7, p. 354 (1958).
[‖] L. M. Branscomb and S. J. Smith, *Journal of Chemical Physics*, vol. 25, p. 598 (1956).

6-4 Atomic Sizes

When we come to look at the chemical behavior of atoms, we often find that a property as simple as the mere size of an atom can have a significant effect on its reactivity. As we have seen in Chapters 4 and 5, the size of an atom has important consequences for the geometry of molecules containing it and for the way it packs in a crystal. (Readers of J. D. Watson's *The Double Helix* will recall the crucial importance of atomic sizes and geometry in the structure of the genetic molecule DNA; these factors appear to be equally important in all aspects of the chemistry of life.)

Just as there is no outer boundary to the earth's atmosphere, there is no outer boundary to the electron cloud surrounding a nucleus. The atmosphere becomes less and less dense as we move away from the earth toward outer space, but we never quite find a perfect vacuum. If we consider the atmosphere to be part of the planet earth, then there is no clear and unambiguous size of the earth. Similarly, the electron density around an atom decreases steadily, becoming vanishingly small as the distance from the nucleus becomes very large, but the electron density never becomes exactly zero. How then can we speak of the size of an atom?

There is no outer surface on an atom.

The answer is that it has no truly definite size in the way a baseball has a size.[3] Nevertheless, atoms do adopt regular spacings in both crystals and molecules, and, as we shall now see, we can divide the distance between the nuclei of neighboring atoms, on the assumption that the atoms are "touching." By this technique, we can arrive at a self-consistent set of radii that are at least valid for comparing one atom with another.

The radii of chemically bound atoms can be thought of in two ways, giving rise to two major kinds of atomic radii, called **covalent radii** and **ionic radii**.

COVALENT RADII. By a number of experimental techniques, it is possible to measure the distance between the nuclei of bonded atoms in a molecule (Chapter 4). The problem of portioning out this distance between the bonded atoms to arrive at an *effective radius* for each is solved by measuring the distance between atoms of the same kind bonded to one another. We shall illustrate this procedure by examining some compounds of bromine and chlorine.

The bond length—the internuclear distance—in the interhalogen compound BrCl is 2.14 Å.[4] Though we may suspect that Br atoms are larger than Cl atoms because the principal quantum level of the outer electrons is the fourth for Br and the third for Cl, there is no way of dividing the bonding electron cloud into "bromine territory" and "chlorine territory," even if such a division were meaningful. But we have some additional experimental information. The internuclear distance in Cl_2 is 1.99 Å, and in Br_2 is 2.28 Å. We

How shall we draw the boundary line?

---

[3] Baseballs and electrons are fundamentally wave phenomena (see Chapter 2), and both a baseball and an atom lack definite size. But because a baseball's wavelength is so short compared to its apparent size, the fuzziness of its outline is not apparent to us.

[4] Recall that the symbol Å signifies 10^{-8} cm, an angstrom unit.

may conclude that a chlorine atom should have a radius of roughly 1.00 Å (that is, $\frac{1}{2}$ of 1.99 Å), and that a bromine atom should have a radius of about 1.14 Å ($\frac{1}{2}$ of 2.28 Å). See Figure 6-6. A confirmation of these guesses is provided by the fact that the sum of these two atomic radii (1.00 + 1.14) is equal to the bond length in BrCl. In other cases, the consistency between independent estimates of covalent radii is less startling, but at least we have the basis for developing a complete series of numbers.

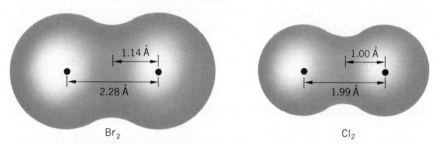

FIGURE 6-6 Illustrations of the calculation of covalent atomic radii. The experimental quantity is the internuclear distance. A reasonable choice for an atomic radius is seen to be half of the internuclear distance.

Here is another example to show how the scheme develops. The bond length in the molecule ICl is 3.32 Å. Combining this information with the covalent radius of 1.00 Å for Cl, we arrive at a guess of 1.32 Å for the covalent radius of I. This guess, in turn, agrees well with half the bond length of I_2 (2.67/2 = 1.33).

When there is *multiple* bonding between atoms, the extra negative charge in the bonding region pulls the nuclei closer together than would be expected from a single-bond covalent radius. For example, the carbon-carbon internuclear distance in ethane (H_3C-CH_3) is 1.54 Å, giving carbon a covalent radius of 0.77 Å, based on that datum alone. But the carbon-carbon internuclear distance in ethene ($H_2C=CH_2$) is only 1.34 Å, and that in ethyne ($HC\equiv CH$) only 1.20 Å. The double-bonded and triple-bonded covalent radii of carbon are smaller in proportion: 0.67 and 0.60 Å, respectively. If the bonded atoms differ greatly in their electronegativities (see Section 6-6), the bonds are *polar* (recall Section 3-5); the result is that the opposite charges at the two ends of the molecule pull the bonded atoms a little closer together, and the apparent radii shrink somewhat. So we conclude that, although a set of "sizes" based on covalent bonding may be deduced, these should be regarded as somewhat conditional numbers depending on the kind of covalent bond (single or multiple, nonpolar or polar) being considered.

Note:
$C-C$ 1.54 Å
$C=C$ 1.34 Å
$C\equiv C$ 1.20 Å

IONIC RADII. X-ray diffraction measurements (Appendix III) on crystals of ionic substances also can give internuclear distances. Since, as we have seen, it is a reasonable assumption that the ions in such crystals are in close contact for efficient ionic bonding, the distances between immediately neighboring ions may again be considered to be the sum of the radii—this time the *ionic* radii—of those ions. The problem of dividing the distance between two unlike, "touching" ions is solved by the happy circumstance that a few ionic substances appear to crystallize in such a way that the larger anions are in contact with one another, in spite of what we might consider the energetic disadvantage of such an arrangement (Chapter 5). Knowing the type of unit cell these substances adopt, it is a simple matter to deduce the effective radius of the larger ion (Figure 6-7), and to combine this number with internuclear distances in other ionic compounds to arrive at a complete set of ionic radii.

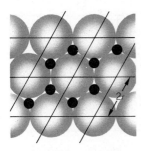

FIGURE 6-7 Calculation of the ionic radius from the spacings in a crystal in which large ions crystallize into virtual contact with one another (r is the radius of the larger ion).

Like the covalent radii, these ionic radii are conditional. For example, they depend on the geometry of counter-ions around a given ion,[5] so for greater accuracy one can list separately the octahedral ionic radii, the tetrahedral ionic radii, and so forth. For metal atoms that form more than one ion, the higher-charged ions are always smaller than those of lower charge. This variation of size with charge is seen among the transition metals where, for example, Fe^{2+} has a radius of 0.76 Å, and Fe^{3+} a radius of 0.64 Å. The reason for this change in radius is not difficult to see—all ions of a given metal possess the same nuclear charge, but different numbers of electrons. Thus, the higher the positive charge, the smaller the number of electrons, and the more tightly these are pulled in.

We must be careful to say what we mean.

PERIODICITY OF ATOMIC SIZES.   Now that we have seen some of the things we can mean (and cannot mean!) by the "size" of an atom, we can turn to the Periodic Table and see how the electronic structure of an atom influences its size. Periodic trends in both covalent and ionic radii are illustrated in Figure 6-8.

In the first place, there is a definite increase in size as we proceed down most groups.[6] This trend is not difficult to understand, since each new period contains atoms whose outer electrons are in successively larger shells. Because the nuclear charge increases rapidly going down a group, the increase in size is not as rapid as would be observed for electrons in successively higher quantum shells around a single given nucleus. The increase in size that we observe is clearly the result of the opposing trends of increasing nuclear charge tending to pull electrons in toward the center, and increasing numbers of screening electrons tending to oppose the nuclear charge.

---

[5] That is, on what kind of lattice hole (Chapter 5) the ion occupies.

[6] The exceptions are the same as those noted before with respect to electropositive character in descending a group, and also are due to lanthanide contraction, taken up later in this section.

(a)

1 H																1 H- 1.5	2 He
3 Li+ 0.68	4 Be^{2+} 0.35											5 B	6 C	7 N	8 O^{2-} 1.40	9 F- 1.36	10 Ne
11 Na+ 0.97	12 Mg^{2+} 0.67											13 Al	14 Si	15 P	16 S^{2-} 1.84	17 Cl- 1.81	18 Ar
19 K+ 1.33	20 Ca^{2+} 0.99	21 Sc^{3+} 0.81	22 Ti^{4+} 0.68	23 V^{5+} 0.59	24 Cr^{6+} 0.52	25 Mn^{7+} 0.46	26 Fe	27 Co	28 Ni	29 Cu	30 Zn	31 Ga	32 Ge	33 As	34 Se^{2-} 1.98	35 Br- 1.95	36 Kr
37 Rb+ 1.47	38 Sr^{2+} 1.12	39 Y^{3+} 0.92	40 Zr^{4+} 0.79	41 Nb^{5+} 0.69	42 Mo^{6+} 0.62	43 Tc^{7+} 0.57	44 Ru	45 Rh	46 Pd	47 Ag	48 Cd	49 In	50 Sn	51 Sb	52 Te^{2-} 2.21	53 I- 2.16	54 Xe
55 Cs+ 1.67	56 Ba^{2+} 1.34	57 La^{3+} 1.14	72 Hf^{4+} 0.78	73 Ta^{5+} 0.68	74 W^{6+} 0.62	75 Re^{7+} 0.56	76 Os	77 Ir	78 Pt	79 Au	80 Hg	81 Tl	82 Pb	83 Bi	84 Po^{2-} 2.3	85 At- 2.2	86 Rn
87 Fr+ 1.8	88 Ra^{2+} 1.43	89 Ac^{3+} 1.18															

Rare earth metals:

58 Ce^{4+} 0.94	59 Pr	60 Nd	61 Pm	62 Sm	63 Eu^{2+} 1.09	64 Gd^{3+} 0.97	65 Tb^{4+} 0.81	66 Dy	67 Ho	68 Er	69 Tm	70 Yb^{2+} 0.93	71 Lu^{3+} 0.85

Uranium metals:

90 Th^{4+} 1.02	91 Pa^{5+} 0.9	92 U^{6+} 0.80	93 Np	94 Pu	95 Am	96 Cm	97 Bk	98 Cf	99 E	100 Fm	101 Mv	102 No	103 Lw

FIGURE 6-8 Periodicity of atomic radii. (a) Ionic radii for selected ions. (b) Covalent radii for selected atoms. [Sources of data: M. C. Day and J. Selbin, *Theoretical Inorganic Chemistry*, 2nd ed., New York: Reinhold, p. 118 (1969); © 1962 and 1969 by Litton Educational Publishing, Inc. Reprinted by permission of Van Nostrand Reinhold Company and R. T. Sanderson, *Chemical Periodicity*, New York: Reinhold, p. 26 (1960). By permission.]

Second, there is a general tendency for the size of an atom to decrease going across a period from left to right. This, too, makes sense when we consider that the quantum number n remains the same in any period, but that the nuclear charge is increasing slowly. This trend is especially clear in the s and p blocks (for the representative elements). Among the transition metals, the size of the atoms goes through a minimum in the course of a general decreasing-size trend across each period. Clearly, this situation is more complex, probably because successive electrons are being added to the shell just within the outer shell (for example, in the fourth period, electrons are going into the $3d$ level). Consequently, each new electron can screen the outer-shell electrons more effectively from the increasing nuclear charge.

Further amplification of this effect occurs between elements 57 and 72, where the $4f$ orbitals are being filled; these are the *rare earth* or *lanthanide*

(b)

1 H 0.37																	2 He 0.93
3 Li 1.34	4 Be 0.90											5 B 0.82	6 C 0.77	7 N 0.75	8 O 0.73	9 F 0.72	10 Ne 1.31
11 Na 1.54	12 Mg 1.30											13 Al 1.18	14 Si 1.11	15 P 1.06	16 S 1.02	17 Cl 1.00	18 Ar 1.74
19 K 1.96	20 Ca 1.74	21 Sc 1.44	22 Ti 1.36	23 V	24 Cr	25 Mn	26 Fe	27 Co	28 Ni	29 Cu 1.38	30 Zn 1.31	31 Ga 1.26	32 Ge 1.22	33 As 1.19	34 Se 1.16	35 Br 1.14	36 Kr 1.89
37 Rb 2.11	38 Sr 1.92	39 Y 1.62	40 Zr 1.48	41 Cb	42 Mo	43 Tc	44 Ru	45 Rh	46 Pd	47 Ag 1.53	48 Cd 1.48	49 In 1.44	50 Sn 1.41	51 Sb 1.38	52 Te 1.35	53 I 1.33	54 Xe 2.09
55 Cs 2.25	56 Ba 1.98	57 La * 1.69	72 Hf	73 Ta	74 W	75 Re	76 Os	77 Ir	78 Pt	79 Au 1.50	80 Hg 1.49	81 Tl 1.48	82 Pb 1.47	83 Bi 1.46	84 Po	85 At	86 Rn 2.14
87 Fr	88 Ra	89 Ac †															

* Rare earth metals

58 Ce	59 Pr	60 Nd	61 Pm	62 Sm	63 Eu	64 Gd	65 Tb	66 Dy	67 Ho	68 Er	69 Tm	70 Yb	71 Lu

† Uranium metals

90 Th	91 Pa	92 U	93 Np	94 Pu	95 Am	96 Cm	97 Bk	98 Cf	99 E	100 Fm	101 Mv	102 No	103 Lw

elements. Here the effect of increasing nuclear charge is emphasized, because the electrons are being added to orbitals "deep down inside" the atom, not making much contribution to atomic size. In fact, after adding 14 positive charges to the nucleus, the atomic radius actually has *shrunk* from 1.69 Å (in La) to 1.44 Å (in Hf). This marked drop in size has its further effect on the elements that follow. Note especially that Zr (Z = 40) has. an atomic radius of 1.45 Å, while its congener Hf (Z = 72), occurring *32 places later* in the Periodic Table, has a radius of only 1.44 Å.[7] Further on, silver and gold have the same atomic radius, 1.34 Å, even though they are separated

---

[7] The nearly identical sizes, coupled with identical electron configuration in the outer shell, make Hf and Zr very similar and extremely hard to separate from each other in a mixture. Hf and Zr are much harder to separate than are adjacent rare earth elements. As might be expected, Hf and Zr occur together, and for a long time Hf eluded discovery as it hid in a much larger amount of Zr. Now that Zr is an important structural material for nuclear reactors, where it must be very pure, a great deal of trouble is encountered in getting out the undesirable neutron-absorbing Hf.

by 31 elements. Hence, if we picture a valence electron on the periphery of a gold atom, we see that it is held by 32 more positive charges than such an electron on a silver atom. That gold electron is removed with more difficulty. Hence gold is *not* more electropositive than its lighter congener, silver, but *less* electropositive and *less* active. Similar conclusions are reached for Pt and Hg versus the elements above them in the Periodic Table. The general effect is called the **lanthanide contraction**.

This is why gold remains lustrous and untarnished.

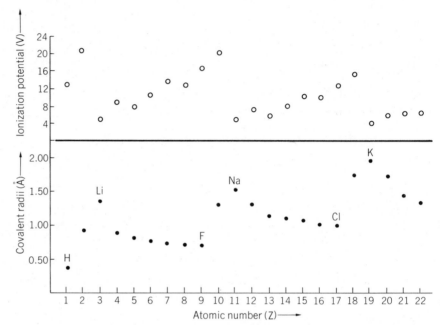

FIGURE 6-9 Plots of covalent radii and of ionization potential, each versus atomic number Z, showing that while ionization potential generally increases with Z across a period, size measured by covalent radius decreases.

Comparing the periodic trends in size and ionization potential, we find an important correlation: *Ionization potential tends to decrease as size increases.* This correlation is made more explicit in Figure 6-9, where the two properties are plotted together. The reason for this antiparallelism of size and ionization potential is not hard to figure out; if the electrons in the outer shell are far from the nucleus because of having a high principal quantum number and/or because they are efficiently screened from the nuclear charge by inner shells of electrons, then the nucleus will have only a relatively small influence on them. These electrons are removed readily in the process of ionization. The immediate chemical consequence of this trend is that elements at the bottom of a group at the right and left sides of the Periodic Table tend to be most ready to give up electrons in a chemical reaction. Thus, in the p block, for instance, the elements at the top (B, C, N, O, and F) tend to bond either covalently or to form negative ions (recall the behavior of fluorine described in Chapter 3), while those at the bottom (Tl, Pb, Bi, Po, and At) are generally metallic, forming positive ions. A possible exception is At, about whose chemistry little is known; it has no stable isotopes. In the center of the Periodic Table, the situation is more complex due to the lanthanide contraction, and the trend may actually be reversed, as we have seen.

The behavior of the positive ions formed by the metals and the negative ions formed by the nonmetals is another (and more complex) story that we shall reserve until later in this chapter.

6-5 Polarizability

On a time average, isolated atoms are always symmetrical. The center of the negative charge due to the electrons coincides with the center of the positive charge of the nucleus; in this way, the attraction between the nucleus and the electrons is best satisfied (that is, the atom finds a minimum of potential energy). But as we have seen in Chapter 5, atoms tend to polarize one another when they approach closely, and one form of van der Waals force—the London dispersion force—arises when the mutually induced dipoles of two atoms align themselves in an electrically favorable way. London dispersion forces exist among all atoms that are near one another, and their variation from atom to atom has an important effect on the different chemical reactivities of the elements. In Chapter 5, we noticed that London dispersion forces appear to be stronger in heavy atoms and molecules than in light ones. We are now in a position to explain that observation and to relate it to the other periodic properties we have discussed.

If isolated atoms tend to adopt a symmetrical spherical structure, it must require an input of energy to distort the electron cloud of an atom from this symmetrical disposition. The London forces resulting will be stronger the more easily the electron cloud of the atom is distorted, since less of the potential energy of the bonding will be lost in overcoming the atom's tendency to return to a symmetrical form. But the influence that leads the electron cloud to be symmetrical is the spherically symmetrical electrical field of the tiny nucleus. Thus, those electron shells that feel the influence of this field least strongly will be polarized most easily.

It takes a push to squash the electronic arrangement.

From what has gone before, we know that nuclear electrical forces decrease as the thickness of the electron cloud between the nucleus and the outer shell of electrons increases. That is, in any given group, atoms (and analogous compounds containing those atoms) exhibit increasing London forces going down the group. Thus, fluorine and chlorine are gases; bromine is a liquid; and iodine and astatine are solids among the elemental halogens, because London forces—the only kind of force possible between symmetrical diatomic molecules of the halogens—increase going down that group. Likewise, CF_4 is a gas; CCl_4 is a liquid; and CBr_4 and CI_4 are solids at room temperature, because the polarizability of the halogens in these compounds increases going down the group.

That the sheer molecular weight of a substance has nothing to do with London forces is illustrated by comparing the solid I_2 (molecular weight 253.8) to the gas IF_7 (molecular weight 259.9). In I_2, the easily polarized I atoms allow large London forces between I_2 molecules. But in IF_7 (Figure 6-10), the small, difficultly polarized fluorine atoms that surround the iodine do not lend much polarizability to the molecule, and at the same time prevent neighboring molecules from approaching the more easily polarized iodine atom.

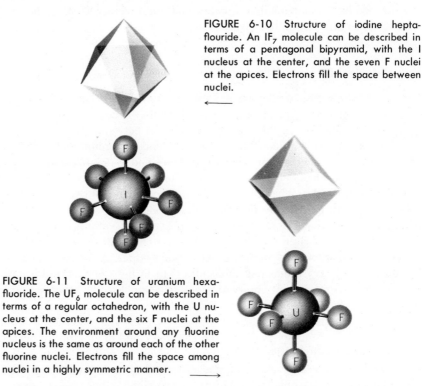

FIGURE 6-10 Structure of iodine heptaflouride. An IF_7 molecule can be described in terms of a pentagonal bipyramid, with the I nucleus at the center, and the seven F nuclei at the apices. Electrons fill the space between nuclei.

FIGURE 6-11 Structure of uranium hexafluoride. The UF_6 molecule can be described in terms of a regular octahedron, with the U nucleus at the center, and the six F nuclei at the apices. The environment around any fluorine nucleus is the same as around each of the other fluorine nuclei. Electrons fill the space among nuclei in a highly symmetric manner.

Precisely the same principle explains the volatility of the historically important compound UF_6. In order to separate the uranium isotope of mass 235 from ^{238}U (see Chapter 10), the fact that light molecules diffuse in gas phase faster than heavier ones can be exploited *if* a gaseous compound of uranium is available. In view of the high atomic weight of uranium, making a gaseous compound seemed to be a tall order. The symmetrical octahedral structure of UF_6 (Figure 6-11) and the unpolarizable nature of the outer skin of fluorine atoms come to the rescue, however, and produce a substance boiling at 56°C. This compound is as volatile as acetone or the lighter fractions of gasoline, and thus there is no difficulty getting reasonable quantities of uranium into the gas phase. Diffusion of the gas through many separation stages eventually yields uranium sufficiently rich in ^{235}U for fission in a reactor of moderate size.

6-6 Scales of Electronegativity

The notion of **electronegativity** was touched upon very briefly in Chapter 3 as a generalization about the Periodic Table. Now that we have taken a closer look at atomic structure and periodicity, we can return to seek a better understanding of this useful concept.

When two atoms are bonded covalently, the bonding electrons are subject to attraction by both of the bonded nuclei. The detailed positions and motions of these electrons are describable only in a statistical way through a wave-mechanical treatment (Chapter 2), but we can say with confidence that *the electrons in a bond will spend a larger fraction of their time nearer the nucleus whose positive electrical field is stronger*, if the two bonded nuclei are non-identical, as in HF, CO, and so forth. This tendency of an atom to make its nuclear charge felt to electrons in and beyond its valence shell is what we mean by electronegativity. We shall describe three different attempts to model this tendency in numerical terms.

A reasonable way to obtain numerical values for electronegativity is to use Coulomb's law (Appendix III) to calculate the force acting on an electron located at a distance, equal to the covalent radius, from the nucleus. This can, in fact, be done by assuming a screening effect exerted by electrons in inner shells, reducing the actual nuclear charge to an "effective nuclear charge," the hypothetical positive charge whose effect is felt in the valence shell of the atom. Not surprisingly, there is some difficulty in deciding how much the actual nuclear charge is to be reduced for purposes of this calculation, but some empirical methods have been developed. With Z_{eff} being the effective nuclear charge, Coulomb's law gives the electronegativity EN as

$$EN = \frac{Z_{eff}}{r^2} \quad (6\text{-}3)$$

where r is the distance from the nucleus to the valence shell of the atom.

An alternative attempt to base an electronegativity scale on purely observable properties of elements, without any assumptions about screening and radii, has been made by calculating EN from the related quantities ionization potential and electron affinity. Ionization potential measures the difficulty with which an electron is removed from the valence shell, and the electron affinity measures the eagerness with which the atom accepts an additional electron, both processes being extremes in electron sharing among atoms. This EN scale simply defines electronegativity as the numerical average of the ionization potential and the electron affinity:

$$EN \equiv \frac{(\text{ionization potential}) + (\text{electron affinity})}{2} \quad (6\text{-}4)$$

The concept of EN is plain enough, but the measurement is not exact.

The difficulty in making use of this scale is that, as we noted in Section 6-3, reliable experimental values for electron affinities are difficult to obtain for most elements.

A third scale—historically the first scale to be developed—is due to Linus Pauling's observation that bonds between atoms of different electronegativities tend to be unexpectedly strong. This may be attributed to the fact that such bonds are always *polar*, and there is a consequent attraction between the positive and negative ends of the molecule, adding to the strength of the bond. Pauling's electronegativity scale relates the difference in electronegativity between two atoms (say X and Y) to the measured strengths of the X-Y bond and the strengths of X-X and Y-Y bonds, reference bonds that involve the same atoms, but do not involve any additional stabilization through polarity. The result is a somewhat roundabout measure of electronegativity, and some ingenuity is required to find compounds containing the proper bonds, and to derive their bond strengths from measurements of energies of reactions in which these bonds are broken.

TABLE 6-2
Electronegativities of the Elements

Element	Z_{eff}/r^2	Pauling EN	Element	Z_{eff}/r^2	Pauling EN	Element	Z_{eff}/r^2	Pauling EN
H	...	2.1	Br	2.74	2.8	Tm	1.11	1.1–1.2
He	2.5–3	...	Kr	3.0		Yb	1.00	1.1–1.2
Li	0.97	1.0	Rb	0.89	2.8	Lu	1.14	1.1–1.2
Be	1.47	1.5	Sr	0.99	1.0	Hf	1.23	1.3
B	2.01	2.0	Y	1.11	1.2	Ta	1.33	1.5
C	2.50	2.5	Zr	1.22	1.4	W	1.40	1.7
N	3.07	3.0	Nb	1.23	1.6	Re	1.46	1.9
O	3.50	3.5	Mo	1.30	1.8	Os	1.52	2.2
F	4.10	4.0	Tc	1.36	1.9	Ir	1.55	2.2
Ne	4.4	...	Ru	1.42	2.2	Pt	1.44	2.2
Na	1.01	0.9	Rh	1.45	2.2	Au	1.42	2.4
Mg	1.23	1.2	Pd	1.35	2.2	Hg	1.44	1.9
Al	1.47	1.5	Ag	1.42	1.9	Tl	1.44	1.8
Si	1.74	1.8	Cd	1.46	1.7	Pb	1.55	1.8
P	2.06	2.1	In	1.49	1.7	Bi	1.67	1.9
S	2.44	2.5	Sn	1.72	1.8	Po	1.76	2.0
Cl	2.83	3.0	Sb	1.82	1.9	At	1.96	2.2
Ar	3.5	...	Te	2.01	2.1	Rn	2.00	...
K	0.91	0.8	I	2.21	2.5	Fr	0.86	0.7
Ca	1.04	1.0	Xe	2.6	...	Ra	0.97	0.9
Sc	1.20	1.3	Cs	0.86	0.7	Ac	1.00	1.1
Ti	1.32	1.5	Ba	0.97	0.9	Th	1.11	1.3
V	1.45	1.6	La	1.08	1.1–1.2	Pa	1.14	1.5
Cr	1.56	1.6	Ce	1.06	1.1–1.2	U	1.22	1.7
Mn	1.60	1.5	Pr	1.07	1.1–1.2	Np	1.22	1.3
Fe	1.64	1.8	Nd	1.07	1.1–1.2	Pu	1.22	1.3
Co	1.70	1.8	Pm	1.07	1.1–1.2	Am	1.22	1.3
Ni	1.75	1.8	Sm	1.07	1.1–1.2	Cm	1.22	1.3
Cu	1.75	1.9	Eu	1.01	1.1–1.2	Bk	1.22	1.3
Zn	1.66	1.6	Gd	1.11	1.1–1.2	Cf	1.22	1.3
Ga	1.82	1.6	Tb	1.10	1.1–1.2	Es	1.22	1.3
Ge	2.02	1.8	Dy	1.10	1.1–1.2	Fm	1.22	1.3
As	2.20	2.0	Ho	1.10	1.1–1.2	Md	1.22	1.3
Se	2.48	2.4	Er	1.11	1.1–1.2	No	1.20	1.3

Sources: E.G. Rochow, *The Metalloids*, Boston: Heath (1966); A.L. Allred and E.G. Rochow, *Journal of Inorganic and Nuclear Chemistry*, vol. 5, p. 264 (1958); F.A. Cotton and G. Wilkinson, *Advanced Inorganic Chemistry*, New York: Interscience, p. 92 (1962); Bing-man Fung, *Journal of Physical Chemistry*, vol. 69, p. 596 (1965).

The fact is that these numbers on these three electronegativity scales are very nicely proportional to one another (we cannot expect them to be identical, since they are measuring different quantities). This proportionality of scales gives us some confidence that there is such a thing as electronegativity, and that we can measure it at least approximately, even if no single scale is free from theoretical objections. Values of the first scale (Z_{eff}/r^2) are given in Table 6-2, numerically adjusted to have values between 0 and about 4 like the Pauling scale, which has been more widely adopted. (Pauling's method, you recall, gives only the *difference* in electronegativity between two atoms, and Pauling chose a scale such that all values would be positive.)

EN is not a precise property of an element, like its weight.

1 H 2.0																	2 He 3.0
3 Li 1.0	4 Be 1.5											5 B 2.0	6 C 2.5	7 N 3.0	8 O 3.5	9 F 4.0	10 Ne 4.4
11 Na 1.0	12 Mg 1.2											13 Al 1.5	14 Si 1.8	15 P 2.1	16 S 2.5	17 Cl 2.9	18 Ar 3.5
19 K 0.9	20 Ca 1.0	21 Sc 1.3	22 Ti 1.4	23 V 1.5	24 Cr 1.6	25 Mn 1.6	26 Fe 1.7	27 Co 1.7	28 Ni 1.8	29 Cu 1.8	30 Zn 1.6	31 Ga 1.7	32 Ge 1.9	33 As 2.1	34 Se 2.4	35 Br 2.8	36 Kr 3.0
37 Rb 0.9	38 Sr 1.0	39 Y 1.2	40 Zr 1.3	41 Nb 1.5	42 Mo 1.6	43 Tc 1.7	44 Ru 1.8	45 Rh 1.8	46 Pd 1.8	47 Ag 1.6	48 Cd 1.6	49 In 1.6	50 Sn 1.8	51 Sb 1.9	52 Te 2.1	53 I 2.4	54 Xe 2.6
55 Cs 0.8	56 Ba 1.0	57 La* 1.1	72 Hf 1.3	73 Ta 1.4	74 W 1.5	75 Re 1.7	76 Os 1.9	77 Ir 1.9	78 Pt 1.8	79 Au 1.9	80 Hg 1.7	81 Tl 1.6	82 Pb 1.7	83 Bi 1.8	84 Po 1.9	85 At 2.1	86 Rn 2.0
87 Fr 0.8	88 Ra 1.0	89 Ac 1.1															

Key: Electronegativity
< 1.0
1.0–1.2
1.2–1.6
1.6–2.0
2.0–2.4
2.4–2.8
2.8–3.2
> 3.2

Rare earth metals:

58 Ce 1.1	59 Pr 1.1	60 Nd 1.1	61 Pm 1.1	62 Sm 1.1	63 Eu 1.1	64 Gd 1.1	65 Tb 1.1	66 Dy 1.1	67 Ho 1.1	68 Er 1.1	69 Tm 1.1	70 Yb 1.0	71 Lu 1.2

Uranium metals:

90 Th 1.2	91 Pa 1.3	92 U 1.5	93 Np 1.3	94 Pu 1.3	95 Am 1.3	96 Cm 1.3	97 Bk 1.3	98 Cf 1.3	99 E 1.3	100 Fm 1.3	101 Mv 1.3	102 No 1.3	103 Lw 1.3

FIGURE 6-12 Periodicity of electronegativity, showing the overall trends of electronegativity values of the elements. Approximate *EN* values are used in this figure, averages of values from Table 6-2. The color of each block indicates the *EN* value of that element, according to the key in the figure.

The periodicity of electronegativity as indicated in Figure 6-12 should be no surprise; electronegativity measures, in its way, the same sort of quality that ionization potential and atomic size do. The fundamental question is: How strong is the influence of the nucleus on electrons located in the outer

On the basis of these numbers, which pairs of elements are most likely to form ionic binary compounds?

electronic shell? The answer is that the strength of this influence, measured by electronegativity (or size, or ionization potential), is greatest for atoms in the upper right-hand corner of the Periodic Table, where there is the largest nuclear charge for the least screening, and least for atoms in the lower left-hand corner, where the opposite conditions apply. Furthermore, the electronegativity of intervening elements, with a few local ripples, varies in a smooth manner between low values for the "supermetals" in the bottom left-hand corner to the "supernonmetals" in the top right-hand. Thus, we find that elements lying on a diagonal perpendicular to this axis (for example, boron, silicon, arsenic, and tellurium in the p block) tend to have similar properties, trading off low nuclear charge and little screening for high nuclear charge and high screening. The diagonal set of elements from boron to tellurium is composed entirely of metalloids, elements whose similar chemical behavior is intermediate between that of metals and nonmetals.

6-7 Periodicity and Chemical Reactivity

We saw in Chapters 3 and 4 how the electronic structure of an atom—and thus its position in the Periodic Table—influences the sorts and shapes of compounds that it is likely to form. We shall conclude this review of periodicity by looking at variations in the "eagerness" with which some of these compounds are formed, and variations in their reactivity once formed. As models for discussion at this time, we shall use reactions in which new bonds are formed through the sharing of electrons between separately stable species. These are *acid-base reactions*. Then later in Chapter 8, we shall consider oxidation-reduction reactions in which electrons are transferred from one chemical species to another. Both kinds of reactions are vitally important in the chemistry that takes place around us and within us.

ACID-BASE REACTIONS: DEFINITIONS. As we have seen, most of the nonmetallic elements tend to form stable compounds in which they have in their valence shells pairs of nonbonding electrons. For example, the oxygen in water has two such "lone pairs." These nonbonding pairs are stable in the neighborhood of one nucleus because of the strength of that nucleus' electrical field; or, what is the same thing, because of the electronegativity of the atom on which they reside. On the other hand, there are electron-poor atoms such as the aluminum in $AlCl_3$, or the many positive ions formed by metals, which can accommodate more electrons in their valence shells, but are not electronegative enough by themselves to hold additional electrons.

An evidently sensible thing for these two classes of compounds to do is to form a new chemical bond by sharing the lone pairs on the electron-rich member of the pair (the *base*), using vacant orbitals on the electron-poor member (the *acid*).[8] In very many cases, the pair of electrons so shared has

---

[8] These definitions of acid and base were proposed by G. N. Lewis, and species so defined are called **Lewis acids** and **Lewis bases**. These are not the only definitions in use by chemists, and we shall encounter another set shortly. All definitions of acid and base are analogous, some being more general than others. The Lewis definitions are among the more general.

lower potential energy in the electrical fields of two nuclei than of one, so that the formation of the bond leads to a more stable situation, and energy is released. As an example, consider the reaction between the Lewis base trimethylamine $(CH_3)_3N$ and the Lewis acid boron trifluoride (BF_3), the structures of which are shown in Figure 6-13. In this reaction, the new compound $(CH_3)_3NBF_3$ is held together by the attraction of both the B and N nuclei for the pair of electrons between them. So stable is this bond that 30 kcal of energy (heat will do) are required to separate a mole of $(CH_3)_3NBF_3$ into the component stable acid (BF_3) and base $(CH_3)_3N$. We might ask whether similar reactions taking place between other acids and bases will lead to equally stable products. Although there are many minor factors that affect the stability of the resulting compound, the answer in general is yes.

ACID-BASE REACTIVITIES. There are at least two ways in which such addition compounds or acid-base adducts can be rendered stable. First, if both atoms involved in the bond are rather small, as in the boron-nitrogen bond described above, then their nuclei are able to come relatively close to the bonding electron pair, and the resulting short bond (all other factors being equal) is unusually strong. Now consider the reaction in solution

$$Hg^{2+} + I^- \longrightarrow HgI^+ \tag{6-5}$$

between ions from the fifth and sixth periods. Here, the product is stabilized not only by the attraction of the mercury nucleus for the electron-rich iodide ion, but also by the relatively large van der Waals forces that inevitably arise between two large, polarizable atoms in close contact.

On the other hand, a reaction between a small, nonpolarizable atom and a large, polarizable atom is less likely to result in so stable a product. The new bond must be long because of the size of the larger atom; but strong van der Waals forces are precluded by the slight polarizability of the smaller reactant. Thus, both the reactions in solution

$$Fe^{3+} + I^- \to FeI^{2+} \tag{6-6}$$

and

$$Hg^{2+} + F^- \to HgF^+ \tag{6-7}$$

result in weaker bonds than the one between Hg^{2+} and I^-; yet the reaction

$$Fe^{3+} + F^- \to FeF^{2+} \tag{6-8}$$

again produces a strong bond. In Equation (6-6), a small Lewis acid (Fe^{3+}) is reacting with a large, polarizable Lewis base (I^-); in Equation (6-7), a larger acid (Hg^{2+}) is reacting with a small, unpolarizable base (F^-); and finally, the stable bond formed in Equation (6-8) is between a small, unpolarizable acid and a similarly small unpolarizable base.

Chemists have begun categorizing acids and bases of the sort we have been considering into **hard** and **soft** groups. As the names imply, soft species are relatively easily polarized; they tend to be atoms or groups with deep and complex electronic structures, and small or zero positive electrical charge.

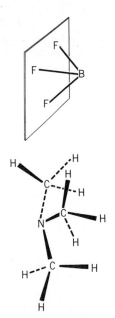

FIGURE 6-13 Structures of BF_3 and $(CH_3)_3N$ molecules.

BF_3 is an irritating and toxic industrial waste. Can you suggest a way to absorb it?

Hard species, on the other hand, are only slightly polarizable, and, among the Lewis acids, number those metal ions with relatively large electrical charge. The reactions (6-5) to (6-8) we have been considering would be classified as soft-soft, soft-hard, hard-soft, and hard-hard, respectively. Of these, the first and last form the most stable bonds.

Of course, any such sweeping generalization about the behavior of many thousands of acids and bases is bound to overlook some counter examples. There are hard-soft pairs which form a more stable bond than some hard-hard or soft-soft liaisons. But on the whole, the soft-hard categorization of reactions is a very useful one and enables us to predict rough orders of reactivity between chemical groups without experimentally trying out every last combination. When a counter example is found, it may set us to looking for the reasons for an exception to such a general rule and thus, sometimes, to the discovery of new effects in bonding. A theory that explains all observations, though the nominal goal of science, would hardly leave much room for argument, creativity, or play—activities that are at least the conscious sources of the scientific endeavor.

TOXICITIES OF LEWIS ACIDS AND BASES. There is a general term—*heavy metal poisoning*—that covers more or less serious disruption of animal metabolism by a variety of metals like lead, barium (alone among the Group II metals), mercury, bismuth, and so forth, all of which are to be found at or near the bottom of their respective periodic groups. Heavy metal poisoning is just one manifestation of the general toxicity of soft Lewis acids and bases. To take another example, among the Group V elements, there are many nontoxic nitrogen and phosphorus compounds, but none containing their heavier (and softer) congeners arsenic and antimony.[9]

Can you predict bad pollutants from their formulas?

Another toxic soft Lewis base is carbon monoxide (see Chapter 3). Carbon monoxide is structurally similar to O_2 and N_2, but it is rendered less electronegative, and thus softer, by the carbon atom in one end. Note that while H_2O is definitely not toxic (taken in moderation), H_2S and H_2Se definitely are. We see that electronegativity decreases, and softness and toxicity increase, as we proceed down the Group VI hydrides.

PROTON TRANSFER REACTIONS. A particularly common and important Lewis acid is the hydrogen ion, H^+. Since hydrogen atoms have only one electron to begin with, hydrogen *ions* are completely devoid of electrons. In the most common hydrogen isotope, a hydrogen ion is a lone proton. A proton would be a particularly hard Lewis acid, having no polarizability at all and a tiny radius; its electrical field is very intense indeed at distances of a small fraction of an angstrom. In fact, this naked proton is never found alone if there are any electron-containing species at hand. Thus, in water solutions, protons are always bound to one of the lone pairs of electrons of a water

---

[9] Thus, one is entitled to skepticism when reassured that the notorious "Agent Blue" defoliant is safe to use in populated areas when one learns that it is an aqueous solution of cacodylic acid, $(CH_3)_2AsO(OH)$.

molecule[10] to form the hydronium ion H_3O^+. In ammonia solutions, the proton is bound to NH_3 to form the ion NH_4^+.

Although free protons are never found in such solutions, a proton can be transferred from one molecule (or ion) to another, as in the reaction

$$H_3O^+(aq) + NH_3(aq) \longrightarrow H_2O(l) + NH_4^+(aq) \qquad (6\text{-}9)$$

This type of reaction—a proton-transfer reaction—is the basis for another definition of acid and base: *a species that has a donatable proton* (such as H_3O^+) *is called a* **Brønsted acid,** and *a species that can accept a proton in such a reaction* (such as NH_3) *is called a* **Brønsted base.** Historically, the Brønsted definitions of acid and base predate the Lewis definitions, but both systems are used by chemists, who specify which they are using if there is any doubt.

Despite all the names, the principles remain the same.

Brønsted acid-base reactivity is also amenable to correlation with the Periodic Table. The reactivity of Brønsted acids and bases is characterized by "acid strength" and "base strength." That is, if a species bonds a proton only rather weakly, then a good Brønsted base can easily remove the proton, in a reaction like Equation (6-9). Such a Brønsted acid is called a **strong acid,** because it protonates bases readily. A **weak acid,** on the other hand, bonds its proton relatively firmly, and consequently does not readily transfer it to a base. In actual fact, such a concept of acid strength is illogical; the acid does not reach out and stick the proton onto the base. Rather, the negative electrical field of the base's nonbonding electron pair attracts the proton away from the acid. The acid plays a passive role in the transaction. Nevertheless, this usage is well entrenched in chemical speech, and we might as well speak the language.

To consider the effect of chemical periodicity on the strength of Brønsted acids, let us begin by thinking about some hydrogen-containing compounds of elements in the second period: LiH, BeH_2, BH_3, CH_4, NH_3, H_2O, and HF. (No electrically neutral hydride of neon is known to exist.) As we recall, the electronegativity of elements in a period increases from left to right. Consequently, the bond between H and the electropositive elements Li and Be is polarized toward the H, which has more than its share of electrons; indeed, these solid compounds are known as "saltlike" hydrides and contain positive metal ions ionically bonded to negative hydride ions, H^-. To remove a proton (H^+) from these, leaving negative metal ions, would require a prohibitive input of energy; indeed, when they do react with other elements, it is usually as hydride-ion donors, rather than proton donors.

Across the Periodic Table again

BH_3 is an electron-deficient compound; that is, the valence shell of the boron has room for more electrons than are available to it in the compound, and BH_3 is always found as a dimer (B_2H_6) in which the deficiency is shared between two BH_3 halves. In any event, since the electronegativities of B and

---

[10] Additional water molecules cluster about the H_3O^+ ion, solvating it to form a variety of species such as $H_5O_2^+$, $H_7O_3^+$, $H_9O_4^+$, and so forth. Conventionally, chemists indicate the solvated hydrogen ion in water as H_3O^+, or sometimes simply as H^+; you will find both notations in this book.

H are very nearly equal, the bond between them is covalent, and only slightly polar. Again, a large input of energy would be required to remove a proton from BH_3 or B_2H_6, and no such reaction has ever been detected. On the other hand, there is no evidence for the existence of H^- in boron hydrides either.

The bond between C and H (for example in CH_4) is also covalent, but this time polarized slightly away from the hydrogen, which is less electronegative than the carbon. It is possible for an exceptionally strong base to remove a proton from a C—H bond; the resulting *conjugate base*[11] (for example, CH_3^-), known as a *carbanion* (carbon-anion), has been detected as a very unstable species transiently existing in some reactions of organic chemistry.

The next element to the right, nitrogen, forms the hydride NH_3, whose conjugate base is NH_2^-, the amide ion. Amide ions are perfectly stable in solid salts and in solutions that are not rich in protons. The relative willingness of NH_3 to part with a proton can be attributed to the fact that nitrogen is nearly one full unit more electronegative than hydrogen, and the electrons in the N—H bond are already pulled away from the hydrogen end of the bond. The more this happens in the prospective acid molecule, the easier it is for a proton-attracting base to pluck the hydrogen nucleus out of its electron-poor environment.

Another element to the right and another jump in electronegativity brings us to oxygen, whose familiar hydride H_2O has the almost equally familiar conjugate base OH^-. Finally, in HF, we have a really willing proton donor whose conjugate base F^- is part of the water supply of thousands of communities.

In the reaction

$$HF + OH^- \rightarrow F^- + H_2O \tag{6-10}$$

we can imagine a stage in which the proton, in the process of being transferred from HF to OH^-, is halfway:

$$\underset{F}{(-)} \ldots \underset{H}{(+)} \ldots \underset{OH}{(-)}$$

Who wins the tug-of-war?

If we imagine that we place a proton in this situation, we can guess that it might move to the left, to produce HF, leaving OH^-, or that it might move to the right, forming H_2O, leaving F^-. A proton will always go where it finds the strongest negative electrical field, and in this case there is no doubt; with a nuclear charge one unit smaller and, consequently, a larger cloud of electrons in its valence shell, OH^- is much more attractive to a positive charge than is F^-. In the language of "acid strength," HF is a stronger acid than H_2O because its conjugate base is less attractive to protons than is the conjugate base of H_2O.

We could perform the same imaginary experiment with any pair of binary hydrogen compounds in this group, and the result would always be the analogous one—the element to the left in the Periodic Table would wind up with

---

[11] Two species that differ only in the number of hydrogen ions they contain are called a *conjugate pair*. CH_4 is the conjugate acid of CH_3^-, and CH_3^- is the conjugate base of CH_4.

the proton. The Brønsted acid strength of hydrides increases uniformly from left to right across a period, because electronegativity increases in the same pattern.

Very well. Then we should expect that the Brønsted acid strength of hydrides should decrease going down a group, by the same reasoning. This is *not* the case, because, as we have found, the *size* of atoms increases rapidly going down Groups VI and VIII, and the proton is a hard Lewis acid that bonds well only when it can form a short bond. Thus, HF is a weaker Brønsted acid than HCl; H_2S is weaker than H_2Se, though stronger than H_2O, and so forth for any pair in a group. The longer the bond between any element and hydrogen, the weaker that bond. With hydrogen, bond length is the dominating factor in determining bond strength, overwhelming any considerations of electronegativity. This effect is not seen going across periods, because there the change in radius is relatively gradual compared to the increase in nuclear charge and thus in electronegativity.

THE OXYGEN ACIDS. Because oxygen is more electronegative than hydrogen, its bond with hydrogen is polar; the oxygen atom is a potential site of proton donation. Oxygen can form two bonds, and so there exist many compounds with the general structure X–O–H, where X may be a single atom (as in Cl–O–H, hypochlorous acid) or a group of atoms (as in CH_3CH_2–O–H, ethanol). If X is a very electronegative element or contains very electronegative atoms near the –O–H group, then the resulting electron-withdrawing effect will polarize the O–H bond even more, and the resulting molecule is likely to be a good proton donor. As examples, let us consider hypochlorous acid and ethanol.

There are thousands of oxyacids, many of which are biologically important.

Hypochlorous acid is an acid because the electronegative chlorine atom withdraws electrons from the O–H bond. On the other hand, ethanol is only faintly acidic (much less acidic than water), because neither C nor H is very electronegative. When wine turns to vinegar, the chemical change is an air oxidation of ethanol to acetic acid,

$$O_2 + \underset{\text{ethanol}}{H_2C-CH_2-O-H} \longrightarrow \underset{\text{acetic acid}}{H_2C-C(=O)-O-H} + H_2O \tag{6-11}$$

The electron-withdrawing effect of that extra oxygen atom on the carbon atom next to the O–H group makes enough difference so that the familiar tang of ethanol turns to the sourness that all acids produce. The disappointed face one makes when the expected bottle of good wine turns out to have gone by to its acetic fate has its real origin in the relatively large nuclear charge of oxygen.

We can see periodicity manifested if we look systematically at the effect of X on the molecule X–O–H, taking various elements in turn for X. Consider, for example, members of the third period from aluminum through chlorine, in the form of the series of compounds Al(OH)$_3$, Si(OH)$_4$, PO(OH)$_3$, SO$_2$(OH)$_2$, and ClO$_3$OH).[12]

In each compound, the oxygen atoms and the O—H groups are bonded directly to the central X atom (Figure 6-14). The electronegativity of X increases uniformly from the metal Al to the strongly electronegative Cl. This increase in electronegativity is reflected in an increase of acid strength in this series of compounds. Let us look at the acid-base behavior of these representative compounds.

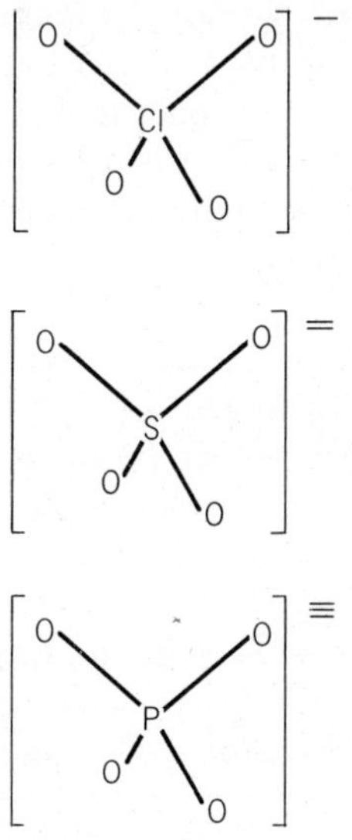

FIGURE 6-14 Structures of the cage of oxygen nuclei surrounding the central atom in phosphoric, sulfuric, and perchloric acids. In each case, the oxygen nuclei are arranged in a tetrahedral configuration in the negative ion and in a distorted tetrahedral configuration in the acid molecule.

The first compound, Al(OH)$_3$, is a solid with almost purely ionic bonds between aluminum ions and hydroxide ions (Al^{3+}, OH$^-$). In one crystalline form, the aluminum ions are packed into octahedral holes in a hydrogen-bonded double layer of OH$^-$ ions. This compound undergoes the reaction

$$\text{Al(OH)}_3(s) + \text{OH}^- \rightarrow \text{AlO(OH)}_2^- + \text{H}_2\text{O} \qquad (6\text{-}12)$$

By thus donating a proton to OH$^-$, it is behaving as a Brønsted acid. On the other hand, it also exhibits Brønsted base behavior:

$$\text{Al(OH)}_3(s) + \text{H}_3\text{O}^+ \rightarrow \text{Al(OH)}_2^+ + 2\text{H}_2\text{O} \qquad (6\text{-}13)$$

Consequently, Al(OH)$_3$ is truly **amphoteric** in being both acidic and basic in the Brønsted system.

Silicic acid, Si(OH)$_4$, is formed in the reaction between finely divided silica and water:

$$\text{SiO}_2(s) + 2\text{H}_2\text{O} \rightarrow \text{Si(OH)}_4 \qquad (6\text{-}14)$$

---

[12] The last three compounds in this series are usually written as H$_3$PO$_4$, H$_2$SO$_4$, and HClO$_4$. These formulas should not fool you; in each compound, the hydrogen atoms are bonded to oxygen (Figure 6-14).

It is a weak Brønsted acid, which can donate a proton to water or to some other Brønsted base:

$$\text{Si(OH)}_4 + \text{H}_2\text{O} \longrightarrow \text{SiO(OH)}_3^- + \text{H}_3\text{O}^+ \quad (6\text{-}15)$$

This reaction is quite feeble and does not go very far toward completion; Si(OH)_4 is a very weak acid.

Moving further to the right within the period, we find increasingly acidic behavior. Phosphoric acid is more acidic than silicic acid, but the sodium salts of both are manufactured in immense quantities for detergents. Sulfuric acid is more so yet; in fact, sulfuric acid is completely transformed into SO_3OH^-, usually written HSO_4^-, in water solutions. Finally, perchloric acid, ClO_3OH, is one of the strongest acids known. The differences within this set of similarly constructed substances all arise from the addition of one positive charge to the nucleus of the central atom each time in going from Al(OH)_3 to ClO_3OH. Such a change in the nucleus, in addition to changing the name and the symbol of the central atom, has the more fundamental effect of increasing the attraction of this central atom for electrons on the periphery of the molecule. We see chemical effects particularly in the O–H bonds. It is this attraction for electrons outside the atom itself that we call electronegativity. This attraction affects the chemical properties of substances in a perfectly regular and reasonable way.

What has this to do with phosphates in lakes and rivers?

The discussion of acid strength we have given here, though a bit long-winded, is still only part of the story. In any reaction involving chemical species (ions and molecules) in a solution, strong forces exist between molecules of the solvent (such as water) and the reactant and product molecules of the reaction. The energy involved in these interactions (**solvation energy**) will, in general, be different for the reactants and products of a reaction. Thus, there will be a change in the total solvation energy when a reaction occurs in solution, and a complete discussion of acid strengths would take differences in solvation energy into account. If you would like to know more, see "Suggestions for Further Reading."

OXIDES AND "EARTHS." Compounds like sulfuric acid, whether written as $\text{SO}_2(\text{OH})_2$ or H_2SO_4, can be thought of as being formed (and they often are) by a reaction between an oxide and water. For instance,

$$\text{SO}_3 + \text{H}_2\text{O} \rightarrow \text{H}_2\text{SO}_4 \quad (6\text{-}16)$$

Sulfur trioxide is spoken of as the **anhydride** of sulfuric acid, a very strong and destructive acid, because it is sulfuric acid *minus* one molecule of water. Because SO_3 is loaded with electronegative atoms, it is also spoken of as an acidic oxide; on reaction with water, it produces an acid in solution. From our earlier discussion, we should expect to find acidic oxides among the oxides of electronegative elements. Unfortunately, most of these are gases, too. What about oxides of metals?

This is why SO_2 and SO_3 are such bad air pollutants.

If we imagine a reaction between, say, magnesium oxide and water to form a possible acid,

$$\text{MgO} + \text{H}_2\text{O} \rightarrow \text{Mg(OH)}_2 \quad (6\text{-}17)$$

we should expect that the resulting hydroxy compound will contain ionic bonds between Mg and O. The difference in electronegativities of Mg and O is 1.8 units, usually sufficient for ionic bonds to exist. Consequently, when magnesium hydroxide dissolves in water (which it does only reluctantly, since it has a quite stable ionic lattice), the parts of the compound separate into independent ions. The reaction is

$$Mg(OH)_2(s) \rightarrow Mg^{2+}(aq) + 2OH^-(aq) \quad (6\text{-}18)$$

The production of the Brønsted base OH^- in solution makes MgO a *basic* oxide. Description of this phenomenon in more old-fashioned language gives the Group II metals their common name of the "alkaline-earth" metals. (*Alkali* means base producing, and *earth* is an archaic name for any metal oxide, especially an insoluble one.)

Although only the Group II metals are called alkaline-earth metals, they are not the only ones to form basic oxides. The "alkali metals" of Group I form even more basic oxides, as one would expect from their position in the Periodic Table. A representative reaction of such an oxide is

$$Na_2O + H_2O \rightarrow 2Na^+ + 2OH^- \quad (6\text{-}19)$$

The dissociation into ions in solution is more nearly complete for sodium oxide than for magnesium oxide, and it is a much stronger base.

6-8 Summary

We brought to this chapter knowledge that the type of bonding—metallic, ionic, or covalent—found in a substance depends primarily on the electronegativity of the atoms of which it is made. Thus, electropositive elements form metallic crystals because of the looseness with which their valence-shell electrons are held, and compounds between elements of widely differing electronegativity are ionic. Covalent crystals are formed by many elements, and by compounds between elements of moderate and nearly equal electronegativity that are capable of forming several bonds per atom to build up a three-dimensional network. Molecular substances result when atoms are not capable of extended covalent bonding, but are too electronegative to form metallic crystals.

Electronegativity is clearly a critical factor in determining the properties of a substance, and in this chapter we have investigated ways in which electronegativity can be expressed in terms of numbers. It was seen that ionization potential and electron affinity are closely related to electronegativity, and that all three quantities correlate well with the Periodic Table. We have seen that both structure and reactivity fit into the grand scheme of chemical periodicity, and we investigated some of the finer details of chemical reactivity by looking at the chemistry of acids and bases.

Bond strengths and chemical reactivity depend on other factors too. Atomic size is such a factor. There are alternative ways of defining atomic size, but we found that both covalent radii and ionic radii showed periodicity. Just about all of the fundamental chemical properties do show periodicity.

The discussion in this chapter, though it may more than satisfy your present curiosity about chemical periodicity, comes nowhere near exhausting the subject and the correlations that can usefully be made. All these correlations hang fundamentally on the simple rules governing the buildup of electronic structure in atoms according to the quantum numbers. Mark Twain's remark is well justified: "There is something fascinating about science. One gets such wholesale returns of conjecture out of such a trifling investment of fact."

GLOSSARY

amphoteric: said of a substance that can react either as an acid or as a base, depending on reaction conditions.

anhydride: compound X and its anhydride A are related by a reaction like X ⇌ yH$_2$O + A.

Brønsted acid: a chemical species that participates in a reaction by donating a proton.

Brønsted base: a chemical species that participates in a reaction by accepting a proton.

covalent radius: an atomic radius derived from internuclear distances in covalently bonded compounds.

electron affinity: the energy released when an electron is added to an atom.

electronegativity: the attraction of an atom for electrons in a compound.

hard acid: a Lewis acid that is essentially nonpolarizable.

hard base: a Lewis base that is essentially nonpolarizable.

ionic radius: an atomic radius derived from internuclear distances in ionic crystals in which the ions are in close contact.

ionization potential (energy): the potential (energy) required to remove an electron from an atom. If expressed in volts, it is a potential; if in electron-volts, an energy. The two are numerically equal here.

lanthanide contraction: a gradual but drastic decrease in the size of atoms in the inner-transition series (atomic numbers 57–70 and 89–102), due to progressive filling of the 4f and 5f orbitals deep within the atom.

Lewis acid: a chemical species (molecule or ion) that participates in a reaction by donating a pair of electrons.

Lewis base: a chemical species (molecule or ion) that participates in a reaction by accepting a pair of electrons.

soft acid: a readily polarizable Lewis acid, typically with many electrons, but with small or zero net charge.

soft base: a readily polarizable Lewis base.

solvation energy: the energy released in forming bonds between an unsolvated molecule or ion, and molecules of the solvent.

strong acid: a Brønsted acid that gives up its proton readily; a Brønsted acid with a large dissociation constant.

weak acid: a Brønsted acid that gives up its proton only in the presence of a strong base; a Brønsted acid with a small dissociation constant.

valence shell: the highest-energy outer electrons; those electrons in an atom that participate in chemical bonding.

PROBLEMS

6-1 Write chemical equations for the formation of Lewis acid-base adducts from the following reactants. Identify the Lewis acid and the Lewis base.

$$(CH_3)_3B + (CH_3)_3N$$

$$Cu^{2+} + NH_3$$

$$Na^+ + H_2O$$

$$Al^{3+} + F^-$$

$$Hg^{2+} + H_3C-\underset{\underset{H}{O}}{CH}-\underset{\underset{H}{S}}{CH}-\underset{\underset{H}{S}}{CH_2}$$

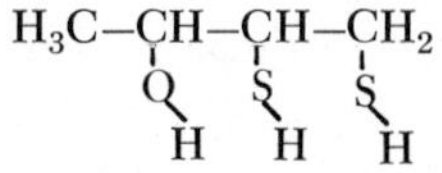

(More than one bond forms in the last reaction. Which atoms on the organic compound are most likely to form strong bonds to the Hg^{2+}? See *Journal of Chemical Education*, vol. 49, p. 28 (January 1972).

6-2 Write chemical equations for proton-transfer reactions between the following reactants. Identify the Brønsted acid and the Brønsted base.

$$H_2SO_4 + NH_3$$

$$HCl + H_2O$$

$$NH_3 + H_2O$$

$$H_2O + H_2O$$

(A discussion of this last reaction is given in Chapter 7, but you should be able to figure out the products for yourself.)

6-3 Write short paragraphs clearly explaining the meaning of the following terms or concepts:

a. screening
b. electronegativity
c. polarizability
d. London force

e. the size of an atom
f. strong (Brønsted) acid
g. weak (Brønsted) acid.

6-4 For Group IV (C, Si, Ge, Sn, Pb), discuss trends in the following parameters, going down the group:

a. electronegativity
b. bonding in the solid crystal
c. tendency to form a Lewis acid

d. toxicity
e. polarizability.

6-5 Discuss the parameters listed in Problem 6-4, going down the group of transition metals Zn, Cd, and Hg.

6-6 Think as far ahead as you can, and write a short statement about the importance of each of the parameters in Problems 6-3 and 6-4 to the toxicity of waste products in air and water.

6-7 The covalent radii of the elements of Groups II, III, and IV are given in the following table:

Period	Group II	Group III	Group IV
2	Be 0.90 Å	B 0.82 Å	C 0.77 Å
3	Mg 1.30 Å	Al 1.18 Å	Si 1.11 Å
4	Ca 1.74 Å	Ga 1.26 Å	Ge 1.22 Å
5	Sr 1.92 Å	In 1.44 Å	Sn 1.41 Å
6	Ba 1.98 Å	Tl 1.48 Å	Pb 1.47 Å

Make a careful plot of these values (covalent radius *versus* period number) on a sheet of graph paper and connect the points with straight lines. Can you think of any reason why the trend in Groups III and IV is not as smooth as in Group II? (*Hint:* The covalent radius does not increase equally rapidly with period number in all groups. Where does the discrepancy between groups begin to be large? What is special about this period?)

6-8 Make a graph of Z_{eff}/r^2 (Table 6-2) versus period number for Groups II, III, and IV and compare it to the graphs you made for Problem 6-7. What statement can you make about the chemical significance of these graphs?

6-9 Predict relative Brønsted acid strengths of the following pairs of acids:

a. H_2Se versus H_2Te
b. H_2S versus HCl
c. H_3AsO_4 versus H_3PO_4 (these are oxygen acids).

6-10 The ionization potentials of F, Cl, Br, and I are, respectively, 17.42, 13.01, 11.84, and 10.45 V.

a. Explain the trend in these numbers.
b. Use Table 6-1 and Equation (6-4) to calculate electronegativities for these elements.
c. Adjust your numbers to the usual scale by dividing them by 2.9 and compare the results to the values given in Table 6-2. For which element does the largest discrepancy arise? Can you find the numerical origin of this discrepancy?

6-11 What should be the relationship between the strength of a Brønsted acid and the basicity of its conjugate base? Explain and illustrate with some examples.

6-12 What acids and bases have you encountered in your daily living, outside of the laboratory? Make a list of them, and designate (if you can) which are strong or weak, hard or soft.

SUGGESTIONS FOR FURTHER READING

Periodicity of chemical properties is the unifying focus for two books recommended for extending the discussions of this chapter: SANDERSON, R. T., *Chemical Periodicity,*

New York: Reinhold (1960); and RICH, R., *Periodic Correlations*, New York: W. A. Benjamin (1965).

Chemists found periodic relationships as they sought to find and explain regularity, order, and system in chemistry. Since there are many periodic relationships and many ways to express the order in systematic ways, it is not surprising that searching for understanding chemical periodicity has led chemists along fascinating paths. If you like intricate detective stories, you will probably like: VAN SPRONSEN, J. W., *The Periodic System of Chemical Elements: A History of the First Hundred Years*, Amsterdam: Elsevier (1969).

Many chemists were surprised, and some were incredulous, when experiments in 1962 revealed that "inert" xenon formed compounds and indeed had a rather rich chemistry. What happened to theories of periodicity, which emphasized the nonreactivity of a whole group of elements, when the nonexistent compounds were made and studied? See HYMAN, H. H., ed., *Noble-Gas Compounds*, Chicago: The University of Chicago Press (1963), a case study of the emergence of a new field of chemistry, written by the chemists who made the discoveries.

A lot of chemistry and a lot of periodic behavior can be encompassed within the realm of acids and bases. An interpretative paperback book, written at the level of this chapter, is: VANDERWERF, C. A., *Acids, Bases, and the Chemistry of the Covalent Bond*, New York: Reinhold (1961). A more advanced book is: LUDER, W. F., and ZUFFANTI, S., *The Electronic Theory of Acids and Bases*, New York: Wiley (1946).

A useful series of compounds to illustrate periodic properties is the hydrides of the elements. We used the hydrides from LiH to HF for just this purpose. You will find lots more information in such books as: SHAW, B. L., *Inorganic Hydrides*, Oxford: Pergamon (1967); HURD, D. T., *An Introduction to the Chemistry of the Hydrides*, New York: Wiley (1952); and STONE, F. G. A., *Hydrogen Compounds of the Group IV Elements*, Englewood Cliffs, N.J.: Prentice-Hall (1962).

7 Chemical Equilibrium

7-1 Abstract

We have examined matter from the atomic and molecular standpoint and have seen how the chemical and physical properties of bulk matter may be explained on the basis of the properties of the atoms and molecules of which it is made.

We are now in need of answers to two questions: (1) How can we express *quantitatively* the relative chemical reactivities we discussed in Chapter 6 as conveniently as we customarily express such bulk properties as density, color, boiling point, and vapor pressure? (2) Do all chemical reactions proceed until at least one of the reactants has completely disappeared?

The answers to both are found in the remarkable phenomenon of chemical **equilibrium**. In this chapter, we shall discuss what a state of chemical equilibrium is, how it comes about (the *why* of it will come in Chapter 9), and review some of the many consequences of chemical equilibrium in solutions. We shall also describe, much more briefly, equilibria that take place in gases and in solids.

7-2 The Phenomenon of Chemical Equilibrium

Your body is steadily metabolizing oxygen and carbohydrates into CO_2 and H_2O to provide the minimal energy you need to stay alive while reading a chemistry book. When you go on to more active pursuits, you burn carbohydrates at a faster rate, producing many more molecules of CO_2 per minute in each active cell. From the discussion in Chapter 6, you might recognize that CO_2—the oxide of a relatively electronegative element—is an **acid anhydride**. The reaction

$$CO_2 + H_2O \rightarrow H_2CO_3 \quad \text{(carbonic acid)}$$

is now taking place in your body.

Carbonic acid, which is $O=C(OH)_2$, is a pretty good proton donor. In fact, it is almost as strong an acid as HF, and a good deal stronger than acetic acid, the compound that gives vinegar its sour taste. It is well established experimentally that an accumulation of as much as 10^{-7} moles of hydrogen ion per liter in the blood causes coma and death; yet the carbonic acid produced by metabolizing a teaspoon of sugar contains a fifth of a mole of loose, ionizable hydrogen. What prevents the mere process of living from loading our bodies with a fatal accumulation of carbonic acid?

As is the case with all potentially harmful metabolic products, CO_2 is excreted from the body, in this case through the lungs. But the process of breathing would not avail if it were not for the fact that the hydration reaction of carbon dioxide and water and the proton-donation reaction of carbonic acid

$$H_2CO_3 + H_2O \rightarrow HCO_3^- + H_3O^+$$

do **not** go completely to the right as they are written here. As the products of these reactions begin to accumulate in the blood, the reverse reaction in each case begins to occur. *Both reactions are reversible.* Just as with the reversible meltings, freezings, evaporations, and condensations examined in Chapter 5, a state of equilibrium is sooner or later achieved when opposite processes are possible. In the present cases, it is *chemical equilibrium.* To indicate that these reactions are reversible and that we are discussing these reactions after equilibrium has been achieved, we shall replace the one-way arrow → with the = sign. Thus, we get

$$CO_2 + H_2O = H_2CO_3 \qquad (7\text{-}1)$$

$$H_2CO_3 + H_2O = HCO_3^- + H_3O^+ \qquad (7\text{-}2)$$

Note the special use of the = sign: $CO_2 + H_2O$ does not equal H_2CO_3; they are in equilibrium.

In this context, equilibrium means that when blood laden with CO_2 and H_2CO_3 is exposed to a gas phase in the lungs, Equations (7-2) and (7-1) together lead to a third equilibrium: **solubility** equilibrium across the gas-liquid phase boundary:

$$CO_2(\text{dissolved}) = CO_2(g) \qquad (7\text{-}3)$$

As your steady breathing pumps out the CO_2-rich gas phase and replaces it with air having a low CO_2 content, the rate at which reaction (7-3) goes to the left decreases, and more dissolved CO_2 enters the gas phase. In other words, the removal of gaseous CO_2 causes the equilibrium (7-3) to "shift" to the right. By a similar argument, an increase in the CO_2 content of the gas phase would cause reaction (7-3) to shift to the left. You can discover this by breathing[1] some CO_2; you will find yourself panting as a CO_2-sensing monitor in your central nervous system detects the increased CO_2 level in your blood. An equilibrium system that responds to change in pressure in such a way as to accommodate the change is following Le Chatelier's principle, just as did the phase equilibria studied in Chapter 5.

Applying Le Chatelier's principle in turn to reactions (7-1) and (7-2), we find that the removal of CO_2 from the blood causes reaction (7-1) to shift to the left, dehydrating carbonic acid, and that the consequent disappearance of carbonic acid from the bloodstream forces reaction (7-2) to shift to the left also (fewer molecules of H_2CO_3 are present, so the rightward rate of (7-2) is temporarily slower than the leftward rate). The leftward shift of reaction (7-2) removes hydrogen ions from the bloodstream, and we are *not* poisoned.

Now that we have settled that, let us spend some time looking at some other aspects and consequences of chemical equilibrium. For one thing, though it is clear that there should be no long-term buildup of hydrogen ions in a healthy bloodstream, we see no reason so far why the particular steady level maintained by reactions (7-1) to (7-3) should be below 10^{-7} M. That it has this low value reflects another equilibrium phenomenon—*buffering*—which we shall examine.

---

[1] This is not a recommended long-term activity, but a few puffs of CO_2 from a cylinder of the gas will demonstrate this behavior of the human respiratory center.

7-3 The Consequences of Equilibrium

It is an experimental fact that when carbon dioxide is dissolved in water, only one molecule in about 300 is hydrated to carbonic acid when equilibrium is reached. Although individual carbon atoms will find themselves now in a molecule of CO_2, and now in a molecule of H_2CO_3, as the rightward and leftward directions of reaction (7-1) proceed at equal rates, there is no further net change in this fraction of hydrated CO_2.

Expressing this fraction mathematically, and supplying the exact numerical quantity,

$$\frac{[H_2CO_3]}{[CO_2]} = 0.0034 \ (25°C) \tag{7-4}$$

where the square brackets signify the number of moles per liter of the indicated substance, and $[CO_2]$ means in particular the number of moles of *dissolved* CO_2 per liter of solution, leaving out of account any gaseous CO_2 that might be present.

In fact, it is a striking and useful fact that the ratio of carbonic acid to dissolved carbon dioxide is the same for any total amount of CO_2 in solution; the value 0.0034 is a *constant* for CO_2 dissolved in aqueous solutions. On the other hand, when reaction (7-1) is carried out in other solvents (for example, in a mixture of water and acetone), the amount of carbonic acid formed from a given content of dissolved CO_2 depends on the amount of water present. More precisely, the numerical value of the equilibrium ratio $[H_2CO_3] / [CO_2]$ is directly proportional to the *fraction* of water molecules in the equilibrium mixture.

The more H_2O present in a solvent, the more CO_2 is hydrated.

The fraction of all molecules in a mixture that is represented by a given kind of molecule is known as the **mole fraction** of that species. The mole fraction is usually symbolized by X, with a subscript labeling the species. For example, in a mixture containing 3 moles of H_2O and one mole of acetone (C_3H_6O),

$$X_{H_2O} = \frac{\# \text{ moles } H_2O}{\# \text{ moles } H_2O + \# \text{ moles } C_3H_6O}$$

$$= \frac{3}{3+1} = 0.75$$

and

$$X_{C_3H_6O} = \frac{\# \text{ moles } C_3H_6O}{\# \text{ moles } H_2O + \# \text{ moles } C_3H_6O}$$

$$= \frac{1}{3+1} = 0.25$$

This dependence of the equilibrium ratio of $[H_2CO_3]$ to $[CO_2]$ on mole fraction of water makes sense. Reflect that, if only a fraction of the molecules that CO_2 encounters is water molecules, then H_2CO_3 will result from collisions of CO_2 with those molecules only a proportionate fraction of the time.

We can take account of this effect if we include the concentration of water molecules in the solution (by mole fraction) in the equation:

$$\frac{[H_2CO_3]}{[CO_2]} = X_{H_2O} \cdot 0.0034 \quad (25°C) \tag{7-5}$$

It is traditional in such equations to keep all the concentration terms on one side, leaving the numerical value on the other. Thus,

$$\frac{[H_2CO_3]}{[CO_2] \cdot X_{H_2O}} = 0.0034 \quad (25°C) \tag{7-6}$$

Note that both Equations (7-5) and (7-6) contain Equation (7-4) as a special case; in pure water as a solvent, with only minor quantities of CO_2 and H_2CO_3 present, X_{H_2O} is practically equal to 1.

A close look at the left side of Equation (7-6) shows that it contains all of the substances present in the chemical equilibrium that it describes:

$$CO_2 + H_2O = H_2CO_3 \tag{7-1}$$

Equation (7-6) as a whole conforms to a convention that places the concentrations of the reactants in the *denominator,* and the concentrations of the product in the *numerator.*[2] This is not a unique result. *For every reaction that goes to equilibrium, the ratio of product concentrations to reactant concentrations [each set of concentrations multiplied as in Equation (7-6)] is a numerical constant.* This numerical constant is called the **equilibrium constant** of the reaction.

Consider, for example, Equation (7-2), which describes an equilibrium reaction. If we multiply together the concentrations of the products and divide by the product of the concentrations of the reactants, we get another numerical constant:

$$\frac{[HCO_3^-][H_3O^+]}{[H_2CO_3] \cdot X_{H_2O}} = 1.32 \times 10^{-4} \quad (25°C) \tag{7-7}$$

This constant is valid for any solution of carbonic acid that has reached equilibrium, regardless of whether one starts with nothing but carbonic acid and water, or with HCO_3^- ions and hydronium ions, or somewhere in between. It is valid regardless of the total concentration of all of these species. **It is important to remember that equations like (7-6) and (7-7) are valid only for chemical systems that have reached equilibrium.** Obviously, if one were to begin with only reactants present, the numerical value of the ratio of products to reactants would initially be zero and would increase to the equilibrium value only as the reaction approached equilibrium, just as we noted in Chapter 5 that the pressure of vapor over a liquid might begin at zero and increase

---

[2] Although it is merely chemical convention that units of moles per liter (molarity) are used for minor solutes (CO_2 and H_2CO_3) and mole fraction for the solvent, we could just as well have used either unit for any component of the solution; the effect would have been to change only the numerical value of the constant, not the principle involved.

until it became constant at the equilibrium value. Different chemical systems approach equilibrium at different rates; for example, reaction (7-1) takes several seconds, but reaction (7-2) takes only a tiny fraction of a second to reach equilibrium. We shall look at some of the reasons for differing chemical rates in Chapter 11.

Before we go on to apply the notion of chemical equilibrium to a variety of chemical reactions, there are a few more things to be said about equilibrium constants in general. First, a minor point; it is chemical convention to omit the solvent term from equilibrium constants of reactions involving the solvent, on the grounds that solvent molecules so dominate the population in a solution (and especially a dilute solution) that $X_{solvent}$ is nearly constant at a value near 1, and so makes no numerical difference.[3] Using this convention, the equilibrium constant for Equation (7-2) would be written

$$K = \frac{[HCO_3^-][H_3O^+]}{[H_2CO_3]} = 1.32 \times 10^{-4} \quad (25°C) \tag{7-8}$$

The symbol K is generally used for equilibrium constants.

Second, some reactions involve more than one particle of some species, as in the gas-phase reaction:

$$3H_2(g) + N_2(g) = 2NH_3(g) \tag{7-9}$$

If we follow the same procedure for writing the equilibrium constant as we have before, we shall write

$$K = \frac{[NH_3][NH_3]}{[H_2][H_2][H_2][N_2]} \tag{7-10}$$

But after all, we have mathematical conventions for expressing a thing multiplied by itself any number of times:

$$K = \frac{[NH_3]^2}{[H_2]^3[N_2]} \tag{7-11}$$

Thus, we see that each concentration term appears in the equilibrium constant raised to a power equal to its stoichiometric coefficient in the chemical equation to which the K applies.[4]

Third, we may note that the numerical value of the equilibrium constant for a reaction gives us a convenient quantitative measure of the "reactivity" which we considered at length, but only qualitatively, in Chapter 6.

For example, we noted a moment ago that carbonic acid is a stronger acid than acetic acid, and not quite as strong as HF. Now we examine proton-donation reactions for each of these acids with water as a base:

---

[3] Perhaps this convention is unwise, since it encourages a chemist to forget that the mole fraction of solvent may be changing unexpectedly as a reaction takes place. Chemists have been known to be baffled by the resulting apparent inconstancy of the equilibrium constant.

[4] As we have seen (Chapter 5), the molar concentration of a substance in the gas phase is proportional to its partial pressure; that implies that equilibrium constants for gas reactions may be written in terms of partial pressures of the products and reactants instead of molarities.

$$H_2CO_3 + H_2O = H_3O^+ + HCO_3^- \tag{7-2}$$

$$\underset{\text{acetic acid}}{H_3CCOOH} + H_2O = H_3O^+ + H_3CCOO^- \tag{7-12}$$

$$HF + H_2O = H_3O^+ + F^- \tag{7-13}$$

The equilibrium constants are, respectively,

$$K_{H_2CO_3} = \frac{[HCO_3^-][H_3O^+]}{[H_2CO_3]} = 1.32 \times 10^{-4} \quad (25°C) \tag{7-8}$$

$$K_{\text{acetic}} = \frac{[H_3CCOO^-][H_3O^+]}{[H_3CCOOH]} = 1.75 \times 10^{-5} \quad (25°C) \tag{7-14}$$

$$K_{HF} = \frac{[F^-][H_3O^+]}{[HF]} = 6.8 \times 10^{-4} \quad (25°C) \tag{7-15}$$

We find that the numerical order of K's reflects the order of acid strengths. [Again, note that the concentration term for the reactant *solvent* molecules is missing from these equilibrium constants, in accord with the convention discussed above. Equilibrium constants for the reaction of an acid with the solvent, like Equations (7-8), (7-14), and (7-15), are called acid "dissociation" constants, and generally given the symbol K_a.]

The stronger an acid is, the more hydronium ions it will produce at equilibrium, and the greater will be the fraction of the substance that has lost its proton. The greater concentration of products compared to reactants leads to a larger numerical value for the K_a. Similar comparisons of reactivity based on the values of equilibrium constants are possible in other kinds of reactions; we shall return to this subject later in the chapter.

A large K implies a large tendency to react.

Finally, in the course of these general remarks about equilibrium constants, we should consider something about the effect of temperature on them. That there **is** an effect of temperature is implied by the fact that we have always mentioned a temperature along with numerical K value; in fact, the latter is meaningless without the former. A full numerical treatment of this subject is complex enough to defer to a later course in chemistry, but we have a very reliable qualitative guide in Le Chatelier's principle, which tells us how equilibrium systems respond to changes in the conditions under which equilibrium is achieved. In the case of phase changes (Chapter 5), we found that equilibria shift in a direction such as to absorb energy at higher temperatures, and to release it at lower temperatures (for example, the ice-water behavior). The same is true for chemical reactions. In the case of phase equilibria, the energy released and absorbed comes from the forces holding together particles in the condensed phases; in chemical reactions, chemical bonds are formed and broken, and the net change in bonding energy shows up as an energy of reaction (usually in the form of heat). Thus, a *rise* in temperature shifts equilibrium and changes the value of its equilibrium constant in such a way that energy is *absorbed* by the reaction; a lowering of temperature has the opposite effect.

For example, we have remarked that water can act both as a Brønsted acid and a Brønsted base; in fact, one water molecule can donate a proton to another, in the **autoprotolysis** reaction:

$$2H_2O = H_3O^+ + OH^- \tag{7-16}$$

Now, since water is not really a very strong acid, nor a very strong base, a good deal of energy is absorbed when reaction (7-16) occurs, and this is reflected in the very small value of the equilibrium constant:

$$K_w = [H_3O^+][OH^-] = 1.01 \times 10^{-14} \quad (25°C) \tag{7-17}[5]$$

If energy is absorbed by reaction (7-16), then it should shift to the right at high temperatures and to the left at lower ones, with a consequent variation of K_w with temperature. This expectation is fulfilled: K_w is 1.15×10^{-15} at 0°C and 2.95×10^{-14} at 40°C.

Having examined some of the consequences of chemical equilibrium, we shall next investigate some further applications to various types of chemical systems.

7-4 Gas-Phase Reactions and the Effect of Pressure

Let us return to reaction (7-9). As we have already observed (Chapter 1), this reaction plays a key role in the present scheme of society, and it is important that the conversion of nitrogen to ammonia in a reactor be as complete as possible. In fact, though, the reaction of Equation (7-9) reaches equilibrium when only part of the available nitrogen has been converted to ammonia at room temperature. The reaction is even less complete at the high temperatures that are required for a reasonably rapid *rate* of conversion of N_2 to NH_3.

But if we examine Equation (7-9), we note that a total of 4 moles of gases on the left become 2 moles of gases on the right. By the gas laws, there should be a consequent decrease in volume in this collection of atoms when reaction (7-9) occurs. Using Le Chatelier's principle as a guide, we should suspect that we can "squeeze" more NH_3 out of the reaction by increasing the pressure. (Remember that equilibrium systems seek the state of minimum volume at high pressures.)

To show that this is so for Equation (7-9), we may recall the relationship between the partial pressure of a gas [in terms of which the equilibrium constant for Equation (7-9) may be written] and its mole fraction in the mixture. We want to maximize the mole fraction of NH_3 that is present at equilibrium. From the law of partial pressures,

$$p_{NH_3} = P_{total} \cdot X_{NH_3} \tag{7-18}$$

where, as before, we use a lowercase p for partial pressure and capital P for total pressure. Now, the equilibrium constant for Equation (7-9), in terms of partial pressure, is

---

[5] The subscript w on the K is a chemical convention that labels K_w the equilibrium constant for the reaction of water with itself.

$$K_p = \frac{p^2_{NH_3}}{p^3_{H_2} p_{N_2}} \qquad (7\text{-}19)$$

Substituting mole fractions for partial pressures for each gas according to Equation (7-18), we find

$$K_p = \frac{X^2_{NH_3} P^2_{total}}{(X^3_{H_2} P^3_{total})(X_{N_2} P_{total})}$$

or

$$= \frac{X^2_{NH_3}}{X^3_{H_2} X_{N_2}} \cdot \frac{1}{(P_{total})^2} \qquad (7\text{-}20)$$

The first term on the right is the equilibrium constant written in terms of mole fractions; since K_p is a constant, K_X must vary with the pressure; in fact,

$$K_X = K_p \cdot P^2_{total} \qquad (7\text{-}21)$$

Thus our Le Chatelier-aided intuition is correct; we may increase the yield of NH_3 from reaction (7-9) (we may increase the mole fraction of NH_3 present at equilibrium) by increasing the total pressure on the system, because by doing so we increase the value of the equilibrium constant K_X, expressed in mole fractions of product and reactants.

Notice that P_{total} would have appeared as a factor an equal number of times in numerator and denominator in Equation (7-20) if there had been an equal number of moles of gas on both sides of Equation (7-9). The consequence would have been that K_X and K_p would have been equal, and equally independent of pressure; squeezing makes no difference if both reactants and products take up the same amount of space. Only if there is a change in the number of moles when a gaseous reaction occurs does pressure affect the position of equilibrium. However, there are many reactions in which the volume of the system *does* decrease on reaction, and it is a common technique in the chemical industry to promote condensations, polymerizations, and similar volume-shrinking reactions with high-pressure reactors.

7-5 Some Solution Equilibria

Nearly three-fourths of the earth's surface is covered with a solution of ionic substances in water; all living things are composed predominantly of similar aqueous solutions; and most of the chemistry carried out in most of the chemistry laboratories of the world takes place between species dissolved in water. We may begin looking at aqueous solution chemistry by considering some of its characteristic reactions from the equilibrium standpoint.

REACTIONS INVOLVING LEWIS ACID-BASE COMPLEXES. The most important and various species to be found dissolved in water are ionic and polar species. But such substances, involving complete or partial separation of electrical charge into positive and negative ions or positive and negative ends of polar molecules, provide a rich field for the association of electron-poor

These are reactions that can be studied in beakers and flasks.

and electron-rich atoms, that is, for Lewis acid-base chemistry (Chapter 6). The most common electron-poor species to be found in water solutions are positive metal ions. Electron-rich species include both negative ions (such as chloride, hydroxide, and so forth) and neutral Lewis bases like NH_3 (with its nonbonding pair of nitrogen electrons) and, preeminently, water itself, with its two nonbonding pairs of oxygen electrons. (Indeed, water molecules are so abundant in water solutions that any metal ion dissolved in water is always associated, more or less tightly, with several water molecules, unless other Lewis bases completely surround it. Such an ion is said to be *aquated*.)

Out of the almost infinite number of reactions we could use as examples for a short mathematical treatment of this situation, we shall choose the reaction of aquated silver ions with thiosulfate ions (Figure 7-1):

$$Ag^+(aq) + S_2O_3^{2-} = AgS_2O_3^- \tag{7-22}$$

$$AgS_2O_3^- + S_2O_3^{2-} = Ag(S_2O_3)_2^{3-} \tag{7-23}$$

FIGURE 7-1 The structure of the Ag^+-$S_2O_3^{2-}$ complex ($AgS_2O_3^-$). When this species is present in aqueous solution, several water molecules are bonded to the silver ion. These solvent molecules are not shown.

The silver ion, with its low charge and abundance of d electrons, qualifies as a soft Lewis acid (Chapter 6), and $S_2O_3^{2-}$, which bonds to metal ions through a sulfur atom, is a soft base, so the association between Ag^+ and $S_2O_3^{2-}$ is quite a firm one:

$$K_1 = \frac{[AgS_2O_3^-]}{[Ag^+][S_2O_3^{2-}]} = 8 \times 10^8 \quad (25°C) \tag{7-24}$$

In fact, the negative charge on $AgS_2O_3^-$ is not enough to discourage a second thiosulfate ion from associating with the silver ion:

$$K_2 = \frac{[Ag(S_2O_3)_2^{3-}]}{[AgS_2O_3^-][S_2O_3^{2-}]} = 4 \times 10^4 \quad (25°C) \tag{7-25}$$

These reactions are used to dissolve and wash unexposed silver ions from exposed and developed photographic film before the developed negative is brought into the light; thiosulfate (formerly "hyposulfate") ions are the active ingredient of photographic fixer, or "hypo."

A detailed algebraic discussion begins.

We might ask, for purposes of economy or curiosity: What quantity of $S_2O_3^{2-}$ must be present in a solution so that practically all of the silver ions present will be tied up in the one-to-one or the two-to-one $S_2O_3^{2-}$—Ag^+ complex? Questions like that are the ones equilibrium chemists like to answer, and the process of answering them rests on the use of equilibrium constants K_1 and K_2.

We shall approach the problem by considering what *fraction* of a given quantity of dissolved silver is in the form of uncomplexed Ag^+, what fraction is in the form of $AgS_2O_3^-$, and what fraction is in the form of $Ag(S_2O_3)_2^{3-}$. We shall find that these fractions depend upon the concentration of $S_2O_3^{2-}$ in the solution in a very regular way.

In mathematical language, we may define the fraction present Ag^+ as

$$\alpha_{Ag^+} = \frac{[Ag^+]}{[Ag^+] + [AgS_2O_3^-] + [Ag(S_2O_3)_2^{3-}]} \tag{7-26}$$

Dividing out the common term in the numerator and denominator gives us

$$\alpha_{Ag^+} = \frac{1}{1 + [AgS_2O_3^-]/[Ag^+] + [Ag(S_2O_3)_2^{3-}]/[Ag^+]} \tag{7-27}$$

Now we are faced with the question of how to evaluate the ratios of concentrations in the denominator, but these follow fairly directly from Equations (7-24) and (7-25), the definitions of K_1 and K_2. Thus,

$$\frac{[AgS_2O_3^-]}{[Ag^+]} = K_1[S_2O_3^{2-}] \tag{7-28}$$

This gives us the first fraction. A similar operation on K_2 leads to

$$\frac{[Ag(S_2O_3)_2^{3-}]}{[AgS_2O_3^-]} = K_2[S_2O_3^{2-}] \tag{7-29}$$

But from K_1, we know that

$$[AgS_2O_3^-] = K_1[S_2O_3^{2-}][Ag^+] \tag{7-30}$$

So we may substitute this bundle in Equation (7-29), where $[AgS_2O_3^-]$ appears, to arrive at the expression:

$$\frac{[Ag(S_2O_3)_2^{3-}]}{[Ag^+]K_1[S_2O_3^{2-}]} = K_2[S_2O_3^{2-}] \tag{7-31}$$

or

$$\frac{[Ag(S_2O_3)_2^{3-}]}{[Ag^+]} = K_1K_2[S_2O_3^{2-}]^2 \tag{7-32}$$

This establishes the identity of the third term in the denominator of Equation (7-27), which now becomes

$$\alpha_{Ag^+} = \frac{1}{1 + K_1[S_2O_3^{2-}] + K_1K_2[S_2O_3^{2-}]^2} \tag{7-33}$$

This equation tells us what fraction of all the silver present in a solution containing $S_2O_3^{2-}$ is in the form of the "free" silver ion, ions complexed only to the solvent water molecules. The fractions present as $AgS_2O_3^-$ and $Ag(S_2O_3)_2^{3-}$ follow more quickly:

$$\alpha_{AgS_2O_3^-} = \frac{[AgS_2O_3^-]}{[Ag^+] + [AgS_2O_3^-] + [Ag(S_2O_3)_2^{3-}]} \quad (7\text{-}34)$$

Using Equation (7-30) in the numerator of this fraction, we find

$$\alpha_{AgS_2O_3^-} = \frac{[Ag^+]K_1[S_2O_3^{2-}]}{[Ag^+] + [AgS_2O_3^-] + [Ag(S_2O_3)_2^{3-}]} \quad (7\text{-}35)$$

But a hard look at Equation (7-35) reveals it to be α_{Ag^+} multiplied by K_1 and $[S_2O_3^{2-}]$:

$$\alpha_{AgS_2O_3^-} = \alpha_{Ag^+} K_1 [S_2O_3^{2-}] \quad 7\text{-}36)$$

The mathematically inclined reader is invited to show, using similar logic, and others are invited to believe, that

$$\alpha_{Ag(S_2O_3)_2^{3-}} = \alpha_{Ag^+} K_1 K_2 [S_2O_3^{2-}]^2 \quad (7\text{-}37)$$

Using Equations (7-33), (7-36), and (7-37), we may calculate what fraction of the total silver dissolved in a solution containing $S_2O_3^{2-}$ is present in each of the three possible forms, knowing only the equilibrium concentration of $S_2O_3^{2-}$ and the values of the two K's, K_1 and K_2.

One way of using this information is to calculate these fractions for a variety of concentrations of $S_2O_3^{2-}$ and plot the results (once and for all) on a graph. The result (Figure 7-2) is rather pretty. Because the equilibrium concentration of $S_2O_3^{2-}$ in the region of interest on the graph ranges from about $0.001\ M$ down to $10^{-10}\ M$ or so, we can display these fractions best if we use an exponential (logarithmic) scale for the horizontal axis rather than a linear one.

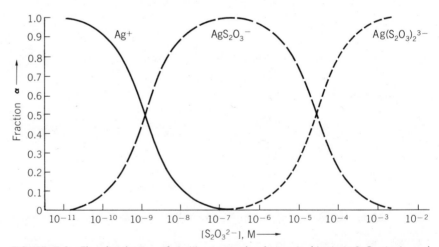

FIGURE 7-2 The distribution of Ag(I) among the forms $Ag^+(aq)$, $AgS_2O_3^-(aq)$, and $Ag(S_2O_3)_2^{3-}(aq)$, as a function of the equilibrium concentration $[S_2O_3^{2-}]$. The use of an exponential (logarithmic) scale for $[S_2O_3^{2-}]$ allows a wide range of numerical values to be displayed clearly on a single graph.

From Figure 7-2, it is clear that uncomplexed Ag^+ ion predominates only at thiosulfate concentrations below $10^{-9}\ M$ and plays no major role in the solution above $[S_2O_3^{2-}] = 10^{-7}\,M$. Between 10^{-9} and $10^{-5}\,M$ thiosulfate, the one-to-one complex $AgS_2O_3^-$ is the predominant form of silver in solution; and above $10^{-4}\,M$, it is $Ag(S_2O_3)_2^{3-}$ that predominates. Since the concentration of thiosulfate ions in fresh photographic fixer is about $1.5\,M$, it is clear from Figure 7-2 that this level of thiosulfate concentration (it would be off the graph to the right) makes $Ag(S_2O_3)_2^{3-}$ the predominant form and Ag^+ practically nonexistent, in photographic fixing baths; the effect of this on the photographic emulsion will be considered later in this chapter.

PROTON-TRANSFER REACTIONS. In looking at a Lewis acid-base reaction from the standpoint of its equilibrium properties, we have found it useful to select a single "master variable" (in our example $[S_2O_3^{2-}]$) and express other variable concentrations as functions of this variable and of the various equilibrium constants. The same approach is widely used in Brønsted acid-base (proton-transfer) equilibria; in this case, it will make sense to use the concentration of solvated protons $[H_3O^+]$ as the master variable.

In water, as in all solvents that undergo **autoprotolysis** (page 208), the concentration of solvated protons is always subject to the autoprotolysis equilibrium

$$2H_2O = H_3O^+ + OH^- \tag{7-16}$$

whose constant is

$$K_w = [H_3O^+][OH^-] = 1.0 \times 10^{-14} \quad (25°C) \tag{7-17}$$

This means that there will be a reciprocal relationship between the concentrations of hydronium ions, H_3O^+, and hydroxide ions, OH^-, in any water solution. The possible values for $[H_3O^+]$ will range from something over 10 M (in a concentrated solution of a strong acid) to very low values in solutions of ionic hydroxides like NaOH. The value of $[H_3O^+]$ can never reach zero, because reaction (7-16) is always going on in any water solution. Just how low $[H_3O^+]$ can go is established by Equation (7-17).

For example, in 10 M solution of NaOH, the concentration of OH^- ions is practically 10 M; so weak is the Lewis acid-base reaction between OH^- and Na^+ that there is practically no association between them in solution. Then, from Equation (7-17),

$$[H_3O^+] = \frac{K_w}{[OH^-]} = \frac{1 \times 10^{-14}}{10^1} = 10^{-15}\,M \tag{7-38}[6]$$

Now, for any variable whose values of interest may range from over 10 M to less than 0.000000000000001 M, it is clearly most convenient to use an exponential scale, as we did in Figure 7-2, to express those values.

---

[6] The numerical value of K_w is somewhat subject to the composition of the solution; if there are many ions present, these exert electrical forces on each other that are practically absent in pure water, and there is an effect on the equilibrium position of Equation (7-16). This effect, which is important when precise calculations are to be carried out, does not alter the principles we are discussing, and need not bother us here.

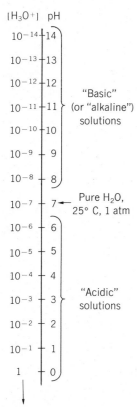

FIGURE 7-3 The pH scale. This figure encompasses the range of pH values ordinarily accessible in water solutions at 25°C. Note that the pH value is just −1 times the power of 10 of the quantity [H_3O^+].

Through most of the range of interest, those exponents are negative, and chemists use a scale which saves ink by changing the sign of the exponent. *The negative of the exponent of the hydronium ion concentration in a solution is known as the pH.* For example, the solution in Equation (7-38) would have a pH of 15, and the range of pH that is observable in water runs from a bit less than −1 to a bit over 15. At the center of this range would be a solution containing equal concentrations of H_3O^+ and OH^-. If these two concentrations *are* equal, as they are in pure water, for example, then by Equation (7-17) each of them must be equal to the square root of K_w:

$$[H_3O^+] = [OH^-] = \sqrt{1.0 \times 10^{-14}}$$

$$= 1.0 \times 10^{-7} \quad (25°C, \text{ pure water}) \quad (7\text{-}39)$$

The pH of such a solution is 7.0. Solutions with this pH are loosely spoken of as "neutral," while those with higher [H_3O^+] (lower pH) are called "acid," and those with higher pH "alkaline." These rough divisions of the pH scale are shown in Figure 7-3.

The values of pH, of course, need not be integers; for example, the optimum level of [H_3O^+] in the blood is 4×10^{-8} M. This value, between 10^{-8} and 10^{-7}, will translate to a pH between 8 and 7; but where? Well, exponents are also called logarithms (Appendix II), and with the aid of a table of logarithms we can make the translation:

$$[H_3O^+] = 4 \times 10^{-8}$$

$$\text{pH} = -\log [H_3O^+] = -\log(4 \times 10^{-8})$$

$$= -\log 4 + (-\log 10^{-8})$$

$$= -0.60 + 8.00$$

$$= 7.40$$

Those readers who have had experience with logarithms should not find this process overly confusing; those who have not may follow the gist of it, and turn to Appendix II (or to your instructor) for further enlightenment.

Now that we have an exponential master variable scale, let us use it to explore some relatively simple and important proton-transfer equilibria. Again, because one picture is worth not only a thousand words, but also several equations, we shall develop a distribution diagram after the style of Figure 7-2 for a Brønsted acid-base system.

A prominent class of Brønsted acid-base conjugate pairs[7] is the indicating dyestuffs. **Indicators** are conjugate pairs whose members are intensely, and differently, colored. For example, consider the indicator *methyl orange* (Figure 7-4). The nitrogen atoms in this compound, with their nonbonding electron pairs, may bond to hydrogen ions. The addition of a proton to methyl orange, as indicated in Figure 7-4, causes a significant change in its electronic structure, and consequently in the electronic energy levels available to the mole-

---

[7] The nomenclature of Brønsted acids and bases, including "conjugate pairs," was introduced on page 191, chapter 6.

cule. We see the result as a change of color (from yellow to red) when the proton is added.

For the reaction

$$HMO + H_2O = MO^- + H_3O^+ \qquad (7\text{-}40)$$

we can write the acid dissociation constant

$$K_a = \frac{[MO^-][H_3O^+]}{[HMO]} \qquad (7\text{-}41)$$

where HMO and MO$^-$, respectively, are the protonated and deprotonated forms of methyl orange.

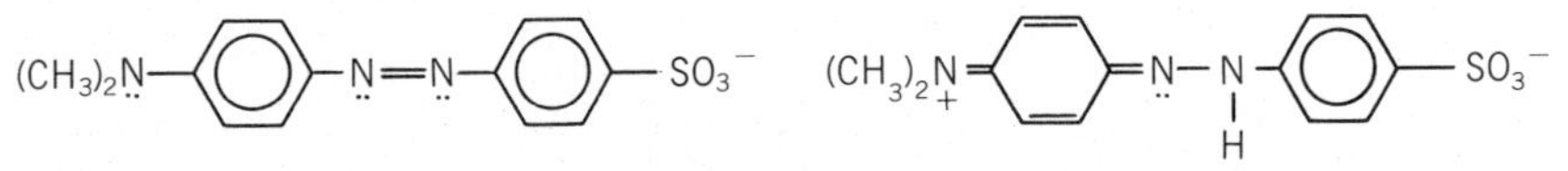

FIGURE 7-4 The molecular structure of methyl orange. (a) The yellow conjugate base form MO$^-$. (b) The red conjugate acid form HMO. A species that, like HMO, has an internal separation of charge is called a **zwitterion** (German: hybrid ion).

The quantity K_a describes the relationship between the hydrogen ion concentration and those of the protonated and deprotonated forms of methyl orange. In that way, it is quite analogous to the equilibrium constants governing the concentrations of $S_2O_3^{2-}$ and the complexed and uncomplexed forms of Ag$^+$ (pages 210–213). We may be confident that algebra analogous to Equations (7-26) through (7-37) will allow us to calculate the fractions of methyl orange present as red HMO and as yellow MO$^-$, as functions of the master variable [H$_3$O$^+$], and that a graph of the distribution of these forms as a function of pH will look similar to Figure 7-2. In fact, it does (Figure 7-5).

Let us take a closer look at the mathematics behind Figure 7-5 for the sake of its future usefulness. We know that in any solution containing methyl orange at a total concentration C_{mo}, part of the methyl orange will be protonated and the rest unprotonated (Figure 7-4):

Organized common sense

$$C_{mo} = [HMO] + [MO^-] \qquad (7\text{-}42)$$

We seek a relationship between the master variable pH, or [H$_3$O$^+$], and the fractions

$$\alpha_{MO^-} = \frac{[MO^-]}{C_{mo}} \qquad (7\text{-}43)$$

and

$$\alpha_{HMO} = \frac{[HMO]}{C_{mo}} \qquad (7\text{-}44)$$

that appear in Figure 7-5.

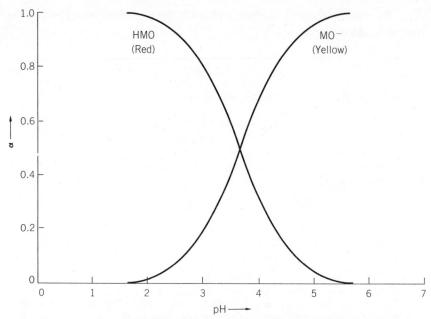

FIGURE 7-5 The distribution of methyl orange among the forms HMO and MO$^-$ as a function of the pH of the solution. Note, as in Figure 7-2, that the use of a logarithmic variable (the pH) allows the display of a wide range of solution compositions on a single graph.

We can obtain α_{HMO} by solving Equation (7-41) for [MO$^-$] and substituting the result in Equation (7-42):

$$[\text{MO}^-] = \frac{K_a[\text{HMO}]}{[\text{H}_3\text{O}^+]}$$

$$C_{\text{mo}} = \frac{K_a[\text{HMO}]}{[\text{H}_3\text{O}^+]} + [\text{HMO}]$$

$$= [\text{HMO}]\left(\frac{K_a}{[\text{H}_3\text{O}^+]} + 1\right) \qquad (7\text{-}45)$$

The fraction α_{HMO}, lurking in Equation (7-45), is revealed by the rearrangement

$$\frac{[\text{HMO}]}{C_{\text{mo}}} = \alpha_{\text{HMO}} = \frac{1}{(K_a/[\text{H}_3\text{O}^+]) + 1}$$

which simplifies somewhat to

$$\alpha_{\text{HMO}} = \frac{[\text{H}_3\text{O}^+]}{K_a + [\text{H}_3\text{O}^+]} \qquad (7\text{-}46)$$

Following the same kind of argument, beginning by solving the K_a equation for [HMO] instead of for [MO$^-$], would bring us to an expression for α_{MO^-}:

$$\alpha_{\text{MO}^-} = \frac{K_a}{K_a + [\text{H}_3\text{O}^+]} \qquad (7\text{-}47)$$

EXERCISE: Since α_{HMO} and α_{MO^-} must sum up to 1 (why?), the same must be true of these algebraic expressions. Is it?

Further exercise: If, indeed, α_{HMO} and α_{MO^-} must add up to 1, what value must each α have at their crossing point in Figure 7-5? Then what must be the relationship between $[H_3O^+]$ and K_a at this point? Finally, verify that this relationship is obeyed in Figure 7-5; the K_a of HMO is about 2×10^{-4}.

The indicating function of substances like methyl orange is readily apparent from Figure 7-5. A drop of this indicator placed in a solution gives an immediate visual signal that tells whether the pH is above about 4.5 (yellow color), is in the range of 3 to 4.5 (shades of orange), or is below 3 (red). Some other indicating dyes are listed in Table 7-1, with the pH range in which they change color. Certain mixtures of these indicators can be prepared so that a continuous "spectrum" of color is displayed over a wide range of pH; such mixtures find extensive use for the rough measurement of pH in chemical laboratories and environmental testing kits.

TABLE 7-1.
Transition Ranges of some Common Acid-Base Indicators

Indicator	Visual Transition Range
Cresol red	pH 0.2 (red) to pH 1.8 (yellow)
	pH 7.2 (yellow) to pH 8.8 (red)
Methyl violet	pH 0.2 (colorless) to pH 1.8 (blue)
	pH 2.0 (blue) to pH 3.2 (violet)
Metacresol purple	pH 1.2 (red) to pH 2.8 (yellow)
	pH 7.6 (yellow) to pH 9.2 (purple)
Thymol blue	pH 1.2 (red) to pH 2.8 (yellow)
	pH 8.0 (yellow) to pH 9.6 (blue)
Bromphenol blue	pH 3.0 (yellow) to pH 4.6 (blue)
Bromcresol green	pH 3.8 (yellow) to pH 5.4 (blue)
Methyl red	pH 4.4 (red) to pH 6.0 (yellow)
Methyl purple	pH 4.8 (purple) to pH 5.4 (green)
Bromcresol purple	pH 5.2 (yellow) to pH 6.8 (purple)
Chlorphenol red	pH 5.2 (yellow) to pH 6.8 (red)
Bromthymol blue	pH 6.0 (yellow) to pH 7.6 (blue)
Phenol red	pH 6.8 (yellow) to pH 8.4 (red)
o-Cresolphthalein	pH 8.2 (colorless) to pH 9.8 (red)
Phenolphthalein	pH 8.6 (colorless) to pH 10.2 (red)
Alizarin yellow GG	pH 10.1 (yellow) to pH 12.1 (lilac)
Sulfo orange	pH 11.0 (yellow) to pH 12.6 (orange)

Reprinted from the *Fisher Chemical Index 64-C* with permission of the copyright holder, Fisher Scientific Company.

THE EXTENT OF DISSOCIATION OF ACIDS AND BASES IN WATER.

Brønsted acids and bases dissolved in water will donate protons to, or accept protons from, water molecules:

$$HA^n + H_2O = H_3O^+ + A^{n-1} \tag{7-48}$$

$$B^m + H_2O = OH^- + HB^{m+1} \tag{7-49}$$

The characteristic equilibrium constants for these reactions are, again,

$$K_a = \frac{[H_3O^+][A^{n-1}]}{[HA^n]} \quad (7\text{-}50)$$

and

$$K_b = \frac{[OH^-][HB^{m+1}]}{[B^m]} \quad (7\text{-}51)$$

Some definitions

As we have seen (page 207), the numerical values of K_a or K_b vary considerably from one acid or base to another. A large value of K_a or K_b indicates that reaction (7-48) or (7-49) lies relatively far to the right when the acid or base is dissolved in water. A **strong acid** produces a number of moles of H_3O^+ essentially equal to the number of moles of strong acid introduced into the solution; a **strong base** similarly produces OH^- equivalent to the quantity of base introduced into the solution.

If the value of K_a for an acid or K_b for a base is not large, then reaction (7-48) or (7-49) will reach equilibrium, while there is still some unreacted acid or base present. In this case, the acid is known as a **weak acid**, or the base as a **weak base**. Some common weak acids, along with the negative logarithm of their K_a's (pK_a's) are listed in Table 7-2.

TABLE 7-2
Dissociation Constants of Brønsted Acids at 25°C

Acid	pK_1*	pK_2	pK_3	pK_4
Acetic	4.7560			
α-alanine (+1)§	2.340	9.870†		
β-alanine (+1)	3.55	10.23		
Ammonium (+1)	9.245			
Anilinium (+1)	4.62			
Arsenic	2.19	6.94	11.50	
Benzoic	4.01			
*iso*butyric	4.86			
n-butyric	4.8196			
Carbonic (H_2CO_3)	−3.88	10.33		
Carbonic ($CO_2 + H_2CO_3$)	6.35	10.33		
Chloric	ca. −2.7			
Chloroacetic	2.861			
Chromic	—	6.49		
Citric	3.128	4.761	6.396	
2,2′-diaminodiethylammonium(+3)	3.64	8.74	9.80	
Diethanolammonium (+1)	9.00			
Dimethylammonium (+1)	10.9			
Ethanolammonium (+1)	9.4980			
Ethylammonium (+1)	10.67			
Ethylenediamminetetraacetic acid (EDTA)	2.0	2.67	6.16	10.27
Ethylenediammonium (+2)	7.18	9.96		
Formic	3.7515			
Hydrazinium (+1)	7.94			
Hydrazoic	4.72			

TABLE 7-2 (Continued)
Dissociation Constants of Brønsted Acids at 25°C

Acid	pK_1*	pK_2	pK_3	pK_4
Hydrocyanic	9.22			
Hydrofluoric	3.17			
Hydrogen peroxide	11.65			
Hydroselenic	3.89	11.0		
Hydrosulfuric	6.99	12.89		
Hydrotelluric	ca. 2			
Hydroxylammonium (+1)	5.98			
Hypobromous	8.62			
Hypochlorous	7.53			
Hypoiodous	10.64			
Iodous	0.77			
Malonic	2.85	5.67		
Mercaptoacetic	3.60	10.55		
Nitric	ca. −1.4			
Nitrilotriacetic (NTA)	1.66	2.95	10.28	
Nitrous	3.29			
Oxalic	1.25	4.285		
Periodic	2.21			
Phenol	9.98			
Phosphoric	2.172	7.211	12.360	
Phthalic	3.14	5.40		
Propionic	4.874			
Pyridinium (+1)	5.18			
Pyrophosphoric	1.52	2.36	6.60	9.24
Pyruvic	2.49			
Selenic	—	1.88		
Sulfuric	—	1.96		
Sulfurous	1.764	7.205		
(±) Tartaric	3.04	4.37		
Thiosulfuric	0.60	1.72		

* $pK = -\log K$
† The presence of more than one entry for an acid indicates that the acid contains more than one ionizable hydrogen. For a discussion of such acids, see Problem 7-5.
§ The electrical charge of the parent acid is indicated for all nonneutral species.

Source: T. R. BLACKBURN, *Equilibrium: A Chemistry of Solutions*, New York: Holt, Rinehart and Winston, pp. 194–195 (1969); two selections from L. G. SILLÉN and A. E. MARTELL, *Stability Constants of Metal-Ion Complexes*, London: Chemical Society (1964).

The extent to which a weak acid or base is "dissociated" (converted, respectively, to its conjugate base or conjugate acid by reaction with water) may, in many cases, be calculated accurately by a simple approximate equation.

Suppose that C_{HA} moles/liter of a weak acid HA are present in a solution, and that we wish to calculate the extent to which HA has dissociated at equilibrium. The reaction[8]

$$HA + H_2O = H_3O^+ + A \qquad (7\text{-}52)$$

---
[8] Throughout this derivation, the electrical charges on HA and A will be omitted for the sake of clarity.

produces A and H_3O^+ in equal quantities. If we assume for the moment that there are no other important sources of A or of H_3O^+ in solution, then

$$[H_3O^+] \simeq [A] \tag{7-53}$$

Also, under the same assumption,

$$C_{HA} = [HA] + [A] \tag{7-54}$$

If HA is truly a weak acid, then in most cases it turns out that the degree of dissociation is very slight, and [HA] is much larger than [A]. In that case,

$$C_{HA} \simeq [HA] \tag{7-55}$$

If we substitute the approximate expressions (7-53) and (7-55) into the equation for the K_a, Equation (7-50), we find

$$\frac{[H_3O^+]^2}{C_{HA}} \simeq K_a$$

Knowing the values of C_{HA} and K_a for a particular situation, we may then calculate

$$[H_3O^+] \simeq \sqrt{K_a \cdot C_{HA}} \tag{7-56}$$

For example, consider a 0.1 M solution of NH_4^+ (perhaps formed by dissolving the ionic substance NH_4Cl in water). Recognizing NH_4^+ as the conjugate acid of NH_3, and finding its K_a (Table 7-2) to be 5×10^{-10} at 25°C, we may calculate the value of $[H_3O^+]$ in this solution and therefore the degree to which the NH_4^+ in it is "dissociated" into NH_3 and H_3O^+ according to Equation (7-48). Using our approximate Equation (7-56), we find

$$[H_3O^+] = [NH_3] \simeq \sqrt{(5 \times 10^{-10})(0.1)}$$
$$= \sqrt{0.5 \times 10^{-10}}$$
$$= 0.7 \times 10^{-5} M = 7 \times 10^{-6} M$$

That is, of the NH_4^+ originally present in the solution, a fraction

$$\frac{7 \times 10^{-6}}{0.1} = 7 \times 10^{-5}$$

or 0.007 percent of it will be found as NH_3 at equilibrium, with 99.993 percent present as NH_4^+. The pH of the solution may be calculated from the value of $[H_3O^+]$ to be 5.15.

We made two assumptions in this calculation that turned out to be well justified in this case. We assumed that there were no other important sources of $[H_3O^+]$ in the solution. Although the autoprotolysis of water

$$2H_2O = H_3O^+ + OH^- \tag{7-16}$$

is always a source of H_3O^+, in the present case it is negligible. We also assumed that most of the NH_4^+ was undissociated at equilibrium; since, in this case, the degree of dissociation was only 0.007 percent, this assumption, too, was quite valid. The corresponding assumptions in other cases may be less accurate, and a more exact calculation may be required. Since such cases constitute a small minority, we may safely leave their treatment to the more advanced books listed in the "Suggestions for Further Reading."

BUFFERS. Of course, the algebra and geometry of Equations (7-40) through (7-47) and Figure 7-5 are independent of whether the acid-base pair is colored or not. Exactly the same principle applies to all such conjugate pairs. Note that there is a region of pH, near the crossing point in Figure 7-5, where the composition of the solution is changing rapidly with changes in pH. Conversely, we might say that, in the region around the crossing point, the pH of the solution changes only rather slowly even if the composition changes from predominantly conjugate acid to predominantly conjugate base. This phenomenon is used, in laboratories and in natural systems such as blood plasma, to stabilize the pH of the solutions. A solution whose pH changes relatively little when an acid or a base is added is called a **buffer**; a mixture of a weak Brønsted acid and its conjugate base is not the only kind of solution with this property, but it is the major kind.

Look at Fig. 7-4 and explain in your own words how buffers work.

Acetic acid (CH_3COOH) and its conjugate base acetate ion (CH_3COO^-) are often used to prepare buffers of pH about 5. Suppose we wish to prepare such a buffer of pH 5.0, knowing that the K_a of acetic acid is 2×10^{-5}:

$$K_a = 2 \times 10^{-5} = \frac{[H_3O^+][CH_3COO^-]}{[CH_3COOH]}$$

From this equation and the desired condition

$$[H_3O^+] = 1 \times 10^{-5} \, M$$

we find that

$$\frac{[CH_3COO^-]}{[CH_3COOH]} = \frac{K_a}{[H_3O^+]}$$

$$= \frac{(2 \times 10^{-5})}{(1 \times 10^{-5})}$$

$$= 2$$

According to this calculation, a solution that contains acetate ions and acetic acid in a 2:1 ratio should have a pH of 5.0. Such a solution might be prepared by adding acetic acid to sodium acetate in the proper ratio, or by partly neutralizing an acetic acid solution with a strong base, or by acidifying a sodium acetate solution with a strong acid.

SOLUBILITY. Solutions are mixtures of substances on the molecular level. To insert particles of one kind between those of another kind requires that the attractive forces between particles in the pure substances be overcome; this will only happen if the forces between unlike particles in the mixture are nearly as large as those in the pure substances to be mixed.[9] This simple reasoning leads to a simple general rule for solubility—nonpolar substances

---

[9] As we shall see in Chapter 9, there is a natural prejudice favoring disorderly situations over orderly ones; this prejudice allows the formation of solutions (with their randomly mixed molecules) even at some disadvantage in energy when the discrepancy is not too great. This leads to the familiar phenomenon of dissolution with cooling. You can observe it by dumping some KNO_3 in a beaker and adding room-temperature water.

dissolve best in nonpolar solvents, while polar and ionic substances dissolve best in polar solvents. Thus, sugar and salt dissolve readily in water, but not in gasoline; carbon tetrachloride and water do not mix, but form a two-layered liquid system. When nonpolar substances like the diatomic halogens (Cl_2, Br_2, and so forth) are present in such a two-liquid system, they collect in the nonpolar phase, while any ionic or strongly polar substances collect preferentially in the water layer (Figure 7-6). Such differential solubilities lie behind *chromatography,* an important technique for separating and analyzing many complex mixtures produced by nature and man in the courses of their activites.

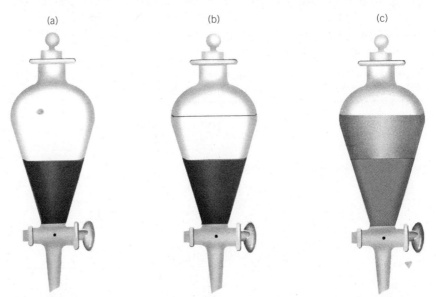

FIGURE 7-6 The separation of two similar substances [triethylamine, $N(C_2H_5)_3$, and triethanolamine, $N(C_2H_4OH)_3$] by partition between two immiscible solvents (water and diethyl ether). (a) A water solution containing both solutes in a separatory funnel. (b) An equal volume of diethyl ether is added. Being immiscible with water and less dense, it floats on top of the water. (c) After equilibrium is established (which may be hastened by shaking the separatory funnel to increase the area of contact between the immiscible liquids), the more polar triethanolamine (color) remains in the polar aqueous solution, while the less polar triethylamine dissolves preferentially in the less polar diethyl ether. The two liquids may now be separated conveniently by draining the more dense water solution out the bottom of the funnel.

IONIC SOLUBILITY. When very polar and ionic substances dissolve in a very polar solvent like water, the large forces in the pure polar solute (for example, the ionic bonding in a crystal of NaCl) are overcome by equally large Lewis acid-base forces of interaction between water molecules and solute ions:

$$NaCl(s) = Na^+(aq) + Cl^-(aq) \tag{7-57}$$

The suffix (aq) indicates that these ions are strongly associated with water molecules and dissolved in excess water. This suffix has been omitted in most previous chemical equations for the sake of clarity.

Notice that the use of the equal sign in Equation (7-57) indicates that the solution and dissolution processes have come to equilibrium; that is, we are talking about a *saturated* solution of salt in water. In writing the equilibrium constant for a solubility equilibrium like Equation (7-57), we take advantage of the fact that the reactant is a pure phase. The mole fraction of NaCl in pure NaCl is exactly 1, so the use of mole fraction for the concentration of NaCl *in the solid* leads to its apparent disappearance from the equilibrium constant:

$$K = \frac{[\text{Na}^+][\text{Cl}^-]}{1} \tag{7-58}$$

The 1 in the denominator is the mole fraction of NaCl in NaCl(s).

Because such expressions lack the denominator usually found in equilibrium constants, they are often called **solubility products** and are designated K_{sp}. The numerical magnitude of a solubility product, like that of any other equilibrium constant, is a statement of how completely the indicated process takes place; in this case, it indicates how soluble the substance is. For NaCl, K_{sp} is about 30 at 25°C, but many salts are much less soluble in water. For AgBr, the solubility product is only 7.7×10^{-13} (25°C).

Solubility products give the mathematical product of the molar concentrations of at least two ionic species in a saturated solution of a salt. Without further information, one cannot say what the value of any of the concentrations might be individually. If we are talking about the **solubility** of a salt composed of two kinds of ions, in pure water, then the dissolution reaction [for example, Equation (7-57)] tells us what the ratio of the two concentrations must be. In the cases of NaCl and of AgBr, the ratio must be 1:1; the concentrations of the two ions must be equal. *In that case*, each concentration must equal the square root of the solubility product; for NaCl, about 5.5 M, and for AgBr, about 9×10^{-7} M.

But only in that case

If one of the ions of the salt is present in the solution from another source, then the solubility equilibrium must shift to accommodate this situation. For example, solid NaCl can be made to crystallize out of salt solutions by the addition of concentrated HCl. The additional Cl$^-$ ions supplied by this strong acid shift Equation (7-57) to the left, removing Na$^+$ from solution until the product [Na$^+$][Cl$^-$] again equals the K_{sp}. The solubility of NaCl in this solution has been decreased, and this effect is known as the "common ion effect." (The fact that HCl has an ion species in common with NaCl makes HCl have an effect on the solubility of NaCl.) Like many other specially named effects, the common-ion effect is just a particular case of LeChatelier's principle operating through a chemical equilibrium.

Another application of Le Chatelier's principle to solubility equilibria is that known as photographic fixing. We referred to this process on page 210, and now we are in a position to finish the discussion.

After a photographic image is registered in atoms of silver by the processes

$$\text{AgBr}(s) + \text{light} \longrightarrow \text{AgBr}°_{\text{photochemically activated AgBr}}$$

$$\text{AgBr}° + \text{developer (reducing agent)} \longrightarrow \text{Ag}(s) + \tfrac{1}{2}\text{Br}_2$$

it is imperative to remove the unused AgBr from the unexposed areas of the developed negative before the negative is brought into the light of day, or else the processes will continue until the entire film is black. The tiny solubility product of AgBr makes the prospect of washing it off with water hopeless, and of course this ionic substance is even less soluble in nonpolar solvents. But what if the exposed film is washed with a solution of $Na_2S_2O_3$? Equations (7-22) and (7-23) combine with the dissolution reactions of AgBr

$$AgBr(s) = Ag^+(aq) + Br^-(aq) \tag{7-59}$$

to produce a scheme of interconnected equilibria:

$$\begin{array}{c} AgBr(s) = Ag^+(aq) + Br^-(aq) \\ + \\ S_2O_3^{2-} \\ \parallel \\ AgS_2O_3^- + S_2O_3^{2-} = Ag(S_2O_3)_2^{3-} \end{array} \tag{7-60}$$

The abundance of $S_2O_3^{2-}$ in the fixing bath drives its Lewis acid-base reaction with Ag^+ toward the formation of $Ag(S_2O_3)_2^{3-}$ (page 212), and the resulting very low concentration of Ag^+ allows AgBr(s) to dissolve freely.

Brønsted acid-base equilibria can also interact with solubility equilibria if the solute is capable of donating or accepting protons. For example, caves are formed in limestone by the dissolution of the solid rock. The principal component of limestone, $CaCO_3$, is ordinarily quite insoluble in water. The reaction

How to make a cave in solid rock

$$CaCO_3(s) = Ca^{2+} + CO_3^{2-} \tag{7-61}$$

has a very small equilibrium constant:

$$K_{sp} = 3 \times 10^{-9} \tag{7-62}$$

But natural waters often have a somewhat acidic pH, because they contain dissolved CO_2 and other naturally occurring Brønsted acids. The reaction

$$CO_3^{2-} + H_3O^+ = HCO_3^- + H_2O \tag{7-63}$$

has a large equilibrium constant:

$$K = \frac{[HCO_3^-]}{[CO_3^{2-}][H_3O^+]} = 2 \times 10^{10} \tag{7-64}$$

The sum of the two chemical reactions, Equations (7-61) and (7-63), is

$$CaCO_3(s) + H_3O^+ = Ca^{2+} + HCO_3^- + H_2O \tag{7-65}$$

and its equilibrium constant

$$K = \frac{[Ca^{2+}][HCO_3^-]}{[H_3O^+]} = 60 \tag{7-66}$$

is the product of Equations (7-62) and (7-64).

EXERCISE: Verify Equation (7-66) for yourself by multiplying the algebraic and numerical parts of Equation (7-62) by the respective parts of Equation (7-64).

We may symbolize the scheme of linked equilibria this way:

$$CaCO_3(s) = Ca^{2+} + CO_3^{2-}$$
$$+$$
$$H_3O^+$$
$$\parallel$$
$$HCO_3^- + H_2O$$
(7-67)

The presence of even a small quantity of H_3O^+ in the water (dissolved CO_2 gives otherwise pure water a pH of about 5) will shift reaction (7-67) toward the production of HCO_3^- and Ca^{2+} and the dissolution of $CaCO_3(s)$. This reaction accounts for both the formation of limestone caves and for the presence of Ca^{2+} in "hard" water.

Or how to dissolve a limestone building

7-6 Equilibria Within Solids

The law of chemical equilibrium is by no means confined in its operations to the gas phase and to aqueous solutions. Semiconducting devices operate because an equilibrium maintains a supply of charge-carrying crystal defects within them. A relatively simple example may show how this works.

We found in Chapter 5 that the crystal structures we were discussing were idealized, in the sense that real crystals always contain imperfections such as missing atoms, dislocations of the lattice, substitutions of one atom for another, and so forth. In particular, such a sobersided compound as NaCl can be made to depart from its ideal crystal structure by heating it. Increased thermal vibration at high temperatures will knock ions out of their ordinary positions, creating an "interstitial" ion (an ion that does not occupy a regular lattice site) and a vacancy at the lattice site from which the ion departed. Both vacancies and interstitial ions can move through the crystal by a sort of leap-frogging process: If an ion next to a vacancy moves into the vacancy, it leaves a new vacancy behind it; similarly, if an interstitial ion knocks another ion out of its lattice site, replacing it there, it creates a new interstitial ion. Finally, an interstitial ion and a vacancy can "collide" and mutually disappear. In other words, the process of forming vacancies and interstitial ions is a reversible one:

$$Na^+ \text{ (lattice)} = Na^+ \text{ (interstitial)} + \text{vacancy} \quad (7\text{-}68)$$

The product of vacancies and interstitial ions is required (and found) to be constant by the law of chemical equilibrium. This and other solid-state equilibria are discussed more fully in the works suggested for further reading.

7-7 The Drift Toward Equilibrium

From what we have been saying, it should be apparent that chemical equilibrium plays a major role in various local situations (for example, in your bloodstream, in chemical flasks, and in single crystals). What we have not really faced is the question of *why* collections of matter should drift toward an equilibrium state. After all, if the electrical potential energy of chemical bonds is the chief driving force of chemical change, there is no apparent reason why one form of energy (potential) should invariably be converted into another (the heat of chemical reactions). But this is almost always the case. Furthermore, spontaneous chemical change is unidirectional; once systems reach equilibrium, they are stuck. An equilibrium system does not change its properties with the passage of time. Of course, equilibria can be affected by the input of energy. For example, hydrogen and oxygen once combined into water can be separated again by electrical energy, in the process of electrolysis. But any source of such energy must itself be on its way toward its own equilibrium state.

Such speculations as these have large consequences. When the full implications of the law of chemical equilibrium were realized in the nineteenth century in the form of the second law of thermodynamics (Chapter 9), they were rejected by many as being unbearably pessimistic; how could God have created a whole universe, just to let it run down?

Whatever the answer to that question is, in Chapter 9 we shall look at the reasons why systems *do* run unidirectionally toward an equilibrium state, and in Chapter 10 we shall have a look at the sources of energy that keep them pumped "uphill" in potential energy so they can keep running downhill for our benefit.

7-8 Summary

When chemical reactions are reversible, a state of chemical equilibrium is eventually reached, in which the rates of the forward and reverse reactions are equal. When this is true, no further change in the physical or chemical properties of the reacting systems is observable. Also, a unique mathematical condition exists: The ratio of the product of concentrations of the reaction products to the product of concentrations of the reactants is equal to a numerical constant, the equilibrium constant. Systems in a state of chemical equilibrium respond to changes in temperature and pressure according to Le Chatelier's principle (as do the phase equilibria discussed in Chapter 5); increased pressure favors the state of lowest volume, and increased temperature induces changes that absorb heat.

We have seen some uses of equilibrium theory in describing acid-base and solubility reactions. In particular, we found that reactivity can be given a quantitative expression in the numerical values of the equilibrium constants for the formation of Lewis acid-base complexes, in the K_a and K_b that govern Brønsted acid and base reactions, and in the solubility product K_{sp} that governs ionic solubility.

Both Lewis and Brønsted acid-base equilibria are most clearly presented when one species' concentration is chosen as a master variable, in terms of which other concentrations or concentration ratios may be calculated. For example, distribution diagrams (Figures 7-2 and 7-5) may be drawn to show the predominant species present in solution as functions of the master variables.

Commonly chosen master variables include the Lewis base concentration and hydronium ion concentration.

We found that the use of the equilibrium constant allowed us to calculate the composition of acid-base buffers required to produce a given pH, the degree of dissociation of a weak acid,[10] and the solubility of a salt. These examples are presented, in fact, only as tokens of the enormous range of equilibrium calculations that are possible. Equilibrium theory (in the form of Le Chatelier's principle) also enabled us to predict the response of solubility equilibria to common ions and to Lewis and Brønsted acid-base interactions. Calculations based on equilibrium theory are very useful.

Finally, we stepped back and found that we have as yet given no explanation for the tendency of chemical systems to seek a state of chemical equilibrium. We shall consider that question in Chapter 9, after we apply the equilibrium principle to electron-transfer reactions in Chapter 8.

GLOSSARY

α (Greek alpha): this symbol is used to represent the fraction of some related group of molecules or ions that is present in a mixture as a particular species. For example, α_{HA} is the fraction that is actually present in the form of the conjugate acid in a mixture of a conjugate pair $\{HA, A^-\}$.

acid anhydride: a substance that reacts with water to form an acid.

autoprotolysis: a reaction in which a substance undergoes a proton-transfer reaction with itself. The autoprotolysis of water is represented by

$$2H_2O = H_3O^+ + OH^-$$

buffer: a solution whose pH changes relatively little when changes are made in its composition (for example, by the addition of acids or bases).

distribution coefficient (partition coefficient): the equilibrium constant for the distribution of a solute between two immiscible solvents.

equilibrium: a state of constant properties brought about by the existence of two opposing processes whose effects exactly compensate for each other. Chemical equilibrium produces a state of constant chemical composition and physical properties in a system as a result of the equal rates of opposing chemical reactions.

equilibrium constant: the numerical constant that results when the product of the concentrations of all products of a reaction at equilibrium is divided by the product of the concentrations of all reactants.

indicator: a substance whose color changes when its structure changes. An acid-base indicator is a conjugate acid-base pair whose colors are different.

K_a: the equilibrium constant for the donation of a proton to water by an acid.

K_b: the equilibrium constant for the donation of a proton to a base by water.

K_{sp}: the equilibrium constant for the dissolution of an ionic substance.

K_w: the equilibrium constant for the autoprotolysis of water.

---

[10] For an extension of this derivation to the case of weak bases in solution, see Problem 7-2.

mole fraction: the number of moles of a given substance present in a solution divided by the total number of moles of all substances in the solution; symbolized by X.

pH: the negative of the logarithm of the hydrogen ion concentration. (In particular circumstances, this symbol may have a slightly different meaning; for discussions of these, see the equilibrium texts listed under "Suggestions for Further Reading.")

polyprotic (acids or bases): substances that are capable of undergoing more than one proton-transfer reaction.

solubility: the number of moles of a given substance that must dissolve to form one liter of a saturated solution in a given solvent.

solubility product: the equilibrium constant for the dissolution of an ionic substance.

strong acid, strong base: an acid or base, respectively, whose proton-transfer reaction with the solvent goes essentially to completion (thus, an acid or base whose K_a or K_b, respectively, is large).

weak acid, weak base: an acid or base whose reaction with the solvent does not go to completion when equilibrium is reached (thus, an acid or base whose K_a or K_b, respectively, is small). There is no sharp distinction between strong and weak acids, or between strong and weak bases. A "weak" acid (with a K_a of 10^{-3}, for example) will be essentially completely dissociated at low enough concentrations. The strength of a particular acid also depends on the solvent in which it is dissolved. Thus, these terms have only relative meaning.

zwitterion (German: hybrid ion): a species in whose structure there is a full positive charge and a full negative charge; merely polar molecules are not considered zwitterions. The example discussed in this chapter is the red conjugate acid form of methyl orange, HMO (Figure 7-4).

PROBLEMS

7-1 A chemist has prepared a buffer of pH 4.5.
  a. What is $[H_3O^+]$ in this buffer?
  b. He places in the buffer one drop of methyl orange solution. Assuming that this addition has no effect on the pH of the buffer, what fraction of the methyl orange is in the acid form HMO? What fraction is present as MO^-? What color is the solution?

7-2 By means of an algebraic argument similar to that on pages 219–220, show that the $[OH^-]$ in a solution of a weak base is given approximately by

$$[OH^-] \simeq \sqrt{K_b \cdot C_B}$$

7-3 a. Write down the equation for the proton-transfer reaction of NH_4^+ with H_2O.
  b. Write the algebraic expression for the equilibrium constant for that reaction.
  c. Write the same two equations for the proton-transfer reaction of NH_3 with H_2O.
  d. Show that the product of the two equilibrium constants you wrote for Parts b and c is K_w, the autoprotolysis constant of water.
  e. Does this relationship hold for any conjugate pair?

7-4 Calculate the pH of the following solutions, assuming each to have a concentration of 0.01 M:

a. HCl (strong acid)
b. NaOH (completely ionized)
c. HOCl (hypochlorous acid, $K_a = 3 \times 10^{-8}$)
d. Pyridinium ion ($K_a = 7 \times 10^{-6}$)
e. Pyridine (conjugate base of pyridinium ion)
f. Pyridinium ion plus 0.005 M pyridine.

7-5 Many Brønsted acids and bases (for example, carbonic acid; see pages 204–206) contain more than one ionizable hydrogen. Such acids are known, in general, as **polyprotic acids** (H_2CO_3 is a diprotic acid). For the triprotic acid H_3PO_4,

a. write the three successive proton-transfer reactions in which H_3PO_4 donates a proton to a molecule of H_2O;
b. write algebraic expressions for the three successive equilibrium constants of K_1, K_2, and K_3 for these equilibria.

7-6 Human blood plasma is buffered at pH 7.4 by numerous weak acid-conjugate base pairs dissolved in it. For $H_2CO_3 + CO_2$, $K_1 = 4 \times 10^{-7}$ and $K_2 = 5 \times 10^{-11}$. Of the three species $\{H_2CO_3 + CO_2\}$ (counted as one), HCO_3^-, and CO_3^{2-}, which is the predominant form in human blood? Which is present in the smallest concentration?

7-7 The indicator thymol blue shows two color changes. Below pH 1.2, it is red; from pH 3 to 8, it is yellow; and above pH 9.6, it is blue.

a. How many ionizable hydrogens must the red form possess?
b. The structure of the red form is

Which hydrogen atoms on this structure are likely to be the ionizable ones? Explain.
c. From the information given above, sketch a distribution diagram (α versus pH) for the three forms of thymol blue. Label the lines with the color and formula of the forms. (*Suggestion:* Use H_2T as a shorthand notation for the red form.)
d. Give approximate numerical values of K_1 and K_2, the stepwise acid dissociation constants of H_2T.

7-8 Write the equilibrium for the Lewis acid-base reaction between Zn^{2+} and pyrophosphate ion, $P_2O_7^{4-}$. Identify the Lewis acid and the Lewis base.

7-9 We noted (page 216) that at the crossing point of a Brønsted acid-base distribution diagram (Figure 7-5), $[H_3O^+] = K_a$. Inspect Figure 7-2 (page 212) to determine what relationship exists between $[S_2O_3^{2-}]$ and the stability constants of the Ag^+–$S_2O_3^{2-}$ complexes at the crossing points in that diagram.

7-10 The stability constant of the complex ion formed by Zn^{2+} and $P_2O_7^{4-}$ is 5×10^8. Sketch a distribution diagram like Figure 7-2 for this system (see Problem 7-9).

7-11 Silver acetate ($Ag^+ CH_3COO^-$) is a slightly soluble salt.

a. Write the equilibrium reaction and the equilibrium constant for the dissolution of $Ag^+CH_3COO^-$ in water.
b. The solubility in pure water at 25° has been measured to be 7×10^{-2} M. Calculate the value of K_{sp}.
c. What effect, if any, will addition of the following substances to the solution have on the solubility of silver acetate?
  i. $Na^+CH_3COO^-$
  ii. $Ag^+NO_3^-$
  iii. HNO_3 (strong acid)
  iv. $Na_2S_2O_3$

7-12 The K_{sp} of $Ba^{2+}CrO_4^{2-}$ is 1.0×10^{-10}.
a. What is the solubility of this salt in pure water?
b. What is its solubility in 0.001 M Na_2CrO_4?

7-13 The K_{sp} of $Ca(OH)_2$ is 1×10^{-5}. Suppose the pH of a natural water sample containing 5×10^{-3} M Ca^{2+} were raised by the addition of NaOH.

a. At what value of $[OH^-]$ would $Ca(OH)_2$ begin to precipitate?
b. What would be the pH?
c. When the experiment was actually done, it was found that $Mg(OH)_2$ precipitated at a pH of 10.2. The K_{sp} of $Mg(OH)_2$ is 2.5×10^{-11}. What must have been the value of $[Mg^{2+}]$ in the water sample?

7-14 Suppose that a solute Z is distributed between two solvents as shown in Figure 7-6. Assuming that the transport of Z molecules across the phase boundary is an equilibrium process:

$$Z(\text{in solvent 1}) = Z(\text{in solvent 2})$$

show that the ratio of concentrations of Z in the two solvents is a constant. (This constant is called the **distribution coefficient,** or the **partition coefficient.**)

SUGGESTIONS FOR FURTHER READING

Equilibrium chemistry can sometimes appear to be only a branch of applied algebra. To dispell this impression, it is worthwhile looking at some of the chemistry behind the equilibrium constant. One short book which attempts to do this is: BELL, R. P., *Acids and Bases: Their Quantitative Behavior*, 2nd ed., London: Methuen (1969).

Equilibria taking place within solid crystals and their analogies to solution equilibrium are discussed by: REISS, HOWARD, "The Chemical Properties of Materials," *Scientific American*, vol. **217**, pp. 210–220 (September 1967).

There are available a number of short textbooks that give unified and consistent treatments of the chemistry and mathematics of equilibrium systems, with strong emphasis on aqueous solutions. Some that you may find helpful as a second-level treatment are:

BLACKBURN, T. R., *Equilibrium: A Chemistry of Solutions,* New York: Holt, Rinehart & Winston (1969).

BUTLER, JAMES N., *Solubility and pH Calculations.* This is a thorough and readable treatment of the topics mentioned in the title and is excerpted from Butler's even more thorough and advanced *Ionic Equilibrium: A Mathematical Approach.* (Both books published by Addison-Wesley, Reading, Mass., 1964.)

FISCHER, R. B., and PETERS, D. G., *Chemical Equilibrium,* Philadelphia: Saunders (1970), is characterized by a notably straightforward and sober writing style and a large number of worked-out examples. This would be a good book to use for practice in working typical equilibrium problems.

Equilibrium chemistry would be a dry subject indeed if it were not for its many applications to the chemistry of natural and technological systems. The following books treat the subject in the context of such applications:

FLECK, G. M., *Equilibria in Solution,* New York: Holt, Rinehart & Winston (1966), especially strong on the application of equilibrium calculations to analytical chemistry.

GUENTHER, W. B., *Quantitative Chemistry: Measurements and Equilibrium,* Reading, Mass: Addison-Wesley (1968), a somewhat longer treatment than Fleck's, even more firmly focused on analysis.

STUMM, W., and MORGAN, J., *Aquatic Chemistry,* New York: Wiley (1970). A thoroughgoing and realistic discussion of chemical species and their equilibria as found in lakes, streams, and oceans.

WILLIAMS, D. R., *The Metals of Life,* New York: Van Nostrand Reinhold (1971). Although only incidentally a book about chemical equilibrium, this discussion of the chemistry of metal ions (particularly their Lewis acid-base interactions) in living systems will make a fascinating and easily followed sequel to this chapter.

8

Oxidation and Reduction: Electrons on the Move

*The cumbrous elements, Earth, Flood, Aire, Fire
Flew upward, spirited with various forms,
Each had his place appointed, each his course,
The rest in circuit walles this Universe.*

JOHN MILTON,
Paradise Lost

8-1 Abstract

The actual movement of electrons from one substance to another in what are called electron-transfer reactions allows us to guide, to count, and to utilize the streams of electrons in many interesting ways. In this chapter, we consider an elementary "bookkeeping" scheme for electrons that allows equations for such reactions to be balanced easily. The measurement of potential for possible reaction is then considered, and from this there follow such handy operations as calculating the equilibrium constant and predicting whether or not a reaction will start up and proceed spontaneously. Manipulation of the electron-transfer reactions also allows us to obtain energy from compact chemical devices (batteries) and to obtain valuable materials (such as metals and industrial chemicals) by conducting electrochemical reactions in what are called electrolytic cells.

8-2 Electron-Transfer Reactions in General

The formation of all chemical compounds and all their varied interactions involve changes in the distribution of electrons about the constituent atomic nuclei, but there the resemblance ends. Some chemical reactions are meek and mild; others are forceful and dramatic. The most spectacular reactions, such as those that blast off spaceships or dazzle the eyes in a fireworks display, are those in which electrons are *transferred* outright from one kind of atom to another. For reasons that will soon appear, these are called **oxidation-reduction** reactions.[1] They constitute a large class of reactions which can be considered as a group, so that we may learn how they take place, how we may write balanced equations for them, how we may put them to use, and how we may predict the amount of energy that can be expected from them.

Some oxidation-reduction reactions, such as the combustion of gasoline or the detonation of miners' blasting powder, produce large amounts of energy and are employed just to obtain that energy; the reagents are only a regrettable expense, and the products are just a tolerated nuisance. On the other hand, some oxidation-reduction reactions that *absorb* much energy are used in a preparative or productive way, on a large scale, just because we want the products (aluminum, magnesium, chlorine, and so forth). Here the energy is

---

[1] Sometimes the slang term "redox" is used instead of writing out "oxidation-reduction." The word *redox* comes from *reduction-oxidation* and has the virtue of being only two syllables long, instead of seven.

simply an expense of production and is endured because of the convenience, or because there is no other feasible method of production.

Since oxidation-reduction reactions involve a physical transfer of electrons from atom to atom, it should be possible to conduct the electrons through a wire to some recipient atoms located in a separate (or even remote) place, instead of allowing a direct local exchange. This can indeed be done as we shall see; the question is merely one of ingenuity in setting up the electron-transport system. If we are successful, the electrons from an exoergic (energy-releasing) reaction may be made to work while they are traversing the external circuit, and so we have a way of transforming the released chemical energy into electrical energy. Conversely, electrons may also be *forced* (against a chemical pressure) into suitably ingenious electrochemical devices so that desirable products of the endoergic (energy-absorbing) reaction can be recovered. Hence we have means of getting energy from chemicals, and vice versa. Such operations constitute the domain of **electrochemistry**, a large and intensely practical subject which can readily be understood in terms of the principles that govern oxidation-reduction reactions. We shall now proceed to those principles.

8-3 The Meaning of Oxidation and Reduction

Although massive castings and forgings of magnesium will not burn, a sufficiently thin sliver or ribbon of magnesium can be ignited in a gas flame, whereupon it burns with a dazzling white light, producing a white powdery oxide of the formula MgO. Similarly, thin iron wires or fine iron powder will burn in air (or, with more enthusiasm, in pure oxygen) to give brown Fe_2O_3 or black Fe_3O_4. A more active metal, such as sodium, can be ignited more easily; sodium burns briskly in air to give a mixture of Na_2O and Na_2O_2, and it burns very vigorously in oxygen to give an almost pure peroxide, Na_2O_2. All these reactions obviously are *oxidations*, because they all are reactions of metallic elements with oxygen. In each one, atoms of a metal give up one or more electrons to become the corresponding positive ions:

$$Mg \rightarrow Mg^{2+} + 2e^- \tag{8-1}$$

(where the symbol e^- again indicates an electron)

$$Fe \rightarrow Fe^{3+} + 3e^- \tag{8-2}$$

$$Na \rightarrow Na^+ + e^- \tag{8-3}$$

and so on. The electrons are taken up by atoms or molecules of oxygen to form the corresponding oxide or peroxide ions:

$$O + 2e^- \rightarrow O^{2-} \tag{8-4}$$

and

$$O_2 + 2e^- \rightarrow O_2^{2-} \tag{8-5}$$

Hence we can recognize two separate processes going on in such experiments: An electropositive element *loses* electrons, forming positive ions, and oxygen *gains* the electrons, forming negative ions.

Compare the terms electropositive *here and* electronegative *in Ch. 6.*

Now we try a different but related experiment. We ignite a piece of sodium in air, and after it has started to burn, we plunge it into a flask of chlorine. The sodium burns just as vigorously, giving forth a bright yellow light and evolving a thick white smoke of sodium chloride. The metallic sodium disappears, and yellow, gaseous chlorine is consumed at the same time. In place of the elements, we have white, solid, water-soluble sodium chloride, made up of sodium ions (Na^+) and chloride ions (Cl^-). Again we recognize two processes:

$$Na \rightarrow Na^+ + e^- \tag{8-6}$$

and

$$e^- + Cl \rightarrow Cl^- \tag{8-7}$$

That is, atoms of sodium have again lost electrons, and an electronegative element (chlorine) has gained the electrons. Since Equations (8-3) and (8-6) are identical, they are both oxidations. Similarly, the loss of electrons from iron, magnesium, aluminum, or any other electropositive element may be termed "oxidation," regardless of what element takes up the electrons; the same process is involved in all such reactions. Therefore, we can define **oxidation** generally as a *loss of electrons*. The opposite process, that of *gaining electrons*, is called **reduction**. The name "reduction" arises from the ancient art of "reducing" ores to metals, a process in which there is a reduction of weight and a large reduction of volume. The fundamental process is seen to involve a gain of electrons by the metal ions in the ore, whether the electrons are supplied electrically or by some electron-rich substance (a "reducing agent"):

The terms oxidation *and* reduction *are much more general than their original meanings. Memorize their definitions.*

$$Al^{3+} + 3e^- \text{ (from electric circuit)} \rightarrow Al \tag{8-8}$$

$$Fe^{3+} + 3e^- \text{ (from carbon as reducing agent)} \rightarrow Fe \tag{8-9}$$

$$Mg^{2+} + 2e^- \text{ (from silicon as reducing agent)} \rightarrow Mg \tag{8-10}$$

Many more examples could be adduced, but the generalizations already are evident. Equation (8-9) is seen to be Equation (8-2) written backward, and Equation (8-10) is seen to be Equation (8-1) written backward. So reductions (8-9) and (8-10) are the opposites of oxidations (8-2) and (8-1). Carbon is used as a reducing agent for reduction (8-9) and ends up as carbon dioxide; silicon may be used as a reducing agent for reduction (8-10), ending up as SiO_2 or $SiCl_4$, depending on whether MgO or $MgCl_2$ was used as starting material. So we see that reducing agents get *oxidized* in the course of their work. Similarly, the *oxidizing agents* O_2 and Cl_2 used in the experiments we described gained electrons in the course of their work [see Equations (8-4) and (8-7)], and so the fate of an oxidizing agent is to become reduced. To summarize

1. *Oxidation* means a *loss* of electrons.
2. *Reduction* means a *gain* of electrons.
3. In the course of oxidation-reduction reactions, oxidizing agents become reduced.

4. Similarly, reducing agents become oxidized.
5. Oxidation and reduction always occur simultaneously and to the same extent, so that no electrons are created or destroyed.

8-4 Half-Reactions

The chemical equation

$$2Na + Cl_2 = 2NaCl \tag{8-11}$$

represents the overall chemical change that occurs in the formation of sodium chloride from the elements, but the equation for the oxidation alone,

$$Na \rightarrow Na^+ + e^-$$

which is Equation (8-3) above, is only half an equation. It represents only half a reaction, the oxidation half. Similarly,

$$e^- + Cl \rightarrow Cl^-$$

which is Equation (8-7), represents only the reduction half. In order to pair off various oxidations with various reductions (since the processes are separate and general, although simultaneous), it is helpful to think of **half-reactions** such as those just written as equations. Obviously, there can be *oxidation half-reactions* [Equations (8-1), (8-2), (8-3), and (8-6)], and there can be *reduction half-reactions* [Equations (8-4), (8-5), (8-7), (8-8), (8-9), and (8-10)], but only these two kinds. We can think of half-reactions in the abstract, and even assign numbers to the driving forces of their electron-transfer processes. The half-reaction concept will be found useful in study and in applications.

8-5 Oxidation Number

The *degree* of oxidation or reduction of any element is best defined in terms of an arbitrary but very useful quantitative concept called the **oxidation number**. For simple ions such as those mentioned earlier in this chapter, the oxidation number is just the *number of electrons lost or gained* by an atom of the element in becoming that ion. The oxidation number is given a positive or negative sign to correspond with the charge on the ion, and the oxidation number of a neutral (unoxidized, unreduced) atom is considered to be zero. Hence the oxidation number (ON) of magnesium in the Mg^{2+} ion is $+2$; the ON of iron in Fe^{3+} is $+3$; the ON of oxygen in O^{2-} is -2; the ON of chlorine in Cl_2 is zero; the ON of chlorine in Cl^- is -1, and so on.

Like most bookkeeping terms, oxidation number is arbitrary but useful.

We have seen that the chemical behavior of many elements in redox reactions is really quite limited, so some conclusions about oxidation number are possible at once:

1. The oxidation number of each element in Group Ia is either zero (in the elementary state) or $+1$ (as ions). It is never anything else, because these elements (Li, Na, K, Rb, Cs, Fr) have only one *s* electron to lose.
2. The oxidation number of each Group IIa element is either zero or $+2$, for similar reasons.

3. The oxidation number of hydrogen in its compounds is usually +1, but in hydrides of electropositive elements (such as NaH) it is −1, corresponding to the H⁻ ion.
4. The oxidation number of oxygen in its compounds is almost always −2, but in peroxides (which contain the O_2^{2-} ion or its equivalent), the oxidation number is −1. That is, *one* electron *per atom* is gained by the element in changing from neutral O_2 to the ion O_2^{2-}.

Since the extent of oxidation must always equal the extent of reduction in any redox process, it follows that we can add another important generalization: *The oxidation numbers of all the atoms in any given compound must add up to zero.* For example, we see that generalizations 2 and 4, above, tell us that in MgO the oxidation number of magnesium is +2, and the ON of oxygen is −2. These add up to zero. Similarly, in sodium peroxide, Na_2O_2, the ON of each oxygen is −1, while that for each sodium is +1.

Exercise: What is the oxidation number of manganese in potassium permanganate, $KMnO_4$?

Answer: The ON of potassium (which is in Group Ia) is +1, and since no peroxide is involved, the ON of oxygen is −2. Since the sum of negative oxidation numbers is 4 × (−2) or −8, and the total ON in the compound must be zero, the situation must be

$$\begin{array}{ll} \text{ON of K} = +1 & \text{ON of four O} = 4(-2) \\ \underline{\text{ON of Mn} = +7} & \phantom{\text{ON of four O}} = -8 \\ \phantom{\text{ON of Mn} =} +8 & \end{array}$$

and the oxidation number of manganese in $KMnO_4$ is +7.

This is all very well for ionic compounds, you may say, but what about covalent compounds, where the oxidation number is not immediately apparent from the charge on an ion? And what about those oxidations and reductions in which electrons merely change position within a molecule or a complex ion, and do not step forth into the wide world? How shall we keep track of the electrons in such cases?

More arbitrariness; more usefulness

It turns out that all the utilitarian purposes of our oxidation-number concept can be served satisfactorily by considering every covalent substance as though it were ionic. This is done by assigning (in a purely formal and arbitrary way) *both* electrons of a shared-pair bond *to the more electronegative element.*[2] For instance, we treat the H_2O molecule as though it consisted of two H^+ ions and one O^{2-} ion. Then the oxidation number of hydrogen in H_2O is +1, and the ON of oxygen is −2. Similarly, in hydrogen peroxide, H_2O_2, the ON of hydrogen is +1, and the ON of oxygen is necessarily −1. If the covalent bond is between atoms of the same element, as in Cl_2 (which is Cl—Cl or Cl:Cl), the electrons are divided equally.

---

[2] Electronegativities of the elements are given in Table 6-2, but often a glance at the Periodic Table (or, better still, some familiarity with the elements) will do as well.

You might object to this convention about covalent compounds as being against the facts, and so arbitrary as to be against the best tradition of science. But oxidation number is not a measured property of an element; it is simply a convenient convention which enables us to balance redox equations, and to think about redox reactions within a simple framework while we put them to work. Oxidation number is a bookkeeping device.

The atoms in most polyatomic ions (such as SO_4^{2-} and NO_3^-) are covalently bonded, so we apply the same convention. The sum of the oxidation numbers of the atoms equals the charge on the ion, so manganese has an oxidation number of $+7$ both in $KMnO_4$ and in the MnO_4^- ion.

8-6 Balancing Oxidation-Reduction Equations

We turn now to some illustrations of how oxidation numbers are of help in balancing equations. Fundamentally, changes in ON allow us to keep track of the electrons that are transferred during a redox reaction. Since the number of electrons gained must equal the number lost, we are provided with a ratio of reactants (relative numbers of molecules of oxidizing agent and of reducing agent) that tells us how to balance the equation.

We start with the simplest kind of redox reaction, in which one element is oxidized and another is reduced:

$$SnCl_2 + HgCl_2 \rightarrow SnCl_4 + Hg_2Cl_2$$

Here tin is being oxidized from an oxidation state of $+2$ (in $SnCl_2$) to $+4$ (in $SnCl_4$). Mercury is being reduced from the $+2$ state (in $HgCl_2$) to the $+1$ state (in Hg_2Cl_2). Since the change in oxidation state is twice as great for tin as it is for mercury, twice as many mercury atoms are required in the reaction to take up the electrons lost by the tin. So we may write in the 2:1 ratio,

$$SnCl_2 + 2HgCl_2 = SnCl_4 + Hg_2Cl_2$$

and the equation is balanced.[3]

We continue with another simple redox reaction: the oxidation of an iodide ion to free iodine, in this case by the use of a dichromate ion as oxidizing agent. The ionic reaction is

$$Cr_2O_7^{2-} + I^- + H^+ \rightarrow Cr^{3+} + I_2 + H_2O$$

Note that the reaction takes place in acid solution, and that the hydrogen ions of the acid end up as water. We must know all the reactants and all the products, or be able to infer them from the information given, or else we cannot balance any equation. The first step, then, is to write down all the reactants and all the products, as shown. The next step is to decide what has been oxidized and what has been reduced. In this case, iodide ions have been oxidized to free iodine molecules, and chromium has been reduced (such is the fate of an oxidizing agent). The oxidation

All reactants and all products must be known first.

---

[3] Many readers will find the solution to this equation-balancing problem so simple that they can write the answer at once, without bothering with oxidation numbers. Excellent! Any balancing that can be done intuitively, or "by inspection," as it is called, is all to the good. Oxidation numbers are for harder cases which do not yield to simple inspection.

number of chromium in Cr^{3+} is obviously $+3$; in the $Cr_2O_7^{2-}$ we reason that the oxidation state of oxygen is -2 and $7 \times (-2) = -14$, so, deducting for the two negative charges on the ion, the two chromium atoms must account for the balance of $+12$. In other words, the oxidation number of chromium in the $Cr_2O_7^{2-}$ ion must be $+6$, for only then can $2(+6) + 7(-2) = 12 - 14 = -2$, the charge on the ion.

The third step is to figure out the electron exchange. The total *increase* in oxidation number must equal the total *decrease* in ON. That is, the number of electrons gained must be equal to the number lost; the overall reaction cannot manufacture electrons. Here the chromium atoms have *each* gained three electrons, for a total of $6e^-$ gained. Hence six electrons must have been lost, and only six iodide ions could furnish them. So the reacting ratio must be $6I^-$ for every $Cr_2O_7^{2-}$, and we write these with confidence:

$$Cr_2O_7^{2-} + 6I^-(+H^+) \to 2Cr^{3+} + 3I_2(+H_2O)$$

The last step is to balance the nonoxidized and nonreduced components of the reaction. The oxidation number of hydrogen has not changed; it is $+1$ in both H^+ and H_2O. Furthermore, the ON of oxygen has not changed, either; it is -2 in $Cr_2O_7^{2-}$ and in H_2O (that is, oxygen does not show up as a peroxide, or as free O_2, or in any other altered oxidation state). With 7 oxygen atoms in $Cr_2O_7^{2-}$ to account for, there must be 7 molecules of H_2O formed, and these will require 14 hydrogen ions. Hence we may write the complete balanced equation:

$$Cr_2O_7^{2-} + 6I^- + 14H^+ = 2Cr^{3+} + 3I_2 + 7H_2O$$

As a check on the validity of the balanced equation, verify that the net number of charges on the left side equals the net number of charges on the right side. Here we add -2 and -6 and $+14$, which equals $+6$, and find that this is the same as $+6$ on the right-hand side. Furthermore, of course, the number of atoms of each kind must be the same on the right and left sides of the balanced equation. As for the $=$ sign, that is optional.

Some who have studied chemistry in high school may be uneasy with the balanced ionic equation just written and may prefer to have every substance appear in neutral or molecular form. If it seems simpler in the neutral or nonionic form, we can just as well write in the formulas for the actual reagents:

$$K_2Cr_2O_7 + 6KI + 7H_2SO_4 = Cr_2(SO_4)_3 + 3I_2 + 7H_2O + 4K_2SO_4$$

The resulting equation looks longer than the ionic equation, because it must provide for the potassium and sulfate ions. To those initiated into ionic equations, it also seems unnecessarily complicated; it includes more elements (such as sulfur) which are neither oxidized nor reduced, but merely go along for the ride, so to speak. An ionic equation comes closer to showing just the essential redox process, without superfluous distractions. However, both types of equation are correct, and the reader may take his choice. Examples of both kinds are included in the problems at the end of this chapter.

Some complexities for which the arbitrariness is useful.

A more complicated situation is encountered when an oxidizing acid is used both as oxidizing agent and to provide an acidic reaction medium. For example, many inorganic sulfides can be oxidized to sulfates by hot, concentrated nitric acid. Thus, the fundamental ionic process for the oxidation of copper sulfide is

$$S^{2-} + 8NO_3^- + 8H^+ \to SO_4^{2-} + 8NO_2(g) + 4H_2O$$

Here sulfur is oxidized from a state of -2 in S^{2-} to $+6$ in SO_4^{2-}, an eight-electron change. Since the reduction of nitrogen (from a $+5$ state in HNO_3 to a $+4$ state in NO_2) is only a one-electron change, eight times as many nitrogen atoms are required as sulfur atoms in order to balance the redox process, and hence the 1:8 ratio shown in the balanced ionic equation. However, when we write a complete *molecular* equation for the oxidation of copper sulfide with nitric acid, giving $Cu(NO_3)_2$ as the copper-bearing product, two more molecules of nitric acid are required to provide the two unchanged, unreduced nitrate ions which appear on the right-hand side:

$$CuS + 10HNO_3 = Cu(NO_3)_2 + H_2SO_4 + 8NO_2 + 4H_2O$$

The ratio of oxidizing agent to reducing agent remains the same, but additional nitric acid must be written into the equation to provide the nitrate ions associated with the copper ion.

Sometimes two or more substances are oxidized by one oxidizing agent, or vice versa (see Problem 8-1c). These are not difficult situations if the separate oxidations are handled in sequence. Perhaps a more puzzling problem occurs when a *single substance is both oxidized and reduced* (a reaction called **autooxidation**). Such an autooxidation occurs when chlorine is passed into a hot, concentrated solution of potassium hydroxide—part of the chlorine is oxidized to chlorate ion, at the expense of a larger part which is reduced to chloride ions:

$$Cl_2 + OH^- \rightarrow ClO_3^- + Cl^- + H_2O$$

Here the oxidation number of chlorine is $+5$ in the ClO_3^- ion, -1 in Cl^-, and zero in Cl_2. Some of the chlorine atoms which start as Cl_2 undergo a five-electron change, to become chlorate ions; others undergo a one-electron change to become chloride ions. Obviously, we shall need five times as many of the latter as the former to make things come out right, so

$$6 \text{ chlorine atoms} + ?OH^- \rightarrow ClO_3^- + 5Cl^- + ?H_2O$$

The oxygen requirement is now balanced by inspection, since oxygen and hydrogen do not change oxidation state in this reaction, and we have

$$3Cl_2 + 6KOH = KClO_3 + 5KCl + 3H_2O$$

Oxidation numbers do not represent real electric charges; they are just a private convenience for use in balancing equations. If we stay with the arbitrary rules, come what may, a balanced equation is sure to result. Take as an illustration the dissolution of black magnetic iron oxide in hot concentrated nitric acid:

$$Fe_3O_4 + NO_3^- + H^+ \rightarrow Fe^{3+} + NO_2 + H_2O$$

Here we find the oxidation number of iron in Fe_3O_4, according to our rules, is $\frac{8}{3}$! A fractional oxidation number may seem strange, but actually it represents a combination of two different oxidation states ($+2$ and $+3$) for iron in the same compound, a situation by no means unique. If we stay with the rules, a change in oxidation number from $\frac{8}{3}$ for iron in Fe_3O_4 to $\frac{9}{3}$ for iron in Fe^{3+} represents a change of $\frac{1}{3}$ in oxidation number for each iron atom, or a total of $\frac{3}{3}$ for the three iron atoms taken together. In other words, the change

$$Fe_3O_4 \rightarrow 3Fe^{3+}$$

is a one-electron change. Since the change

$$NO_3^- \rightarrow NO_2$$

is also a one-electron change (with nitrogen changing ON from +5 to +4), it follows that one nitrate ion can oxidize one Fe_3O_4 "molecule." The rest is just a matter of adding enough hydrogen ions from the acid to combine with the oxygen. Hence the balanced equation is

$$Fe_3O_4 + NO_3^- + 10H^+ = 3Fe^{3+} + NO_2 + 5H_2O$$

or, if the neutral form of equation is preferred,

$$Fe_3O_4 + 10HNO_3 = 3Fe(NO_3)_3 + NO_2 + 5H_2O$$

We conclude that the oxidation-number method for balancing equations always works, if we give it a chance. Some students will find it an unnecessary bother, however, because they grasp the concept of half-reactions so well that the exchange of electrons becomes obvious immediately. That is, they see the preceding redox process in terms of an oxidation half-reaction

$$Fe_3O_4 \rightarrow 3Fe^{3+} + 4O^{2-} + e^-$$

and a reduction half-reaction

$$NO_3^- + e^- \rightarrow NO_2 + O^{2-}$$

Since both of these are one-electron changes, they must take place on a one-to-one basis. The student then need only add 10 H^+ ions to combine with the 5 O^{2-} ions, giving $5H_2O$, and the operation is finished. If you are one of those who can discern half-reactions easily, there is no need to devote time and thought to oxidation numbers. The half-reactions will work just as well for you. Probably even better!

8-7 Chemical Energy from Oxidation-Reduction Reactions

If the transfer of electrons in redox processes is indeed a physical reality, it should be possible to observe the actual flow of electrons by suitable means. It should also be possible to measure the number or quantity of electrons, and, if we are clever enough, to measure the tendency for reagent A to give up electrons versus the tendency for reagent B to do the same thing. Such measurements would give us a quantitative comparison of oxidizing and reducing agents, and would allow us to assign numbers to what we now call vaguely "the tendency to react." In short, there is promise that oxidation-reduction processes can be put on a clear, firm, numerical basis, without any ambiguity. Then we shall be able to make predictions about reactions and put some to use for the benefit of all of us. In order to achieve this high promise, however, we must be willing to use physical concepts and electrical units to make the measurements and to do the comparisons. This brings us to the combination of chemistry and electricity, which is called **electrochemistry**. Naturally electrochemistry has much to do with *energy*, for it is through electrochemical principles that we can get energy from chemicals, and use energy to produce chemicals.

Energy is simply *the ability to do work* (see Appendix III, if necessary). The kind of energy we shall deal with in this chapter is the so-called **free** energy, or *freely convertible* energy: electrical, mechanical, nuclear, chemical, and radiated energy. All these kinds of energy may be converted freely from one to another, by means of suitable devices (such as an electric motor) which introduce only frictional losses. Heat energy is something else again; it is different from all the forms of free energy listed above, in that it is not interconvertible with just frictional losses. Heat energy is fundamentally different, in a way that will be taken up first in Section 8-8 and then treated more thoroughly in Chapter 9.

We have discussed this before, and it will come up again in Ch. 9.

UNITS AND DEFINITIONS. A discussion of any form of energy, be it thermal, electrical, mechanical, or chemical, requires the independent consideration of the *pressure* or *potential* of the energy, on the one hand, and the *quantity* (or flow of energy over a period of time) on the other. Thus, the energy available from a mountain lake or reservoir impounded behind a dam may be described in terms of the *pressure* (the "head" of water, or the vertical distance it may fall as it seeks a lower level) and the *quantity* or volume of water that flows under this pressure. If we think of energy as the ability to do work, a thin trickle of water could not be expected to deliver much energy even though it might be under high pressure, and even a very large volume of water would be of little use for power if it were down near sea level and hence had little potential. It is the product of *potential times quantity* that tells us how much water power is available. Similarly, electrical energy is measured by the product of the *potential* (electromotive force or voltage) times the *quantity* of electricity (number of electrons). Chemical energy can be measured in similar terms, and in this section we seek means for determining both the chemical potential and the quantity of electrons from the reacting material. The *potential* is really the *tendency to react*, or the pressure of electrons as they leave a reducing agent and are gained by an oxidizing agent in an oxidation-reduction reaction. The *quantity* of reagents used, or of products formed, may once again be stated conveniently in *moles* or in *gram-equivalents*. One **gram-equivalent** of any substance is that weight of substance that involves Avogadro's number (6.023×10^{23}, or one mole) of electrons in a chemical change.

Energy = potential times quantity.

The units in which we shall express chemical and electrical pressures and quantities have the following definitions and interrelations:

1. The *coulomb* (C) is the practical unit of quantity of electricity, defined in electrochemical terms as the amount of electricity that will deposit 0.001118 g of silver from a solution containing Ag^+ ions, without regard for time. This particular weight of silver is chosen for historical reasons of correspondence with earlier units (such as those in the electrostatic system) and has no simple relationship with the atomic weight of silver or any other chemical unit of quantity. In terms of practical electrochemistry, the coulomb is inconveniently small.
2. The *faraday* is a larger unit of quantity and is the amount of electricity that will deposit one mole (107.880 g) of silver. It follows that one faraday is equal to 96,494 C (usually rounded off to 96,500 C), and that one faraday also is equivalent to 6.023×10^{23} electrons.

ELECTRONS ON THE MOVE

All these units, except hp and BTU, come from names of scientists.

E is potential, I is current, and R is resistance.

3. The *ampere* (A) is a unit of flow of electricity, or current. One coulomb of electricity passes through a point in a circuit each second when the flow is one ampere.
4. The *ohm* is the unit of *resistance* to the flow of electricity, defined in practice as the resistance to the flow of electrons offered by a thread of mercury 106.3 cm long and weighing 14.4521 g, at 0°C. Secondary standards of resistance, in the form of solid metallic conductors carefully calibrated by the U.S. National Bureau of Standards, are much more convenient to use.
5. The *volt* (V) is the practical unit of electrical pressure or potential (often called electromotive force), defined as that potential which will cause a current of one ampere to flow through a resistance of one ohm. Hence the volt is defined in terms of Ohm's law, which says that potential (in volts) is equal to the product of current (in amperes) times resistance (in ohms), or $E = IR$.
6. *Energy* (the ability to do work) is expressed in many units, of which the following two will be most useful to us:

$$\text{joules (J)} = \text{volts (V)} \times \text{coulombs (C)}$$

and

$$1 \ cal = 4.185 \ J$$

7. Since *power* is the *rate of expenditure* of energy, we have as units

$$\text{watts (W)} = \text{volts (V)} \times \text{amperes (A), or J/sec}$$

and

$$1 \ horsepower \ (hp) = 745.7 \ W$$

Furthermore, since power integrated over a period of time represents total energy, we have

1 W-sec	= 1 J
1 W-hr	= 3600 J
1 kilowatt-hour (kWhr)	= 3,600,000 J

and

1 horsepower-hour (hp-hr)	= 0.7457 kWhr
	= 2545 British Thermal Units (BTU)
	= 2.6845×10^6 J

These units now allow the preceding paragraphs to be summarized in the following relationships:

a. Electrical *free energy* in joules = *potential in volts* × *quantity in coulombs*.

b. Chemical *free energy of reaction* in joules per equivalent weight = potential in volts × quantity in coulombs per equivalent weight = *potential in volts* × *96,500 C per equivalent weight*.

c. Total chemical energy in *joules* = volts × 96,500 C per equivalent weight × no. of equivalent weights.

d. Total chemical energy in *calories* = volts × 23,100 C per equivalent weight × no. of equivalent weights.

THE MEASUREMENT OF CHEMICAL POTENTIAL FOR REACTION. We now seek to measure the *potential* of a chemical system for reaction. This chemical potential is the system's tendency to react, its pressure to undergo reaction. We shall consider only oxidation-reduction reactions, knowing that electrons are being removed from a substance being oxidized and transferred to a substance being reduced. We shall attempt to carry the electrons through an external circuit of wires and meters in order to measure the *pressure* and the *current* electrically before returning the electrons to the chemical system. In order to collect or deliver electrons, we shall make use of rods or sheets of conducting material inserted into the reagents, and these will be called **electrodes**.

Pressure *is used figuratively here. The proper term is introduced below.*

We can imagine an oxidation half-reaction going on in one container (which we shall call a **half-cell**), and a reduction half-reaction going on in another container (the other half-cell). There will, of course, be no reaction unless electrons can get from the one container to the other. If we connect the half-cells with a copper wire, electrons will travel for a short time from the substance being oxidized to the substance being reduced. The experimental arrangement is shown in Figure 8-1.

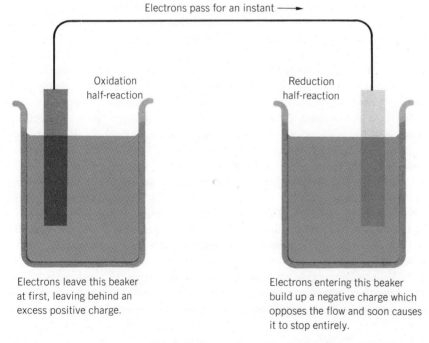

FIGURE 8-1 Two half-reactions, isolated in space from each other in two separate beakers, but connected with a copper wire attached to electrodes in the beakers. Electrons can flow freely along the copper wire, and the whole system quickly achieves electrostatic equilibrium.

The flow of electrons will not continue indefinitely, however, because electrons will accumulate in the right-hand container and build up a negative charge there. This charge will oppose the flow of more electrons by repelling

This liquid pathway is (aptly) called a salt bridge.

Electron donation and electron acceptance proceed at the same rate at equilibrium.

their negative charges, and the flow will come to a stop when the repulsion in the right-hand beaker becomes equal to the tendency of electrons to leave the left-hand beaker. The entire system then stands in electrostatic equilibrium, and the reaction cannot proceed. We can obtain a continuous flow of electrons to allow the reaction to proceed by providing a path for ions in the liquid to pass from one half-cell to the other. For example, if the oxidation half-reaction produces positive ions that accumulate in the left-hand beaker while electrons travel through the wire, we shall need a path by which these positive ions can reach the right-hand beaker in numbers corresponding to the charge transferred by electrons. Being massive bodies, the ions cannot push aside the atoms of a strong metal and make their way through a wire, but we know that they *can* pass through a liquid such as water.[4] Those ions should be able to migrate along the liquid pathway under the influence of the electric field that develops, and a continuous electric current should flow. The completed circuit is shown in Figure 8-2. By the same reasoning, if negative ions were left behind as a result of the reduction half-reaction, they should also be able to traverse the liquid path and help neutralize the surplus positive charge in the left-hand beaker. If ions are produced in both reactions, both will migrate, and the exact proportion of electric current carried by each will depend upon their relative mobilities in the liquid medium.

To sum up, if we were to divide an oxidation-reduction reaction physically into an oxidation half-reaction and a reduction half-reaction, and if we were to provide one suitable path for electrons and another suitable path for ions, then we should observe a flow of electrons from the substance oxidized to the substance reduced as the reaction proceeds. Now the terms *oxidizing agent* and *reducing agent* are relative;[5] therefore, an instrument inserted in the external circuit measures the tendency of one substance to give up electrons *relative* to the tendency of the other substance to take them up. If, by strange chance, two different reagents had equal tendencies, there would be no net pressure of electrons to escape from one reagent to the other, and there would be zero potential for reaction. In all other cases, where the tendencies were unequal, there would be a definite potential, which should be reproducible under specified conditions. The greatest potential for reaction should be observed at the instant reaction begins; then as products accumulate and reagents disappear, there should be a steady diminution of potential, because the reacting system has a steadily decreasing tendency to react and less capacity for reaction. When the reaction reaches *chemical equilibrium*, the tendencies to lose and gain electrons are equal, and the potential is zero. That is, *at equilibrium all capacity for further net reaction is gone;* the concentrations of reagents

---

[4] As an analogy, consider the difficulty a boat would have trying to make its way through a brick wall by pushing aside the bonded bricks, versus the comparative (although not perfect) ease with which it makes its way through water. The difference is exactly that which differentiates the liquid from the solid state. As for how the electrons can make their way through copper, see H. Sisler, *Electronic Structure, Properties, and the Periodic Law*, New York: Van Nostrand Reinhold, pp. 93–97, paperback (1963).

[5] For example, hydrogen peroxide may be either an oxidizing agent or a reducing agent, depending upon the other substances available for reaction.

and products do not change; the unidirectional electron transfer has ceased; and all pressure to react has disappeared.

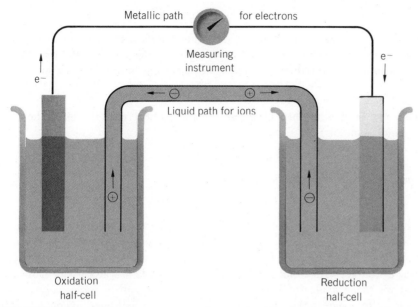

FIGURE 8-2 Two half-reactions, separated in two beakers, but coupled in a completed electrical circuit. Pathways are provided for flow of electrons and ions. Each beaker in this couple is called a half-cell.

This general discussion has narrowed itself down to oxidation-reduction reactions that involve ions and that take place in liquids. This limitation is unnecessarily restrictive; the reagents could be any of the elements (except some of the Group 0 gases) or any of their compounds, and the liquid could be anything from a molten silicate at bright red heat[6] to liquid ammonia at −33°C. Under specified conditions, standard oxidation-reduction potentials for hundreds of reagents could be measured in any one of these liquids. A few actually have been measured in nonaqueous systems, but a vastly greater amount of effort has been expended on *aqueous* solutions at room temperature. This solution chemistry in water under ambient conditions is a field of immense interest, having great practical value in industrial and biological chemistry. For these reasons, and in order to permit an examination of some principles and applications in a reasonable space, we shall confine our discussion to aqueous solutions at room temperature and one atmosphere pressure, and to the solids or gases reacting with such solutions under these conditions.

---

[6] See R. Didtschenko and E. G. Rochow, "Electrode Potentials in Molten Silicates," *Journal of the American Chemical Society*, vol. **76**, p. 3291 (1954), in which the authors describe the measurement of potentials in systems of metal oxides, using the oxygen-covered platinum electrode at 800°C. See also H. A. Laitinen, C. J. Nyman, and C. E. Shoemaker, "Polarography in Liquid Ammonia," *Journal of the American Chemical Society*, vol. **70**, pp. 2241, 3002 (1948) and vol. **72**, p. 663 (1950), where electrochemical measurements made with a mercury electrode in liquid ammonia are reported.

STANDARD ELECTRODE POTENTIAL. To apply the principles that we have developed to a specific oxidation reaction, let us consider the common reaction of zinc with dilute hydrochloric acid:

$$Zn + 2HCl = ZnCl_2 + H_2$$

Since the zinc will react with dilute sulfuric acid in the same manner, and indeed with any strong acid in aqueous solution, we see that the reaction is really an oxidation-reduction exchange of electrons with hydrogen ions:

$$Zn + 2H^+ = Zn^{2+} + H_2$$

This may be divided into an oxidation half-reaction (with the usual thermochemical symbols)

$$Zn(s) \rightleftharpoons Zn^{2+}(aq) + 2e^-$$

and a reduction half-reaction

$$2H^+(aq) + 2e^- \rightleftharpoons H_2(g)$$

where the double arrows emphasize that both the zinc ions and the hydrogen ions may be reduced, and that both elements may be oxidized to their ions.[7] Following the strategy of the preceding section, we place a rod of metallic zinc in contact with a molar solution of zinc chloride in water in one half-cell, and we place a film of hydrogen gas in contact with a molar solution of hydrochloric acid in the other half-cell. The film of hydrogen in contact with the acid is readily obtained by saturating a clean platinum electrode with hydrogen gas, whereupon an adsorbed film of hydrogen completely covers its surface.[8] We now connect the two half-cells with a *salt bridge* to provide the necessary path for positive and negative ions. The salt bridge may be filled with some of the solution of zinc chloride; its only purpose is to serve as a low-resistance pathway for the transport of equivalent numbers of ions. (The ends of the salt bridge hold wads of glass wool to prevent gross mixing of the solutions. If the zinc were to come into contact with hydrochloric acid, it would dissolve directly in the acid, and we should not be able to observe the passage of electrons in an external circuit.)

In this system, the zinc has a certain tendency to give up electrons. We are trying to measure this tendency. As soon as a few atoms have given up electrons, however, the zinc will come to equilibrium, because the electrons stay in the metal while the Zn^{2+} ions wander off into the solution; the accumulation of extra electrons inhibits the ionization of more atoms of zinc. Similarly, elementary hydrogen comes to equilibrium with hydrogen ions of the acid solution in the other half-cell, as indicated by our equations for the half-reactions. If the zinc and hydrogen electrodes now are connected by a

Most corrosion of metals is an electrochemical process. How is it similar to the zinc's behavior here?

---

[7] Indeed, such reversibility is absolutely essential to the measurement of oxidation-reduction potentials, because the measurements depend on a balance of equilibria.

[8] The platinum is seen to be merely a device to make electrical connection to gaseous hydrogen; the platinum does not react with the solution. If clean platinum is exposed to chlorine and immersed in a solution of chlorine, it has the properties of a chlorine electrode, and so on.

short, stout copper wire, electrons will be able to flow freely from the metallic zinc to the hydrogen ions, and the reaction

$$Zn + 2H^+ = Zn^{2+} + H_2$$

may take place rapidly, reaching equilibrium in the shortest possible time. We desire to measure the maximum potential of the reaction and do not want it to proceed far toward equilibrium because the potential will drop. Hence we insert in the electron circuit the highest possible resistance compatible with accurate measurement[9] and allow the electron pressure to build up. Only relatively few electrons pass while the measurement is being made, and so we approach the ideal situation as closely as possible. Having taken these precautions, we arrive at the completed zinc-hydrogen cell depicted in Figure 8-3. With the cell set up in this way, a potential difference of 0.763 V between the electrodes is observed, and this is the potential for reaction.

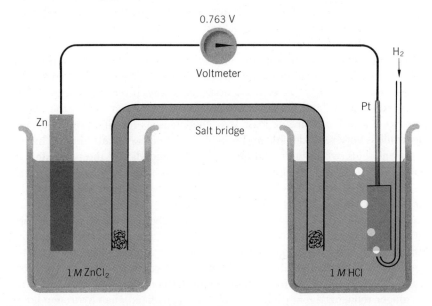

FIGURE 8-3   A completed zinc-hydrogen cell, measuring the electron-releasing tendency of zinc, relative to hydrogen, in an aqueous system. The high-resistance voltmeter allows only a few electrons to pass through the wire, and the measurement of $+0.763$ V gives the maximum potential for the chemical reaction $Zn + 2H^+ = Zn^{2+} + H_2$.

We must now pay particular attention to the *sign* of this potential. Formerly, there was a great deal of controversy over this point, but fortunately, the International Union of Pure and Applied Chemistry has now resolved the

---

[9] That is, a voltmeter of highest possible resistance (at least 100,000 ohms/V of scale deflection) is used. Vacuum-tube voltmeters draw still less current and are to be preferred for this reason. Any current flowing through electrical resistance in the solutions and in the salt bridge will reduce still further the observed voltage, providing an additional reason why the external resistance should be high compared with the internal resistance.

controversy in an admirably simple way: *The electrode potential is given the same sign which the electrode itself exhibits in the laboratory, relative to a hydrogen electrode.* In our example, a dc voltmeter must have its positive terminal clipped to the hydrogen electrode and its negative terminal to the zinc electrode, or else it will deflect off scale below the zero mark. Similarly, any other device for indicating polarity, be it chemical[10] or electrical, shows the zinc to be *negative* with respect to the hydrogen electrode.

Indeed, this was to be expected, for the zinc is the reducing agent and is giving up electrons as it dissolves; these electrons accumulate in the zinc and follow the metallic path out of it, making the zinc a rich source of electrons and hence a source of negative electricity. The hydrogen electrode *consumes* electrons, and so is always somewhat deficient in them in a flow system; that is, the hydrogen electrode has a positive charge with respect to the zinc. If we take the hydrogen electrode as agreed-upon reference point, then the electrode potential for zinc is -0.763 V (always with a negative sign).

A cell such as that shown in Figure 8-3 measures the electron-releasing tendency of zinc *relative to hydrogen,* and this only as it appears in an aqueous system. A different metal, such as copper, would manifest its different chemical behavior in a different voltage relative to hydrogen. Similarly, a different voltage would be expected for zinc versus copper instead of zinc versus hydrogen. For simplicity, half-cell potentials are always listed versus a particular accepted standard, and the internationally established standard is the hydrogen electrode. There are at least two good reasons for this: The hydrogen electrode is very reproducible and reliable; most of the metals liberate hydrogen from acids, while the metalloids, the noble metals and the nonmetals do not, so that a natural division ensues which often is helpful.

For the same reason, altitudes are measured relative to the arbitrary standard sea level.

With this method of measurement in mind, we may now define **standard electrode potential (E^0)** as *the electromotive force* (in volts) *developed by an oxidation or reduction half-cell* relative to a hydrogen half-cell taken as zero, all solutions being at unit activity and the cell operating at 25°C and one atmosphere pressure. For our purposes, "unit activity" may be considered the same as unit molarity, leaving the distinction to advanced courses in chemistry. If the substance involved in the half-cell under consideration turns out to be a stronger reducing agent (that is, more readily oxidized) than hydrogen under the stated conditions, the standard electrode potential E^0 will be negative for the reasons given. If the substance turns out to be a weaker reducing agent (that is, is more readily reduced) than hydrogen, the potential will be observed and listed as positive. The International Union of Pure and Applied Chemistry recommends that electrode potentials for all half-reactions be tabulated as though the reactions were *reductions,* as in

---

[10] A favorite indicator of polarity in the old days of cells and batteries was a strip of filter paper moistened with a solution of NaCl and phenolphthalein (an acid-base indicator which is colorless in acidic solution and pink in alkaline solution) and connected to the terminals. Since H^+ ions are discharged at the negative pole and given off as H_2, residual OH^- ions accumulate around this electrode and color the paper pink. Similarly, starch-iodide paper will turn black in the vicinity of a wire connected to the positive electrode of a cell or battery.

$$\text{MnO}_2 + 4\text{H}^+ + 2e^- = \text{Mn}^{2+} + 2\text{H}_2\text{O} \qquad E^0 = +1.23 \text{ V}$$

or

$$\text{Zn}^{2+} + 2e^- = \text{Zn} \qquad E^0 = -0.763 \text{ V}$$

and that the voltage be given its natural electrostatic sign. Because of the convention of writing all half-reactions as though they were reductions, the electrode potentials are sometimes called *reduction potentials*. When written in the reverse direction, they are oxidations, and the voltages are *oxidation potentials*, with opposite sign.

Table 8-1 gives some standard electrode potentials selected from the hundreds known.[11] Since any two half-reactions may be combined,[12] one being reversed to serve as reducing agent, it is seen that the table provides information about more than 2300 different possible reactions.

Even the briefest examination of Table 8-1 raises some questions which need to be answered at this point. There are some general trends which were to be expected. The strongest reducing agents, comprising the alkali metals of Group I and then the alkaline-earth elements of Group II, are found at the top of the list with the highest negative electrode potentials, indicating the greatest driving force to reduce other substances, and in turn to undergo oxidation and form positive ions. The strongest oxidizing agents, such as ozone, fluorine, and chlorine, are found at the bottom of the list with the greatest positive potentials.

But why should lithium precede cesium, cesium supposedly being a far more electropositive element? An answer is to be found in our commitment to the chemistry of dilute aqueous solutions. A measurement of the ionization potential of metal atoms *in a vacuum* does indeed show that cesium (ip 3.87 V) loses its one bonding electron much more readily than lithium (ip 5.36 V), and in dry reactions of the metals with a gas such as oxygen, this difference shows up in far greater reactivity for cesium. When we measure the tendency to form *hydrated* unipositive ions in dilute aqueous solution, however, as we do in measuring the standard electrode potential, we combine the *energy of ionization with the energy of hydration* of the resulting ion; the two cannot be separated. In the hydration reaction lithium ion readily excels cesium because its small size (0.60 Å) gives it a much more intense electric field than that of the cesium ion, which has the same charge but much larger radius (1.69 Å). Similarly, the intermediate element potassium loses its one electron less readily than cesium but hydrates *more* readily, and so shows an aqueous solution potential which is comparable with that of cesium. It is this combination of innate atomic reactivity *and* subsequent ionic hydration effects that

The tendency for reaction depends on the properties of reactants *and of* products.

---

[11] The values given in Table 8-1 are from W. M. Latimer, *Oxidation Potentials*, 2nd ed., Englewood Cliffs, N.J.: Prentice-Hall (1952). Mostly reactions in acid or neutral solutions are shown; many more, which occur in basic solution, are listed separately by Latimer.

[12] Notice that it is *not* necessary to select one half-reaction that stands above hydrogen in the series and one from below; any two may be used, for one substance is always a stronger oxidizing agent than the other.

determines the place of an element on the list of standard electrode potentials, and the merging of the two energies is unavoidable.[13]

TABLE 8-1
Selected Electrode Potentials

Half-Reaction	Standard Potential, E^0
$Li^+ + e^- = Li$	-3.045 V
$K^+ + e^- = K$	-2.925 V
$Rb^+ + e^- = Rb$	-2.925 V
$Cs^+ + e^- = Cs$	-2.923 V
$Ba^{2+} + 2e^- = Ba$	-2.90 V
$Sr^{2+} + 2e^- = Sr$	-2.89 V
$Ca^{2+} + 2e^- = Ca$	-2.87 V
$Na^+ + e^- = Na$	-2.714 V
$La^{3+} + 3e^- = La$	-2.52 V
$Mg^{2+} + 2e^- = Mg$	-2.37 V
$\frac{1}{2}H_2 + e^- = H^-$	-2.25 V
$Th^{4+} + 4e^- = Th$	-1.90 V
$U^{3+} + 3e^- = U$	-1.80 V
$Al^{3+} + 3e^- = Al$	-1.66 V
$Ti^{2+} + 2e^- = Ti$	-1.63 V
$Zn(NH_3)_4^{2+} + 2e^- = Zn + 4NH_3$	-1.04 V
$Zn^{2+} + 2e^- = Zn$	-0.763 V
$Cr^{3+} + 3e^- = Cr$	-0.74 V
$Au(CN)_2^- + e^- = Au + 2CN^-$	-0.60 V
$H_3PO_3 + 2H^+ + 2e^- = H_3PO_2 + H_2O$	-0.50 V
$Fe^{2+} + 2e^- = Fe$	-0.440 V
$Cd^{2+} + 2e^- = Cd$	-0.403 V
$PbSO_4 + 2e^- = Pb + SO_4^{2-}$	-0.356 V
$Co^{2+} + 2e^- = Co$	-0.277 V
$PbCl_2 + 2e^- = Pb + 2Cl^-$	-0.268 V
$Hg_2Cl_2 + 2e^- = 2Hg + 2Cl^-$	-0.2676 V
$Ni^{2+} + 2e^- = Ni$	-0.250 V
$N_2 + 5H^+ + 4e^- = N_2H_5^+$	-0.23 V
$AgI + e^- = Ag + I^-$	-0.151 V
$Pb^{2+} + 2e^- = Pb$	-0.126 V
$2H^+ + 2e^- = H_2$	0.0000 V
$Ag(S_2O_3)_2^{3-} + e^- = Ag + 2S_2O_3^{2-}$	$+0.01$ V
$S + 2H^+ + 2e^- = H_2S$	$+0.141$ V
$Sn^{4+} + 2e^- = Sn^{2+}$	$+0.15$ V
$SO_4^{2-} + 4H^+ + 2e^- = H_2SO_3 + H_2O$	$+0.17$ V
$PuO_2(OH)_2 + e^- = PuO_2OH + OH^-$	$+0.26$ V
$Cu^{2+} + 2e^- = Cu$	$+0.337$ V

---

[13] Neither is it objectionable, for electrode potentials have a very pragmatic purpose; we want to use them to bring order into *solution* chemistry and to serve practical electrochemistry as well.

TABLE 8-1 (Continued)
Selected Electrode Potentials

Half-Reaction	Standard Potential, E^0
$Fe(CN)_6^{3-} + e^- = Fe(CN)_6^{4-}$	+0.36 V
$H_2SO_3 + 4H^+ + 4e^- = S + 3H_2O$	+0.45 V
$MnO_2 + NH_4^+ + H_2O + e^- = NH_3 + Mn(OH)_3$	+0.5 V
$I_2 + 2e^- = 2I^-$	+0.5355 V
$Au(CNS)_4^- + 3e^- = Au + 4CNS^-$	+0.66 V
$O_2 + 2H^+ + 2e^- = H_2O_2$	+0.682 V
$Te + 2H^+ + 2e^- = H_2Te$	+0.70 V
$Fe^{3+} + e^- = Fe^{2+}$	+0.771 V
$Hg_2^{2+} + 2e^- = 2Hg$	+0.789 V
$Ag^+ + e^- = Ag$	+0.7991 V
$ClO^- + H_2O + 2e^- = Cl^- + 2OH^-$	+0.89 V
$PuO_2^{2+} + e^- = PuO_2^+$	+0.93 V
$NO_3^- + 3H^+ + 2e^- = HNO_2 + H_2O$	+0.94 V
$NO_3^- + 4H^+ + 3e^- = NO + 2H_2O$	+0.96 V
$HNO_2 + H^+ + e^- = NO + H_2O$	+1.00 V
$Br_2 + 2e^- = 2Br^-$	+1.0652 V
$ClO_4^- + 2H^+ + 2e^- = ClO_3^- + H_2O$	+1.19 V
$O_2 + 4H^+ + 4e^- = 2H_2O$	+1.229 V
$MnO_2 + 4H^+ + 2e^- = Mn^{2+} + 2H_2O$	+1.23 V
$Cl_2 + 2e^- = 2Cl^-$	+1.3595 V
$PbO_2 + 4H^+ + 2e^- = Pb^{2+} + 2H_2O$	+1.455 V
$Au^{3+} + 3e^- = Au$	+1.50 V
$MnO_4^- + 8H^+ + 5e^- = Mn^{2+} + 4H_2O$	+1.51 V
$HClO + H^+ + e^- = \frac{1}{2}Cl_2 + H_2O$	+1.63 V
$NiO_2 + 4H^+ + 2e^- = Ni^{2+} + 2H_2O$	+1.68 V
$PbO_2 + 4H^+ + SO_4^{2-} + 2e^- = PbSO_4 + 2H_2O$	+1.685 V
$MnO_4^- + 4H^+ + 3e^- = MnO_2 + 2H_2O$	+1.695 V
$H_2O_2 + 2H^+ + 2e^- = 2H_2O$	+1.77 V
$S_2O_8^{2-} + 2e^- = 2SO_4^{2-}$	+2.01 V
$O_3 + 2H^+ + 2e^- = O_2 + H_2O$	+2.07 V
$F_2 + 2e^- = 2F^-$	+2.87 V
$F_2 + 2H^+ + 2e^- = 2HF$	+3.06 V

As might be expected from this heavy dependence upon solution effects, the electrode potential of an element is influenced markedly by the presence of complexing or coordinating agents. This is one reason why the potentials have different values in basic solution, the other reason being the frequent participation of hydrogen ion in the reactions. Ions and molecules that are better donors of electron pairs than hydroxyl ion have an even more marked effect on the potential. For example, the oxidation of iron(II) to iron(III) ordinarily requires a quite potent oxidizing agent:

$$Fe^{3+} + e^- = Fe^{2+} \quad E^0 = +0.771 \text{ V}$$

but the hexacyanoferrite ion oxidizes much more easily:

$$Fe(CN)_6^{3-} + e^- = Fe(CN)_6^{4-} \quad E^0 = +0.36 \text{ V}$$

Or consider the noble metal gold. Putting gold into solution with the help of only the usual ionic hydration effects requires a very strong oxidizing agent,

$$Au^{3+} + 3e^- = Au \qquad E^0 = +1.50 \text{ V}$$

but in the presence of thiocyanate ion, it is much more easily oxidized:

$$Au(CNS)_4^- + 3e^- = Au + 4CNS^- \qquad E^0 = +0.66 \text{ V}$$

In fact, in the presence of cyanide (CN^-) ion, metallic gold is actually a strong reducing agent and is readily oxidized just by air:

$$Au(CN)_2^- + e^- = Au + 2CN^- \qquad E^0 = -0.60 \text{ V}$$

This is why gold dust can be extracted from its ore by the cyanide process.

Compare this to the effect of thiosulfate ion on the solubility of silver salts, Sec. 7-5.

Other examples of coordination effects that show up in measurements of standard electrode potential are the easy oxidation of silver in the presence of thiosulfate ion (+0.01 V versus +0.7991 V without thiosulfate), and the oxidation of plutonium (V) to plutonium(VI) in separation processes which takes place more easily in strongly basic solution (+0.26 V with OH^- ion present, +0.93 V without).

A similar consideration of reaction potential as the result of competition or collaboration between solution, hydration, and complexing effects shows why the potential changes considerably when an insoluble substance is involved. Thus, the silver versus silver-iodide half-reaction given in Table 8-1 has a negative potential (−0.151 V), whereas *dissolved* silver ion versus metallic silver shows a positive potential of +0.7991 V. Further, the potentials of metallic lead and lead dioxide as they change to insoluble lead sulfate in the lead storage battery also differ considerably from the corresponding potentials for Pb and PbO_2 versus dissolved Pb^{2+} ion, for the same reason. Since the formation of a precipitate substantially alters any ionic equilibrium, we must expect that the presence of a precipitate will have a substantial effect on the balance of half-cell equilibria that determines the measured reaction potentials.

A final remark about the position of hydrogen in Table 8-1 may be in order. Hydrogen was given a value of 0.0000 V in the list, *not* because it has no tendency to oxidize or reduce, but because it is a suitable reference electrode substance[14] that separates the "active" metals from the historically "noble" metals and the nonmetals. This arbitrary choice neatly divides the list between one group comprising the substances that are predominantly reducing agents (with potentials extending to approximately −3 V) and another group comprising the substances that are predominantly oxidizing agents (with potentials extending approximately to +3 V). These mere 6 V constitute the whole scale of electrochemical potential, showing how limited is the range of chemical force available in dilute aqueous solutions. How remarkable that all the electron rearrangements encompassed by thousands

---

[14] Whatever its chemical virtues, hydrogen does not lend itself to a convenient *physical* arrangement for a reference electrode because of the problem of gas supply. The saturated calomel electrode, using solid Hg_2Cl_2 and a saturated solution of KCl in contact with metallic mercury, is much more convenient, and often is used as a working standard. It has a potential of +0.2444 V versus the hydrogen electrode, at 25°C.

upon thousands of reactions in solution should be made possible by such a modest electromotive force!

CHANGE OF POTENTIAL WITH CONCENTRATION. We have seen that chemical reagents in a chemical system develop an electromotive force (emf) if they have a tendency to react by transfer of electrons. It seems logical that this emf should decrease as the concentration of reactants decreases, for the system then is under less stress to react and has lost part of its capacity for reaction. Similarly, as the products of the reaction accumulate, the reverse reaction should begin to operate against the initial tendency to react, decreasing the observed potential. We want now to inquire about the quantitative relation between observed potentials and the concentrations of reagents and products.

For any general reaction of the type

$$a\text{A} + b\text{B} \rightleftarrows c\text{C} + d\text{D}$$

(where the lowercase letters represent coefficients or molar proportions of the reagents A and B and the products C and D), the expression relating observed voltages to standard electrode potentials is

$$\text{observed voltage} = \text{E}^0 - \frac{0.059}{n} \log \frac{[\text{C}]^c[\text{D}]^d}{[\text{A}]^a[\text{B}]^b} \qquad (8\text{-}12)$$

where n is the number of electrons transferred in the redox reaction; it is 2 in the reaction

$$\text{Zn} + \text{Cu}^{2+} \rightleftarrows \text{Zn}^{2+} + \text{Cu}$$

or 3 in the reaction

$$\text{Al} + 3\text{H}^+ \rightleftarrows \text{Al}^{3+} + \tfrac{3}{2}\text{H}_2$$

The figure 0.059 comes from a combination of constants which are inserted in a more general equation, called the Nernst equation,[15] which relates chemical energy to chemical "activities" (which, in turn, are related to concentrations). This equation allows us to determine the effect of a nonstandard concentration (not 1 M) on the observed electrode potential. It also allows us to calculate equilibrium constants, and it points out the way to manipulate concentrations to obtain electrical energy. These three aspects will be considered separately.

First, let us inspect Equation (8-12) closely. We see that if the reactants A and B and the products C and D are all at unit concentration (1 M), and the coefficients a, b, c, and d, are all 1 (for simplicity), then the fraction

A few manipulations

$$\frac{[\text{C}]^c \, [\text{D}]^d}{[\text{A}]^a \, [\text{B}]^b}$$

---

[15] For details about this application of the Nernst equation, see T. R. Blackburn, *Equilibrium: A Chemistry of Solutions*, New York: Holt, Rinehart & Winston, especially pp. 152–168 (1969).

becomes unity. Its logarithm then becomes zero, for

$$\log 1 = 0$$

and the observed voltage is the standard potential. On the other hand, if the products are present at high concentrations, let us say 10 M, while the reactants stay at 1 M, then we have

$$\log \frac{10 \times 10}{1 \times 1} = \log 100 = 2.00$$

and the correction term becomes

$$-\frac{0.059}{n} \times 2 = -\frac{0.118}{n}$$

For a one-electron change, the result would be

$$\text{observed voltage} = \mathbf{E}^0 - 0.118 \text{ V}$$

Similarly, if the concentrations of reactants were high compared with those of the products, the opposite effect would occur. At 10 M concentrations for the reactants, and only 1 M concentrations for the products,

$$\log \frac{1 \times 1}{10 \times 10} = \log 0.01 = -2$$

so the correction term becomes

$$-\frac{0.059}{n} \times (-2) = +\frac{0.118}{n}$$

and for a one-electron change, the result would be

$$\text{observed voltage} = \mathbf{E}^0 + 0.118 \text{ V}$$

Both results are predicted by logic, for a high concentration of reactants (relative to products) means a high pressure to react, and hence a high reaction potential, while a low concentration of reactants (relative to products) means a diminished pressure to react in the face of an augmented reverse reaction, and hence a decreased potential.

Since different voltages are expected at different relative concentrations of reactants and products in an electron-transfer reaction, it should be possible to construct a cell utilizing the same half-reaction (but at two different concentrations) in both the half-cell compartments. Such an arrangement is called a **concentration cell**. Here the tendency to react is greater in one half-cell than in the other, and so a current of electrons flows in the wire connecting the electrodes, while an equal current of ions flows through a salt bridge connecting the two half-cells. Such concentration cells are seldom used to provide energy, but may very well be used to measure concentration by measuring the potential generated between the half-cell of unknown concentration and a similar half-cell operating at known concentration. Indeed, this is the principle of the well-known *pH meter*, an instrument widely used in chemical, medical, and industrial practice to measure the acidity of a solution. We can picture two hydrogen-coated platinum electrodes, one immersed in a solution of 1 M HCl as the standard half-cell and the other immersed in the solution of unknown concentration being tested. If we provide a salt bridge or suitable equivalent as an ionic pathway between the solutions, and then compare the potentials on the platinum electrodes by means of a sensitive potentiometer, we should be able to use the Nernst equation to calculate the concentration difference from the potential difference. The actual pH meter saves us the work and is much more ingenious in other ways too. A very thin membrane of glass takes the place of a salt bridge; it is permeable to Na$^+$ and H$^+$ ions and at the same time prevents mixing of the two solutions. Within the hemispherical glass membrane is the reference half-cell of fixed potential, usually the Hg$_2$Cl$_2$ cell or one just as compact and reliable, and not the bulky H$_2$/Pt/1 M HCl half-cell. The potentiometer, which indicates or records the potential difference, has a scale graduated directly in pH (see Chapter 7) instead of in millivolts, and its output may be recorded or even used to control operations in a chemical plant. (For more about pH meters, see Section 16-4.)

A further application of the principles outlined in this section is the calculation of *equilibrium constants* from observed cell potentials or from a table of standard electrode potentials. Suppose we had the reactants and products initially at standard conditions, so that the observed voltage was equal to E^0. We then connect the half-cells with a stout wire and let the reaction proceed. As the reactants are used up and the products accumulate, the observed voltage will fall. When the voltage dwindles to zero, the reverse reaction has become so prominent (due to increased concentration of products) that it balances all tendency of the reaction to continue in the forward direction. All capacity for further change is gone, and the chemical system stands at equilibrium. At this point

E^0 and K are two different numbers that tell the same story.

$$\text{observed voltage} = 0 = E^0 - \frac{0.059}{n} \log K$$

and so

$$\frac{0.059}{n} \log K = E^0$$

and

$$\log K = \frac{E^0 n}{0.059}$$

The equilibrium constant can be calculated for any reaction for which we can obtain an E^0 value. If a table of standard electrode potentials is used, the same table will indicate the value of n, the number of electrons transferred in the unit process; if not, the equation for the reaction must be written in order to determine n.

EXERCISE: Determine the equilibrium constant for the reaction of metallic zinc with a solution of a copper salt, as used in the ancient Daniell cell:

$$Zn + CuSO_4 = ZnSO_4 + Cu$$

or

$$Zn + Cu^{2+} = Zn^{2+} + Cu$$

or

$$Zn = Zn^{2+} + 2e^- \quad \text{and} \quad 2e^- + Cu^{2+} = Cu$$

Answer: The two half-cell standard potentials, 0.763 V and 0.337 V, add to give a total cell voltage E^0 of 1.10 V. (The basis for the additivity will be treated in the next section.) Hence we have

$$\log K = 1.10 \frac{n}{0.059}$$

and since $n = 2$ in the half-reaction above,

$$\log K = \frac{2.20}{0.059} = 37.3$$

This means that the equilibrium constant for the reaction is in the neighborhood of 10^{37}; from a table of logarithms, the value is found to be 2×10^{37}. Since the concentration of product (Zn^{2+}) is 2×10^{37} times as great as the concentration of reactant (Cu^{2+}) at equilibrium, we are justified in saying that this reaction is essentially complete.

EXERCISE: Calculate the equilibrium constant for the oxidation of lead ions (Pb^{2+}) to lead dioxide by the action of potassium permanganate in acid solution, making use of the standard potentials given in Table 8-1.

Answer: $\log K = 9.32$ and $K = 2.09 \times 10^9$.

DETERMINING THE FEASIBILITY OF A REACTION. Since standard electrode potentials give us a measure of the potential for reaction (and potential times quantity of electricity equals energy), we can find out how much energy to expect from the reaction of a given quantity of oxidizing agent and reducing agent. We simply add the standard electrode potentials for the two half-reactions, multiply the sum by the number of moles of reactant in shortest supply times the appropriate constant, and we have the quantity of free energy available. That is,

$$\text{volts} \times \text{moles} \times (96{,}500 \text{ C/mole}) = \text{energy in joules}$$

and

$$\text{volts} \times \text{moles} \times (23{,}100 \text{ cal/mole-V}) = \text{energy in calories}$$

If the calculated amount of energy has a positive sign, the reaction will start up spontaneously and go toward completion (as far as the equilibrium constant allows) until the reagent in shortest supply is used up. If the energy has a negative sign, the reaction will not proceed by itself; energy would have to be poured into the reaction from another source.

Just what is the physical meaning of combining two half-reactions or their half-cells? Why should the potentials be additive? To make the answers clear, we shall consider first (and rather slowly) an example that is simple. We shall also find the answer to a question central to all of chemistry: "Will this reaction go?" Then we shall look at some examples where the answers are not so obvious.

Since all other parts of this calculation have positive signs, the sign of E^0 is the same as that of the energy as calculated here.

EXERCISE: We begin with the zinc and copper half-cells

$$Zn^{2+} + 2e^- = Zn \quad E^0 = -0.763 \text{ V}$$

$$Cu^{2+} + 2e^- = Cu \quad E^0 = +0.337 \text{ V}$$

and we ask whether the reaction of zinc metal with dissolved copper ion will start up and proceed spontaneously. We are asking whether the reaction

$$Zn + Cu^{2+} = Zn^{2+} + Cu$$

will proceed spontaneously in the indicated direction. It will. The outmoded Daniell cell used this very reaction. But here we want to know what can be predicted from the standard electrode potentials in Table 8-1.

Answer: We find first that the Zn^{2+}, Zn half-cell has a potential of -0.763 V relative to the hydrogen half-cell, and so the Zn, Zn^{2+} half-reaction has a potential of $+0.763$ V. (Remember that reversing a stated half-reaction reverses its sign; here we want to start with *metallic* zinc.) The Cu^{2+}, Cu half-cell has a potential of $+0.337$ V. The zinc and copper cells compare as shown diagrammatically in Figure 8-4. Notice that electrons flow *out* of the zinc but *into* the copper.

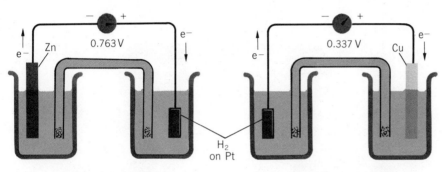

Zinc dissolving to form zinc ions:

$$Zn = Zn^{2+} + 2e^-$$

Copper ions reduced to metal:

$$Cu^{2+} + 2e^- = Cu$$

FIGURE 8-4 · A Zn, Zn^{2+} oxidation half-cell coupled to a hydrogen half-cell, and a Cu^{2+}, Cu reduction half-cell coupled to a hydrogen half-cell. Coupling is via copper wires for electron flow, and salt bridges for ion flow.

The Cu^{2+}, Cu versus hydrogen cell would operate just as well if we used the same hydrogen half-cell as reference electrode for both metals (see Figure 8-5). Obviously, a single voltmeter connected from the zinc electrode directly to the copper electrode would read $0.763 \text{ V} + 0.337 \text{ V} = 1.10 \text{ V}$. Furthermore, since equal numbers of electrons are going into and out of the hydrogen electrode,[16] the net current down the wire to this electrode is zero. *We could dispense with it entirely.* Removing the whole middle beaker, and connecting the two end beakers with one salt bridge,[17] we get the simpler arrangement shown in Figure 8-6.

Removing the middle electrode cannot change the meter reading.

---

[16] Remember that oxidation and reduction must not only proceed simultaneously, but also to equal extents. For every electron given up by the zinc, one must be taken up by copper.

[17] The beaker containing the hydrogen electrode is, in fact, serving only as part of the salt bridge.

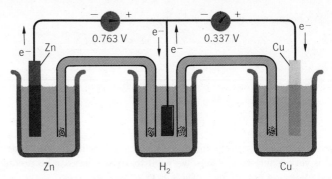

FIGURE 8-5 A Zn Zn^{2+} oxidation half-cell and a Cu^{2+}, Cu reduction half-cell, both coupled to the same hydrogen half-cell. The voltmeter readings are the same as in Figure 8-4.

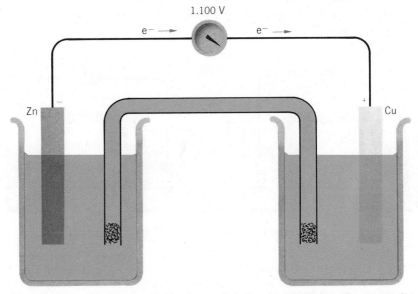

FIGURE 8-6 Zinc and copper half-cells coupled directly, eliminating the intermediate hydrogen reference electrode. The observed voltmeter reading is the sum of the two voltmeter readings in Figure 8-4, or in Figure 8-5.

This demonstration of additivity, arranged by physically coupling two actual electrochemical cells, reveals by manipulation of electrodes and wires the mathematical operations we need to perform in this and in every other case:

A principle and some examples

1. The oxidation half-reaction is rewritten in the direction in which it is to be used, with its corresponding standard electrode potential necessarily changing sign:

$$\text{Zn} = \text{Zn}^{2+} + 2e^- \qquad E^0 = +0.763 \text{ V}$$

2. The reduction half-reaction also is written in the direction in which it is to be used, which is the same way as it appears in the table:

$$Cu^{2+} + 2e^- = Cu \quad E^0 = +0.337 \text{ V}$$

Be sure that all the electrons released in reaction 1 are used in reaction 2.
3. The resulting potentials are then added to get the overall voltage:

$$+0.763 \text{ V}$$
$$\underline{+0.337 \text{ V}}$$
$$+1.100 \text{ V total}$$

Since the voltage is positive, free energy will be produced by the reacting system, and the reaction can be expected to proceed of its own accord. We cannot tell anything about the *rate* of reaction, but the higher the total potential, the greater will be the driving force of the reaction and the further it will go before it reaches equilibrium.

EXERCISE: Let us consider a reaction which is not so well known:

$$H_3PO_2 + CoCl_2 + H_2O = H_3PO_3 + Co + 2HCl$$

Here we should like to know whether or not Co^{2+} ions can be reduced to metal by hypophosphorous acid.

Answer: To find out, we follow the sequence of operations just developed:

1. Write the oxidation half-reaction in the direction in which it is to be used (the reverse of what appears in the table):

$$H_3PO_2 + H_2O = H_3PO_3 + 2H^+ + 2e^- \quad E^0 = +0.50 \text{ V}$$

2. Write the reduction half-reaction in the direction in which it is to be used (which happens to be the same as that given in the table):

$$Co^{2+} + 2e^- = Co \quad E^0 = -0.277 \text{ V}$$

3. Add the half-cell potentials:

$$+0.50 \text{ V}$$
$$\underline{-0.277 \text{ V}}$$
$$\text{total } +0.223 \text{ V for the } H_3PO_2\text{--}Co^{2+} \text{ cell}$$

4. Since the resulting potential is positive, the reaction will proceed spontaneously. Cobalt will be precipitated; free energy will be released; and the reaction will go on until a condition of chemical equilibrium is reached.

EXERCISE: Now let us consider whether a similar reduction of Cr^{3+} ion by hypophosphorous acid can occur:

$$3H_3PO_2 + 2CrCl_3 + 3H_2O = 3H_3PO_3 + 2Cr + 6HCl$$

Answer: We proceed by the same steps as before:

1. Write the oxidation half-reaction:

$$H_3PO_2 + H_2O = H_3PO_3 + 2H^+ + 2e^- \quad E^0 = +0.50 \text{ V}$$

(This is necessarily the reverse of what appears in Table 8-1.)

2. Write the reduction half-reaction:

$$Cr^{3+} + 3e^- = Cr \quad E^0 = -0.74 \text{ V}$$

3. Add the half-cell potentials:

$$+0.50 \text{ V}$$
$$\underline{-0.74 \text{ V}}$$
$$-0.24 \text{ V total}$$

4. This reaction will *not* go spontaneously; it is endoergic in terms of free energy. Hence chromium(III) ion is *not* reduced to metal by hypophosphorous acid the way nickel, cobalt, thallium, indium, and many other ions are reduced.

Significance of the sign of the cell potential

Other reactions of solution chemistry can be tested in the same way for feasibility. The result is definitive: A positive net potential always indicates a reaction that will proceed of its own accord. A negative net potential, on the other hand, tells us that the reaction is incapable of proceeding of its own accord. There mere fact that a reaction is possible tells us nothing about its rate, however, nor anything about how to conduct the reaction. These are questions that must be answered by experiment and by experience.

8-8 The Concept of Entropy

Or—what's the difference between free energy and heat?

Having learned how to determine the potential and energy of a redox reaction, we now turn to a favorite lecture-demonstration experiment in which zinc is oxidized in one half-cell and free iodine is reduced in another. The two half-cells are arranged so that ions may move readily from one half-cell to the other through a porous diaphragm, and the two electrodes are connected by copper wires to a small lamp, as shown in Figure 8-7. An electric circuit thereby is completed, and as electrons flow from the zinc (as it is oxidized) to the iodine (which is being reduced), the pressure and volume at which they flow through the tiny tungsten wire cause the lamp to light up and to glow as long as the reagents last. This is a simple form of electrochemical **cell** to produce electrical energy from chemical reaction, and an assembly of two or more such cells is called a battery. (In colloquial usage, even a single cell is called a battery.)

The standard electrode potentials for the zinc and iodine half-cells sum to 1.30 V; a somewhat higher voltage is achieved in practice by raising the concentration of iodine above 1 M. Using the relation given in the previous section, the free energy liberated by the consumption of one mole of zinc or iodine is 60,014 cal. This represents the free energy of formation of one mole of zinc iodide

$$Zn + I_2 = ZnI_2$$

as carried out in an electrochemical cell.

It is even easier, of course, to allow zinc to combine with iodine without benefit of an electrochemical cell, and to obtain zinc iodide directly. When this is done under laboratory conditions in a calorimeter (an instrument which measures heat output), and the resulting zinc iodide is dissolved in water in the same calorimeter, it is found that the direct formation and solution of one mole of zinc iodide produces 61,400 cal of heat, or 1386 cal more than the electrochemical preparation. Why is there a difference?

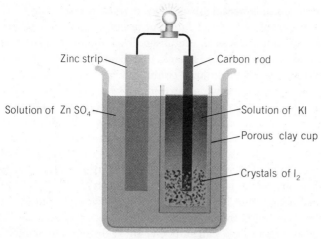

FIGURE 8-7 Zinc metal is converted to Zn^{2+} ions by oxidation of the zinc strip electrode, while at the same time solid I_2 is reduced to I^- ions. Electrical contact with I_2 crystals is made with a carbon rod. Electrons, flowing from zinc strip to carbon rod, do work in the high-resistance filament of the light bulb, heating the filament to incandescence. The electrical circuit is completed by passage of ions through the porous walls of the clay cup that holds the KI solution and the I_2 crystals.

We inquire first whether the reaction between zinc and iodine is unique, and we find that it is not. Practically all electrochemical reactions produce an amount of chemical or electrical energy per mole which differs from the amount of heat energy liberated by direct combination of the reagents. The difference between the two results illustrates the fundamental difference between free energy and heat energy, a difference which will be explored in more detail in Chapter 9. The difference involves the concept of **entropy**, a concept which may be explained in many ways, but one which emerges clearly from the present example.

There are many forms of energy. The various forms of so-called *free energy* are free in the sense of being entirely available for conversion into the other forms. Electrical energy may be converted into mechanical energy in an electric motor, for example, or the reverse change can occur in a generator or dynamo, and the only losses will be minor ones due to the electrical resistance of copper wire and magnetic losses in the iron. The better we make the electrical machine (using heavier copper wire or even silver wire for the conductors and improved silicon-steel laminations in the magnetic circuit, for example), the more we can reduce these losses, so that with further improvement in materials and design, the completely quantitative interconversion of electrical and mechanical energy becomes an approachable reality. Similarly, in a well-designed electrochemical cell, the interconversion of chemical and electrical energy becomes nearly quantitative, and so we have considered the two forms of energy as completely equivalent in the calculations just carried out. Similar arguments apply to the conversion of electricity to light (and vice versa), the conversion of light and other forms of radiant energy to chemical change, and so on. Only when we come to thermal energy do we

What's free about free energy?

Some value judgments about kinds of energy

meet up with a fundamental block. *The various forms of free energy are completely convertible into heat, but heat is not completely convertible into free energy.* This point has been established by a hundred years of strenuous effort; the answer always comes back that heat is a lower-grade form of energy, and that not all the caloric content of heat can be recovered in the form of free energy. Hence we must think not only of a *quantity* of energy but of its *quality* as well, and heat is a low-quality or degraded form of energy. This is why the thermal change that accompanies a chemical reaction (the change of enthalpy, to use a term from thermochemistry, which is covered in Chapter 9) is not, in itself, a sufficient guide as to what reactions will go and what ones will not, and this is why a consideration of *free-energy change* becomes necessary to the understanding of chemistry.

If thermal energy and free energy are both measurable in exact units, is not the difference between them measurable? Indeed so, and the difference is proportional to what we call **entropy**. In Webster's Dictionary, entropy is defined as "a measure of that portion of heat energy which is not recoverable in a heat engine," and so it measures exactly the loss or the difference we have been considering. We may write the definition as an equation,

$$\Delta H - \Delta G = T\Delta S \qquad (8\text{-}13)$$

where ΔH is the change in heat content or enthalpy associated with a particular transformation (such as we are considering in the formation of zinc iodide from zinc and iodine); ΔG is the corresponding free-energy change; T is the absolute temperature; and ΔS is the change in entropy, expressed simply in "entropy units." The equation usually is written in textbooks as

$$\Delta G = \Delta H - T\Delta S \qquad (8\text{-}14)$$

but you may remember it better written as a literal translation of the definition, as in Equation (8-13).

Entropy and disorder

The concept of entropy is new to most readers of this book. To give it concrete meaning, let us consider entropy to be a measure of the *degree of disorder* in a chemical system. A pack of playing cards all neatly arranged in order of suit and sequence has mass and can acquire momentum; it can knock over an empty flower vase in one well-aimed shot. But if the cards become scattered and disordered in the process, a great deal more work is required to put them back in their previous well-ordered condition. Here entropy is a measure of the irrecoverable work associated with the increased disorder. In the same way, if the myriad atoms in an orderly crystal of metal are scattered to random positions in space by chemical reaction, there is an increase in entropy. To restore the system to its original condition, the atoms of metal must be separated from their partners in chemical combination (as in dissociating the Zn^{2+} and I^- ions of ZnI_2), and the metal atoms must be neatly arranged in rows and layers corresponding to their original lattice. This is why the degree of order and disorder (that is, entropy) in a chemical system must be considered together with changes associated with the making and breaking of chemical bonds.

But why the term $T\Delta S$ rather than just ΔS in Equation (8-13)? Reflecting now on the difference between heat energy and free energy, we see that heat energy involves a more disordered condition of matter. Heat consists in *random* motion of particles such as atoms or molecules, and temperature measures the degree of this random thermal agitation. Randomly moving molecules of a gas, for example, have an average kinetic energy which is proportional to the absolute temperature T. If we cool the gas, the random motion decreases. If we continue cooling the gas, eventually it condenses and freezes to a solid, certainly a more ordered form of the substance. Nevertheless, the lattice components of the solid continue in random motion about their mean lattice positions; further cooling toward absolute zero reduces such motion. A perfect[18] crystal at 0°K has an entropy content of zero, but at any temperature above that, it has a definite entropy content which increases with the absolute temperature. This is why the difference between heat energy and free energy appears in our equation as $T\Delta S$, instead of just ΔS.

Thinking of entropy as a measure of disorder reveals just why heat energy is not quantitatively convertible to free energy. All the forms of *free* energy may be thought of in terms of orderly motion: the unidirectional motion of translation of a train, the unidirectional rotational motion of a crankshaft or a jet-engine rotor, the unidirectional motion of electrons in a copper wire carrying an electrical current, or the unidirectional motion of light rays through space. Heat energy, on the other hand, consists of *random* motion of atoms and molecules. In order to convert heat energy to free energy, certain components of the random motion must, in effect, be selected, while other components must be thrown away. Hence every heat engine in actual use must reject some low-grade heat while it is transforming another component of heat to nice, interconvertible free energy, and that is why an automobile throws away low-grade heat from its radiator or cooling fins while it is turning the wheels, and why a central electric power station must throw away low-grade heat in its condenser cooling water while it generates electricity.

What's wrong with disorder?

Recyclable energy versus throw-away heat

Now let us return to our chemical reaction of zinc and iodine. We saw that if the reaction were carried out in an appropriate electrochemical cell, the accompanying change in chemical free energy could be measured by measuring the completely equivalent and convertible amount of electrical energy moving through the external circuit. This energy, ΔG, amounted to 60,014 cal. On the other hand, if we allowed the electrons from zinc to move directly to iodine atoms without benefit of half-cells and wires and instruments, simply by carrying out the reaction in a single vessel called a calorimeter, all of the chemical energy was transformed into a measureable amount of heat called ΔH, found to be 61,400 cal. Then the difference

$$\Delta H - \Delta G = 61{,}400 \text{ cal} - 60{,}014 \text{ cal} = 1386 \text{ cal}$$

is $T\Delta S$, the entropy change associated with the reaction. How can this entropy change be related to changes in molecular order accompanying the reaction in which zinc iodide is formed and dissolved? The zinc crystal is an orderly array of zinc atoms closely packed together. Iodine is a rather open but very orderly structure of oriented I_2 molecules, each made up of two iodine atoms

Entropy and structure

---

[18] An *imperfect* crystal will have some inherent disorder even at 0°K, of course, and so has an inherent entropy content associated with its state.

bound together. Some of the chemical energy of the oxidation of zinc (and reduction of iodine) must be consumed in pulling atoms of zinc out of their crystal positions, and in dissociating I_2 molecules pulled from the iodine lattice. The solid cubic crystals of zinc iodide formed in the dry reaction also are orderly, but they are less dense and hence less tightly ordered than those of zinc metal (density = 4.74 g/cm^3 versus 7.14 g/cm^3 for the metal) or even of iodine (density = 4.94 g/cm^3). So far, then, there has been a decrease in order as zinc iodide is formed. In the solution process, we start with liquid water, the molecules of which are in a state of considerable disorder, and allow the molecules to arrange themselves about zinc and iodide ions in hydrates such as $Zn(H_2O)_4^{2+}$. This orientation is due to the intense electric field of the small but doubly charged Zn^{2+} ion. Similar orientation occurs in the field of each negative iodide ion. Considering the combination of processes, noting the very large number of water molecules oriented around ions, the increase in degree of disorder has not been large, but it has been sufficient to make the heat change somewhat greater than the change in free energy. Hence as the reaction proceeds electrochemically, *the cell gets warm* at the same time free energy is furnished to the external circuit.

In other reactions, the free-energy change in an electrochemical reaction may differ much more from the thermal change accompanying direct reaction, and the obtainable heat may even be less than the obtainable free energy. In one example,[19] where lead is caused to undergo an electron exchange with mercury (I) in the reaction

$$Pb + Hg_2Cl_2 = PbCl_2 + 2Hg$$

the net potential is −0.536 V, and the change in free energy per mole is

$$\Delta G = 0.536 \times 2 \times 23{,}100 = 24{,}720 \text{ cal}$$

Cooling off by doing work

However, the heat evolved when the reaction takes place in a calorimeter instead of an electrochemical cell is only 22,730 cal. Here the free energy for doing work *exceeds* the combined heats from reaction and hydration by 1990 cal, and therefore a Pb–Hg_2Cl_2 cell turning out this free energy *absorbs* heat from its surroundings and gets *colder* during operation. This is no contradiction of thermodynamics, but a demonstration of the generality of its laws. In such cases, where ΔG exceeds ΔH, the electrochemical cell acts as a transformer, changing some of the heat of the surroundings (that is, thermal motion) into useful work. The close coupling of the reacting atoms and molecules with the vibrating molecules of their surroundings enables them to do this. And once again we are reminded that it is the *change of free energy* during a reaction that is the fundamental criterion to be considered in predicting the course of that reaction, not just whether it is exothermic or endothermic. If free energy is released, a reaction can start off spontaneously and continue, even though the system may get colder because of the reaction.

---

[19] From G. N. Lewis and M. Randall, *Thermodynamics*, 2nd ed., revised by K. S. Pitzer and L. Brewer, New York: McGraw-Hill, pp. 158–160 (1961). The entire Chapter 15 in this book will prove to be a helpful reference.

8-9 Practical Considerations about Electric Power from Chemical Reactions

Since oxidation-reduction reactions are capable of generating an electrical potential when carried out under suitable conditions, and since the oxidation or reduction of a particular number of equivalent weights of reagents gives a corresponding quantity (that is, the same number of faradays) of electricity, it becomes possible to obtain practical quantities of electrical energy, as well as to make calculations and predictions. This process of converting chemical free energy to electrical free energy has been exploited for over a century and a half in "primary" cells, storage cells, and batteries of such cells for the production of electricity. The principles of operation of such power-producing cells are readily understood in terms of the preceding sections of this discussion.

Getting down to practical cases (more or less)

Since electrical energy is measured in terms of joules or their equivalent, such that

$$\text{energy in joules} = \text{volts} \times \text{coulombs}$$

it follows that the most productive reactions for the generation of electrical energy (or power, which is energy per unit of time) would theoretically be those that furnish the highest oxidation potentials and also involve the maximum quantity of electricity in the conversion of a given weight of starting reagent. Very practical application of standard electrode potentials leads us to a choice of the more active metals, like magnesium and aluminum, as desirable reducing agents for cells or batteries. In terms of *quantity* of electrons, those reagents that have the lowest equivalent weight (that is, which can furnish the largest number of faradays per pound) would be best, and here the light metals again come to the fore. What we should like on theoretical grounds, then, is a cell which oxidizes lithium, aluminum, magnesium, or carbon with oxygen or hydrogen peroxide, for example.

Unfortunately, these principles soon come to grief when they are directed toward practical use in working cells and batteries, because of some other special requirements which are imposed. A battery is expected at times to give large amounts of power immediately upon demand, and also at other times to lie idle for long periods of time without loss of reagents. It should not involve gases, which are bulky and awkward, and preferably should not be subject to spilling or freezing. The voltage generated by the battery should not decrease steadily (as it inevitably would do if only dissolved reagents were used), but should be maintained at a constant level until all the reacting materials are gone. Finally, the internal resistance of the cell must be low, so that no large portion of the generated electrical energy will be lost as heat when large currents flow. It is especially necessary that no high-resistance films of gases or of insoluble solids be formed directly upon the electrode surfaces as a result of reaction (that is, no "polarization"). These conditions rule out gaseous oxidizing agents like oxygen or fluorine,[20] and they discourage

---

[20] For special reasons, gases are used in the special electrochemical devices known as fuel cells; see Section 8-11.

use of the light metals because these polarize readily through formation of adherent insulating films of products. The cells that have been most successful through many years of use are those that use solid reagents in a minimum of liquid electrolyte and that make special provisions for the prevention of polarization. These reagents are incorporated in *primary cells,* which are intended to use up their reagents in one reaction and are then discarded, or in *storage cells* ("accumulators" in Europe), which are rechargeable through reversal of their energy-producing reactions.

8-10 Some Practical Batteries

The best-known type of portable primary cell is the Le Clanche dry cell, which uses metallic zinc as the reducing agent and manganese dioxide as oxidizing agent. The mechanical arrangement is shown in Figure 8-8. A carbon rod is immersed in a paste of ammonium chloride (to provide ions), manganese dioxide (oxidizing agent), and powdered carbon (to decrease the internal resistance), contained in a porous paper cup within a zinc cylinder. The paper separates the oxidation and reduction half-reactions, but provides a low-resistance path for the ions. The reactions are

oxidation: $\quad Zn + 4NH_3 = Zn(NH_3)_4^{2+} + 2e^-$

reduction: $MnO_2 + NH_4^+ + H_2O + e^- = NH_3 + Mn(OH)_3$

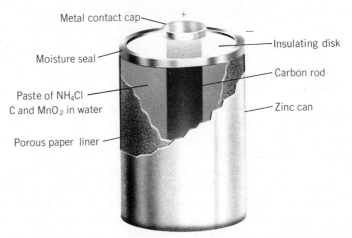

FIGURE 8-8 The LeClanche dry cell, a portable 1.48 V electrochemical cell commonly used for flashlights. Electrical energy is supplied by oxidation of the zinc container by manganese dioxide.

Since zinc is being oxidized, the zinc can has a surplus of electrons and therefore is the negative electrode. The carbon rod is the positive terminal. A potential of 1.48 V is developed, maintained fairly well during most of the life of the cell although dropping considerably in very cold weather. The long-standing advantages of the dry cell are its portability, its reliability under changing loads, its good shelf life, and its inexpensive simplicity. As disadvantages, the cell consumes its own container and therefore becomes leaky if the zinc is used efficiently, and its voltage is not as constant under high load as might be desired. As for its output, each equivalent of zinc (32.7 g) gives 96,500 C, or about 27 A-hr of electrical energy at 100-percent efficiency. At 75-percent efficiency, which is more likely, each equivalent of zinc would produce 72,000 C, which is 72,000/(60 × 60) = 20 A-hr. A flashlight cell probably provides only a tenth of one equivalent of zinc for reaction, producing 2 A-hr at 1.48 V. This is about 3 W-hr of electrical energy, and, at 5 cents per 1000 W-hr for domestic electrical energy, is worth about 0.015 cent, or about 1/2000 of the cost of the dry cell. This is not a chemical performance to be proud of.

How "good" are your flashlight batteries?

Another type of primary cell widely used in instruments is the sealed mercury cell, shown in Figure 8-9. In this cell, zinc is oxidized in an alkaline medium to zinc hydroxide (which dissolves in the solution of potassium hydroxide used as electrolyte) at the negative electrode, and mercury(II) oxide is reduced to mercury at the positive electrode. The cell maintains its potential of 1.32 V extraordinarily well throughout its life. It is capable of furnishing large currents without polarizing, because any hydrogen formed is oxidized rapidly by the mercury oxide. One variation of the cell uses amalgamated zinc (that is, zinc covered with a layer of mercury) as the highly reactive reducing agent and incorporates powdered silver with the mercury oxide to reduce its electrical resistance.

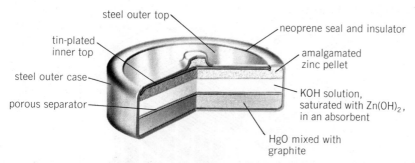

FIGURE 8-9 A mercury dry cell, a very stable source of a 1.32 V potential, used in electronic equipment. At the zinc electrode, zinc metal is oxidized to Zn^{2+} ions. At the other electrode, HgO is reduced to metallic mercury. Excellent electrical contact with the mercury is achieved by using a mercury-iron alloy (mercury alloys are called *amalgams*) as the electrode.

There is no fundamental distinction between *primary cells* and *storage cells*, but there is an important practical distinction. Storage cells are mechanically designed for repeated reversals of the oxidation-reduction reaction that furnishes the electrical energy. Such ready reversibility also requires a favorable selection of reactants, of course, but once choice has been made, the principal requirement is to keep the products of reaction in close proximity to the electrodes so that they may readily be converted back to reagents. Since the reagents must be solids (for maintenance of the potential), this mechanical requirement poses some difficulty. It has been met most successfully through the years by the lead-acid storage battery, which we shall consider here in detail.

Storage cells

The mechanical arrangement of a single lead storage cell is shown in Figure 8-10. The oxidizing agent is lead dioxide, compressed into thin slabs supported by an open grid made of lead. A thin sheet of wood or woven glass fiber separates this

PbO_2 electrode from the reducing agent, a spongy form of metallic lead compressed into a thin slab on a similar supporting grid of metallic lead. Another separator keeps this from another slab of oxidizing material, and so on. The alternate plates or slabs are connected together, the spongy lead plates in parallel forming the negative electrode of the cell, and the lead dioxide plates the positive electrode. The assembly is immersed in a solution of moderately concentrated sulfuric acid contained in a plastic or hard-rubber jar, and in this solution the following half-reactions take place:

$$\text{Oxidation:} \qquad Pb(s) + SO_4^{2-} = PbSO_4(s) + 2e^- \qquad + 0.356 \text{ V}$$

$$\text{Reduction:} \quad PbO_2(s) + SO_4^{2-} + 4H^+ + 2e^- = PbSO_4(s) + 2H_2O \quad \underline{+ 1.685 \text{ V}}$$

$$+ 2.041 \text{ V}$$

The metallic lead furnishes electrons as it is converted to lead sulfate (which is insoluble), and the lead dioxide absorbs the electrons as it is reduced to lead sulfate. Thus, lead appears in three different oxidation states in the cell: the $+4$ state in PbO_2 (which decreases to the $+2$ state in $PbSO_4$), and the zero state in Pb (which increases to $+2$ in $PbSO_4$). Sulfuric acid is consumed as the reaction proceeds and is replaced by water.[21] The potential of 2.04 V is maintained fairly well, with some drop due to decreasing concentration of sulfuric acid, until almost all of the lead dioxide and spongy lead are converted to the sulfate, and then the potential drops rapidly as the chemical system approaches equilibrium. At equilibrium, the battery is completely discharged, or "dead."

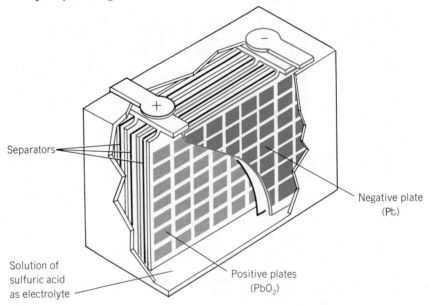

FIGURE 8-10 A lead-acid storage cell, with reactants and products of the redox reactions arranged so that the discharging reactions can be reversed by connecting the cell terminals to a source of electric power. Each cell has a potential of about 2V, and six cells wired in a series give a battery that delivers a potential of 12V, adequate for the electrical system of an auto.

---

[21] This is why a discharged lead storage battery will freeze more readily in cold weather than will a fully charged battery, and also why the degree of charge can be measured by measuring the density of the acid solution with a hydrometer.

Since all of the lead sulfate produced in both half-reactions is held in place on the plates by the separators, it is in a position to be reconverted to the original reactants by reversal of the reactions that produced it. Such reversal requires (1) that electrons be forced through the cell by an external source, such as a generator, (2) that a voltage *higher* than the cell potential be applied, and in the opposite direction, to overcome the electromotive force generated by the reactants and also to overcome the electrical resistance of the cell materials, and (3) that at least one faraday of electricity be forced through the cell in this reverse direction for each equivalent weight of lead dioxide to be regenerated.[22] When these conditions have been fulfilled, the battery is recharged and is ready once again to produce electrical energy upon demand. The successive processes are the generation of electrical free energy by a chemical reaction (discharge cycle), followed by reversal of the reaction by forcing electrical free energy into the chemical system (charging cycle). These processes may be repeated alternately many times, until such time as a large proportion of the active reagent is dislodged from the plates, and the battery consequently is unable to perform its function. At this point, the addition of soluble sulfate salts (proprietary battery additives or "restorers") may succeed in decreasing the internal resistance slightly by increasing the concentration of ions, but such additives cannot replace the active reagent dislodged from the plates and so cannot restore the battery to its original condition.

The convenient reversibility of the PbO_2–Pb cell and the ability of its common embodiment to deliver very large currents for a limited time have made it a favorite for starting automobile engines and for propelling small vehicles. Nevertheless, a critical assessment of it shows that it leaves much to be desired. It is too heavy; if we could rely purely on standard potentials and on equivalent weights, we would certainly never choose so inactive and so very heavy a metal as lead for the reducing agent (or lead dioxide for the oxidizing agent, for that matter). One equivalent of lead weights 104 g, and so about a *pound* of lead must be converted to lead sulfate in order to supply the quantity of electricity normally required from an automobile battery for starting a cold engine. The actual calculation is easily made:

120 A-hr (usual rating) × 3600 sec/hr = 432,000 C required

$$\frac{432,000 \text{ C}}{96,500 \text{ C/equivalent}} = 4.48 \text{ equivalents of Pb required}$$

4.48 equivalents × $\frac{207.2}{2}$ g/equivalent = 465 g of Pb required at 100 percent efficiency

Even though a typical lead storage battery weighs 40 to 50 lb, and occupies one-third of a cubic foot or more, its output in terms of horsepower-hours is very limited. It will scarcely propel an automobile for 100 ft while it converts its several pounds of lead, whereas a pound of gasoline converted to energy in the engine will propel the car 3 miles at a good speed. In terms of electrical energy from a central power station, the battery's output becomes

The problem of power density

120 A-hr × 3600 sec/hr = 432,000 C

432,000 C × 12 V = 5,200,000 J (or W-sec) = 1442 W-hr = 1.442 kWhr

---

[22] One faraday suffices to regenerate *both* one equivalent weight of PbO_2 and one equivalent weight of Pb, for the same electrons do for both half-reactions. Here Pb^{2+} gives up two electrons to form Pb^{4+}, and another Pb^{2+} receives them to form Pb^0.

At 5 cents per kWhr for domestic electricity, 1.442 × $0.05 = $0.072 worth of energy. So this large, expensive, and heavy object produces only 7 cents' worth of electrical energy. Clearly, there is room for a great deal of improvement, and indeed one of the great needs of our time is a way to store electrical energy more economically than the storage battery does. Great rewards await invention in this area.

The nickel-iron battery

The only serious competitor of the lead-acid storage cell has been the nickel-iron-alkaline cell, which uses the half-reactions

$$\text{Oxidation:} \qquad Fe + 2OH^- = Fe(OH)_2 + 2e^-$$

$$\text{Reduction:} \quad NiO_2 + 2H_2O + 2e^- = Ni(OH)_2 + 2OH^-$$

This generates a potential of only 1.35 V, compared with 2.04 V for the lead cell, so more cells are required to furnish 6 or 12 V. The battery still weighs less than a lead-acid battery, however, because of the much lower equivalent weights of the metals. It is also said to be more rugged and durable than the lead-acid battery.

8-11 Fuel Cells

In the preceding section, the production of electrical free energy from the oxidation of zinc by manganese dioxide in a dry cell was readily explained in terms of the electron transfer in redox reactions. As a practical source of electricity, however, such a cell was found to be expensive, heavy, and bulky. The same drawbacks were present in greater measure in the lead storage cell. Part of the disappointment in the practical utility of these cells may be blamed on the large masses of material that must be converted in order to obtain appreciable current—a pound of zinc is used up in order to obtain 2 cents' worth of electrical energy, or several pounds of lead to get 7 cents' worth of electricity. Unfortunately, the metals consumed in primary cells do not occur naturally, but must be obtained by reduction from natural ores at considerable expense; the pound of zinc, which produces 2 cents' worth of energy, costs 32 cents itself. Nor would this situation be much improved if the lighter active metals, such as aluminum and magnesium, were to be used in primary cells instead of zinc; these metals must be produced by electrolysis (see Section 8-12), and they cannot give back more electricity than was consumed in reducing them from their compounds in the first place.[23] All we would really accomplish, if we were to make aluminum for use in primary cells on a large scale, would be to store electrical energy in the form of potential chemical energy of the solid metal. This might someday be desirable for convenience, but it cannot result in any net gain in useful energy. Indeed, there must be considerable loss because both electrolytic processes (production and use) are far from 100 percent efficient.

It has occurred to many people that if some natural product which is oxidizable, like coal or petroleum or wood, could be oxidized in a compartmented electrochemical cell the way zinc is oxidized in a dry cell, it should be possible

---

[23] Although the equivalent weight of aluminum is only 9 g, compared with 32.7 g for zinc, a faraday of electricity is needed to *produce* this one equivalent weight just as certainly as it is given up in a cell.

to convert the chemical free energy of the fuel directly into electrical free energy. Such a process would be highly desirable, for at present these fuels are simply burned without benefit of electrochemistry, and the heat of combustion is used to generate electricity by way of a steam turbine. This practice of degrading all the chemical free energy to heat and then attempting to recover the usable portion of the heat as mechanical free energy is very wasteful. This waste is inherent in the difference between thermal energy and free energy. The maximum theoretical proportion of heat energy convertible into free energy is known and is given by the curve in Figure 8-11, calculated from equations for the work done by a theoretical heat engine operated by the expansion and cooling of a gas.[24] It is obvious that the higher the temperature of the steam supplied to a steam engine or turbine, the more free energy can be extracted. The high pressure and corrosive chemical activity of water vapor at temperatures above 400°C place the upper limit in a practical sense.

The wastefulness of thermal uses of fuels

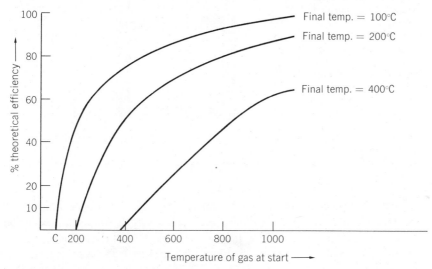

FIGURE 8-11 Maximal theoretical efficiency of a heat engine, the maximum percentage of heat energy that can be converted into free energy.

The best modern power plants operate in the region of 500°C maximum temperature, and here the theoretical maximum efficiency is 48 percent; the actual realized efficiency is in the neighborhood of 40 percent.[25] This means

---

[24] This is the Carnot cycle of thermodynamics, a general consideration of work done by an expanding gas in *any* heat engine. See G. N. Lewis and M. Randall, *Thermodynamics*, 2nd ed., revised by K. S. Pitzer and L. Brewer, New York: McGraw-Hill, chap. 9 (1961), or F. T. Wall, *Chemical Thermodynamics*, San Francisco: Freeman (1958).

[25] To the credit of the electric power companies, further improvement in the efficiency of very large generating stations has been realized by using a mercury vapor cycle at much higher temperatures ahead of the steam cycle. They also try to minimize thermal pollution by the use of cooling towers and other devices, but they cannot circumvent the natural laws of thermodynamics (see Chapter 9).

that from every 100 tons of coal used in a power station, electrical energy equivalent to the chemical energy from burning only 40 tons of coal is generated and later sent out over the wires, and low-grade heat equivalent to burning 60 tons of coal is lost to the cooling water which circulates in the condensers beyond the last turbine. The rejection of waste heat is necessary but objectionable; it gives rise to the so-called thermal pollution of our rivers. So a heat engine is *inherently* wasteful as a converter of chemical energy to electrical energy, and objectionable in other ways. It would be highly desirable to convert the chemical free energy directly to electricity without going through a heat cycle, if this could be accomplished.

See the discussion of the thermo-dynamics of power plants in Ch. 18.

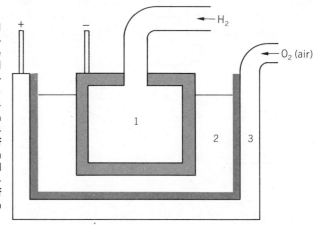

FIGURE 8-12 A fuel cell for the direct conversion of chemical free energy into electrical energy, utilizing the reaction of $H_2 + \frac{1}{2}O_2 = H_2O$. 1. Hollow electrode of porous carbon through which H_2 diffuses. 2. Electrolyte of fused salts such as a mixture of Na_2CO_3 and K_2CO_3. 3. Hollow container with inner wall of porous carbon through which air diffuses.

Partial success in the search for direct conversion has been attained by the electrochemical device called a **fuel cell,** which is simply an oxidation-reduction cell adapted for using gaseous or solid *fuels* as the reducing agents. One form of fuel cell is shown in Figure 8-12. Hydrogen passes through the porous walls of the inner electrode and is oxidized on the platinized carbon surface. Oxygen is supplied (as air) through the porous lining of the outer electrode, which has silver deposited on it. The cell is kept hot partly by its own ohmic resistance under load, and partly by a portion of the chemical energy which is not converted into electrical energy. This cell operates well on hydrogen, but is not satisfactory for other gaseous fuels (such as carbon monoxide or methane), which require expensive rare-metal catalysts for their activation. Since hydrogen is not yet abundantly available as a fuel, the practical use of such a cell is rather limited. Nevertheless, compact lightweight hydrogen-oxygen fuel cells, which use a moist ion-exchange membrane as electrolyte and pure gaseous hydrogen and oxygen as reactants, are of great interest as power sources in space vehicles, where the pure water produced by the cell serves as drinking water. It will probably turn out that liquid fuels will be converted to hydrogen by preliminary reactions just before use, or that hydrogen will be produced electrolytically by nuclear power plants, and the hydrogen used in fuel-cell-powered vehicles.

Maybe H_2 will be our portable fuel of the future.

Attempts have been made to use powdered coal as fuel in a fuel cell. In one design, the coal is mixed with the oxides of tungsten and cesium (WO_3 and CeO_2, both solids) and is oxidized by air admitted through a granular mass of Fe_3O_4. This cell operates at a potential of several volts and at 70-percent efficiency for conversion of chemical to electrical energy, but trouble is experienced with the accumulation of ash from the coal.

None of these designs has solved all the problems of the fuel cell, so there is room for much invention in this important area. A fuel cell that would extend our coal reserves by giving twice as much electric power per ton of coal as the present boiler-plus-heat-engine designs, without waste heat and thermal pollution, would have a far-reaching effect on our economy and our history. Undoubtedly, it would be easier to design and perfect a fuel cell to use natural gas or petroleum as fuel, but in view of the increasingly important use of these irreplaceable resources for organic synthesis (of fibers, plastics, rubber, chemicals, and even basic foodstuffs), such a development would not serve our long-term interests as well as a coal fuel cell. A cell which uses combustible waste, such as garbage or agricultural waste, would be still more welcome.

Fuel cells for garbage?

8-12 Chemicals from Electricity: Electrolysis

We have seen that chemical oxidation-reduction reactions, which liberate free energy, may be used to solve the problem of portable production of electrical energy. Sometimes the reverse problem must be solved—a reaction for the production of some useful material *consumes* free energy, and a way must be found to force energy into the chemical system to carry out the reaction. Often this can be done satisfactorily and with good control electrochemically, and we call the process **electrolysis.**

Electrolysis is the direct use of electricity to carry out chemical change. An example has already been met in the recharging of the lead storage battery. The requirements were described there; in summary, the reaction must be adaptable mechanically to the desired oxidation and reduction at electrodes, and electrical energy at a potential greater than the spontaneous potential of the cell (and in a quantity corresponding to the number of moles to be produced) must be supplied by an external source. Ways in which these requirements are met in present practice are seen in the few examples that follow.

OXIDATION THROUGH THE USE OF ELECTRICAL ENERGY. The preparation of fluorine offers a clear-cut example of the rationale and the methodology of electrolysis. Free fluorine cannot be prepared by the chemical oxidation of fluoride ion; there is no oxidizing agent stronger than fluorine, and hence none to do the job. Fluoride ion can only be oxidized electrolytically. To accomplish the oxidation, we must first find a liquid **electrolyte** which contains fluoride ions as the *only* negative ions (since all other negative ions would be oxidized preferentially in the cell), and which has low electrical resistance at a moderate temperature. Electrolysis of this liquid in a cell of appropriate design should then produce the free halogen.

Electrolytic production of F_2

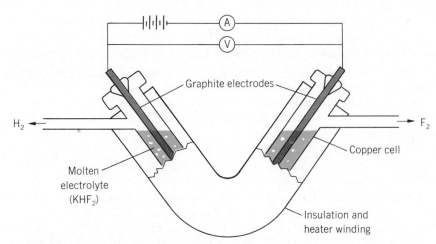

FIGURE 8-13 An electrolysis cell for the small-scale production of fluorine by the oxidation of F⁻ ion. The free energy required to force electrons to leave F⁻ ions is supplied by the external source of electricity, ┤│┤│┤├ , monitored by the ammeter A and the voltmeter V.

A small laboratory cell for the preparation of fluorine is shown in Figure 8-13.[26] Here about a kilogram of melted potassium hydrogen fluoride, KHF_2, is contained in a V-shaped copper vessel fitted with insulated caps that support graphite electrodes extending into the molten salt. A current of 8 to 10 A is passed through the electrolyte, and the two gases appear at the electrode surfaces:

oxidation
(at positive electrode) $\quad 2F^- = F_2(g) + 2e^-$

reduction
(at negative electrode)[27] $\quad 2H^+ + 2e^- = H_2(g)$

These gases bubble out of the electrolyte and leave the cell through the side arms; the cell design must be such that they never mix, for violent explosions would result. The fluorine may be purified by passing it through a copper U-tube filled with granular potassium fluoride, where hydrogen fluoride is removed and retained according to the reaction

$$KF + HF = KHF_2$$

The copper cell resists the action of fluorine at the operating temperature (about 280°C), because an adherent protective film of copper fluoride forms on it. Nickel and its alloy with copper may similarly be employed at 300°C or less, since they

---

[26] L. M. Dennis, J. M. Veeder, and E. G. Rochow, *Journal of the American Chemical Society*, vol. **53**, p. 3263 (1931).

[27] The potassium ions are not reduced because they require a much higher voltage for reduction than do hydrogen ions. As long as there are many hydrogen ions about, they will undergo reduction in preference to the potassium ions. The latter simply accumulate in the electrolyte, and the overall chemical change is seen to be an electrolysis of HF, converting the original KHF_2 gradually to KF.

are protected by fluoride films, but all such metals would be consumed if they were heated to 1000°C in fluorine. Even platinum reacts with fluorine at red heat.

Much larger quantities of fluorine are produced in commercial cells constructed of low-carbon steel, in which an electrolyte of the composition KH_3F_4 is electrolyzed at about 100°C with nickel electrodes. The fluorine is purified as before and may be stored at moderate pressure in cylinders of nickel. Some fluorine is liquified (it boils at −188°C) and transported in this condition. It is used as an oxidant in experimental rocket engines, and to prepare a variety of anydrous fluorides, of which volatile uranium hexafluoride (for separation of the isotopes of uranium) is one example.

Chlorine is prepared most conveniently and economically by electrolytic oxidation of chloride ion, either in a bath of molten sodium chloride (mp 801°C) or in an aqueous solution. In molten anydrous sodium chloride, the reactions are simply

$$\text{oxidation:} \quad 2Cl^- = Cl_2(g) + 2e^-$$

$$\text{reduction:} \quad 2Na^+ + 2e^- = 2Na$$

Gaseous chlorine and liquid metallic sodium are obtained. The chlorine is cooled, compressed, liquified, and shipped in tank cars; it is used for the preparation of many plastics, elastomers, and industrial organic and inorganic chemicals, as well as for bleaching and for the purification of water. The sodium is cast in drums or cans and is used at present for the manufacture of tetraethyllead (for ethyl gasoline), as a heat-transfer agent, and as an industrial reducing agent for the preparation of detergents and drugs.

The preparation of chlorine by electrolysis of a *water solution* of sodium chloride also is possible, because the hydroxide ions of water require a higher electrode potential for their discharge than do the chloride ions. Hence, as brine is electrolyzed in separated electrode compartments, the half-reactions which take place are

$$\text{oxidation:} \quad 2Cl^- = Cl_2(g) + 2e^-$$

$$\text{reduction:} \quad 2H^+ + 2e^- = H_2(g)$$

and sodium hydroxide is left in solution. Alternatively, if both electrode reactions are allowed to take place in one compartment at an appropriate potential, chloride ion may be oxidized directly to hypochlorite ion in the basic solution:

$$Cl^- + 2OH^- = ClO^- + H_2O + 2e^-$$

The resulting solution of sodium hypochlorite (household bleach) is widely used as a germicide and bleaching agent.

As a final example of electrolytic oxidation, consider the preparation of *hydrogen peroxide* by the persulfate method. Examination of oxidation potentials shows that water cannot be oxidized to hydrogen peroxide by lead dioxide or by potassium permanganate, but it could be oxidized by fluorine, by ozone, or by persulfuric acid, $H_2S_2O_8$. Hence a cold concentrated solution of sulfuric acid may be oxidized electrolytically to persulfuric acid, which may later be allowed to react with water to form a solution of hydrogen peroxide and sulfuric acid. The hydrogen peroxide may be isolated from the solution by distillation, and the residual sulfuric acid may be reused in the oxidation process.

It is illuminating to look at the power requirements for preparing hydrogen peroxide this way:

UF_6: see Ch. 6.

Bleach power

Current to make one mole (194 g) of $H_2S_2O_8$ (at 100 percent efficiency) in one minute is

$$2 \times 96{,}500 \text{ C in one minute} = \frac{2 \times 96{,}500}{60} = 3217 \text{ A}$$

At an operating potential of 5 V, the power required is

$$3217 \text{ A} \times 5 \text{ V} = 16{,}085 \text{ W} = 16.09 \text{ kW required}$$

or

$$16.09 \times 14.34 = 231 \text{ kcal of energy per minute}$$

It is apparent even from this one simplified example that electrolysis on a practical scale requires a very large expenditure of power, and so its use as a method of industrial production usually is governed by the availability and cost of electric power in the vicinity.

REDUCTION THROUGH THE USE OF ELECTRICAL ENERGY. It is incorrect to think of a particular electrolysis purely as an oxidation or as a reduction, since both types of reaction must occur simultaneously and in equivalent proportions. However, when the principal objective is to prepare a particular substance by electrolytic oxidation, the design of a cell necessarily emphasizes that aspect at the expense of reduction, and vice versa.[28] So it happens that the term *electrolytic reduction* is applied to electrolyses designed primarily for the "winning" of metals from their ores or compounds. Such processes are very common.

Aluminum production

Perhaps the best known electrolytic reduction is the Hall process for the production of aluminum considered in Chapter 3. An anhydrous electrolyte composed of aluminum oxide dissolved in molten sodium aluminum fluoride, Na_3AlF_6 (which once was mined as a natural mineral but now is synthesized), is maintained at red heat in a large furnace lined with graphite, and heavy graphite electrodes are lowered into the melt from above (see Figure 8-14). The graphite lining serves as the negative electrode, at which aluminum ions from the aluminum oxide are reduced:

$$Al^{3+} + 3e^- = Al(l)$$

Since the temperature of the electrolyte is above the melting point of aluminum, the metal collects as a liquid layer underneath the electrolyte (which is less dense) and is drawn off from time to time. The vertical (positive) electrodes attract the oxide ions, which are oxidized to oxygen and then combine with the carbon of the electrode:

$$2O^{2-} = O_2(g) + 4e^-$$

$$2C + O_2(g) = 2CO(g)$$

---

[28] Thus, a cell designed for the preparation of fluorine throws away the hydrogen (the product of reduction) because it has negligible value in comparison with that of the fluorine. Present fuel problems will surely change this soon. Some commercial processes, on the other hand, turn out two or even three valuable products.

The process consumes carbon, aluminum oxide, and electric power; it produces metallic aluminum of a purity sufficient for most uses and carbon monoxide, and it conserves the solvent, sodium aluminum fluoride.

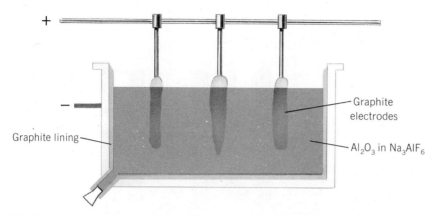

FIGURE 8-14 Production of aluminum by electrolysis, using dipping graphite electrodes which are consumed in the process by reaction with oxygen. The negative electrode is the graphite lining of the reaction vessel. The molten salt electrolyte is maintained at red heat, and the reduced aluminum metal is collected from the bottom of the vessel as a liquid.

Magnesium is produced by a similar electrolytic process in which melted anhydrous magnesium chloride is the electrolyte. Magnesium ions are reduced at the negative electrode, forming a layer of molten metal which floats on the more dense electrolyte. Chlorine is evolved at the positive (carbon) electrode and is collected for reuse in the conversion of magnesium hydroxide to chloride. In most installations, the source of magnesium hydroxide is sea water, which contains 0.13 percent by weight of magnesium as Mg^{2+} ion. The sea water is treated with low-cost alkali in the form of calcium hydroxide from the calcination of shells, and magnesium hydroxide precipitates. This is filtered off, converted to magnesium chloride by the action of hydrochloric acid, and then dehydrated for use in the electrolytic cells.

Magnesium from the sea

The production of sodium, beryllium, and calcium metals by electrolysis of their fused halides follows a similar pattern, except for the sources of starting material; we have already seen that the electrolysis of fused sodium chloride provides sodium by reduction as well as chlorine by oxidation. The plating of silver, nickel, and chromium on other metals is another example of electrolytic reduction of these metals from aqueous solutions of their salts.

As a final example of electrolytic oxidation and reduction, we may consider the purification of copper as it is carried out on a large scale. The high-temperature metallurgy of copper follows a long path from dispersed particles of copper sulfide ore in the ground through mining, concentration of the ore, separation of the Cu_2S, and oxidation of this sulfide by blowing air through a molten mass of it. The resulting copper contains varying amounts of iron, arsenic, zinc, silver, and gold, as well as residual sulfides, oxides, and phosphides of these elements. Some of these impurities interfere seriously with the electrical conductivity of the copper and hence spoil it for its major use, so some purification of the copper is necessary. The most effective purification is accomplished by oxidizing the copper to Cu^{2+} ions electrolytically and then reducing these ions to metal at another site; control of the conditions for

Copper refining

both half-reactions by application of the principles outlined earlier in this discussion allows these processes to be specific, favoring copper over the other metals and excluding the nonmetals.

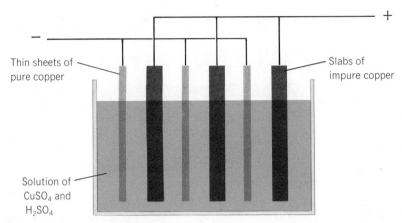

FIGURE 8-15 Electrolytic refining of copper. Copper atoms from slabs of impure copper (the positive electrodes) are oxidized, passing into solution as Cu^{2+} ions and migrating through solution to the negative electrodes to be reduced to copper metal. With careful control of voltage, extremely pure copper metal can be prepared by this method.

The arrangement of the electrolytic cell for copper is simple in principle, as is shown in Figure 8-15. Cast slabs of the impure blister copper are alternated with thin sheets of pure copper hung in an aqueous solution of copper sulfate and sulfuric acid. A direct current of closely controlled voltage is passed through the cell, in a direction which removes electrons from the impure copper and supplies them to copper ions at the surface of the pure copper. The oxidation reaction at the positive electrode,

$$Cu = Cu^{2+} + 2e^-$$

Gold and silver are by-products.

is not confined to copper alone; iron, zinc, and other metals more active than copper also go into solution from the slab of blister copper. The less active metals (gold, silver, and the platinum metals) do not go into solution, but fall to the bottom as small particles and are recovered later. At the negative electrode, the copper, having a lower electrode potential than the more active metals, is more readily reduced:

$$Cu^{2+} + 2e^- = Cu$$

The zinc and iron, therefore, accumulate in the electrolyte as soluble salts, and the solution must be replaced periodically. Exceedingly pure copper results from the process, and the more valuable metals are recovered on such a scale as to pay for the entire refining process. The principal cost is for electric power consumed in overcoming resistance (that is, in transporting the ions within the cell and forcing the electrons through the external circuit). Since the cell uses copper for both electrodes, it generates no appreciable potential that must be overcome, and it does not consume free energy for conducting the principal chemical process.

8-13 Summary

Oxidation is a general process involving a loss of electrons. Reduction is its opposite, involving a gain of electrons. Oxidation and reduction take place simultaneously and to an equal extent, so electrons are transferred in the process. The number of electrons lost or gained may be followed by using a convenient formalism called the oxidation number, and by balancing the increases and decreases of oxidation number we can balance chemical equations which would otherwise prove difficult. Examples were worked out, and additional equations to be balanced (together with the correct answers or results) are given at the end of the chapter.

This further bit of chemical symbolism and arithmetic is just a minor (but necessary and helpful) aspect of oxidation-reduction processes. The major emphasis is upon the actual transfer of electrons from reducer to oxidizer in external electrical circuits, opening up the broad practical area of *electrochemistry*. Here, there are three main aspects of immediate importance:

1. A physical basis is found for measuring the push or potential of an electron-transfer process, giving rise to a table of standard electrode potentials which can be used to predict the likelihood of thousands of reactions and to calculate their equilibrium constants.
2. Means are devised for the direct conversion of chemical energy to electrical energy in batteries and fuel cells, which are described in some detail.
3. Electrochemical devices are described for applying electrical energy to the production of desirable materials which would be difficult to obtain in any other way.

In the course of these matters, the difference between free energy and heat energy became apparent, a difference measureable in terms of the concept of entropy. We shall have much more to say about this concept soon. Furthermore, an inevitable source of thermal pollution was found in the operation of any heat engine, be it an automobile engine, a coal-fired power station, or a nuclear power station. The desirability of direct conversion of chemical energy to electricity thereby was emphasized, leading to a critical assessment of present-day batteries and an awareness of the need for much new ingenuity here.

We conclude that if ever we are to have electric automobiles and pollution-free sources of power, a great deal of research and inventing will have to be done. Dreams and rumors of battery-operated cars that will cruise for hundreds of miles abound, but the promises are not yet realized. The realities of electrochemistry show us why so many such dreams fail, and at the same time point out the possible directions for making firm progress.

GLOSSARY

autooxidation: a reaction in which a single element is both oxidized and reduced, some of the atoms of that element gaining electrons at the expense of other atoms of the same element.

cell: in electrochemical usage, an arrangement of electrodes, reagents, and solution(s) in which an oxidation-reduction reaction may be exploited to force an electric current through an external circuit. *Primary* cells are one-shot devices; *secondary* cells or storage cells can be recharged. *Electrolytic cells* produce materials.

concentration cell: an electrochemical arrangement in which two half-cells contain the same kind of electrode and the same active reagent, but at different concentrations. The dependence of half-cell potential on concentration of reagent causes the arrangement to generate a voltage and to furnish energy.

electrochemistry: the art and science of obtaining electrical energy from chemicals, and vice versa; the domain of reactions at the surfaces of electrodes in suitably arranged chemical systems.

electrode: a rod, wire, or sheet (usually metallic) inserted into a reagent in order to conduct electricity into or out of the reagent.

electrolysis: an electrochemical process in which electrical energy is used to decompose a chemical compound, or to produce economically desirable materials, by a process of oxidation and reduction within a cell.

electrolyte: an ion-containing solution, or a melted ionic compound, used as the ionically conducting material in an electrochemical cell.

entropy: a measure of that portion of heat energy which is not recoverable in a heat engine; a measure of degree of disorder or randomness on the molecular level. (See also Chapter 9 and its glossary.)

free energy: freely convertible energy (mechanical, electrical, chemical, nuclear, or radiant energy), as distinct from thermal energy or heat.

fuel cell: an electrochemical primary cell in which the substance oxidized is not a metal, but a substance normally considered to be a combustible fuel, such as coal or natural gas. The oxidant is usually air, but sometimes pure oxygen or H_2O_2.

gram-equivalent: that weight of substance which will lose or gain 6.023×10^{23} electrons (a mole of electrons) in a chemical change.

half-cell: a container of reagent, with associated electrode, in which a half-reaction may take place.

half-reaction: one-half of an electron-transfer reaction, consisting of the oxidation half or the reduction half.

oxidation: a chemical process in which a substance loses electrons.

oxidation number: the number of electrons lost or gained by one atom of an element in attaining the state of combination under consideration.

reduction: a chemical process in which a substance gains electrons.

standard electrode potential: the potential registered by a given half-cell versus a standard hydrogen electrode under standard conditions of concentration (1 M), temperature (25°C), and pressure (1 atm); symbolized by **E°**.

PROBLEMS

8-1 Oxidation-reduction equations to be balanced.

a. $Cl_2 + FeCl_2 \rightarrow FeCl_3$

b. $I_2 + H_2S \rightarrow S + HI$

c. $HNO_3(conc) + FeS \rightarrow Fe(NO_3)_3 + NO_2 + H_2SO_4 + H_2O$

d. $Na_2O_2 + CrCl_3 + NaOH \rightarrow Na_2CrO_4 + NaCl + H_2O$

e. $PbO_2 + Mn(NO_3)_2 + HNO_3 \rightarrow Pb(NO_3)_2 + HMnO_4 + H_2O$

f. $KMnO_4 + MnSO_4 + H_2O \rightarrow MnO_2 + K_2SO_4 + H_2SO_4$

g. $FeSO_4 + K_2Cr_2O_7 + H_2SO_4 \rightarrow Fe_2(SO_4)_3 + Cr_2(SO_4)_3 + K_2SO_4 + H_2O$

h. $SnCl_2 + HgCl_2 \rightarrow Hg_2Cl_2 + SnCl_4$

i. $H_2S + HNO_3 \rightarrow S + NO + H_2O$

j. $Na_2SO_3 + H_2O + I_2 \rightarrow NaI + H_2SO_4$

k. $Na_2S_2O_3 + I_2 \rightarrow NaI + Na_2S_4O_6$

l. $H_2C_2O_4 + KMnO_4 + H_2SO_4 \rightarrow MnSO_4 + K_2SO_4 + CO_2 + H_2O$

m. $Cr_2O_7^{2-} + I^- + H^+ \rightarrow Cr^{3+} + I_2 + H_2O$

n. $Cr_2O_7^{2-} + Sn^{2+} + H^+ \rightarrow Cr^{3+} + Sn^{4+} + H_2O$

o. $MnO_4^- + I^- + H^+ \rightarrow Mn^{2+} + I_2 + H_2O$

p. $MnO_4^- + H_2S + H^+ \rightarrow Mn^{2+} + S + H_2O$

q. $H_2S + Cr_2O_7^{2-} + H^+ \rightarrow S + Cr^{3+} + H_2O$

r. $NaClO + As + NaOH \rightarrow NaCl + Na_3AsO_4 + H_2O$

s. $KMnO_4 + Na_2SnO_2 + H_2O \rightarrow MnO_2 + Na_2SnO_3 + KOH$

t. $Au + HNO_3 + HCl \rightarrow AuCl_3 + NO + H_2O$

8-2 a. Calculate the free-energy change in joules per mole for the reaction

$$FeCl_2 + Zn = Fe + ZnCl_2$$

b. What is the equilibrium constant for the reaction?

8-3 What is the standard potential E^0 for the following reactions? Which ones will start up spontaneously?

a. $Fe^{2+} + Sn^{2+} = Fe + Sn^{4+}$
b. $H_2O_2 + HClO = O_2 + HCl + H_2O$
c. $2Fe^{3+} + Sn^{2+} = 2Fe^{2+} + Sn^{4+}$
d. $PbO_2 + Mn^{2+} = MnO_2 + Pb^{2+}$

8-4 Pure sodium chloride is melted and electrolyzed, passing a current of 200 A for a 20-hr period.

a. What products are formed?
b. How much energy was put into the electrolytic cell?
c. How many moles of each product were formed?

8-5 An electrochemical cell is set up to exploit the reaction

$$Zn + CuSO_4 = Cu + ZnSO_4$$

How much electrical energy (in joules and in horsepower-hour) has been produced by this cell when it has operated long enough for 10.3 g of copper to be deposited?

8-6 Either sodium or magnesium may be used as a reducing agent in a particular synthesis (the products of oxidation being Na^+ or Mg^{2+}, of course). If sodium costs $0.22 per pound, and magnesium costs $0.31 per pound, which metal will be the less expensive to use?

8-7 Arrange the following substances in the order of their strength as oxidizing agents in 1 M solution, placing the strongest one first: I_2, K_2SO_4, Cl_2, $KMnO_4$, S, and Br_2.

8-8 A pair of platinum electrodes is immersed in a solution of a salt of a metal M, and the cell is connected in series with a similar cell made up of platinum electrodes immersed in a solution of $AgNO_3$. An electric current is passed through both cells, depositing silver and the metal M. The following experimental data are obtained:

$$\text{weight of silver deposited} = 215.6 \text{ g}$$

$$\text{weight of M deposited} = 241.4 \text{ g}$$

$$\text{current} = 0.1 \text{ A}$$

$$\text{voltage} = 3.5 \text{ V}$$

What is the equivalent weight of M? If its ions are M^{2+}, what is the atomic weight of M?

8-9 Recent publicity about new batteries being tested for possible use in electric automobiles concerns a sodium-sulfur cell and a lithium-chlorine cell. Which would give the higher potential? Which would give the more energy per pound of metal used?

8-10 A fuel cell which utilizes natural gas as a reducing agent and hydrogen peroxide as oxidizing agent is proposed as a source of domestic electric power instead of a nuclear power station. Is such a plan

  a. possible?
  b. logical?
  c. wise from the standpoint of ecology?

8-11 Manganese as a metal is an active element (standard electrode potential for $Mn^{2+} + 2e^- = Mn$ is -1.05 V). Indeed, it is so active that at one point in history it was confused with magnesium. But black MnO_2 is not a basic oxide, so we conclude that manganese in the $+4$ oxidation state is quite different from manganese of lower oxidation number. Then we turn to compounds in which manganese has a still higher oxidation number, and find they are acids: green H_2MnO_4 (manganic acid, a moderately strong acid) and purple $HMnO_4$ (permanganic acid, a strong acid). So we find manganese exhibiting a wide range of chemical behavior, from decidedly basic to decidedly acidic. Now the questions:

a. Why does manganese have so many oxidation states?
b. What explanation do you find for the increasingly acidic behavior of manganese with rising oxidation number?
c. Is manganese alone in this kind of behavior? From the information given in Table 8-1, what other elements change from reducing agents to oxidizing agents as their oxidation numbers rise considerably?

ANSWERS TO OXIDATION-REDUCTION EQUATIONS

a. $Cl_2 + 2FeCl_2 \rightarrow 2FeCl_3$

b. $I_2 + H_2S \rightarrow S + 2HI$

c. $12HNO_3(conc) + FeS \rightarrow Fe(NO_3)_3 + 9NO_2 + H_2SO_4 + 5H_2O$

d. $3Na_2O_2 + 2CrCl_3 + 4NaOH \rightarrow 2Na_2CrO_4 + 6NaCl + 2H_2O$

e. $5PbO_2 + 2Mn(NO_3)_2 + 6HNO_3 \rightarrow 5Pb(NO_3)_2 + 2HMnO_4 + 2H_2O$

f. $2KMnO_4 + 3MnSO_4 + 2H_2O \rightarrow 5MnO_2 + K_2SO_4 + 2H_2SO_4$

g. $6FeSO_4 + K_2Cr_2O_7 + 7H_2SO_4 \rightarrow 3Fe_2(SO_4)_3 + Cr(SO_4)_3 + K_2SO_4 + 7H_2O$

h. $SnCl_2 + 2HgCl_2 \rightarrow Hg_2Cl_2 + SnCl_4$

i. $3H_2S + 2HNO_3 \rightarrow 3S + 2NO + 4H_2O$

j. $Na_2SO_3 + H_2O + I_2 \rightarrow 2NaI + H_2SO_4$

k. $2Na_2S_2O_3 + I_2 \rightarrow 2NaI + Na_2S_4O_6$

l. $5H_2C_2O_4 + 2KMnO_4 + 3H_2SO_4 \rightarrow 2MnSO_4 + K_2SO_4 + 10CO_2 + 8H_2O$

m. $Cr_2O_7^{2-} + 6I^- + 14H^+ \rightarrow 2Cr^{3+} + 3I_2 + 7H_2O$

n. $Cr_2O_7^{2-} + 3Sn^{2+} + 14H^+ \rightarrow 2Cr^{3+} + 3Sn^{4+} + 7H_2O$

o. $2MnO_4^- + 10I^- + 16H^+ \rightarrow 2Mn^{2+} + 5I_2 + 8H_2O$

p. $2MnO_4^- + 5H_2S + 6H^+ \rightarrow 2Mn^{2+} + 5S + 8H_2O$

q. $3H_2S + Cr_2O_7^{2-} + 8H^+ \rightarrow 3S + Cr^{3+} + 7H_2O$

r. $5NaClO + 2As + 6NaOH \rightarrow 5NaCl + 2Na_3AsO_4 + 3H_2O$

s. $2KMnO_4 + 3Na_2SnO_2 + H_2O \rightarrow 2MnO_2 + 3Na_2SnO_3 + 2KOH$

t. $Au + HNO_3 + 3HCl \rightarrow AuCl_3 + NO + 2H_2O$

SUGGESTIONS FOR FURTHER READING

For a next step in learning about electrochemical cells, try Chapter 6, "Oxidation-Reduction Equilibria and Electrochemical Cells," in BLACKBURN, T. R., *Equilibrium: A Chemistry of Solutions,* New York: Holt, Rinehart & Winston (1969). Then you will be ready for: LINGANE, J. J., *Electroanalytical Chemistry,* 2nd ed., New York: Wiley, Interscience Division (1958).

The Nernst equation is derived in all of the standard physical chemistry textbooks. See, for instance, DANIELS, F., and ALBERTY, R. A., *Physical Chemistry,* 3rd ed., New York: Wiley, p. 248 (1966). Here use is made of thermodynamics in the derivation.

A rich storehouse of information about redox reactions of the elements and the relationship between redox potential and pH is: POURBAIX, M., *Atlas of Electrochemical Equilibria in Aqueous Solutions,* London: Pergamon (1966). The single most influential American book dealing with electrochemistry has surely been LATIMER, W., *The Oxidation States of the Elements and Their Potentials in Aqueous Solutions,* 2nd ed., Englewood Cliffs, N.J.: Prentice-Hall (1952), both a complete source of electrochemical data and an excellent discussion of the methods required for obtaining and using these data.

Metabolism in any living cell requires redox reactions to permit the utilization of light energy (in photosynthesis) or chemical energy (from nutrients). Many aspects of such reactions are treated in detail by 20 authors in: SINGER, T. P., ed., *Biological Oxidations,* New York: Wiley, Interscience Division (1968). A more general and more philosophical approach is found in: MOROWITZ, H. J., *Energy Flow in Biology: Biological Organization as a Problem in Thermal Physics,* New York: Academic (1968).

9
Energy and Chaos: Why Chemical Reactions Happen

*The force that through the green fuse drives the flower
Drives my green age; that blasts the roots of trees
Is my destroyer.
And I am dumb to tell the crooked rose
My youth is bent by the same wintry fever.*

DYLAN THOMAS

The Poems of Dylan Thomas, Copyright 1939 by New Directions Publishing Corporation. Reprinted by permission of New Directions Publishing Corporation.

9-1 Abstract

Throughout this book, so far, we have considered the electrical forces between particles (between protons and electrons, between positive and negative ions, and so forth) as the ultimate explanation of chemical behavior. There are other reasons for chemical change, some of them having to do with heat and other forms of energy, and some based subtly on statistics. In this chapter, the ways in which chemical forces manifest themselves are examined in a systematic way. The principles of **thermochemistry** are introduced as a way of exploring a great deal of chemistry with pen and paper, instead of by tedious experiment. The concept of entropy, encountered briefly in Chapter 8, is developed on the basis of probability as a far-reaching principle governing not just the behavior of chemicals in a laboratory, but also the lives and growth of people, plants, and animals—governing, indeed, the entire universe. The whole chapter is about the domain of thermodynamics and its inexorable laws, fundamental laws which are simple statements of far-reaching truth.

9-2 Heat, Work, and Energy in Chemical Change

Thermodynamics of a simple reaction

Let us begin with a reasonably simple chemical example. The brown gas NO_2 is a frequent sight in chemical laboratories (as a reduction product of nitric acid; see Section 8-6) and on the industrial landscape (sometimes issuing wholesale from factory flues). Nitrogen dioxide undergoes a reversible *dimerization* (doubling) reaction to colorless dinitrogen tetroxide,

$$2NO_2 \rightleftarrows N_2O_4 \tag{9-1}$$

in the course of which a new chemical bond is formed between the two nitrogens of N_2O_4, and the nitrogen-oxygen bonds are rearranged (Figure 9-1). The force that drives this dimerization reaction is the electrical attraction between the electrons in the new bond and the positive nitrogen nuclei, somewhat modified by a rearranged bonding between N and O in the dimer.

When one mole of N_2O_4 is formed by the reaction of 2 moles of NO_2 at 25°C, in a container of constant volume,[1] the system becomes warmer.

---

[1] The specification of constant volume allows us to ignore the energetic effects of a surrounding pressure, which would tend, according to Le Chatelier's principle, to drive reaction (9-1) to the right. For a discussion of the effect of pressure on the reactions of gases, see Chapter 7 and Section 9-4.

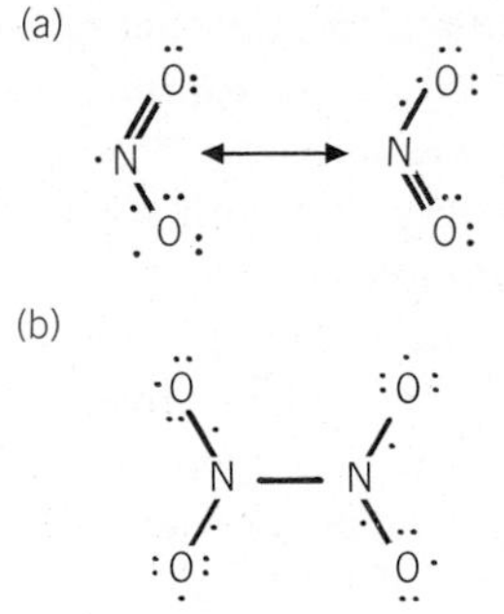

FIGURE 9-1 Structures of (a) NO$_2$ and (b) N$_2$O$_4$. The NO$_2$ molecule contains 17 electrons from outermost atomic shells, an odd number. There must be at least one unpaired electron, so the simple Lewis dot-type structure with shared pairs of electrons will not suffice for NO$_2$. The structures shown here were adapted from suggestions of J. W. Linnett in *The Electronic Structure of Molecules*, London: Methuen (1964). In NO$_2$, the N-O bond length is 1.19 Å, and the

$$N<\genfrac{}{}{0pt}{}{O}{O}$$

angle is 134°. The N$_2$O$_4$ molecule is planar.

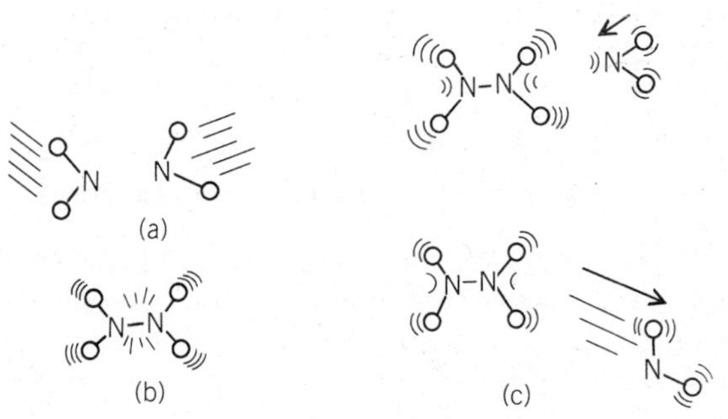

FIGURE 9-2 (a) Two NO$_2$ molecules collide. Because of the attractive force between them, the collision is more forceful than would otherwise have been the case. After the collision, a new bond results from this force. (b) The newly formed N$_2$O$_4$ molecule vibrates with extra energy because of the forcefulness of the collision. (c) This extra energy is shared with all of the other molecules in the gas through collisions.

The rise in temperature results because the attractive force between electrons and nuclei in N$_2$O$_4$ has been converted (by reactive collisions of NO$_2$ molecules) into vibrations, rotations, and straight-line motions of the molecules in the system (Figure 9-2). One kind of energy (the potential energy of separated NO$_2$ molecules relative to the more stable N$_2$O$_4$) has been converted into that microscopic collection of jiggles, tumbles, and velocities that we call thermal energy or, loosely, heat.

We can conveniently measure the amount of energy involved in this transaction by measuring how much energy (heat) would have to be withdrawn from the collection of N$_2$O$_4$ molecules to return their temperature to 25°C. This quantity turns out to be 13.0 kcal (kilocalories) of energy for every mole of N$_2$O$_4$ formed. We can attribute this quantity of energy directly to the satisfaction of electrical forces between nuclei and electrons in the process of forming N$_2$O$_4$ out of NO$_2$. An analogous process occurs in nature when a meteorite, obeying an attractive gravitational force, smashes into the earth. The potential energy that this body had while far from the earth is converted first into its own kinetic energy, as it falls faster and faster, and then into heat; the heat can be seen first in the glowing fireball that plunges through the atmosphere and then in the explosive effects of the collisions of the mete-

Measuring the energy of a reaction

orite with the earth. Of course, some of the meteorite's kinetic energy is converted into chemical potential energy, as chemical bonds are ruptured in the physical disintegration of the meteorite at the point of impact.

Exactly the same kind of process is carried out every day with smaller objects involving lesser quantities of energy, as for example when you drop a suitcase, a baseball, your watch, or a hair on the floor. We simply fail to notice the heat effects in these cases because the quantity of heat involved is much smaller. Finally, when two molecules collide and stick together because electrical potential energy is thereby lowered, it takes inconceivable numbers of such events to generate a detectable amount of heat. In the case of Equation (9-1), the formation of 6×10^{23} molecules of N_2O_4 generates 13 kcal of heat. How much heat is 13 kcal/mole? Let us compare this 13 kcal/mole of N_2O_4 to the quantity of energy involved in a drastic rearrangement of the electrons in atoms; for example, to the ionization potentials that we discussed in Chapter 6. Making the conversion in units (Appendix III), we find that the formation of one *molecule* of N_2O_4 releases only 0.56 eV of energy, or one-tenth of the quantity required to ionize an atom of lithium (Section 6-1). Clearly, the formation of N_2O_4 is a relatively mild process, since it involves not the removal, but only the minor rearrangement, of electrons in atoms.

9-3 Conservation of Energy: The First Law of Thermodynamics

If the formation of a mole of N_2O_4 from NO_2 is allowed to take place in a container of variable volume, but at the constant *pressure* of one atmosphere, the heat effect turns out to be not 13.0 kcal, but 13.6 kcal. The extra 0.6 kcal is the heat equivalent of the work done by an external pressure of one atmosphere squeezing 2 moles of gas into the space of one; since the surrounding atmosphere has done some work in thus squeezing the system, the energy of that work appears as heat. In the same way, when you rub your hands together briskly, the work that you do against friction appears as heat that you can feel. In these cases, and in the conversion of potential energy into heat that we discussed in the last section, it has invariably been found that *energy can be converted from one form into another without detectable loss*, but with the limitation of direction encountered in Chapter 8.

For some time, it was thought that this conservation of energy during its conversion from one form to another was an absolute law. Since the advent of relativistic physics, it develops that the true conservation is a combined one: the equivalence and conservation of mass *and* energy referred to in Chapter 2. The mass equivalent to the energy of ordinary chemical reactions is very small, by the standard of ordinary molecular masses. Einstein's famous equation

$$E = mc^2 \tag{9-2}$$

gives the relationship (E is energy in ergs; m is mass in grams; and c is the velocity of light in centimeters per second). Since c is very large, 3×10^{10}

cm/sec, and c^2 that much larger, it follows that only a very tiny amount of mass had to disappear to compensate for the materialization of 13 kcal of "potential energy" as heat; from Equation (9-2) we can calculate that the lost mass is 3.5×10^{-14} g, out of the 92 g of nitrogen and oxygen involved in the reaction. Thus, with discrepancies much smaller than can be detected with the most sensitive balance, chemists have felt free to consider matter and energy as separately conserved in chemical reactions. So just as we saw (in Chapter 2) that it is possible to give an accounting of the changes in mass in a chemical reaction using the idea of gram-molecular mass, we can give equally valid accountings of energy changes if we can find analogous molar energy functions.

It is impossible to sum up the multitudes of potential and kinetic energies that constitute the molar energy of a substance, except in particularly simple cases. And there is no instrument that will tell us the chemical energy content of a chunk of matter in the way a balance can tell us its mass.[2] What we can observe is the *change in energy* in any chemical transaction, manifested either as heat or (in many cases like automobile engines and electrical batteries) as heat and work. The *first law of* **thermodynamics** simply states that the energy change in any event is composed of these heat and work effects:

We can't measure the chemical energy in matter.

But we can balance an energy budget.

$$\Delta E = Q - W \qquad (9\text{-}3)$$

where, by thermodynamic convention of long standing, ΔE signifies the change in energy of a particular system (a chunk of matter, a machine, a chemical reaction); Q indicates energy flowing into the system in the form of heat; and W indicates energy flowing *out* of the system in the form of work.[3]

9-4 The Energy of Formation of Chemical Compounds

Equation (9-3) gives us confidence that we can measure the energy *change* in a chemical reaction by measuring the heat and work effects it produces. Furthermore, the notion that energy is conserved makes it reasonable to suppose that the energy of any given substance is always the same under specified conditions of temperature and pressure, even though we cannot conveniently measure that energy. But if every substance has a definite energy under definite conditions, then the difference in energy between any two substances must also have a constant and characteristic value (Figure 9-3). All that needs to be done is to define an arbitrary reference level of energy, and measure differences from that reference level. In exactly the same way, we talk about mountain heights and airplane positions by using a scale of altitude which takes the average level of the sea as an arbitrary reference; we do not bother about measuring altitudes above the center of the earth.

Using the energy balance

---

[2] Of course, we can always compute the total energy equivalent of any given mass through Equation (9-2). But, as we have noted, we are dealing only with a tiny fraction of this total in a chemical reaction, a tinier fraction than we can weigh.

[3] The peculiar reversal of the sign convention for heat and work reflects early thermodynamicists' concern with heat engines, for which heat normally flows in, and work normally flows out.

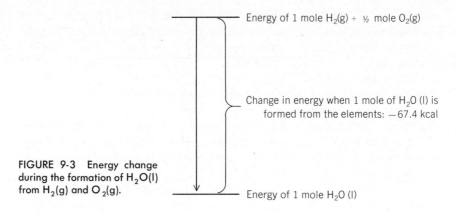

FIGURE 9-3 Energy change during the formation of $H_2O(l)$ from $H_2(g)$ and $O_2(g)$.

The convenient arbitrary reference level that is chosen is the energy of the individual chemical elements in their usual states at room temperature and one atmosphere pressure. Then the energy of formation of any compound is defined in terms of a measurable quantity of energy, the *difference* in energy between that compound and the elements of which it is composed. Thus, for example, the energy of formation of water is defined as the energy change in the reaction

$$H_2 \text{ (g, 1 atm)} + \tfrac{1}{2} O_2 \text{ (g, 1 atm)} \to H_2O \text{ (l, 1 atm)} \tag{9-4}$$

9-5 The Enthalpy Function

We remarked that during the dimerization of NO_2 the heat effect of that reaction depends noticeably on whether it is carried out under conditions of constant volume or of constant pressure. A container of constant volume (such as the strong steel bombs that are sometimes used to measure such heat effects) does not allow the pressure of the surrounding atmosphere to do any work on the reacting system by compressing it as the number of moles of gases present changes. On the other hand, a container of variable volume, like a cylinder with a piston, or like the open beakers used to carry out chemical reactions in liquids and solids, does allow this work to be done. If the reaction *is* carried out in an open container, the atmospheric pressure on the reacting mixture is constant, and the work may be calculated readily (see Appendix III) to be given by that pressure multiplied by the change in volume of the reacting system (Figure 9-4). We can use the first law [Equation (9-3)] to calculate the heat effect under these circumstances:

$$Q = \Delta E + W \tag{9-5}$$

$$= (E_{\text{products}} - E_{\text{reactants}}) + P(V_{\text{products}} - V_{\text{reactants}})$$

Now the function on the right-hand side of Equation (9-5) contains the difference in energy between products and reactants and the difference in their volumes. That is, it depends on the properties of the reactants and products of the reaction. Let us collect together terms referring to the reactants, separating them from those referring to the products:

$$Q = (E_{products} + PV_{products}) - (E_{reactants} + PV_{reactants})$$

This difference in the properties (E, P, and V) of the reactants and products can be expressed more compactly:

$$Q = (E + PV)_{products} - (E + PV)_{reactants} \qquad (9\text{-}6)$$

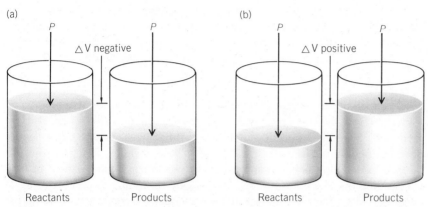

FIGURE 9-4 Work done by a reacting chemical system under a constant external pressure P. The work done by the system in both cases is $P\Delta V$. In (a), the products occupy less volume than the reactants; ΔV is negative, and Q ($= \Delta E + W$) is more negative than ΔE. In (b), the products occupy more volume than the reactants, so by the same reasoning, Q is more positive than ΔE.

Equation 9-6 shows that the heat effect in a reaction carried out at constant pressure still depends only on properties of the reactants and products of the reaction, just as in the case of reactions carried out at constant volume; but in this case, more than just the energies of the products and reactants is involved. Consequently, the heat effect in reactions carried out at constant pressure is not such a direct measure of the change in potential energy on a molecular scale as is the heat effect of a constant-volume reaction, which gives ΔE directly.

The significance of $(E + PV)$

Nevertheless, it is very often more convenient to carry out reactions under constant-pressure conditions than under those of constant volume. In particular, in the case of reactions taking place in solids and liquid phases, enormous forces develop when reactions occur at constant volume. Because the molecules of solids and liquids are in contact, there is no room to accommodate new molecules in the system if there is an increase in the number of molecules during the reaction. If there is a decrease in the number of molecules during the reaction, the opposite problem arises: Van der Waals forces strive mightily to fill in any holes left by the disappearance of molecules. Furthermore, open containers like beakers, flasks, and test tubes are convenient to use, and necessarily, reactions in them are constant-pressure processes.

Fortunately, the discrepancy between ΔE and $\Delta(E + PV)$ for most reactions is very small. Even in the gas-phase dimerization of NO_2, the difference is only 0.6 kcal, less than 5 percent of the very modest ΔE value. For reactions involving larger energies, and for reactions having solid or liquid reactants and products with their much smaller molar volumes, the effect of the PV term is even more insignificant. In any event, the ΔPV term is readily calculated, and chemists can use heats of reaction measured either at constant pressure or at constant volume as an indication of the change of molecular potential energy during the reaction.

With constant-pressure heats of reaction playing such a prominent role in chemical energetics, it has been found convenient to give a special name and symbol to the function $(E + PV)$, whose change is equal to the constant-pressure heat of reaction. It is called the **enthalpy**, and given the symbol H. Thus,

$$H \equiv E + PV \qquad (9\text{-}7)$$

Enthalpy: a name for a useful function

It is possible to measure and tabulate for any substance its *enthalpy of formation* from the elements, just as energies of formation are known. Because E, P, and V are all properties of substances that have definite values under definite conditions, the enthalpy change in any chemical reaction (the difference in enthalpy between products and reactants) is independent of how we imagine the reaction to be carried out. By imagining the reactants to be decomposed to their elements, and the elements recombined to form the products,[4] we may use enthalpies of formation to calculate the enthalpy change in any reaction for which the enthalpies of formation of the products and reactants are known. The enthalpy of formation is often called the **heat of formation**. Perhaps this principle will be clearer if we put it to work at once on an example.

The dimerization of NO_2 is, in fact, not a reaction whose heat effect can conveniently be measured directly. A container of NO_2 will begin to dimerize, but the reaction reaches equilibrium

$$2NO_2 = N_2O_4 \qquad (9\text{-}1)$$

long before all of the NO_2 present has reacted. Thus, any heat that is observed results from the reaction of only part of any sample of NO_2 taken. On the other hand, the value of ΔH for this reaction may be *calculated* from enthalpies of formation by following this circuitous (but not circular!) reasoning:

$$\tfrac{1}{2}N_2(g) + O_2(g) \rightarrow NO_2(g): \ \Delta H \ (= \Delta H \text{ of formation of } NO_2)$$

$$= +8.091 \text{ kcal/mole of } NO_2 \ [5]$$

$$N_2(g) + 2O_2(g) \rightarrow 2NO_2(g): \ \Delta H = +(2)(8.091) \text{ kcal} \qquad (9\text{-}8)$$

---

[4] Such an imaginary process carries with it the assumption that a roundabout route to the products does not affect the energy change involved, and this indeed is true. Experiment shows it to be so. Sometimes the principle is stated as Hess's law: The enthalpy change of a reaction is independent of the number of steps.

[5] It is implied by the sign conventions for heat and work that a positive value of ΔH means an increase in the enthalpy of the chemical system; that is, that heat flows *into* the system at constant pressure. A negative sign, of course, means that heat flows out, accompanying a decrease in the enthalpy of the system.

But the reverse of Equation (9-8) would have the opposite heat effect:

$$2NO_2(g) \rightarrow N_2(g) + 2O_2(g): \Delta H = -(2)(8.091) \text{ kcal}$$
$$= -16.182 \text{ kcal} \quad (9\text{-}9)$$

The heat of formation of N_2O_4 is

$$N_2(g) + 2O_2(g) \rightarrow N_2O_4(g): \Delta H \; (= \Delta H_{formation} \text{ of } N_2O_4)$$
$$= +2.54 \text{ kcal/mole} \quad (9\text{-}10)$$

We can now sum up Equations (9-9) and (9-10), eliminating terms which appear on both sides, to obtain

	ΔH, kcal
$2NO_2(g) \rightarrow N_2(g) + 2O_2(g)$	-16.182
$N_2(g) + 2O_2(g) \rightarrow N_2O_4(g)$	$+2.54$
$2NO_2(g) \rightarrow N_2O_4(g)$	-13.64

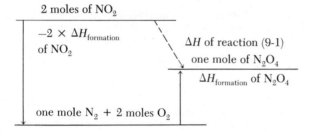

The same reasoning, shown in a diagram

Thus, we have used enthalpies of formation to calculate the magnitude of a heat effect that we cannot readily observe directly, and we have used the result of that calculation (Section 9-2) to say something about molecular energy changes in the dimerization of NO_2. There are available in the chemical literature the enthalpies of formation of many thousands of chemical substances (and unstable species such as $O_2{}^+$), allowing thermochemistry to be done on paper much more quickly and conveniently than it can be done in a laboratory, given reliable values of the data. This correlation and distillation of the data from millions of thermochemical experiments would be impossible without the fact (or, as we have seen, the near-fact) of the conservation of energy in chemical reactions. Such correlations and distillations are one of the primary values of comprehensive theories like the first law of thermodynamics.[6] We shall make use of these published values in two illustrative examples.

More examples

For the first example, we draw upon the terrible and tragic World War II history of London during the "blitz," when fire bombs rained down nightly. Fire extinguishers that use carbon dioxide were a familiar household item. The CO_2 smothers flames from burning wood and paper, serving very well to put out accidental fires.

---

[6] The other primary value is aesthetic.

But CO_2 extinguishers were found to be totally ineffective against fire bombs which employed burning magnesium powder as the incendiary agent; the magnesium appeared to burn more vigorously in CO_2 than in air!

The same reaction is demonstrated easily on the lecture table. A mixture of powdered magnesium and powdered dry ice (solid CO_2) can be ignited by a piece of magnesium ribbon as a fuse, and the mixture blazes vigorously despite the low temperature of the dry ice. The reaction leaves a mass of black carbon and white magnesium oxide, showing that the carbon dioxide has been reduced:

$$2Mg + CO_2 = 2MgO + C$$

We want to know the **heat** of this **reaction**, per mole of CO_2. Starting with the reactants at standard state, for the moment, we find in a chemical handbook that the standard enthalpy of formation of CO_2 is -94 kcal/mole, while that of MgO is -146 kcal/mole. We add the enthalpies of formation of the products

$$2MgO = 2 \times (-146 \text{ kcal}) = -292 \text{ kcal}$$
$$+ 1C \text{ (zero by definition)} = \underline{ 0}$$
$$\text{sum } H_{\text{products}} = -292 \text{ kcal}$$

Remember that enthalpies of formation of the elements are zero by definition.

and then we add the enthalpies of formation of the reactants

$$2Mg \text{ (zero by definition)} = 0$$
$$+ 1CO_2 = \underline{-94 \text{ kcal}}$$
$$\text{sum } H_{\text{reactants}} = -94 \text{ kcal}$$

Since

$$\Delta H_{\text{reaction}} = \text{sum } H_{\text{products}} - \text{sum } H_{\text{reactants}}$$

the heat of reaction is

$$\begin{array}{r} -292 \text{ kcal} \\ \underline{-(-94) \text{ kcal}} \\ -198 \text{ kcal/mole of } CO_2 \text{ involved} \end{array}$$

This is a tremendous amount of heat, and it shows that as far as this oxidation-reduction reaction is concerned, CO_2 is a richer source of oxygen than air.

In the lecture-demonstration experiment, the low temperature of the CO_2 snow does not change matters much. The heat required to vaporize one mole of CO_2 at $-78°C$ is 2.64 kcal, and the heat required to raise the temperature of the resulting one mole of gas from $-78°C$ to $+25°C$ is only 0.86 kcal, so by the expenditure of only $2.64 + 0.86$ kcal, or 3.50 kcal, we have one mole of CO_2 as gas at room temperature. This 3.50 kcal must come out of the heat of reaction, but the subtraction leaves 194.5 kcal net heat of reaction to provide the brilliant flame.

Our other example arises from a matter of laboratory practicability. We can determine the heat of combustion of an inflammable substance quite easily by burning a weighed sample in pure oxygen within the closed vessel of a calorimeter and measuring the rise in temperature of the calorimeter system. So heats of combustion are found in every handbook. Determining the heat of formation of a substance is much harder, especially if it is an endothermic process that takes place only at high temperature. Making use of the principles developed in this section, it becomes possible to *calculate* enthalpies of formation instead of measuring them. For example, what is the enthalpy of formation of carbon disulfide, CS_2?

We imagine CS_2 taken apart into its elements (the *opposite* of formation), and then the oxidation of carbon and sulfur separately. The sum of the two operations must be the same as the heat of combustion directly from CS_2 to SO_2 and CO_2. That is, the upper cycle of reactions A → B → C in the diagram

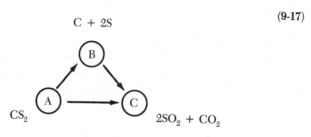

(9-17)

must produce the same end result as the lower pathway, A → C, or else we should have the workings of a perpetual-motion machine (see also Hess's law, footnote page 294). But the Process A → C represents the heat of combustion CS_2, which is −265 kcal/mole (negative because the heat is evolved), and the process B → C represents the heat of combustion of carbon (−94 kcal/mole) and of sulfur (−70 kcal/mole). So the cycle A → B → C comprises $-\Delta H_{formation}(CS_2) + (-94 \text{ kcal}) + (2)(-70 \text{ kcal})$, or $-\Delta H_{formation}(CS_2) - 234$ kcal. The process A → C gives −265 kcal. Hence the difference between −265 kcal and −234 kcal, or −31 kcal, must be the heat of dissociation of CS_2, and we conclude that the opposite (the heat of formation of CS_2, $\Delta H_{formation}$) must be +31 kcal/mole. We see also that the substance CS_2 is an endothermic compound; its formation absorbs 31 kcal/mole from the surroundings.

9-6 The Ongoing Nature of Things

Light a match. Dive into a swimming pool. Drop a brick, or your watch. Each of these happenings involves a transformation of potential energy (arising from electrical or gravitational forces) into heat. Each is an *irreversible* event; the precise opposite does not occur. You never see watches and bricks rising from the floor, swimmers leaping up to diving boards, smoke and oxides of carbon, hydrogen, sulfur, and phosphorus reconstituting an unburned match, while absorbing from the surroundings the heat released during the "forward" process. Not, of course, that it is impossible to lift watches and swimmers, or even to reconstitute matches. All of these things can be done if they are coupled to other more forceful, irreversible events. For example, a swimmer can be plucked out of a swimming pool by attaching him to a rope, on the other end of which is either a pulley and another heavier weight, or a crane powered by an engine that uses the energy of burning fuel. But the larger weight, once dropped, the fuel once burned, can only be restored to *their* original states by the conversion of still greater quantities of potential energy into heat. Thus, our common experience seems to indicate that there is a natural tendency for potential energy to be converted first to kinetic energy (the colliding atoms of fuel and oxygen, the falling swimmers and bricks) and then into heat, where most of the energy is inaccessible for further use.

It is possible to convert *some* of the energy of heat into potential energy: A steam-operated crane can lift weights, and so forth. But it has been the universal experience of mankind without exception that it is impossible to convert *all* of the energy of heat into potential energy. The well-known impossibility of perpetual-motion machines is a consequence of this rule; if it were not so, then an engine could be devised that would use the heat produced by a falling body to lift that body to its original height and release it for another fall, in an endless cyclic process. Or a chemically operated perpetual-motion machine could use the heat generated by a chemical reaction to power a chemical plant that would reconstitute the original reactants (Figure 9-5).

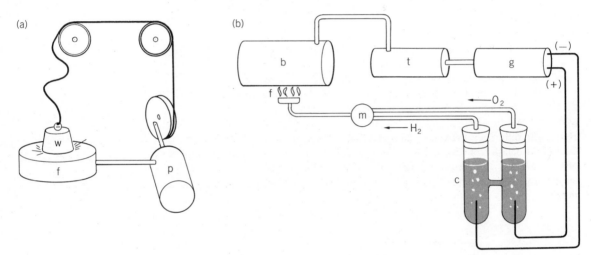

FIGURE 9-5 (a) An (impossible) perpetual-motion machine. Heat generated by the falling weight (w) causes a fluid (in f) to expand, operating piston (p), which raises the weight to its original height. (b) Another (impossible) perpetual-motion machine. A hydrogen-oxygen flame (f) heats a boiler (b), creating steam which drives a turbine (t), and a dc electrical generator (g). The electrical power is used to operate an electrolysis cell (c) which decomposes water into hydrogen and oxygen, which are mixed at (m) and burned at (f). A condensing system (not shown) catches the water vapor produced in the flame and condenses it to liquid water, which is returned to the electrolysis cell.

Perpetual motion

That such machines have never operated is not the fault of the many who have spent lifetimes trying. The creation of an immortal machine would give our Appolonian culture not only a free source of energy for all its works, but a symbolic victory over death. But the very same law of energy that gives each life an end as well as a beginning forces the hard lesson that perpetual-motion machines are impossible. Every realizable process is irreversible, and the great majority of them convert energy from the accessible (sunlight, molecular, or macroscopic potential energy) to the inaccessible (uniform and all-pervading heat).

9-7 Chaos, Probability, and Equilibrium States of Matter

There are counterexamples to the conversion of potential energy to heat. Any natural process that causes a *drop* in temperature (that is, the absorption of heat) must, according to the first law of thermodynamics, accomplish the conversion of the absorbed heat to potential energy. Familiar examples abound: the expansion of a nonideal gas into a vacuum; the dissolving of NaCl or, more strikingly, of KNO_3 in water; and the contraction of a stretched rubber band. (You can observe the last of these by stretching a rubber band, holding it stretched for a few seconds, and then letting it snap back to its original shape. The drop in temperature is rather subtle and is best felt at some temperature-sensitive spot on your body, such as the wrist or lips.)

These heat-absorbing spontaneous processes are just as irreversible as those described in Section 9-6. That is, they can be reversed only at the cost of greater changes elsewhere in the world—gases may be recompressed, rubber bands restretched by falling weights; salt solutions can be decomposed into pure water and pure salt by distilling off the water over an irreversible fire.

What single law can account plausibly for the unidirectionality of natural change and for the preponderance among these changes of heat-releasing ones? Certainly the first law of thermodynamics is no help to us here. Each of the impossible reversed processes and perpetual-motion machines described in Section 9-6 proceeds with conservation of total energy. We need a principle that will describe in what ways today must be different from yesterday, thus putting directionality in time into the energy balancing of the first law.

Time and chemistry

Atoms and molecules behave according to the timeless principles of quantum mechanics. In the quantum-mechanical description of stable atoms and molecules, chemists specifically eliminate the time dependence of the electronic wave function, and properly so, since a single atom left to its own devices will remain stable "forever."[7] But atoms and molecules are much smaller than people, so that any event that we can see involves unimaginably large numbers of them. Since we cannot possibly follow in detail the course of each of these multitudes of atoms, we are limited, in our descriptions as in our perceptions, to considering them as groups exhibiting certain average characteristics. Directionality in time arises, not from the behavior of any single particle, but from the statistical behavior of many. This is true not only of processes, but of unreacting matter. When we put a finger over an open gas jet we feel, not the impact of a single gas molecule, but the pressure generated by billions of such impacts. We feel, not the vibrational energy of one iron atom, but the average energy of septillions in the temperature of a hot iron skillet. If, then, we see around us the statistical, mob behavior of multitudes of particles, we should expect that the laws of probability will govern the ways of the world.

Statistics and chemistry

You may object that the world that we see does not show very obvious probabilistic features. Events, far from being random seeming, are inexorable. Falling objects *always* generate heat; matches *always* burn, never "unburn";

---

[7] Provided its nucleus is stable with respect to decay into a simpler one.

leaves and men change color and die, each at characteristic rates. The answer is that the behavior we see is statistical all right, but that the statistics are stacked heavily in favor of *increasing microscopic disorder*.

To see that this is so, let us consider in detail the simplest kind of spontaneous process, the expansion of a gas to fill all of the space available to it. From consideration of this, we should be able to see why the other spontaneous processes we have been considering are so reliably one-directional.

Suppose, then, we have a cylinder containing a gas under pressure, and we open this cylinder in a telephone booth, whose volume just happens to be nine times that of the cylinder (Figure 9-6). After the valve is opened, the gas has available to it just 10 times as much volume as it did before. (We shall assume, for simplicity, that there was no gas in the telephone booth before ours was introduced.) "Naturally," the gas rushes out until the pressure in the larger volume is equal to the pressure of the gas remaining in the cylinder. If we could follow the careers of individual molecules, we could readily see that the chances of any particular molecule being inside the cylinder is 1 in 10 as long as we leave the valve open, so that the whole volume of the cylinder and the telephone booth is available to each molecule.

"Naturally." What are the chances that the gas would *not* rush out of the cylinder? Or what are the chances (what is the probability) that this natural, spontaneous process will reverse itself, so that all of the gas in the booth rushes back into the cylinder? It turns out that we cannot answer that question until we know how many molecules of gas we have; so let us consider various numbers of molecules.

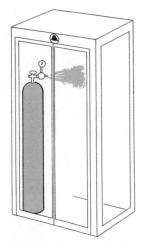

FIGURE 9-6 The irreversible expansion of gas into a telephone booth. The volume of the telephone booth plus the cylinder is just 10 times the volume of the cylinder (in color) alone.

1. If there is just *one* molecule, and the valve is opened, then the "rush" of gas out of the cylinder is going to be a pretty mild thing, but that one molecule is bound eventually to occupy all of the volume open to it in its ceaseless caroming about. The chance that we shall happen to find it outside the cylinder at any moment is 9 out of 10, or 0.9, and the chance that it will be inside the cylinder is 0.1. The probability, that is, of finding any *one* molecule in a given space is just proportional to the size of that space.
2. What if we have two molecules in our gas? The chance of finding them both *inside* the cylinder at once (that is, of reversing the spontaneous outflow) is the chance of finding molecule 1 in the cylinder while molecule 2 happens also to be there. It will happen 10 percent of 10 percent of the time, or with a total probability of 0.01.
3. From the above consideration, we see that the probability of reversing the spontaneous direction of a gas expansion is just the probability of a single molecule reversing the expansion, raised to a power equal to the number of molecules. If we have three molecules, the probability is (0.1) (0.1) (0.1), or 0.001—one chance in a thousand.

At this point, it should be clear that a really gross reversal of a gas expansion has a probability that is low enough for only one molecule, but completely hopeless even for a handful of molecules. Perhaps we are being too ambitious. We can detect much more subtle departures from equilibrium than a tenfold one. Suppose we specify that we can detect, within our telephone booth, the pressure rise caused when all of the molecules in a gas in it crowd themselves into 99.9 percent of the space, rather than 100 percent. That would be a reversal of the spontaneous expansion by only 0.1 percent, a mild enough miracle to occur. Let us calculate the likelihood of this miracle:

1. The probability of finding *one* molecule within 99.9 percent of the space within the telephone booth is 0.999, and the probability of finding it outside that space is 0.001.
2. The probability of finding each of *10* molecules within 99.9 percent of the available volume is, then, $(0.999)^{10} = 0.991$.
3. The probability of finding each of *100* molecules within 99.9 percent of the available space is $(0.999)^{100} = 0.912$. That is, there is still a 91.2-percent chance (or about a 10-to-1 chance) that a gas composed of 100 molecules will *spontaneously* shrink into only 99.9 percent of the space available to it! Let us carry on for two more calculations.
4. By the same reasoning, the probability of a 0.1-percent shrinkage of a gas of *1000* molecules is 0.4, or 40 percent. The odds are now against the shrinkage.
5. For a gas of 10,000 molecules, the probability of such a shrinkage (that is, of finding each molecule within 99.9 percent of the space available) is $(0.999)^{10,000} = 0.000002$.

This time, we find a real unlikelihood only for much larger numbers of molecules. But let us remember that 10,000 is still absurdly tiny compared to the number of molecules we encounter every day in telephone booths. A reasonable-sized telephone booth might have a total volume of about 2 m³, or 2×10^3 liters, and so contain roughly 100 *moles* of gas at one atmosphere pressure. This amount of gas is comprised of some 6×10^{25} molecules, or more than 10,000 by a factor of 6×10^{21}. Where now is the probability of even a trivial spontaneous shrinkage of this gas into only part of the space available to it? What was a perfectly expectable thing for a single, timelessly, aimlessly wandering molecule to do has become an improbability of horrendous dimensions for the amount of air in a telephone booth. On the contrary, the likelihood that a gas, compressed ever so slightly, will expand to fill *all* of the space available to it becomes an overwhelming, dead certainty, not because of any push on each little molecule to spread out from the others, but simply from the crushing weight of statistics operating on Avogadro's number of particles.

It now remains to us to make the connection between this statistical result and the flow of heat, on the one hand, and molecular disorder, on the other.

When most gases expand, they become cooler, as their molecules become farther apart. To separate the molecules of a gas, work must be done against the van der Waals forces (Chapter 5) among them; in the process, heat energy is converted into molecular potential energy in the van der Waals electrical fields. But, as we have seen, they expand only because they "probably" will, with a probability that is overwhelming. Thus, thermal energy *can* be converted into potential energy if the odds are right. Expanded gases have the right odds; why do dissolved salts and contracted rubber bands? In each of these cases, the high-probability state is one that we may anthropocentrically[8]

Statistics and time

---

[8] It turns out to be nearly impossible to give a clear definition of randomness or chaos except as absence of order, and what is "order" but a human choice of one or a few of an infinite number of arrangements of large numbers of things? Consider the arbitrariness of the distinction between an ordered and a shuffled deck of cards, an alphabetized group of people and one not so ordered.

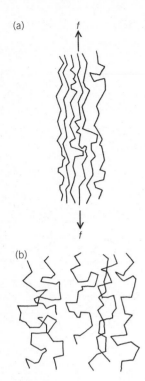

FIGURE 9-7 Entropy changes when stretched rubber relaxes. (a) Stretched state. Polymeric rubber molecules are pulled into an orderly and somewhat close packing by a vertical force f. (b) Relaxed state. When the vertical force is removed, random thermal collisions "instantly" knock the polymer molecules into random coils, which are both shorter and farther apart laterally.

characterize as chaotic or random. Chaotic states have a high statistical probability because they may be realized in many ways, while ordered states have fewer possibilities of realization. Consider, for example, our gas. Before it expands, *all* of its molecules are segregated into *part* of the space. Afterward, any individual molecule can be either inside or outside of the cylinder. Only *one* arrangement of a deck of cards is called ordered; the rest of the arrangements of the 52 cards is "shuffled."

The science of *statistical mechanics* concerns itself with computing the probabilities of various configurations of atoms and molecules. In every case, it develops, as in the case of our simple gas, that the most chaotic arrangements of molecules have the highest probability. In the case of dissolving salts, this means the conversion of an orderly crystal and a pure liquid into a random mixture of hydrated ions and excess water molecules. The case of rubber bands is more subtle—long, roughly parallel polymeric molecules (described in Chapter 14) are pulled into orderly alignment when the rubber is stretched. The pull of the rubber band is the result of random thermal vibrations trying to knock these orderly arrays into disorderly and shorter random coils (Figure 9-7). Because of sideways van der Waals forces between the parallel molecules, the stretched state is actually lower in potential energy than the unstretched state; in the thinner stretched rubber band, the molecules approach one another more closely than in the fat and random relaxed state. This accounts for the rubber band warming up when it is stretched, and cooling when it relaxes.

Homework assignment: Blow up a rubber balloon (inside a telephone booth, if you like), let it go, and watch entropy increase.

If randomness must prevail, then why do the forces between particles have any effect at all? Why is the universe not composed of uniformly distributed gas? The answer lies in the statistical behavior of energy. Remember, from our earlier discussion, that the formation of stable chemical bonds leads to the conversion of molecular potential energy into heat. The *random* energy of heat has a high statistical probability, as the following simple example shows.

Let us return to our falling meteorite, and this time follow the motions of its component atoms as it falls and strikes the earth. As it is falling, each atom has kinetic energy in a path that is approximately parallel to that of all the other atoms in the meteorite (Figure 9-8). The motion of these atoms is what we might call *orderly*. The meteorite strikes the earth, and the blow sets its atoms into violent, random motion (let us suppose that it is strong enough, or falls gently enough, not to break apart). Now the kinetic energies of its atoms are directed *randomly* in all directions, and we observe the effect not as motion of the whole meteorite, but as heat. The probability that these atoms will *by chance* all move in the same direction again (so that heat is converted back to kinetic energy of the whole meteorite) is surely seen to be terribly small. Such an event is of an unlikelihood commensurate with that of the spontaneous self-compression of a gas.

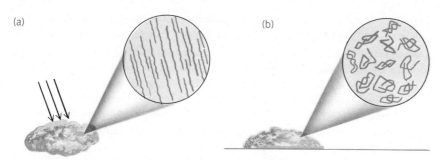

FIGURE 9-8 Changes in randomness of molecular motion when an object falls and strikes the earth. (a) The cold meteorite falls through space toward the earth. Each atom vibrates relatively little within the crystal lattices of the meteorite, but all have a common, parallel motion (magnified about $10^7 \times$ in the inset). (b) After collision with the earth, the ordered energy of motion has been converted to the greatly increased random motions of thermal energy.

When the bonding forces between particles are strong enough, they allow the generation of much heat when a bond is formed. The randomized energy of this heat creates a new state (hot but bonded), whose overall probability can compete with that of the separate atoms (cool but free). The bond energy, once lost to the surroundings as heat, can "never" spontaneously flow back into the system again. So matches burn, but never unburn. It is not that matches cannot unburn, but that *probably* no match in your lifetime will unburn.

9-8 The Second Law of Thermodynamics: The Entropy Function

We can now see that the law of nature that drives spontaneous changes always in one direction is a statistical one. A possible way of stating this *second law of thermodynamics* is this: Every possible process is one in which the molecular disorder of the world increases. Stating it this way, we take care of both heat-absorbing (endothermic) and heat-releasing (exothermic) processes. In the former, the increased disorder in the changing system (expanding gas, dissolving salt) predominates in spite of cooling effects; in the latter, the heat released to the surroundings compensates for increased bonding and decreased randomness in the changing system (for example, in the freezing of water into beautiful ordered crystals).

It is useful to have some numerical measure of the molecular disorder of substances under various circumstances, so that we can compare the influence of disorder with that of potential energy in causing chemical changes. We can find such a measure by relating disorder to probability; remember how the increase in spatial disorder in an expanded gas makes the expansion highly probable. It would be nice also if our disorder-measuring function were additive, in the sense that 2 moles of some substance should contain twice as much disorder as one, just as the energy and enthalpy functions are additive. But

Statistics, disorder, heat, chemistry, and time: the second law of thermodynamics

Don't confuse entropy with enthalpy.

we found while thinking about our gas that probabilities are multiplicative rather than additive; the probability of finding two molecules in some state is the square of that of finding just one. We can make a multiplicative function additive by taking its logarithm;[9] the logarithm of XY is $\log X + \log Y$. The disorder-measuring function that chemists use is the **entropy** (symbolized by S) which we encountered in the previous chapter (Section 8-8) and which may be now defined as being related to the statistical probability that a given collection of atoms will be found in the form of a given substance:

$$S \equiv 2.303\, R \log \omega \qquad (9\text{-}11)$$

The quantity ω measures the number of ways that the collection of atoms can combine to form a given substance in a given state; thus, ω is the statistical weight of that state. Its evaluation is a problem in statistical mechanics, into which you are invited to delve some other time (see "Suggestions for Further Reading"). The R in Equation (9-11) is the same constant as is found in the ideal gas law (Chapter 5). It has dimensions of energy divided by temperature and number of moles of substance; the usual units used by chemists are calories per mole-degree.[10] For our purposes, we can get a good feel for entropy by looking at the calculated molar entropies of a number of substances and relating them to what we know about the internal structure of those substances.

Molar entropy

As we examine the numbers in Table 9-1, we should see a pattern emerging. *Most strikingly, the molar entropies of most gases are larger than those of other substances.* This is perfectly reasonable on order-disorder grounds; what could be more disordered than a substance, no two of whose molecules have any particular relationship to each other, and any molecule of which is free to move throughout the whole mass? The entropies of liquids are, by and large, smaller than those of most gases, and the entropies of solids are smaller yet. Again, this observation simply reflects the increasing ordering of molecules. In a liquid, there is no rigid structure, but molecules are much more constrained spatially, and temporary structures held together by van

---

[9] See Appendix II.

[10] It is not immediately evident that such a unit is an appropriate measure of randomness, and a complete proof of the relationship is best left to the reader's second acquaintanceship with entropy. We can, though, give some rationalization of this relationship, which will make sense in light of our earlier discussion of order and disorder. Whenever heat flows into a substance, the energy so gained causes an increase in disorder in one of three ways: either by expansion of the substance in space (as in an expanding gas), by loosening or breaking bonds within the substance (as in melting ice), or by an increase in the violence of random internal motions (that is, by an increase in temperature). The *increase* in disorder caused by such an influx of heat will be larger, the greater the quantity of heat flowing in; and smaller, the greater the thermal disorder already present. Thus, an increase in entropy should be proportional to the heat inflow (in calories), and inversely proportional to the temperature of the substance (in degrees Kelvin). But an increase in a quantity must have the same units as the quantity itself; thus, entropy must also have units of energy divided by temperature. The disorder within a substance is also proportional to the quantity of substance present, so that characteristic entropies are put on a molar basis. Thus, the full proper unit for the entropy of a substance should be energy divided by temperature *per* (divided by) quantity: cal/mole-deg.

der Waals forces come and go within the bulk of the liquid. And in solids, of course, each molecule has even less freedom and more ordering. The rigidly structured diamond crystal has a spectacularly low entropy.

TABLE 9-1
Molar Entropies of Selected Substances

I. Monatomic gases		IV. Simple liquids	
He	30.126	CS_2	36.10
Ar	36.983	Br_2	36.4
Xe	40.53	V. Hydrogen-bonded liquids	
II. Diatomic gases		H_2O	16.716
H_2 ($^1H_2$)	31.211	CH_3OH	30.3
D_2 ($^2H_2$)	34.602	VI. Metals	
N_2	45.767	Hg	18.5
O_2	49.003	Pb	15.51
F_2	48.6	W	8.0
Cl_2	53.286	VII. An ionic substance	
		NaCl (crystal)	17.30
III. Polyatomic gases		NaCl	27.6
CH_4	44.50	(1 M solution in water)	
C_2H_6	54.85	VIII. Covalent solids	
SiH_4	48.7	SiO_2 (quartz)	10.00
GeH_4	51.21	SiO_2 (glassy)	11.2
NO_2	57.47	C (diamond)	0.5829
N_2O_4	72.73	Si	4.51

All values are for the conditions of one atmosphere pressure and 25°C. The units are cal/mole-deg. These data were selected from the very useful National Bureau of Standards Circular 500, F.D. Rossini, D.D. Wagman, W.H. Evans, S. Levine, and I. Jaffe, *Selected Values of Chemical Thermodynamic Properties*, Washington, D.C.: U.S. Government Printing Office (1952).

Second, we note that *the entropy of similar substances is larger, the larger the molecular weight*. Thus, we observe a steady increase of molar entropy within the monatomic gases with increasing molecular weight. A similar effect is shown by the hydrogen molecules of different isotopic composition. The fundamental reason for this effect is quantum mechanical. Like all energies of confined motion, the kinetic energy of a gas molecule is quantized; that is, it can take on only certain values. But the number of energy states so allowed is larger for heavier molecules than for light ones, so that a given collection of heavy molecules can spread its energy content over more states than can a light one; the energy is thus more randomly distributed, and the molar entropy is higher.

Third, *the entropy is higher, the more complex the molecular structure*. Complex molecules have modes of motion and thus opportunities for disorder that are not available to simple ones—diatomic molecules can rotate in two planes, while single atoms cannot, and polyatomic molecules can rotate in three planes. C_2H_6 has an internal rotation (of one-half of the molecule with respect to the other half) that CH_4 has not (recall Chapter 4). Further, complex molecules have internal vibrations of one atom or group of atoms against the

Internal vibrations and rotations increase entropy.

others; the more bonds, the more vibrations. Thus, all of the polyatomic gases listed have higher molar entropies than the monatomic gases, even though one of the latter (xenon) is heavier.

Fourth, *the molar entropy reflects internal ordering resulting from association between molecules.* The nonpolar liquids have higher entropies than the hydrogen bonded ones (though in this case, part of the effect reflects difference in molecular weight as well). The strength of metallic bonding is reflected in mercury, lead, and tungsten, whose entropies decrease, and melting points increase, in that order.

Fifth, *the molar entropy of a given substance varies with the disorderliness of its state.* Thus, crystalline sodium chloride has a much lower entropy than dissolved sodium chloride,[11] and SiO_2 has an entropy that depends on how it solidifies. In quartz, a regularly crystalline substance (whose structure is discussed in Chapter 6), there is a long-range order in the arrangement of Si and O atoms that holds throughout the crystal. In glassy or vitreous SiO_2, each Si atom is properly surrounded by oxygen atoms, but the relationship of these units to neighboring units changes *randomly* throughout the mass; there is short-range order, but no long-range order, and the entropy is correspondingly higher.

The Second Law again

Now that we have, in the entropy, a function that seems to be a faithful reflection of our hitherto qualitative ideas about order and chaos in matter, we can give the second law of thermodynamics yet another statement, equivalent to those we have already discussed—*changes of the form of matter will always occur in such a direction that the total entropy of the world increases.*

9-9 The Entropy of the World and Chemical Equilibrium

Let us apply the entropy function to some familiar one-directional processes, to see how it controls the direction which those processes adopt. First, we must note that the second law refers to the total entropy of the world, not to that of any particular chemical system in the world. The entropy of any part of the world may decrease as long as there is an associated and larger increase in entropy elsewhere. For the purpose of this discussion, then, we divide the world into two parts: the particular system of matter and energy that we are interested in, and its surroundings—the rest of the world. From the standard molar entropies of substances like those given in Table 9-1, we can always calculate the change in entropy for any chemical reaction, just as we calculate heats of reaction from enthalpies of formation. Thus, for

System + surroundings = "world."

---

[11] One cannot measure the thermodynamic properties of a single substance in a solution, but only its apparent contribution to the properties of the whole solution. The entropy tabulated for dissolved NaCl is, in fact, the rate at which the entropy of the solution increases as the NaCl content increases. It reflects not only the sodium and chloride ions, but their interactions with the solvent water molecules. Nevertheless, dissolved sodium and chloride ions are surely in a more disordered state than those in an ionic crystal, and the entropy contribution of NaCl to the solution reflects that fact. See also the remarks about changes in entropy when a metal dissolves electrochemically (Section 8-8).

example, the conversion of 2 moles of NO_2 into one mole of N_2O_4 must require a decrease in the entropy of the chemical system, as follows:

$$2NO_2(g, 1\text{ atm}) \longrightarrow N_2O_4(g, 1\text{ atm}) \qquad (9\text{-}12)$$
$$(S = 2 \times 57.47 \text{ cal/deg}) \qquad (S = 72.73 \text{ cal/deg})$$
$$\Delta S = S_{\text{products}} - S_{\text{reactants}} = -42.21 \text{ cal/deg}$$

Note that the entropies in this calculation are those of the pure gases at one atmosphere pressure, but, of course, when NO_2 does dimerize, the result is a mixture of N_2O_4 and as yet unreacted NO_2. We shall look at the consequences of this fact in a moment.

We should be able, then, to calculate the entropy change in the *world* caused by any chemical reaction if we had a method of calculating the entropy change in the surroundings of the chemical system. There will certainly be some such change, because chemical reactions release or absorb heat, and this heat will have an effect on the random thermal motions of molecules in the surroundings of the system, that is, an effect on the entropy of those surroundings. Calculation of this effect is not difficult; it is equal to the heat released by the process in the system divided by the temperature of the surroundings. At constant pressure,

$$\Delta S_{\text{surroundings}} = \frac{-\Delta H_{\text{system}}}{T_{\text{surroundings}}} \qquad (9\text{-}13)$$

A proper derivation of this result should wait for a thorough course in thermodynamics. We shall use it here without such derivation, only pointing out that Equation (9-13) does indicate the proper units for ΔS, energy per degree Kelvin. Also, see page 304, footnote.

With this result, we are in a position to calculate the entropy change in the *world* for any proposed chemical change and use the criterion of increasing world entropy to determine whether such a change is a possible one.

We can expect chemical changes to occur in the direction either of increased entropy *of the chemical system,* or of the release of much heat, leading to an increase in the entropy of the surroundings. Some chemical reactions create entropy in both ways (for example, the burning of gunpowder converts low-entropy solids to high-entropy gases with the release of heat); others must trade off lower entropy in the system for high entropy in the surroundings, or vice versa.

The entropy balance

To take a simple and concrete example, we know that at any temperature above 0°C, ice spontaneously melts to liquid water:

$$\text{ice(c)} \rightarrow \text{water (l)} \qquad (T > 0°C) \qquad (9\text{-}14)$$

This is a unidirectional change, as long as the temperature is above the melting point. We could easily assert that ice melts because when it does, a low-entropy solid is converted to a higher-entropy liquid. This is certainly so, but does nothing to account for the *freezing* of water at lower temperatures. Let us calculate the entropy change in the system (the H_2O) and its surroundings in both cases:

1. At 25°C, the molar entropy of liquid water is 16.7 cal/deg, and that of ice is 10.6 cal/deg.[12] Then the entropy change in the system when reaction (9-14) goes to the right is +6.1 cal/deg. The heat of fusion of ice is 1.7 kcal/mole, so that the entropy change in the surroundings, from Equation (9-13), must be

$$\Delta S_{\text{surroundings}} = \frac{-1.7 \text{ kcal}}{298.2 \text{ deg}}$$

$$= -5.6 \text{ cal/deg}$$

A positive world entropy change when ice melts

The total change in the entropy of the universe is $+6.1 - 5.6 = +0.5$ cal/deg. Thus, we have confirmation from the second law of thermodynamics (if we needed it) that warm ice melts. What about the freezing of cold water?

2. At 25° *below* the freezing point (248°K), the molar entropy of ice is 8.9 cal/deg, and that of water is 13.4 cal/deg.[13] (Note that both values are lower at the lower temperature; the colder substances are, the less their random thermal energy.) Thus, if reaction (9-14) were to go to the right, the system would again increase in entropy, this time $13.4 - 8.9$ cal/deg, or 4.5 cal/deg. The surroundings would decrease in entropy as before, but the lower temperature makes a difference:

$$\Delta S_{\text{surroundings}} = \frac{-1.7 \text{ kcal}}{248.2 \text{ deg}}$$

$$= -6.7 \text{ cal/deg}$$

A negative one when subzero ice melts

This time, summing up the entropy changes shows us that the total entropy change in the world is $+4.5 - 6.7 = -2.2$ cal/deg. But a negative value for the entropy change of the world is a signal that the proposed process cannot happen, any more than a gas can spontaneously compress itself back into a container from which it has escaped. Ice does *not* melt at 248°K; on the contrary, water freezes! In summary, we may say that the ice-water system seeks a state of high entropy at high temperatures, but a position of low entropy and low potential energy (more hydrogen bonding) at low temperatures, where the entropy change in the surroundings is relatively large. These relationships, generally true of chemical systems, are summarized graphically in Figure 9-9.

Systems "seek" high entropy at high temperature, low energy at low temperature.

Very well, then. We have seen that matter will always drift to a position of maximum world entropy. How can that account for the phenomenon of chemical equilibrium, in which a collection of atoms stops in the middle of a process of chemical change? If we are to believe the second law, then we must believe that a chemical system in a state of equilibrium belongs to a maximum in world entropy, a value of entropy that is *higher* than the entropy of either the products or the reactants of that reaction. This is indeed the case, and the reason for it is a phenomenon known as the *entropy of mixing*.

The notion of an entropy of mixing should not be surprising to you. Clearly, a molecular mixture (that is, a solution) of two or more substances

---

[12] This value is extrapolated from the entropy of ice at 0°C.
[13] Extrapolated from the entropy of liquid water at 0°C.

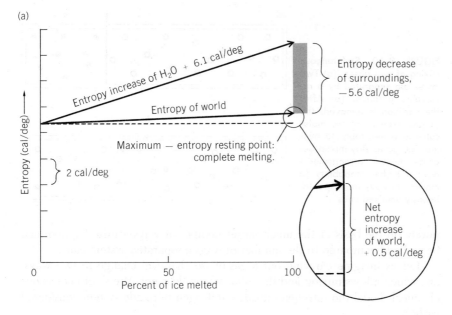
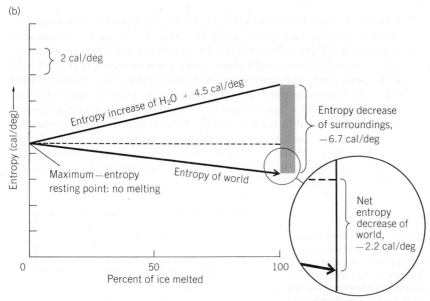

FIGURE 9-9 (a) World entropy changes for melting one mole of ice at 25°C (298.2°K). (b) World entropy changes for melting one mole of ice at −25°C (248.2°K).

is a more random (and thus more probable) state than a self-sustained separation. Suppose we have two gases in a container separated by a partition (Figure 9-10). Now we remove the partition. The molecules of the two gases will spontaneously mingle in a process that is driven (as is the expansion of gases) by the statistics of randomly arranged systems as opposed to orderly ones. The entropy change that occurs when the two gases mix is calculable

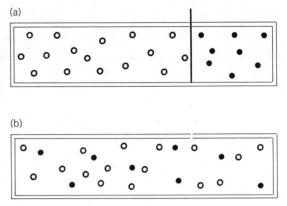

FIGURE 9-10 Spontaneous mixing of two gases. (a) Two gases are separated by a removable partition. (b) After the partition is removed, the random motions of the molecules lead inevitably to mixing. The gases *stay* mixed because mixed arrangements have a higher probability (a higher entropy) than spontaneously unmixed ones.

purely on the basis of the much larger number of ways a mixed system can be realized compared to the number of ways a separated system can.

Let us imagine carrying out a gas-phase chemical change in two ways: one which allows mixing and the other which does not. We shall take as our chemical model the nitrogen dioxide–dinitrogen tetroxide system considered earlier.

1. Pure NO_2 is converted chemically to pure N_2O_4 in a special (in fact, imaginary) chamber such that each molecule of N_2O_4 is placed in a separate compartment as soon as it is formed (Figure 9-11). The system resembles the ice-water one, with its boundary between the phases, and there is no mixing of product N_2O_4 with reactant NO_2. The entropy changes in the world are summarized in Figure 9-12(a). It happens in this system that the entropy changes in the system and surroundings are very nearly equal, but there is a slight balance in favor of dimerization, caused by the fact that the entropy gained by the surroundings in absorbing the familiar 13.6 kcal enthalpy of reaction overbalances the decease in entropy suffered by the system when 2 moles of gas become one. The maximum entropy resting point for the world is reached in this case when all of the NO_2 is converted to N_2O_4.

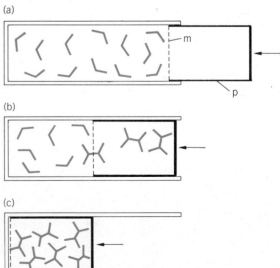

FIGURE 9-11 Dimerization of pure NO_2 ($<$) to form pure N_2O_4 ($\rangle$—$\langle$), without mixing of products and reactants. The reactor has a sliding piston (p) fitted with a membrane (m) that is permeable only to N_2O_4 molecules. (a) Initial state: 12 NO_2 molecules in volume V, at a pressure of one atmosphere. (b) As each N_2O_4 molecule forms, it passes through the membrane into the interior of the piston, where it remains without dissociating. (c) Final state: 6 molecules of N_2O_4 in volume V/2, at a pressure of one atmosphere.

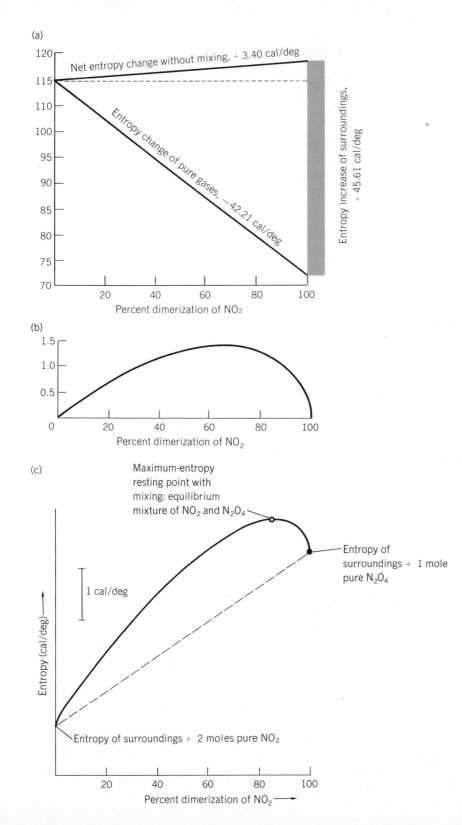

FIGURE 9-12 World entropy changes during the dimerization of NO_2. (a) Reaction entropy without mixing. (b) Entropy of mixing (cal/deg). (c) Total entropy change.

2. In any reaction in solution, the products and reactants must mingle. The entropy of mixing generated by the dimerization is shown in Figure 9-12(b), calculated statistically as a function of the percent of NO_2 dimerized to N_2O_4. Note that this entropy of mixing must be zero at both 0 percent dimerized and at 100 percent dimerized, since there is only one substance present, and no mixing possible, at both of those points. Any function (see Appendix II) that is zero at each of two points and has positive values between *must* exhibit a maximum.

The total entropy effect in the world is the sum of what we might call the inherent reaction entropy shown in Figure 9-12(a) and the entropy of mixing shown in Figure 9-12(b). These two contributions are added together graphically in Figure 9-12(c). The reacting system stops short of a complete reaction, because the entropy of the world reaches a maximum somewhere between pure reactants and pure products. Any change by the system away from this maximum-entropy resting point would necessarily involve a decrease in the entropy of the world, which is contrary to the second law of thermodynamics.[14]

For a longer discussion of reaction rates, see Ch. 11.

This result—the influence of the entropy of mixing on the final state of a reacting system—is a general one, and accounts for the fact that every reaction occurring among substances in solution has an equilibrium point that is short of completion. For some reactions, this point is reached virtually instantly; for example, the reaction between H_3O^+ and OH^- to form water (Chapter 7) is complete within a billionth of a second. Others drift toward equilibrium with great slowness, and indeed appear not to be reacting at all, though far from a position of maximum world entropy. For example, diamonds are unstable with respect to graphite at ordinary pressures. They are thus unfaithful, though lingering, friends to working girls. Furthermore, since there is no mixing of diamond and graphite crystal lattices (Chapter 5), this is a nonequilibrium reaction that will not stop short of complete conversion of diamond to plain old graphite. Fortunately, the rate at which this consummation is sought by diamonds is undetectably slow.

9-10 The Second Law and Living Systems

Early philosophical opposition to the second law of thermodynamics was based primarily on its pessimism with regard to the utilization of energy for useful work (the limitations on conversion of heat to potential energy that we discussed earlier). But after all, machines are only projections of the human body, and underlying the bad news about perpetual motion machines we may now discern some more fundamental statements about the thermodynamics of life. Living systems are beautifully organized, orderly chemical systems that daily sustain their order through the conversion of simple substances into complex ones, and also reproduce themselves (with only rare slips) from generation to generation through great stretches of time. We shall have a close

---

[14] Remember that the second law of thermodynamics is statistical in nature. Someday a mixture of N_2O_4 and NO_2 *might* just react to completion, but the chance is as remote as that of a camel passing through the eye of a needle.

look at some of the mechanisms by which these functions are accomplished in Chapter 17, but let us now take a thermodynamic overview of this apparent creation and preservation of order in the teeth of the randomizing dictates of the second law. The miraculous ability of life to so buck the odds has led some to believe that living systems are somehow outside of the prescriptions of thermodynamics; in fact, there is no reason to think that this is the case. We shall here consider three aspects of life: the daily continuation of order, the long-range preservation (and evolution) of species, and the mortality of individuals.

METABOLISM: ORDER OUT OF CHAOS. Every living thing builds its marvelously complex tissues out of relatively simple substances. Plants, of course, begin most simply, breeding lilacs out of the air and water, with the help of sunlight. The conversion of a gas (CO_2) and a liquid (H_2O) into a solid polymeric sugar (starch), to name just one of their achievements, represents an enormous decrease in entropy. The accompanying entropy increase in the surroundings results from the conversion of the energy of sunlight ultimately to heat, through the pathways of photosynthesis and plant metabolism. Even plants, though, generally need food—the nitrates and phosphates, for example, that we pour so liberally into our sewers, to the great benefit of the algae in our rivers and lakes. Molecular potential energy in these substances is converted partly into heat and partly into potential energy in more complex molecules. Once plants are cut off from the available energy input of sunlight and nutrients, they quickly cease this production of complexity out of simplicity, and they soon die.

Where do we come from?

Animals eat plants, and some of them eat other animals. In these cases, too, order is selected out of chaos, with the production of more chaos elsewhere. The first chemical move an animal makes after ingesting the complex tissues of its plant or animal food is to begin separating them into their component molecules, and to simplify those molecules by means of digestive chemicals (as when starch is converted to sugars by enzymes). (Humans pre-simplify their food still more by cooking, thus randomizing its structure further by heating.) Thus animals, too, must re-create complex molecules out of simple ones, and the necessary entropy production appears in the surroundings as heat. A dietetic "calorie" is in fact a kilocalorie, and we produce thousands of them per day.

Thus, all living things increase order within their own structures by exporting disorder (heat and the simple chemical residues of metabolism) to the surroundings. The total disorder so exported must still be larger than the order created in the tissues of the organism.

EVOLUTION: EXTERNAL GENETICS. You resemble your parents not only in the subtleties of face, hair, and eye color, but primarily in being a recognizable human being. How can this specificity of structure be transmitted across the abyss of nonbeing from one generation to another? As we shall see in Chapter 17, physical and chemical characteristics are transmitted in code by deoxyribonucleic acid (DNA) molecules, which are *ordered* arrange-

Who are we?

ments of phosphate, sugar, and nitrogen base groups into a single polymeric array. The information on how you are to be constructed and how the assembled person is to function, both physically and chemically, is all contained in the *sequence* in which groups are joined in the polymer. A different sequence would have led to a different kind of person, a different organism, or no organism at all. Here it is quite clear that the genetic information is closely related to order; the more the sequence of groups is randomized, the more genetic information is lost.[15] But highly ordered substances are inherently improbable. Any joggling or disruption of the structure of DNA should lead to loss of order and thus loss of the genetic information.

Fortunately, the sugar-phosphate-base groups of DNA are held in place by strong covalent bonds. Because the electrons in those bonds have quantized energies, they cannot be nudged out of them a little bit at a time by a succession of gentle thermal collisions; either they stay where they are, or the bond is totally disrupted by a violent influx of energy. Such an energy input "cannot" arise from thermal vibrations and collisions at the body temperature of organisms; to collect the randomized thermal energy present in even a small region of your body at 98.6°F into one or a few bonds within your genetic material would require a gross violation of the second law, which in this context operates, paradoxically, as a protector of information and order.

Genetic material is, of course, damaged by exposure to high-energy radiation. Mutations have occurred throughout the history of life, and one source of them has surely been the natural radioactivity of the earth and the high-energy particles of "cosmic rays." In such cases, enough energy to disrupt chemical bonds is concentrated in a single event, or set of events.

A familiar result of the occasional mutations in genetic material through the ages has been the phenomenon of evolution. Again, here is a seeming contradiction of the second law in that very complex organisms have evolved from simple ones, both the organisms themselves and their genes becoming more complexly organized with the passage of time. More information is encoded in the DNA of a human being than in that of an amoeba. Do we see in the history of species a reversal of the law of increasing entropy and decreasing information? A little thought will convince you that we do not. No single individual increases the complexity of his genetic material during his lifetime; random changes do occur, and sometimes those which lead to greater complexity also lead to higher survival characteristics. But if such changes do not benefit the organism, we never hear of them, since the process of natural selection artificially (from a physicochemical standpoint) isolates "beneficial" mutations out of the multitude wrought through the hundreds of

---

[15] The order-information correlation is present in simple cases as well. When a gas is compressed inside a cylinder, we know that every molecule of it is in there. Open the valve, and the information that we have about the location of any molecule decreases as the space available to it expands. Indeed, an entire theory of information transfer and storage has been developed and has been found to contain, as an implication, the second law of thermodynamics. See "Suggestions for Further Reading."

millions of years of evolution. Thus, evolution is as wasteful a process as any other governed by the second law; decreases in entropy are still achieved only locally, and at the expense of increases elsewhere.

AGING AND DEATH: INTERNAL GENETICS. Every organism replenishes the matter out of which it is made many times in the course of its life. If you feel like a different person from the one who left high school, it is quite true in a material sense that you are. All of the new atoms that enter your body to replace those sloughed off in the endless cycle of cell life and death must be organized into viable patterns of functioning and useful cells, each in its proper place. The information required to achieve this continuity of your person is encoded in the nucleus of each of your cells, so that the amino acids you ingested last week can be properly organized into the proteins of, for example, new hair growth, and so that this hair can be the same color, texture, curliness, and so forth, as what is already there. The same kind of breaking-in of new materials goes on all through your body, and it depends on the retention intact of genetic material. We have seen that the chances of disruption of a DNA molecule are small (in the absence of high-energy radiation). However, a small chance taken over a long time interval eventually becomes a certainty. We age because our cells lose the ability to regenerate themselves, and even if we can avoid gross accidental destruction of our bodies, they must inevitably bow to the statistical improbability of keeping so much genetic information intact for much more than three score years and ten. The internal message from cell to cell becomes garbled and inoperative; entropy creeps in inevitably, not because it must, but because it *probably* will.

Where are we going?

Artists from Coleridge to Bergman have used the common figure of death playing statistical games for men's souls; and Sigmund Freud and Norman O. Brown have written convincingly of the deep-seated unconscious drive toward death, dissolution, and chaos. These are not the first manifestations in history of poets, mystics, and speculative thinkers arriving at significant truths about the physical world. It would be too bad if, in our time, intellectual polarization should prevent our sharing both kinds of insight.

9-11 Summary

We have seen that chemical changes are accompanied by heat effects and by a certain amount of work done on or by the surroundings. The total energy change embraces both effects and fits within the first law of thermodynamics: the conservation of energy. Refinements of this principle enable us to relate the heat of formation (enthalpy content) and heat of reaction, so that accumulated data on these aspects of thermochemistry may be used to calculate the results of chemical reaction without always doing the experiment. Then the second law (every possible process is one in which the molecular disorder of the world increases) was introduced on the basis of the statistical behavior of immense numbers of molecules. The quantitative measure of degree of disorder (change in entropy, ΔS) was brought into the framework of the second law, enabling us to calculate the degree of randomness and to find in this the reason why some reactions have a strong "push" to take place, even though they evolve little heat, or even get colder. Finally, the concept and consequences of entropy change were applied to the growth of living systems, to their reproduction along genetic lines, and to their statistically inevitable death. Here thermodynamics, poetry, and psychology find common ground.

GLOSSARY

enthalpy: the heat of reaction (or the heat of formation of a compound) at constant pressure, defined by the relation $H \equiv E + PV$, where H is the enthalpy; E is energy; and PV (pressure times volume) represents work.

entropy: the disorder-measuring function in a chemical system related to the statistical probability that a given collection of atoms will be found in the form of a given substance by

$$S \equiv 2.303\ R \log \omega$$

where S is the entropy function; R is the universal gas constant; and ω measures the number of ways that the collection of atoms can combine to form a given substance. (See also the definition of entropy given in Chapter 8 and its glossary.)

heat of formation: the enthalpy change, or the amount of heat evolved or absorbed at constant pressure, by the formation of one mole of a compound from its elements, all being considered to be at their standard states (normal condition at 25°C and one atmosphere pressure).

heat of reaction: the enthalpy change, or the amount of heat evolved or absorbed at constant pressure, during the completion of a stated reaction. (Sometimes heat of reaction is given per mole of reactant, but more usually in terms of the numbers of moles of reactants and products defined by a specific equation.)

thermochemistry: the area of chemistry that has to do with the enthalpy content of chemical substances, and with the exact amount of heat energy evolved or absorbed during chemical change.

thermodynamics: formerly, the science of heat engines; here, the application of the classical laws of conservation of energy and statistical increase of disorder to chemical change in general, resulting in guidelines for such change and in explanations for even the most complicated chemical behavior.

PROBLEMS

9-1 When pulverized ammonium nitrate and zinc are mixed, the dry powders do not react (because the ions of NH_4NO_3 have no mobility and the granules touch at only a few places). When a drop of water is placed on the mixture, however, a vigorous reaction ensues:

$$Zn + NH_4NO_3 = ZnO + N_2 + 2H_2O$$

This is but one of many examples of fires started by water. In this case, given the enthalpies of formation

$\Delta H_{formation}$ of NH_4NO_3 = −87.93 kcal/mole

$\Delta H_{formation}$ of ZnO = −84.35 kcal/mole

$\Delta H_{formation}$ of $H_2O(g)$ = −57.83 kcal/mole

calculate the heat of reaction. (Ans.: −112.07 kcal.)

9-2 "Gas" for household heating and cooking usually means methane, CH_4, the chief constituent of natural gas. How much heat is given up by the combustion of one mole of methane? *Hint:* Write the balanced equation for the oxidation of methane to CO_2 and H_2O, taking the enthalpies to be

$\Delta H_{formation}$ of CH_4 = −18.0 kcal/mole

$\Delta H_{formation}$ of CO_2 = −94.0 kcal/mole

$\Delta H_{formation}$ of $H_2O(g)$ = −57.8 kcal/mole

Then calculate the heat of reaction. (Ans.: −191.6 kcal.)

9-3 How much heat must be added

a. to 8.064 g of crystalline hydrogen at 13.9°K to convert it to liquid at the same temperature?
b. to 8.064 g of liquid hydrogen at 20.3°K to convert it to hydrogen gas at the same temperature?
c. to 72.064 g of ice at 0°C to convert it to steam at 100°C?

Make use of the following data from a handbook

	H_2	H_2O
ΔH_{fusion}	14 cal/g	80 cal/g
$\Delta H_{vaporization}$	108 cal/g	540 cal/g

9-4 The normal melting points of chlorine and bromine, both composed of diatomic molecules, are −100.98°C and −7.2°C, respectively. Predict which has

a. the higher normal boiling point;
b. the larger molar heat of fusion;
c. the smaller molar heat of vaporization;
d. the lower critical temperature;
e. the higher equilibrium vapor pressure for the liquid at 0°C.

9-5 In the design of a rocket engine, an important limitation may be imposed either by restricted volume or by restricted weight. How much energy would either of the following reactions produce if (a) the size of the chambers for the oxidizing and reducing agents is limited to 500 liters, or (b) the weight of oxidizing agent is limited to 700 kg? The equations are:

$$C_2H_5OH + 3O_2 \rightarrow 2CO_2(g) + 3H_2O(g) + 295.6 \text{ kcal}$$

$$H_2 + F_2 \rightarrow 2HF(g) + 120 \text{ kcal}$$

The other necessary data are: density of liquid O_2 = 1.14 g/ml; density of alcohol (C_2H_5OH) = 0.789 g/ml; density of liquid H_2 = 0.070 g/ml; density of liquid F_2 = 1.11 g/ml.

9-6 By comparison with Equation (8-14) (page 264), show that the "entropy change of the world," discussed in Section 9-9, is given by

$$\Delta S_{world} = \frac{-\Delta G}{T}$$

when a change occurs at constant temperature and pressure. Must the free energy G of a system increase, or decrease, in a spontaneous process?

9-7 There is a rule of thumb that matter seeks the state of lowest energy (at constant pressure, this is lowest enthalpy) at low temperatures, but the state of highest entropy at high temperatures. Use Equation (8-14) and the result of Problem 9-6 to show why this is true.

9-8 Given below are some standard entropies at 298°K. Comment briefly on the significance of the numerical values, and calculate and comment on the significance of ΔS^0_{298} for the reaction

$$CaCO_3(s) \rightarrow CaO(s) + CO_2(g)$$

Compound	S^0_{298}, cal-deg^{-1}-mole^{-1}
CaCO$_3$	22.2
CaO	9.5
CO$_2$	69.09

9-9 The standard entropy of carbon in the form of graphite at 298°K is 1.3609 cal-deg^{-1}-mole^{-1}.

a. Compare this value to those for C(diamond) and any other pure element in Table 9-1, and comment.
b. From the data in Table 9-1, calculate ΔS^0_{298} for the reaction

$$C(graphite) + 2H_2(g) \longrightarrow CH_4(g)$$

and comment on the significance of this result.

9-10 Use the data in Problems 9-2 and 9-9 to calculate the entropy change in the world when one mole of $CH_4(g)$ is formed from C and H_2 at 298° K and one atmosphere pressure. What is the significance of this result? How do you account for the fact that CH_4 is present in abundance in the atmosphere of Jupiter?

SUGGESTIONS FOR FURTHER READING

Classical thermodynamics has been treated in a number of short paperback books, any of which will give you references to full-dress treatments. Two paperbacks that might provide a good starting place are:

NASH, L. K., *Elements of Chemical Thermodynamics*, Reading, Mass.: Addison-Wesley (1962).

VAN NESS, H. C., *Understanding Thermodynamics*, New York: McGraw-Hill (1969).

A full-length treatment which is couched in an original, thoughtful, and readable style is: BENT, H. A., *The Second Law*, New York: Oxford University Press (1965).

Thermodynamics and quantum chemistry inevitably raise fundamental philosophical speculations about our world and ourselves. If you would like to read more in this vein, try Schrödinger, E., *What is Life? And Other Scientific Essays,* Garden City, N.Y.: Doubleday (1956). The author is also responsible for Equation (2-3) (Chapter 2) and so combines creative scientific insight with his philosophy. Our discussion about the thermodynamic stability of genetic material derives from this book.

Blum, H. F., *Time's Arrow and Evolution,* 3rd ed., Princeton, N.J.: Princeton University Press (1968). Sir Arthur Eddington coined the phrase, "Entropy is time's arrow," because the increase in entropy determines which of two states must follow the other in time. Blum gives a thorough discussion of the role of the second law in biological evolution, with attention to the question whether such life processes might violate it.

Brown, N. O., *Life Against Death: The Psychoanalytical Meaning of History,* Middletown, Conn.: Wesleyan University Press (1959), is more a critique of our present culture than a thermodynamic treatise, but just as there are philosophical overtones to thermodynamics, there are thermodynamic undertones to the Freudian view of man that Brown so tellingly elaborates here. You are sure to be profoundly influenced by this book in your lifetime, and fairly certain to read it eventually. Try it with the second law fresh in your mind.

For contemporary bibliographies and a behind-the-scenes view of professors struggling with the teaching of thermodynamics, look at a collection of papers on "The Teaching of Thermodynamics" in the *Journal of Chemical Education* (May 1970). Titles of contributions range from "Chemical Equilibrium As a State of Maximal Entropy" to "Our Freshmen Like the Second Law."

For an introduction to the relationships between thermodynamics and information, see Tribus, M., and McIrvine, E., "Energy and Information" in *Scientific American,* vol. 225, no. 3, p. 179 (September 1971).

10

The Ultimate Source of Energy

Our living is a little heat.
For this we house, we clothe, we eat.

In winter burn the tree, the coal.
In summer sun us body and soul.

Yet all the heat we have is one
Essentially, the summer sun.

Only the summer sun burns deep.
Only the summer sun will keep.

We live to learn new ways to hold
Summer sun through winter cold.

ROBERT FRANCIS[1]

10-1 Abstract

Chemistry is the study of matter and its interrelations with energy. The opening chapters of this book have shown how close that relationship is—we cannot consider the structure of matter without bringing in the concept of energy, and we cannot talk about energy for long without bringing in forms of matter which generate, contain, or furnish the energy. Having considered some of the formal aspects of the matter-energy relationship in the previous chapter, we turn now to the different manifestations of energy and their various sources, in particular the ultimate source, **nuclear reactions.**

This chapter is about the atomic **nucleus:** its behavior, its apparent structure, and some of the aspects of its reactions. We shall go into more detail in these matters than is traditional in introductory chemistry texts, because we think that nuclear reactions and their technology are of overwhelming importance to everyone who hopes to live through the end of this century and the beginning of the next.

We shall begin by discussing some of the phenomena on which our present understanding of nuclear structure is based. This will be followed by an introduction to the nuclear **isotope chart,** which plays a role for nuclear reactions similar to that played by the Periodic Table for chemical reactions. We shall find that patterns of nuclear stability have their origin in a system of quantized nuclear energy levels, just as the periodicity of chemical behavior is based on a system of quantized electronic energies.

We shall then turn to the technology of nuclear power: nuclear fission and fusion as energy sources, the structural design of a fission reactor, and the nature and effects of nuclear radiation, including its effect on living tissue.

---

[1]"Summer Sun." Copyright and permission by Robert Francis.

10-2 Energy Sources and Energy Flows

It was pointed out in Chapters 8 and 9 that a very real distinction can be drawn between free energy and heat energy, because free energy can be transformed quantitatively into heat, but heat cannot be transformed quantitatively into free energy. Suppose we now take up common examples of these transformations of energy, following chains of transformation back to an ultimate source.

An automobile obtains its mechanical energy of revolving parts and forward motion from the chemical energy contained in the fuel. That chemical energy is not converted directly into motion of the car, of course; the chemical energy is first transformed into heat by burning the fuel in the cylinders, and then a part of that energy is abstracted by allowing the hot gas in the cylinder to expand against a retreating piston. The rest of the heat is thrown away, by way of the radiator and the exhaust. Part of the piston's mechanical energy sets the car in motion, and part generates electricity for the ignition system (and the radio). The electricity eventually becomes heat, and ultimately the kinetic energy of the car becomes heat, too, as its momentum is checked by friction in the brakes. So all the conversions reduce to one major one—the fuel gives up chemical energy in the combustion reaction and produces heat. And where did the fuel get its energy? From sunlight acting on plants in the world of long ago. Petroleum and coal are simply fossilized sunshine, stored all those millions of years in the materials derived from living matter. We can get no more energy out of the fossil fuels than the sun's radiant energy put there when the ancient plants were growing.

Specifically, the mechanism for the fixing of solar energy is photosynthesis (see Chapters 1 and 17). In the terms introduced in the last chapter, simple substances of low energy and high entropy (H_2O and CO_2) are converted, using the energy of sunlight, to complex high-energy compounds, the carbohydrates. Conversion of these through geologic time to coal and oil does not greatly alter the thermodynamic energy content. When the coal and oil are burned to give low-energy CO_2 and H_2O again, the trapped energy is released as heat.

By the same reasoning, electric power generated in a conventional steam-power plant also represents stored-up solar energy, for the steam boiler burns coal or oil to generate the power. Electricity from water power does not draw upon fossil fuel, but it still is derived from solar energy; heat from the sun evaporated the water and lifted its warm vapors to a place where the rain could fall onto mountains and run down in streams. Wood burned in a fireplace represents solar energy from recent years, and electricity from solar batteries on a satellite represents immediate solar energy. All these manifestations show how dependent we are on the sun's radiation for all aspects of life on earth.

The sun is indeed generous to us. Each day the earth receives on every $1\frac{1}{2}$ square miles of its surface an amount of energy equivalent to an atomic bomb of the Hiroshima type. If we could convert all the sunshine falling on

Remember that energy is always conserved in any transformation.

THE ULTIMATE SOURCE OF ENERGY

Compare this to the monetary value of electricity in batteries (Sec. 8-9).

the roof of a house directly to electricity, we would get from each 15 ft² of roof enough electricity to run all the appliances in the house. The earth is a tiny speck 90 million miles away from the sun, receiving only one two-billionth of the sun's output, but nevertheless the earth's share (if converted all to electricity) is equivalent to $2,000,000,000 worth of power *each minute*. No wonder people feel challenged to employ all this wealth of energy more efficiently!

Is there any earthly source of energy besides the sun? Yes, we have nuclear energy. The disintegration of radioactive elements deep within the earth heats the interior enough to cause volcanos and serves to keep the regions of the earth's surface warm even when the sun is shining elsewhere. Furthermore, a much higher rate of emission of such nuclear energy can be achieved by conducting a controlled fission reaction in uranium or plutonium, so electrical energy (by way of heat energy) comes from this source also. In the future, it will be inevitable that more and more heat for making steam, for distilling sea water, and for processing chemical materials will come from nuclear reactors rather than from fossil fuels.

Comparing nuclear energy and solar energy more closely, we see that they are really one and the same thing. The sun derives its energy from a nuclear fusion reaction, in which hydrogen is converted continuously to helium and to heavier elements, with consequent conversion of some of the sun's mass to energy (which is radiated). In the earthbound nuclear reactors, the reaction is one of fission of heavy elements rather than fusion of light elements, but again a portion of the mass of the reactants is liberated as energy. So upon analysis, *we find only one ultimate source of energy* in our universe, and that is *nuclear energy*.

Inasmuch as nuclear reactions have not been treated before in this book, and indeed are so distinctly different from conventional electron-rearrangement reactions, it is important to devote some space to the general matters of nuclear stability, nuclear structure, and nuclear reactions. They will be considered in that order.

10-3 Stable versus Unstable Nuclei

Some elements, such as silicon and oxygen, have existed unchanged in rocks that are three billion years old, while other elements, such as radium and uranium, disintegrate spontaneously and so change irreversibly and steadily into other elements. Our first inquiry is into this matter of stability: Which elements have the most stable nuclei, and *why* are they the most stable?

Since all the elements are being formed from hydrogen by nuclear fusion reactions in the stars, and since matter reaching us from outer space (in the form of meteorites) has a composition about like that of the earth and the moon, we may think of a genesis of all the chemical elements, starting with hydrogen and neutrons. If all the other elements can be made from hydrogen, and are so made, then *the most stable and unreactive nuclear configurations should persist*, while the less stable ones perish or are transformed.

TABLE 10-1
Abundances of Elements in the Earth's Crust

Atomic Number Z	Symbol	Grams per metric ton	Atomic Number Z	Symbol	Grams per metric ton
1	H	present	50	Sn	40
2	He	0.003	51	Sb	1
3	Li	65	52	Te	0.0018 (?)
4	Be	6	53	I	0.3
5	B	3	54	Xe	—
6	C	320	55	Cs	7
7	N	46.3	56	Ba	250
8	O	466,000	57	La	18.3
9	F	600–900	58	Ce	46.1
10	Ne	0.00007	59	Pr	5.53
11	Na	28,300	60	Nd	23.9
12	Mg	20,900	61	Pm	—
13	Al	81,300	62	Sm	6.47
14	Si	277,200	63	Eu	1.06
15	P	1180	64	Gd	6.36
16	S	520	65	Tb	0.91
17	Cl	314	66	Dy	4.47
18	Ar	0.04	67	Ho	1.15
19	K	25,900	68	Er	2.47
20	Ca	36,300	69	Tm	0.20
21	Sc	5	70	Yb	2.66
22	Ti	4400	71	Lu	0.75
23	V	150	72	Hf	4.5
24	Cr	200	73	Ta	2.1
25	Mn	1000	74	W	1.5–69
26	Fe	50,000	75	Re	0.001
27	Co	23	76	Os	present
28	Ni	80	77	Ir	0.001
29	Cu	70	78	Pt	0.005
30	Zn	132	79	Au	0.005
31	Ga	15	80	Hg	0.077–0.5
32	Ge	7	81	Tl	0.3–3
33	As	5	82	Pb	16
34	Se	0.09	83	Bi	0.2
35	Br	1.62	84	Po	0.0000000003
36	Kr	—	85	At	present
37	Rb	310	86	Rn	present
38	Sr	300	87	Fr	present
39	Y	28.1	88	Ra	0.0000013
40	Zr	220	89	Ac	0.0000000003
41	Nb	24	90	Th	11.5
42	Mo	2.5–15	91	Pa	0.0000008
43	Tc	0	92	U	4
44	Ru	present	93	Np	probably present
45	Rh	0.001	94	Pu	present
46	Pd	0.010	95	Am	probably present
47	Ag	0.10	96	Cm	probably present
48	Cd	0.15	97	Bk	?
49	In	0.1	98	Cf	?

SOURCE: T. Moeller, *Inorganic Chemistry: An Advanced Textbook*, New York: John Wiley & Sons, Inc., pp. 30–32 (1952).

THE ULTIMATE SOURCE OF ENERGY

What is stable should be reflected by what is here now.

With this thought in mind, we can look at a table of terrestrial abundances and note which elements are prominent. Table 10-1 gives such information about the solid crust of the earth, insofar as it has been sampled and analyzed; the hydrosphere and the atmosphere are not included in this survey, nor is the core of the earth. A glance at the table reveals no obvious relationship between atomic number and abundance. It seems definite that light elements are more abundant than heavy elements, in general, but there are scarce light elements (B, Sc, As), and there are moderately plentiful heavy elements (Ba, Ce, Pb). The so-called rare earths ($Z = 58$ to 71) are more abundant than the transition metals that come before (45 to 54) or the transition metals that come after (72 to 80), so they are misnamed. Among the heavier elements, barium, lanthanum, cerium, neodymium, lead, and thorium stand out. That is all that can be said.

**TABLE 10-2
Cosmic Abundances of the Elements**

Atomic Number Z	Symbol	Atoms per 10,000 atoms of Si	Atomic Number Z	Symbol	Atoms per 10,000 atoms of Si
1	H	3.5×10^8	18	Ar	130
2	He	3.5×10^7	19	K	69
6	C	8×10^4	20	Ca	670
7	N	16×10^4	21	Sc	0.18
8	O	21×10^4	22	Ti	26
9	F	90	23	V	2.5
10	Ne	10,000	24	Cr	95
11	Na	462	25	Mn	77
12	Mg	8870	26	Fe	18,300
13	Al	882	27	Co	99
14	Si	10,000	28	Ni	1340
15	P	130	29	Cu	4.6
16	S	3500	30	Zn	1.6
17	Cl	190			

Thinking that a broader view might reveal something more, we can turn to a table of *cosmic abundances* (Table 10-2). These numbers have been determined by analysis of meteorites, and by quantitative examination of atomic spectra obtained from light coming from the sun and stars. The preponderance of light elements is striking; the universe is made up mostly of hydrogen and helium, with appreciable amounts of carbon, nitrogen, oxygen, magnesium, neon, silicon, and iron thrown in. The other elements are much less prominent. *All* of the elements heavier than hafnium ($Z = 72$), taken together, constitute less than one-tenth of one percent of the universe.

It seems that the process of building up elements from hydrogen results in an accumulation of some stable nuclear species which usually have *even* numbers of protons and *even* numbers of neutrons.

Some species having neutron and proton numbers[2] that are some multiple of those of $^4_2$He are especially prominent: $^{12}_6$C, $^{16}_8$O, $^{20}_{10}$Ne, $^{24}_{12}$Mg, $^{28}_{14}$Si, and $^{32}_{16}$S. Other abundant elements, such as N, Al, Fe, and Ni, however, do not fit this description. And there we have exhausted the information and must leave the question of abundance, for the present.

RADIOACTIVITY. Another source of information about atoms and their nuclei is the observed instability of some elements which we call *radioactive*. The phenomenon of radioactivity was discovered by Henri Becquerel in 1896; before that, no one had the faintest idea that some of the well-known elements (such as uranium and thorium) gave off radiations. Becquerel was studying the phenomenon of fluorescence, and particularly the phosphorescence of some minerals which keep on glowing after exposure to sunlight. He recorded the phosphorescence photographically, and like every careful photographer, he always developed the first plate in a box, unexposed, to see whether the plates were in good condition to use. He was surprised one day to find black spots developing on such an unexposed first plate from a new box, and he traced the event back to some uranium minerals which he had placed on top of the unopened box some days before. He repeated the sequence of events several times, and came to the conclusion that some minerals emitted mysterious penetrating radiation which could go right through black paper, and yet could affect a photographic emulsion like light. He called the phenomenon **radioactivity,** and he traced it to the heavy elements thorium and uranium. Others took up the study of this new phenomenon, and it soon was found out that the emission of penetrating radiation was unaffected by heat or pressure and was independent of the state of combination of the element. Later, other radioactive elements were found; as methods for detection improved, their number grew. Today we recognize 28 elements that have naturally occurring radioactive isotopes, including such familiar elements as potassium, vanadium, platinum, and lead. Some elements, such as radium and polonium, have been discovered only by observing their radiations. The art of making new elements and new isotopes of old elements, most of them also radioactive, has also developed in the intervening years.

The penetrating radiations from the naturally radioactive elements have been found to be of only three kinds: (1) alpha rays (α rays), composed of a stream of α particles, which were found to be ionized helium atoms, $^4_2$He^{2+}, (2) beta rays (β rays), made up of β particles, which are ordinary electrons, e$^-$, with high kinetic energies, and (3) gamma rays (γ rays), which are electromagnetic radiation of wavelength shorter than that of X-rays. The α and β particles, being electrically charged, may be deflected from their paths by electric and magnetic fields, but γ rays are unaffected by such fields. Alpha particles are exceedingly massive compared to β particles, so they cause severe

---
[2] The number of protons is equal to the atomic number Z. It is customary to write the mass number (the atomic weight to the nearest whole number) as A. Then the number of neutrons, N, is $A - Z$. (See Chapter 2.)

ionization of nearby matter as they collide with its atoms and are slowed down. For the same reason, α particles are easily stopped; a few sheets of paper or a thin sheet of metal will shield against them. Beta particles penetrate farther if they have high kinetic energy, and they, too, cause ionization in the matter that stops them. Several centimeters of water or plastic or aluminum suffice to shield against low-energy β rays ("soft betas"). Gamma rays are much more penetrating and require correspondingly more shielding materials. The more mass of material between source and observer the better, so lead, cast iron, and barium-containing concrete are favorites for shielding.

A kind of radiation not recognized by early experimenters (because it was very rare then; it has since become a persistent danger in this age of nuclear technology) is a stream or shower of *neutrons*. The neutron was discovered by Chadwick in 1932 in the course of experiments on artificial radioactivity, and it was evident from the start that neutrons (being uncharged) penetrate matter very easily. At the same time, when a neutron is captured by a nucleus, the usual result is emission of an ionizing γ ray and conversion of the target nucleus into that of another element. So neutrons destroy the structure of proteins and damage living tissue extensively, just as α, β, and γ rays do, and their ease of penetration makes neutrons that much more dangerous. The best way to stop a neutron is to slow it down by multiple collisions with particles nearly its own weight, such as nuclei of hydrogen or carbon or oxygen, until the slowed-down neutron can be captured. Hence great thicknesses of water, paraffin, and wood make the best neutron shields.

Information about which isotopes give off α, β, γ, or neutron rays appears in tables of nuclei and on complete isotope charts. Such information about unstable nuclei will come into this discussion repeatedly, where needed.

CROSS SECTION. A third (and totally different) source of information about the stabilities of nuclei is the study of what the nuclear physicist calls "**cross section.**" A source of neutrons is arranged in such a way that a collimated (parallel) beam of neutrons (of adjustable but known kinetic energy) enters a detecting and recording device, so that the intensity of the beam (number of neutrons per second) is recorded. A sample of the element under study is then introduced into the beam, and the absorption of neutrons by the sample is studied over a range of neutron energies. The physicist thinks of the nuclei of the sample as though they presented a certain frontal area to the neutron beam, stopping a corresponding proportion of the neutrons. He knows how the atoms are packed, and therefore he knows the spacing of the nuclei. From the Rutherford experiments (see Chapter 2), he also knows that most nuclei have radii of about 10^{-12} cm, hence having cross-sectional areas of about 10^{-24} cm². If the nuclei were opaque and would let no neutrons pass by, they would be acting like little shields 10^{-24} cm² in area. Considering such action normal, the cross section area of 10^{-24} cm² is called one **barn** (meaning the nucleus acts like the broad side of a barn, and a neutron cannot miss it). But nuclei of some elements seem quite transparent to neutrons; helium as $^{4}_{2}$He presents no barrier at all to neutrons of low kinetic energy, and carbon as $^{12}_{6}$C has a cross section of only 0.0037 barns to neutrons moving with only the velocity imparted by thermal agitation[3] at 25°C. That is, ^{12}C acts as though its nuclei presented only 0.0037 as much barrier to the neutron stream as would be expected from its known nuclear diameter. On the other hand, cadmium as $^{113}_{48}$Cd has a cross section for thermal neutrons of 27,000 barns. Here the nuclei of ^{113}Cd act so aggressively in capturing neutrons that they *appear* to have frontal areas 27,000 times as great as expected from the nuclear radius. Since it is hard to picture literally what this means, it seems better to think of the measurements of cross section as indicating *probability of reaction*, and nothing more.

Table 10-3 lists the cross section for **thermal neutrons** of a number of elements (each element consisting of its natural mixture of isotopes, if any) and of a number of isolated isotopes. A complete survey would be prohibitive here, because several hundred such cross sections are known; the interested student should consult an up-to-date isotope chart.[4] Even the few data in Table 10-3 show the great variation in cross section among the various elements, and also among the various isotopes of an element. These variations tell a great deal about comparative stabilities of different nuclei, suggesting that there must be corresponding distinctions in nuclear *structure*.

TABLE 10-3
Some Cross Sections for Thermal Neutrons

Elements as Naturally Occurring Mixtures of Isotopes		Single Isotopes	
H	0.33 barn	^{1}H	0.35 barn
He	0.007	^{2}H	0.00057
Li	71	^{3}He	5400
Be	0.010	^{4}He	0.0000
B	755	^{14}N	1.83
C	0.0037	^{15}N	0.00
N	1.88	^{24}Mg	0.033
O	0.0002	^{25}Mg	0.28
F	0.009	^{26}Mg	0.03
Ne	0.035	^{106}Cd	1
Na	0.53	^{124}Xe	74
Mg	0.063	^{128}Xe	5
Al	0.23	^{129}Xe	45
Si	0.16	^{130}Xe	5
P	0.20	^{131}Xe	120
S	0.52	^{132}Xe	0.2
Cl	34	^{197}Au	98.8
Ar	0.63	^{235}U	678

---

[3] Neutrons that are moving only in a random manner under the influence of thermal energy at room temperature are called **thermal neutrons**.

[4] Isotope charts may be bought from several publishing companies and may also be obtained free from a few sources such as the General Electric Co., Educational Relations Dept., Schenectady, N.Y. The student is urged to get his own copy of the isotope chart.

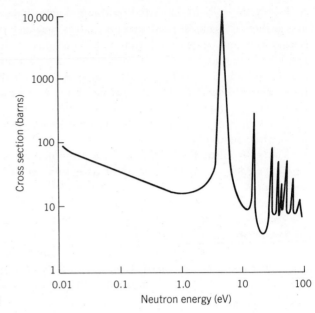

FIGURE 10-1 Neutron absorption by silver. (*Source:* AECU-2040.)

More important to this discussion is the variation of cross section with neutron energy. Figure 10-1 is a graph of cross section (in barns) versus energy of the bombarding neutrons (in electronvolts, both on logarithmic scales) for silver, showing very little change in the probability of reaction with neutrons over the range 0.01 to 1.0 eV, but many peaks at higher energies. A gradual drop in cross section with increasing neutron energy can be explained by saying that the faster a neutron moves through a nucleus, the less time there is for reaction; very fast neutrons might go through with little probability of any reaction at all. But a sharp peak in the curve cannot be rationalized away; it indicates an interaction that requires a special set of circumstances, involving subtleties of the structure not foreseen in simple theories about atomic nuclei. Nor is Figure 10-1 an extreme example; similar curves for platinum, tungsten, and dozens of other metals show a succession of such peaks (called **resonances**). The interpretation of these features is best left to advanced books on nuclear structure.[5] The point to be made here is that the stability of some nuclei cannot be expressed in a simple way, being a function of the nature and the energy of a bombarding particle.

Can you think why the term resonance is used?

ENERGY OF REACTION. A fourth source of information comes from measurements of the energy evolved during specific nuclear reactions. If some nuclei are capable of absorbing neutrons, as is indicated in the experiments on cross section, then we could also learn something about their relative reactivities by exposing the nuclei to neutrons of appropriate energy and measuring the amount of energy given off. Presumably those nuclei with stable

---

[5] For example, the excellent treatise by G. Friedlander, J. W. Kennedy, and J. M. Miller entitled *Nuclear and Radiochemistry*, 2nd ed., New York: Wiley, pp. 337–341 (1964).

configurations will not show much enthusiasm (as it were) for combination with an additional neutron, but those nuclei that can attain stability by adding a neutron will react more energetically. Table 10-4 summarizes the results of a few such tests. In it we see that $^{3}_{2}$He reacts violently with an additional neutron to capture it and attain the $^{4}_{2}$He structure, but $^{4}_{2}$He is completely "disinterested" in combining with another neutron. The isotopes $^{12}_{6}$C, $^{16}_{8}$O, $^{24}_{12}$Mg, and $^{28}_{14}$Si (all of them with even numbers of neutrons and protons, and with $Z = N$) release only modest amounts of energy as they combine with additional neutrons. By contrast, the nuclei with an odd number of neutrons or protons, or both ("odd-even" and "odd-odd" nuclei) such as $^{14}_{7}$N, $^{25}_{12}$Mg, and $^{29}_{14}$Si, release substantially more energy, indicating that they have become more stable after absorbing another neutron. Among the several isotopes for any one element, we see that the even-even isotopes are quite "disinterested" in additional neutrons, but the intervening even-odd isotopes ($^{25}_{12}$Mg, $^{27}_{14}$Si, $^{29}_{14}$Si, $^{53}_{26}$Fe, $^{55}_{26}$Fe, $^{61}_{128}$Ni) invariably release more energy as they absorb a neutron. This apparent preference for even numbers is taken to mean that there must be *pairing* of neutrons and *pairing* of protons in atomic nuclei, reminiscent of electron pairing in atomic orbitals.

TABLE 10-4
Energy of Reaction With An Additional Neutron

^{1}H	2.19 MeV		^{27}Si	17.12 MeV
^{2}H	6.16		^{28}Si	8.47
^{3}He	20.51		^{29}Si	11.5
^{4}He	−0.8		^{53}Fe	13.80
^{12}C	4.81		^{54}Fe	8.80
^{14}N	10.72		^{55}Fe	11.30
^{16}O	4.12		^{52}Cr	8.9
^{17}O	7.9		^{60}Ni	4.4
^{24}Mg	7.1		^{61}Ni	12.4
^{25}Mg	12.0			

The energies are expressed in million electron volts (MeV), meaning *millions of electronvolts*. As noted before, an electronvolt is the energy acquired by an electron as it falls through a potential difference of one volt. One million electronvolts is equal to 3.83×10^{-14} cal.

BINDING ENERGY. A fifth (and very important) source of information on the relative stabilities of nuclei comes from a consideration of their **binding energies**. We do the imaginary experiment of constructing a particular nucleus by putting together the required numbers of neutrons and protons, comparing the sum of the masses of these neutrons and protons with the observed mass of the actual nucleus. If the two are not equal, then some mass must have been lost or gained in the actual cosmic generation of the atom. By converting any mass difference to the equivalent energy according to the Einstein equation $E = mc^2$, we get an estimate of the binding energy involved. From comparative studies of such heats of reaction, we may draw conclusions about the relative stabilities of nuclei, just as we compare the stabilities of ordinary chemical compounds.

For example, we may look upon the hypothetical reaction

$$2 \text{ neutrons} + 2 \text{ protons} \longrightarrow {}_2^4\text{He}^{2+}$$

and inquire, "What is the heat of reaction?" With the method given in Chapter 2, the simple calculation goes this way:

Explain this calculation in your own words.

$$2p = 2 \times 1.007825 = 2.015650 \text{ atomic mass units (amu)}$$
$$2n = 2 \times 1.008665 = \underline{2.017330} \text{ amu}$$
$$\text{sum} = 4.032980 \text{ amu}$$
$$\text{subtract } observed \text{ mass} \quad \underline{4.002600 \text{ amu}}$$
$$\text{so "mass defect"} \quad = 0.030380 \text{ amu}$$

This "mass defect" represents the mass equivalent of energy lost during the combination of the four nucleons[6] to make an α particle. The alert reader may object that the mass of the ^{1}H *atom* was used, not the mass of a proton. True, and we also used the mass of the helium *atom*. Thus the reaction being considered really is

$$2 \text{ neutrons} + 2 \text{ electrons} + 2 \text{ protons} \longrightarrow {}_2^4\text{He atom}$$

But adding two electrons to the left side of the equation and also to the right side does not change the *difference* in masses between the two sides, so that mass defect is still 0.030380 amu for each α particle constructed.

In ordinary chemical language, we express heat of reaction in calories *per mole*, but in nuclear chemistry, it is customary to express the energy of a nuclear reaction in calories or million electron volts *per event*. So we do not need to multiply the mass defect by Avogadro's number to convert it to a molar basis; we leave it as the result of one event. To express the result in energy units, instead of mass units, we use the Einstein expression $E = mc^2$ with cgs units, and find that

$$1 \text{ amu} \equiv 931 \text{ MeV of energy}$$

So the energy for our imaginary reaction is

$$0.03038 \text{ amu} \times 931 \text{ MeV/amu} = 28.3 \text{ MeV}$$

For comparison with other reactions of this sort, it is customary to divide the result by the number of nucleons involved, getting the binding energy per nucleon and considering the matter without giving false advantage to heavy nuclei. In this case,

$$\frac{0.03038 \text{ amu} \times 931 \text{ MeV/amu}}{4 \text{ nucleons}} = 7.07 \text{ MeV per nucleon}$$

Similar calculations for other isotopes indicate that the binding energy for ^{3}He is only 2.57 MeV per nucleon, and that for ${}_3^6$Li (the next heavier stable isotope after ^{4}He) is 5.33 MeV per nucleon. Obviously, the binding energy for ^{4}He represents a maximum in this little sector of the isotope chart,

---

[6] "Nucleon" is an inclusive name given to protons and neutrons in a nucleus without distinction between the two.

as we might have expected from the other properties of ^{4}He and α particles. The calculations may be continued by the reader; you will find that maxima occur at ^{12}C (7.66 MeV per nucleon), ^{16}O (7.95 MeV), and ^{28}Si (8.45 MeV). From there on, there are fewer fluctuations, but a slow rise continues, reaching a broad maximum around ^{56}Fe (8.73 MeV). Afterward, the binding energy declines slowly to 7.7 MeV for ^{238}U. A complete graph of binding energy per nucleon versus mass number A, therefore, exhibits early peaks and then a broad maximum in the vicinity of iron and nickel, as shown in Figure 10-2. By analogy with conventional heats of formation for compounds, and in keeping with the inexorable laws of thermodynamics, it follows from Figure 10-2 that the middleweight elements have the most stable nuclei, and that all the heavy nuclei above $A = 120$ are progressively unstable with respect to the middleweight nuclei. Moreover, certain light-element nuclei also could give up energy if they could be converted to middleweight nuclei. So the curve of Figure 10-2 will prove very useful when we come to the subject of releasing nuclear energy, and the student should bear it in mind for that reason.

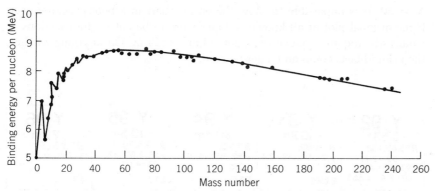

FIGURE 10-2 Binding energy per nucleon as a function of A, the mass number of the isotope.

MAGIC NUMBERS. A sixth source of information about nuclear stability and instability is the set of "**magic numbers**." These are really summations of experience with cross sections, energies of reaction, abundances of isotopes, and related observables; they are definitely empirical, like Mendeleev's periods of 2, 8, 18, and 32 elements. They have to do solely with *nuclear* properties, however.

The accepted magic numbers are 2, 8, 20, 28, 50, 82, and 126. Any isotope that has a magic number of protons, or neutrons, has an extra measure of stability relative to its neighbors. For example, we have seen how $^4_2$He distinguishes itself in the tests so far described in this chapter, and we note that $^4_2$He has two protons *and* two neutrons. Further, $^{16}_8$O has a very stable and unreactive nucleus with 8 neutrons and 8 protons, and oxygen is the most plentiful element on earth. We find a larger number of stable isotopes for tin ($Z = 50$) than for any other element, and we note the far greater abundance of lead (Z = 82) than of the nearby elements platinum and gold. Moreover, the principal isotope of lead is $^{208}_{82}$Pb, which has 82 protons and 126 neutrons (and so has *two* magic numbers in its favor). And so it goes, throughout any tabulation of nuclear properties; the magic numbers continue to have special significance.

For more magic numbers and superheavy elements, see Suggestions for Further Reading.

10-4 The Isotope Chart

All nuclear properties of the many known isotopes, stable and unstable, may be organized and displayed in a very useful way by setting up a rectilinear plot of the atomic number Z versus the number of neutrons N. When this is done on a large enough scale, which requires a piece of paper about 7 ft × 7 ft, each isotope can be allotted space about $\frac{1}{2}$ in. × $\frac{1}{2}$ in. in which to print the important characteristics: atomic mass, relative abundance, and neutron cross section for stable isotopes; mode of disintegration, energy of radiations, and half-life for radioactive isotopes. Figure 10-3 shows a portion of such a chart, a section cut out from a complete chart in the vicinity of $N = 50$. It is impossible to show the entire chart in a book, because even if the original plot of all known isotopes were to be cut up into sections, it would still require a piece of paper 50 in. × 29 in. On anything smaller, the printed facts could no longer be read.

Y 92 $^{2-}$ 3.53h β^- 3.63,1.32, 1.59-2.71 γ.93,.47,1.4, .07-2.4 E3.63	Y 93 $^{(1/-)}$ 10.2h β^- 2.89,·· γ.27,.94,.49-2.4 E2.89	Y 94 20.3m β^- 5.0,·· γ.92,.55,1.14,1.67·· 3.5 E5.0	Y 95 10.5m β^- 4.3,.76,1.17,·· γ.95,2.18,1.32, .43-4.07 E4.3	Y 96 2.3m β^- 3.5 γ.7,1.0,·· E7
Sr 91 $^{5/+}$ 9.67h β^-1.09,1.36, 2.67,·· γ(.551),.65-1.65 E2.67	Sr 92 2.69h β^-.55,1.5 γ1.37,.44,.23 E1.9	Sr 93 7.5m β^- 2.9,2.6,2.2 γ.60,.80,.3-2.1 E4.8	Sr 94 1.29m β^- 2.1 γ1.42 E3.5	Sr 95 26s β^-
Rb 90 4.28m|2.7m β^- | β^-6.6, 5.8,2.2,·· γ.83,.53-5.2 E6.6	Rb 91 57.9s β^- 4.6 γ.094,.35 E6	Rb 92 4.48s β^- (n) E8	Rb 93 5.87s β^- (n)	Rb 94 2.67s β^- (n)
Kr 89 3.18m β^- 4.9,4.6,3.8,·· γ.22,.59,.086-4.7 E5	Kr 90 32.3s β^- 2.8,·· γ.1215,.540,1.12, 1.538,.11-3.6 E4.6	Kr 91 8.6s β^- 3.6,·· γ.51,.11 E5.7	Kr 92 1.84s β^- (n)	Kr 93 1.29s β^- (n)

FIGURE 10-3   A portion of the isotope chart in the vicinity of $N = 50$.

Examination of an isotope chart, not just the piece of one shown in Figure 10-3, leads to some general conclusions which are listed here, subject to verification by the student:

1. With all the vast field exposed in a plot of Z versus N, with N varying from 1 to 158 (that is with $103 \times 158 = 16{,}274$ spaces available), the stable isotopes occupy only about 280 spaces. These are all clustered in a rather narrow band, which starts out along the diagonal $Z = N$ line, but afterward curves off toward the horizontal. Obviously, very few of the possible nuclear compositions are stable.
2. The naturally radioactive species (shown in black) are concentrated among the heavier elements, but naturally radioactive isotopes of H, C, K, V, Rb, In, Te, La, Ce, Nd, Sm, Gd, Dy, Lu, Hf, Ta, Re, Pt, Pb, Tl, and Bi also are known.
3. Spread along both sides of the band of stable isotopes (shown in gray) are the artificially radioactive species, shown in white. These isotopes have been "manufactured" by bombarding other stable or unstable isotopes in such a way as to cause nuclear reactions, producing new species with the properties indicated. Notice from their short lives that the isotopes *below* the stable band are too neutron rich to be stable.

Fig. 10-3 shows some of these. Just to the left of the sample, in the table, are Y 89 and Sr 88 (gray, stable) and Rb 87 (black, naturally radioactive).

Any isotope, natural or artificially prepared, which is off to the *neutron-rich* side of the stability curve, is unstable, but could revert to the stable range if it could exchange one or more of its neutrons for protons. This is not as imaginary a procedure as it seems, for isolated neutrons by themselves are unstable particles, which decay rather rapidly into protons and electrons. Those neutrons that are thoroughly involved in the exchange forces of nuclear bonding seem to go on indefinitely in that condition, like the nitrogen atoms bonded together in the N_2 molecule by strong electron-exchange forces. A neutron that is superfluous to any extent, and hence is not completely involved in the nuclear bonding processes, is reactive in the sense that an isolated nitrogen atom is reactive, and may revert to a lower-energy condition by decaying into a proton and an electron. The result is that the electron comes flying out of the nucleus at high velocity, while the slow proton is "recaptured" and is retained within the nucleus by the now-improved bonding situation. Such an emission of an electron, historically called a β particle, actually constitutes a common form of nuclear reaction, and many of the elements have heavy isotopes which react in this way.

The process of β emission leaves the nucleus with *one more positive charge* than it had originally, and with a place for one more electron in the extranuclear atomic orbitals. That is, the original nucleus is transformed into an isotope of the element of next higher atomic number, and the enhanced positive charge enables the new nucleus to hold one more electron. A neutron-rich isotope may be so far to the right of the stability curve that it is not stabilized by a single emission. In that case, it may undergo a second or third β decay, increasing its atomic number by one unit each time. When successive β particles are emitted, they usually come off with progressively less energy as the capacity for further reaction is decreased by each event.

An isotope with a neutron deficiency (one which lies above the band of stable nuclei in the table) can become more stable only by losing positive charges. This particular condition sometimes results in emission of a **positron,** a positively charged particle with the electron mass. The actual mechanism by which such a particle is emitted is more difficult to imagine than that for β emission, but the net result fits into the framework of situation and reaction within which we have been considering nuclear changes.

4. The elements of even Z (42, 44, 46, 48, and 50, for example) have more stable isotopes than the elements of odd Z. Those of odd Z either are anisotopic (Y, Nb, Rh near Figure 10-3; F, I, and Au among the common elements) or have only two stable isotopes. When there are two isotopes, they are separated by one space (note Ag, Ar, and Cl).[7]
5. Extra large numbers of stable isotopes occur along the lines corresponding to a magic number for Z or N, as illustrated by the 10 stable isotopes of tin.
6. Many isotopes have **isomers** (that is, nuclei containing the same numbers of neutrons and protons, but in two or more different configurations). Isomers appear, for example, at ^{87}Sr, ^{89}V, ^{93}Nb, ^{103}Rh, ^{105}Pd, ^{107}Ag, ^{109}Ag, and ^{111}Cd, to name only the spaces where one of the isomeric pairs is a stable isotope. Close examination of the chart shows that isomers are very common. Usually a less-stable isomer decays to a more-stable one by γ emission.

It is evident from these matters that a wealth of information is available in an isotope chart, and that a great deal of understanding about nuclear structure and nuclear reactions can be gained by poring over one. Such information is not contained at all in a Periodic Table of the elements.

10-5 Theories of Nuclear Structure

The earliest view of nuclear composition was that nuclei were made up of 2Z protons and Z electrons and that these electrons *within the nucleus* neutralized half the charge, leaving Z net positive charges to cope with the extranuclear electrons. We found in Chapter 2 that this view is untenable, because electrons, whether considered as particles or waves, cannot fit into the tiny nucleus. After neutrons were discovered, and it was realized that nuclei are made up of neutrons and protons, it was thought that these nucleons might be close packed at random, and that some close-range interaction between neutrons and protons kept the agglomerate together. There was good reason to believe in the close-packing part of this concept, for the radius of a ^{4}He nucleus is 2×10^{-13} cm, while the radius of a ^{238}U nucleus is only 8×10^{-13} cm. Hence the ratio of diameters is 4:1, and so the ratio of volumes is 4^3:1, or 64:1. Since the mass of the helium nucleus is 4, the mass of the uranium nucleus should be 4×64, or 256, if the same kind of packing occurs. The actual mass, 238, is close enough to emphasize the crowding together of the nucleons. However, the arrangement cannot be random; the existence of isomers, the strong indications of neutron and proton pairing inherent in

---

[7] There are seven exceptions to this last rule, and they occur at the seven places where odd-odd nuclei are found (as at $^{10}_{5}$B and $^{14}_{7}$N).

the odd-versus-even phenomena, and the importance of the magic numbers all militate against that.

The existence of some kind of short-range interaction between protons and protons remains an indisputable fact, however. So far as can be found out, Coulomb's law of electrical repulsion (Appendix III) is valid for distances that are only a small fraction of a nuclear diameter, so the protons in a nucleus repel one another strongly. The nuclear binding force is much stronger than the Coulomb force of repulsion, however, and it swamps out that repulsion with something hundreds of times as great. The nuclear-bonding force is of very short range; a nucleon is bonded only to its nearest neighbors, and the nuclear forces do not extend appreciably beyond the nuclear radius. They are even noticeably diminished at the nuclear surface.

Another theory of nuclear structure, which arose quite early in the study of nuclei, is the α-particle theory. This view held that the α particle is the epitome of nuclear stability and is the one true pattern; all the most stable nuclei contain an integral number of α particles. This might be so for $^{12}_{6}C$, $^{16}_{8}O$, $^{20}_{10}Ne$, $^{24}_{10}Ne$, $^{24}_{12}Mg$, $^{28}_{14}Si$, and $^{32}_{16}S$, all of which are stable and abundant nuclei, but it cannot be so for any isotope heavier than $^{40}_{20}Ca$, because that is the last stable isotope with an equal number of neutrons and protons. There are many more stable isotopes heavier than $^{40}_{20}Ca$, and all of them have more neutrons than protons (to an ever-increasing extent), so they cannot be made up purely of α particles.

After the highly successful development of the theory of electronic structure in atoms, by applications of quantum mechanics and wave mechanics, much effort was devoted to applying the same principles to nuclear structure. It turned out, however, that none of the conclusions about electron orbitals or energy levels applied to nucleons. Even with extensive modification, there was no satisfactory quantum-mechanical explanation for nuclear composition or structure of any isotope above $^{40}_{20}Ca$.

Beginning in the late 1930s, a comprehensive shell-structure theory of the nucleus began to evolve, based on a system of energy levels or "closed shells." This scheme, similar to that for the extranuclear electrons, is based on the quantization of spin and orbital momentum of the nucleons.[8] Again, a pattern of quantized energies emerges. These are shown in Figure 10-4 and may be compared to those of Figure 2-12. Because of the great strength of intranuclear forces, the difference in energy between levels is much greater than that for electronic shells. Thus it is that chemical reactions, involving rearrangement of electrons, have energies measurable in electronvolts (see Chapter 6), while nuclear reaction energies are measured in *millions* of electronvolts (MeV).

In the shell theory, each energy level has the capacity for neutrons *or* of protons indicated at the right in parentheses, and the cumulative number of

---

[8] For a general discussion of nuclear shell structure, see B. H. Flowers, *Journal of Chemical Education*, vol. 37, p. 610 (December 1960). Maria Mayer, who received the Nobel Prize for her work in this area, describes the theory in *Physical Review*, vol. 78, p. 16 (1950). See also the book by M. G. Mayer and J. H. D. Jensen, *Elementary Theory of Nuclear Shell Structure*, New York: Wiley (1955).

neutrons *or* of protons appears in square brackets. Notice that complete filling of the first "shell" or level corresponds to the isotope $^4_2$He, and complete filling of the second level to $^{12}_6$C. The third level is filled at $^{16}_8$O, and the fourth at $^{28}_{14}$Si. Successive levels become filled at the points corresponding to the elements O, Ca, Ni, Se, Sr, Zr, Sn, Ce, Gd, Er, Yb, Pb, and U. All of these elements have high nuclear stabilities, and all are relatively abundant. The magic numbers come in at the intervals indicated in the column at far right to emphasize the extra importance of *particular* combinations of filled shells.

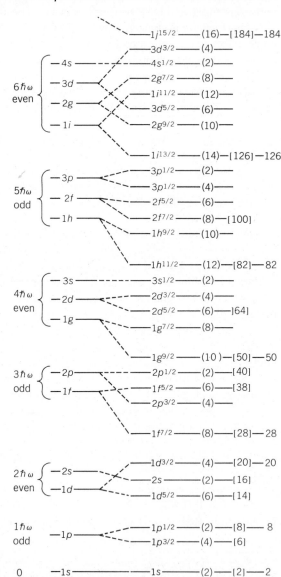

FIGURE 10-4 Splitting of energy levels by spin-orbit coupling in nuclei, showing derivation of the magic numbers. [Source: G. Friedlander, J. W. Kennedy, and J. M. Miller, *Nuclear and Radiochemistry*, 2nd ed., New York: Wiley (1964).]

So it turns out that *nuclei have a compact but ordered structure;* that *there can be two or more distinct structures for the same Z and N* resulting in isomers; that *the nuclear forces are of very short range* but much more potent than the electrical forces that hold together atoms; and that *the details of structure fit in with the measurable properties of nuclei.* The current theory of nuclear bonding, although far from perfect, accounts well enough for the observed facts to allow us to go on to the more practical matters of nuclear reactions and nuclear energy.

10-6 Nuclear Reactions

In conventional chemistry, chemical reactions are carried out either to prepare new substances or to obtain energy. The same can be said about nuclear reactions; they are conducted to release energy as heat (in nuclear power plants, Project Plowshare explosions, bombs, and such), or to produce new elements or scarce materials. Consider the small-scale production of neutrons for laboratory or lecture-demonstration use, for example. A tiny amount of radium sulfate is mixed with a half gram of beryllium oxide, and the mixture is sealed in a nickel capsule. The radium disintegrates steadily, with a half-life of 1620 years, emitting α particles:

$$^{226}_{88}\text{Ra} \xrightarrow{1620 \text{ yr}} {}^{222}_{86}\text{Rn} + {}^{4}_{2}\text{He}$$

For more about half-life, see Sec. 11-3.

The α particles strike beryllium nuclei, producing carbon and neutrons:

$$^{4}_{2}\text{He} + {}^{9}_{4}\text{Be} \rightarrow {}^{12}_{6}\text{C} + {}^{1}_{0}\text{n}$$

The neutrons penetrate the nickel capsule readily and come flying off to be used in further nuclear reactions. The reaction goes on silently and steadily, as long as the radium lasts.

How could you calculate the energy change in these processes?

Notice that *equations* can be used to summarize the two reactions and that the notation is simple but concise. In a balanced equation, the sum of all the mass numbers of the left side of the equation must equal the sum of all the mass numbers of the right side, and the same is true of the atomic numbers. Hence if we were given the incomplete equation

$$^{226}_{88}\text{Ra} \rightarrow {}^{222}_{86}\text{Rn} + \text{?}$$

we could figure out that the other product is ${}^{4}_{2}\text{He}$. There may also be emission of γ rays, of course, but they have no A or Z notation.

A balanced nuclear equation allows us to do something that a balanced conventional chemical equation will not allow—we can calculate the *energy change* (heat of reaction) just from the equation and a table of isotopic masses. In the reaction of α particles with beryllium given above, the diminution of mass is calculated and converted to energy units:

for the products ^{12}C = 12.000000 amu

       n = <u>1.008665 amu</u>

      sum = 13.008665 amu

for the reactants ^{9}Be = 9.01219 amu

      ^{4}He = <u>4.00260 amu</u>

      sum = 13.01479 amu

so the difference is

$$13.014790 \text{ amu}$$
$$-13.008665 \text{ amu}$$
$$0.006125 \text{ amu}$$

and the energy is 0.006125 amu × 931 MeV/amu = 5.7 MeV. This is the energy per event. To convert this to kilocalories per mole, for comparison, we multiply by Avogadro's number and the conversion factor[9] for million electron volts to calories:

$$5.7 \text{ MeV/atom} \times 6.023 \times 10^{23} \text{ atoms/mole} \times 3.82 \times 10^{-14} \text{ cal/MeV}$$
$$= 1.32 \times 10^{11} \text{ cal/mole or } 1.32 \times 10^{8} \text{ kcal/mole}$$

This is seen to be about a million times greater than the highest energies for ordinary chemical reactions and is typical for nuclear reactions. The energy appears as kinetic energy of neutrons, recoil kinetic energy of carbon (dissipated as heat at the source), and some γ rays emitted simultaneously.

Nuclear reactions in general can be understood in terms of the *Bohr theory*, proposed in 1936 when nuclear experiments were done with particles of 10 to 30 MeV energy, but just as useful today if a few modifications are made. Niels Bohr considered that nuclear reactions take place in five distinct stages:

This is the same man who gave us the Bohr theory of atomic structure.

1. The incident particle hits a target nucleus, penetrates, and is bound after several collisions with nucleons.
2. The *compound nucleus* contains all of the mass and energy of projectile and target, including the kinetic energy and the binding energy. This total energy is distributed over all the nucleons by a multitude of collisions.
3. The active compound nucleus exists for a long time (ca. 10^{-12} sec, 1,000,000,000 times as long as it would take a 0.5 MeV neutron to traverse a single nucleus).
4. Random fluctuations eventually concentrate a sizable fraction of the energy in one particle, from time to time.
5. Such a temporarily activated particle sometimes escapes, especially if it is near the surface, where the nuclear-binding forces are diminished.

Where have we seen statistical behavior before?

The Bohr theory is seen to depend on random and statistical behavior, suiting nuclear phenomena very well. Perhaps the most important corollary of the theory is that the *formation* and *disintegration* of a compound nucleus *are independent events*, separated by a comparatively long lapse of time.

---

[9] The conversion factors are

$$1 \text{ MeV} = 1.6 \times 10^{-6} \text{ erg}$$
$$1 \text{ MeV} = 3.82 \times 10^{-14} \text{ cal}$$
$$1 \text{ amu} = 931 \text{ MeV} = 3.55 \times 10^{-11} \text{ cal}$$
$$1 \text{ g/mole} = 6.023 \times 10^{23} \times 3.55 \times 10^{-11} \text{ cal} = 2.14 \times 10^{13} \text{ cal}$$
$$1 \text{ gram of mass} = \$400{,}000 \text{ of energy at current electric rate}$$

An illustration will make this clear. The bombardment of aluminum by neutrons gives a compound activated nucleus, ^{28}Al, which decays by four different paths,

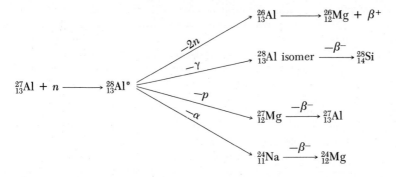

Such multiple disintegration indicates statistical governance of the mechanism. Moreover, the same compound nucleus, ^{28}Al°, can be made by at least two other methods,

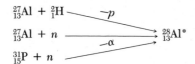

Here the asterisk indicates a "hot" nucleus which will disintegrate.

but *the disintegration pattern is always the same,* regardless of the preparative reaction. The Bohr theory holds for all these related reactions.

Bohr did not envision the phenomenon of nuclear fission when he proposed his theory, and yet the theory holds quite well for this radically different mode of decay. The splitting of a compound nucleus into two unequal portions, at the same time that it evolves neutrons, is a more complicated reaction, but one which can be understood in terms of the principles and examples just considered.

10-7 Fission

THE FISSILE SPECIES. It is well known that the scarce isotope of uranium, ^{235}U, is caused to undergo **fission** by thermal neutrons, whereas the much more plentiful isotope, ^{238}U, is not. More energetic neutrons *will* fission in ^{238}U, especially if they are within a particular resonance range (see page 330). What makes the difference? And are there any other isotopes subject to fission by thermal neutrons?

These questions can be answered by considering the energy content of a nucleus just before it undergoes fission. Some nuclei, especially those of the heavy transuranium elements, are unstable and undergo *spontaneous* fission. Every once in a while, a nucleus of such an element (for example, ^{254}Cf or ^{256}Fm) simply flies apart into two unequal pieces, emitting a shower of neutrons in the process. Such statistical behavior is shown by all nuclei of mass greater than 250. Since no other source of energy seems necessary, we assume that the energy content of the heavy nucleus is already sufficient to bring about the reaction. This state of affairs can be indicated on a diagram, Figure 10-5, where curve *a* shows no energy hump to be surmounted before fission occurs. At low mass numbers, however, the half-life for spontaneous fission increases; for ^{235}U and ^{238}U it is extremely large, so that these isotopes have persisted in natural minerals for billions of years. Curve *b* of Figure 10-5 illustrates this situation, where a small energy hump exists, and some additional input is necessary in order to bring about fission. It has been found that even lead and bismuth (mass numbers around 200) undergo fission reactions if they are given sufficiently large activation energy, as is shown by curve *c* of Figure 10-5. When we consider the 6 MeV barrier to fission for uranium, and then inquire into the binding energy for one additional neutron added to the even-odd $^{235}_{92}$U nucleus, we find that it is about 6.5 MeV—just enough to activate the ^{235}U nucleus for fission. On the other hand, the even-even nucleus $^{238}_{92}$U has a considerably lower binding energy for an additional neutron, as would be expected, and *this is insufficient to surmount the 6 MeV barrier.* Instead, additional activation must be supplied in the form of kinetic energy of the neutron. Similarly, about 20 MeV additional energy is necessary for the fission of lead. The more rapidly a neutron is moving, the faster it traverses a nucleus and the lower the probability of capture, so fission with fast neutrons is a low-yield process.

Compare the concept of "energy barrier" for fission to "activation energy" for chemical reaction (Sec. 11-8).

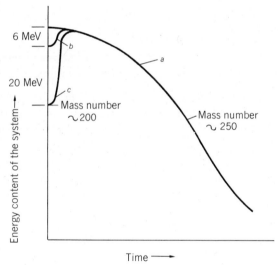

FIGURE 10-5 Spontaneous fission versus induced fission in heavy isotopes. Curve a describes any unstable nucleus which can undergo spontaneous fission without any input of activation energy. Curve b describes a nucleus such as ^{235}U which undergoes fission if given a small amount of energy. Isotopes of lead and bismuth and other elements of mass number ca. 200 require even more energy (curve c) to become sufficiently activated for fission.

From this analysis, it is apparent that other even-odd isotopes with mass numbers in the region 230 to 240 should be fissionable with slow neutrons. Since there is not much ^{235}U in the world, this is comforting, provided that suitable candidates can be found. There are no natural ones in sufficient quantity, but $^{233}_{92}$U and $^{239}_{94}$Pu can be made in useful quantities by ancillary reactions that go on during the operation of a fission reactor, so these have become practicable fuels along with ^{235}U. Their production is described below.

REACTIONS AND PRODUCTS. The fission of ^{235}U can be brought about by several agencies besides slow neutrons (by γ rays, α particles, or protons, for example), but the fission pattern is always the same, in accordance with Bohr's theory. The actual fission pattern is very complicated, for 35 elements are represented, and over 100 isotopes have been identified as fission products. Almost all the fission products are violently radioactive, so that they decay again and again and complicate the picture further. Hence a great many equations may be written for the fission reaction, each representing its own part of the pattern. As an example, which is fairly typical, consider

$$n + {}^{235}_{92}U \rightarrow {}^{82}_{35}Br + {}^{151}_{61}Pm + \gamma + 4\beta^- + 3n$$

These two fission products were chosen because they are among the most common. The position of these two isotopes on the fission yield curve (Figure 10-6) shows why they are among the easiest to identify; they are among the most abundant products. Figure 10-6 shows that no one isotope accounts for more than 7 percent of the total yield and that there is a million-fold difference in abundance between the two ends of the curve and the two peaks. Moreover, the isotopes falling under the peaks are *hundreds* of times more abundant in the fission mixture than the isotopes around mass 115. So the overwhelming majority of uranium nuclei split into two *unequal* pieces. The fission of ^{229}Th also gives a bimodal pattern of products, but interestingly, the fission of lead gives a yield pattern with only a single symmetrical peak.

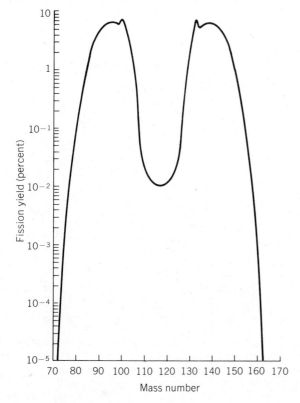

FIGURE 10-6 Yields of fission-product chains as a function of mass number for the slow neutron fission of ^{235}U. [Reprinted from G. Friedlander, J. W. Kennedy, and J. M. Miller, Nuclear and Radiochemistry, New York: Wiley (1955), by permission.]

A glance at the binding-energy curve (Figure 10-2) shows how much energy to expect from the fission of ^{235}U. There is a difference of about 1 MeV between the curve's maximum and the position of ^{235}U, so the splitting of ^{235}U into two middleweight fragments (both near the maximum in binding energy) causes about 1 MeV per nucleon to be evolved (the actual measured amount is 200 MeV per fission event or a bit less than 1 MeV per nucleon). Reference to a complete isotope chart will show that the N/Z ratio for uranium is 1.5, while the ratio for $^{82}_{35}$Br is 1.34, reflecting a lower neutron excess required for stability of the lighter nucleus. Hence the splitting of a ^{235}U nucleus requires that some of the excess neutrons be given off, and the average number emitted per event is 2.3. Of these, 1 is necessary to continue the chain reaction, but the remaining 1.3 (average) are available for generating new fissionable material by way of the **breeding** reactions. By "breeding" is meant the synthesis of fissile isotopes from the excess neutrons, which are absorbed in a layer of depleted uranium[10] or a layer of natural thorium surrounding the core of the fission reactor. There the breeding reactions take place. From ^{238}U the fissile plutonium is synthesized:

$$n + {}^{238}_{92}\text{U} \rightarrow {}^{239}_{92}\text{U}^\circ \xrightarrow[23m]{-\beta-} {}^{239}_{93}\text{Np} \xrightarrow[2.3d]{-\beta-} {}^{239}_{94}\text{Pu}$$

The product $^{239}_{94}$Pu is an α emitter, with a half-life of 24,000 yr; its high α activity makes it less desirable as a reactor fuel than ^{233}U, which is synthesized from natural thorium:

$$n + {}^{232}_{90}\text{Th} \rightarrow {}^{233}_{90}\text{Th}^\circ \xrightarrow[25m]{-\beta-} {}^{233}_{91}\text{Pa} \xrightarrow[27d]{-\beta-} {}^{233}_{92}\text{U}$$

The ^{233}U is also an α emitter, but with a half-life of 162,000 yr, and hence with lower α intensity. Since there is much more thorium on earth than uranium, this breeding reaction allows fissile fuel to be made from a plentiful raw material.

Mass and energy balances:
$E = mc^2$

MATERIAL AND ENERGY BALANCES. When ^{235}U is used as a fuel in a nuclear reactor, only 0.1 percent of its mass is converted to energy. The rest remains as matter, according to the following balance sheet:

1 kg ^{235}U gives 989.0 g of fission fragments

10.2 g of neutrons

0.7 g of kinetic energy[11]

0.1 g of radiation
_____
1000.0 g

---

[10] In the United States, it is customary to "enrich" uranium in ^{235}U by methods of isotopic separation, leaving part of the natural uranium depleted in ^{235}U (more nearly all ^{238}U).

[11] This is the mass equivalent of the kinetic energy possessed by the fission fragments and the neutrons. This kinetic energy (and almost all of the radiation) is converted to heat.

A further detailing of the energy balance shows that:

$$\text{fission of 1 atom of } {}^{235}\text{U yields 200 MeV}$$
$$= 5 \times 10^{10} \text{ cal/g}$$
$$= 60{,}000 \text{ kWhr/g available as}$$

kinetic energy of fission fragments	160 MeV
kinetic energy of neutrons	5 MeV
energy of γ rays released	5 MeV
from later radioactive decay	20 MeV
energy from neutrons absorbed	10 MeV
total energy	200 MeV

REQUIREMENTS FOR FISSION REACTORS. Every nuclear reactor, whether used for research or for generating electric power, has six necessary components:

1. nuclear "fuel" (a fissile isotope, fabricated into suitable shape to transfer the heat generated within it to the cooling medium);
2. a **moderator** to slow down the neutrons (which come off from the fission reaction at kinetic energy of about 1 MeV) to thermal velocities of about 0.02 MeV, so they can be absorbed by ^{235}U and cause more fission; the best moderators are helium, heavy water, graphite, CO_2, beryllium, and ordinary water, in that order;
3. controls (adjustable means of absorbing a few neutrons to keep the chain reaction from going at too fast a rate, or stopping);
4. a cooling medium (a fluid, often acting also as moderator, which will carry off the heat developed within the fuel elements and transfer it to an external heat exchanger, where steam is produced to run the turbogenerators);
5. a breeding blanket (of ^{238}U or ^{232}Th) where the synthesis of more fissile material is carried on, and from which the fissile isotopes are separated at long intervals by chemical processing;
6. a reflector and shield, in four parts: a reflecting layer of graphite to turn back half the errant neutrons, a thermal shield (perhaps of cast iron) to absorb much of the γ radiation, a main shield of concrete in layers 4 to 8 ft thick, and a biological shield of cellulose fiber to absorb neutrons, β rays, and X-rays that otherwise would injure people.

The actual design of the reactor is a complicated engineering matter determined by the size and the intended purpose of the device. There also are regional preferences; in England there are many natural-uranium power reactors moderated by graphite and cooled by CO_2 under pressure, while in the United States most power reactors use uranium enriched in ^{235}U and are moderated and cooled by ordinary water kept under high pressure. Research reactors often are moderated and cooled by heavy water, D_2O, and are set up so that samples can be exposed to the neutron flux. Figure 10-7 is a schematic drawing of a research-type reactor. The fuel rods FR contain the fissile material, and these are suspended in a tank of heavy water. The fission reaction heats the water, which is cooled by being pumped through a heat exchanger HE. P is a platinum-black catalyst, which reconstitutes that heavy water which was decomposed into D_2 and O_2 by radiolysis within the core tank. CR is the control rod, which regulates the intensity of the fission reaction. Fast neutrons from the reaction are moderated by the heavy water and by the graphite shell G. The graphite also acts as a neutron reflector in that neutrons are scattered by multiple collisions, so that half of them are scattered right back into the tank. A breeding blanket of thorium or ^{238}U surrounds the graphite. Slowed-down neutrons can wander through a column of pure graphite G', and come out at port P_1, for use in experiments. Fast neutrons are available for experiments through port P_2. A concrete shield surrounds the assembly and supports the reactor. Within the concrete block, there is a sump S, which is intended to receive the heavy water and the debris if there should be an accidental overheating of a fuel rod and a rupture of the tank. Experimenters work at the floor level F, and the working space is monitored continuously by instruments which are capable of shutting down the reactor automatically if the radiation level exceeds a safe level.

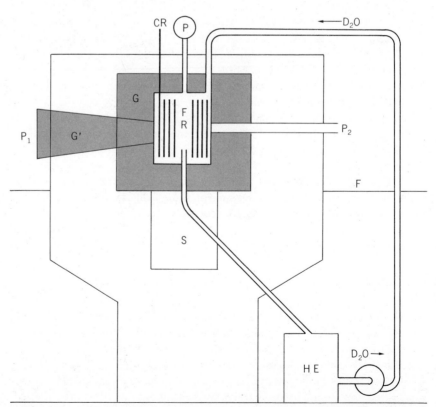

FIGURE 10-7 A research reactor cooled by heavy water. Fuel rods FR and a control rod CR are suspended in a tank of D_2O coolant, which circulates through a heat exchanger HE. Thermal neutrons (moderated by graphite G') exit at port P_1; fast neutrons come out at port P_2.

It should be emphasized that a nuclear reactor for research (as in Figure 10-7) or for generating steam for tubogenerators, *is not a nuclear bomb*. It cannot explode like a bomb. If it should overheat accidentally, moderator water would be boiled off, and the reaction would stop for lack of enough slow neutrons. Similarly, any expansion of the fuel-rod spacing will stop the reaction, for the assembly always stands on the threshold of criticality,[12] and any increase of dimensions makes it subcritical. Reactors are designed with adequate shielding to protect the nearby personnel and the public.

Reactor systems are designed to minimize the escape of radioactive matter from the reactor core, and thus to minimize contamination of the environment. It seems clear that early reactor designs, particularly including some boiling-water reactors now in operation, allow needlessly large emission of radiation, for example, releases of the order of one million curies[13] per year in the form of the radioactive gas ^{85}Kr. Smaller in scale, but of more long-term concern, are emissions of chemically active elements such as $^{90}Sr^{2+}$, which can be taken up in biological cycles (see Chapter 1) because of its chemical similarity to Ca^{2+}. The major cause of radioactive emissions is the seemingly inevitable presence of pinhole leaks in the zirconium-alloy containers, which separate the fission fuel elements from the primary cooling water, and in the heat exchanger.

Emissions from the presently operating and planned power reactors are, and should remain, well below the guidelines established by the Atomic Energy Commission (which has the dual responsibility of promoting and regulating the generation of nuclear power in the United States). Yet it is also clear that reactor emissions are larger than they might be, based on present reactor design and construction technology. Newer designs, not yet operational at the time of this writing, promise greatly reduced emission rates. This would be highly desirable in view of the great increase in power consumption forecast for the future.

The other major environmental impact of nuclear reactors is their production of waste heat, which can be absorbed in cooling towers (as is done in England) or put to other good use (such as the desalting of sea water by distillation, to increase the supply of fresh water). Unfortunately, at present the waste heat often is discarded as warm water into cooling ponds or into streams. Such waste heat is of course not peculiar to nuclear power plants; all heat-power stations must export some heat to their surroundings in order to convert the rest to free energy. This is true whether the heat be supplied initially by coal, or oil, or a nuclear reactor. Waste heat is always the price to be paid for getting free energy from a heat engine, even in an automobile (see Chapters 8 and 9), and nuclear power plants are no exception to this general rule of thermodynamics. However, far from exporting more waste heat, nuclear power plants actually put out *less* waste heat than a coal-fired power plant, because they have no furnace and hence no flue gas. They avoid heating up all the air that would be required for combustion of coal or oil, and they put out no smoke or cinders.

The authors of this book disagree about the environmental dangers of nuclear power plants, but we agree on this: you must *inform yourself!*

---

[12] A reactor that is just on the point of sustaining a self-continuing nuclear fission reaction is said to be *critical*.

[13] See references cited in Section 10-8.

ADVANTAGES AND DISADVANTAGES OF FISSION. As a means of deriving electric power from the release of nuclear energy, fission reactors have the advantages of compactness, long operation on one charge of fuel, and no demand on fossil fuel. Moreover, the technology of building and operating fission reactors has reached a high level, due to long experience. The *disadvantages* are these: (1) the amount of ^{235}U in the world is decidedly limited, making it necessary to breed new fissile material to replace the ^{235}U used; (2) the fission products include many strong absorbers of neutrons, making it necessary to interrupt the reaction in order to dissolve the fuel rods and conduct a chemical separation of the still useful uranium; and (3) the fission products are highly radioactive, and so the safe disposal of this waste is a troublesome and expensive operation.

10-8 Fusion

By nuclear **fusion** is meant the combination of nuclei of light elements to produce heavier nuclei and emit the corresponding binding energy. A glance at the binding-energy curve (Figure 10-2) shows that any nuclear fusion which produces ^{4}He or ^{12}C or ^{16}O liberates binding energy, and that if the reactants are chosen carefully, more energy (per gram of reactant) can be obtained than is realized from fission. A few suitable reactions are

$$^3\text{H} + {}^2\text{H} \rightarrow {}^4\text{He} + n + 17.6 \text{ MeV per event}$$

$$^2\text{H} + {}^2\text{H} \rightarrow {}^3\text{He} + n + 4.0 \text{ MeV per event}$$

$$^3\text{H} + {}^3\text{H} \rightarrow {}^4\text{He} + 2n + 11.4 \text{ MeV per event}$$

$$^3\text{He} + {}^2\text{H} \rightarrow {}^4\text{He} + {}^1\text{H} + 18.3 \text{ MeV per event}$$

$$^7\text{Li} + {}^1\text{H} \text{ (as LiH, a salt)} \rightarrow 2{}^4\text{He} + 17.3 \text{ MeV per event}$$

There are also the reactions used in the French H-bomb of August 24, 1968:

$$^6\text{Li} + {}^2\text{H} \rightarrow 2{}^4\text{He} + 22.4 \text{ MeV per event}$$

$$^6\text{Li} + {}^3\text{H} \rightarrow 2{}^4\text{He} + {}^1\text{H} + 16.1 \text{ MeV per event}$$

All of these fusion reactions require that the reactants be heated in compact form to millions of degrees in order to start the fusion, and the only way known so far to accomplish this (at will, by man) is to ignite the mass with a fission-type bomb. No sustained controlled fusion reaction has yet been achieved for useful periods of time in a laboratory. There are schemes for confining hot ionized gases (H, ^{2}H, ^{3}H, and so on) by means of surrounding magnetic fields, and for exciting small regions of this "plasma" by continuous input of energy, but the temperature necessary for continuous nuclear fusion has not yet been reached in such laboratory experiments.[14] The nearest really contin-

---

[14] See report on fundamental nuclear energy research for 1969 of the United States Atomic Energy Commission (Washington D.C.: U.S. Government Printing Office, Jan. 1970), pp. 177–183. See also preceding annual reports of the AEC, available each year as supplements to the Annual Report to Congress. See also D. J. Rose, "Controlled Nuclear Fusion: Status and Outlook," *Science*, vol. **172**, p. 797 (May 21, 1971).

uous fusion reactor is the sun, which derives its energy from the fusion of hydrogen to helium and radiates huge quantities of that energy to us. It is unfortunate that we do not know how to collect and use that energy more efficiently.

The *advantages* of fusion reactions over fission reactions for the generation of power are considerable: Fusion requires only light elements, available in very large quantities everywhere, and it produces no dangerously radioactive by-products to dispose of. It also produces four or five times as much energy per gram of reactant as fission does. The main *disadvantage* is that we do not know how to do it yet. Fusion holds the key to unlimited electric power, for all nations, if only the methods and technology become available.

10-9 Effects of Radiation

Any discussion of nuclear energy should include a consideration of what radiation can do to materials. All matter is affected by radiation, but the effects differ widely, according to the kind of matter and the kind of radiation. For example, metals are affected very little by γ rays, which excite emission of X-rays and some loss of electrons, but do not damage the crystal structure. Fast neutrons, on the other hand, collide with metal nuclei and displace them from their lattice positions (and at the same time produce new elements which do not fit the original lattice). Fission fragments cause even more damage to the physical structure of metals and can turn some strong metals into useless powders. Nonmetals, such as organic plastics, suffer much more from intense radiation, for rays cause ionization of elements formerly united by covalent bonds by knocking electrons out of bonding orbitals, and that destroys the molecular structure. Neutrons and fission fragments cause even worse ionization, with consequent very heavy damage. Relatively few materials can be used in the actual construction of nuclear reactors, and these materials are those that are affected least by radiation and that have low cross section for neutrons (such as graphite, metallic beryllium and beryllium oxide, aluminum, and zirconium).

Living organisms suffer just as much physical damage as inanimate organic matter, because they are composed of the same elements. In addition, there are some extra considerations which involve special problems in living organisms:

Also see Sec. 9-10.

1. Since protein material is vastly more complicated in structure than wood or nylon or rubber, a few disruptions by ionization can cause a complete collapse of the protein structure and loss of all its activities. Worse than that, the disintegrated structure is just so much waste material for the body to eliminate, putting extra strain on the kidneys and (in quantity) causing fever and nausea.
2. A still more sensitive tissue is the genetic material contained within an egg cell or sperm. Since the protein material there contains the entire genetic code for the reproduction of the species, *any* change induced by radiation, however small, causes an irreversible alteration of the reproductive pattern (a mutation). Almost all mutations are

lethal, and most of the rest are undesirable, so it is important to avoid the irradiation of genetic material to the very greatest extent possible.

3. The different parts of a highly complex organism are not independent units capable of survival by themselves, but are highly dependent on other parts for supply of oxygen and nutrients (as well as for chemical control messengers and coordination impulses). Any disruption of a part of this complicated network has an immediate effect on the functioning of the rest, so the extent of the original radiation damage may be less important than the multiplication of its effect as it shows up in the impaired activity of the organism. It is not just a question of change of physical properties, as in irradiation of a plastic, but of possible immediate or long-term lethal effects due to disruption of even a small but important part.

With these specific matters of vulnerability in mind, it must be the personal concern of everyone to protect himself in a rational way from the dangers of radiation. Further, he should know enough about the subject to advise others who have no access to information. In this case, ignorance is not bliss; ignorance often gives rise to cavalier disregard of safety precautions, or on the other hand to fear and even to panic. By contrast, dependable information from instruments can lead the informed person to rational acts of protection, just as in case of fire or flood or any other emergency. Radioactive materials will be used and shipped throughout the country in ever-increasing amounts as the nuclear age unfolds, and no one ever knows when or where an accident might occur. So the prudent person will give some thought to his protection. He will also see to it that proper safeguards are maintained, through popular agreement and insistance, so that pollution of the environment with radioactive material will be held to an absolute minimum.

The first fact to remember is that the human senses do not respond at all to low or even lethal levels of radiation. True, massive bursts of radiation may be felt as heat, but it is perfectly possible for a person to receive a fatal dose of radiation without his being aware at all of what was happening at the time. Light and heat and noise we can detect, and we respond with instinctive protective measures, but nuclear radiation is silent and unfelt. This is why instruments are so important.

Where is the nearest radiation detector to where you are now?

Early experience with X-rays (and then with the use of radium in luminescent paints) led to the development of simple dependable instruments for recording the level of radiation. Mostly these are instruments for detecting γ rays, since these are given off as by-products of most nuclear changes, but specific detectors for α, β, and neutron radiation also are available.

Some detectors, such as the little portable ionization chamber dosimeter commonly carried by those who work with radioactive materials, respond to all kinds of radiation to a useful degree. The ionization chamber is a thin aluminum tube about the size of a fountain pen containing a tubular air capacitor charged to 100 or 200 V and connected to a tiny internal silica-fiber electroscope. The extent of charge can be read in terms of lateral displacement of the fiber on a scale, by means of light coming through a small lens at the end of the tube. If there is no environmental radiation, the electrostatic charge will take a long (but measurable) time to leak out of the capacitor, but any incoming radiation causes ionization of the gas within the tube, and the ions carry away part of the charge. So the rate of loss of charge becomes the important measurement. The scale can be read at intervals of one minute, or one hour or day, and the scale is calibrated to show the corresponding density of radiation that causes the loss of charge. Such instruments are inexpensive, durable, and harmless, and can be worn all the time. They can (and should) be supplemented by other devices: film badges, such as are worn by dental technicians; "squawkers" or "sparrows," which utter a warning chirp whenever the radiation level rises beyond a preset amount (as detected by a miniature Geiger counter and electronic amplifier circuit); and nonportable instruments that can flash an alarm automatically as well as record the level continuously on a chart for the continuous monitoring of a particular space. See any textbook on radiochemistry for details on the operation of the instruments.

The unit of radiation measurement, which grew out of the early use of X-rays, is the **roentgen** (r), which is the dose of X-rays or γ rays that will produce one electrostatic unit of electricity in one gram of air or water without limit of time. The roentgen corresponds to 1.6×10^{12} ion pairs produced in one gram of air, and the amount of ionization energy it represents is 83 ergs. When used in connection with other kinds of radiation which might harm people, the roentgen is changed to the **rem** (roentgen, equivalent, man) according to a scale relating the relative harmfulness of the radiation, so that

$$1 \text{ rem} = 1.0 \text{ r for } \gamma \text{ rays}$$
$$= 0.1 \text{ r for } \beta \text{ rays}$$
$$= 0.2 \text{ r for fast neutrons}$$
$$= 0.05 \text{ r for thermal neutrons}$$
$$= 0.02 \text{ r for protons}$$
$$= 0.01 \text{ r for } \alpha \text{ emitters which can get into the body}$$

Sometimes the *rad* is used instead of the *rem;* the *rad* is similar except that it is based on 100 ergs of ionization energy instead of 83 ergs.

The unit of radioactivity (as distinct from radiation) is the **curie** (c), which is the amount of radiation activity of all kinds delivered continuously by one gram of radium in association with all its products. The curie corresponds to 3.7×10^{10} nuclear disintegrations *per second*. It is a large unit, because one gram of radium would be a dangerously large amount to have collected in one place. In radiochemical work, the more usual units are the millicurie—mc (one-thousandth of a curie, and about the level of a stock solution of a radioisotope) and the microcurie—μc (one-millionth of a curie, and about the level used in a tracer experiment involving radioisotopes). Monitoring and surveying instruments usually are calibrated in microcuries.

The *natural background level* of radiation represents unavoidable radiation from all natural sources. It is there all the time, and we can do nothing about it. There are naturally radioactive elements in rocks, concrete, wood, and even in our bodies (as radioactive potassium and carbon). Moreover, we receive radiation from outer space all the time in the form of cosmic rays and the earthly radiations they produce as they hit the air and earth around us. The sum of radiations from all natural sources amounts to about 6.4 rem per 70-year lifetime at sea level and to more at higher altitudes where there is less shielding: 8.1 rem at Pittsburgh and 14.5 rem at Colorado Springs. The fact that the people of Colorado Springs live just as long as those in the coastal cities and have just as many healthy offspring is taken to mean that the difference between 6.4 and 14.5 rem is not significant in human experience.[15]

For most of us, diagnostic X-rays and exposure to luminous-dial watches and other "cultural" sources of radiation add another 7 or 8 rem in a lifetime. Fallout from the testing of nuclear weapons adds about 0.1 rem per year at the time of this writing; what may happen if more nations insist on developing their own nuclear weapons remains to be seen. It is estimated that it would require a lifetime dose of 35 to 100 rem above the unavoidable natural background to double the number of human mutations. To double the number of defective children would certainly be a tragedy, so anything that increases the environmental radiation by a factor of 10 is certainly to be avoided. Twenty-five years of experience with large-scale nuclear reactors in the United States (for power and to make fissile materials) has resulted in no detectable increase in the general background radiation. However, the nuclear generating capacity being built or definitely planned represents more that 16 times that operating at the time of this writing (95 million kilowatts versus 5.8 million kilowatts). Since the technology exists to reduce reactor emissions to a small fraction of their present magnitude, there is no reason to expect that the total radiation contamination of the environment should increase over its present level. And, as we have noted, there is every reason to insist that it does not, even at the price of an estimated one-percent increase in the cost of each generating plant.

10-10 Summary

All of the energy that we observe and use in the world has its origin in nuclear reactions, although energy may pass through many forms and locations before finally appearing as heat.

Nuclear energy arises from the transformation of unstable nuclei into stable ones, with a concomitant conversion of matter into energy. The question

---

[15] Much larger amounts of radiation are required to produce substantial damage. A single whole-body exposure dose of about 400 rad would be fatal to a man, rather quickly. A whole-body dose of 100 to 200 rad would cause severe radiation sickness, a debilitating fever, and nausea produced by the toxic effects of sloughed-off decomposed protein within the body (from which recovery is very slow and intermittent). For people who work with radioactive materials, the usual maximum allowable dose is 3 rem for any period of 13 consecutive weeks for people over the age of 18, with a total of not more than 5 rem per year. Radiations from sources or reagents usually are limited (by shielding) to no more than 100 μc at a distance of 10 cm, and no more than 0.1 μc of α emitter if there is any danger of its being accidentally ingested or inhaled. The student should consult more specific rules which apply in his own college.

of which nuclei are more stable than others, and why, may be approached from six points of view. (1) Examination of terrestrial and cosmic abundances of the elements should reveal which kinds of atoms are stable enough to appear and persist for cosmic lengths of time. The striking feature of cosmic abundances is the predominance of light elements over heavy ones. (2) When unstable nuclei decay, they emit "radiation"—the matter and energy debris left over from the formation of stable nuclei. Four kinds of radiation are commonly observed: α particles, which are $^{4}_{2}$He nuclei; β particles (electrons); γ rays (very-short-wavelength electromagnetic radiation); and neutrons. (3) The probability that a nucleus will react with a passing neutron is a clue to the structure and stability of the nucleus. (4) The energy released when a nucleus reacts with another neutron is a measure of the change in stability on adding the neutron. (5) The mass defect (difference in mass between an actual nucleus and the sum of the masses of its component particles) is a direct measure of the stability of a nucleus, because it is the energy that would be required to decompose the nucleus into its components. Binding energies, equivalent to the mass defect, are best compared by calculating the binding energy *per nucleon*. When these are plotted versus the nuclear mass numbers for all nuclei (Figure 10-2), it is evident that some nuclei represent local islands of stability and that a broad maximum in stability covers the region from mass numbers of approximately 20 to approximately 200. (6) Finally, it is empirically found that certain numerical combinations of nucleons are especially stable; those with even numbers of both neutrons and protons have an edge over even-odd and odd-odd combinations, and of the even numbers, certain ones (2, 8, 20, 28, 50, 82, 126, and 184) appear to confer extra stability and have earned the facetious title "magic numbers."

All of these features of nuclear stability, including the "magic numbers," have their origin in the system of nuclear energy levels (Figure 10-4). The theory of this energy-level chart, like that of electronic energy levels, involves mathematical and physical concepts beyond the range we assume you bring to this book. However, the qualitative notion that such a scheme exists and underlies nuclear reactivity, as atomic orbital structure underlies chemical reactivity, is itself a significant insight.

Nuclear reactions may be represented by equations exactly like chemical equations, in which the atomic number and mass number are shown with the atomic symbol. Because the energy change in nuclear reactions is so large, it results in a measurable mass change and may, in fact, be calculated from the total masses of reactants and products. According to the Bohr theory of nuclear reactions, a nuclear reaction begins with formation of a compound nucleus (for example, by the bombardment of a nucleus with a neutron) that lasts long enough to distribute its excess energy among all of its nucleons. Only when one or several of these happen to accumulate enough of this randomly distributed energy does the unstable compound nucleus disintegrate.

Fission is a nuclear reaction in which an unstable heavy nucleus flies apart into smaller fragments, plus several neutrons. Nuclei of very heavy elements may undergo spontaneous fission; more stable nuclei can be activated to fission by bombardment with neutrons. Uranium 235, especially, will undergo fission if activated by about 6 MeV—just the energy it acquires by binding a thermal neutron.

Since the fission of ^{235}U produces two (unequal) nuclei whose masses are well within the region of a maximum binding energy per nucleon (Figures 10-6 and 10-2), a considerable quantity of energy is evolved during the process: about 200 MeV per event. Also, since lighter nuclei generally require fewer neutrons per proton for stability, some "excess" neutrons are emitted during fission, and these may serve to activate other ^{235}U nuclei to fission, in a chain reaction. Alternatively, the excess neutrons may "breed" other fissile elements.

A fission reactor has five necessary components: (1) fuel, (2) a moderator, which absorbs enough of the kinetic energy of neutrons to increase the probability (cross section) of their reaction with new fissile nuclei, (3) control rods, (4) a coolant, and (5) a neutron reflector and shielding. In addition, a blanket of "breeding material" for the synthesis of new fissile fuel is also generally present to take advantage of the neutron output of the main reaction. These components are shown in Figure 10-7. Every reactor now operating is designed with elaborate and generally effective safety measures built in, to minimize the loss of radioactive material to the environment and to prevent catastrophic failure (which could result from excess heating, not a nuclear explosion).

Although fission reactors are practical energy sources (for example, for the generation of electricity), they have some disadvantages which are avoided by nuclear fusion. The latter has not yet been made a practical source of continouous energy release, except for the energy we get free from the sun. There is reason to hope that controlled, continuous nuclear fusion may become a reality, since it would use hydrogen, an abundant element, as fuel and would produce far less dangerous radioactivity.

We have reviewed the damaging effects of nuclear radiation on simple crystals and on covalent polymers. The principal mechanism of damage to the latter is through the disruption of covalent bonds, with formation of ions. Living matter is much more susceptible to radiation damage, since not only the bulk matter, but also the patterns of molecular organization, may be disrupted. Protection against radiation begins with the availability of simple and reliable instruments for its detection, since even lethal doses of radiation have no effect on the human senses. Radiation is measured in roentgens, a unit based on its ability to produce ions in air. Biological susceptibility to various kinds of radiation is correlated through the *rem* (*r*oentgen, *e*quivalent, *m*an). We have compared the level of naturally occurring (background) radiation to "cultural" sources and have seen that the two are about equal at the present time. Nuclear power reactor emissions represent a very small fraction of the present background radiation. However, nuclear power generation will increase by a large factor in the next few years, and improved reactor design will be required to keep this contribution to the background radiation at a negligible level. Since more efficient and leakproof reactors are possible with existing technology, any increase in total reactor emissions would be needless and inexcusable.

GLOSSARY

barn: a unit of cross section of atomic nuclei, defined specifically as 10^{-24} cm^2.

binding energy: the energy (usually deduced from calculated changes in mass) evolved or absorbed when a given nucleus is formed from protons and neutrons.

breeding: the process by which extra neutrons from fission reactions are absorbed and employed to produce new fissile material.

cross section: probability of a specified nuclear reaction, expressed in barns; probability of capture of a particular particle, expressed as apparent area presented to an oncoming stream of particles.

curie: unit of radioactivity; level of radioactivity corresponding to presence of one gram of radium; 3.7×10^{10} disintegrations per second.

fission: the cleavage of an atomic nucleus into two almost equal parts, spontaneously in some very heavy elements, but usually by bombardment with neutrons or by equivalent activation.

fusion: the combination of two light nuclei to form a heavier one, with consequent emission of a small portion of the total mass as energy.

isomers: in the nuclear sense, nuclei which have the same charge and mass, but differ in structure (and hence in stability).

isotope chart: an exposition of all the known isotopes, both stable and unstable, of all the elements, together with information on mass, half-life, mode of decay, and so forth. There are various ways of organizing the exposition, but the usual arrangement is a plot of atomic number Z (that is, proton number) versus neutron number N, assigning a small square to each isotope.

magic numbers: certain even numbers (2, 8, 20, 28, 50, 82, and 126) of protons or of neutrons associated with exceptional stability of nuclei.

moderator: a substance which slows down fast-moving neutrons by repeated collisions, preferably with little or no absorption of the neutrons.

nuclear reactions: reactions that take place wholly within the nucleus of an atom, hence involving only the elementary particles within the nucleus (protons, neutrons, mesons, and so on), rather than the extranuclear electrons (the rearrangements of which constitute ordinary chemical reactions).

nucleus: that portion of an atom (about 10^{-12} cm in diameter) that contains all of the positive charge and almost all of the mass of the atom. See also Chapter 2.

positron: an elementary particle with the mass (and many other characteristics) of an electron, but with a unit *positive* charge.

radioactivity: the spontaneous disintegration of active or excited atomic nuclei, resulting in the emission of energetic particles or radiation and the formation of a nucleus of different properties.

rem: unit of radiation (*r*oentgen *e*quivalent, *m*an) on a scale designed specifically in terms of radiation damage to human beings.

resonance: a sharp peak in the absorption curve for cross section versus neutron (or other) velocity, indicating greatly enhanced and specific interaction of the target material with the projectile particles.

roentgen: dose of radiation which produces 1.6×10^{12} ion pairs in one gram of air ($\equiv 83$ ergs).

thermal neutrons: neutrons that have not been imparted extra velocity or energy; neutrons that contain only the kinetic energy of thermal motion at room temperature, which is 0.02 to 0.04 eV.

PROBLEMS

10-1 In 1919 Rutherford accomplished the first transmutation of a chemical element: He made oxygen out of nitrogen. He did this by passing a stream of α particles (from natural radioactive material) through a small sample of nitrogen, obtaining ^{17}O and protons:

$$^{14}_{7}\text{N} + ^{4}_{2}\text{He} \rightarrow ^{17}_{8}\text{O} + ^{1}_{1}\text{H}$$

How much energy was associated with this change? Was it evolved or absorbed? (The isotopic masses are: ^{14}N = 14.0030744, ^{4}He = 4.0026036, ^{17}O = 16.999133, and ^{1}H = 1.0078252.)

10-2 Obtain a complete isotope chart and look it over carefully. Inspection of the entire isotope chart reveals six generalizations about the number of stable isotopes of an element and their relative abundance. These generalizations are known as the *stability rules*. Write down as many of these "rules" as you can deduce, listing any exceptions. (Three rules will be considered par; there should be a bonus for any extra "rules" discovered.)

10-3 The iodine ^{131}I used in a laboratory experiment as NaI has a half-life of 8.05 days. When the supply was received on March 3, one drop of the solution sufficed to give 7500 counts per minute. How many drops would be needed to provide the same activity on the date of the last laboratory period, March 17?

10-4 Consider the portion of Table 10-1 between $Z = 34$ and $Z = 83$. By reference to an isotope chart and any appropriate theory of nuclear structure, explain the remarkably high amounts of Rb, Sr, Zr, Sn, Ba, and Pb found in the earth's crust.

10-5 Sulfur has four stable isotopes of mass numbers 32 (the most abundant), 33, 34 (the next most abundant), and 36. The specific isotopic masses of ^{32}S and ^{33}S are 31.972074 and 32.971460. Calculate and compare the binding energies of these two nuclei. Is the result what you would have expected? What does this indicate about the "state of the art" concerning the theory of nuclear structure at present?

10-6 The radioactive decay of natural uranium 238 takes place in many steps, but eventually produces lead 206. The overall reaction may be written

$$^{238}_{92}\text{U} \rightarrow ^{206}_{82}\text{Pb} + 8\,^{4}_{2}\text{He} + 6e^{-}$$

From the masses of the ^{238}U and its products, calculate the energy (in million electron volts) liberated during the reaction. (The isotopic masses are: ^{238}U = 238.0508, ^{206}Pb = 205.97446, ^{4}He = 4.0026036; the mass of the electron is 0.000548597.) The equation seems to imply that negative charge is produced by the reaction; show that in fact the reaction results in overall conservation of electrical neutrality.

10-7 Moeller gives 4 g per ton as the terrestrial abundance of uranium. Considering the distribution to be uniform throughout the earth's mass of 6.6×10^{21} tons, and remembering that one-half of the ^{238}U disappears every 4.5×10^9 yr, how much heat (in kilocalories) is liberated by this particular process each year? (For this purpose, consider the decay to be linear with time, so that a $(0.693)/(14.5 \times 10^9)$th part is used every year. The answer to Problem 10-6 gives the energy in million electron volts per event; the conversion factor from *electron volts per event* to kilocalories per mole is 23.05 kcal/mole/eV.) What do you think about the resulting quantity of energy as a source of heat for activating volcanoes and for keeping the interior of the earth molten? Compare this quantity of energy with that received from the sun each year, and with the energy requirement of all of the people in the world, which is 22×10^{12} horse-power hours per year.

SUGGESTIONS FOR FURTHER READING

TSIVOGLOU, E. C., "Nuclear Power: The Social Conflict," *Environmental Science and Technology*, vol. **5**, pp. 404–410 (May 1971) is a balanced and cool-headed article, reviewing briefly the main sources of radioactive emissions from existing power plants, the means presently used to minimize them, and the prospects for greatly reduced emissions in the future.

SEABORG, G. T., and BLOOM, J. L., "The Synthetic Elements: IV," *Scientific American*, vol. **220**, no. 4, pp. 57–67 (April 1969). Generally speaking, very heavy elements tend to be unstable with respect to spontaneous fission or other decay processes, as we have seen. However, certain nuclei containing magic numbers of neutrons and protons can be imagined that would be heavier than any presently known element. This article describes the predicted stability of these "superheavy" elements and the techniques that may be used to make them. The possibilities for producing such superheavy nuclides in a yet-to-be-constructed heavy-ion accelerator are discussed in: HAMILTON, J. H., and SELLIN, I. A., "Heavy Ions—Looking Ahead," *Physics Today*, vol. **26**, no. 4, pp. 42–49 (April 1973).

The significance of the data in Tables 10-1 and 10-2 is discussed from a geochemical point of view in the paperback: AHRENS, L. H., *Distribution of the Elements in Our Planet*, New York: McGraw-Hill (1965).

The following is a selection of textbooks, beginning with a book written for college freshmen, and continuing with more-advanced works:

OVERMAN, R. T., *Basic Concepts of Nuclear Chemistry*, New York: Reinhold (1963)—a paperback in the series *Selected Topics in Modern Chemistry*.

CHOPPIN, G., *Nuclei and Radioactivity*, New York: W. A. Benjamin (1964)—an introduction to nuclear chemistry at the undergraduate level.

GLASSTONE, S., *Sourcebook on Atomic Energy*, 3rd ed., Princeton, N.J.: Van Nostrand Reinhold (1967).

FRIEDLANDER, G., KENNEDY, J. W., and MILLER, J. M., *Nuclear and Radiochemistry*, 2nd ed., New York: Wiley (1964). Has excellent discussions of nuclear forces and nuclear shell structure.

11
Mechanisms of Reactions

11-1 Abstract

Material change is central to chemistry. Questions about the rates of change and about relative rates of competing changes are asked throughout all of chemistry. These chemical changes are of extraordinary diversity. In this chapter, we shall focus on ways to talk about change while it is occurring. We shall discuss some aspects of change itself, and then concentrate on *rates of chemical reactions.*

Chemical change involving a chunk of matter large enough to see is never an all-or-nothing affair, nor an all-at-once event. Time is always required for the many molecular changes needed to transform a substantial mass of matter. The time may indeed be very short, as in an explosion. The time may be frustratingly long, as in the aging of wine. Or the time may be measured in millennia, as in the metamorphosis of rocks.

Each chemical reaction has its own unique features. Yet there are some common aspects of change in general, and of chemical change in particular, that are worth knowing about. We shall begin our discussion with an analysis of changes of human population on the planet earth. We shall make a model of this population in an abstract way by means of an **exponential function**, examining some of the significant features of the model and noting some of its limitations. Then we shall look at an example of radioactive decay, modeling it also with an exponential function. Finally, we shall examine rates of change in a chemically reacting system, modeling the system both with an exponential function and with an arithmetic model well suited for use with a computer.

Studies of reaction rates can give clues to ways for achieving control and mastery over a chemical system. Interpretations of reaction-rate data can lead to insights into fundamental mechanisms of chemical change. The chapter closes with some current ideas about the molecular mechanisms of chemical transformations. The key words for this chapter are *change, rate, time, model,* and *mechanism.*

11-2 A Model to Simulate Global Population Change

The human population of the earth doubled between 1920 and 1970. Judging from current trends, what population changes are to be expected during the next 50 years? Census data, such as those plotted in Figure 11-1, are not sufficient by themselves to yield numbers for the probable population in some future year. To extend Figure 11-1 into the future, and thus to make forecasts about global population, a *model* is needed. We shall develop a **quantitative** model that yields predictions about populations, both past and future. We shall then see whether this model conforms to the general requirements that must hold for any closed dynamic system. If flaws in the model are detected, we shall look for ways to mend the model.

Does this seem to be a reasonable set of assumptions?

The human population of the earth is increased only by births and is decreased only by deaths, disregarding the comparatively few people arriving and leaving on spaceship trips. Expressing these facts in terms of *rates*, we can write

rate of change of population (11-1)
　　= *rate of increase by births* − *rate of decrease by deaths*

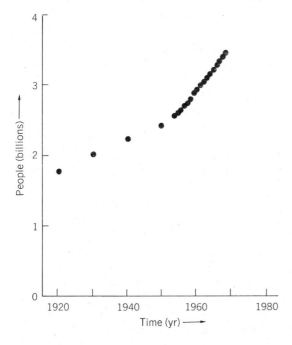

FIGURE 11-1 Population of the planet earth throughout the middle of the twentieth century. (Source: *United Nations Demographic Yearbook,* New York: United Nations, published annually.)

We shall symbolize the number of people on earth at any instant by P, and write the rate of change of P as $\mathscr{R}P$. When the rate $\mathscr{R}P$ is positive, the population is increasing; when $\mathscr{R}P$ is negative, the population is decreasing. Using this shorthand notation, Equation (11-1) becomes

$$\mathscr{R}P = \text{rate of increase by births} - \text{rate of decrease by deaths} \quad (11\text{-}2)$$

You have seen vital statistics given in the form of a number such as "33 births per year per 1000 people." When this number is divided by 1000, the result is 0.033 births per year per person, and we shall denote such a number by the symbol k_{birth}, the number of births per year per person. We shall use the symbol k_{death} for the number of deaths per year per person. We can then write the rates of increase and decrease as

$$\frac{\text{births}}{\text{year}} = \frac{\text{births}}{(\text{year})(\text{person on earth})} \times (\text{number of persons on earth})$$

and

$$\frac{\text{deaths}}{\text{year}} = \frac{\text{deaths}}{(\text{year})(\text{person on earth})} \times (\text{number of persons on earth})$$

That is the same as saying

$$\text{rate of increase by births} = k_{\text{birth}} \times P \tag{11-3}$$

$$\text{rate of decrease by deaths} = k_{\text{death}} \times P \tag{11-4}$$

Putting this all together, that is, combining Equations (11-2), (11-3), and (11-4), we have

$$\mathscr{R}P = \left(k_{\text{birth}} \times P\right) - \left(k_{\text{death}} \times P\right)$$

or

$$\mathscr{R}P = \left(k_{\text{birth}} - k_{\text{death}}\right)P \tag{11-5}$$

Equation (11-5) is a quantitative model for population change on this planet. Let us examine some ways in which we can use this model to simulate the population changes that have occurred in the past and that may be expected in the future.

During the period 1963 to 1969, k_{birth} for the whole earth had the value 0.033 births per year per person, and k_{death} was 0.014 deaths per year per person.[1] Thus, Equation (11-5) becomes

$$\mathscr{R}P = 0.019 \cdot P \tag{11-6}$$

The rate of population change is proportional to the population itself. Thus, if population doubles, the rate of change also doubles, and the population increase accelerates. This is one example of the widespread phenomenon of **exponential change**. Equation (11-6) implies the exponential equation

$$P = P_0 e^{0.019t} \tag{11-7}$$

See Appendix II.

where t is time; P_0 is the population at the time we choose to call "zero time"; and $e = 2.71828\ldots$, the base of natural logarithms. Because Equations (11-6) and (11-7) may look unfamiliar to many readers of this book (and because such pairs of equations are of great importance in studying change, in chemistry and many other fields), it is important to answer some questions: Why use the base e? When is zero time? What are the proper units of time for the variable t?

WHY USE THE BASE e?   The base e was used so that there would be a neat and obvious relationship between the proportionality factor in Equation (11-6) and the exponent in Equation (11-7). Thus, if $\mathscr{R}P = mP$, then $P = P_0 e^{mt}$. Use of any base other than e obscures the fundamental role of m in linking the paired **rate equation** and exponential equation. For instance, written in terms of base 10, we have the pair $\mathscr{R}P = mP$, $P = P_0 10^{(0.43429\ldots)mt}$; this equation appears to be more complicated than when e was used. For additional information, see Appendix II.

---

[1] *United Nations Demographic Yearbook*, New York: United Nations (1970).

WHEN IS ZERO TIME? You can start the clock at zero whenever you choose. Since $e^{m \cdot 0} = e^0 = 1$, the choice of zero time determines the value of P_0. If we decide to count time from 1970, letting 1970 be 0, 1971 be $+1$, 1969 be -1, and so forth, we obtain the equation

$$P = (3.63 \times 10^9)e^{0.019t} \tag{11-8}$$

since the population of the earth at time zero (midway in the year 1970) was 3.63 billion people. When mid-1970 is chosen as time zero, P_0 becomes equal to 3.63×10^9 people.

WHAT ARE THE PROPER UNITS FOR t? Any time units will do, although some units seem especially appropriate for certain cases. Talk about population changes usually focuses on the year (rather than the day, or the century) as a useful time unit. Thus, both k_{birth} and k_{death} are in time units of years. It is important that consistency be observed; both m and t must have the same units for time.

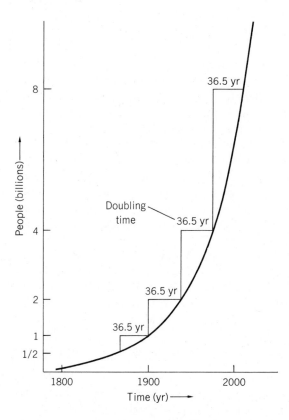

FIGURE 11-2 Simulation of the population of the earth—a plot of Equation (11-7). Pick any point on the graph and read the population; 36.5 yr before, the population was half that value; 36.5 yr later, the population will be twice that value. This simulated graph approaches zero faster than the actual population estimates indicate. The simulation implies no upper limit to population, even though the finite resources of the planet require such a limit. The data plotted in Figure 11-1 fit quite well on this graph.

Let us examine the shape of the simulated population curve. Figure 11-2 is a plot of Equation (11-8), showing population as a function of time during the two centuries between 1800 and 2000. The equation gives a population figure for any year, no matter how far in the past or how far in the future;

the time span chosen is about the longest period that can be conveniently plotted on a graph with a linear population scale. There is no indication on the graph of the time chosen for "time zero" in the calculations; happily, this arbitrary choice has no effect on the graph.

For our model, there is no upper limit to population; according to Equation (11-8), the population increases each successive year at an increasing rate. The time needed for the population to double is 36.5 yr, no matter where on the curve one starts. Indeed, one informative way to characterize the rate of exponential growth is to state the **doubling time**. Even without looking at any census data, we can see that the lack of an upper limit to P is a serious flaw in our population model. The earth is essentially a closed system with respect to exchange of matter; there must be an upper limit to the growth of *any* material aspect of the earth. As an absurdly high upper limit, the requirement of conservation of matter demands that the population stop growing when all carbon, hydrogen, oxygen, or nitrogen on the surface of the earth has been converted into human flesh. Clearly, other factors will limit population long before that time.[2] *In any closed system, whether composed of men or molecules, appreciable material growth cannot continue forever.* We shall have to modify the model so that an upper limit on population is predicted.

What happens when the predictions of this model are confronted with some population data from the past? The agreement as shown in Table 11-1 between data and model is excellent for the two-decade period between 1950 and 1970. However, the model yields numbers that are much too small for 1830 and 1650, and we might guess that the discrepancy would be even greater for earlier times. What can be done to reconcile the model with the population estimates for earlier centuries? Equation (11-2) is probably valid for any conceivable situation. The deficiency in our model arises when, in writing Equations (11-3) and (11-4), we assumed that k_{birth} and k_{death} are time-independent constants, numbers that are unchanging year after year. In earlier centuries, the difference ($k_{birth} - k_{death}$) was certainly smaller than the value of 0.019 that was appropriate for the 1950 to 1970 period.

What can we say about projections into the future? Since growth cannot continue unabated forever, eventually the difference ($k_{birth} - k_{death}$) must become zero or negative. Our model is not sufficiently detailed to give any information about when or how this difference is to change from a positive number, although we could incorporate all sorts of ideas into the model. Each such idea of how either k_{birth} or k_{death} might be expected to change with time would be a proposed *mechanism* for population change on the earth.

---

[2] Population growth can be sustained only so long as living conditions permit a high birth rate and a low death rate. Many factors can affect these two numbers. The conclusion of one team of researchers is that the earth's interlocking resources probably cannot support the rates of economic and population growth of 1970 much beyond the year 2100. Their report—D. H. Meadows, D. L. Meadows, J. Randers, and W. W. Behrens, *The Limits to Growth*, New York: Universe Books (1972)—is a very readable supplement to this chapter.

TABLE 11-1
Confrontation of Predictions of Model With Census Data

Year	Population of the Earth (billions of people)	
	According to the Model Equation (11-7)	According to Census Figures and Other Estimates
1650	0.0083	0.50°
1830	0.25	1.00°
1950	2.48	2.52†
1955	2.72	2.69†
1960	3.00	3.00†
1965	3.29	3.29†
1970	3.63	3.63†
2070	24.3	
2170	162	
2231	515§	
2270	1070	

° W. D. Borrie, *The Growth and Control of World Population*, London: Weidenfeld and Nicholson, p. 19 (1970).
† *United Nations Demographic Yearbook*, New York: United Nations, published annually.
§ This population is equivalent to a population density of the entire land area of the planet equal to the 1969 population density of Hong Kong, 3859 people per square kilometer.

Some conclusions can be drawn. In earlier centuries, there was only a marginal difference between the numbers of births and deaths, and population growth then was relatively slow. In the twentieth century, most dramatically at mid-century, medical advances made possible marked increases in life expectancy throughout the world, reflected in a substantial difference between k_{birth} and k_{death} and therefore a sharp rise in population. If worldwide average fertility and mortality stay the same as at present, then global population can be predicted with Equation (11-8), and the consequences are staggering: With a population doubling time of 36.5 yr, a planet already hard pressed to meet the needs for food, shelter, and clothing for its people would have extraordinary problems within the lifetimes of many of the readers of this book. The only way to slow the population growth and postpone crises is to reduce the difference $(k_{birth} - k_{death})$. The only way to achieve population stability is for this difference to be substantially zero for a long period of time. The social implications of various means of achieving a zero or negative value of $(k_{birth} - k_{death})$ are vast; the quality of life would be very different indeed if k_{birth} were decreased than if k_{death} were increased.

Explain this in your own words.

We have built a model that allows us to talk quantitatively about global population change. The model is based on a rate equation that predicts an exponential dependence of population on time. We saw that the main feature of exponential growth is the doubling time, and that an exponential growth curve with a constant doubling time has no upper limit. We found that the critical number in discussing population statistics in terms of this model is the difference ($k_\text{birth} - k_\text{death}$), and we concluded that a realistic model must allow for this difference to change with time. To predict how this difference depends on time, we would have to assume many details about the factors that influence both births and deaths; these details would constitute a mechanism for population change. We shall see, in the following sections, that analyses of many other types of changing systems involve markedly similar considerations.

11-3 The Production and Decay of Radioactive Carbon: A Key to Dating the Past

Cosmic rays—radiation (including some energetic neutrons) from outer space—give rise in the earth's upper atmosphere to the nuclear reaction

$$^{1}_{0}n + {}^{14}_{7}N \rightarrow {}^{14}_{6}C + {}^{1}_{1}H$$

The carbon thus produced by bombardment of nitrogen nuclei is not the common, abundant isotope ^{12}C (only one in about 10^{12} carbon nuclei in the atmosphere is ^{14}C). The rate of this process of carbon-14 production depends upon the intensity of cosmic radiation and the number of nitrogen nuclei being bombarded. Since both of these factors might be expected to remain reasonably constant, year after year, it would seem likely that the rate of production of ^{14}C has been constant for a long time.

Why does ^{14}C not accumulate, and its concentration steadily increase, if this isotope is being manufactured in the atmosphere at a constant rate? The answer is that production of ^{14}C is opposed by a destructive process. A feedback control mechanism is operative, bringing the processes of production and destruction into a self-adjusting balance. This destruction occurs because the isotope ^{14}C is radioactive, undergoing the nuclear reaction

$$^{14}_{6}C \rightarrow \beta^{-} + {}^{14}_{7}N$$

at a rate that is proportional to the number of ^{14}C atoms in existence. That is, every ^{14}C nucleus has a definite probability (during any particular period of time) of emitting a β particle and transmuting into a ^{14}N nucleus. If there are ten ^{14}C nuclei, then the probability is 10 times greater that there will be a transmutation during the same time period, and so on. Writing these statements in terms of rates, we get an equation for the net production of ^{14}C on the planet:

$$\mathscr{R}[^{14}C] = \text{rate of production of } ^{14}C \text{ by neutron bombardment of } ^{14}N \text{ nuclei} \quad (11\text{-}9)$$
$$- \text{ rate of destruction of } ^{14}C \text{ by transmutation}$$

or

$$\mathscr{R}[^{14}C] = k_1[^{14}N] - k_2[^{14}C] \quad (11\text{-}9)$$

where $[^{14}N]$ is the concentration of ^{14}N nuclei in the atmosphere; $[^{14}C]$ is the concentration of ^{14}C nuclei on the earth; and k_1 and k_2 are proportionality constants. The overall rate $\mathscr{R}[^{14}C]$ is almost exactly zero, and therefore the rate of destruction of ^{14}C (a rate dependent on the intrinsic properties of the unstable ^{14}C nucleus) is balancing the rate of production of ^{14}C (a rate dependent on the composition of the upper atmosphere and the intensity of cosmic radiation). How is this balance achieved and maintained?

The balance is achieved by the operation of a simple control mechanism. The rate of destruction of ^{14}C depends on the amount of ^{14}C already present. The more ^{14}C, the faster the destruction; the less ^{14}C, the slower the destruction. We have, then, a self-regulating mechanism for maintaining a **steady-state** concentration of ^{14}C in the atmosphere. Maintenance of this steady-state composition is a dynamic affair, involving two opposing rates. In our model, we have one constant rate and another variable rate controlled by **negative feedback**.

How is this negative feedback like Le Chatelier's principle (Sec. 11-5)? How is it different?

Transmutation of ^{14}C to ^{14}N can be monitored with radiation detectors[3] that respond to β radiation. The number of β particles emitted within a sample is proportional to the number of ^{14}C nuclei in that sample, giving a convenient and accurate method for determining extremely small percentages of ^{14}C in a small sample. Archaeologists have taken advantage of the ease of detection of ^{14}C radiation, and of the predictable rate of self-destruction of ^{14}C nuclei, in a method for determining the age of ancient artifacts made of wood or other once-living materials. Next, we shall examine some aspects of this method of radiochemical dating.

For more on radiation detectors, see Ch. 10.

Living plants convert carbon dioxide which contains ^{14}C into carbohydrates, proteins, and fats which contain ^{14}C. Since the minute amounts of ^{14}C are carried along in all chemical processes with the predominant ^{12}C isotope, all living cells on earth that obtain their carbon directly from the atmosphere will have nearly the same ratio of atoms, $^{14}C/^{12}C$, as occurs in the atmosphere. If the fraction $^{14}C/^{12}C$ in the atmosphere has been constant during past centuries, then the $^{14}C/^{12}C$ ratio in plant cells alive a thousand years ago must have been the same as the $^{14}C/^{12}C$ ratio in plant cells alive today. When a cell dies, the metabolic incorporation of carbon from CO_2 in its environment ceases; the ^{14}C atoms in a dead but intact cell transmute to nitrogen, and the $^{14}C/^{12}C$ ratio decreases with the passage of time. The decrease in the number of atoms of ^{14}C would be expected to be given by Equation (11-9) without the rate-of-production term. That is, *in a dead cell*,

$$\mathscr{R}[^{14}C] = -k_2[^{14}C] \qquad (11\text{-}10)$$

If this rate equation [Equation (11-10)] faithfully represents the rate of change of $[^{14}C]$, then the concentration itself can be written as the exponential equation

$$[^{14}C] = [^{14}C]_0 \, e^{-k_2 t} \qquad (11\text{-}11)$$

where $[^{14}C]_0$ is the value of $[^{14}C]$ at time zero. The currently accepted value for k_2 is 1.21×10^{-4} yr^{-1}.

---

[3] Commonly used detectors include Geiger-Müller counters and liquid-scintillation counters. An authoritative source of information about the assay of ^{14}C is V. F. Raaen, G. A. Ropp, and H. P. Raaen, *Carbon-14*, New York: McGraw-Hill, chap. 11 (1968).

A bristlecone pine tree provides an independent source of extensive data for evaluating the radiocarbon dating method. Bristlecone pines more than 4500 yr old are alive in California's White Mountains, and in those trees is some ^{14}C that has been trapped for 4500 yr. After each year of tree growth, a new layer of wood remains—an annual ring of newly dead plant cells. Imagine such an annual ring produced in the year 2500 B.C. In 2500 B.C., the $^{14}C/^{12}C$ ratio should have been 1.0×10^{-12}, the same in the living plant cells as in the atmosphere. Thereafter, ^{14}C in the dead cells would have been expected to disappear at the rate given by Equation (11-10)—to disappear with a time dependence given by Equation (11-11). Figure 11-3 is a plot, calculated from Equation (11-11), that simulates the predicated decay of ^{14}C and therefore the decay of the radioactivity due to ^{14}C in that tree ring. By the year 3245 A.D., half of the original ^{14}C will have disappeared, and one **half-life** of the radioactive decay process will have passed. The half-life of ^{14}C is 5745 ± 50 yr, meaning that each passage of 5745 yr halves the number of ^{14}C atoms. *Half-life* is an informative number for characterizing the rate of exponential decay. If each tree ring began with a $^{14}C/^{12}C$ ratio of 1.0×10^{-12}, then we can tell its age today by determining its present $^{14}C/^{12}C$ ratio, and using Equation (11-11). The lower the $^{14}C/^{12}C$ ratio, the older the tree ring.

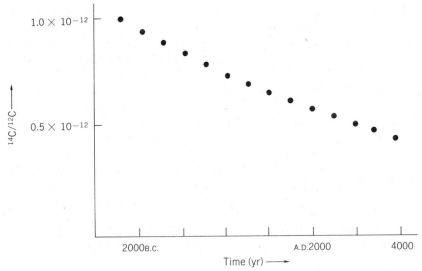

FIGURE 11-3 Prediction of the radioactive decay of ^{14}C trapped in an inner growth ring (wood in a cell that died in 2000 B.C.) of a bristlecone pine tree still alive—a plot of Equation (11-10).

The fantastic life span of bristlecone pines makes possible a check on the reliability of the carbon-14 dating technique, and thus on the model that we have been developing. By counting annual growth rings, the exact age of samples of wood from within the trunk can be determined; the same sample can then be dated by determining its $^{14}C/^{12}C$ ratio. Rather good agreement has been found[4] all the way back to 1500 B.C., but for earlier times there is a significant discrepancy that amounts to 700 yr by 2500 B.C. Present indications are that the intensity of cosmic radiation (and consequently the rate of production of ^{14}C in the atmosphere) has fluctuated in the past and that our model could be improved by including time-dependent variations of k_1 in Equation (11-9).

This model for radioactive decay is not a special model applicable to ^{14}C alone. Equations like (11-10) and (11-11) apply to the radioactive decay of ^{7}Be (half-life 53.6 days), ^{8}Be (half-life 10^{-16} sec), ^{10}Be (half-life 2.5×10^6 yr), and to all other radioactive isotopes. Because of the wide applicability of the model, it makes sense to write it in a general way. For instance, we can write a general reaction

$$A \xrightarrow{k} B \qquad (11\text{-}12)$$

We might call this a radioactivity decay law.

where A and B are unspecified isotopes, and k is a proportionality constant, which we shall call the **rate constant** for the nuclear disintegration. The rate equation for the reaction is

$$\mathscr{R}[A] = -k[A] \qquad (11\text{-}13)$$

Paired with this rate equation is an equation that gives the concentration $[A]$ as a function of time:

$$[A] = [A]_0 e^{-kt} \qquad (11\text{-}14)$$

The value of $[A]$ at time zero is $[A]_0$.

The experimental decay curve is characterized by a particular value of the half-life (here symbolized by $t_{1/2}$), and $t_{1/2}$ is closely related to the value of the rate constant k. As you can verify by working through Problem 11-6, the half-life is related to the rate constant by the equation

$$t_{1/2} = \frac{0.693}{k} \qquad (11\text{-}15)$$

Thus, we see that an experimental determination of the half-life for the radioactive decay of an isotope is a way to find the value of the rate constant. It also turns out[5] that the average life expectancy of a nucleus is equal to $1/k$. Equation (11-14) is plotted in Figure 11-4 for four half-lives to show the characteristic shape of an exponential decay curve. For this illustration, $t_{1/2}$ was chosen to be 1.00 hr, and therefore k has the value 0.693 hr^{-1}. Each nucleus, at any instant, has a life expectancy of 1.44 hr.

---

[4] Methods of correcting carbon-14 dates to match the bristlecone scale, as well as a discussion of how accurate dating of ancient artifacts is contributing to our view of prehistoric Europe, are given in C. Renfrew, "Carbon 14 and the Prehistory of Europe," *Scientific American*, vol. **225**, no. 4, p. 63 (October 1971). See also W. F. Libby, "The Radiocarbon Clock," *Chemical Technology*, vol. **3**, p. 142 (1973).

[5] See G. Friedlander, J. W. Kennedy, and J. M. Miller, *Nuclear and Radiochemistry*, 2nd ed., p. 69, New York: Wiley (1964).

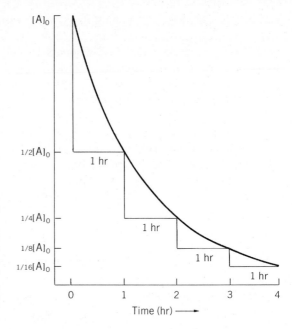

FIGURE 11-4 An exponential decay curve—a plot of Equation (11-13) with $k = 0.693$ hr^{-1}. Pick any point on the graph and read the value of [A]; one hour later, the concentration will be half that value; one hour earlier, the concentration was twice that value.

Just as with global population growth, our discussion of the production and decay of radioactive carbon has involved the formulation of an algebraic model with a rate equation whose solution is an exponential function. If we had known only that ^{14}C was produced by cosmic radiation, but had not realized that ^{14}C is radioactive, then our model would have included only the production process, and we would have predicted unrestrained exponential growth until the nitrogen content of the atmosphere became depleted. (Such a curve, when plenty of nitrogen is present, would look like the population curve in Figure 11-2.) If we had considered only the destruction of ^{14}C by its spontaneous radioactive decay, ignoring the process of production, then we would have predicted exponential decrease in [^{14}C], with a decay curve that would have the same shape as the plot in Figure 11-4. In fact, the destruction of ^{14}C via radioactive disintegration restrains growth and allows this isotope to be maintained in the atmosphere at a low but nearly constant steady-state concentration. The spontaneous decay of ^{14}C occurs continuously, but the supply of ^{14}C is steadily replenished by the production of ^{14}C in the upper atmosphere.

Thus, we have seen that one way to achieve long-term stability in a dynamic system is to have two opposing processes,[6] with the rate of at least one process controlled by negative feedback. We shall find in the next section that this is the way that equilibrium is achieved and maintained in chemically reacting systems. As in our earlier model of global population, we set up a model on the basis of limited data, and then compared the predictions of the model with extensive observations. The model was found to be imperfect,

---

[6] It is tempting to call these two processes "birth of carbon-14" and "death of carbon-14," and to speak of the feedback control as a mechanism for stabilizing "carbon-14 population."

but this discovery was just another step along the pathway to additional discovery and toward perfecting the model. The model gives us a picture of some fundamental physical and chemical processes, while at the same time being very useful for assigning ages to ancient objects. Finally, we observed that the overall model that describes superimposed production and destruction of ^{14}C has parts that can be studied separately. The decay of isolated samples of radioactive isotopes can be studied in the laboratory, and there is presently great confidence in the validity of the generalized radioactive decay model presented in Equations (11-12) to (11-15). Excellent agreement has been found between the predictions of this model and laboratory data, particularly for isotopes with short half-lives so that reliable data can be accumulated in reasonable lengths of time.

11-4 A Model for a Chemical Reaction

How do molecules of a substance react with molecules of a second substance to produce molecules of yet a third substance? Chemists have speculated about the mechanisms of such molecular transformations and have devised experiments to test the predictions of various alternative mechanisms in order to find evidence for choosing among such alternatives. Many of these experiments involve measurements of the *rates of chemical change*. The speculation about a particular reaction is formalized in terms of a chemical model called the **reaction mechanism**. We shall examine ways in which a chemical reaction mechanism, used as a model for the real system, can be used to simulate the time behavior of the reacting system. We shall also see how to test this proposed mechanism against the reality of experimental data. And we shall see the exponential function again!

Distinguish between a reaction rate and a reaction mechanism.

The study of reaction mechanisms—the field of *chemical kinetics*—is the study of chemical systems while they are reacting. Time is an important factor. A fundamental goal is the description of chemically reacting systems in terms of elementary chemical processes that are simple and believable. Often such an analysis reveals how a complex chemical system can be studied in terms of simple subsystems, reducing a complicated research problem to a set of manageably small projects. We shall look next at ways to tackle one such small project.

Consider the reaction in a solution[7] in which the overall chemical change is

$$D + E = X$$

where D, E, and X symbolize chemical substances. We are not going to be specific and say which particular ones; let us say only that it is *known* experimentally that quantities of substances D and E react to form quantities of

---

[7] Chemical reactions also occur in gases, solids, mixtures of immiscible liquids, and in systems containing solids, liquids, and gases. Important and interesting reactions take place in a baking loaf of bread, in a living cell, in a burning flame, and in lake-bottom mud. Given a choice, a chemist will probably try to find conditions so that a reaction can be studied in the most convenient laboratory situation. Reaction-rate studies in solution, without precipitation of solids or evolution of gases, are especially neat and convenient.

X. There must surely be some mechanism whereby individual molecules of D and E are transformed into molecules of X. *Suppose* that the actual molecular events in this transformation involve a meandering molecule of species D encountering a molecule of E. During this chance encounter, the closely associated D and E molecules get involved in some electronic rearrangements, and the result is the formation of a molecule of X. This **elementary chemical process** can be written as

$$D + E \rightarrow X \qquad (11\text{-}16)$$

Also, distinguish between the overall change and the proposed mechanism.

Process (11-16) is a model for what *might* be the reaction mechanism. If process (11-16) really were the actual mechanism, then we could make some definite and reliable predictions about how the reaction would proceed, and how the reaction would approach equilibrium. We shall make some predictions and then see how to arrange a confrontation between these predictions and the data from experiments. Right now, as a working hypothesis, we shall assume process (11-16) and examine the quantitative consequences of this model for the reaction.

Not every encounter of D and E molecules results in formation of a molecule of X. But under a particular set of reaction conditions, there is a definite value for the probability that a particular encounter will be a reactive encounter. The rate of formation of X is proportional to the number of encounters per second. For random movement of molecules of D and E through the solution, there will be molecular encounters of D and E at a rate such that

The number of encounters per second is proportional to [D]

and

The number of encounters per second is proportional to [E]

where the quantities [D] and [E] are the concentrations of species D and E expressed in moles of D or E per liter of solution. The pair of proportionality conditions can be replaced by the single statement that

The number of encounters per second is proportional to [D] × [E] *(that is, to* [D] *multiplied by* [E])

We have set two requirements for the actual occurrence of this process: (1), there must be an encounter between a molecule of D and a molecule of E in order that there be possibility of a reaction, and (2), after the molecules have gotten together, there must be an appropriate (perhaps very complicated and intricate) rearrangement of electrons in order for the collision pair to become a product molecule X. No matter how involved and complicated the details of the conversion of a collision pair to a molecule of X, there is a definite probability that a given encounter will lead to formation of X. These requirements together demand that

Rate of formation of X is proportional to the number of encounters per second

In order to satisfy all of these proportionality requirements, we write the equation

$$\mathscr{R}[X]_{process\ (11\text{-}16)} = k_1[D][E] \quad (11\text{-}17)$$

The proportionality constant that relates the rate $\mathscr{R}[X]_{process\ (11\text{-}16)}$ to the two concentrations [D] and [E] is called a **bimolecular rate constant**[8] and is symbolized by k_1. The subscript "process (11-16)" indicates that the rate is associated with the single elementary process and is included only for clarity.

The rate constant k_1 is the number required to make Equation (11-17) a valid equation for a particular reaction. Scrambled together in this one number are all the details that make one bimolecular process different from all the countless others: processes with different reactants, in different solvents, at different temperatures. If we were to try to predict the value of the number, we would need to consider a myriad of details about the movement of reactants through solution, about the many possible geometries of a two-molecule encounter, about the ways in which an encounter can result in sharing and rearrangement of electrons so that old bonds can be cleaved, new bonds formed, and a new molecular species actually produced. Chemists' present knowledge of these various details of an elementary process in solution is exceedingly incomplete. We do know that rate constants get larger (and the corresponding elementary processes get faster) when temperature increases. Often the effect is impressively large, with many reaction rates doubling when temperature is increased by only about 10 degrees.

Equation (11-17) describes numerically how the rate of the forward process depends upon concentrations. This forward process gets D and E converted into X. We have wrapped up all the details about *how* the process takes place into a single number represented by the symbol k_1 and given it the name "rate constant." We shall return later to this rate constant and considerations of what we can do with it (Section 11-5), how we can measure it (Section 11-6), and what it means (Section 11-8). But first we need to complete our picture of the reaction of D, E, and X by including the reverse process, the conversion of X into D and E.

Process (11-17), if continued long enough, would convert *all* of either D or E (whichever is present in smaller concentration) into X. But we know (recall Chapter 7) that at equilibrium, there will be *some* of each of the

One can write down (and even measure) a rate constant in complete ignorance of the reasons for its particular value.

---

[8] Process (11-16) is called a *bi*molecular process because there are two reactant molecules participating in the process prior to the chemical transformation. Thus, we can imagine various elementary processes:

$A + A \rightarrow B$	bimolecular process
$A + B \rightarrow C$	bimolecular process
$A + B + C \rightarrow D + E$	termolecular process
$A + B + B \rightarrow C$	termolecular process
$C \rightarrow D$	unimolecular process
$C \rightarrow D + E$	unimolecular process

reactant and product molecules. In fact, at equilibrium the concentrations of D, E, and X must satisfy the equation

$$K = \frac{[X]_\infty}{[D]_\infty [E]_\infty} \tag{11-18}$$

where K is the equilibrium constant for the reaction, and the subscript ∞ reminds us that these are "infinite-time" or equilibrium concentrations. The proper equilibrium state can be achieved by including in the model the reverse process[9]

$$D + E \leftarrow X \tag{11-19}$$

The rate of consumption of X via process (11-19) is proportional to the number of molecules of X in solution, and therefore is proportional to the molar concentration [X]. Thus, for the reverse process, we may write the rate equation

$$(-\mathscr{R}[X]_{\text{process (11-19)}}) = k_{-1}[X] \tag{11-20}$$

where k_{-1} is the **unimolecular rate constant** for the process. Note that this rate (including the minus sign) is the rate of *consumption* of species X; if the sign were changed, it would be the rate of *formation* of X. The subscript on the rate is a reminder that this rate considers only the single elementary process.

Now we need to put it all together. Neither elementary process can occur by itself in isolation. In the real microscopic world in which molecules D, E, and X are moving through solution, colliding with solvent molecules and with one another, and occasionally reacting, the two elementary processes are occurring simultaneously. The actual **elementary chemical reaction** is a superposition of the two opposing elementary processes. These two elementary processes are usually written together with a pair of reaction arrows, and each rate constant is then written with the appropriate arrow. Thus, the chemical model—the chemical reaction mechanism—is written as

$$D + E \underset{k_{-1}}{\overset{k_1}{\rightleftharpoons}} X \tag{11-21}$$

The overall net rate of change of the concentration of species X via this elementary reaction is found by superimposing the two opposing processes: the formation and the consumption of X. The net rate of these two simultaneous processes is the sum of the rates for the separate processes. Thus,

$$\mathscr{R}[X]_{\text{reaction (11-21)}} = \mathscr{R}[X]_{\text{process (11-16)}} + \mathscr{R}[X]_{\text{process (11-19)}} \tag{11-22}$$

---

[9] The assertion that there is a reverse process corresponding to each elementary chemical process is called the *principle of microscopic reversibility*. The forward and reverse elementary processes together comprise a reversible elementary chemical reaction that can bring about chemical equilibrium between reactants and products. Early in a reaction, the forward and reverse processes may have very different rates. At equilibrium, though, the elementary reaction must balance itself, so that the average rate of the forward process is equal to the average rate of the reverse process.

This means that the rate of formation of X via the elementary reaction is found by substituting expressions for the rates of the individual processes (rates with subscripts indicating *processes*) from Equations (11-17) and (11-20) into Equation (11-22). The result is an equation in which the only rate has a *reaction* subscript. This overall reaction rate for the formation of species X is

$$\mathscr{R}[X]_{\text{reaction (11-21)}} = k_1[D][E] - k_{-1}[X]$$

Since this rate applies to the entire mechanism, there is no longer any need (in the name of clarity) to use the qualifying subscript. Without the subscript on the rate, we have

$$\mathscr{R}[X] = k_1[D][E] - k_{-1}[X] \tag{11-23}$$

Following the same procedure, but focusing on the rates of change of [D] and [E], we can get

$$\mathscr{R}[D] = -k_1[D][E] + k_{-1}[X] \tag{11-24}$$

$$\mathscr{R}[E] = -k_1[D][E] + k_{-1}[X] \tag{11-25}$$

We now need to find out what this quantitative model implies.

11-5 Computer Simulation of Our Chemical-Reaction Model

How would the concentration of X be expected to change with time if our model were operative? This prediction (and other related predictions) can be made, using only pencil, paper, and elementary-school arithmetic. Actually, the arithmetic is repetitive and boring, and the task is badly suited to the temperament of most people, but ideal for a small digital computer. We shall look at these calculations later in this section. First, though, let us recall some facts about rates and velocities that apply to chemical reactions, but that you may have learned first when planning a hike or a bicycle trip.

A way to calculate the velocity of a bicycle that is moving at a constant velocity is to divide the distance traveled in some time interval by the length of that time interval. That is, to calculate the rate of change of the bicycle's position, you perform the division $\delta x/\delta t$, where δx is the change of position during the time interval δt (δ is the Greek letter delta, here used to mean "difference"). If the bicycle's velocity is changing, this procedure gives a good estimate of its velocity at any instant if the time interval is sufficiently small.

Similarly, for sufficiently small time intervals, it is an excellent approximation to write

$$\mathscr{R}[X] = \frac{\delta[X]}{\delta t}$$

where $\delta[X]$ is the change in the value of [X] that occurs during the time interval δt. The model of the reaction [Equations (11-23) to (11-25)] can thus be written as

$$\delta[D] = \{-k_1[D][E] + k_{-1}[X]\}\delta t \tag{11-26}$$

$$\delta[E] = \{-k_1[D][E] + k_{-1}[X]\}\delta t \tag{11-27}$$

$$\delta[X] = \{k_1[D][E] - k_{-1}[X]\}\delta t \tag{11-28}$$

Each equation gives the change in a concentration that occurs during a short time interval. After this time interval, each concentration has a new value given by

$$[D]_{new} = [D]_{old} + \delta[D] \tag{11-29}$$

$$[E]_{new} = [E]_{old} + \delta[E] \tag{11-30}$$

$$[X]_{new} = [X]_{old} + \delta[X] \tag{11-31}$$

These new concentration values are the concentrations at the new time,

$$t_{new} = t_{old} + \delta t \tag{11-32}$$

The computation procedure is straightforward. We start with initial values for the concentrations [D], [E], and [X]. We then calculate the three changes $\delta[D]$, $\delta[E]$, and $\delta[X]$, which are in turn used to calculate three new values of [D], [E], and [X]. With these updated new concentrations, we calculate another set of concentration changes, in turn used to calculate new values of concentrations, and so on, and so on.

TABLE 11-2
Simulation of a Chemical Reaction

t	[D]	$\delta[D]$	[E]	$\delta[E]$	[X]	$\delta[X]$
0.010	0.999971	−0.000030	0.009970	−0.000030	0.000030	0.000030
0.020	0.999941	−0.000029	0.009940	−0.000029	0.000059	0.000029
0.030	0.999911	−0.000029	0.009910	−0.000029	0.000089	0.000029
0.040	0.999882	−0.000029	0.009881	−0.000029	0.000118	0.000029
0.050	0.999852	−0.000029	0.009852	−0.000029	0.000147	0.000029
0.060	0.999823	−0.000028	0.009823	−0.000028	0.000176	0.000028
0.070	0.999795	−0.000028	0.009794	−0.000028	0.000205	0.000028
0.080	0.999766	−0.000028	0.009765	−0.000028	0.000234	0.000028
0.090	0.999738	−0.000028	0.009737	−0.000028	0.000262	0.000028
0.100	0.999709	−0.000028	0.009709	−0.000028	0.000290	0.000028
0.110	0.999681	−0.000027	0.009681	−0.000027	0.000318	0.000027
0.120	0.999654	−0.000027	0.009653	−0.000027	0.000345	0.000027
0.130	0.999626	−0.000027	0.009626	−0.000027	0.000373	0.000027
0.140	0.999598	−0.000027	0.009598	−0.000027	0.000401	0.000027
0.150	0.999571	−0.000027	0.009571	−0.000027	0.000428	0.000027
0.160	0.999544	−0.000026	0.009544	−0.000026	0.000455	0.000026
0.170	0.999517	−0.000026	0.009517	−0.000026	0.000382	0.000026
0.180	0.999491	−0.000026	0.009491	−0.000026	0.000508	0.000026
0.190	0.999464	−0.000026	0.009464	−0.000026	0.000535	0.000026
0.200	0.999438	−0.000026	0.009438	−0.000026	0.000561	0.000026
0.210	0.999412	−0.000026	0.009412	−0.000026	0.000587	0.000026
0.220	0.999386	−0.000025	0.009386	−0.000025	0.000613	0.000025
0.230	0.999360	−0.000025	0.009360	−0.000025	0.000639	0.000025

The first few numbers from a computer simulation of a chemical reaction proceeding via mechanism (11-21). The algebraic equations used were Equations (11-26) to (11-32). The initial concentrations were as follows: [D] = 1.000 mole/liter; [E] = 0.0100 mole/liter; [X] = 0.0000 mole/liter. Values of the rate constants used for this simulation were: k_1 = 0.30 sec^{-1} · (mole/liter)$^{-1}$ k_{-1} = 0.40 sec^{-1}.

If you had time enough and patience enough, you could do these calculations a thousand times with pencil and paper, and you would have a thousand values for each of the concentrations at a thousand successively larger (but larger by only very small steps) values of t. A better method, less tiring and boring, less subject to human mistakes, and much more likely to get attempted and completed, is to let an electronic computer do the repetitive arithmetic. The first few numbers from one such computer calculation are presented in Table 11-2. Plots of a thousand pairs $\{t, [E]\}$ and another thousand pairs $\{t, X]\}$ are shown in Figure 11-5. All the computer did was the tedious, repetitive arithmetic. It took our algebraic equations and our numerical values for the rate constants and initial concentrations, and it left the chemical interpretations for us.

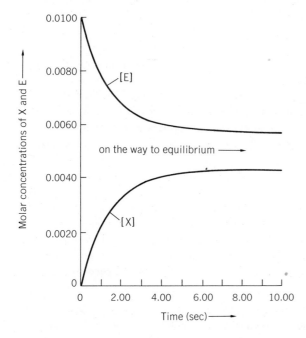

FIGURE 11-5 Computer simulation of a chemical reaction. Plots of a thousand pairs $\{t, [E]\}$ and a thousand pairs $\{t, [X]\}$, calculated from the mathematical model for the mechanism $D + E \rightleftarrows X$. The first few plotted numbers are given in Table 11-2.

To make the calculations, we first had to be specific about the numerical values of the rate constants and the numerical values of the initial concentrations. Later in this chapter, we shall see how values of rate constants can be obtained for a real chemical system from experimental data; right now, as we investigate the sorts of behavior that would be expected for a mechanism, we are free to try all sorts of numbers to see what will happen. For our simulation here, we chose an initial concentration of D much greater than the initial value of [E], so that [D] will undergo less than one-percent change during the reaction.

If this paragraph is puzzling, review Ch. 7.

Now for some chemical interpretations of the two plots in Figure 11-5. Each concentration approaches an equilibrium value. In fact, the whole collection of chemical species is headed for an equilibrium state in which, even though two elementary processes continue to transform molecules, all concentrations have constant values. Matter is conserved in this chemical reaction; at the same time that D and E are being consumed, X is being produced. If the rate constants had been larger, the approach to equilibrium would have been faster. If the ratio k_1/k_{-1} had been larger, the equilibrium state would have had more X and less E.

Initially, there was E in solution, but no X. Such a state is unstable, and the chemical system will change to a more stable state if there exists a mechanism for relieving the stress of nonequilibrium concentrations. For our mechanism, as for any chemically acceptable mechanism, the change must be in the direction of the relaxed state of equilibrium concentrations. The steady, continuous, monotonic, sure change of the concentrations [X] and [E] toward their equilibrium values is called a **relaxation.** After 10 sec, the system is quite close to equilibrium, and the values of [X] and [E] have almost reached their permanent, unchanging equilibrium values. Even though the reaction is reversible at the microscopic level of molecular transformations, the macroscopic change from nonequilibrium state toward equilibrium state is irreversible. Molecules continue forever to interconvert by the two-way molecular reaction mechanism, but the system evolves along a one-way pathway toward equilibrium. This system can never spontaneously move backward away from equilibrium; such a change would be a movement backward in time.

Negative feedback operates in this mechanism to counteract any displacement from equilibrium. Equation (11-23), for instance, shows that more X is produced whenever $k_1[D][E]$ is greater than $k_{-1}[X]$, but that X is consumed whenever $k_1[D][E]$ is less than $k_{-1}[X]$. The equilibrium concentrations —the concentrations that result in no change in [X]—are the concentrations that make $k_1[D][E]$ equal to $k_{-1}[X]$. Any change from these concentration values produces an imbalance in the rates of forward and reverse processes that restores equilibrium.

Exercise: Show that the requirement that $k_1[D][E]$ be equal to $k_{-1}[X]$ at equilibrium means that the ratio k_1/k_{-1} is equal to the equilibrium constant K in Equation (11-18).

There are many relaxation phenomena in the world, and most relaxations can be described by using the exponential function. A hot cup of coffee cooling toward room temperature is a nonequilibrium system relaxing toward equilibrium. The progress of the relaxation can be followed with a thermometer. A plot of temperature versus time represents this relaxation, and that plot is an exponential curve (see Figure 11-6). Consider also another nonequilibrium system: two upright pipes, connected by a small orifice at the bottom, with water initially in one pipe only (see Figure 11-7). As time passes, water flows through the orifice, leaving the left side and entering the right side, until

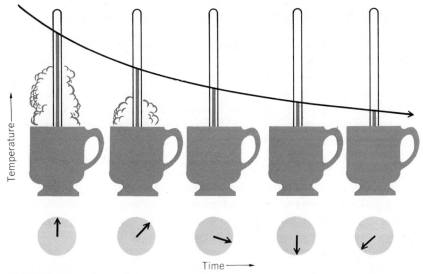

FIGURE 11-6 An exponential relaxation phenomenon: a hot cup of coffee cooling toward room temperature.

an equilibrium state is achieved. This relaxation—this approach to equilibrium—can also be described by an exponential function. We shall take our cues from these relaxing systems and from our earlier experience with changing systems and *try* an exponential function for describing our chemical relaxation. This will be part of our strategy as we see solutions for rate equations in Section 11-6.

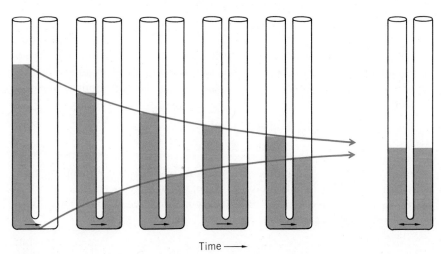

FIGURE 11-7 Another exponential relaxation phenomenon: water falling and rising in two interconnected pipes, finally reaching a state of equilibrium in which the two levels are the same.

11-6 Evaluation of Rate Constants from Experimental Data

There is no way to know the appropriate numbers to assign to k_1 and k_{-1} unless some rate-of-reaction data from experiments have been obtained and analyzed. But if we did have some experimental data, perhaps concentrations determined at various times and plotted like the graphs in Figure 11-5, what would we do with them? How could we find the values needed for the two k's in the computer simulation if we had available only the assumed quantitative model and the experimental plots? Our strategy here is to find an algebraic solution to the equations of the quantitative model, to utilize some of the properties of that solution to give us a graphical method for analyzing plots of concentration versus time, and thereby to evaluate the k's.

A solution to the set of rate equations is an equation (without rates) giving the *concentration* of some species as a function of time. A graph of such a solution must look like one of the plots in Figure 11-5. In fact, Figure 11-5 displays graphs of our desired solutions for a particular set of k's and initial concentrations. But we are also looking for the solutions written out in algebraic form, so that we can work on and with them.

To illustrate this procedure, we shall obtain results for one special case: the case in which the initial concentration of D (signified by $[D]_0$, the subscript denoting zero time) is much greater than $[E]_0$, thus requiring that $[D]$ be almost unchanged in value from the beginning of the reaction all the way to equilibrium. This condition was met in the simulation. With this approximation, the three equations of the quantitative model [Equations (11-23) to (11-25)] become

If $[O]_0$ is much greater than $[E]_0$, then nearly all of it would be left at the end of the reaction.

$$[D] = [D]_0, \text{ always constant} \tag{11-33}$$

$$\mathscr{R}[E] = -\{k_1[D]_0\}[E] + k_{-1}[X] \tag{11-34}$$

$$\mathscr{R}[X] = \{k_1[D]_0\}[E] - k_{-1}[X] \tag{11-35}$$

Because of the marked similarity between the computer-calculated curves of the simulation and the exponential function, we might expect an exponential equation as a solution. In fact, the equations giving concentrations as functions of time are

$$[E] = E_\infty + E_m e^{-mt} \tag{11-36}$$

$$[X] = X_\infty + X_m e^{-mt} \tag{11-37}$$

where X_∞, X_m, E_∞, E_m, and m are constants (they do not depend on time, although they all do depend on the values of the initial concentrations), whose particular values give a rate curve its own distinctive shape. The value of m is related to the values of the k's by the equation

$$m = k_1[D]_0 + k_{-1} \tag{11-38}$$

Equations (11-36) to (11-38) have important significance as we seek ways of relating changes in macroscopic properties of a chemically reacting system to the details of microscopic transformations of reactant molecules and product molecules. Equations (11-36) and (11-37) tell how macroscopic properties (molar concentrations of E and X) would change with time if the assumed reaction mechanism were operative. The key quantity in these macroscopic equations is m, called the **macroscopic rate constant.** Equation (11-38) tells how the macroscopic rate constant, a quantity obtained by analysis of the experimental rate data, is related to k_1 and k_{-1}, the two **microscopic rate constants** of the proposed mechanism. We anticipate that an experimental rate curve of [E] versus t will be a graph of Equation (11-36), because that is the prediction of the model that provides the framework for our interpretation of the data. [If Equation (11-36) is not a good representation of the data, then the validity of the proposed mechanism is brought into question.]

If indeed the experimental data do fit Equation (11-36), then there is a neat method[10] for obtaining a value of m. We define the concentration of E at time t by the symbol $[E]_t$, and the concentration at a later time $(t + \tau)$ by the symbol $[E]_{t+\tau}$. (τ is the Greek letter tau.) It turns out that a plot of $[E]_t$ versus $[E]_{t+\tau}$ will be a straight line with slope equal to $e^{m\tau}$. The concentration values used are for a series of different values of t, but always the same τ. In using this method, data are arranged in a table such as

The way to obtain rate constants from laboratory data

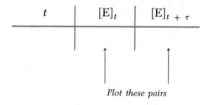

Plot these pairs

The natural logarithm (see Appendix II) of the slope is $m\tau$. Since we already know τ, we can calculate m. This graphical method is a way to obtain a numerical value for m from the changing values of concentration observed experimentally during a reaction. Because we plot two numbers, one lagging the other, this method is called the *time-lag method.*

If we now perform a series of rate experiments, identical except that the value of $[D]_0$ is different in each case, and analyze each experiment by making a plot according to the time-lag method, we can obtain a table of numbers

Equation (11-38) predicts that a plot of m versus $[D]_0$ will be a straight line of slope k_1. The intercept of the line (the intersection of the line with the $[D]_0 = 0$ axis) is k_{-1}. We have worked out a procedure for using macroscopic data from the laboratory to obtain numerical values for both microscopic rate constants of the assumed reaction mechanism.

---

[10] F. J. Kezdy, J. Jaz, and A. Bruylants, *Bulletin des Sociétés Chimiques Belges,* vol. **67**, p. 687 (1958); P. C. Mangelsdorf, *Journal of Applied Physics,* vol. **30**, p. 442 (1959); E. S. Swinbourne, *Journal of the Chemical Society* (London), vol. **1960**, p. 2371.

All results in Section 11-6 depend on one critical aspect of experimental design. All the rate experiments were performed under conditions such that the value of [D] could legitimately be considered constant. If [D] had been a time-dependent variable, then Equations (11-36) and (11-37) would not have resulted, and data analysis would not have been as straightforward. Thus, careful experimental design made possible mathematical simplification, which, in turn, led to straightforward ways of analyzing data. From this set of experiments, we see how to complete the model by finding numbers for k_1 and k_{-1}. Predictions by computer simulations could then be made for any set of initial concentrations of reactants and product. If the model is valid, the predictions will be good predictions.

11-7 Testing a Reaction Mechanism

Interpretation of experimental rate data cannot proceed unless someone has first proposed a mechanism as a working hypothesis. This mechanism then serves as the conceptual framework for analyzing the data, and for thinking and talking about the reaction. Imaginative chemists often can think of several mechanisms that seem to be reasonable possibilities. Then the predictions of each mechanism are compared in turn with the observed data. To be an acceptable mechanism, the chemical and mathematical model considered in this chapter would have to pass several tests of consistency with experimental data. The concentration-versus-time curves must have the shape predicted by Equations (11-36) and (11-37). The time-lag plot must be a straight line. And a plot of Equation (11-38) must be a straight line.

If reliable experimental data are inconsistent with these predictions, then a new or modified mechanism must be tried. One goal is to design experiments that will distinguish among rival mechanisms, permitting a chemist to discard all but one of the mechanisms that have been proposed. Hopefully, one would end up with a single chemical model that is faithful to all available data. This would be a reliable, useful model with a high probability of yielding accurate predictions for experiments as yet untried.

But although a mechanism may be faithful to the data, that mechanism cannot be *proved* to be real or true. There are indeed very few chemical reactions whose mechanisms are confidently said to be established beyond a reasonable doubt. And even one of these established mechanisms may fall victim to a series of carefully planned, carefully executed experiments, experiments that yield data inconsistent with the predictions of the mechanism.

11-8 Little Rate Constants, Big Rate Constants, and Activation Energies

Why are some reactions fast, and others slow? Why does a change in temperature change the rate of reaction? We would know the answers to these and many related questions if we understood all the factors that determine the numerical value of a rate constant, for fast reactions have large rate constants, and slow reactions have small rate constants.

See Eq. 11-17.

Many of these factors can be recognized by following the path of reactant molecules on a trip along a reaction path over a reaction landscape. As an example, let us return to reaction (11-21), and trace the progress of an E molecule and a D molecule that are fated to become an X molecule. We shall begin the trip with independent, separated reactants, and follow them through encounter and eventual transition to product molecule. Just as a trip planner can lay out an automobile trip along winding roads as a schematic straight-line strip map, so can we represent the meanderings through reaction space as a straight-line path—a line on paper. Thus, we represent the reaction potentialities for D and E as the positions along the line in Figure 11-8.

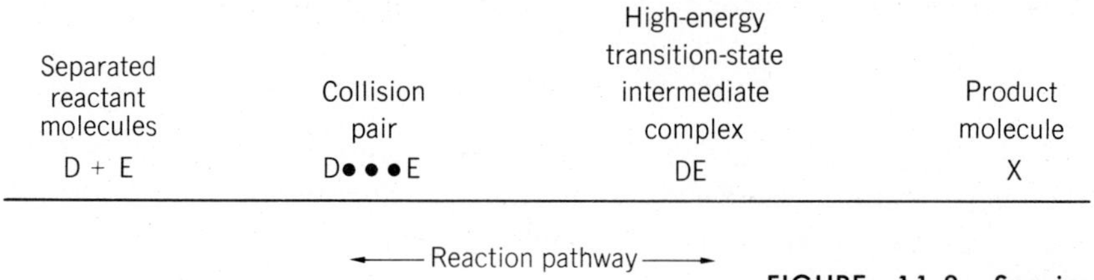

Separated reactant molecules	Collision pair	High-energy transition-state intermediate complex	Product molecule
D + E	D●●●E	DE	X

←——— Reaction pathway ———→

FIGURE 11-8 Species encountered along a reaction pathway leading between reactants and product.

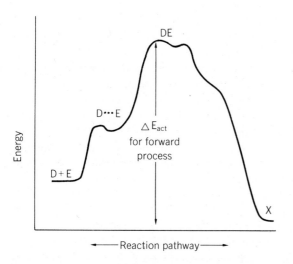

FIGURE 11-9 A two-dimensional plane-of-inquiry slice of the reaction landscape, showing that energy is required to produce an activated transition-state complex required before reaction to give product can occur.

A significant feature of this reaction path is the high energy of the transition-state complex, much higher than either the separated reactants or the products. This feature is displayed in the plot of energy versus distance along the reaction pathway given in Figure 11-9. To make the transition from collision pair to product, enough energy must be obtained so that the transition-state complex can be formed. If a lot of energy is needed to activate the collision pair, then only very seldom (only by a particularly fortuitous series of collisions with solvent molecules that just happen individually to have more than the average amount of kinetic energy) will a collision pair get enough energy to form a transition-state complex. Such high-energy transitions will occur very seldom, and the overall observable reaction rate will be very slow. And so we have come to an important generalization: A high energy of the transition-state complex—a high *energy of activation* for the process—implies a small value for the rate constant of the process.

Large activation energy means slow reaction.

The way to get more energy for transition is to get more energy distributed among those solvent molecules that every once in a while—"in a series of fortuitous collisions"—transfer enough energy to the collision pair to activate it. The way to get more thermal energy distributed among the solvent molecules is to raise the temperature. In fact, increasing the temperature does result in larger values of rate constants. The relationship turns out to be

$$k = Ae^{-\Delta E_{\text{activation}}/RT} \tag{11-39}$$

where $\Delta E_{\text{activation}}$ is the energy of activation; R is the gas constant; and T is the temperature on the Kelvin scale. And A? This constant includes factors such as geometry (that is, how the molecules happen to collide) and factors also that nobody yet understands!

11-9 Summary

Much of the interesting chemistry in our world involves compounds and processes on the way to equilibrium, *but not there yet*. The final end products of the living biological world will probably be simple molecules in a then-uninteresting equilibrium system, uninteresting because there will be no one around to take any interest. We are alive because the rates of many simultaneous chemical reactions are delicately balanced to maintain steady states, with some chemical concentrations dynamically driven far away from equilibrium. If we are to understand the chemistry of any biological system, we must know about the rates of many individual reactions; we must know about the factors influencing these rates; and we need insight into the ways in which the many simultaneous individual reactions interact. This chapter has been an introduction to ways of learning about and ways of talking about mechanisms of reaction and the dynamics of change.

We have examined changing systems by means of models that have certain features in common. Each model considered contains one or more rate equations. Each model, whether describing unrestrained growth, a steady state, continuous decay, or relaxation toward equilibrium, could be modeled with an exponential function. Although exponential functions cannot describe all changing systems, they are very useful in chemical rate studies, and the examples chosen are representative. Of special significance are models in which stability is achieved by one or more negative feedback processes. In every elementary chemical reaction, there are two processes involving negative feedback, and this fact guarantees that change will be in the direction of the equilibrium state, and that the equilibrium state will always be a stable state.

Chemical models are molecular descriptions of the mechanisms of chemical reactions, and much of this chapter was devoted to an examination of one such model. Associated with the chemical model is a mathematical model. Mathematical models have the advantage that unique numerical predictions can be made in such a way that data from laboratory rate experiments can be confronted directly with the predictions of a mechanism. A simulation method was discussed and then used for gaining insight into a reaction mechanism. A method was developed whereby an experimental rate curve can be analyzed, in terms of an assumed mechanism, to give numerical values of microscopic rate constants.

A model does not say all that there is to say about a changing system. It is always incomplete, and it may be partially incorrect. But is is an aid to careful and rational analysis, directing thought and providing a means for thinking about, talking about, and dealing with a complex system.

GLOSSARY

bimolecular rate constant: the proportionality constant between the rate of concentration change of either reactant in a two-reactant elementary process and the arithmetic product of the concentrations of the two reactants, symbolized by the letter k (and with an identifying subscript to distinguish it from all other rate constants of the mechanism). The rate constant is defined by imagining a hypothetical situation in which the elementary process is occurring in isolation, without competition from any other chemical process.

doubling time: the time required for an increasing quantity to double in value. For the special case of the function e^{mt}, where m is any positive number, the doubling time has the same value in all regions of the curve, from small values of t to large values.

elementary chemical process: a chemical transformation described as the *irreversible* conversion of the individual reactant molecule(s) into the individual product molecule(s); one of the components of an *elementary chemical reaction* (q.v.).

elementary chemical reaction: a chemical transformation described as the reversible interconversion of the individual reactant molecule(s) and the individual product molecule(s); a pair of forward and reverse elementary chemical processes considered as an inseparable couple. The designations reactant/product and forward/reverse are arbitrary, and conventionally indicate the direction [forward is left to right, reactant(s) → product(s)] in which the reaction has been written on paper.

exponential function: in general, a function of a variable that appears in an exponent. Thus, y is an exponential function of x if $y = a^x$. In this chapter, the exponential function is always of the form $y = e^{mt}$, where y is a variable such as population or concentration; t is time; m is a *macroscopic rate constant* (q.v.); and e (the base of natural logarithms) is 2.71828.... See Appendix II.

half-life: the time required for a decreasing quantity to change to a value halfway between its present value and its equilibrium value. In radioactive decay of an isotope that is not being replenished, none of the isotope is present at equilibrium, and for this case the half-life is the time required to halve the number of radioactive atoms. For the case of a change described by the function e^{-mt}, where m is any positive number, the half-life has the same value in all regions of the curve, from small values of t to large values.

macroscopic rate constant: a quantity, symbolized by the letter m, that characterizes the rate of a chemical relaxation. Called a *macroscopic* quantity because its value is obtained by analysis of time changes in observable properties of a laboratory-size chemical system.

microscopic rate constant: a proportionality constant in the algebraic rate equation that describes an elementary process, thus having meaning only in terms of a particular molecular reaction mechanism. Called a *microscopic* quantity because it is not directly observable, but must be deduced from macroscopic quantities. Symbolized by the letter k.

negative feedback: a form of control of a process in which the products of the process act to reduce the rate of accumulation of those products.

quantitative: expressed in the form of numbers.

rate constant: A proportionality constant that characterizes the rate of a process or reaction; symbolized by k or m.

rate equation: an equation that expresses the relationship of the rate of change of a quantity to some of the variables on which that rate depends.

reaction mechanism: a model describing a chemical reaction in terms of molecular interactions and molecular transformations.

relaxation: a monotonic approach to equilibrium, typically exponential in form.

steady state: a condition of a chemically reacting system in which some, but not all, of the species concentrations have values that do not change with time.

unimolecular rate constant: the proportionality constant between the rate of concentration change of the reactant in a one-reactant elementary process and the concentration of that reactant, with the elementary process considered independent of all other processes of the mechanism.

PROBLEMS

11-1 Write the mathematical model (five rate equations) for the reaction mechanism

$$A + B \underset{k_{-1}}{\overset{k_1}{\rightleftharpoons}} C + D$$

$$C + B \underset{k_{-2}}{\overset{k_2}{\rightleftharpoons}} E$$

The rate equation with $\mathscr{R}[C]$ has four terms, as does the rate equation with $\mathscr{R}[B]$.

11-2 The equilibrium constant for the reaction

$$X = Y$$

is defined by the equation

$$K = \frac{[Y]_{eq}}{[X]_{eq}}$$

where $[X]_{eq}$ and $[Y]_{eq}$ are the equilibrium concentrations of the species X and Y. Find the relationship between K and the two microscopic rate constants, assuming the mechanism of the reaction to be

$$X + H^+ \underset{k_{-1}}{\overset{k_1}{\rightleftharpoons}} Y + H^+$$

Here H^+ is serving as a catalyst for the reaction, facilitating the reaction without being consumed. Remember that at equilibrium, the time rate of change of each of the concentrations is zero.

11-3 According to an ancient legend, a king was offered an exquisite chessboard by one of his clever subjects, with the request that the king give in return 1 grain of rice for the first square on the board, 2 grains of rice for the second square, 4 grains of rice for the third, and so forth, doubling the number of grains of rice for each square. The king agreed, but was unable to meet his obligation in the bargain. Show why. Calculate the number of grains of rice required for the 20th square. If there are 1000 grains of rice in an ounce, find the square that requires a ton of rice. What square requires a million tons of rice? (A standard chessboard has 64 squares.)

11-4 To grow vegetables in a garden, a continuing battle must be waged with weeds. Let us suppose that, during a period of favorable growth conditions, the area of soil covered by weeds doubles each day. If the weeds are allowed to grow unrestrained, all the soil will be covered in four weeks, choking off all the desired plants. For several days the weeds seem insignificant, so you decide not to worry about having to hoe the garden until half the soil is covered by weeds. On what day do you hoe? How many days do you have to save the garden?

11-5 Suppose that there is sufficient radioactive ^{35}S in a sample so that the amount can be determined with a precision of one percent. That is, a measurement of 1.00 unit is uncertain because of the limitations of the radiation detector, and the number should be written as 1.00 ± 0.01. After a half-life has passed, the amount of ^{35}S present is only 50 units, but the uncertainty of measurement is still ± 0.01. How many half-lives must pass until the uncertainty in measurement is as great as the expected value of the measurement itself?

11-6 a. Consider a system in which the concentration of A changes with time according to the equation

$$[A] = [A]_0 e^{-kt}, \quad (k \text{ being a positive number})$$

Write this equation for the particular time $(t = t_{1/2})$ when $[A] = \frac{1}{2}[A]_0$. Making use of the facts that $\log_e(e^x) = x$, and that $\log_e(\frac{1}{2}) = -0.693$, show that $t_{1/2} = 0.693/k$.

b. Consider a system in which the concentration of B changes with time according to the equation

$$[B] = [B]_0 e^{kt}, \quad (k \text{ being a positive number})$$

Find the relationship between the doubling time t_2, and k ($\log_e 2 = 0.693$).

c. Consider a system in which the concentration of C changes with time according to the equation

$$[C] = C_\infty + C_m e^{-mt}, \quad (m \text{ being a positive number})$$

The constant C_∞ is the value of [C] at equilibrium, and $t_{1/2}$ is the time for the total remaining concentration change $\{[C] - C_\infty\}$ to be halved. Find the relationship between $t_{1/2}$ and m.

11-7 Using the following experimental rate data, obtain a value for m by plotting the data according to the time-lag method.

t	concentration
0 min	0.1000 mole/liter
3	0.0911
6	0.0823
9	0.0743
12	0.0670
15	0.0605
18	0.0547
21	0.0494
24	0.0446
27	0.0403
30	0.0364
33	0.0330
36	0.0298
39	0.0270
42	0.0244
45	0.0221
48	0.0200
51	0.0182
54	0.0165
57	0.0149
60	0.0136
63	0.0123
66	0.0112
69	0.0102
72	0.0093
75	0.0085

If you do not have access to a table of natural logarithms, take advantage of the fact that

$$\log_e e^{m\tau} = m\tau = 2.303 \log_{10} e^{m\tau}$$

11-8 Suppose that the following values of macroscopic rate constants have been determined from rate experiments involving the reaction of D and E, performed under conditions such that [D] did not change appreciably with time.

$[D]_0$	m
1.00 mole/liter	13.6 min^{-1}
0.75	11.8
0.50	10.2
0.30	8.6
0.10	7.1

Assuming mechanism (11-21), evaluate both microscopic rate constants.

11-9 Write and execute a computer program to simulate the reaction

$$D + E \underset{k_{-1}}{\overset{k_1}{\rightleftharpoons}} X$$

Choose whatever values for the k's and initial concentrations that you think are appropriate, or that you think might be interesting.

SUGGESTIONS FOR FURTHER READING

Two very readable introductions to chemical kinetics are: CAMPBELL, J. A., *Why Do Chemical Reactions Occur?*, Englewood Cliffs, N.J.: Prentice-Hall (1965); and KING, E. L., *How Chemical Reactions Occur*, New York: W. A. Benjamin (1963). Both are available in paperback.

For additional examples of chemical mechanisms, the associated mathematical models, and methods of evaluating rate constants from experimental data, see: FLECK, G. M., *Chemical Reaction Mechanisms*, New York: Holt, Rinehart & Winston (1971). For an introduction to numerical methods of analyzing kinetic data, see: SWINBOURNE, E. S., *Analysis of Kinetic Data*, London: Thomas Nelson and Sons (1971).

A powerful approach to the formulation of mathematical models for biological systems involves the use of compartment system models. For example, in simulating drug action in a human, the plasma of the blood constitutes one compartment, and each of the organs or tissues in which the drug is distributed is another compartment. The important rates are considered to involve transfer of the drug from one compartment to another. The mathematics is identical to the mathematics discussed in this chapter. An excellent book devoted entirely to such models is: RESCIGNO, A., and SEGRE, G., *Drug and Tracer Kinetics*, Waltham, Mass.: Blaisdell (1966).

Since 1955, the number of mathematical models of physiological systems described in the research literature has nearly doubled every year. An introduction to the modeling of biochemical rate processes, extending many of the concepts introduced in this chapter, is: RIGGS, D. S., *Control Theory and Physiological Feedback Mechanisms*, Baltimore: Williams and Wilkins (1970).

Insight into the concept of a chemical model can be gained by examining the use of a mathematical model in a quite different context. In *The Mathematician and the River* (a Horizons of Science film, 19 min, 16 mm sound, color, Princeton: Educational Testing Service), flood control on the Mississippi River is presented as an important but complex problem that can be investigated with a concrete scale model of the river, and with a mathematical model of the river. Use of animation very effectively illustrates much of the material of Section 11-8 in *An Introduction to Reaction Kinetics* (CHEM Study Film No. 4121, 13 min, 16 mm sound, color). This film explores, for a gas-phase reaction, the effect of temperature, molecular geometry, and the presence of a catalyst on the rate of an elementary process.

Radioisotopes have been used as tagged atoms to trace species through complex chemical reactions. This method of elucidating reaction mechanisms is described in "Introduction to the Kinetics of Tracers in Reaction Systems," chap. 8 in BRAY, H. G., and WHITE, K., *Kinetics and Thermodynamics in Biochemistry*, 2nd ed., New York: Academic (1966).

Many environmental problems involve large-scale chemical reactions in lakes, rivers, and oceans, and chemical *rates* are often critical factors in controlling the effects of these reactions. To understand such complicated systems, and perhaps to be able to exercise some control over the important chemical reactions, reaction-rate models have been devised. These models, based on principles set forth in this chapter, use computer calculations to simulate effects expected when various alternative actions of man and his technology are introduced. Three such illustrative chemical-simulation studies are: RANDERS, J., "DDT Movement in the Global Environment"; ANDERSON, A. A., and ANDERSON, J. M., "System Simulation to Identify Environmental Research Needs: Mercury Contamination"; and ANDERSON, J. M., "The Eutrophication of Lakes." Each study is a chapter in the book, MEADOWS, D. L., and MEADOWS, D. H., eds., *Toward Global Equilibrium: Collected Papers*, Cambridge, Mass.: Wright-Allen Press (1973). A large-scale simulation depends on reliable information about the mechanisms of the constituent chemical processes; for an example of how information about such subsystems is obtained and reported, see: LEAN, D. R. S., "Phosphorus Dynamics in Lake Water," *Science*, vol. **179**, pp 678–680 (1973).

12
Molecular Orbitals

12-1 Abstract

This chapter is about electrons in molecules. Our objective is to understand more about the nature of chemical bonding in molecules and something about the principles that quantum-minded chemists use in developing theories about the distribution of electrons in molecules. This is a chapter about quantum theory and about geometry and symmetry. Like the field of molecular quantum chemistry itself, it will seem like a curious blend of exactness and approximation. The key word throughout is *orbital*.

Functions: See Appendix II.

We shall look at a collection of mathematical functions that describe the cloud of electrons spread throughout an atom. This collection serves as the basis for discussing a multitude of more complex atoms and molecules, by a procedure called the *linear combination of atomic orbitals*. You are already familiar with linear combinations of various sorts, although the odds are against your ever having used that phrase to describe ordinary processes of mixing. We shall examine some commonplace examples of linear combinations, and then proceed to the special case of mixing atomic orbitals to get molecular orbitals.

It is a tricky business trying to map the interior of a cloud in the sky by drawing lines on a sheet of paper. For exactly the same reasons, it takes some special tricks to draw a map of the electronic cloud of a molecule. We shall adopt a type of map well known to hikers and mountain climbers: a contour map. Using some contour maps and taking advantage of the language of symmetry, we shall examine some of the properties of atomic and molecular orbitals.

Molecular orbitals are made by combining atomic orbital functions that have the same symmetry and that have similar energies. We shall devote some time to sorting out atomic orbital functions according to these criteria, using a molecule of water as an example. Then, by combining similarly symmetric atomic orbitals, we shall arrive at two bonding molecular orbitals which, when occupied by electrons, provide the electronic glue that holds together the oxygen and hydrogen nuclei in H_2O. Other electrons in water will be found to occupy orbitals that do not contribute to bonding, and we shall call them nonbonding orbitals. Finally, we shall examine the concept of antibonding orbitals and show how one such orbital might be constructed for a water molecule.

A review of the elementary discussion of wave mechanics in Chapter 2, of symmetry in Chapter 4, and of mathematical functions in Appendix II will be helpful to your understanding of this chapter.

12-2 Orbitals

A hypothetical molecule formed of just nuclei would be extremely unstable. The mutual repulsions of the positively charged nuclei would cause self-destruction of any arrangement of naked nuclei. Stability in a real molecule is achieved by electrons which attract the nuclei and which shield the nuclei from one another. This electronic glue bonds the nuclei and stabilizes the whole molecule. We say that *electrons form chemical bonds.*

We have talked about the nuclei as points in space, and we have described the collection of nuclei by pinpointing the location of each of the nuclei. In each molecule, there is an arrangement in space of the individual nuclei. We cannot locate the chemical bonds in the same way, and we cannot talk about the arrangement of individual electrons in space. It often is not very helpful even to consider the electrons to be distinctly individual. In many cases, electrons in molecules exhibit cooperative, communal behavior, and we need to use an appropriate language for this swarm of electronic charge; we shall turn from words like "arrangement" to words like "distribution." We shall keep track of the number of electrons in a region of space, but we shall not try to pinpoint a location of any particular electron.

Electrons in molecules are delocalized.

The electrons of a molecule are dispersed throughout the space that surrounds the nuclei, forming a cloud of negative charge that is dense near each nucleus and dense between pairs of nuclei that are chemically bonded together. This fog of charge is diffuse in the outer reaches of the molecule. Just as the atmosphere of the earth has no clearly defined outer boundary (the air just gets thinner and thinner as a spaceship moves farther and farther away from the earth), so the electronic atmosphere of a molecule has no outer surface. But this electronic atmosphere, a wispy mist around the outside, a dense fog in portions of the interior, has definite form and shape. *The symmetry of the electron-density distribution is the same as the symmetry of the arrangement of nuclei, and each of these symmetries is the same as the symmetry of the molecule as a whole.* However vague our picture may be of the smeared-out electrons that constitute the electronic atmosphere of a molecule, we can be precise and explicit about the symmetry of the spatial smear.

Happily, we do not have to think about all of the electrons at the same time. It is fruitful to consider the molecule to be divided into overlapping but independent regions called **orbitals**. Each orbital is the domain of zero, one, or two electrons, and chemists speak of empty, half-filled, and filled orbitals. An orbital has shape and form, and together all of the occupied orbitals have the shape and form of the overall electronic distribution of the molecule. That is, the symmetries of the individual orbitals must combine to give the symmetry of the electronic distribution of the whole molecule.

Orbital: A region in space and a mathematical model of that region

Electrons are not distributed uniformly throughout an orbital. A mathematical function that describes the distribution is called a **state function.** Every orbital has associated with it a particular state function, symbolized by the Greek letter ψ (psi). Each state function is a solution of the Schrödinger equation. As we found in Chapter 2, there are many mathematical functions that are solutions to the Schrödinger equation, and the first task confronting chemical quantum mechanics is to find a function that describes how ψ varies as one moves about in space around the nucleus in an atom, or between nuclei in a molecule. We shall be spending most of our time in this chapter manipulating state functions—adding and subtracting them (forming linear combinations of them), investigating their symmetry properties, and finally reading

A function may be the solution to an equation just as a number may be.

some chemical meaning into the values of their energies. We shall pay close attention to the values of the squared[1] state function ψ^2, which we assert is proportional to the electron density in the orbital at each point in space.

12-3 The Three-Body Dilemma and the Necessity for Relying on Hydrogen-Atom Calculations

We shall develop a picture of the electronic distribution in molecules that is based on the results obtained by solving the Schrödinger equation for the simplest of chemical systems, the hydrogen atom.[2] We shall not present any exact solutions for chemical systems more complicated than the two-body (one electron and one proton) hydrogen-atom system. In fact, with three particles instead of two, the problem becomes too tough for the best present-day mathematicians to solve exactly. There is no real choice but to concentrate on the hydrogen atom as a model for all chemical systems. This has been the approach of chemical theoreticians throughout the mid-twentieth century, and it is to be our approach in gaining an understanding of chemical bonding.

The three-body dilemma is not unique to quantum chemistry. The problem caused trouble earlier when mathematicians working in the field of celestial mechanics tried to write down an algebraic description of the motions of the planets in our solar system. It turned out that there was a straightforward solution to the problem of the dynamics of two bodies which attract each other by gravitational force. But the next more difficult problem—the three-body problem of three mutually attractive bodies moving in space—has no exact solution that can be written down as alebraic equations. So analysis of the solar system, or even the sun-moon-earth system, has had to be made by way of various approximation methods. Approximation methods can give very good answers to certain questions. For instance, approximation methods utilizing high-speed computers can be precise enough so that men can navigate to the moon, land, and return to earth.

Approximations are not exact—but they may be very good.

The algebra in the sun-moon-earth analysis gets complicated because each of the bodies attracts both of the other two. For the same reason, chemists have found that no exact solution can be found for the dynamics of the particles (always more than two) that constitute any molecule. Anything more complicated than the hydrogen atom requires approximations. Since complete analysis without any approximation appears impossible for any molecular system of electrons and nuclei, chemists have paid particularly close attention

---

[1] When the state function is a complex number (a number containing $i \equiv \sqrt{-1}$), then instead of squaring the state function to give ψ^2, we multiply ψ by its complex conjugate ψ° to give $\psi^\circ \psi$. The function ψ° is the same as ψ, except that i has been replaced by $-i$. Such complications are all sidestepped in this text.

[2] The details of obtaining these results are beyond the scope of this book. With some perseverance, though, you can come to grips with the mathematical model and the mathematical techniques without being a mathematician. An excellent first book for a chemistry student who wants to learn how to solve the Schrödinger equation is J. W. Linnett, *Wave Mechanics and Valency*, London: Methuen (1960).

to the results obtained for the hydrogen atom. They have developed an extensive vocabulary and a system of approximation methods, based on hydrogen as a prototype for all chemical systems.

If we imagine a rectilinear xyz coordinate system with its origin at the nucleus of a hydrogen atom, we can locate all points in the atom by specifying a value for each of the x, y, and z coordinates. When we consider an electron moving in the sphere of influence of a nucleus, it is often helpful and informative to use in addition the radial coordinate r, the distance from the nucleus.[3] A collection of solutions of the Schrödinger equation for an isolated hydrogen atom is presented in Table 12-1, using a combination of the coordinates x, y, z, and r. All angular dependence of these solutions is contained in the dependence on x, y, and z. There is one solution that corresponds to the ground state (the lowest-energy state), and it has the quantum numbers $n = 1$, $l = 0$, $m_l = 0$. All the solutions with $n = 2$ have identical energy values, energy greater than for $n = 1$. It turns out that any of these solutions can be multiplied by a constant, and the result will still be a solution.

Quantum numbers: See Sec. 2-8.

TABLE 12-1
Hydrogen Atomic Orbitals

Quantum Numbers			Name of Orbital	Orbital State Function ψ
n	l	m_l		
1	0	0	$1s$	10^{-r}
2	0	0	$2s$	$(2.303r^{-2}) \cdot 10^{-r/2}$
2	1	0	$2p_z$	$z \cdot 10^{-r/2}$
2	1	± 1	$2p_x$	$x \cdot 10^{-r/2}$
2	1	± 1	$2p_y$	$y \cdot 10^{-r/2}$

In the absence of any hope of obtaining similar solutions for any atom with more than one electron, we need some approximation that will permit plausible guesses for such polyelectronic atoms. We shall assume that *in a polyelectronic atom, each electron or pair of spin-paired electrons occupies an orbital that is described by a hydrogenlike state function.* This hydrogenlike state function may have an altered radial dependence, but the angular dependence will be assumed to be the same as in the hydrogen atom. The significant consequence of these assumptions is that each atomic orbital in a polyelectronic atom is to have the same symmetry properties as one of the hydrogen-atom orbitals. We also assume that in a polyelectronic atom, the first two electrons will occupy the lowest-energy orbital, and that the aufbau principle introduced in Chapter 2 is followed in placing all subsequent electrons. This bookkeeping procedure is the reason why Table 2-2 and Table 12-1 look so much alike.

Exact algebraic solutions for multielectron atoms do not exist.

---

[3] If x, y, and z are known, then r can be calculated by means of the three-dimensional Pythagorean formula

$$r = \sqrt{x^2 + y^2 + z^2}$$

12-4. Pencil-and-Paper Mapping of the Interior of an Orbital

A map of the interior of an electron cloud might be a representation of the density of the cloud at each point in space. This would have to be a four-dimensional graph, giving a value of ψ^2 at each point in space. Another useful map would be a representation of ψ at each point.

The orbital state function for the $2p_x$ orbital is

$$\psi = x \cdot 10^{-r/2} \tag{12-1}$$

For each set of numbers (x, y, z), Equation (12-1) allows us to calculate a number ψ. To plot or to graph this state function, we need a way to display graphically a table of numbers with four columns:

x	y	z	ψ

When we recall that a table with two columns of numbers is plotted as a line on a sheet of graph paper, and that a table with three columns of numbers is plotted as a three-dimensional surface, we see that our table with four columns of numbers may not be easy to plot. In order to get some guidance about what to do, we shall examine the methods used when mapmakers try to represent the three-dimensional surface of the earth by drawing lines on a sheet of paper.

Suppose that a mapmaker has the table of numbers

latitude	longitude	elevation
30″	60″	610 ft
30	50	646
30	40	738
30	30	773
30	20	652
30	10	485
30	0	449
40		
⋮	⋮	⋮

These numbers give the height above sea level at various points on Davis Hill and Shingle Hill in the little New England town of Williamsburg, Massachusetts. How can the mapmaker represent this collection of numbers on a piece of flat paper? One common representation is a *contour map*. On such a map, points of equal elevation are connected with **contour lines**. A hiker, or a surveyor, or a logger who is familiar with these hills has a mind's-eye picture of the lay of the land; any other person who is accustomed to looking at contour maps can translate the contour lines of Figure 12-1 into a similar mind's-eye picture of the three-dimensional landscape. Since the elevation points did not happen to have the values of 600, 650, 700, and so forth, at which we wished to draw the contour lines, it was necessary to estimate the position of the contour elevations between adjacent known values. Then the

Contour lines can help you construct a reliable picture in your mind.

contour lines were drawn to connect these estimated points. Many more elevation points were actually used in drawing the map; the more elevation data available, the more accurate is the resulting map.

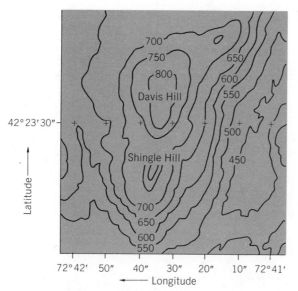

FIGURE 12-1 A contour map of the region around Davis Hill (elevation 841 ft) and Shingle Hill (elevation 767 ft). Lots of elevation data were required in order to draw this detailed map; typical of the data used are the numbers given in Table 12-3 and indicated on the map as + symbols along the latitude 42°23′30″ line. Contour lines connect points of equal elevation. Note that two contour lines can never cross (why?), that each contour line is a continuous, unbroken line (why?), and that each contour line on a large map is a closed curve (why?). Locate the tops of Davis Hill and Shingle Hill. Find some other hilltops. Where are the steepest slopes? Where do you suppose Nungee Swamp is? This contour map is a plot of three variables—longitude, latitude, and elevation—on a two-dimensional book page. [Map redrawn from the *Williamsburg (Massachusetts) Quadrangle Topographical Map,* Washington, D.C.: U.S. Geological Survey (1948).]

Now we return to the problem of representing a $2p_x$ orbital in a hydrogen atom by plotting Equation (12-1). We can reduce the complexity of the problem from four columns of numbers to only three columns by deciding to look first at just a planar slice through the hydrogen atom. For instance, we can sample the orbital by looking at ψ in a plane perpendicular to the z axis and thus in a region of space in which z is constant. We shall pick the plane in which z is equal to zero. Our strategy will be to choose a reasonably large number of points in this plane, and then to calculate the value of ψ at each of the points by using Equation (12-1). The plane and the points are shown in Figure 12-2.

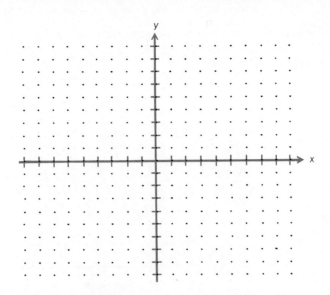

FIGURE 12-2 An array of points in the xy plane. We shall sample the function ψ by calculating the value of ψ at each of these points, replacing each dot by the value of ψ at that point. The result is Figure 12-3.

The numbers arrayed in Figure 12-3 are the result of 361 calculations, each using Equation (12-1) together with different values of x and y corresponding to the positions of the points in Figure 12-2. Combining the information in these two figures, we can draw the contour diagram shown in Figure 12-4. For example, the locations of points having $\psi = +0.6$ were estimated from the known values of ψ at the selected calculation points, and then these estimated points of constant ψ were connected with a contour line. We then labeled the line "+0.6."

The array of numbers in Figure 12-3, the contour lines in Figure 12-4, the state function, and the orbital itself all have the same symmetry properties. Each (the pattern of numbers, the pattern of contour lines, Equation (12-1), and the $2p_x$ orbital) is symmetric with respect to reflection in the xz mirror plane, antisymmetric (each changes sign) with respect to a C_2 rotation around the y axis, and antisymmetric with respect to reflection in the yz mirror plane. Because of the spherical symmetry of the factor $10^{-r/2}$ (that is, because there is no angular dependence in the factor $10^{-r/2}$), we would have obtained precisely the same array of numbers and thus the same contour lines for any plane that contains the x axis. Thus, there is an infinity of σ planes containing the x axis, making the x axis a C_∞ axis. The $2p_x$ orbital is symmetric with respect to reflection in any of these σ planes and is also symmetric with respect to any rotation about the C_∞ axis.

Review the names and symbols for symmetry operations in Ch. 4.

The page is a table of numerical values of $\psi(x, y, z)$ at $z = 0$, oriented sideways on the page. The rows (read along the page's long axis) give values as a function of one coordinate, and labels within the array spell out "ALL NUMBERS IN THIS QUADRANT ARE NEGATIVE", "ALL NUMBERS IN THIS QUADRANT ARE POSITIVE" (twice, for the two right-hand quadrants), and "ALL NUMBERS IN THIS QUADRANT ARE NEGATIVE" / "POSITIVE" in the four quadrants defined by the x and y axes.

0.016	0.019	0.023	0.027	0.029	0.029	0.026	0.020	0.011	0.000	0.011	0.020	0.026	0.029	0.029	0.027	0.023	0.019	0.016				
0.022	0.028	0.034	0.040	0.045	0.046	0.042	0.032	0.018	0.000	0.018	0.032	0.042	0.046	0.045	0.040	0.034	0.028	0.022				
0.030	0.039	0.050	0.060	0.068	0.071	0.067	0.053	0.029	0.000	0.029	0.053	0.067	0.071	0.068	0.060	0.050	0.039	0.030				
0.040	0.054	0.070	0.086	0.101	0.109	0.105	0.085	0.048	0.000	0.048	0.085	0.105	0.109	0.101	0.086	0.070	0.054	0.040				
0.052	0.072	0.095	0.121	0.146	0.163	0.163	0.135	0.078	0.000	0.078	0.135	0.163	0.163	0.146	0.121	0.095	0.072	0.052				
0.065	0.091	0.124	0.163	0.203	0.236	0.246	0.214	0.127	0.000	0.127	0.214	0.246	0.236	0.203	0.163	0.124	0.091	0.065				
0.078	0.112	0.155	0.210	0.270	0.328	0.360	0.330	0.206	0.000	0.206	0.330	0.360	0.328	0.270	0.210	0.155	0.112	0.078				
0.090	0.130	0.184	0.254	0.339	0.428	0.495	0.486	0.327	0.000	0.327	0.486	0.495	0.428	0.339	0.254	0.184	0.130	0.090				
0.097	0.142	0.204	0.287	0.391	0.509	0.617	0.654	0.493	0.000	0.493	0.654	0.617	0.509	0.391	0.287	0.204	0.142	0.097				
0.100	0.147	0.211	0.299	0.410	0.541	0.669	0.736	0.607	0.000	0.607	0.736	0.669	0.541	0.410	0.299	0.211	0.147	0.100				
0.097	0.142	0.204	0.287	0.391	0.509	0.617	0.654	0.493	0.000	0.493	0.654	0.617	0.509	0.391	0.287	0.204	0.142	0.097				
0.090	0.130	0.184	0.254	0.339	0.428	0.495	0.486	0.327	0.000	0.327	0.486	0.495	0.428	0.339	0.254	0.184	0.130	0.090				
0.078	0.112	0.155	0.210	0.270	0.328	0.360	0.330	0.206	0.000	0.206	0.330	0.360	0.328	0.270	0.210	0.155	0.112	0.078				
0.065	0.091	0.124	0.163	0.203	0.236	0.246	0.214	0.127	0.000	0.127	0.214	0.246	0.238	0.203	0.163	0.124	0.091	0.065				
0.052	0.072	0.095	0.121	0.146	0.163	0.163	0.135	0.078	0.000	0.078	0.135	0.163	0.163	0.146	0.121	0.095	0.072	0.052				
0.040	0.054	0.070	0.086	0.101	0.109	0.105	0.085	0.048	0.000	0.048	0.085	0.105	0.109	0.101	0.086	0.070	0.054	0.040				
0.030	0.039	0.050	0.060	0.068	0.071	0.067	0.053	0.029	0.000	0.029	0.053	0.067	0.071	0.068	0.060	0.050	0.039	0.030				
0.022	0.028	0.034	0.040	0.045	0.046	0.042	0.032	0.018	0.000	0.018	0.032	0.042	0.046	0.045	0.040	0.034	0.028	0.022				
0.016	0.019	0.023	0.027	0.029	0.029	0.026	0.020	0.011	0.000	0.011	0.020	0.026	0.029	0.029	0.027	0.023	0.019	0.016				

Labels within the array: In the upper-left quadrant, "ALL NUMBERS IN THIS QUADRANT ARE NEGATIVE". In the upper-right quadrant, "ALL NUMBERS IN THIS QUADRANT ARE POSITIVE". In the lower-left quadrant, "ALL NUMBERS IN THIS QUADRANT ARE NEGATIVE". In the lower-right quadrant, "ALL NUMBERS IN THIS QUADRANT ARE POSITIVE".

(axes: y horizontal, x vertical)

FIGURE 12-3 Equation (12-1), displayed for selected values of x and y, with $z = 0$. This array of numbers is one way of plotting $\psi(x, y, z)$ on a two-dimensional book page. These numbers do not make much graphic impact as presented here. A better way to display the essence of this function is shown in Figure 12-4.

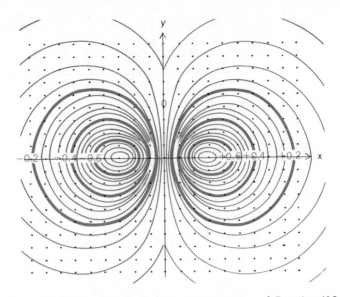

FIGURE 12-4 The salient feature of this contour mapping of Equation (12-1) is the symmetry of the state function and thus the symmetry of the orbital itself. This contour map was made by examining the numbers in Figure 12-3 and estimating the places where the function must have values such as +0.200, +0.400, and so forth. Then smooth curves were sketched through points having the same value of ψ.

A contour map of an appropriate planar slice of an orbital is probably the clearest way to represent accurately that orbital on the printed page. A related way, a sort of shorthand contour-map description, is better suited for use on a lecturer's blackboard, in a student's notes, and for schematic representations in a textbook. The shortcut is to use just one contour line, preferably a line that surrounds most of the orbital. In Figure 12-5, such a picture is drawn, using only the $\psi = \pm 0.08$ contour line. In this picture, a change of color is used to indicate that the two lobes of the orbital have different signs. We can imagine a surface, generated by revolving the picture in Figure 12-5 around the x axis. Such a surface would contain most of the orbital and would have the full symmetry of the orbital.

Our technique of picking a two-dimensional plane to sample the properties of a three-dimensional entity is called the **plane-of-inquiry** method. It is the mathematical analogue of the use of thin sections to reveal the anatomy of a biological specimen. In that case, the plane of inquiry is the plane exposed by the sectioning knife. (The wary shopper who buys a cut-in-half watermelon is trusting a particular plane of inquiry to have revealed a representative sample of that melon.) In the present case, we have in effect sliced open a $2p_x$ orbital along the xy plane. Our procedure was to restrict attention to a limited range of variables (we restricted z to the value zero), ignoring for the moment the overall complexity of the total problem. After inquiring about the nature of this particular sample, we tried some other planes of inquiry. In the case of the $2p_x$ orbital, we found there to be an infinity of planes of inquiry that are identical to the one we originally chose. Since this infinity of planes sweeps out all space and thus all regions of the orbital, we were able to complete the analysis of the orbital with a single well-chosen *plane-of-inquiry* sample. A single slice suffices because of the cylindrical symmetry of the orbital.

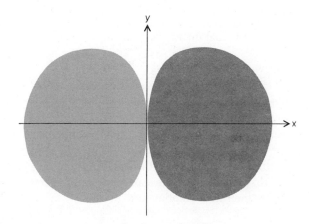

FIGURE 12-5 A schematic representation of the electron density of the $2p_x$ orbital. The lines are contour lines, connecting points in the xy plane with equal values of ψ^2. Color is used to distinguish between regions in which the state function ψ (but, of course, not the electron-density function ψ^2) is positive and regions in which it is negative. The yz plane is for the state function a mirror that reflects and changes sign; that same plane is for this picture a mirror that reflects and changes color.

We could repeat the procedure used for analyzing the $2p_x$ orbital, and we would find that the results for the $2p_y$ and $2p_z$ orbitals are identical except for the labeling of the coordinate axes. The $2p_y$ orbital has a C_∞ axis that coincides with the y axis. The $2p_z$ orbital has a C_∞ axis that is the z axis.

At this point, it makes sense to refer to figure 2-6 to see the relationship between the notions that have just been developed and the artist's perspective drawings used to depict the electronic distribution in orbitals. Note that the geometric shapes representing the p_x, p_y, and p_z orbitals have the symmetry elements C_2, C_∞, σ (perpendicular to the C_∞ axis), and an infinity of σ's (each containing the C_∞ axis). Because the shapes in Figure 2-6 represent ψ^2 (the electron density) instead of the state function ψ itself, there are no negative regions in the pictures. These figures are contour *surfaces*, each surface having been generated by connecting points of equal electron density. Such a contour surface can also be generated by revolving a contour line around a C_∞ axis.

12-5 Linear Combinations

Mixing sometimes results in a combination whose properties are the sums of the properties of the individual constituents. A mixture of oxygen and nitrogen (for instance, the air we breathe) is a mixture of N_2 and O_2 molecules that do not interact chemically to any appreciable extent. The pressure P of the gas mixture is simply the sum of individual pressures, p_{O_2} and p_{N_2}. That is,

$$P = p_{O_2} + p_{N_2} \qquad (12\text{-}2)$$

We say that the total pressure is a **linear combination** of the pressures of the individual constituents. Likewise, the mass M of the gas mixture is a linear combination of the individual masses m_{O_2} and m_{N_2}:

$$M = m_{O_2} + m_{N_2} \qquad (12\text{-}3)$$

In Equations (12-2) and (12-3), we quite properly *added* the individual contributions. Sometimes, however, mixing involves a subtraction when the linear combination is formed. The mixing of two waves, for example, may result in either reinforcement (addition) or cancellation (subtraction), depending on the extent to which the waves are in phase or out of phase (in step or out of step) with each other. Thus, the height H (the amplitude) at each point of the wave that results from the superposition of two waves of heights h_1 and h_2 is given by the linear combination

$$H = f_1 h_1 + f_2 h_2 \qquad (12\text{-}4)$$

where f_1 and f_2 are some numbers between -1 and $+1$. The mixing of two waves can give rise to a resultant wave that has a height equal to the sum of the heights of the two waves, or to the difference between the heights of the two waves, or to any intermediate height.

We shall use an analogous method for mixing atomic orbitals to give linear combinations that we shall call *molecular orbitals*. Since there is no way in sight to use the Schrödinger equation to obtain exact state functions for a molecule which contains not two or three but many particles, we must use approximation methods. We shall assume that electrons in a molecule can be described in terms of a collection of independent molecular orbitals, each of which can serve as the domain for zero, one, or two electrons. State functions for these orbitals will be thought of as superpositions in space of atomic orbitals, as linear combinations of atomic orbitals. To a certain extent, the appropriate linear combinations are arrived at by guess, retaining those combinations that best describe the properties of the actual molecule. As in any such game, experience is the best guide to guessing. However, since we do not assume any such experience on your part, we shall rely on two guidelines that are also used by professional chemists in constructing molecular orbitals.

GUIDELINE I. For lack of a better technique, we shall use hydrogenlike atomic-orbital functions (Table 12-1) as a basis for the construction of molecular orbitals. We shall combine them into molecular-orbital functions by the simple process of linear combination (that is, by adding or subtracting them).

Hydrogen-like orbitals form the basis; and then we mix them.

Some linear combinations make more sense than others, and part of the job is to select combinations that appear to offer the best bonding.[4]

GUIDELINE II.   The symmetry of the molecular orbital is a very important consideration. The final electron distribution must show the full symmetry of the molecule, as we noted in Chapter 4. Since many linear combinations of atomic orbitals do not meet this requirement, the use of symmetry allows us to reject many combinations, thus greatly simplifying our search for realistic molecular-orbital functions. We shall test the symmetry of each atomic orbital by the straightforward geometric process of applying the actual symmetry operations of the molecule to the spatial orbital domains.

The electron distribution, proportional to ψ^2, has the full symmetry of the molecule. We can be sure that ψ^2 has the complete symmetry of the molecule if the overall state function ψ for the whole assemblage of electrons in the molecule is related to the state functions for the individual orbitals by an equation of the form

$$\psi^2 = \psi_1^2 + \psi_2^2 + \psi_3^2 + \psi_4^2 + \cdots \qquad (12\text{-}5)$$

where ψ_1, ψ_2, ψ_3, and so forth, are state functions of the individual orbitals.[5] There is a straightforward bookkeeping device that will assure us that the total electron distribution has the proper symmetry; we shall assume that for each orbital, ψ_i^2 has the full symmetry of the molecule (ψ_i being the state function of that orbital). There are two ways that ψ_i^2 can be symmetric with respect to each of the symmetry operations of the molecule—the state function ψ_i itself may be either symmetric or antisymmetric with respect to each of the symmetry operations of the molecule. Thus, Guideline II can be stated in very precise terms by requiring that the application of a symmetry operation of the molecule to any molecular-orbital state function must leave the state function unchanged (and thus the orbital is symmetric with respect to that symmetry operation) or must change the sign of the state function at every point in space, but otherwise leave it unchanged (thus showing that the orbital is antisymmetric with respect to the symmetry operation). Fundamentally, this guideline works for the same reason that there are two different square roots of the number one, the distinctly different numbers -1 and $+1$.

The molecular orbitals must be "symmetry correct" (see Ch. 4).

12-6 Electron Distribution in the Water Molecule

Now let us apply these principles to a specific molecule, an isolated molecule of water. We continue with consideration of the symmetry of the water molecule, resuming just at the point where we left that molecule in Chapter 4.

---

[4] A full discussion includes determining the orbital that implies the lowest electron energy. Such a treatment is beyond the scope of this textbook. Information on this subject can be found in most of the references listed in "Suggestions for Further Reading" at the end of this chapter.

[5] We are assuming that each of these orbitals is fully occupied by two electrons, and we are not paying attention to the proportionality constant that relates ψ^2 and electron density. Similar equations in other books may look more complicated, but the *form* is the same.

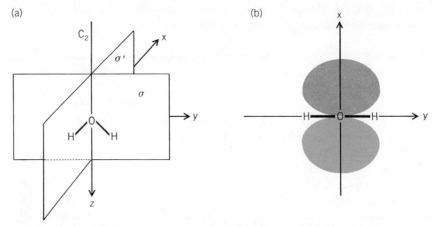

FIGURE 12-6(a) Symmetry elements of the water molecule. Because we shall need them in the following discussion, we have superimposed the rectilinear xyz coordinates. The reflection planes σ and σ' are identical, respectively, to the yz and xz planes; the z axis is identical to the C_2 axis; and the positive z direction is chosen in the direction from the oxygen nucleus toward the hydrogen nuclei. The oxygen nucleus is at the origin of the coordinate system. (b) The same system as shown in Figure 12-6(a), but viewed along the z axis in the positive direction (toward the hydrogen end of the molecule). The oxygen $2p_x$ orbital, whose symmetry in this molecular system we shall first discuss, is added to the picture. XXblueXX: positive lobe; YYgreenYY: negative lobe (compare Figure 12-5).

Figure 12-6 shows a water molecule located at the center of the xyz coordinate system, with the symmetry elements σ and σ' (two perpendicular reflection planes) and C_2 (twofold rotational axis) indicated. Now we propose to construct molecular orbitals for the 10 electrons of H_2O by linear combination of hydrogenlike atomic-orbital functions centered on the oxygen and hydrogen nuclei. Because orbitals with the same symmetry properties will combine to form a symmetry-correct molecular orbital, we shall begin by considering the symmetry of these atomic orbitals under the symmetry operations of the water molecule.

Consider again the state function for the $2p_x$ orbital (Figure 12-4). We shall successively apply the symmetry operations to it, and check its response:

1. IDENTITY OPERATION I. As always, this operation makes no difference to anything to which it is applied. Thus, the $2p_x$ function is symmetric with respect to the identity operation.

2. REFLECTION OPERATION σ'. This operation is a reflection through the xz plane (Figure 12-6). Consulting Figure 12-4, we find that each point in the plane of inquiry (and in fact each point in the orbital) is reflected onto a point identical in value. Thus, the $2p_x$ state function is symmetrical also with respect to the σ' operation. To take just one of the 361 examples from Figure 12-3, the extreme upper-right-hand value (+0.016) is reflected onto the extreme lower-right-hand value (+0.016) by the σ' operation (by reflection through the xz mirror plane). The same relationship applies for all of the numbers in Figure 12-3, as well as for the contour lines of Figure 12-4.

3. REFLECTION OPERATION σ. As shown in Figure 12-6(a), the σ operation results in reflection in a plane perpendicular to the σ' plane; for our figure the σ plane is the yz plane. Returning to Figures 12-3 and 12-4, we find that reflection through the yz plane interchanges numerical values that are equal in magnitude but opposite in sign. The $2p_x$ orbital is **antisymmetric** with respect to the σ operation. Taking the same arbitrary example as before, the upper-right-hand value of $+0.016$ is interchanged by reflection through the yz plane with a value of -0.016 (the number in the upper-*left*-hand corner). Application of the σ operation changes the sign of the $2p_x$ state function at each point in space, without altering its magnitude.

Remember: antisymmetric = symmetric, with a change of sign.

4. ROTATION OPERATION C_2. When the entire plane of inquiry in Figure 12-4 is rotated 180° (half of a full revolution) around the z axis, the contour lines of the rotated plane superimpose on contour lines of the original plane. The contour-line labels all have changed sign. The $2p_x$ state function is antisymmetric with respect to the C_2 rotation.

We may summarize the symmetry properties of the oxygen $2p_x$ orbital with respect to the symmetry operations of the water molecule by means of the table

	I	σ	σ'	C_2
$2p_x$	$+$	$-$	$+$	$-$

where a $+$ sign stands for a symmetric result, and a $-$ sign for an antisymmetric result, when the $2p_x$ state function is subjected to the indicated operation. You may verify (and you are urged to do so) that the symmetry properties of the other oxygen atomic orbitals can also be expressed in the same way. The results are presented in Table 12-2.

TABLE 12-2
Symmetry Properties of the Oxygen Orbitals in H_2O

Oxygen Orbital	I	σ	σ'	C_2
$1s$	$+$	$+$	$+$	$+$
$2s$	$+$	$+$	$+$	$+$
$2p_x$	$+$	$-$	$+$	$-$
$2p_y$	$+$	$+$	$-$	$-$
$2p_z$	$+$	$+$	$+$	$+$

If there is to be a combination of atomic orbitals into a molecular orbital, the symmetry rule (Guideline II, Section 12-5) requires that all orbitals to be combined show the same results in a table like Table 12-2; each orbital in the combination must be symmetric (or antisymmetric) with respect to the same symmetry operations. In forming molecular orbitals that are bonding, we must include atomic orbitals centered on the hydrogen nuclei. Therefore, our next move should be to categorize similarly those hydrogen-centered orbitals in the water molecule. However, when we try to do so, we quickly discover that these orbitals are neither symmetric nor antisymmetric *as individuals*. For purposes of combining into molecular orbitals, each hydrogen-centered orbital is useless *by itself*. For example, consider a 1s orbital function centered on one of the hydrogen nuclei (which we shall label nucleus A). Applying the C_2 operation to this orbital, we find that $1s_A$ is transformed into $1s_B$ (where $1s_A$ is a 1s orbital centered on nucleus A, and $1s_B$ is an identical 1s orbital centered on the other nucleus, labeled B). *This is not a proper symmetry relationship;* neither $1s_A$ nor $1s_B$ *alone* partakes of the full symmetry of the water molecule, and these orbitals may not be used alone in the construction of a symmetry-correct molecular orbital. We can wiggle out of this problem neatly, however, by using the linear combinations $\{1s_A + 1s_B\}$ and $\{1s_A - 1s_B\}$. When both hydrogen orbitals are used in the combination, the symmetry rule is obeyed, as we see in Table 12-3. For example, the σ' operation interchanges the 1s state functions centered on nuclei A and B, with the following results:

Linear combinations are introduced in Sec. 12-5.

TABLE 12-3
Symmetry Properties of Combined Hydrogen 1s Orbitals in H_2O

Combined Orbital	I	σ	σ'	C_2
$\{1s_A + 1s_B\}$	+	+	+	+
$\{1s_A - 1s_B\}$	+	+	−	−

The σ' reflection of $\{1s_A + 1s_B\}$ gives $\{1s_A + 1s_B\}$, which is equal to the original function $\{1s_A + 1s_B\}$. This linear combination is symmetric with respect to the σ' operation, and so + is entered in Table 12-3.

The σ' reflection of $\{1s_A - 1s_B\}$ gives $\{-1s_A + 1s_B\}$ which is equal to $-\{1s_A - 1s_B\}$, the negative of the original function. This linear combination is antisymmetric with respect to the σ' operation, and so − is entered in Table 12-3.

Comparing Tables 12-2 and 12-3, we find several ways to collect together the atomic orbital functions that have the same symmetry properties. These collections are given in Table 12-4. We could go on to characterize higher-energy orbitals such as the 3s, 3p, and 3d orbitals on oxygen and the 2s and 2p orbitals on hydrogen. But we expect that the lowest-energy molecular orbitals will result from combination of the lowest-energy atomic orbitals, and we already have parts at hand to construct more than enough molecular orbitals for our 10 electrons.

Now that we have collected together the atomic orbitals having identical symmetry properties under the symmetry operations of the water molecule, we can make symmetry-correct molecular orbitals by combining atomic orbitals that are in the same collection[6] in Table 12-4. To choose symmetry-correct combinations that will lead to good bonding, we shall use two further criteria:

---

[6] These collections are technically known as "representations."

TABLE 12-4
Classification of Oxygen and Hydrogen Atomic Orbitals Into Collections of the Same Symmetry

Orbitals	I	σ	σ'	C_2
Oxygen $1s$ Oxygen $2s$ Oxygen $2p_z$ Hydrogen $\{1s_A + 1s_B\}$	+	+	+	+
Oxygen $2p_y$ Hydrogen $\{1s_A - 1s_B\}$	+	+	−	−
Oxygen $2p_x$	+	−	+	−

GUIDELINE III. The superimposed (added or subtracted) atomic-orbital functions should lead to a high electron density between oxygen and hydrogen nuclei. It is this concentration of electron density between nuclei, after all, that leads to the electrical attraction of bonding.

GUIDELINE IV. The atomic orbitals to be combined should be similar in energy. If this guideline is violated and, as a consequence, part of a molecular orbital represents a deep well of potential energy relative to the other parts, electrons in that molecular orbital will simply spend almost all of their time in the low-potential-energy region near the nucleus that contributed the low-energy atomic orbital. Then the electrons will not be between the nuclei, the region where they must be for good bonding, and the molecular orbital will not contribute much to holding the nuclei together.

Because of Guideline IV, we shall not combine hydrogen $1s$ orbitals with an oxygen $1s$ orbital, even though such combinations meet the symmetry requirement. The $+8$ nuclear charge of oxygen gives the oxygen $1s$ orbital a much lower energy than any of the others we have considered. The oxygen $1s$ orbital will be left alone, to be filled with two electrons; the combination of the oxygen nucleus and a filled oxygen $1s$ orbital is known as the oxygen **core** or **kernel**. Most approximate molecular-orbital calculations involve only the valence-shell electrons, with atomic cores considered to be isolated units. However, the oxygen valence-shell orbitals (the $2s$ and the three $2p$ orbitals) and the hydrogen $1s$ orbital are all about equal in energy; recall (Chapter 6) that the ionization energies of oxygen and hydrogen are each about 14 eV.

This is simply an abstract way of saying that a charge of $+8$ attracts electrons much more strongly than a charge of $+1$.

What does the ionization energy measure?

Let us now get down to the business of constructing molecular orbitals for H_2O from these atomic orbitals.

BONDING ORBITALS. The combination of orbitals from Table 12-4 that gives the best overlap in the bonding region between nuclei is

$$\psi_1 = \{1s_A + 1s_B\}_{\text{hydrogen}} + \{2s + 2p_z\}_{\text{oxygen}} \tag{12-6}$$

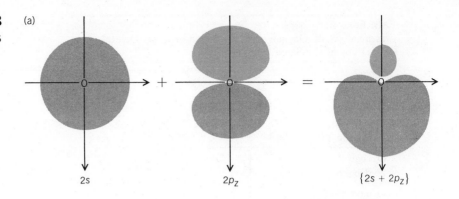

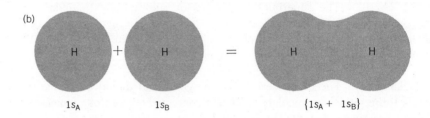

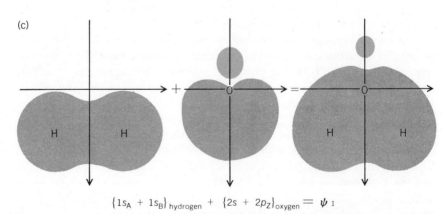

FIGURE 12-7 The construction of the bonding orbital ψ_1. (a) Combination of the oxygen $2s$ and $2p_z$ orbitals. Notice that the positive part of the $2s$ orbital combines with the $2p_z$ orbital to increase the positive lobe (XXblueXX), and, by cancellation, to shrink the negative lobe (YYgreenYY). (b) Combination of the hydrogen $1s$ orbitals. Because the signs of the orbital functions are everywhere positive, there is an increase in the electron density where they overlap. (c) Combination of the hydrogen and oxygen contributions to ψ_1. Overlap of the positive part of the oxygen contribution with the positive hydrogen contribution makes for large values of electron density in the regions between the oxygen and hydrogen nuclei, and thus makes for good bonding.

Note that in this combination, we follow Guideline II, which requires that the atomic-orbital ingredients have the same symmetry characteristics. This combination of orbitals was chosen because it places lots of electron density in the region between the oxygen and hydrogen nuclei (Figure 12-7). It is a **bonding** orbital.

Another bonding orbital can be made from the combination

$$\psi_2 = \{1s_A - 1s_B\}_{\text{hydrogen}} + \{2p_y\}_{\text{oxygen}} \tag{12-7}$$

All of these atomic orbitals are taken from the second collection in Table 12-4. The construction of this molecular orbital is shown schematically in Figure 12-8.

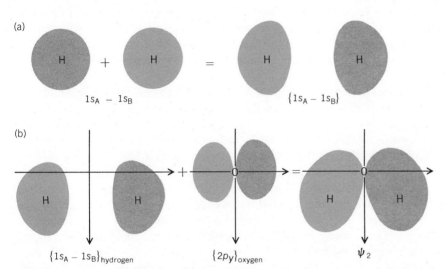

FIGURE 12-8 Construction of the bonding orbital ψ_2. (a) Combination of the hydrogen 1s orbitals. Since these orbitals are now combined with opposite signs, there is some cancellation in regions where they overlap, leading to a decrease in electron density between the nuclei and a compensating increase outside this internuclear region. (b) Combination of hydrogen and oxygen orbitals to form ψ_2. Overlap of orbital functions identical in sign on either side of the z axis leads to an increase in electron density between the oxygen and hydrogen nuclei. But since both atomic contributions to ψ_2 have a plane of zero electron density (called a *nodal plane*), ψ_2 itself also has a nodal plane; this nodal plane is coincident with the xz (σ') plane.

We now have, counting the 1s atomic orbital of the oxygen core, enough orbitals to accommodate 6 of the 10 electrons of a water molecule. However, we have used up all of the *additive* combinations of compatible atomic-orbital functions in Table 12-4. There are other possibilities for combination, but we shall find either that they will have no effect on bonding (they are **nonbonding** orbitals), or in fact are destabilizing because they have the effect of removing electrons from the region between nuclei (they are **antibonding** orbitals). Since we need at least two more symmetry-correct orbitals for the four remaining electrons, we shall now turn to nonbonding combinations.

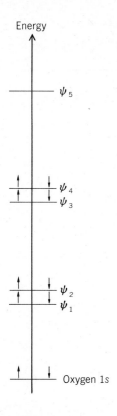

FIGURE 12-9 Energies and occupation of orbitals in the water molecule. Two electrons occupy each of five orbitals, with spins opposed in accordance with the Pauli principle. Note that the energy of the oxygen 1s orbital is considerably below that of any other orbital, and that the energies of the bonding orbitals ψ_1 and ψ_2 are below those of the nonbonding orbitals ψ_3 and ψ_4. (Electrons in nonbonding orbitals have about the same energy as they would in isolated O and H atoms.) The high-energy orbital ψ_5 is an antibonding orbital; its significance is discussed in Section 12-7. Since this pattern of orbital occupation places electrons as low in energy as possible, it is a model of the *ground state* of the water molecule.

NONBONDING ORBITALS. The isolated $2p_x$ orbital on oxygen has no possibility of entering into a bonding molecular orbital by combination with the hydrogen orbitals. This atomic orbital is itself a nonbonding molecular orbital:

$$\psi_3 = \{2p_x\}_{\text{oxygen}} \qquad (12\text{-}8)$$

Other nonbonding molecular orbitals arise from symmetry-allowed combinations of oxygen and hydrogen atomic orbitals that do not place electrons between the oxygen and hydrogen nuclei. Consider, for example, the function

$$\psi_4 = \{1s_A + 1s_B\}_{\text{hydrogen}} + \{2p_z - 2s\}_{\text{oxygen}} \qquad (12\text{-}9)$$

The negative sign in the last part of this linear combination has the effect of increasing the state function in the region away from the hydrogen atoms, just as the positive sign in the corresponding part of ψ_1 directs the state function toward the hydrogens. Since there is little or no added electron density between the oxygen and hydrogen nuclei when this orbital is occupied, it contributes nothing to the bonding. It is another nonbonding orbital.

THE COMPLETE MOLECULE. We have now assembled 5 orbitals for the 10 electrons of H$_2$O. They are summarized in an energy diagram in Figure 12-9. The resulting model of the water molecule, in somewhat more descriptive words, is this: Two electrons (in the oxygen 1s orbital) are practically unaffected by the fact that the oxygen atom happens to be in a water molecule. They and the oxygen nucleus constitute the oxygen core. Another four electrons are located in molecular orbitals labeled 1 and 2 (with state functions ψ_1 and ψ_2) that place them between the oxygen core and the hydrogen nuclei. The electrical attraction between the positive nuclei and these electrons is the force that holds the molecule together, so these molecular orbitals are called *bonding orbitals*. Finally, four electrons occupy orbitals (with state functions ψ_3 and ψ_4) that place them near the oxygen core, but out of the internuclear region. Since these electrons are in an energetic environment not much different from that of an unreacted oxygen atom, the electrons and the orbitals they occupy are given the discriptive tag *nonbonding*.

You are urged to return to Chapters 2 and 4 and compare the models of the water molecule discussed there. You should find that each model accounts for some of the properties of H$_2$O molecules (and thus of water itself), and that with each successive model some detail has been added to the picture.

12-7 Antibonding Orbitals and the Photodissociation of Water

High over your head,[7] water vapor in the upper atmosphere is being bombarded by high-energy light from the sun. As a result, some of it is decomposed into atoms by the reactions

$$H_2O \longrightarrow H + OH$$

---

[7] Or, if you are reading this at night, "high" under your feet.

and (12-10)

$$OH \longrightarrow H + O$$

If you wonder how this process can come about, you might recall (Chapter 5) that photons can be absorbed by matter if they have the right energy to promote electrons from low-energy, occupied orbitals to high-energy, unoccupied orbitals. In the case of the water molecule, we might speculate on what high-energy orbitals there may be, and what electron distributions they would produce. We have not exhausted the possibilities for symmetry-correct linear combinations of atomic orbitals from Table 12-4 (although we have run out of *bonding* combinations). Consider, for example, the combination

$$\psi_5 = \{1s_A - 1s_B\}_{\text{hydrogen}} - \{2p_y\}_{\text{oxygen}} \quad (12\text{-}11)$$

State function ψ_5 is just like ψ_2 except for the minus sign joining the hydrogen and oxygen contributions. The effect of that minus sign is shown in Figure 12-10. The opposing signs of the hydrogen and oxygen contributions to the state function make the state function small in the internuclear region, where the hydrogen and oxygen orbitals overlap. With relatively little electron density in this region, the nuclei are left to confront (and repel) each other across a nearly empty region of space. This is a region where bonding would occur if electrons were present. Instead, a pair of electrons in orbital 5 actually destabilizes the molecule; this is an *antibonding* molecular orbital.

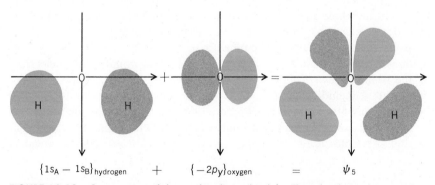

FIGURE 12-10 Construction of the antibonding orbital ψ_5. The orbitals used to construct the bonding orbital ψ_2 are combined with opposing signs. The resulting partial cancellation of the wave function in the region of overlap causes a small electron density in the region between the oxygen and hydrogen nuclei. Compare Figure 12-8.

Also, because an electron in orbital 5 is kept outside the positive electrical field between the nuclei, it is not held to those nuclei very tightly; it has a relatively high potential energy, and thus ψ_5 is a high-energy orbital state function (see Figure 12-9). It is not occupied in the ground state (Chapter 2) of the water molecule. However, when a photon of the proper energy interacts with a water molecule, electrons may be promoted from bonding or nonbonding up to antibonding orbitals. When that happens, the nuclei are no longer held together, and their mutual repulsions tear the water molecule apart—**photodissociation** occurs.[8]

12-8 Summary

In this molecular model of chemical bonding, electrons hold together the nuclei by occupying orbitals called bonding molecular orbitals whose domains lie mainly between nuclei. These molecular orbitals, like their nonbonding and high-energy antibonding counterparts, are described mathematically by molecular-orbital state functions. Since correct state functions cannot be deduced by solving the Schrödinger equation for more than two interacting particles (that is, we cannot get an exact solution for any chemical entity more complex than the hydrogen atom), approximate state functions are constructed as guesses at the undiscoverable "true" functions. The most commonly used method for constructing trial molecular-orbital functions is by adding and subtracting hydrogenlike atomic-orbital functions.

To prevent this combination process from being simply a series of haphazard guesses, we take advantage of the reasonable postulate that the overall electron distribution in a molecule has the same symmetry as the arrangement of nuclei. The state function that produces this symmetry of electron density (by being squared) must itself be either symmetric or antisymmetric with respect to the symmetry operations of the molecule. Also, the orbitals to be combined should be similar in energy.

Application of these principles to the water molecule produces an array of atomic and molecular orbitals that may be assumed to be occupied by zero, one, or two electrons. Two molecular-orbital functions (ψ_1 and ψ_2) were found to place electronic charge predominantly in the area between the oxygen and hydrogen nuclei, and so these are bonding molecular orbitals. Only 4 electrons of the 10 in H_2O can occupy these two orbitals, and the rest occupy nonbonding orbitals, including the oxygen $1s$ orbital, too low in energy to combine well with any other orbital function of either oxygen or hydrogen.

While discussed here in detail only for the H_2O molecule, the molecular-orbital principles apply to every molecule and to every case of covalent chemical bonding in nature.

GLOSSARY

antibonding: having the effect of weakening or preventing bonding in a molecule; said of an orbital, and of electrons in antibonding orbitals.

antisymmetric: having the property of reversing direction or sign as the result of a symmetry operation; said of a vibration (Chapter 4), a mathematical function, or an orbital.

---

[8] Sometimes the integrity of an excited molecule can be preserved by the emission of light with the resulting restabilization of the molecule. Such light emission gives rise to the phenomena of fluorescence, phosphorescence and bioluminescence, all of which are discussed in P. G. Seyhold, "Luminescence," *Chemistry*, vol. 46, no. 2, p. 7 (Feb. 1973).

bonding: having the effect of holding together nuclei in a molecule through electrical attraction; said of electrons and of those orbitals that place electrons between nuclei so as to stabilize a molecule.

contour line: on a map or other diagram, a line connecting points with equal values of some numerical variable; on a representation of an orbital, a line connecting points with equal values of the state function ψ, or with equal values of the electron density, ψ^2.

core: the nucleus plus inner shells of electrons—an atomic core.

kernel: an atomic core (q.v.).

linear combination: a combination produced by adding or subtracting, including combinations resulting when the variables are multiplied by constants before addition (or subtraction). The following two expressions are linear combinations of the variables x and y: $(x + y)$; $(x - y)$; $(Ax + By)$, where A and B are constants.

nonbonding: having no effect on the bonding in a molecule; said of orbitals and of electrons in such orbitals.

orbital: a domain in space, described by a mathematical function of the space variables (the electronic state function, ψ), which may be occupied by zero, one, or two electrons.

photodissociation: the breakage of chemical bonds by input of energy from the absorption of light.

plane of inquiry: A plane created (physically or mathematically) by choosing to fix the value of one of the space coordinates. In effect, the plane exposed by a slicing operation.

state function: a mathematical function that describes the properties of an object or collection of objects when in a particular state; that is, a function that specifies the state of a system when particular values of variables (such as energy, or quantum numbers) are chosen.

PROBLEMS

12-1 Consider the electron distribution in an isolated H_2 molecule:

a. Sketch the electron distribution (draw contour lines) that would result from the molecular orbital

$$\psi_I = \{1s_A + 1s_B\}$$

for the H_2 molecule. ($1s_A$ is a $1s$ orbital centered on atom A, and $1s_B$ is a $1s$ orbital centered on atom B.)

b. Does this molecular orbital partake of the full symmetry of the H_2 molecule?

c. The function ψ_I is a bonding molecular orbital for H_2. How could an antibonding orbital be constructed?

d. Sketch the electronic distribution that should result from your antibonding orbital. Explain briefly why electrons occupying this orbital could not prevent the hydrogen nuclei from repelling each other and dissociating the molecule.

12-2 Suppose the positive sign between the contributions from hydrogen and oxygen in the expression for ψ_1 [Equation (12-6)] were changed to a negative sign.

a. Using Figure 12-7 for comparison, sketch the electron distribution that would result.
b. Would this molecular orbital be bonding or antibonding? Explain.
c. Determine whether this orbital is symmetric or antisymmetric with respect to each of the symmetry operations of the water molecule.

12-3 Calculate the wavelength of the photon that has just enough energy to break an O—H bond in water, if the bond energy is 117.8 kcal/mole.

12-4 The process of "pumping" on a swing is an everyday example of the addition of two periodic (wave or state) functions (the overall motion of the swing and the pumping motion of the swinger's body) in a symmetry-correct combination to produce a larger wave function (higher swinging).

a. What are the symmetry elements of the curved path swept out by a swinging child?
b. Is it possible to swing while holding on with only one hand? Is it possible to "pump" this way? (If you are not sure, go out and try it.) Why, or why not? Is symmetry involved in this problem? How?
c. Is there an "antipumping" mode of swinging whose effect is to reduce the overall motion? (Again, if you are not sure, go try it.) If so, how is it related to the pumping mode?

12-5 a. Give a verbal summary of the general locations in space of the ten electrons in H_2O, according to the molecular orbital description of this chapter.
b. Give an informal account of what holds the water molecule together, according to this description.
c. Compare your answers to parts a. and b. of this problem to those implied by the Lewis "dot diagram" for H_2O introduced in Section 3-5.

SUGGESTIONS FOR FURTHER READING

The ideas, concepts, and methods of this chapter are extended in a very readable paperback; URCH, D. S., *Orbitals and Symmetry,* Harmondsworth, Middlesex, England: Penguin Books (1970). This book contains some excellent illustrations depicting molecular orbitals.

Three clearly written and authoritative books about molecular orbitals are: ORCHIN, M., and JAFFÉ, H. H., *Symmetry in Chemistry,* New York: Wiley (1965) (paperback); *The Importance of Antibonding Orbitals,* Boston: Houghton Mifflin (1967) (paperback); and *Symmetry, Orbitals, and Spectra (S.O.S.),* New York: Wiley-Interscience (1971).

Symmetry considerations play a central role in discussions of molecular orbitals, and a systematic treatment of symmetry, using the methods of mathematical group theory, has proved a powerful aid to understanding molecular bonding. Two recommended texts are: HALL, L. H., *Group Theory and Symmetry in Chemistry,* New York; McGraw-Hill (1969); and COTTON, F. A., *Chemical Applications of Group Theory,* 2nd ed., New York: Wiley-Interscience (1971).

A large group of computer-drawn representations of molecular orbitals is presented in the book: STREITWIESER, A., and OWENS, P. H., *Orbital and Electron Density Diagrams,* New York: Macmillan (1973) (paperback).

13
Around the Coordination Sphere

I described a bond, a normal simple chemical bond; and I gave many details of its character (and could have given many more). Sometimes it seems to me that a bond between two atoms has become so real, so tangible, so friendly that I can almost see it. And then I awake with a little shock: for a chemical bond is not a real thing; it does not exist: no one has ever seen it; no one ever can. It is a figment of our imagination.

C. A. COULSON[1]

13-1 Abstract

Theories about the way molecules are put together and how they stay together are essential to the understanding of chemistry. Yet these theories, particularly if they lead to predictions that conflict with experimental observations, can stand in the way of our understanding of molecules and their reactions. A new theory, even though imperfect, can carry new glimpses of truth, bring fresh insights, and stimulate research activity; the same theory, years later as a part of textbook dogma, can freeze thinking, stifle imagination, and impede research progress. We shall look at three times in the history of molecular theory when a young man with fresh ideas started and helped carry off an intellectual revolution. These men are August Kekulé (1839–1896), Alfred Werner (1866–1919), and Gilbert Newton Lewis (1875–1946). The story, as we shall tell it, revolves about compounds of the transition-metal cobalt. We shall learn about some of the chemistry of cobalt, about some chemistry shared by most of the transition metals, and some more about the general principles of chemical bonding. We begin in Europe, in the middle of the nineteenth century.

13-2 Kekulé, His Models, and Carbon Compounds

In 1850, chemistry had become stalled on the central question of the structure of individual molecules. Continued progress in chemistry seemed to require some general agreement, however speculative and tentative, about the structures of the most common molecules. There really were no adequate laboratory data for establishing such structures, but at least some plausible structures would have served better than no structures at all. To facilitate communication among chemists, some simple and reasonable rules were needed. These rules were needed to provide guidance for predicting the structure of a molecule in the absence of detailed experimental data. Such rules should yield very few predictions in direct conflict with experimental observations.

---

[1] C. A. Coulson, Tilden Lecture: "The Contributions of Wave Mechanics to Chemistry," *Journal of the Chemical Society* (London), vol. **1955**, p. 2084.

And these rules needed rather general acceptance throughout the scientific community. The time was ripe for such new ideas.

The conceptual breakthrough came in the late 1850s when a new generation of chemists decided to proceed as if carbon compounds were composed of three-dimensional molecules, with the arrangement of nuclei around each carbon atom ordinarily tetrahedral, and with each carbon atom having a **combining capacity** of four.

August Kekulé was one of the first proponents of structural chemistry in three dimensions. He persuasively advocated and persistently used models for carbon compounds in which each carbon atom had four tetrahedrally oriented bonds. His work was an important factor in the dramatic rise of organic chemistry in the 1860s and 1870s. Kekulé not only used spatial models; he also acted as if he believed the consequences of his models. Thus, when his models permitted chains of carbon atoms linked together, he proposed such structures. His models permitted a closed ring of carbon atoms, and he proposed a ring structure for the benzene molecule. His models had four bonds for each carbon atom, and he proposed a combining capacity of four for every carbon atom.

One of the most significant simplifying propositions advanced by Kekulé was the idea of constant combining capacity of elements. In 1857, the young man wrote[2]

Three-dimensional models for molecules are described in Sec. 4-4.

In a molecule of a compound, the number of atoms of an element or of a radical bound to an atom of another element (or radical) is dependent upon the basicity or affinity numbers of the constituents of the molecule.

Elements fall in this respect into 3 principal groups:

1. Unibasic   (I) H, Cl, Br, K
2. Dibasic   (II) O, S
3. Tribasic (III) N, P, As

Carbon is easily shown to be tetrabasic: that is, 1 atom of carbon is equivalent to 4 atoms of H.

Kekulé's rules, giving hydrogen a combining capacity of one, oxygen a combining capacity of two, and carbon a combining capacity of four, are consistent with the experimentally observed structures for a multitude of compounds. The rules are adequate for millions of compounds. Although the rules fail for carbon monoxide, they will seldom lead you astray if you confine your attention to the compounds of carbon, hydrogen, and oxygen. This concept of constant combining capacity, often rigidly applied, was extremely successful in bringing order into the rapidly expanding field of synthetic carbon chemistry, the domain of the organic chemists of the late 1800s. The rules still work well for most compounds of carbon, as we shall see in Chapter 14.

---

[2] A. Kekulé, *Liebig's Annalen der Chemie und Pharmacie*, vol. **104**, p. 129 (1857); trans. from German in W. G. Palmer, *A History of the Concept of Valency to 1930*, Cambridge: Cambridge University Press, p. 52 (1965).

13-3 Werner: Expansion of the Notion of Combining Capacity

A generation after the innovative ideas of Kekulé had been proposed and used so effectively, some of his ideas were causing trouble. The constant-combining-capacity concept, so fruitful in carbon chemistry, was being applied rigidly to all elements and was blocking progress in understanding a large number of new compounds involving metals. Twenty-six-year-old Alfred Werner had spent a great deal of time thinking about the metal ammines, a group of compounds which contemporary bonding theory could not explain.

With typical romantic fervor, Werner awoke one night at 2 A.M. with the solution to the riddle, which had come to him like a flash of lightning. He arose from his bed and wrote so furiously and without interruption that by 5 P.M. the following day, he had finished his most important paper, in which he proposed his now famous coordination theory.[3]

He considered a confusing series of compounds of cobalt:

$Co(NH_3)_4Cl_3$ green or violet, depending on how the compound is prepared

$Co(NH_3)_5Cl_3$ purple

$Co(NH_3)_6Cl_3$ yellow

which he decided to formulate as

$[Co(NH_3)_4Cl_2]^+Cl^-$

$[Co(NH_3)_5Cl]^{2+}(Cl^-)_2$

$[Co(NH_3)_6]^{3+}(Cl^-)_3$

Werner suggested that not all the chlorine atoms are directly bound to cobalt.

Werner then began to speculate about the structure of the cobalt-containing species written above within the brackets. Werner's paper presented revolutionary ideas with little experimental evidence. Bitter attacks came from the traditionalists who demanded evidence, not conjecture. Werner prevailed, however, after the accumulated evidence of a quarter century of prolific investigation in his laboratory, research that culminated in the Nobel Prize in 1913. During these years, Werner and his students prepared and characterized thousands of compounds; there still exists a collection of over 8000 substances prepared in his laboratory. We shall now examine Werner's coordination theory in some detail.

Werner postulated a general scheme of bonding in which a cluster of atoms or molecules surrounds a central atom. Each such atom or molecule (we now call these atoms or molecules *ligands*) in the cluster is directly bonded to the central atom. In methane, there are four hydrogen ligands bound to the central

---

[3] G. B. Kauffman, *The Selected Papers of Alfred Werner*, New York: Dover, p. 5 (1968). Werner's paper, "Contribution to the Constitution of Inorganic Compounds," is given in English translation on pp. 9–88; the original appeared in *Zeitschrift für anorganische Chemie*, vol. 3, p. 267 (1893).

carbon. In $Co(NH_3)_3Cl_3$, there are three ammonia ligands and three chloride ligands bound to the central cobalt. The possibilities for different geometrical arrangements with six ligands in the cluster are great, and Werner exploited these geometrical possibilities to the fullest. Each arrangement of ligands yields a distinct compound; these compounds are called **isomers** of one another. We shall now examine the geometry of such complexes in which six ligands are coordinated to a central atomic nucleus, drawing most of our examples from the work of Werner.

13-4 *cis* and *trans* Isomerism in Octahedral Coordination Complexes

Six ligands, arranged in a cluster around a central metal nucleus so that the ligand-metal distance is the same for all ligands and so that the ligands are as far apart as possible from one another, lie at the apices of an imaginary regular octahedron (see Table 4-3). For example, the six nitrogen nuclei in the cobalt(III)hexammine ion, $Co(NH_3)_6^{3+}$, are in symmetry-equivalent (and therefore chemically equivalent) positions in the octahedral configuration

An octahedron has 8 faces but only 6 apices.

If we arbitrarily pick out one nitrogen nucleus and say that it is in position 1, then there are four nearest-neighbor nitrogens said to be *cis* to nitrogen 1, and a single nitrogen which is farther away and said to be *trans* to nitrogen 1. (*trans*-: from Latin, across; *cis*-: from Latin, on this side.)

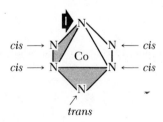

An imaginary line passing through any two nuclei *trans* to each other also passes through the central cobalt nucleus. There are three σ reflection planes containing four nitrogen nuclei and the cobalt nucleus, reflecting the remaining two nitrogen nuclei into each other. There are also planes containing just two nitrogen nuclei and the cobalt nucleus. Each nitrogen is itself located within an imaginary pyramid consisting of the central cobalt nucleus, one of the nitrogen nuclei, and the three hydrogens associated with that nitrogen. Thus, each nitrogen is enclosed in a cage like this:

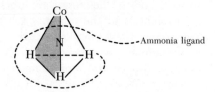

Note that all the lines in these figures have been drawn as aids for visualizing the three-dimensional aspects of this complex; none of the lines indicates the location of a chemical bond.

Now consider the cobalt(III)dichlorotetrammine ion, $Co(NH_3)_4Cl_2^+$, a chemical species in which the four nitrogen nuclei of the ammonia ligands, and the two chlorine nuclei, lie at the apices of a slightly distorted regular octahedron. *There are two different ways for these ligands to be arranged.* If the two chloride ligands occupy positions *trans* to each other, then the ion has the structure

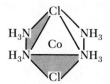

trans-Co $(NH_3)_4Cl_2^+$

which is exactly the same as

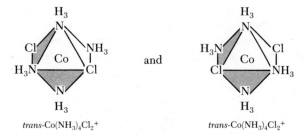

trans-Co$(NH_3)_4Cl_2^+$ *trans*-Co$(NH_3)_4Cl_2^+$

But if the chlorides are in positions *cis* to each other, the ion is

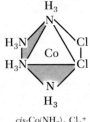

cis-Co$(NH_3)_4$ Cl_2^+

How many molecular symmetry elements (Sec. 4-5) can you find in these complexes?

Note that there are several ways of writing this structure on two-dimensional paper, but only one way of constructing a three-dimensional molecular model.

Salts of the *trans* ion, such as the chloride *trans*-[Co(NH$_3$)$_4$Cl$_2$]Cl, have an intense green color, whereas salts of the isomeric *cis* ion, such as *cis*-[Co(NH$_3$)$_4$Cl$_2$]Cl, have an intense cobalt-blue color. Werner[4] prepared the chloride, bromide, iodide, nitrate, dithionate, and sulfate salts of the *cis* ion to demonstrate that different compounds (they must be different compounds, because they have different colors) result when the arrangement in space of substituents around a central metal ion is changed. He felt that the existence of two different ions with the same overall chemical composition but with different properties was strong evidence for the octahedral configuration of ligands around the central cobalt. Certainly, *if* the actual arrangement is octahedral, there would be two possible isomers. There is, however, a more stringent test that we shall now investigate.

13-5 Optical Rotation and Isomerism

At meetings of the Institut de France on October 15 and 30, 1815, Jean Baptiste Biot reported[5] that polarized light, passing through turpentine, laurel oil, lemon oil, or a camphor solution, was rotated. An explanation proposed by Augustin Fresnel was that this rotation

> may result from a peculiar constitution of the refracting medium or of its molecules, which produces a difference between the directions right to left and left to right; such, for instance, would be helicoidal arrangement of the molecules of the medium. . . .[6]

A helix is either clockwise or counterclockwise in its spiraling; thus, it is either right-handed or left-handed. Therefore, a helix is not identical with its own mirror image. Louis Pasteur focused attention on superposability of mirror images:

Are your left and right hands superimposable?

> When we study material things of whatever nature, as regards their forms and the repetition of their identical parts, we soon recognize that they fall into two large classes of which the following are the characters. Those of the one class, placed before a mirror, give images which are superposable on the originals; the images of the others are not superposable on their originals, although they faithfully reproduce all the details. A straight stair, a branch with leaves in a double row, a cube, the human body—these are of the former class. A winding stair, a branch with the leaves arranged spirally, a screw, a hand, an irregular tetrahedron—these are so many forms of the other set. The latter have no plane of symmetry.[7]

Does your hand have a plane of symmetry?

---

[4] Werner reported these syntheses and discussed the significance of this series of compounds in a paper which was published in *Berichte der Deutschen Chemischen Gesellschaft*, vol. **40**, p. 4817 (1907). For an English translation, see G. B. Kauffman, *The Selected Papers of Alfred Werner*, New York: Dover, p. 144 (1968).

[5] Published in *Bulletin de la Société Philomatique*, p. 190 (1815).

[6] A. Fresnel, *Bulletin de la Société Philomatique*, p. 147 (1824); trans. from French by T. M. Lowry, *Optical Rotatory Power*, New York: Dover, p. 19 (1964).

[7] Delivered on January 20 and February 3, 1860, before the Société Chimique de Paris; trans. from French in *Alembic Club Reprints*, vol. **14**, p. 26.

Pasteur attributed the rotation of polarized light by a liquid or by a solution to molecular dissymmetry—to the handedness of individual molecules.[8] This interpretation is also the modern interpretation. A solution containing dissymmetric molecules (molecules that are not superimposable on their own mirror images) will rotate polarized light, unless there happen to be equal numbers of the left- and right-hand forms. The solution and the molecules are said to be *optically active*. The left- and right-hand forms of dissymmetric molecules are called *optical isomers*.

For a tetrahedral arrangement of ligands surrounding a central nucleus, optical rotation will be observed if each of the four ligands is different. This fact has been exploited in many investigations of the structure of carbon-containing compounds. (For discussion of isomerism in carbon compounds, see Section 15-6.)

If experiments show that solutions containing a particular cobalt complex rotate polarized light, then the only acceptable structures for that complex are those that are nonsuperimposable on their own mirror images. Werner synthesized a series[9] of complexes containing the ligand ethylenediamine,

$$
\begin{array}{c}
\text{H} \text{H} \\
\text{H} | | \text{H} \\
\diagdown | | \diagup \\
\text{N}\text{---}\text{C}\text{---}\text{C}\text{---}\text{N} \\
\diagup | | \diagdown \\
\text{H} | | \text{H} \\
\text{H} \text{H}
\end{array}
$$

ethylenediamine, en

Both nitrogen atoms of an ethylenediamine molecule can bond to a central cobalt atom. Werner prepared two different compounds with the formula [Co en$_2$ NH$_3$ Br]Br$_2$. Most properties of the two compounds were found to be identical. For instance, both compounds are dark, red-violet, needlelike crystals. However, Werner reported that a solution of one compound (labeled with the prefix *l* for **levo**: left) rotates polarized light $-196.2°$ per mole, whereas a solution of the other (labeled *d* for **dextro**: right) rotates polarized light $+201.65°$ per mole. If these numbers are considered of equal magnitude (within experimental uncertainties) although of opposite sign, then it makes sense to consider molecules of these two isomers to be mirror images of each other. Werner proposed the following two structures for the [Co en$_2$ NH$_3$ Br]$^{2+}$ cations:

---

[8] Pasteur used the phrase "une dissymetrie dans les molecules" as early as 1848 when he was 24 years old; see L. Pasteur, *Annales de Chimie et de Physique*, ser. 3, vol. **24**, p. 443 (1848).

[9] A. Werner, "Toward an Understanding of the Asymmetric Cobalt Atom," *Berichte der Deutschen Chemischen Gesellschaft*, vol. **44**, p. 1887 (1911); English trans. in G. B. Kauffman, *The Selected Papers of Alfred Werner*, New York: Dover, p. 159 (1968).

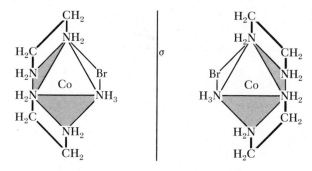

Neither mirror-image molecule has a mirror plane within itself.

Neither mirror-image structure can be superimposed on the other. These are two different molecular species. The two structures bear the same relation as left-hand and right-hand gloves, and so could properly be designated as the right and left forms. Here we do not know which is which, but if one is called right-handed, then the other is left-handed.

Werner's great influence on chemistry was in the opening of new vistas. Chemistry, he demonstrated, is not confined to molecules with just one, two, three, or four groups around an atom. He showed that a lot can be learned about the structures of rather complicated compounds, especially with the combination of imaginative minds and productive hands in the laboratory. But the *nature* of the chemical bonds remained a mystery. Werner's legacy was a multitude of compounds whose bonding needed to be explained. And that also was a legacy of nineteenth-century organic chemistry. Insight into the nature of bonds themselves had to await the birth of the electron into science and the new generation of chemists who were to grow up with the electron.

13-6 G. N. Lewis: Bonding by Sharing Electrons

The nineteenth century closed without any explanation for the stability of covalent bonds. The electron had been isolated and studied during the period 1895 to 1897 by J. J. Thomson, and as we know, the twentieth-century explanation was to be that covalent chemical bonding results from sharing of these electrons. Yet it was not enough to say in 1900 simply that electrons bound together atoms. A chemically useful theory had to embody the tetrahedral geometry and fixed combining capacity of carbon elucidated by Kekulé and exploited by a host of successful organic chemists. A chemically acceptable theory also had to account for variable combining capacity of other elements, and for the octahedral geometry of complexes being prepared in Werner's laboratory. And a chemically plausible theory had to be consistent with the periodic behavior of the elements as so clearly set forth by Mendeleef.

G. N. Lewis, a 26-year-old instructor at Harvard in 1902, began to combine notions about electrons, geometry, and chemical bonding. He recalls how it was:

A good 20th-century theory has to deal with 19th-century facts and ideas.

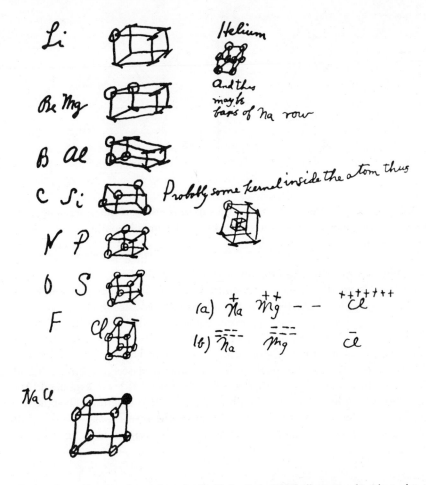

FIGURE 13-1 Memorandum written by G. N. Lewis in 1902, illustrating his ideas about electrons in atoms. Note the kernel of electrons buried inside the atom, and the outer shell that contains from one to seven valence electrons. Compound formation is illustrated with a completed octet of valence-shell electrons. Lewis was wrong in his electronic structure of helium, and he had probably not yet realized the significance of shared electron pairs in describing bonding in covalent compounds. [Reproduced from G. N. Lewis, *Valence and the Structure of Atoms and Molecules,* New York: The Chemical Catalog Co. (1923), by permission of Litton Educational Publishers, copyright owner.]

In the year 1902 (while I was attempting to explain to an elementary class in chemistry some of the ideas involved in the periodic law) becoming interested in the new theory of the electron, and combining this idea with those which are implied in the periodic classification, I formed an idea of the inner structure of the atom which, although it contained certain crudities, I have ever since regarded as representing essentially the arrangement of electrons in the atom. In Figure 13-1 is reproduced a portion of my memorandum of March 28, 1902, which illustrates the theory.

The main features of this theory of atomic structure are as follows:

1. The electrons in an atom are arranged in concentric cubes.
2. A neutral atom of each element contains one more electron than a neutral atom of the element next preceding.

3. The cube of 8 electrons is reached in the atoms of the rare gases, and this cube becomes in some sense the kernal about which the larger cube of electrons of the next period is built.
4. The electrons of an outer incomplete cube may be given to another atom, as in Mg^{2+}, or enough electrons may be taken from other atoms to complete the cube, as in Cl^-, thus accounting for "positive and negative valence."[10]

And so Lewis began to develop a working model for the chemical atom. He did not base this model on the properties of electrons, nor on the quantitative physical properties of charged particles. In the chemical tradition of Kekulé, Werner, and their contemporaries, Lewis was trying to find a model for chemical bonding to provide a language for talking about chemistry, to suggest new and fruitful experimental investigations, and to expose more fundamental questions about molecules and their reactions. As he thought, talked, and taught, Lewis put together a carefully stated and deceptively simple model for chemical bonding. In a paper[11] sent to the *Journal of the American Chemical Society* during the 1915 Christmas recess, he proposed that of the many electrons in most atoms, only a few are available for bonding; that a chemical bond consists of a *shared pair* of electrons; that in a molecule most atoms share *eight* electrons; and that a convenient way to describe bonding is to use electron-dot diagrams (recall Section 3-5).

Lewis says that a chemical bond is a shared pair of electrons.

The idea of chemical bonds as shared electron pairs has proved to be an extraordinarily fruitful concept, both as a descriptive device and as a theoretical tool in guiding the development of chemical quantum theory. The rule of eight turned out to do very well for compounds of carbon, oxygen, nitrogen, hydrogen (where it becomes the rule of two), and so forth, but it must be modified for transition-metal compounds with perhaps six ligands surrounding a single metal nucleus. We shall devote the remainder of this chapter to some of these transition-metal compounds, usually talking about the bonding in terms of shared electron pairs. We shall continue to focus on compounds of cobalt.

13-7 Isomerization of an Octahedral Complex

How could a cis complex become a trans complex?

Pairs of electrons, if they hold molecules together, must surely also hold together the high-energy, unstable intermediates that make possible the transition from reactant to product molecules in a chemical reaction. The difference is that the electrons do a better job of holding together a stable molecule than they do in holding together an unstable intermediate. We shall consider a reaction of interconversion of Co(III) ions having the formula [Co

---

[10] G. N. Lewis, *Valence and the Structure of Atoms and Molecules,* New York: The Chemical Catalog Company, pp. 29–30 (1923); paperback reprint, New York: Dover (1966). This book grew out of a graduate seminar on **valence** conducted by Lewis at the University of California at Berkeley for many years.

[11] G. N. Lewis, "The Atom and the Molecule," *Journal of the American Chemical Society,* vol. **38**, p. 762 (1916).

en$_2$ Cl$_2$]$^+$, and see how many of the fundamental aspects of this reaction can be exposed by keeping track of electron pairs.

Recipes have been developed for synthesis of the *trans* isomer of [Co en$_2$ Cl$_2$]$^+$, as well as for each of the mirror-image *cis* isomers.[12] Under appropriate conditions, each isomer is inert.[13] Yet chemical changes do occur. In solution, at rates that depend on the solvent, the temperature, and the presence of acids or bases, these isomers interconvert. We shall examine a mechanism that describes such interconversions.

The positions of the bonding atoms of the ligands define a **coordination sphere,** a distorted sphere[14] with a cobalt nucleus at its center, and four nitrogen nuclei and two chlorine nuclei lying on its surface. Each ligand bonding atom occupies a region of space such as indicated in Figure 13-2. This space around the cobalt atom is filled with electrons, many of which contribute directly to chemical bonding. Each ligand is constrained to a particular region of the coordination sphere because four other ligands fill the adjacent space. If the equilibrium internuclear distances between the cobalt nucleus and each of the six bonding ligand nuclei are maintained, there is no way to change the relative positions of the ligands.

There are two ways in which changes in the coordination sphere might occur. One of the ligands might dissociate from the complex, taking along an electron pair and leaving behind an intermediate with just five ligands bound to the cobalt atom. Then either the same or a different ligand could reenter the coordination sphere to form an octahedral complex again. Such a reaction mechanism is called a **dissociative mechanism.** Another possibility might involve the entry of a seventh ligand into the coordination sphere, forming a high-energy, unstable, and very labile intermediate that quickly loses one of its ligands and returns to octahedral geometry. This is called an **associative mechanism.** We shall discuss the bonding in the intermediate complexes for these two mechanisms, following Lewis by emphasizing electron pairs, while also using quantum principles that we have been accumulating in the preceding chapters.

The cobalt kernel contains 18 electrons, and there are available atomic orbitals in the outer bonding shell for 9 more electron pairs, even though an atom of cobalt does not have that many electrons. The cobalt nucleus in a Co(III) complex can be said to contribute nine atomic orbitals (one 4s, five 3d, and three 4p) and three electron pairs to the coordination sphere. Each

First, a ligand comes off, or ...

... first, another ligand goes on.

---

[12] J. C. Bailar and C. L. Rollinson, *Inorganic Syntheses*, vol. **2**, p. 222 (1946).

[13] If ligands of a complex or isomer are replaced rapidly by others, or if the geometrical arrangement of the ligands changes rapidly, the complex or isomer is said to be **labile**. When the ligands are replaced only slowly, the complex is called **inert**. Although there is not a clear-cut distinction between these two categories, it is often considered that a complex must maintain its integrity for at least a few minutes under particular reaction conditions to be called inert. This allows time for many sorts of measurements to be made using ordinary laboratory techniques.

[14] The coordination "sphere" is not exactly spherical, because the Co-N distance is 1.99 Å, whereas the Co-Cl distance is 2.22 Å.

of the bonding nuclei of the ligands can be imagined to contribute an electron pair and an appropriately directed atomic orbital. A molecular orbital description of the bonding in this coordination sphere is straightforward (although the details are too involved for us here). In essence, we would form symmetry-correct linear combinations (Chapter 12) of the various atomic orbitals, obtaining nine orbitals of bonding or nonbonding type, and by changing signs in the linear combinations, an antibonding orbital from each of the bonding orbitals. Further calculation would allow us to arrange these molecular orbitals in order of increasing energy.

The bonding electrons are smeared out in molecular orbitals that envelop the whole coordination sphere.

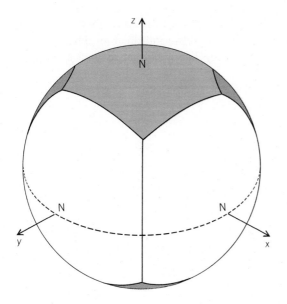

FIGURE 13-2 A view of the coordination sphere of a cis-[Co en$_2$ Cl$_2$]$^+$ complex, showing each nitrogen atom of the ethylenediamine ligands occupying a region of space bounded by four other ligands. Two of the atoms shown (and two other atoms that are out of sight) lie on an imaginary equator of this sphere (in the xy plane), and the remaining two atoms lie at the poles of the sphere (on the z axis).

An electron pair can go into each molecular orbital. Or sometimes two electrons will separate and go into two different orbitals. To pair two electrons requires energy, so if two empty orbitals of about the same energy are available, one electron will go into each. But if the two available orbitals have energies differing by more than the electron-pairing energy, then the electrons take the low-energy road and the pair goes into the lowest-energy orbital.

A complex with all electrons paired is **diamagnetic** (and therefore slightly repelled by a magnetic field); a complex with unpaired electrons is **paramagnetic** (and thus attracted by a magnetic field) because the electrons with unpaired spins act as if they were little magnets. Changing the geometry of the complex changes the relative energies of the molecular orbitals, changing the energy required to promote an electron to the lowest unfilled orbital, thereby changing the color of the complex. Changing the geometry may also cause a change between the electron-paired and the electron-unpaired

Magnetic properties and color depend on the electronic configuration.

A way for left to become right, or for both to become trans

situations.[15] Thus we see that information about the bonding of these isomers may be obtained by analyzing the spectra, and by determining the magnetic properties, of each of the isomers. But what about questions concerning the mechanisms of the interconversions? The answers depend on the relative energies of the various conceivable intermediate complexes. Seldom does the concentration of intermediate become great enough so that its spectrum or its magnetic properties can be measured experimentally. The answers may come, in the future, by precise quantum calculations of the energies of the various molecular orbitals, and even today some shrewd guesses can be made. Information about the mechanism, and about the nature of the reaction intermediates, often can be obtained most readily and directly by studying the rates of the reaction. This is a kinetic test of a proposed mechanism.

Let us examine a particular dissociative mechanism. Suppose that one end of an ethylenediamine ligand dissociates from the coordination sphere. Those five bonding nuclei of the five remaining ligands might assume a trigonal bipyramidal arrangement such as

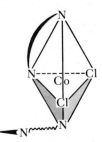

Then the uncomplexed ethylenediamine nitrogen atom might return to the coordination sphere, entering at any of the three edges of the triangle

As the nitrogen atom enters the coordination sphere, the other three atoms in that triangle move closer together, and the square arrangement of an octahedral complex results. Depending on which edge of the triangle was entered, the resulting complex will be *d-cis*, *l-cis*, or *trans*. Figure 13-3 shows how one *cis* isomer might be converted via this mechanism into either of the other isomers, and also how it could be transformed into the intermediate and then emerge unchanged in configuration from the original *cis* isomer. A similar scheme can be drawn, starting with the mirror-image *cis* isomer, or with the *trans* isomer. All three possibilities can be summarized compactly by a mechanism written in terms of the elementary chemical reactions as

---

[15] These fascinating and informative relationships among molecular geometry, spectra, and magnetism are explained in detail by R. S. Drago, *Physical Methods in Inorganic Chemistry*, New York: Reinhold, especially chap. 3 (1965).

$$d\text{-}cis \underset{k_{-1}}{\overset{k_1}{\rightleftharpoons}} X \qquad (13\text{-}1)$$

$$l\text{-}cis \underset{k_{-1}}{\overset{k_1}{\rightleftharpoons}} X \qquad (13\text{-}2)$$

$$trans \underset{k_{-2}}{\overset{k_2}{\rightleftharpoons}} X \qquad (13\text{-}3)$$

where the species X is the trigonal bipyramidal intermediate. The rate constants for reaction (13-1) and (13-2) have the same numerical values, because the two mirror-image *cis* isomers are identical chemical entities in all respects except handedness. The rate constants for (13-3) are different.

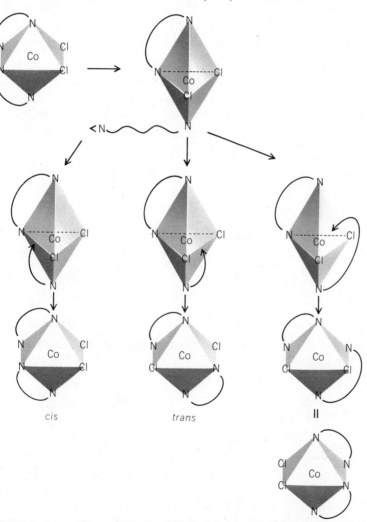

FIGURE 13-3 A dissociative mechanism for the interconversion of three isomers of [Co en$_2$ Cl$_2$]$^+$, showing how a common reaction intermediate can be transformed into any of three different isomers by reentry of the dissociated nitrogen atom back into the coordination sphere.

Predictions of the mechanism can be made using the methods of Sec. 11-5.

The kinetic test of whether this mechanism is an adequate description of the behavior of $[Co\ en_2\ Cl_2]^+$ at a particular temperature and in a particular solution requires that the predictions of the mechanism be compared with the observations of experiment. This mechanism turns out to be incomplete. Both acids and bases play important roles in these isomerizations, probably being involved in dissociating the nitrogen atom from the coordination sphere. In some solvents, there is the additional factor that solvent molecules are potential ligands. In water, one must include a collection of competing reactions of the type

$$[Co\ en_2\ Cl_2]^+ + H_2O \rightleftharpoons [Co\ en_2\ H_2O\ Cl]^{2+} + Cl^-$$

in which, for each isomer, the two chloride ligands are successively replaced by water molecules. With these changes, the mechanism provides a satisfactory picture.

13-8 Reaction of Cobalt Ions with EDTA: Complexation with a Polyfunctional Ligand

Ethylenediamine, because it is a single molecule with two groups available for coordination (it has two *functional groups* that can coordinate with a single metal atom), gives different possibilities for isomerism in $[Co\ en_2\ Cl_2]^+$ than exist in the similar ion $[Co(NH_3)_4Cl_2]^+$. Another favorite polyfunctional ligand is EDTA (*e*thylene*d*iamine*t*etra*a*cetate)

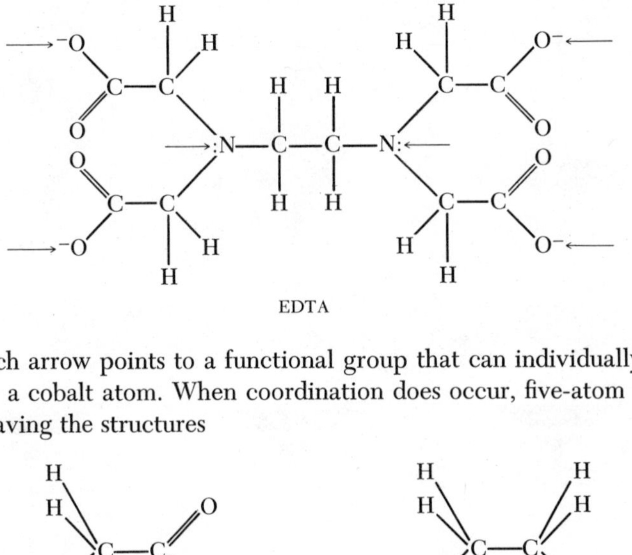

EDTA

where each arrow points to a functional group that can individually coordinate with a cobalt atom. When coordination does occur, five-atom rings are formed having the structures

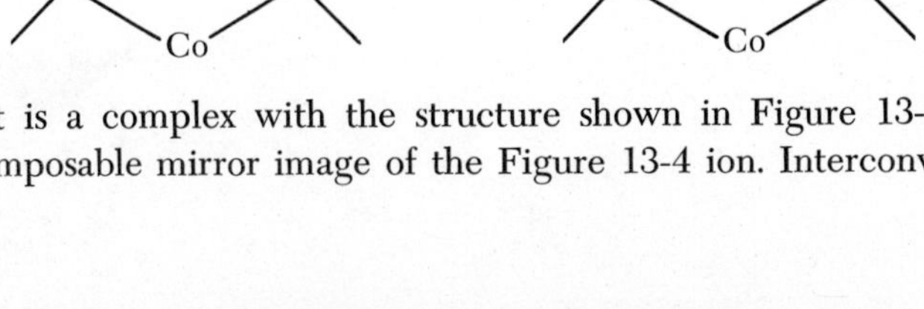

The result is a complex with the structure shown in Figure 13-4, or the nonsuperimposable mirror image of the Figure 13-4 ion. Interconversion of

left- and right-hand EDTA-Co(III) complexes cannot proceed by just the breaking and reestablishing of one or two Co-O bonds.

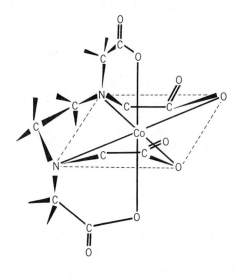

FIGURE 13-4 Structure of a complex in which EDTA is coordinated to a cobalt(III) ion. The various hydrogen atoms on the EDTA ligand have been omitted from this figure to show the geometry of the complex more clearly. Locate the four oxygen nuclei and the two nitrogen nuclei that are octahedrally arranged in the coordination sphere of the cobalt(III) ion. [From A. M. Sargeson, "Conformations of Coordination Chelates," *Transition Metal Chemistry*, vol. **3**, Fig. 17, p. 331 (1966). Used by permission of Marcel Dekker, Inc.]

EXERCISE: Draw structures for a series of EDTA-Co(III) complexes that would constitute a mechanism for the conversion of one isomer into its mirror image. Can you find any way to do this without breaking all four Co-O bonds?

These EDTA-Co(III) complexes, like most other Co(III) complexes, are relatively inert compared to the corresponding Co(II) complexes. The EDTA-Co(II) complex is comparatively labile, but it is certainly **stable.** For instance, consider the reaction in water solution of EDTA with the hydrated Co(II) ion

$$EDTA^{4-} + Co(H_2O)_6^{2+} = CoEDTA^{2-} + 6H_2O \tag{13-4}$$

The equilibrium constant (recall Chapter 7) for reaction (13-4) is given by the equation

$$K = \frac{[CoEDTA^{2-}]}{[EDTA^{4-}][Co(H_2O)_6^{2+}]} = 10^{16} \tag{13-5}$$

A large equilibrium constant means a very stable complex. Equation (13-5) can be rewritten as

Stability of a complex is enhanced when several bonds are formed between the metal and the same ligand.

$$\frac{[CoEDTA^{2-}]}{[Co(H_2O)_6^{2+}]} = 10^{16} \times [EDTA^{4-}]$$

which says, for instance, that at a concentration of $EDTA^{4-}$ of 0.001 M, the ratio of complexed Co(II) to free Co(II) has the extraordinarily large value of $10^{16} \times 10^{-3} = 10^{13}$. EDTA is an effective complexing agent for Co(II), and because of the lability of the complex, equilibrium in solution is established quickly.

EDTA is also an excellent complexing agent for just about all transition-metal ions, and for many other metal ions as well. The EDTA-Fe(II) complex, called "iron **chelate**" and sold to gardeners, is used to provide iron for plants that cannot meet their requirements for iron from the soil in which they are growing. This problem is said to arise for such plants as azalea, rhododendron, and mountain laurel when they are grown in basic soil, because OH^- then competes effectively for Fe^{3+} and Fe^{2+} ions, precipitating almost all iron as the various iron hydroxides. The iron chelate, when applied to soil, gets iron into solution, allowing this iron complex to pass across the cell membrane where, apparently, iron can be captured from the EDTA complex.

The use of EDTA as a complexing agent has been widely exploited in chemical analysis. One illustrative example of this use is given in Section 16-3. EDTA has also been used as a scavenger to pick up trace amounts of contaminating radioactive metal ions.

13-9 Vitamin B_{12}: Another Cobalt Complex

One of the most recently discovered vitamins (isolated as red crystals from liver in the late 1940s) is vitamin B_{12}, a six-coordinate complex of cobalt(II). The structure of this compound was established by the methods of X-ray crystallography by Dorothy Crowfoot Hodgkin and her co-workers in England.[16] The entire molecular structure is depicted in Figure 13-5; we shall focus our attention on just the portion of the molecule in the vicinity of the cobalt atom.

Four nitrogen nuclei are held, by a rather rigid molecular framework of carbon-carbon double and single bonds, into a plane, with space between just about right for a cobalt atom. A fifth coordination position is filled by a nitrogen in the imidazole group

Focus your attention on the region around the coordination sphere of the metal atom.

And the sixth ligand position is occupied by a cyanide ion. The cobalt atom is in an octahedral environment, with the geometry determined largely by the geometry of the four coplanar nitrogen atoms.

Vitamin B_{12} has been used in the treatment of pernicious anemia, and the compound is considered to be essential for normal growth of rats, pigs, chicks, and humans. However, the molecular details of how vitamin B_{12} is involved in metabolism remain subjects of active biochemical research.

Vitamin B_{12} bears a striking resemblance to chlorophyll, whose structure is given in Figure 13-6. Chlorophyll is an essential photosensitive component

---

[16] See C. Brink-Shoemaker, D. W. J. Cruickshank, D. C. Hodgkin, M. J. Kamper, and D. Pilling, "The Structure of Vitamin B_{12}," *Proceedings of the Royal Society* (London), ser. A, vol. **278**, p. 1 (1964).

of the chemical system in plants that permits the energy of sunlight to reduce the carbon in CO_2 to the carbon in carbohydrates. The role of magnesium in the functioning of chlorophyll is not yet clearly understood.

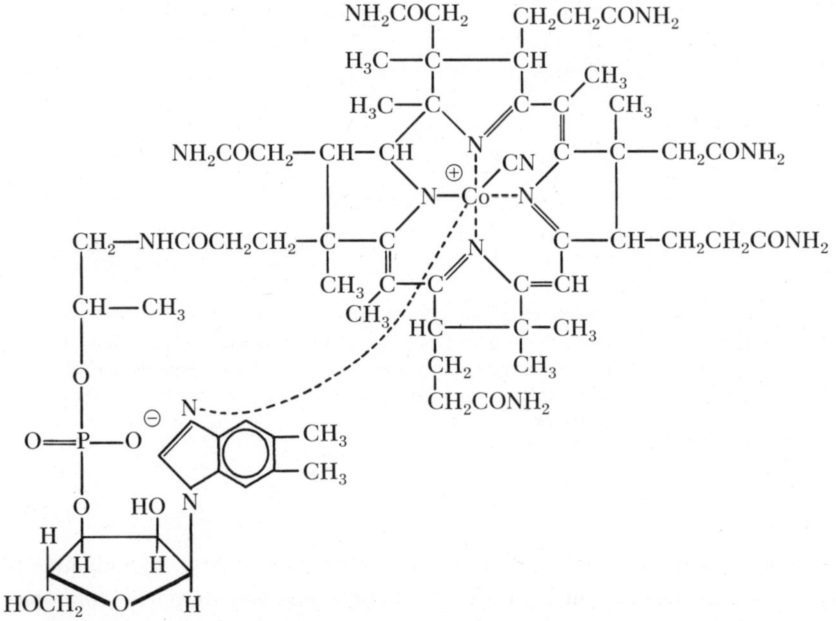

FIGURE 13-5 Structure of a molecule of vitamin B_{12}. Included in the octahedral coordination sphere around the cobalt nucleus are four coplanar nitrogen atoms, a cyanide ion, and the nitrogen of an imidazole group.

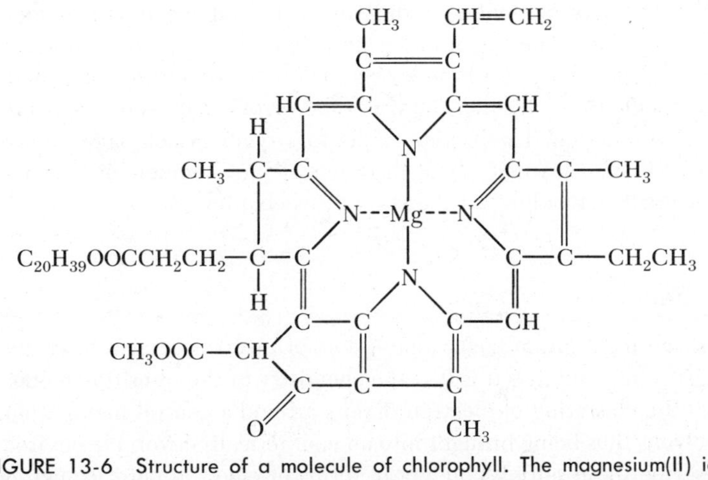

FIGURE 13-6 Structure of a molecule of chlorophyll. The magnesium(II) ion is coordinated within a square-planar configuration of four nitrogen atoms. Absorption of light by this molecule is the first step in the production of carbohydrates from CO_2 and H_2O in a plant leaf. From the point of view of human economics, absorption of sunlight by this molecule is the first step in the chemical production of most food, fiber, and fuel.

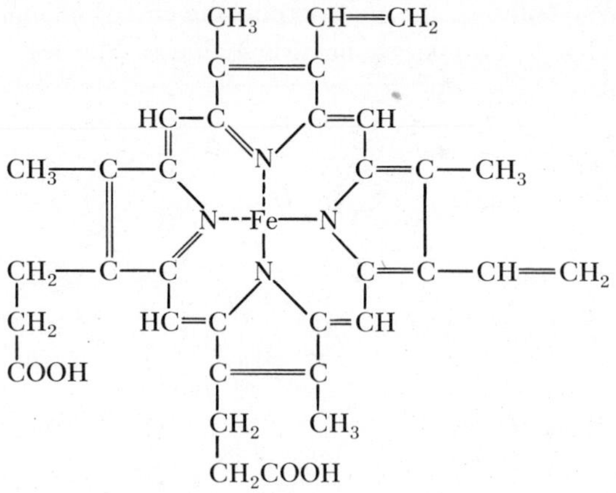

$$\begin{array}{c}
\text{heme structure with Fe center coordinated to four N atoms of porphyrin ring,}\\
\text{with methyl, vinyl, and propionic acid side chains}
\end{array}$$

FIGURE 13-7 A portion of a hemoglobin molecule. Attachment of this heme molecule to a globular protein molecule is by coordination to the iron atom. An oxygen molecule can be coordinated to the iron atom. Other ligands can also bond to the iron atom, replacing oxygen and inactivating heme as the oxygen carrier in this vital life support system. Two such poisoning ligands are cyanide ion and carbon monoxide.

Both vitamin B_{12} and chlorophyll bear striking structural resemblance to heme (see its structure in Figure 13-7), the oxygen-carrying portion of hemoglobin. The requirements for an oxygen carrier in blood are quite stringent. Oxygen must be bound securely, but oxygen must also come off readily, and so the strength of the oxygen-hemoglobin bond must be just right. And it seems that the details of geometry within the coordination sphere of the iron atom are such that the proper energy requirements for making and breaking a bond to oxygen are met. The oxgen molecule is bound to the iron, while the sixth octahedral coordination position is occupied by a bond to the protein portion of the hemoglobin molecule. The trials and errors of evolution seem to have yielded a well-designed molecular carrier for oxygen. Much is now known, and more still is being learned, about the chemistry of hemoglobin; see "Suggestions for Further Reading" at the end of this chapter.

13-10 Summary

A transition-metal atom has available lots of electrons and orbitals for even more electrons. And thus it is that the chemistry of the transition metals often involves the clustering of electron donors around a central metal atom, with the electrons thus being brought into an energetically favorable environment. We describe the resulting bonding in terms of electron pairs (sometimes the electrons are not *all* paired, but then the magnetic properties of the compound are different) residing in molecular orbitals. The chemical nature of the metal and the ligands and the geometry of the ligand arrangement around the central atom together determine the nature of the resulting compound.

The many beautiful and striking colors of transition-metal compounds are interesting to most people. The geometrical aspects of these compounds are intriguing to others. The transition-metal compounds have played central roles during the past century in the development of our ideas about chemical bonding.

Yet there is much more of potential interest and value within the domain of coordination chemistry, and there is much more to be learned in what is probably still an infant science. In the short Section 13-9, we saw how two important biochemical processes—photosynthesis and oxygen transport—involve transition-metal complexes. Can man design such transition-metal compounds for needs that he specifies? Many chemists today are working in just this area. There has been success in designing specific catalysts by placing just the right ligands around just the right metal atom in just the right geometrical configuration. Here is an area where man's ingenuity and creativity can find useful and meaningful and perhaps dramatic expression in the near future.

GLOSSARY

associative mechanism: a reaction mechanism in which the first elementary process involves the association of two reactants to form a reaction intermediate.

chelate: a coordination compound or ion formed by a polyfunctional ligand with multiple bonds to the central metal atom (from Greek *chele:* claw).

cis: prefix denoting a geometrical configuration of a molecule in which two groups are near each other; opposed to *trans* (q.v.).

combining capacity: the number of combining units that can be bound by a particular atom, with the combining capacity of hydrogen set at unity, and oxygen at two. Then carbon is usually four.

coordination sphere: a region of space around an atom in which the primary bonding to that atom occurs.

dextro: one of the nonsuperimposable mirror-image forms of an asymmetric compound (from Latin *dexter:* right), denoted by the prefix *d*.

diamagnetic: said of a substance that is repelled by the stronger part of a nonuniform magnetic field. Most substances are weakly diamagnetic.

dissociative mechanism: a reaction mechanism in which the first elementary process involves either a dissociation of a reactant into two fragments, or the breaking of a bond in the reactant.

inert: said of a complex in which bonds, once formed, stay formed for significantly long periods of time.

isomers: two compounds whose molecules have the same overall composition, but different structures.

labile: said of a complex in which bonds are being broken at a rapid rate.

levo: one of the nonsuperimposable mirror-image forms of an asymmetric compound (French *lévo-:*left), denoted by the prefix *l*.

ligand: an ion or molecule that coordinates to a central atom to form a complex.

paramagnetic: said of a substance that is attracted in a magnetic field. Substances with unpaired electrons are paramagnetic.

stable: said of a complex with a large equilibrium constant for the formation reaction. Nothing is implied in the term about rates of reaction.

trans: prefix denoting a geometrical configuration of a molecule in which two groups are far from each other; opposed to *cis* (q.v.).

valence: synonym for combining capacity.

PROBLEMS

13-1 Demonstrate that left-handed and right-handed helices are nonsuperimposable mirror images by making two mirror-image helices. Take two pipe cleaners, and wrap each into a spiral around a pencil. Wrap one clockwise, and the other counterclockwise. If you cannot convince yourself that the two are really different, make a third, and compare the two that are the same with the different one.

13-2 There are two different mirror-image forms of the EDTA-Co(III) complex. In EDTA, the two bonding nitrogens and the four carboxyl groups are all connected to the same ligand. Now consider the related situation in which the nitrogens and carboxyls are freed from each other. How many isomers are possible for the complex $[Co(III)(CH_3COO^-)_4(NH_3)_2]^-$? Identify which, if any, of these isomers have nonsuperimposable mirror images.

13-3 A reading-research problem. Write a short paper, describing one recent theory about a role of vitamin B_{12} in metabolism. Look up "vitamin B_{12}" in the subject indexes of *Chemical Abstracts* to get a lead on sources of information.

13-4 Consider the seven-coordinate and five-coordinate reaction intermediates discussed in Section 13-7. There are a great many conceivable geometries possible for each. For instance, you can write planar structures, in which five or seven ligands are arranged in a circle around the central cobalt nucleus. Make models for several of these arrangements (marshmallows and toothpicks can be used). Use a ruler to measure the various distances. What arrangements seem to fill the space in the coordination sphere most evenly?

13-5 A reading-research problem. Look up a procedure for synthesizing a coordination compound. Excellent sources for such procedures are the series: *Inorganic Syntheses,* New York: McGraw-Hill; and *Preparative Inorganic Reactions,* New York: Interscience. Describe the procedure in your own words, writing the structure of each starting material and each intermediate compound. Where would you obtain the starting materials? Are they listed in chemical catalogs? Are they available in your laboratory stockroom? Speculate on mechanisms for producing the chemical transformations that occur.

13-6 Consider the ions $[Co(NH_3)_6]^{3+}$, $[Co(NH_3)_6]^{2+}$, $[Cu(NH_3)_6]^{2+}$, $[Cr(NH_3)_6]^{3+}$, $[Cr(NH_3)_6]^{2+}$, and $[Ni(NH_3)_6]^{2+}$. For each species, determine the number of electrons involved in the coordination sphere of the metal atom. Which of these compounds must be paramagnetic?

13-7 Why are HCN and CO such lethal gases? Find information in the library about the relative bonding effectiveness of O_2, HCN, and CO to hemoglobin. Is the competition for oxygen dominated by stability-instability considerations, or is it a lability-nonlability matter? Write a short paper on the molecular explanation of poisoning by either HCN or CO. Document your statements.

SUGGESTIONS FOR FURTHER READING

Biographical information about many chemists is found in: FARBER, E., ed., *Great Chemists,* New York: Interscience (1961). See chapters about Kekulé (pp. 697–702), Pasteur (pp. 641–660), and Werner (pp. 1233–1243). A full-length biography of Werner is KAUFFMAN, G. B., *Alfred Werner, Founder of Coordination Chemistry,* Berlin: Springer-Verlag (1966). For Lewis, see: HILDEBRAND, J. H., "Gilbert Newton Lewis," *Biographical Memoirs of the U.S. National Academy of Sciences,* vol. 31, pp. 210–235 (1958).

Two well-written and authoritative books about the development of our ideas about chemical combination are: PALMER, W. G., *A History of the Concept of Valency to 1930,* Cambridge: Cambridge University Press (1965); and RUSSELL, C. A., *The History of Valency,* Leicester: Leicester University Press (1971).

An excellent presentation of coordination chemistry written for freshman chemistry students is: BASOLO, F., and JOHNSON, R. C., *Coordination Chemistry,* New York: W. A. Benjamin (1964). Some more-specialized books include: SUTTON, D., *Electronic Spectra of Transition Metal Complexes: An Introductory Text,* New York: McGraw-Hill (1968); LANGFORD, C. H., and GRAY, H. B., *Ligand Substitution Processes,* New York: W. A. Benjamin (1965); and TAUBE, H., *Electron Transfer Reactions of Complex Ions in Solution,* New York: Academic (1970).

For lots more about the uses of EDTA in analytical chemistry, see: SCHWARZENBACH, G., and FLASCHKA, H., *Complexometric Titrations,* 2nd ed., London: Methuen (1969). For more about the structures of EDTA complexes and related chemical species, see: SARGESON, A. M., "Conformation of Coordination Chelates," *Transition Metal Chemistry,* vol. 3, p. 303 (1966).

Much has been written about photosynthesis. You may find one of the following helpful and interesting: CALVIN, M., and BASSHAM, J. A., *The Photosynthesis of Carbon Compounds,* New York: W. A. Benjamin (1962); RABINOWITCH, E., and GOVINDJEE, *Photosynthesis,* New York: Wiley (1969); and FOGG, G. E., *Photosynthesis,* New York: Elsevier (1968).

14
Organic Chemistry

14-1 Abstract

The distinction between organic and inorganic chemistry is purely historical and is now anachronistic. There is no line of demarcation between the two. Yet sometimes—as in this chapter, for instance—it is convenient to consider the compounds of carbon as a special category. Early in the development of chemistry, it was thought that the mysterious vital force of living organisms endowed all products of such organisms with a special quality that could not be achieved in any other way. Such products, virtually always compounds of carbon, were called *organic compounds*. We shall explore some of the reasons why compounds containing carbon occupy such an important place in the chemistry of living organisms.

Our first discussions of organic chemistry involve rather simple compounds, but we cannot go far without using the conventional names for more complicated organic compounds, and we shall spend some time with rules for naming these substances.

Structure and reactivity are the woof and warp of the fabric of organic chemistry. We shall weave these threads together as we discuss diamonds and graphite, the properties and reactions of isomers, the organization of organic chemistry in terms of functional groups, and finally the unifying concepts of reaction mechanisms.

The chapter concludes with some practical examples of the application of the facts and the principles of organic chemistry to the design of gasolines, and to the synthesis of polymeric plastics.

14-2 The Myth of Vital Force

Can some compounds be made only by living cells?

One rather appealing, romantic view of the world places man at the center of all life and places the community of living organisms at the center of the world. This world view easily embraces a chemical mysticism, a belief held by many rational and informed people in earlier centuries and still current among superstitious men today, that imagines a hierarchy of forms of matter, with *organic* substances created by living organisms having a special status, set apart from the inanimate stuff of the *inorganic* realm. This mysticism asserts that the vital force of living organisms gives special qualities to the chemicals of life; only living organisms can create such organic compounds. It is asserted that there are benefits to man, to other animals, and, indeed, to plants when these organic materials of the living realm are used in preference to inorganic substances. So very distinctive are these substances of life that chemists are *incapable* of making compounds defined as organic by these folk.

The intellectual foundations for this theory of chemical vitalism began to crumble early in the nineteenth century, accompanied by bitter and sustained controversy. A crushing blow to these foundations came in 1828, when Wöhler synthesized urea,

$$H_2N-\underset{\underset{\displaystyle}{\overset{\displaystyle O}{\|}}}{C}-NH_2$$

a clearly organic compound, which is the end product of protein metabolism by animals. Wöhler made this compound in the laboratory, from the inorganic compound ammonium cyanate, without the assistance of plants or animals.[1] In 1845, Kolbe synthesized acetic acid by a series of reactions that started with the uncombined elements as initial reactants. Then and now, man-made acetic acid is indistinguishable from acetic acid made from apple cider by *Acetobacter aceti*.

Vitalists clung to the hope that living cells could bestow *some* special property on chemical compounds. One favorite property was optical activity (see Sections 13-5 and 14-6). Even into the twentieth century, there were reputable chemists who held out for the proposition that only compounds containing carbon, and therefore having a special kinship with life, could be optically active. Then in 1914 Alfred Werner reported the synthesis of the carbon-free dissymetric compounds, d- and l-$\{Co[Co(OH)_2(NH_3)_4]_3\}Br_6$. The demonstration that each salt gave an optically active solution removed any factual basis for this remnant of vitalism.[2]

Today some important methods for large-scale commercial preparation of organic compounds use inorganic substances like sodium cyanide (from sodium carbonate) and calcium carbide (from calcium oxide and coke) as sources of carbon. Chemists no longer attempt to distinguish between the mineral forms or compounds of carbon (diamond, graphite, silicon carbide, and so on) and the plant or animal products (glucose, urea, and so on). It is generally agreed that the hydrocarbons and their derivatives constitute a large and somewhat separate area of chemistry that blends with biochemistry on one side and with inorganic chemistry on another. This area we shall call *organic chemistry*. We describe the domain of organic chemistry in terms of the composition of the compounds studied, without reference to the origin and past history of those compounds.

Organic chemicals are routinely made from inorganic substances.

But do not assume, simply because the rational foundations of chemical vitalism have been demolished by experimental facts, that the myth of chemical vital force is dead. Far from it! The myth lives on as a point of view, as a prejudice, as a belief. It shows itself in devotion to the view that plants are poisoned by inorganic sources of nitrogen, phosphorus, and potassium, and that the only appropriate sources of plant fertilizers are animal manures and composted vegetable matter.[3] The myth appears also in a popular distrust of synthesized vitamins versus "natural" ones, and in a layman's wariness of "chemicals" in general versus "natural" products. In assessing the validity of such attitudes, you may find a consideration of the manufacture and the use of the compound urea to be relevant.

Not all the authors of this book share the philosophical stance behind this debunking; however, these are arguments that must be considered.

---

[1] Reported in *Poggendorf's Annalen der Chemie*, vol. **12**, p. 253 (1828).

[2] See A. Werner, "On Optical Activity among Carbon-free Compounds," *Berichte der Deutschen Chemischen Gesellschaft*, vol. **47**, pp. 3087–3094 (1914); trans. from German by G. B. Kauffman, *The Selected Papers of Alfred Werner*, New York: Dover, pp. 175–184 (1968).

[3] Thousands of laboratory tests have shown that plants do not distinguish between the source (manufactured versus animal produced) of the same compound, when applied in the same manner, in the same concentration.

Urea is manufactured on a very large scale from ammonia and carbon dioxide via the reaction

$$CO_2 + 2NH_3 \longrightarrow (NH_2)_2C=O + H_2O$$

Half the manufactured urea goes into urea-formaldehyde plastics. The other half is used in various forms as a lawn, garden, and farm fertilizer, competing effectively in the marketplace with urea-containing fertilizers derived from animal manures. But what about this organic compound, man-made from inorganic substances? Are vegetables raised with factory-made urea different than vegetables raised with animal-produced urea? Can a grower claim to have produced "organically grown" vegetables if he has used the factory-made urea as a fertilizer? Is urea different in any way when made by different methods? Can there be an *unnatural* urea? Or an *inorganic* urea? Because pure urea is a chemical compound, its properties are invariant, no matter how it was made, no matter what the history of each particular molecule. The results of the use of pure urea have always been found to be the same, no matter what its source. And chemists assert that the same is true of all other compounds.

Throughout this chapter and book, organic compounds are considered simply as interesting and useful chemical compounds, and nothing more.

How can you define an organic food? A natural food?

14-3 Is Carbon Different?

Known organic compounds far outnumber compounds without carbon.[4] This is due, in part, to the fact that chemists have given much more attention to organic compounds. Compounds of other elements have seemed of lesser interest to organic man. But are there chemical reasons for this abundance of carbon-containing compounds?

It is often claimed that carbon is unique among the elements in being able to link to itself, thus being able to form chains of carbon atoms. The process of forming chains of identical atoms is called *catenation* (compare the catenary curve of mathematics). While catenation is a significant factor in accounting for the multitude of different compounds containing carbon, catenation is certainly not unique to carbon. Long chains of sulfur atoms are joined together in plastic sulfur, and even in the most stable crystalline forms of sulfur, the structural unit is a ring of eight sulfur atoms linked together by covalent bonds. Silicon, germanium, and boron also form many compounds, in which identical atoms of the element are linked together.

Carbon is just one of the elements that bonds to itself, forming chains.

It is fruitful to compare catenation of carbon atoms with comparable instances involving silicon atoms. Catenation of silicon atoms gives rise to a series of silicon hydrides

---

[4] About 2 million organic compounds are listed, together with their properties and methods of synthesis, in the organic chemist's lexicon, *Beilsteins Handbuch der Organischen Chemie*, published in 31 volumes of several parts each, plus many supplements.

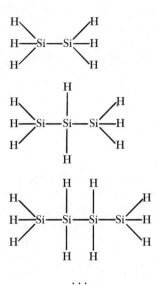

These compounds are entirely analogous to the saturated hydrides of carbon, the hydrocarbons

Physical properties of members of the silicon hydride series parallel those of the hydrocarbons. Moreover, among the corresponding **binary** chlorides, such as the two series

$$\text{Cl}_3\text{Si}-\text{SiCl}_2-\text{SiCl}_3 \qquad \text{Cl}_3\text{C}-\text{CCl}_2-\text{CCl}_3$$

... ...

silicon is known to form much longer chains than carbon. Among the binary fluorides, compounds in the fluorosilane series Si_nF_{2n+2} are known up to $Si_{14}F_{30}$, and rubbery high **polymers** of the compositions $(-SiF_2-)_x$, with x very large, have been made. Far fewer of the analogous compounds of carbon have been made. It appears that catenation is favored for halides in silicon chemistry, but it is favored for hydrides in carbon chemistry. Phosphorus, tin, lead, and many transition metals also form catenation compounds. Catenation is a widespread phenomenon, clearly not unique for carbon.

We must turn to other aspects of carbon chemistry to find the reasons for the large number and rich variety of organic compounds. We are likely to find reasons in the areas of reaction rates and reaction mechanisms. Let us compare the chemical behavior of various compounds with water. Carbon compounds such as CCl_4, C_2Cl_6, CH_4, and C_2H_6 do not react with water at a detectable rate. By contrast, all the halides of silicon hydrolyze rapidly; even the hydrides of silicon hydrolyze readily. So in our watery world, the catenated compounds of silicon perish, but those of carbon persist.

Hydrolysis of silicon compounds proceeds readily because tetracovalent silicon can easily coordinate to water molecules by accepting electrons in its d orbitals. Formation of such a fifth bond to silicon sets the stage for cleavage of the Si-Si bond when the silicon atom returns to its more stable tetravalent state. Hydrolysis occurs, according to the picture presented in this reaction mechanism, because there is an easily formed intermediate compound whose formation represents a halfway state between original compound and the products of hydrolysis.

Most carbon compounds persist because they react only very slowly with air and water.

Such a mechanism will not work for carbon. Carbon atoms have no d orbitals available for attaching water or oxygen. Thus, tetracovalent carbon can react with water or air only by slower and more roundabout mechanisms. Since the individual carbon atoms are so small, the surrounding atoms in a carbon compound often can block off access to the carbon; any attack by an invading reactant molecule is impeded. Both electronic structure and geometry combine to keep many reactions very slow. So it can be said that most of those 2 million organic compounds are stable for lack of any mechanism by which they can be readily destroyed by water or air.[5] We know that in air, most organic compounds are thermodynamically unstable with respect

---

[5] Most of these compounds are stable, at room temperature, for rather long periods of time. But heat almost any organic compound to the temperature of a flame, and oxidation will occur. Most organic compounds will burn, or can be burned, in air.

to conversion to carbon dioxide and water, but without any mechanism available to facilitate the oxidation, the compounds stay together.

The bonds between carbon and other common elements are moderately strong, as is shown in Table 14-1. The ability of carbon to supplement its σ bonds with one or two strong π bonds appears very clearly from the bond-dissociation values for multiple bonds, numbers considerably higher than corresponding values for single bonds, to carbon, nitrogen, and oxygen. So the simple compounds of carbon with these elements would be expected to be strongly bonded, as well as stabilized by the mechanistic considerations just discussed. Thus, the existence of the large number of stable organic compounds is plausible. We now proceed to examine the significance of these compounds.

TABLE 14-1
Average Bond-Dissociation Energies for Carbon

C–C	83 kcal/mole	C–O	82 kcal/mole
C=C	148 kcal/mole	C=O	169 kcal/mole
C≡C	194 kcal/mole	C–S	62 kcal/mole
C–H	99 kcal/mole	C–F	105 kcal/mole
C–N	70 kcal/mole	C–Cl	79 kcal/mole
C=N	147 kcal/mole	C–Br	66 kcal/mole
C≡N	210 kcal/mole	C–I	57 kcal/mole
C–Si	69 kcal/mole	C–P	63 kcal/mole

Source: L. Pauling, *Nature of the Chemical Bond*, 3rd ed., Ithaca, N.Y.: Cornell University Press, p. 85 (1960); and F.A. Cotton and G. Wilkinson, *Advanced Inorganic Chemistry*, 2nd ed., New York: Interscience, p. 100 (1966).

14-4 Some Simple Compounds of Carbon

Carbon exists as an element in two crystalline forms: *diamond*, which has the carbon atoms closely bound tetrahedrally in a cubic lattice of high density and hardness; and *graphite*, which has hexagonal rings of carbon atoms more loosely bound in extended, flat, parallel sheets that are separated from one another by a relatively large distance (see Figure 14-1). All the other black varieties of elemental carbon (charcoal, carbon black, and coke) contain minute crystallites of the graphite lattice, but no other crystal structure. None of these substances is pure carbon except graphite and diamond. The density of diamond is 3.51 g/cm³, and that of graphite is 2.25 g/cm³. Both burn at 800°C in oxygen.

Graphite is the stable form of carbon at room temperature and one atmosphere pressure. All gem diamonds are doomed to become graphite eventually, but the transformation is immeasurably slow at temperatures below 2000°C. The reverse process—conversion of graphite to diamond—has been achieved by raising the temperature to 3000 or 4000°C and applying a pressure of one million pounds per square inch, using materials and machines which had to be devised especially for this purpose. Rapid cooling keeps the newly synthesized diamond crystals from changing back into graphite. Any source of carbon will do for this preparation—diamonds have even been made from peanut butter! The resulting crystals of synthetic diamond are small and imperfect, but they serve very well for grinding, polishing, and boring operations, extending our meager supply of naturally occurring industrial diamonds.

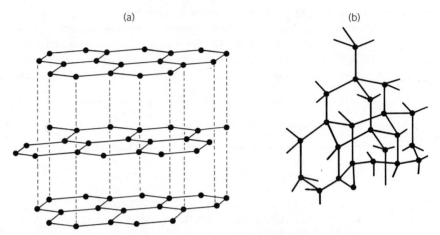

FIGURE 14-1 Crystal structures of (a) graphite and (b) diamond. The relatively more open graphite structure consists of parallel sheets in which the carbon nuclei are arranged in hexagonal rings. The nuclei are closer together in the diamond structure, with each atom bonded to four other tetrahedrally located carbon atoms.

Carbon oxidizes to CO (combustible, poisonous gas; mp −207°C, bp −190°C) and to CO_2 (noncombustible, suffocating gas; mp 56.6°C at 5 atm, sublimation point −78.5°C). The monoxide is as insoluble in water as is oxygen, but the dioxide dissolves readily to the extent of 1.71 v/v of water at 0°C. When CO_2 dissolves in water, carbonic acid is formed. Carbonic acid dissociates in water according to the equilibria

For a geochemical interpretation of these equilibria, see Sec. 18-3.

$$H_2O + H_2CO_3 = H_3O^+ + HCO_3^-$$

$$K_1 = \frac{[H_3O^+][HCO_3^-]}{[H_2CO_3]} = 4 \times 10^{-7}$$

$$H_2O + HCO_3^- = H_3O^+ + CO_3^{2-}$$

$$K_2 = \frac{[H_3O^+][CO_3^{2-}]}{[HCO_3^-]} = 5 \times 10^{-11}$$

When ammonia is dissolved in a solution of carbonic acid, the salt NH_4HCO_3 precipitates. When ammonia is dissolved in a solution of carbonic acid that is also saturated with sodium chloride, equilibrium considerations point to the precipitation of the least soluble compound that can be formed from the ions H_3O^+, OH^-, NH_4^+, HCO_3^-, CO_3^{2-}, Na^+, and Cl^-. That compound is sodium hydrogen carbonate, $NaHCO_3$, commonly called sodium bicarbonate. These facts from the laboratory are the basis for the Solvay process for industrial manufacture of sodium bicarbonate in huge quantities. Most of this $NaHCO_3$ is converted to sodium carbonate to be used in making glass and in many other applications where an inexpensive soluble alkali is required.

The Solvay process may be summarized in these equations:

$$CO_2 + 2H_2O = H_3O^+ + HCO_3^-$$

$$NH_3 + H_2O = NH_4^+ + OH^-$$

$$CO_2 + NH_3 + H_2O + Na^+ + Cl^- \longrightarrow NaHCO_3 + NH_4^+ + Cl^-$$

$$2NaHCO_3 \xrightarrow{270°C} Na_2CO_3 + CO_2 + H_2O$$

The carbon dioxide liberated in the last step is returned to the process. If calcium carbonate (in the form of limestone or oyster shell) is the original source of the carbon dioxide,

$$CaCO_3 \xrightarrow{700°C} CaO + CO_2$$

then the residual calcium oxide can be used to recover the ammonia for recycling through the process:

$$CaO + 2NH_4Cl \longrightarrow CaCl_2 + 2NH_3$$

The only by-product is calcium chloride, which has no major use except for melting ice and snow in the wintertime and keeping down road dust (by absorbing water from the air) in the summertime.

In addition to use in making window glass, sodium carbonate is also used for a purpose closer to organic chemistry: the manufacture of sodium cyanide and hydrocyanic acid. This route to the cyanides begins by heating calcium carbonate to give calcium oxide. Then the calcium oxide is heated with coke in an electric furnace to make calcium carbide:

$$CaO + 3C \xrightarrow{2000°C} CaC_2 + CO$$

This reaction is a gateway to many possible organic compounds. Calcium carbide absorbs nitrogen from pure N_2 at red heat to form calcium cyanamid via the reaction

$$CaC_2 + N_2 \longrightarrow CaCN_2 + C$$

Much of the calcium cyanamid produced commercially is used to make organic compounds of nitrogen for plastics and fibers. Some of the mixture with carbon (obtained as just shown) is treated with sodium carbonate to make sodium cyanide and regenerate calcium carbonate:

$$Na_2CO_3 + CaCN_2 + C \longrightarrow 2NaCN + CaCO_3$$

Sodium cyanide reacts with various strong acids to yield HCN gas, which can, in turn, add to ethyne by a procedure soon to be considered in this chapter:

$$HCN + HC{\equiv}CH \longrightarrow H_2C{=}CHCN$$

From oyster shells to an Orlon sweater

The product, acrylonitrile, is the compound from which acrylic fibers for carpets and textiles are made.

We have just seen examples of how organic material can be made by a sequence of operations beginning with carbon dioxide or with coke. In principle, virtually any organic compound can be synthesized from these sources of carbon. In practice, the would-be maker of compounds also has available a multitude of materials in the fossil deposits of petroleum and coal, and in the produce of forests and farms, to serve as raw products for a chemical synthesis.

Carbon reacts with sulfur vapor to form the disulfide, CS_2, a volatile and highly flammable substance important in the manufacture of rayon. Chlorination of carbon disulfide leads to carbon tetrachloride, a decidedly nonflammable liquid. Carbon tetrachloride was popular as a cleaning solvent until it was discovered that serious liver damage is caused by inhalation of CCl_4 vapors. The same CCl_4 is also used in compact fire extinguishers, where it serves well in an emergency. However, the smelly vapors from the just-extinguished fire should not be breathed, because they contain not only harmful CCl_4, but also sometimes the oxidized compound phosgene, $O=CCl_2$. Phosgene dissociates to carbon monoxide and chlorine and is exceedingly poisonous.

14-5 The Naming of Organic Compounds

At first, organic compounds were given individually descriptive names that told something of their origin or their properties. Heliotropine, $C_8H_6O_3$, smells like heliotrope. Fluorene, $C_{13}H_{10}$, fluoresces in ultraviolet light. Thyroxine, a complicated iodine-containing amino acid, is produced by the thyroid gland. Putrescine, $H_2N(CH_2)_4NH_2$, has a "characteristic" odor. Because there are so many organic compounds, it soon became impossible to assign such descriptive or "trivial" names to all of them. What is worse, the descriptive names gave no indication of the structure, nor even of the empirical formula, so that it was necessary to commit to memory the formula of every compound along with its associated name. Eventually, the need for a systematic scheme of nomenclature was recognized, and such a system was devised at an international congress of chemists at Geneva in 1892.

Today's extended and internationally accepted "Geneva System" is reliable and unambiguous. The name specifies the structural formula, and from a structural formula, one can deduce the name. Not everyone uses the Geneva System all the time, through reasons of laziness, brevity, familiarity with older trivial names, or, more often (as with every other triumph of the human intellect based on unemotional logic), disdain. You should be prepared for a generous sprinkling of trivial names, and a good many derivative names, in listening to chemists' talk. Nevertheless, we present the essentials of the Geneva System as an all-embracing, memory-saving scheme, pointing out that a Geneva name is always socially acceptable and correct usage, whether it appears on a label, an examination paper, or the pages of a scientific journal.

SYSTEMATIC NOMENCLATURE OF ORGANIC CHEMISTRY: THE GENEVA SYSTEM

1. All compounds are named as derivatives of the hydrocarbons, which include
    a. Aliphatic hydrocarbons

methane	CH_4	heptane	C_7H_{16}
ethane	C_2H_6	octane	C_8H_{18}
propane	C_3H_8	nonane	C_9H_{20}
butane	C_4H_{10}	decane	$C_{10}H_{22}$
pentane	C_5H_{12}	undecane	$C_{11}H_{24}$
hexane	C_6H_{14}	dodecane	$C_{12}H_{26}$, and so forth

    Hydrocarbon groups as part of a molecule are named with the same root, but with the suffix *–yl*, as in *methyl*, $-CH_3$; *ethyl*, $-C_2H_5$; *propyl*, $-C_3H_7$; *butyl*, $-C_4H_9$; and so forth.

    b. Aromatic hydrocarbons
       benzene C_6H_6 ($-C_6H_5$ is called the *phenyl* group, a name that is an exception to the rules.)
       naphthalene $C_{10}H_8$
       anthracene $C_{14}H_{10}$
       biphenyl $C_6H_5-C_6H_5$
       terphenyl $C_6H_5-C_6H_4-C_6H_5$

    All 12 nuclei of a benzene molecule lie in the same plane; all 6 C—C bonds are equivalent; all 6 C—H bonds are equivalent; and the molecule has a C_6 axis of symmetry. A shorthand designation for benzene, representing all these structural features, is a hexagon with an inscribed circle.

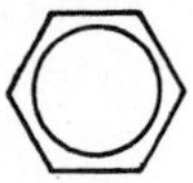

    Each corner of the hexagon represents a carbon atom, and it is assumed that a hydrogen atom is bonded to each carbon unless something else is shown as a replacement for hydrogen. The circle indicates the cloud of delocalized π electrons above and below the ring.[6] Other aromatic structures may be written in an analogous fashion:

    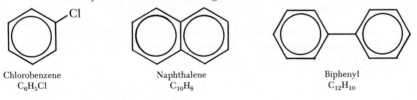

    Chlorobenzene C_6H_5Cl

    Naphthalene $C_{10}H_8$

    Biphenyl $C_{12}H_{10}$

---

[6] Sometimes a benzene molecule is depicted as

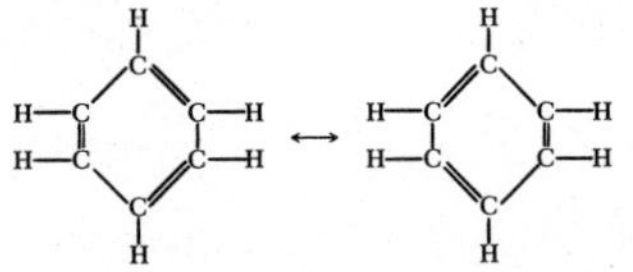

where four bonds are explicitly indicated for each carbon atom. However, any picture with alternating single and double bonds around the ring is a formalism, since all these C—C bonds are identical. Such pictures are cumbersome and misleading.

2. Procedure for naming organic compounds:

   a. Name the longest carbon chain as a hydrocarbon. For example, the carbon-atom skeleton

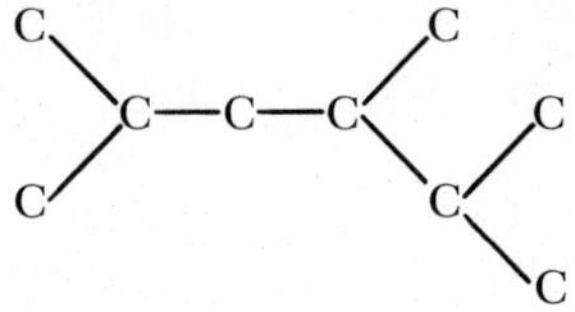

   $$\begin{array}{c} C \\ C-C-C-C \\ C \end{array} \begin{array}{c} C \\ C \\ C \end{array}$$

   is a *hexane* and can be written as

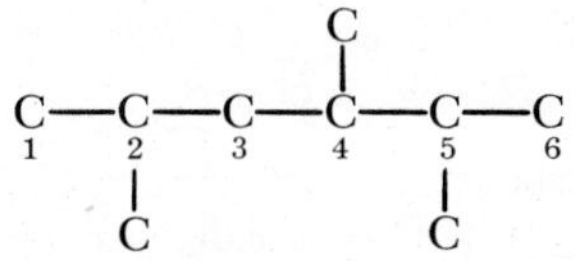

   $$\underset{1}{C}-\underset{2}{C}-\underset{3}{C}-\underset{4}{\overset{\overset{C}{|}}{C}}-\underset{5}{\overset{\overset{}{|}}{C}}-\underset{6}{C}$$
   $$||$$
   $$CC$$

   where the molecule is described as a six-carbon chain with three substituents. Each carbon is assumed to have four bonds, with all unindicated bonds to hydrogen atoms.

   b. Number the carbon atoms from the end that gives the smallest numbers in the final name.

   c. Show quantity and position of substituent groups by prefix and position number. By convention, hydrogen is not considered a substituent. Thus,

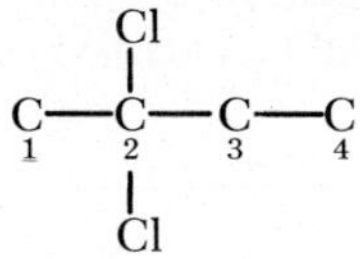

   $$\underset{1}{C}-\underset{2}{\overset{\overset{Cl}{|}}{\underset{\underset{Cl}{|}}{C}}}-\underset{3}{C}-\underset{4}{C} \quad \text{or} \quad C_4H_8Cl_2 \text{ is 2,2-dichlorobutane}$$
   (*not* 3,3—)

   Another example:

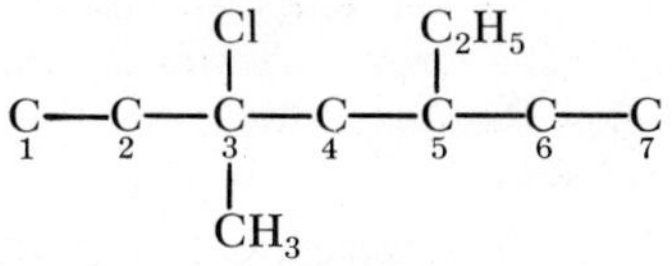

   $$\underset{1}{C}-\underset{2}{C}-\underset{3}{\overset{\overset{Cl}{|}}{\underset{\underset{CH_3}{|}}{C}}}-\underset{4}{C}-\underset{5}{\overset{\overset{C_2H_5}{|}}{C}}-\underset{6}{C}-\underset{7}{C} \quad \text{is 3-chloro-3-methyl-5-ethylheptane}$$

   Whenever there is a choice, arrange to give the smallest numbers.

   d. Indicate position on a ring by numbering the carbon atoms clockwise, starting at the top:

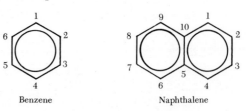

   Benzene   Naphthalene

3. Names for **functional groups** (see Section 14-7):

   a. Multiple bonds: C = C and C ≡ C are indicated by the suffixes *—ene* and *—yne*, as in ethene, $H_2C = CH_2$; and ethyne, $HC \equiv CH$. The corresponding

These names are part of the international scientific vocabulary that transcends national language differences.

(A rule that is broken in the structure above. If only people were as rational as their systems!)

three-carbon compounds are propene and propyne. When there is a possibility of ambiguity, use numbers to indicate positions of multiple bonds. Thus

$$C=C-C-C \quad \text{1-butene}$$
$$C=C-C=C \quad \text{1,3-butadiene}$$
$$HO-C-C\equiv C-C-OH \quad \text{2-butyne-1,4-diol}$$

b. Alcohols: use the suffix —*ol*, as in methanol, CH_3OH; or 2-butyne-1,4-diol.
c. Amines: use the prefix *amino—*, as in amino acid. Note, however, the usage: methylamine, CH_3NH_2; dimethylamine, $(CH_3)_2NH$; and so forth.
d. Aldehydes: use the suffix —*al*. Thus,

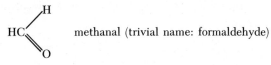   methanal (trivial name: formaldehyde)

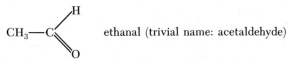   ethanal (trivial name: acetaldehyde)

e. Ketones: use the suffix -*one*. Thus,   -one *is pronounced "own."*

$$H_3C-\overset{\overset{O}{\|}}{C}-CH_3 \quad \text{2-propanone (trivial name: acetone)}$$

f. Acids: use hydrocarbon root plus the suffix -*oic*, as in

HCOOH   methanoic acid (trivial name: formic acid)
CH_3COOH ethanoic acid (trivial name: acetic acid)
C_2H_5COOH propanoic acid (trivial name: propionic acid)

g. Ethers: name the two groups bonded to the oxygen. Thus

$$H_3C-O-CH_3 \quad \text{dimethyl ether}$$

For more detailed rules of naming organic compounds, see any textbook of organic chemistry. Facility comes with use, and cannot be expected at once.

14-6 Isomerism in Organic Compounds

There is only one structural form of the methane molecule that can be written. Similarly, there is only one form of ethane and only one form of propane. True, we can write the three-carbon-atom chain of propane as

$$H_3C-CH_2-CH_3$$

and we can also write the same structure as

$$H_3C\diagup^{CH_2}\diagdown CH_3$$

to emphasize the tetrahedral bond angle and remind ourselves that the

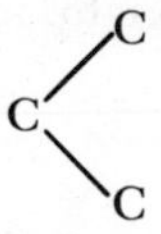

$$C\diagup^{C}_{\diagdown C}$$

angle is not 180°. But the hydrocarbon is still the same propane. There is only one way to connect three carbon atoms and eight hydrogen atoms. With four atoms of carbon, however, there is a crucial difference.

We can write two distinct structures for butane, each with the formula C_4H_{10}, but with the atoms connected differently. These structures are

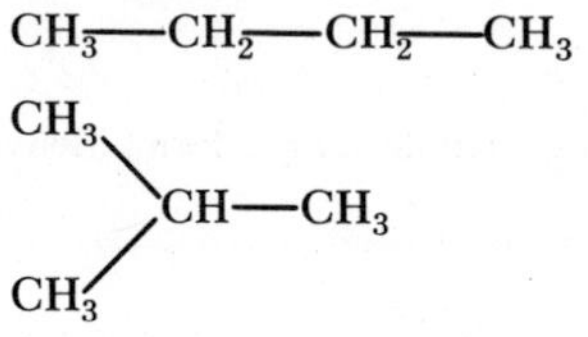

CH_3—CH_2—CH_2—CH_3 normal butane, or *n*-butane

$$\begin{matrix} CH_3 \\ \diagdown \\ CH—CH_3 \\ \diagup \\ CH_3 \end{matrix}$$ isobutane, or 2-methylpropane

These structural formulas represent two distinct compounds. The first one melts at −135°C and boils at −0.3°C; the second one melts at −145°C and boils at −10°C. These compounds are both C_4H_{10}, but there are **structural isomers** of C_4H_{10}.

Isomerism accounts for much of the diversity in carbon chemistry.

Structural isomerism is very prevalent in organic chemistry. The more atoms in a compound, the more structural variations that are possible. So it is that there are 3 isomeric pentanes, 5 hexanes, 9 heptanes, 18 octanes, 75 decanes, and 62,491,178,805,831 tetracontanes (all $C_{40}H_{82}$). The advisability of using Geneva System names for these hydrocarbons is evident: It is much more helpful and specific to write "2,3-dimethylbutane" than to write "one of the branched-chain hexanes" or just "a hexane."

Still more structural isomers are possible in ternary compounds, because the third kind of atom can take up different positions. Thus, there are only two hydrocarbons of the simple formula C_4H_8, but when we introduce an oxygen atom, we get several distinct structural isomers of C_4H_8O. Among the possibilities are the compounds

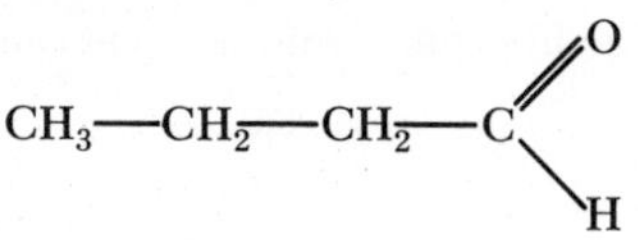

$$CH_3—CH_2—CH_2—C\diagup^{O}_{\diagdown H}$$

butanal, an aldehyde
mp −99.0°C, bp 75.7°C

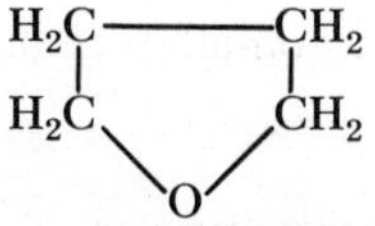

$$\begin{matrix} H_2C\!\!-\!\!-\!\!-\!\!-\!\!-\!\!-\!\!CH_2 \\ || \\ H_2CCH_2 \\ \diagdown\diagup \\ O \end{matrix}$$

tetrahydrofuran, a cyclic ether
mp −65°C, bp 64.5°C

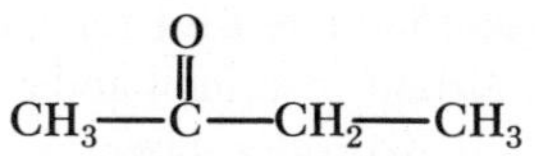

$$CH_3-\underset{\underset{\text{2-butanone, a ketone}}{}}{\overset{\overset{O}{\|}}{C}}-CH_2-CH_3$$
2-butanone, a ketone
mp −86.4°C, bp 79.6°C

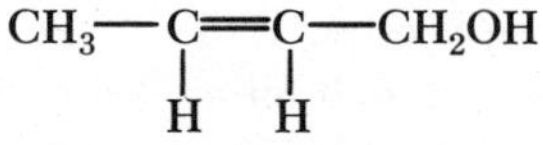

$$CH_3-\underset{H}{\overset{}{C}}=\underset{H}{\overset{}{C}}-CH_2OH$$
2-buten-1-ol, an unsaturated alcohol
mp <−30°C, bp 118°C

Not only are these different compounds, but they are different *types* of compounds. Structural isomerism becomes still more important when we are dealing with four, five, or six kinds of atoms in the same molecule. Perhaps the ultimate in such isomerism is reached in the proteins, polymers containing oxygen, nitrogen, and sulfur in addition to carbon and hydrogen, and in the nucleic acids, which also contain phosphorus. These molecules, with very complicated yet very specific structures, control the functioning of the living cells in all organisms (see Chapter 17).

With two substituents of the same kind, more structural isomerism becomes possible. There are, for instance, two different dichloroethanes:

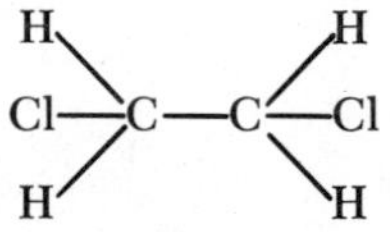

1,2-dichloroethane 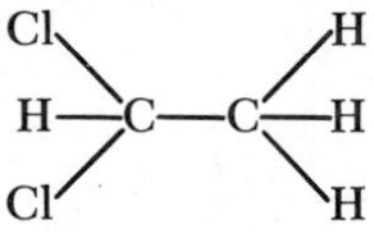   1,1-dichloroethane

There is only one isomer (at room temperature) of 1,2-dichloroethane. Although an infinite number of conformations can be achieved by rotating the two $-CH_2Cl$ groups with respect to each other around the C—C axis (see Section 4-10), the thermal energy available at room temperature is sufficient to interconvert these conceivable isomers. Thus, an actual sample of 1,2-dichloroethane on the laboratory reagent shelf is an equilibrium mixture of all the various conformational isomers; we *say* that there is only one isomer.

The energy barrier to rotation about a σ bond is modest compared to the energy available from ordinary molecular collisions at room temperature, and there is virtually free rotation. Things change drastically, however, when there is a π bond in addition to the σ bond. There is a strong electrostatic barrier to rotation about such bonds, because the electronic distribution of a π orbital does not have cylindrical symmetry about the bond axis. This situation leads to geometrical isomerism in certain compounds containing a C=C bond. Thus, there exist *two* isomeric 1,2-dibromoethenes:

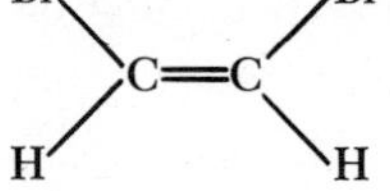

cis-1,2-dibromoethene 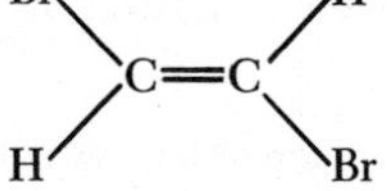   *trans*-1,2-dibromoethene

This *cis-trans* isomerism (see Section 13-4 for another example of *cis-trans* isomerism) is especially important in animal and plant products. For example, the only difference between natural rubber and gutta percha (the tough, horny, inelastic material used for golf-ball covers) is that rubber is the *cis* form of the hydrocarbon polyisoprene, while gutta percha (sometimes called balata) is the *trans* form (see Section 15-5).

A third type of isomerism that often arises in organic chemistry, and especially in biochemistry, is **optical isomerism** (see Section 13-5). Optical isomerism occurs whenever there are four demonstrably different atoms or groups attached to one tetrahedral carbon atom. Designating the substituents as W, X, Y, and Z, we see (Figure 14-2) that there are two different spatial arrangements possible about the carbon atom. These two arrangements are actually mirror images of each other. Interchange any two groups, and you have interconverted the two arrangements. The two arrangements cannot be superimposed on each other.[7] An atom with four different groups tetrahedrally arranged around it is an *asymmetric* atom, because that assemblage has no symmetry elements passing through the central atom. Most (but not all) organic molecules that are optically active have one or more asymmetric carbon atoms.

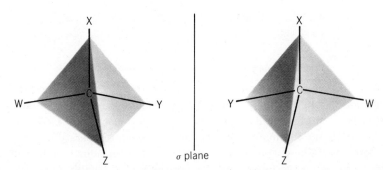

FIGURE 14-2 Optical isomerism about an asymmetric carbon atom. The two arrangements in space of four different substituents around a carbon atom are nonsuperimposable mirror images of each other.

Ordinarily, as we make dissymmetric compounds in the laboratory, we get a 50-50 mixture, of the D and L isomers. The task of separating them is often very difficult. Living organisms are very partial, however, and will build one or the other, rather than both. For example, all protoplasm is made up of proteins constructed with a variety of amino acids, but only L-amino acids appear. No living creature on earth is made of dextrotype amino acids. This may have been a matter of chance at the very beginning of life here, all

---

[7] Set up two tetrahedra; mark the four corners differently; and convince yourself of these properties of a tetrahedral arrangement of bonded atoms. Better still, buy an inexpensive set of molecular models and use them to explore this point and to illustrate all the rest of structural organic chemistry.

subsequent organisms having followed the original pattern. We do not know. Perhaps creatures on other planets will prove to be made of dextrotype amino acids.

14-7 The Functional Groups of Organic Chemistry

Since there are so many organic compounds, we need a way to classify them other than in terms of their many different structures and isomers. In particular, we should like to think about organic compounds functionally in terms of *what they do*, rather than just in terms of what they are. Fortunately, we can frequently identify a particular cluster or group of atoms that always exhibits general reactions characteristic of the group, wherever it may occur in analogous compounds. These characteristic reactions indicate the presence of the particular reactive group, and help us to detect and recognize that group in a substance of unknown structure. Such a group is also depended upon to react in predictable ways when we synthesize new compounds.

These common reactive groups are called *functional groups*, and our aim is to find out how they function. We shall consider 11 of the most important such groups.

1. MULTIPLE BONDS, ESPECIALLY C=C AND C≡C.

Any multiple bond is an assemblage of σ and π bonds, and so it contains some regions of high electron density. It is not surprising that electron-seeking reagents attack this high concentration of electrons, effecting a rearrangement of the overall electronic distribution and therefore a rearrangement of the bonds. We would expect that such reactions would reduce the multiplicity[8] of the carbon-carbon bond, and that the electron seeker would often become attached to a carbon atom, because carbon does not readily transfer electrons to form a permanently charged carbon ion.

As an example, consider the action of the electronegative (electron-seeking) element bromine on a carbon-carbon double bond. Let us distinguish between the σ and π bonds by writing the σ bond in the usual way as C—C, then showing the π bond by indicating an off-center pair of electrons, C$\stackrel{..}{-}$C. Now we might depict the collision of a molecule of ethane and a molecule of bromine, focusing attention on the fate of two particular pairs of electrons:

$$\underset{\text{reactants}}{H_2C\stackrel{..}{-\!\!\!-}CH_2 + Br\!:\!Br} \longrightarrow \underset{\text{intermediate}}{H_2C\overset{Br\,:\,Br}{-\!\!\!-\!\!\!-}CH_2} \longrightarrow \underset{\text{product}}{H_2\overset{Br}{C}-\overset{Br}{C}H_2}$$

(The actual molecular mechanism is more complicated, and the details depend on the nature of the solvent. The bromine molecule may first dissociate into the ions Br^+ and Br^-, or into the free radicals $Br\cdot$ and $Br\cdot$. Then the two fragments may add to the double bond in two distinct reaction steps.) The

---

[8] The C≡C bond has a multiplicity of 3; C=C has a multiplicity of 2; and C—C has a multiplicity of 1.

result is 1,2-dibromoethane, and we say that bromine has **added** across the double bond. The net reaction, conventionally written, is

$$H_2C=CH_2 + Br_2 \longrightarrow \underset{\underset{Br}{|}}{H_2C}-\underset{\underset{Br}{|}}{CH_2}$$

In the same way, one molecule of bromine could add to ethyne, and then another molecule of bromine to the resulting 1,2-dibromoethene until the organic molecule is **saturated:**

$$HC \equiv CH + Br_2 \longrightarrow \underset{H}{\overset{Br}{>}}C=C\underset{Br}{\overset{H}{<}}$$

and

$$\underset{H}{\overset{Br}{>}}C=C\underset{Br}{\overset{H}{<}} + Br_2 \longrightarrow H-\underset{\underset{Br}{|}}{\overset{\overset{Br}{|}}{C}}-\underset{\underset{Br}{|}}{\overset{\overset{Br}{|}}{C}}-H$$

The other halogen elements also add to multiple bonds, as do hydrogen halides, hydrogen cyanide, and hydrogen itself. Typical reactions include

$$HC \equiv CH + HCl \xrightarrow{HgCl} H_2C=CHCl$$
chloroethene
"vinyl chloride"

$$H_2C=CH_2 + H_2 \xrightarrow{Ni} H_3C-CH_3$$
ethane

Since alkenes (the aliphatic compounds with one or more double bonds) are plentiful by-products of petroleum refining, and ethyne is produced readily from calcium carbide and water, the reagents for such addition reactions are abundant and cheap. Thus, these reactions figure prominently in industrial organic chemistry.

Provided that petroleum remains abundant

2. THE HYDROXYL GROUP IN ALCOHOLS, R—OH. The symbol R is used here to represent any unspecified aliphatic hydrocarbon group, be it methyl or ethyl or anything higher. The **alcohols** R—OH are analogues of water, and they are waterlike in many ways. For instance, they react with metallic sodium to evolve hydrogen, just as water does:

$$2ROH + 2Na \longrightarrow 2RONa + H_2 \uparrow$$

$$2HOH + 2Na \longrightarrow 2HONa + H_2 \uparrow$$

The resulting RONa is an ionic compound, like NaOH, and the RO$^-$ ion is a strong base, like the OH$^-$ ion.

Intermolecular association involving hydrogen bonding occurs with alcohols, but to a lesser extent than with water. Methyl, ethyl, and propyl alcohols are miscible with water in all proportions, and so is tertiary butyl alcohol

$$\text{HO}-\underset{\underset{\text{CH}_3}{|}}{\overset{\overset{\text{CH}_3}{|}}{\text{C}}}-\text{CH}_3$$
t-butyl alcohol

It is as though a *t*-butyl alcohol molecule had been designed for maximum solubility of a C_4 alcohol in water; the hydrocarbon portion is almost spherical, presenting the smallest possible hydrocarbon surface for interaction with water molecules. Significantly, *n*-butanol

$$\text{CH}_3-\text{CH}_2-\text{CH}_2-\text{CH}_2-\text{OH}$$

with its long hydrocarbon chain and high hydrocarbon surface area, is only moderately soluble in water.

Alcohols are dehydrated to ethers in the presence of sulfuric acid, as for example

$$2\text{C}_2\text{H}_5\text{OH} \xrightarrow{\text{H}_2\text{SO}_4} \underset{\text{diethyl ether,}\atop\text{an anesthetic}}{\text{C}_2\text{H}_5-\text{O}-\text{C}_2\text{H}_5} + \text{H}_2\text{O}$$
ethanol

Under much more drastic conditions, a molecule of water may be expelled from a *single* molecule of an alcohol, forming the corresponding alkene. Thus,

$$\text{C}_2\text{H}_5\text{OH} \xrightarrow[\text{Al}_2\text{O}_3]{350°\text{C}} \text{H}_2\text{C}=\text{CH}_2 + \text{H}_2\text{O}$$

Do not think of alcohols as hydroxides, for they are not basic. They are acidic in the way water is, as shown by the displacement of protons by sodium. All alcohols show acidic behavior in other ways as well. The acidity is much more prominent, however, in the aromatic (benzenoid) alcohols, which have measurable acid dissociation constants in water. The simplest aromatic alcohol (C_6H_5OH, phenol) had the earlier name of *carbolic acid*.

All alcohols are toxic, but to varying degrees. Methanol is very poisonous. It acts on the nervous system and particularly the optic nerve, causing blindness, coma, respiratory failure, and eventually death. Since methanol is only slowly eliminated from the body, and since its metabolic products are also toxic, cumulative poisoning can occur even from the inhalation of the vapor. Ethanol is less toxic because it is much more quickly oxidized to carbon dioxide and water and does not act on the nervous system in the way methanol does. Ethanol does have narcotic effects, of course, and large doses will also cause coma. The vapor is about one-fifth as toxic as that of methanol. The propanols, butanols, and pentanols cause severe gastric disturbances and also are difficult for the body to eliminate. Isopropanol, $(CH_3)_2CH-OH$, is the alcohol most commonly used for body rubs, for it is not potable, and its vapor is only mildly toxic.

3. THE CARBONYL GROUP, $>C=O$. Carbonyl compounds are of three types—aldehydes, ketones, and carboxylic acids—and in each type the $C=O$

group acts somewhat differently. If a hydrogen atom is bonded to the carbonyl group, the compound is called an **aldehyde**:

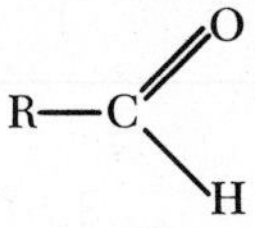

$$R-\overset{\displaystyle O}{\underset{\displaystyle H}{C}} \qquad H-\overset{\displaystyle O}{\underset{\displaystyle H}{C}}$$

an aldehyde 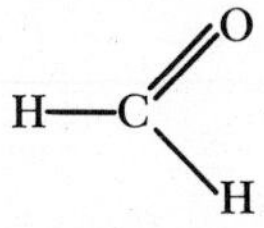   Formaldehyde
                   Geneva name: methanal

Aldehydes serve as reducing agents and are easily oxidized to the corresponding organic acids. Note the sequence

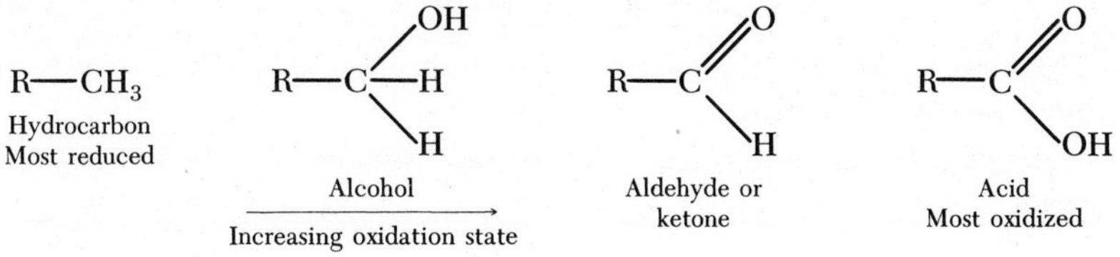

R—CH$_3$	R—C(OH)(H)(H)	R—C(=O)(H)	R—C(=O)(OH)
Hydrocarbon Most reduced	Alcohol	Aldehyde or ketone	Acid Most oxidized

Increasing oxidation state →

If there are *two* hydrocarbon groups bonded to the C=O group, then the compound is a **ketone**

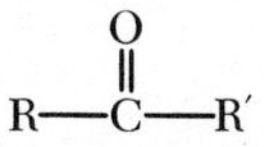

$$R-\overset{\displaystyle O}{\underset{\displaystyle \|}{C}}-R'$$

a ketone

The two R groups may be the same, or different

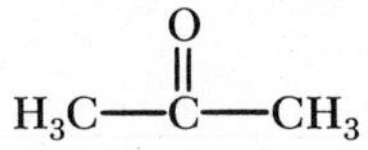

$$H_3C-\overset{\displaystyle O}{\underset{\displaystyle \|}{C}}-CH_3 \qquad H_3C-\overset{\displaystyle O}{\underset{\displaystyle \|}{C}}-CH_2-CH_3$$

Acetone   Methyl ethyl ketone
Geneva name: 2-propanone   Geneva name: 2-butanone

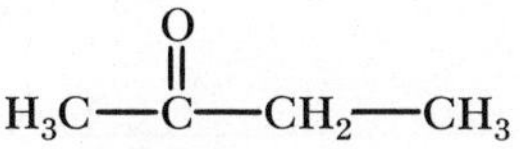

Ketones have a polar C=O group and a high concentration of electrons in the double bond, so they are good electron donors and good solvents. Their low boiling points and high flammability make them a bit dangerous to use, however. Having no hydrogen in a position to act as an electron acceptor, they are not associated in the liquid state; hence they are volatile. All ketones are toxic when ingested. Methyl ethyl ketone ("MEK" in slang) is an excellent solvent for many plastics and for paint films.

4. THE CARBOXYLIC ACID GROUP, —COOH. Every organic acid has a carboxyl group, and some acids have two or more such groups. An organic acid can be written as

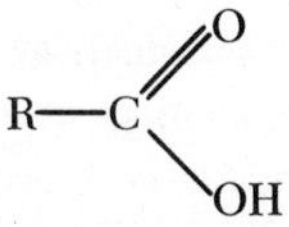

$$R-\overset{\displaystyle O}{\underset{\displaystyle OH}{C}}$$

Because the —OH group permits hydrogen bonding, the lower acids (the acids with just a few carbons), like the lower alcohols, are associated and are very soluble in water. The chief effect of the carbonyl group is to intensify greatly the acidity of the hydroxyl hydrogen by withdrawing electrons from the O—H bond, leaving the hydrogen freer to dissociate as H^+. The simplest acid, methanoic (formic) acid, H—COOH, is a moderately strong acid. The second in the series, ethanoic (acetic) acid, H_3C—COOH, is moderately weak. Replacement of the methyl hydrogen atoms in ethanoic acid by fluorine or chlorine atoms substantially enhances the acid strength by further electron withdrawal from the O—H bond; Cl_3C—COOH is almost as strong an acid as H_2SO_4.

Organic acids are effective solvents, combining the properties of alcohols and ketones. However, they also react with most of the substances they dissolve! Organic acids react with alcohols reversibly to form **esters**,

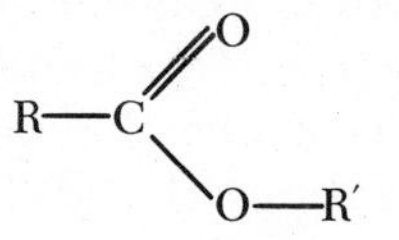

$$R-C\begin{smallmatrix}\nearrow O \\ \searrow O-R'\end{smallmatrix}$$

One such esterification reaction is

$$H_3C-C\begin{smallmatrix}\nearrow O \\ \searrow OH\end{smallmatrix} + HO-C_2H_5 \longrightarrow H_3C-C\begin{smallmatrix}\nearrow O \\ \searrow O-C_2H_5\end{smallmatrix} + H_2O$$

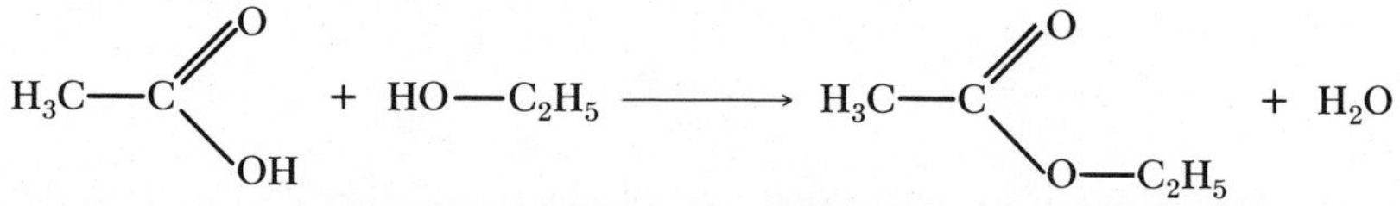

acetic acid + ethanol ⟶ ethyl acetate + water

Here the alcohol loses only its hydrogen atom, and the acid loses its —OH group. This fact was shown by performing a tracer experiment with ^{18}O-labeled alcohol. After reaction, the isotope ^{18}O was found in the ester, but not in the water. Thus, the actual reaction mechanism must be consistent with the observation

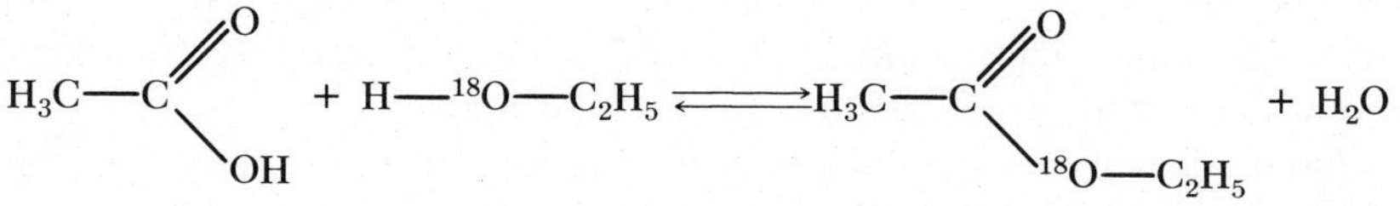

$$H_3C-C\begin{smallmatrix}\nearrow O \\ \searrow OH\end{smallmatrix} + H-{}^{18}O-C_2H_5 \rightleftharpoons H_3C-C\begin{smallmatrix}\nearrow O \\ \searrow {}^{18}O-C_2H_5\end{smallmatrix} + H_2O$$

The C—O bond in the alcohol is not cleaved during the reaction. The esterification reaction is markedly accelerated by the presence of a little strong acid.

The reverse reaction, really a hydrolysis of the ester, is facilitated by using a mole of base to combine with each mole of acid that is formed. This operation is called *saponification* (soap making). Soap (a mixture of the sodium salts of long-chain carboxylic acids) is in fact produced when the esters used are the animal-fat or vegetable-oil esters of the trihydric alcohol glycerol:

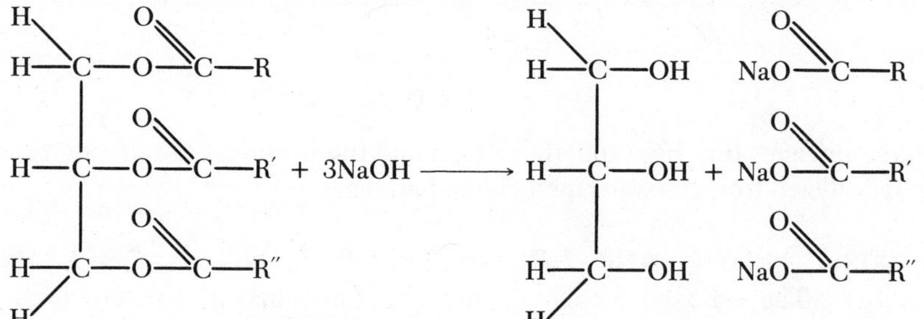

$$\begin{array}{c} HO \\ H-C-O-C-R \\ | \\ H-C-O-C-R' + 3NaOH \\ | \\ H-C-O-C-R'' \\ | \\ H \end{array} \longrightarrow \begin{array}{c} H \\ H-C-OH \\ | \\ H-C-OH + \\ | \\ H-C-OH \\ | \\ H \end{array} \begin{array}{c} O \\ NaO-C-R \\ O \\ NaO-C-R' \\ O \\ NaO-C-R'' \end{array}$$

vegetable oil or animal fat + base ⟶ glycerol + soap

The R groups represented here are principally those of palmitic acid, $CH_3(CH_2)_{14}COOH$, and stearic acid, $CH_3(CH_2)_{16}COOH$. The vegetable oils differ from animal fats, containing more unsaturated R groups, each with one or more $C=C$ bonds.

Esters, like ketones, are good solvents and are rather nonreactive. Although all the volatile organic acids have unpleasant odors, most esters of these same acids have quite pleasant odors; often these esters are used as constituents of fruit and flower fragrances.

5. **THE AMINO GROUP, $-NH_2$.** Organic **amines** are substituted ammonias and are basic to varying degrees. They bear the same relation to ammonia as alcohols do to water. Amines can be of three types: primary, of the type $R-NH_2$; secondary, of the type $RR'-NH$; and tertiary, of the type $RR'R''-N$.

$$
\begin{array}{ccc}
R-\ddot{N}-H & R-\ddot{N}-H & R-\ddot{N}-R'' \\
| & | & | \\
H & R' & R' \\
\text{Primary} & \text{Secondary} & \text{Tertiary} \\
\text{amine} & \text{amine} & \text{amine}
\end{array}
$$

In each case, there is an unbonded pair of electrons, indicated by two dots, available for reaction with electron acceptors. The R groups may be the same or different. In the series CH_3NH_2, $(CH_3)_2NH$, and $(CH_3)_3N$, the base strength decreases markedly as additional methyl groups are added. All three methylamines have the odor of decayed fish; all amines have unpleasant smells. Aromatic amines, such as $C_6H_5NH_2$ (phenylamine or aniline), are less basic than their aliphatic counterparts. Amines form coordination compounds with transition-metal ions, and so the long-chain amines adhere strongly to the oxidized surfaces of copper, nickel, and stainless steel. Amines are good electron-donor solvents, but often are too reactive to enjoy much use.

Just as alcohols react with carboxylic acids to form esters, so amines react with carboxylic acids to form **amides.** Thus,

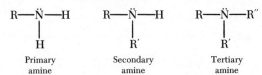

Such amide bonds

$$-C{\overset{\displaystyle O}{\underset{\displaystyle NH-}{\diagup\!\!\!\diagdown}}}$$

are the linkages that hold together nylon and the proteins. Amide bonds can be hydrolyzed to regenerate the acid and amine.

6. **NITRO, NITRITE, AND NITRATE GROUPS: $-NO_2$, $-ONO$, AND ONO_2.** The oxidized nitrogen-containing functional groups are far less stable than the reduced (amino) groups, readily losing their oxygen to form

N_2 and oxidation products of the R group. This instability has been exploited on a massive scale; all conventional (that is, nonnuclear) explosives except gunpowder are organic nitro and nitrate compounds. A nitro compound has an $-NO_2$ group attached directly to a carbon atom by a C—N bond, whereas a nitrite has a C—O—NO configuration, and a nitrate has a C—O—NO_2. Two examples of nitro compounds may suffice to show how they are made and what they are like:

$$\underset{\text{benzene}}{C_6H_6} + \underset{\text{nitric acid}}{HNO_3} \xrightarrow{H_2SO_4} \underset{\substack{\text{nitrobenzene,} \\ \text{a solvent for waxes}}}{C_6H_5NO_2} + H_2O$$

$$\underset{\text{toluene}}{C_6H_5CH_3} + 3HNO_3 \xrightarrow{H_2SO_4} \underset{\substack{\text{trinitrotoluene,} \\ \text{TNT, an explosive}}}{C_6H_2(NO_2)_3CH_3} + 3H_2O$$

TNT requires a strong detonator and booster charge to make it explode, and when it does, it lacks enough oxygen to oxidize all its carbon

$$2C_6H_2(NO_2)_3CH_3 \longrightarrow 5H_2O + 7CO + 3N_2 + 7C$$

For this reason, ammonium nitrate may be mixed with it to furnish extra oxygen and hence more energy:

$$NH_4NO_3 \xrightarrow{1000°C} N_2 + 2H_2O + [O]$$

The most famous organic nitrate is glycerol trinitrate, popularly (and erroneously) called nitroglycerine. This compound is made by esterification of glycerol (obtained by saponification of fats, or from petroleum-produced propene) with nitric acid via the reaction

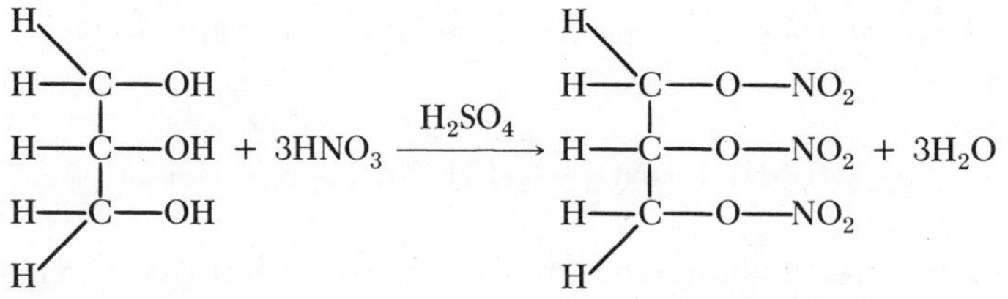

<!-- glycerol + 3HNO3 → glycerol trinitrate + 3H2O -->

Pure liquid glycerol trinitrate is very dangerous to handle, because it is sensitive to shock and detonates so readily. When soaked up by a porous solid (clay, sawdust, or nitrated cellulose), it is called *dynamite* and is safer to handle. In contrast to TNT, the nitrate groups in glycerol trinitrate furnish a surplus of oxygen when detonation occurs:

$$4C_3H_5(ONO_2)_3 \longrightarrow 10H_2O + 12CO + 6N_2 + 7O_2$$

Note that the stoichiometry allows all of the liquid explosive to be transformed into gases; the volume change upon explosion is impressively large.

Organic nitro compounds and nitrates are good electron-donor solvents, but seldom are they safe ones to use.

7. THE NITRILE GROUP, —C≡N.

Organic cyanides, R—C≡N, are called **nitriles.** They are closely related to organic acids. The nitrile group hydrolyses in the presence of acid to yield the corresponding organic acid (or the ammonium salt of that acid), the carbon of the —C≡N group staying with the R group. Thus,

$$R-C\equiv N + 2H_2O \xrightarrow{H^+} R-C\underset{OH}{\overset{O}{\diagup\!\!\!\diagdown}} + NH_3$$

Since HCN can be added to **alkenes** and **alkynes** (as indicated earlier in this section in the discussion of reactions of multiple bonds), this reaction provides a way to add one more carbon atom to an R group. This is also a way to make a carboxylic acid out of a hydrocarbon. The hydrolysis of a nitrile is a very handy reaction in chemical synthesis and manufacture.

The other important type of reaction which nitriles may undergo is *reduction* to an organic amine. This can be accomplished with hydrogen (in the presence of a catalyst like nickel), or by certain other reducing agents:

$$R-C\equiv N + 2H_2 \xrightarrow{Ni} R-CH_2-NH_2$$

Nitriles by themselves are quite stable compounds and are not as poisonous as the lethal inorganic cyanides.

8. THE DIAZO GROUP, —N=N—.

Organic diazonium compounds are rather rare. However, the diazo group is a useful functional group for syntheses where it is desired to attach a chlorine atom to carbon without the destructive oxidation that accompanies reaction with elementary chlorine. To accomplish this, an amine (here 1-methyl-2-aminobenzene, called toluidine) is treated with an ice-cold acidified solution of sodium nitrite to give a diazonium salt:

$$CH_3C_6H_4NH_2 + HNO_2 \xrightarrow[HCl]{0°C} CH_3C_6H_4N=N^+Cl^- + 2H_2O$$

The unstable diazonium chloride readily splits out nitrogen when warmed with a solution of copper (I) chloride, leaving chlorine attached to carbon:

$$CH_3C_6H_4N=NCl \xrightarrow[CuCl]{50°C} CH_3C_6H_4Cl + N_2$$

The specific 1-methyl-2-chlorobenzene is obtained, without any by-product isomers. This selective chlorination cannot be achieved by direct reaction with chlorine, because the methyl group would be chlorinated extensively, and the benzene ring would also become chlorinated at several different positions. A mixture of products would result.

The diazo group also figures prominently in a series of dyes, where intense yellow, orange, and red colors are developed as a result of the intense absorption of blue and green light by the configuration —C=C—N=N— (see Section

15-6). Such an arrangement of alternating single and double bonds is called a *conjugated system* and represents a sort of pathway (or pipeline or wave guide) for easy transfer of electrons in organic molecules. Whenever conjugated systems are encountered, it may be expected that absorption of light and enhanced reactivity of adjacent functional groups will be found.

9. CARBON-HALOGEN BONDS. Carbon-fluorine bonds are strong and decidedly polar, but C—Cl and C—Br are progressively weaker and much less polar. Carbon-iodine bonds have almost zero polarity, and the C—I bonds in aliphatic iodides are so weak as to be destroyed photochemically by daylight. Aromatic halides of the simple RX type (X = F, Cl, Br, or I) have stronger and less reactive C—X bonds than their aliphatic analogues.

Methyl and ethyl chlorides are somewhat soluble in water. They react extremely slowly with water to form the corresponding alcohols. For example,

$$C_2H_5Cl + H_2O \rightleftharpoons C_2H_5OH + HCl$$

This hydrolysis reaction is reversible under proper conditions, and so one of the ways of preparing methyl and ethyl chlorides involves that reverse reaction. Thus, C_2H_5Cl can be prepared by reaction of C_2H_5OH with anhydrous HCl gas. A single chlorine atom is selectively placed at the former site of the —OH group. By contrast, direct chlorination of methane or ethane always gives a mixture of monochlorinated, dichlorinated, and polychlorinated products.

The isolated carbon-fluorine bond is decidedly reactive toward water and toward many ionic reagents, although the lack of an effective mechanism for a fast reaction provides some kinetic stability. There is also a high probability of splitting out hydrogen fluoride (because of its high heat of formation) if there is a hydrogen atom nearby. But if *all* the hydrogen atoms of a hydrocarbon group or molecule are replaced by fluorine atoms, the C—F bond distance shrinks, the fluorine is held very strongly, and there is no HF to split out. Hence fluorocarbons are very stable and unreactive substances, unaffected by water and oxygen and even by aqueous acids and bases. Teflon is a polymeric compound of repetitive —CF_2— units, renowned for its chemical inertness to reagents and solvents.

Carbon-chlorine bonds are decidedly more reactive than C—F bonds, and sometimes will react with alcohol-soluble sodium compounds to split out alcohol-insoluble NaCl and introduce a new group into the molecule. Thus,

$$-CH_2Cl + NaCN \xrightarrow{C_2H_5OH} -CH_2CN + NaCl$$

When the C—Cl bond is insufficiently reactive to permit reaction with sodium compounds, silver compounds can often be used, because AgCl is less soluble than sodium chloride:

$$R'C(=O)OAg + RCl \longrightarrow R'C(=O)OR + AgCl$$

Chlorine or bromine atoms attached to the carbon atom of a carbonyl group are in a special class because their reactivity is greatly enhanced by the C=O group. An example is acetyl chloride, CH_3COCl, which is a potent **lachrymator** and always smells of hydrochloric and acetic acids because it hydrolyses so readily to produce both these acids:

$$CH_3-C\overset{O}{\underset{Cl}{\diagup\!\!\!\diagdown}} + H_2O \longrightarrow CH_3-C\overset{O}{\underset{OH}{\diagup\!\!\!\diagdown}} + HCl$$

Such carbonyl chlorides are termed acid chlorides because they can so easily produce the corresponding acid, and of course they serve as excellent reagents for introducing the formyl and acetyl groups:

$$H-C\overset{O}{\diagup\!\!\!\diagdown} \qquad H_3C\overset{O}{\diagup\!\!\!\diagdown}$$
Formyl group   Acetyl group

10. SULFUR-CONTAINING GROUPS. There are sulfur analogues of alcohols and ethers, all malodorous substances. Compounds of the type R—SH, analogous to alcohols, are called *mercaptans*. Some are added to natural gas to warn the householder of escaping gas; others serve in the skunk's arsenal. Compounds of the type R—S—R′ are *thioethers*. Proteins contain sulfur in the form of amino acid residues having C—S—S—C bonds, and in the bacterial disintegration of proteins, this sulfur usually is evolved as hydrogen sulfide. Since petroleum is derived from plant and animal tissue, it also contains sulfur; some of this is present in gasoline and fuel oil, and especially in "residual oil," a heavy, undistilled fuel oil used in power stations and ships. This sulfur eventually becomes oxidized to the air pollutant sulfur dioxide when the fuel is burned.

For further word on SO_2, see Ch. 18.

11. INORGANIC ESTER GROUPS: $-OPO_3H_2$, $-OSO_3H$, AND THE LIKE. Phosphoric acid groups are important constituents of plant and animal tissue and are intimately concerned with the formation of sugars, fats, and proteins from carbon dioxide and water (and ammonia) in the very complicated sequence of reactions we call photosynthesis. Similarly, phosphate esters are intimately involved in the stepwise oxidation of glucose[9] in muscle tissue, where the chemical energy of the glucose is converted to heat and to free energy of locomotion. The phosphate groups in organic molecules have the structure

$$-\overset{|}{\underset{|}{C}}-O-P\overset{OH}{\underset{OH}{\diagup\!\!\!\diagdown}}=O$$

---

[9] Glucose is a sugar of the composition $C_6H_{12}O_6$, a simple carbohydrate derived from foods by the hydrolysis of starch or cane sugar and the principal fuel for the oxidation in the human body.

They may be considered to be derived by the reaction of phosphoric acid (acting as a monoprotic acid) with the —OH group of an alcohol. Carbohydrates and glycerol have many such —OH groups available for esterification.

Esters of sulfuric acid play key roles in some organic syntheses, particularly of dyes and drugs. A simple example of how sulfate esters react is provided by the use of concentrated[10] sulfuric acid in an outmoded preparation of industrial ethyl alcohol from ethene:

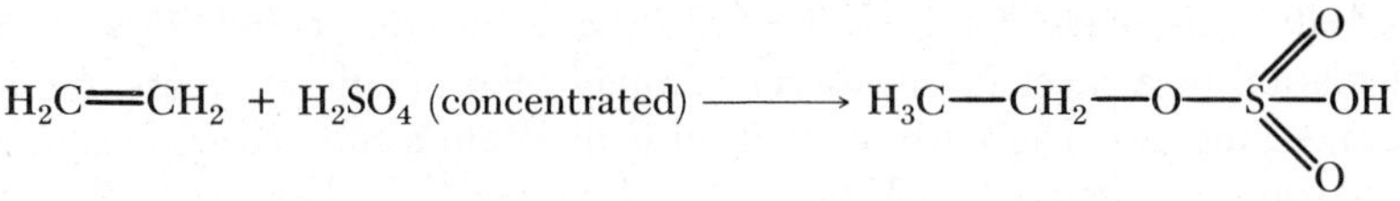

$$H_2C=CH_2 + H_2SO_4 \text{ (concentrated)} \longrightarrow H_3C-CH_2-O-\overset{\overset{O}{\|}}{\underset{\|}{S}}-OH$$

$$H_3C-CH_2-OSO_3H + H_2O \longrightarrow CH_3CH_2OH + H_2SO_4$$

In some aromatic compounds, the sulfur atoms of sulfuric acid are attached directly to carbon atoms, rather than attached through oxygen atoms as in the above ethylsulfuric acid. For example, when benzene is treated with fuming sulfuric acid, there is formed a compound of the composition $C_6H_5SO_3H$:

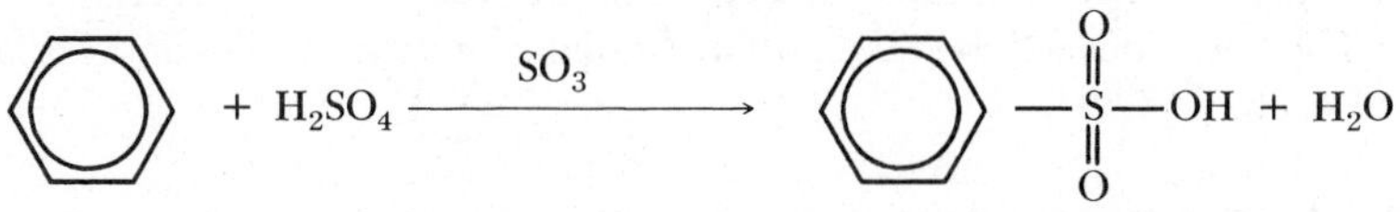

$$\text{C}_6\text{H}_6 + H_2SO_4 \xrightarrow{SO_3} \text{C}_6\text{H}_5-\overset{\overset{O}{\|}}{\underset{\underset{O}{\|}}{S}}-OH + H_2O$$

benzene sulfonic acid

Such substances are called *sulfonic* acids to distinguish them from the sulfate esters. The usual reasons for introducing a sulfonic acid group are to increase the solubility in water and to provide a reactive functional group for attaching something else at that site later. Sometimes the subsequent reactions lead to the final inclusion of —SO_2— groups or —SO_2NH_2 groups, as in these well-known compounds:

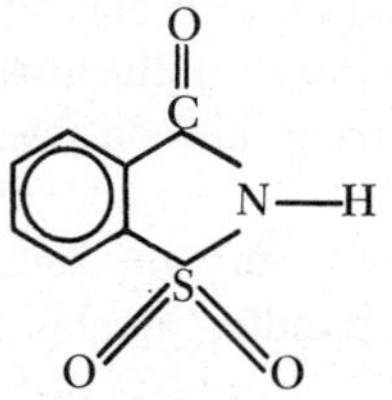

saccharin,
a sweetener

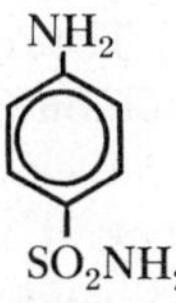

sulfanilamide, first
of the sulfa drugs

Further acquaintance with the functional groups will best be provided by practical examples of organic syntheses and by considering the large-scale production of common materials. In order to understand just *how* the reactions of functional groups take place, however, it will be necessary to examine a few mechanisms.

---

[10] Pure "concentrated sulfuric acid" has a composition corresponding to equal numbers of moles of SO_3 gas and H_2O. Fuming sulfuric acid is produced by dissolving additional SO_3 in concentrated H_2SO_4.

14-8 Reaction Mechanisms in Organic Chemistry

Classification of reactions according to mechanisms brings order to organic chemistry.

A reaction mechanism is a complete description of just how the reactant molecules become the product molecules in a particular chemical reaction. We should like to identify the individual steps (the elementary reactions) of the reaction, and we hope to identify the intermediate compounds (the transition states) that are involved. To do all this, a thorough study of the kinetics of the reaction becomes necessary, for the choice among several speculated possibilities must be in accordance with these rates-of-reaction data. Actual elucidation of a mechanism, therefore, may take months or years, for it is a painstaking job. The purpose of it all is to attain a rational understanding of the ways in which reactions take place, followed by a classification of many reactions according to type of mechanism. This would put a great deal of order into organic chemistry, making unnecessary the remembering of a huge catalog of individual reactions.

Experimental study of hundreds of reactions by the methods of physical organic chemistry during the past 20 or 30 years has indeed led to some systematic understanding. The most important point to emerge is that *the general types of organic reactions are surprisingly few*. Five principal types of reactions, in fact, embrace almost all of the reactions considered in this book. The five types will be taken up in sequence, with examples and explanation of each.

1. **SUBSTITUTION OR DISPLACEMENT REACTIONS.** In a **heterogeneous** reaction, such as the reaction of a lump of metallic sodium with liquid alcohol, we can easily conceive of molecules of alcohol striking the surface of the sodium. The sodium can then displace the active (the acidic) hydrogen atom of the —OH group as the oxygen becomes temporarily coordinated to the metal atom. Hydrogen atoms then combine on the metal surface, becoming H_2 molecules that eventually form a gas bubble and escape from solution. The sodium alkoxide produced at the metal surface diffuses away into the alcohol. This replacement reaction can take place only at the interface between the two phases. The reaction rate depends on the area of sodium surface exposed to the alcohol, on the rate of stirring, and on the temperature. The molecular picture is rather simple.

But now let us look at a **homogeneous**[11] reaction, such as the reversible hydrolysis of an alkyl halide to the corresponding alcohol and hydrohalogen acid:

$$RX + H_2O \rightleftharpoons ROH + HX$$

Here a hydroxyl group replaces (or substitutes for) the halogen atom of the halide RX. The halogen in turn substitutes for the hydroxyl group of water.

---

[11] If we are to study this reaction in homogeneous solution, we need a reaction solvent. Liquid alkyl halides and water are generally immiscible; only for a few alkyl halides is there an appreciable range of mutual solubility with water. Thus, for a reaction medium, we need a liquid that is a solvent for water and the halide, as well as for the alcohol and hydrogen halide formed as products. An ether or a ketone usually will do.

How does this reaction proceed? What is the molecular mechanism? Chemists have envisioned two contrasting mechanisms for this solution reaction: an *associative mechanism* and a *dissociative mechanism*. We shall examine both possibilities.

Reaction according to the associative mechanism begins with the collision of an RX molecule and a water molecule. These two reaction partners bump, adhere in some way, rearrange their bonds, and then separate as transformed products ROH and HX. The mechanism can be written in terms of the elementary reactions, beginning with *association* of the two reactants to give a transition-state complex.

$$RX + H_2O \underset{k_{-1}}{\overset{k_1}{\rightleftharpoons}} \begin{array}{c} X \\ R \diamond H \\ O \\ H \end{array}$$

$$\begin{array}{c} X \\ R \diamond H \\ O \\ H \end{array} \underset{k_{-2}}{\overset{k_2}{\rightleftharpoons}} ROH + HX$$

reactants   transition-   products
             state
             intermediate

Formation of the unstable intermediate (by either the forward or the reverse reaction) is via a *bimolecular* elementary process.

Now compare the same overall reaction proceeding by the *dissociative* mechanism. Each reactant molecule begins the reaction by *dissociating* into ions:

$$RX \underset{k_{-1}}{\overset{k_1}{\rightleftharpoons}} R^+ + X^-$$

$$H_2O \underset{k_{-2}}{\overset{k_2}{\rightleftharpoons}} H^+ + OH^- \quad \text{very fast}$$

$$R^+ + OH^- \underset{k_{-3}}{\overset{k_3}{\rightleftharpoons}} ROH$$

$$X^- + H^+ \underset{k_{-4}}{\overset{k_4}{\rightleftharpoons}} HX$$

reactants   intermediates   products

The two different mechanisms have different intermediates.

Here the alkyl halide RX dissociates to give two ionic intermediates that react with the ions of water. All elementary processes leading to intermediates are *unimolecular*. Any reaction condition likely to facilitate ionization of the reactants is likely to lead to increased concentrations of reactants and thus to faster reaction rates.

How could we design experiments to distinguish between these alternative mechanisms? One way takes advantage of the mathematical rate equations associated with each mechanism (see Section 11-4). These equations predict different dependence of the reaction rates on values of the concentrations of reactants. This distinguishing concentration dependence arises because in one case the intermediate is formed by a bimolecular process, whereas the intermediates in the other mechanism are formed by unimolecular processes. Another experimental method involves studying the reaction in a variety of different solvents; if the associative mechanism were operative, then only a small change in rates would be expected when solvents were changed, but the major role of ionization played in the dissociative mechanism means that a large increase in reaction rates would be expected when ionizing solvents are used, and that a decrease in rates would be anticipated for nonpolar solvents in which ionization is suppressed.

Recall for comparison the discussion of mechanism of octahedral substitution in Sec. 13-7.

We can now turn to the results, published in the chemical research journals, asking which mechanism agrees with experimental facts. The answer is that *both* mechanisms are encountered! For simple halides like CH_3Br and C_2H_5Cl, the bimolecular, associative mechanism prevails. But the ionic, unimolecular, dissociative mechanism prevails when the R group is large and bulky (hence capable of fending off an approaching water molecule from the halogen atom), and especially when the R group is also a branched structure that can distribute a charge and so stabilize the positive ion, as in the solvated ion

$$H_3C-\underset{\underset{CH_3}{|}}{\overset{\overset{CH_3}{|}}{C^+}}-CH_3$$

2. ADDITION REACTIONS. The **addition** of chlorine to the π-electron cloud of ethene is an example of an addition reaction. So are the related additions

$$\underset{H_3C}{\overset{H_3C}{\diagdown}}C=O + HCN \longrightarrow \underset{H_3\diagup}{\overset{H_3C\diagdown}{\diagdown}}\underset{\diagdown CN}{\overset{\diagup OH}{C}}$$

and

$$CH_3-CH=CH_2 + HCl \longrightarrow CH_3-\underset{\underset{}{|}}{\overset{\overset{Cl}{|}}{CH}}-CH_3$$

We can picture Cl_2 attacking the localized electron density of a $C\!\!\stackrel{..}{=}\!\!C$ bond in ethene to give dichloroethene

$$H_2C\!\stackrel{..}{-}\!CH_2 + Cl_2 \longrightarrow H_2\ddot{C}\overset{Cl}{}\quad\overset{Cl}{}\ddot{C}H_2$$

but the addition of gaseous hydrogen chloride to propene to give $CH_3CHClCH_3$ is more puzzling. Why does the chlorine atom end up on the middle carbon atom? Why not on the terminal carbon atom half the time?

These questions have been settled by a thorough investigation of the rate of reaction versus conditions and concentrations, leading to these conclusions:

  a. With an ionizing solvent present, or with some suitable metal halide to stabilize the chloride ion by coordination, the hydrogen chloride dissociates to H^+ and Cl^- ions.
  b. The hydrogen ion, being an electron seeker, attaches itself at the site of the π bond:

$$CH_3-CH=CH_2 + H^+ \longrightarrow CH_3-\overset{+}{C}H-\overset{\overset{H}{|}}{C}H_2$$

  The positive charge becomes distributed more easily over the rest of the molecule if it resides mostly on the *center* carbon atom, and so the ion is said to be stabilized by delocalization of charge. As the attacking hydrogen ion transfers its positive charge to the center carbon atom, it becomes a hydrogen atom bound to the terminal carbon atom by the electrons of the former π bond.
  c. Lastly, the chloride ion is attracted to the positively charged central carbon atom and neutralizes that charge by donating a pair of electrons to form a covalent bond:

$$CH_3-\overset{+}{C}H-CH_3 + Cl^- \longrightarrow CH_3-\overset{\overset{Cl}{|}}{C}H-CH_3$$

The result is 2-chloropropane, not 1-chloropropane. A similar result could be expected from the addition of any acid or other polar substance to a double bond; the positive ion becomes attached to the double-bond partner which bears the greater number of hydrogen atoms (here the terminal carbon atom). This is known as *Markownikoff's rule* for such additions, a rule that allows prediction of the structure of the product.

Addition and condensation mechanisms are often exploited for making polymers; see Sec. 14-9 and 15-4.

3. CONDENSATION-ELIMINATION REACTIONS. When ethanol is heated with some sulfuric acid, two molecules of the alcohol combine, splitting out water and leaving diethyl ether:

$$C_2H_5OH + HOC_2H_5 \longrightarrow C_2H_5OC_2H_5 + H_2O$$

This is a **condensation** reaction, for two small molecules have condensed to make a larger one. It also is an **elimination** reaction, because a very small molecule (water) has been eliminated as a result of the interaction of two functional groups. The eliminated molecules are usually ones that have a high heat of formation, such as HCl or H_2O. Elimination reactions are the opposite of addition reactions, as is clear from two equations:

$$\text{addition:} \quad H_2C=CH_2 + Cl_2 \longrightarrow \underset{\underset{Cl}{|}}{H_2C}-\underset{\underset{Cl}{|}}{CH_2}$$

$$\text{elimination:} \quad \underset{\underset{Cl}{|}}{F_2C}-\underset{\underset{Cl}{|}}{CF_2} + Zn \longrightarrow F_2C=CF_2 + ZnCl_2$$

Condensation reactions are particularly useful in making polymers, where the aim is to make a few very large molecules out of many small ones. The building of proteins from amino acids is a process of condensation.

4. OXIDATION-REDUCTION REACTIONS. Oxidation and reduction have the same meaning here as they had in the chapters on inorganic chemistry and are just as useful as they were there. The situation is somewhat simpler in that organic oxidation reactions generally involve the introduction of oxygen into the molecule, and reduction generally involves the introduction of hydrogen. For example, alcohols may be oxidized by oxygen itself to give aldehydes or ketones, and aldehydes are oxidized to acids:

$$2CH_3OH + O_2 \xrightarrow{CuO} 2HC\begin{smallmatrix}O\\H\end{smallmatrix} + 2H_2O$$

$$2HC\begin{smallmatrix}O\\H\end{smallmatrix} + O_2 \xrightarrow{CuO} 2HC\begin{smallmatrix}O\\OH\end{smallmatrix}$$

Only primary alcohols (that is, alcohols that have the hydroxyl group attached to a CH_2 group) can undergo these two oxidation reactions. Secondary alcohols (which have the hydroxyl group attached to a CH group within a carbon chain) have insufficient hydrogen on the hydroxyl-bearing carbon atom to form aldehydes, and so end up as ketones:

Remember that an oxidation must always be coupled with a reduction.

$$6CH_3\overset{OH}{\underset{|}{C}}HCH_3 + K_2Cr_2O_7 + 4H_2SO_4 \longrightarrow 6CH_3\overset{O}{\underset{|}{C}}CH_3 + Cr_2(SO_4)_3 + K_2SO_4 + 7H_2O$$

2-propanol (isopropyl alcohol) → 2-propanone (acetone)

The last three equations are not practically reversible. However, with suitable choice of reducing agents, all three reactions can be reversed. Reduction of aldehydes gives primary alcohols, and by reduction of ketones we get secondary alcohols. Tertiary alcohols which have the configuration

$$\begin{array}{c} C \\ | \\ C-C-OH \\ | \\ C \end{array}$$

are difficult to oxidize at all, and oxidize only by fragmentation.

A further example of reduction occurs in the conversion of a nitrile to an amine:

$$R-C\equiv N + 2H_2 \xrightarrow{Ni} R-CH_2-NH_2$$

This reaction is important in the manufacture of nylon, which is a polymer formed by condensing a diamine with a diacid (see Section 15-4).

5. REARRANGEMENT REACTIONS.
In rearrangements, there is no oxidation, no reduction, no condensation, no addition—just isomerization. The various components of the molecule change position, but there is no loss or gain of atoms. For example, when 3-butenoic acid is heated, it rearranges to give 2-butenoic acid, which has a conjugated system of bonds:

$$CH_2{=}CH{-}CH_2{-}COOH \xrightarrow{heat} CH_3{-}CH{=}CH{-}COOH$$

Rearrangements can be furthered by heterogeneous catalysts, as in the isomerization of n-octane to isooctane to make high-octane gasoline:

$$CH_3{-}CH_2{-}CH_2{-}CH_2{-}CH_2{-}CH_2{-}CH_2{-}CH_3 \xrightarrow{\text{Pt on } Al_2O_3 + SiO_2}$$

n-octane, C_8H_{18}

$$CH_3{-}\underset{CH_3}{\overset{CH_3}{\underset{|}{\overset{|}{C}}}}{-}CH_2{-}\underset{}{\overset{CH_3}{\underset{}{\overset{|}{CH}}}}{-}CH_3$$

2,2,4-trimethylpentane, "isooctane"

Here the n-octane is broken into fragments by the very high temperature employed, but the fragments are held on the surface of the platinum catalyst until they can regroup in another more stable configuration. Aluminum chloride is also a catalyst for conducting rearrangements, but it acts by encouraging *ionization* of the reactant to form (positive) carbonium ions and (negative) stabilized carbanions, which later may recombine in a different way to give an isomeric product.

What other rearranged octanes can you imagine?

14-9 Some Practical Examples

The importance and the utility of the various type reactions and the functional groups can best be appreciated by considering some actual processes. The processes discussed in this section have been selected because they are relevant to our current industrial culture and also relevant to the principles that have been outlined in the previous section.

GASOLINE HYDROCARBONS.
There was a time when "gasoline" meant only the volatile fraction of straight-chain hydrocarbons distilled from crude oil. As early as 1925, the supply of this "straight-run" gasoline became inadequate, however, and something had to be done to increase the yield of gasoline from each barrel of crude oil. The "something" was thermal cracking, a process which split up large (nonvolatile) molecules of hydrocarbon into smaller (volatile) molecules that could be put in with the straight-run gasoline. Later, cracking was done more efficiently by using a suspended clay catalyst in a cyclic operation. An example of a cracking reaction is

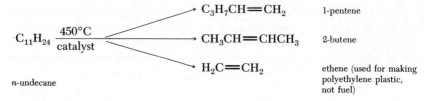

Saturated paraffins are aliphatic compounds of composition C_nH_{2n+2}

The hydrocarbons in crude oil are mostly **alkanes** (saturated paraffinic hydrocarbons with no double bonds), and especially unbranched alkanes such as the *n*-undecane shown above. The products of cracking, however, are **alkenes** (unsaturated hydrocarbons with double bonds). It soon became evident that the new fuel from the cracking operation did not act like the older straight-run fuel for automobiles. In one way, it was worse, because it formed gummy deposits in the tanks and fuel lines, due to oxidation and oxygen-induced polymerization at the double bonds. In another way, it was better, for it did not cause as much knocking in the engine at full throttle.

An intensive study[12] of the actual combustion flames in internal combustion engines revealed that the straight-chain hydrocarbons tended to detonate in the engine cylinders if the compression ratio were greater than 5:1. The detonation resulted from an initial slow oxidation of the hydrocarbon, heating the air-hydrocarbon mixture, followed by a very rapid oxidation of the mixture by another mechanism (the detonation). The slow preliminary oxidation takes place only at the ends of the straight-chain hydrocarbon, and no way has been found to hurry the process or to make such hydrocarbons burn smoothly at higher compression ratios. At the same time, higher compression ratios are highly desirable from the standpoint of efficiency, because the more the hot mixture of gases can be made to expand in the engine cylinder, the more work they can be made to do, and the less heat must be thrown away. The key to higher-compression-engines obviously lay in an improved gasoline, and ways had to be found to redesign the hydrocarbon molecules that occur naturally in crude oil in order to make that improved gasoline.

Three types of hydrocarbon were found to work better than straight-chain alkanes in experimental engines with high compression ratios: alkenes (which could oxidize at the double-bond sites as well as at the ends of the chain), aromatics like benzene (which also have plenty of π electrons and can oxidize throughout the molecule), and highly branched alkanes (which have several points for attack by oxygen because of the branching). One of the worst offenders in causing knocking was *n*-heptane, $CH_3CH_2CH_2CH_2CH_2CH_2CH_3$. The best fuel at the time of the study was an isomer of octane, and specifically 2,2,4-trimethylpentane, called "isooctane,"

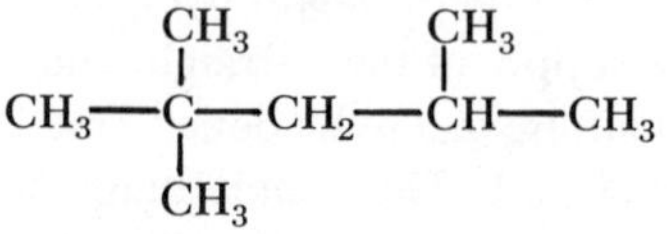

$$CH_3-\underset{\underset{CH_3}{|}}{\overset{\overset{CH_3}{|}}{C}}-CH_2-\underset{}{\overset{\overset{CH_3}{|}}{CH}}-CH_3$$

---

[12] Started by Thomas Midgley, who not only invented antiknock gasoline (with tetraethyl lead in it to prevent knocking), but also invented the Freon nontoxic refrigerants and one of the earliest types of synthetic rubber.

This hydrocarbon has about the same vapor pressure as *n*-heptane, so mixtures of the two could be expected to evaporate without change of composition. A performance scale was devised, based on these two hydrocarbons; *n*-heptane was assigned a rating of zero, and isooctane a rating of 100 on the "octane scale." Commercial gasolines are mixtures of a great many compounds, of course, but all can be rated against the two standards of the octane scale. If a particular gasoline causes an experimental engine to knock as badly as *n*-heptane does, then no matter what its composition may be, it is said to have an octane rating of zero. Similarly, a sample of gasoline that behaves the same way *in an engine* as a mixture of 50-percent *n*-heptane and 50-percent isooctane is said to have an octane rating of 50, and a gasoline that behaves exactly the same as isooctane itself has an octane rating of 100. Through the years, the octane rating of regular gasoline has been pushed upward steadily by improvements in refining operations and by the introduction of tetraethyl lead as an antiknock catalyst. In 1930 an octane rating of 60 was common, but by 1970 the octane ratings of 93 for regular gasoline and 97 for premium gasoline were considered about par. This drastic improvement has allowed the introduction of more efficient engines of high-compression ratio, with consequent saving in the fuel consumption per horsepower-hour.

It is interesting to consider a few of the reactions by which refiners now produce high-octane gasoline, because these reactions refute the often-expressed view that alkanes are so nonreactive that not much can be done with them. One of the earliest reactions was *alkylation*, which allowed the synthesis of branched-chain alkanes from *n*-alkanes produced in cracking:

$$2C_2H_6 + CH_3CH\!=\!CHCH_3 \xrightarrow{\text{HF} \atop \text{BF}_3} CH_3\!-\!\underset{\underset{C_2H_5}{|}}{\overset{\overset{C_2H_5}{|}}{CH}}\!-\!CH\!-\!CH_3 + H_2$$

ethane   2-butene   3,4-dimethylhexane

Another reaction used now is *hydroforming*, a way of making benzene[13] from *n*-hexane:

$$CH_3CH_2CH_2CH_2CH_2CH_3 \xrightarrow[\text{high } T \text{ and } P]{Cr_2O_3\ +\ Mo_2O_5} C_6H_6 + 4H_2$$

n-hexane   benzene

A third method is *catalytic reforming*, by which smaller cycloalkanes can be aromatized to yield benzene and hydrogen:[14]

---

[13] Benzene has a high density and a high octane rating, so it gives more energy per gallon than *n*-hexane.

[14] All these refinery operations produce hydrogen as a by-product. Some of this hydrogen is used for desulfurizing petroleum products, but most of it is available for making ammonia to be converted to agricultural fertilizer. Hence oil refining and nitrogen fixation go together.

methyl cyclopentane → (10% MoO₃ on Al₂O₃) → cyclohexane → benzene + 3H₂

Still other operations give highly branched alkanes by isomerization, as has been shown in the discussion of rearrangement reactions (Section 14-8). Some of the products have been found to have octane numbers over 100 (that is, they give better high-compression performance than pure isooctane itself). By linear extrapolation of the scale, one compound can be said to have an octane rating of about 200:

$$CH_3-C(CH_3)(CH_3)-CH(CH_3)-CH_3 \quad \text{2,2,3-trimethylbutane, "triptane"}$$

This is the hydrocarbon of highest octane rating known so far.

The development of high-octane gasoline in the United States made it possible to build smaller, lighter, more efficient engines, and hence smaller and lighter cars with as much power as those of the 1930s, but with greatly reduced gasoline consumption. Unfortunately, the American public chose to take its dividend in the form of more power rather than reduced gasoline consumption, and so engines became even more powerful and cars even larger. This choice, made very clearly in the marketplace, greatly increased the pollution problem, for a high-compression engine produces more oxides of nitrogen than a low-compression one, due to the higher combustion temperature. Moreover, the emission of hydrocarbons and carbon monoxide is much worse from a large engine than from a small one.[15] So from the standpoints of conservation of resources and alleviation of air pollution, the public choice (perhaps an uninformed choice) of the 1950s and 1960s proved a bad one. A greater awareness of the chemical facts, coupled with a more enlightened demand for cars that will treat our atmosphere and our resources better, could lead to considerable improvement in the latter 1970s.

SOME COMMON PLASTICS. The reactions by which polymers of very high molecular weight can be made are of two types: addition polymerization, by which alkenes add to the double bonds of similar or identical molecules, and condensation polymerization, by which large molecules are made from simple substances containing functional groups by an interaction that splits out small molecules such as water. A good example of addition polymerization is provided by the manufacture of polyethylene, the most popular plastic in

---

[15] The 1970 United States standards exempted all new 4-cycle engines of 50 cm³ displacement or less from emission controls, because they do not produce enough pollutants to bother about.

terms of weight produced. Polyethylene of random structure is made by compressing the gas ethene (ethylene) to several thousand atmospheres and heating it in the presence of some oxygen. Partial oxidation produces a few **free radicals** of unknown composition, which we shall represent by R·. An R· radical can attack the π electrons of a neighboring molecule of ethene, seizing an electron:

$$R\cdot + H_2C=CH_2 \longrightarrow R-CH_2-\dot{C}H_2$$

The product, being also a free radical because it has an unpaired electron left from the π bond, attacks another molecule of ethene, and so on:

$$R-CH_2-\dot{C}H_2 + H_2C=CH_2 \longrightarrow R-CH_2-CH_2-CH_2-\dot{C}H_2$$

A free radical is a nonionic molecular fragment with an unpaired electron.

The process goes on until two free radicals happen to meet and join, thereby eliminating the active sites. It follows that addition polymerization can lead to high molecular weights only if very few free radicals are introduced in the first place, and these few are given a long time to grow in the presence of monomer.

Polyethylene with a more orderly structure (and hence with a greater density and some crystallinity) is made by using a heterogeneous catalyst derived from an organometallic coordination compound. The ethylene bonds itself temporarily to an octahedral site on the coordination complex, and the ordered polymerization takes place while the ethene is held in this constrained position. Polypropylene is made from propene by a similar method. Many other analogues or derivatives of ethene also produce useful plastics by addition polymerization, but usually by the free-radical method. Some are listed in Table 14-2.

TABLE 14-2
Polymerizable Derivatives of Ethene

Compound	Formula	Name	End Use of Polymer
Vinyl chloride	$H_2C=CHCl$	PVC	rigid pipes and ducts°
Vinyl acetate	$H_2C=CH-O\overset{\overset{O}{\|}}{C}CH_3$	PVA	adhesives°
Styrene	$H_2C=CHC_6H_5$	poly-styrene	foamed insulation, copolymers†
Acrylonitrile	$H_2C=CHCN$	Orlon, Acrilan, and so on	fibers, copolymers†
Methyl methacrylate	$H_2C=\overset{\overset{CH_3}{\|}}{C}-COOCH_3$	PMMA	transparent organic glass
Tetrafluorethene	$F_2C=CF_2$	Teflon	unreactive plastic

° A copolymer of vinyl chloride and vinyl acetate is vinylite, or common vinyl plastic for upholstery, floor tiles, and so on.
† A copolymer of acrylonitrile, butadiene ($H_2C=CH-CH=CH_2$), and styrene is ABS, a plastic with high impact strength.

ORGANIC CHEMISTRY

Brown or black Bakelite is found in households as pan handles and in electrical equipment.

Condensation polymers are of many kinds (nylon, silicones, polyesters, phenolics, and so on), but the reactions used for making them are basically simple. We may take the preparation of Bakelite-type phenol-formaldehyde polymers as an example, and follow all the steps:

1. Phenol and formaldehyde are made from coal, air, and water

$$H_2O + C \text{ (coke from bituminous coal)} \xrightarrow{800°C} CO + H_2$$

$$CO + H_2O \xrightarrow[\text{catalyst}]{450°C} CO_2 + H_2$$

$$CO + 2H_2 \xrightarrow[\text{200 atm, 350°C}]{Cr_2O_3 + ZnO} CH_3OH$$

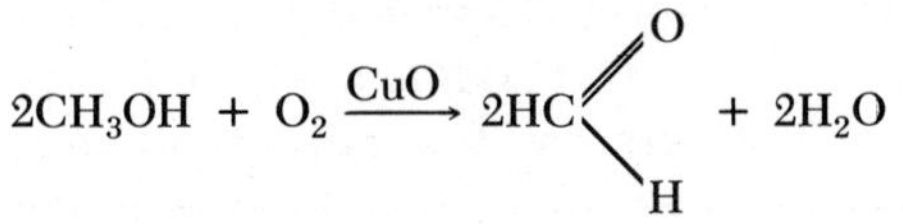

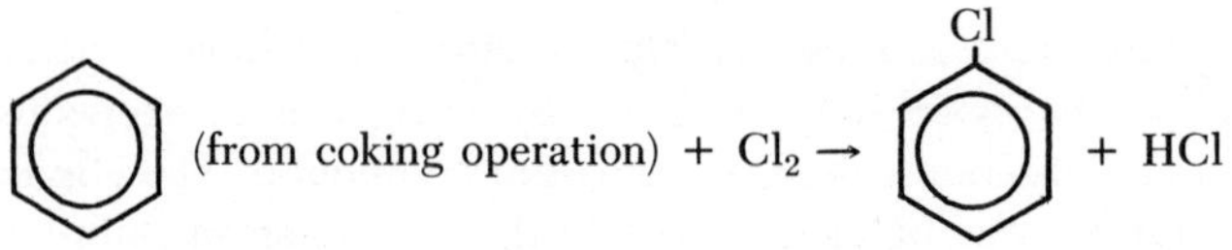

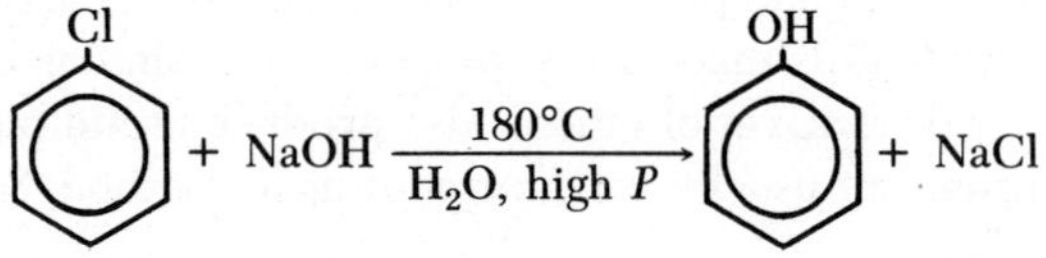

$$2CH_3OH + O_2 \xrightarrow{CuO} 2HC\!\!\begin{array}{c}\diagup O\\ \diagdown H\end{array} + 2H_2O$$

[benzene] (from coking operation) + Cl$_2$ → [chlorobenzene] + HCl

[chlorobenzene] + NaOH $\xrightarrow[\text{H}_2\text{O, high }P]{180°C}$ [phenol] + NaCl

2. The phenol and formaldehyde combine to form an alcohol

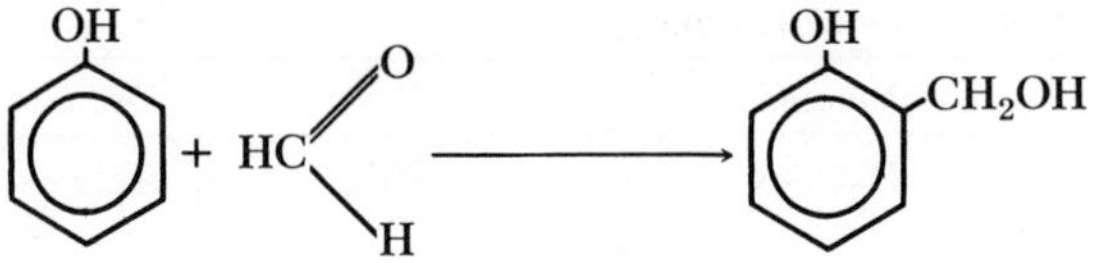

[phenol] + HC(=O)H ⟶ [o-hydroxybenzyl alcohol: phenol with CH$_2$OH]

3. This alcohol undergoes condensation with one of the ring-position hydrogen atoms

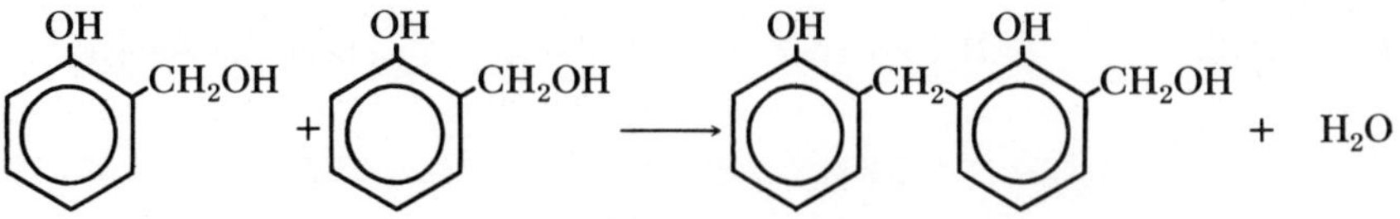

[phenol-CH$_2$OH] + [phenol-CH$_2$OH] ⟶ [phenol-CH$_2$-phenol-CH$_2$OH] + H$_2$O

and the operation is repeated many times to give a polymer of the type

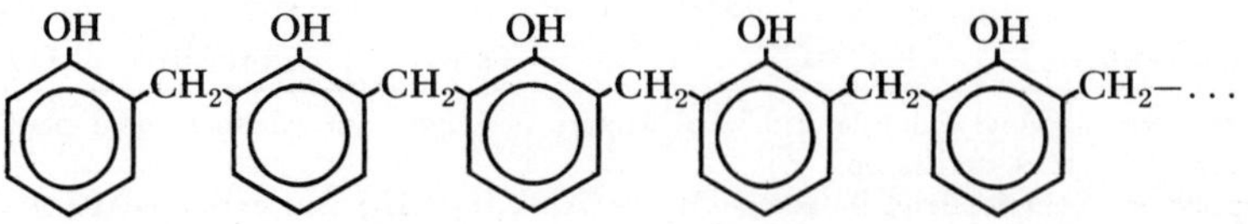

[phenol-CH$_2$-phenol-CH$_2$-phenol-CH$_2$-phenol-CH$_2$-phenol-CH$_2$-...]

The polymer is by no means as regular nor as linear as shown. Other hydrogen atoms on the ring positions also react, giving rise to cross-links and branched chains which make the final product rigid. The last stages of the condensation take place in the mold during fabrication, in the presence of fillers and reinforcing fibers, and with the aid of an acid catalyst.

OTHER MATERIALS FROM ETHENE, ETHYNE, AND PROPENE. Ethanol is used in large quantities as a solvent and as a starting material for making esters and synthetic rubber. Formerly it was produced by fermentation of grain or potatoes or molasses, putting it in competition with the food supply. Now it is made from petroleum (by hydration of ethene), or from coal (by way of calcium carbide and ethyne). Starting with coal and limestone, the reactions are

$$CaCO_3 \xrightarrow{750°C} CaO + CO_2$$

$$CaO + 3C \xrightarrow{1000°C} CaC_2 + CO$$

$$CaC_2 + 2H_2O \longrightarrow Ca(OH)_2 + HC\equiv CH$$

$$HC\equiv CH + H_2O \xrightarrow[H_2SO_4]{HgSO_4} CH_3-C{\overset{O}{\underset{H}{\diagdown}}}$$
ethanal
(acetaldehyde)

Reduction of the aldehyde gives ethanol

$$CH_3CHO + H_2 \xrightarrow{Ni} CH_3CH_2OH$$

and oxidation gives acetic acid

$$2CH_3CHO + O_2 \xrightarrow{CuO} 2CH_3COOH$$

Furthermore, addition of ethene to the aldehyde gives 2-butanone, a very useful solvent:

$$CH_3C{\overset{O}{\underset{H}{\diagdown}}} + H_2C=CH_2 \longrightarrow CH_3-\overset{O}{\underset{\|}{C}}-C_2H_5$$
2-butanone
(methyl ethyl ketone)

Hence the production of three useful substances is linked to one initial reaction, the hydration of ethyne.

Ethylene glycol for radiator antifreeze also is made from ethene by oxidation, followed by hydration of the epoxide:

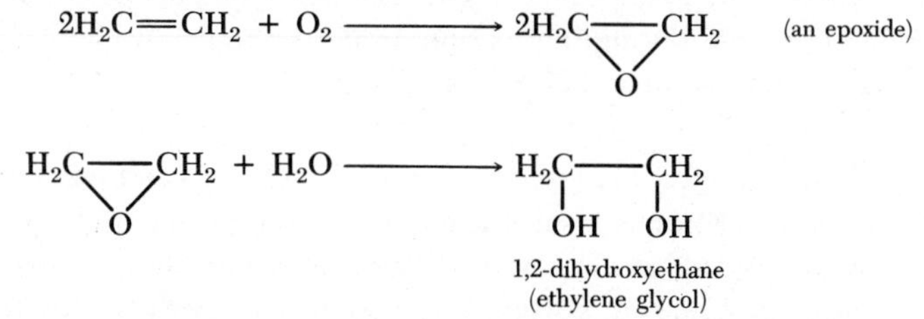

$$2H_2C=CH_2 + O_2 \longrightarrow 2H_2C\underset{O}{\overset{}{-\!\!\!-\!\!\!-}}CH_2 \quad \text{(an epoxide)}$$

$$H_2C\underset{O}{\overset{}{-\!\!\!-\!\!\!-}}CH_2 + H_2O \longrightarrow \underset{\underset{OH}{|}}{H_2C}-\underset{\underset{OH}{|}}{CH_2}$$

1,2-dihydroxyethane
(ethylene glycol)

Styrene for plastics and synthetic rubber also is made from ethene, by addition of hydrogen chloride and then condensation of the product with benzene:

$$H_2C=CH_2 + HCl \xrightarrow{CuCl} CH_3CH_2Cl$$

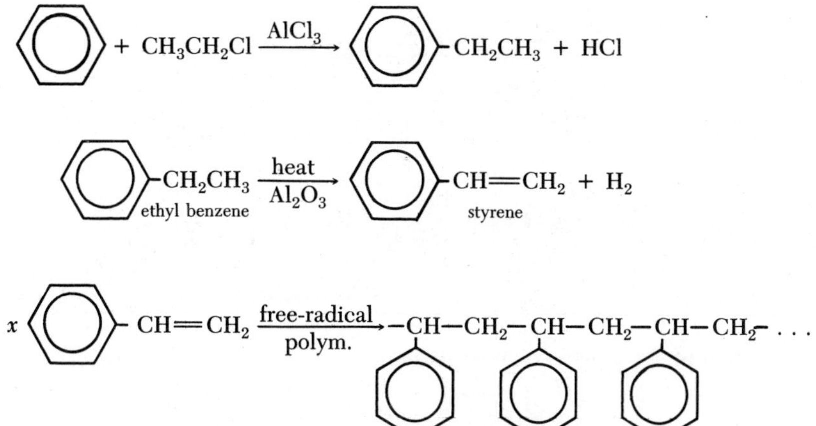

$$C_6H_6 + CH_3CH_2Cl \xrightarrow{AlCl_3} C_6H_5-CH_2CH_3 + HCl$$

$$\underset{\text{ethyl benzene}}{C_6H_5-CH_2CH_3} \xrightarrow[Al_2O_3]{\text{heat}} \underset{\text{styrene}}{C_6H_5-CH=CH_2} + H_2$$

$$x\, C_6H_5-CH=CH_2 \xrightarrow[\text{polym.}]{\text{free-radical}} -CH(C_6H_5)-CH_2-CH(C_6H_5)-CH_2-CH(C_6H_5)-CH_2- \ldots$$

Glycerol formerly was obtainable only as a by-product from making soap, but when it came into greater demand for making condensation polymers (particularly with terephthalic acid, $C_6H_4(COOH)_2$), a more plentiful supply was needed, and one not linked to the supply of edible fats. Since propene is a product of petroleum cracking operations, a way was developed to make glycerol from propene by chlorination, extraction of HCl, and hydration:

$$\underset{\underset{CH_2}{\|}}{\overset{\overset{CH_3}{|}}{CH}} + Cl_2 \xrightarrow{500°C} \underset{\underset{CH_2}{\|}}{\overset{\overset{CH_2Cl}{|}}{CH}} + HCl$$

(Note that at very high temperatures, chlorine does not add permanently to the double bond, but engages in a substitution reaction instead. At room temperature, it would be quite the reverse.)

$$\begin{array}{c} CH_2Cl \\ | \\ CH \\ \| \\ CH_2 \end{array} + HOCl \xrightarrow[20°C]{H_2O} \begin{array}{c} CH_2Cl \\ | \\ CHCl \\ | \\ CH_2OH \end{array} \xrightarrow{Ca(OH)_2} \begin{array}{c} CH_2Cl \\ | \\ CH \diagdown \\ O \\ CH_2 \diagup \end{array} + HCl$$

an epoxide

$$\begin{array}{c} CH_2Cl \\ | \\ CH \diagdown \\ O \\ CH_2 \diagup \end{array} + H_2O \xrightarrow{NaOH} \begin{array}{c} CH_2OH \\ | \\ CHOH \\ | \\ CH_2OH \end{array} + NaCl$$

glycerol

Many other examples could be given to illustrate the types of reaction we have covered, but this chapter already is too long. Some further exercises of the principles considered here will be found in the next chapter, on the design of materials.

14-10 Summary

We have sampled the rich variety of carbon-containing compounds, inquiring about the molecular structures of these compounds, asking what these substances can do and what they can be used for.

Organic compounds can be used for many things. Gasoline provides power for our automobiles. Plastics serve a steadily increasing number of roles in our daily lives. Explosives make possible grandiose construction projects, extensive mining of minerals, and alas, also destructive warfare. Other organic compounds, some from the farms and some from the factories, comprise the food and fiber that feed and clothe us. A primary goal of this book is to help you read, talk, and think about such molecules. Yet to talk about gasoline or about plastics requires knowledge about isomers, about reactions of functional groups, and about those arbitrary yet essential conventions involved in naming organic compounds. That is what this chapter has been all about.

Eleven functional groups have been described. We showed how to give names to compounds containing these groups. We examined some of the properties conferred on substances by the presence of these functional groups. And, moving to the heart of organic chemistry, we explored some of the reactions of these functional groups with one another. We saw, in discussing gasolines and plastics, that care and creativity in using these reactions can give rise to new materials with properties very useful to mankind. We shall explore these possibilities for the design of materials in the next chapter. We also saw, in those same examples, that in providing new materials for society, some unwanted (and perhaps unforeseen) side effects have also been produced. There is much more to say about this in the next chapter.

This has just been a beginning. For more, see the "Suggestions for Further Reading," and the last four chapters of this book.

GLOSSARY

addition: a type of reaction that results in bonding a new group to a molecule without removal of an existing group; typically, the bonding of two new atoms or groups to the two carbon atoms of a multiple bond.

alcohol: a compound of the type R—OH.

aldehyde: a compound of the type $R-C{\overset{\displaystyle O}{\underset{\displaystyle H}{}}}$

alkane: a saturated hydrocarbon; a compound containing only the elements carbon and hydrogen and having no multiple bonds.

alkene: a hydrocarbon with one or more double bonds.

alkyne: a hydrocarbon with one or more triple bonds.

amine: a compound of the type R—NH (a primary amine), RR'—NH (a secondary amine), or RR'R''—N (a tertiary amine), with all the R groups bonded directly to the nitrogen atom.

amide: a compound containing the linkage $-C{\overset{\displaystyle O}{\underset{\displaystyle NH}{}}}-$.

binary: said of a compound that contains atoms of two elements.

***cis-trans* isomers:** two compounds that differ only in the relative positions of two like groups; when involving substituents on the carbon atoms of a double bond, the *cis* isomer has the two like substituents on different carbon atoms but on the *same* side of the molecule, whereas the *trans* isomer has the substituents on different carbon atoms and on *opposite* sides of the molecule.

condensation: a type of reaction in which two molecules combine to form a larger molecule, usually with reaction of two functional groups and often with the splitting out of a small molecule such as water.

elimination: a type of reaction in which the reaction of two molecules is accompanied by the splitting out of a small molecular fragment, such as water, ammonia, or a hydrogen halide.

ester: a compound of the type $R-C{\overset{\displaystyle O}{\underset{\displaystyle O-R'}{}}}$

free radical: a highly reactive molecular fragment formed by the removal of an atom or group, leaving with the free radical an unpaired electron; symbolized by R·.

functional group: a collection of atoms, often transferable as a unit from one molecule to another, that has a particular set of chemical characteristics; thus one can meaningfully speak of the chemical reactions of a functional group, reactions that are to a large extent independent of the molecule in which the group appears.

heterogeneous: consisting of two or more phases, as, for instance, a system consisting of both a liquid and a solid, or two immiscible liquids, or a gas and a liquid. A heterogeneous reaction requires the presence of at least two phases.

homogeneous: consisting of a single phase, as, for instance, a solution without precipitate.

ketone: a compound of the type $\underset{R'}{\overset{R}{>}}C=O$.

lachrymator: a substance that irritates the eyes and causes copious flow of tears; a tear gas.

monomer: small molecules with two or more functional groups, capable of bonding together to form long chains or other large molecular structures; the constituent repeating units of a polymer (q.v.).

nitrile: an organic cyanide; a compound of the type $R-C\equiv N$.

optical isomers: compounds identical in molecular structure in every way except that the molecules of one isomer are the mirror image of molecules of the other isomer; said only of molecules whose mirror images are not superimposable.

polymer: a substance, often of very high molecular weight, consisting of molecules made by joining many small (similar or identical) monomer molecules.

saturated: containing the full complement of hydrogen atoms or their equivalent; said of a molecule when there are no double or triple bonds; said of a bond when the bond multiplicity is one.

structural isomers: compounds whose molecular structures are different, but whose formulas are the same.

PROBLEMS

14-1 How much volume change actually occurs when an industrial explosive detonates, moving large quantities of earth and rock? Consider the detonation of glycerol trinitrate in dynamite, the equation for which is given in Section 14-7. The density of glycerol trinitrate is 1.60; for convenience, take one mole of the compound and calculate the volume change when this one mole is transformed entirely to gases at 1000°C and one atmosphere pressure. Express the result as a ratio of volume of gases to volume of glycerol trinitrate at the start.

14-2 Consider two straight-chain compounds, both of the composition C_4H_9Br. For convenience, we shall call them X and Y. When X is warmed with silver oxide and a little water, it becomes $C_4H_{10}O$. When this is oxidized, it gives C_4H_8O, an excellent solvent for paints and resins. No further oxidation with potassium dichromate is possible. On the other hand, compound Y yields a different $C_4H_{10}O$ upon treatment with water and silver oxide, and when *this* is oxidized, it gives first C_4H_8O and then $C_4H_8O_2$, a vile-smelling liquid which gives a pH of 5 when stirred with water. When this same malodorous $C_4H_8O_2$ is warmed with ethanol and a drop of concentrated sulfuric acid, a substance $C_6H_{12}O_2$ (having a sweet, flowery odor) is produced. Identify X and Y and trace all the steps they undergo, writing names and structural formulas for all the compounds involved.

14-3 In the preparation of aminobenzene (aniline) from benzene by the two reactions

$$\underset{}{C_6H_6} \xrightarrow{HNO_3, H_2SO_4} C_6H_5NO_2 \xrightarrow{Fe, HCl} C_6H_5NH_2$$

132 g of aminobenzene were obtained from 156 g of benzene. What was the yield, as percent of the theoretical?

14-4 Write structural formulas for the following organic compounds:

a. 2,2-dimethylpropane
b. 2-butyne-1,4-diol
c. 1-hydroxy-2,4,6-trinitrobenzene (picric acid, an explosive)
d. 1-amino-2,2-dichloro-5-heptene.

14-5 Write names for the following compounds in the Geneva System:

a. $(H_3C)(C_2H_5)CH-CH_2-CH=CH-CH_2OH$

b. 1-chloro-4-vinylbenzene (Cl–C$_6$H$_4$–CH=CH$_2$)

c. hydroxy-chloro-dinitro naphthalene (OH, Cl, O$_2$N, NO$_2$ substituted naphthalene)

14-6 Write equations for the following:

a. the reaction of ethanol with 1,1,1-trifluoroacetic acid
b. the synthesis of 1,2-ethanediol ("ethylene glycol") from coal
c. the synthesis of hydrogen cyanide from coal, air, water, and salt
d. the synthesis of methyl alcohol from coal and water
e. the preparation of aspirin

[Structure: salicylic acid (C$_6$H$_4$(OH)(COOH)) → aspirin (C$_6$H$_4$(COOH)(O–CO–CH$_3$))]

from salicyclic acid (2-hydroxybenzoic acid) (benzoic acid is C_6H_5COOH).

f. the preparation of ethyl alcohol from petroleum products.

SUGGESTIONS FOR FURTHER READING

The student who is interested is urged to read not only textbooks of organic chemistry, but also in the peripheral areas such as organometallic chemistry and biochemistry. A selection of such books follows:

HENDRICKSON, J. B., CRAM, D.J. and HAMMOND, G. S., *Organic Chemistry*, 3rd ed., New York: McGraw-Hill (1970).

GOULD, E. S., *Mechanism and Structure in Organic Chemistry*, New York: Holt, Rinehart & Winston (1960).

BENFEY, O. T., *Introduction to Organic Reaction Mechanisms*, New York: McGraw-Hill (1970) (a *Chemistry-Biology Interface* paperback).

ROCHOW, E. G., *Organometallic Chemistry*, New York: Reinhold (1964) (a *Selected Topics in Chemistry* paperback).

MAHLER, H. R., and CORDES, E.H., *Biological Chemistry*, 2nd ed., New York: Harper & Row (1971).

15
The Design of Materials

15-1 Abstract

Having been introduced to the principles of inorganic and organic chemistry, you are now invited to consider the relation between *structure* and *function* in synthetic materials. By "synthetic materials" we mean those substances or compositions that are purposefully designed and then manufactured to meet a particular need. Usually there is no naturally occurring counterpart of the material, although sometimes a natural product is too expensive or too variable in its properties, and so a synthetic material takes over. Buildings are now constructed mostly of synthetic materials. Metals (such as iron, steel, copper, aluminum), ceramics (bricks, cement or concrete, glass, tile), and organics (plastics, rubber, paints, roofing) are manufactured building materials; only wood and stone survive as naturally occurring products useful in building. Similarly, clothing is largely made from manufactured fibers and colored with synthetic dyes. Automobiles are made entirely of manufactured materials. So is most other machinery.

The composition and the structure of each synthetic material must be specified by some creative person with training in chemistry, and economical processes for their manufacture must be devised by some inventive person who also has training in chemistry. Often rational decisions about purchase, care, and eventual disposal of such materials require knowledge of chemical facts and chemical principles. How the manufacturing is done, how well it is done, and how the material is used have large effects on the environment and on the future of our world. This entire matter deserves considerable attention in any introductory course in chemistry, and hence this chapter.

We shall be concerned here chiefly with the *design* of materials, with the *molecular architecture* that is responsible for their properties, and with the *natural resources* that enter into the manufacture of these materials. Space allows only the consideration of a few representative examples, chosen from five rather separate fields. In an industrial society, each one of these fields involves an enormous amount of applied chemistry, along with other technologies. We shall consider metals, ceramics, plastics, elastomers, and dyes.

15-2 Metals

Metals derive their strength from the unique and highly characteristic type of bonding described in Section 5-5. Metals conduct electricity readily, always by transfer or transport of electrons, not ions. The actual conductivity varies from one metal to another, being highest for copper and silver[1] and much lower for chromium and mercury. The conductivity is greatest at low temperatures, where there is the least vibration of the atomic cores and hence least

---

[1] On a volume basis. If we choose equal *weights* instead of volumes, then sodium and aluminum rival copper and silver as conductors. Thus, copper is used for electrical wiring where thin wires are required, but aluminum (being less expensive) is often used for outdoor overhead wires where space is no problem.

resistance to the passage of electrons. The conductivity of a metal always becomes poorer as the temperature is raised, because increased thermal agitation of the atoms interferes more with the motion of electrons. Those metals with the highest electrical conductivity also have the highest heat conductivity, because heat also is conducted by the freely moving electrons, which can transmit vibrational energy as well as carry charge.

Many metallic properties derive from free electrons.

TABLE 15-1
Tensile Strengths of Some Metals At 20°C

Metal	Tensile Strength (lb/in^3)
Aluminum wire	35,000
Brass (Cu + Zn) wire	100,000
Bronze (Cu + Sn) wire	67,000
Copper wire	65,000
Gold wire	20,000
Iron wire	110,000
Magnesium wire	33,000
Nickel wire	90,000
Platinum wire	50,000
Silver wire	42,000
Sodium wire	250
Steel (Fe + C) wire	330,000
Tantalum wire	130,000
Tin wire	5000
Tungsten wire	590,000
Zinc wire	26,000
Cast cobalt	33,000
Cast iron	23,000
Cast tin	4000
Cast zinc	9000
Aluminum sheet	32,000
Stainless steel (Fe + Cr + Ni) sheet	112,000
Titanium (pure) sheet	104,000
Titanium alloy (Ti + C + N) sheet	140,000

   This picture of metallic bonding leads us to expect that metals furnishing only one electron per atom to the all-pervading electron "gas" within the metal are mechanically weak, and those metals contributing more electrons per atom to the electron gas are stronger. This is found to be true; the Group I metals are weak and have low melting points, while the strongest metals are found in the middle of the transition series. Specifically, the strongest metals per unit volume are titanium, chromium, manganese, iron, cobalt, and nickel in the first transition series, and zirconium, tantalum, molybdenum, and tungsten among the heavier metals (see Table 15-1). Some alloys of these metals are much stronger than the pure elements, because **intermetallic compounds** form. Further increase in tensile and flexural strength can be achieved by *precipitation hardening*, a process in which an insoluble phase is precipitated in finely divided condition throughout the mass of metal under controlled thermal conditions. The earliest example[2] of such hardening is the change from weak and soft pure iron to much harder and stronger *steel* when minute particles of iron carbide, Fe_3C, are caused to precipitate within the iron by thermal treatment. The most advantageous distribution of carbide can be stabilized by alloying the iron with chromium and nickel at the same time, yielding *stainless steel*, a particularly tough, strong, and useful combination of metals.

Let us utilize these few facts in considering the design of alloys for *aircraft* use, particularly for fast jet transports. As for all aircraft, the maximum strength per unit weight (*not* per unit volume) is essential. In addition, this strength must be maintained at high temperatures, for the skin of a fast aircraft becomes heated by atmospheric friction, and the engines and their mountings are heated by burning fuel. We should like to depend on only abundant and inexpensive raw materials that can be transformed to the finished structural metals with a minimum of pollution of air and water during the manufacturing processes. We shall examine these requirements, considering the properties of elements that are abundantly available on earth.

High-speed aircraft require light-weight structural materials that retain strength when heated.

TABLE 15-2
The Twenty Most-Abundant Elements on Earth

Element	Percent of Earth's Crust (by weight)	Density at 20°C (g/cm³)*
Oxygen	49.5	
Silicon	25.7	2.33
Aluminum	7.50	2.70
Iron	4.71	7.87
Calcium	3.39	1.55
Sodium	2.63	0.97
Potassium	2.40	0.86
Magnesium	1.93	1.74
Hydrogen	0.87	
Titanium	0.58	4.50
Chlorine	0.19	
Phosphorus	0.12	1.83
Manganese	0.09	7.43
Carbon	0.08	2.25
Fluorine	0.08	
Sulfur	0.06	2.07
Barium	0.04	3.5
Chromium	0.03	7.19
Strontium	0.03	2.60
Zirconium	0.02	6.49

* For elements that are solids at 20°C. For comparison, recall that water has a density of 1.00 g/cm³ at 20°C.

---

[2] This venerable art of heat treating steel is described by C. S. Smith and R. J. Forbes, "Metallurgy and Assaying," vol. 3, chap. 2, in C. Singer, E. J. Holmyard, A. R. Hall, and T. I. Williams, *A History of Technology*, London: Oxford University Press (1957). Other chapters about the beginnings of man's manufacture of materials are also interesting reading.

The most plentiful structural metals,[3] in order of decreasing abundance, are aluminum, iron, magnesium, and titanium. Turning to the requirement of light weight, we check (Table 15-2) the density of each of these metals. Aluminum and magnesium have the clear advantage in lightness. Moreover, each can be produced from readily available natural supplies of raw material without much contamination of the atmosphere. The production of aluminum has already been described in Chapters 3 and 8; production of magnesium is accomplished by electrolyzing molten anhydrous magnesium chloride at 708°C:

$$Mg^{2+} + 2e^- \rightarrow Mg$$

$$2Cl^- \rightarrow Cl_2 + 2e^-$$

The chlorine produced is used for chlorinations, converting it to HCl that is returned to the process to make $MgCl_2$ from the magnesium hydroxide precipitated from *sea water*. This process of obtaining $MgCl_2$ from sea water uses CaO derived from *oyster shells* to provide a base for precipitating $Mg(OH)_2$. Thus, starting with oyster shells and sea water, we utilize the following chemical reactions to obtain $MgCl_2$ for the electrolysis:

$$CaCO_3 \text{ (in oyster shells)} \xrightarrow{750°C} CaO + CO_2$$

$$H_2O + CaO + MgCl_2 \text{ (in sea water)} \longrightarrow Mg(OH)_2 + CaCl_2$$

$$Mg(OH)_2 + 2HCl \text{ (returned from process)} \longrightarrow MgCl_2 + 2H_2O$$

Even though sea water is a very dilute solution of magnesium salts, the volume of the oceans is so vast that we have a literally inexhaustible source of supply of magnesium, available to all nations having access to an ocean. It has been estimated that if we produced magnesium at the present rate of steel production (100 million tons per year), and kept this up for one million years, allowing none of the magnesium to return to the sea, the magnesium content of the seas would drop from the present 0.13 percent only down to 0.12 percent. Thus, the only irreplaceable resource consumed in making magnesium is the fuel used to generate electric power, and as we have seen, there is the possibility of generating most of this from nuclear energy in the future. As for aluminum, methods have been devised for extracting the necessary Al_2O_3 from clays, so we are not likely to run out of raw material. We need only electric power, and carbon for electrodes. It would be nice if we could find a way to avoid consumption of carbon.

However, pure aluminum is rather soft and weak, and pure magnesium is not much better. Hence much effort has gone into devising stronger alloys

Soft and weak metals are often hardened and strengthened by alloying.

---

[3] That is, the most abundant elements (see Table 15-2) that also are metals, and that are of sufficient chemical nonreactivity to stand up against air, moisture, sea water, and other potential reactants.

of the two light metals. Aluminum is best strengthened by alloying with copper (1 to 5 percent), magnesium (0.5 to 5 percent), silicon (0.4 to as much as 12 percent in casting alloys), and manganese (0.2 to 1 percent). It is further hardened by precipitation of $CuAl_2$ and Mg_2Si within the metal matrix; both copper and silicon dissolve in red-hot molten aluminum, but upon proper treatment at 120°C to 250°C, the indicated compounds precipitate and harden the metal in much the same way as Fe_3C hardens iron to make steel. Similarly, magnesium is alloyed and hardened with aluminum, manganese, zinc, thorium, and zirconium. The technology has advanced to the point where excellent aircraft alloys are made from these two light metals, aluminum being used for spars and skin, and magnesium for parts such as forged landing gear and landing wheels. The rapid advances in aircraft construction during the 1950s and 1960s were made possible by the development of these materials.

The advent of supersonic transports and the huge "air bus" makes necessary an abrupt change in materials for constructing aircraft for one critical reason: Aluminum and magnesium lose much of their strength when heated, and some of the functional parts of modern aircraft run much hotter than their earlier counterparts. Pure aluminum loses strength rapidly above 150°C, and its alloys weaken above 200°C.[4] What materials can be substituted for those which have been so successful with slower aircraft? One answer has been that future high-speed aircraft should be made of stainless steel. Two very different sorts of objections have been raised: Making stainless steel creates substantial atmospheric pollution, and stainless steel also loses strength when heated. We shall explore both of these objections.

In making iron and steel by present methods, enormous volumes of preheated air are blown through blast furnaces, generating a great deal of dust. This preheated air conducts the high-temperature oxidation-reduction reactions

$$O_2 + 2C \text{ (coke)} \xrightarrow{1500-2000°C} 2CO$$

$$3CO + Fe_2O_3 \text{ (from iron ore)} \longrightarrow 2Fe + 3CO_2\uparrow$$

Because air is only one-fifth oxygen, it is unfortunately necessary to heat not only the required oxygen but also four times as much nitrogen, which serves no useful purpose at all, but only wastes fuel. The only desired chemical change is the reduction of iron ore to metal; the high temperature unfortunately is needed for the traditional oxidation-reduction reactions given above, and also is necessary to convert the siliceous earthy impurities in the iron ore to calcium silicate slag, which can then be removed from the molten iron

---

[4] Therein lies the great danger from any fire which develops aboard a flying aircraft; the wing spars and other structural members weaken when heated, and will no longer sustain their loads.

metal. This removal of impurities involves the use of calcium oxide (obtained by thermal decomposition of limestone in the blast furnace), which reacts with silica:

$$CaCO_3 \text{ (limestone)} \xrightarrow{750°C} CaO + CO_2$$

$$CaO + SiO_2 \xrightarrow{1000-1500°C} CaSiO_3$$

The hot gases that must be discharged to the atmosphere (N_2 and CO_2) carry vaporized Fe_2O_3 and $CaSiO_3$; a flight over a steel plant will show the great reddish plumes of this dust-laden air. Because of this air-pollution problem, it seems undesirable to make twice as much, or ten times as much steel in the future, *if the same methods must be used.*

Present methods of steel production waste fuel and add dust to the air.

The other objection to stainless steel is that it, too, loses strength too rapidly when heated. A popular stainless steel, called alloy 302, weakens at 300°C and drops to one-fifth its original strength at 400°C (see Figure 15-1). So any aircraft structural parts that become heated to redness could not safely be made of stainless steel, even if the weight could be tolerated (remember that steel is about three times as dense as aluminum).

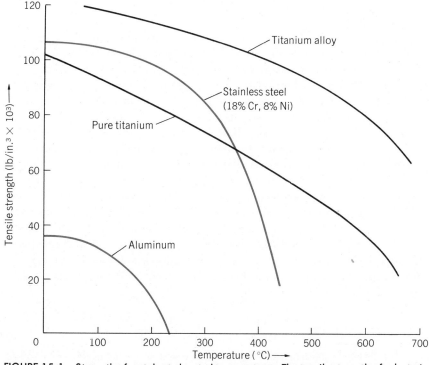

FIGURE 15-1 Strength of metals at elevated temperatures. The tensile strength of selected metals is plotted versus temperature, showing the superiority of titanium as a structural material at temperatures over 400°C.

Titanium ore requires unconventional reduction methods.

There is a way out of these difficulties. The way is to use the heretofore-neglected metal *titanium*. Although abundant (comprising 0.58 percent of the earth's crust; see Table 15-2) and widespread, titanium is uncommon because it cannot be obtained by the classical procedures of metallurgy. It cannot be reduced from its oxide by carbon in a blast furnace because it combines readily with carbon to form a brittle carbide, and because it also combines with nitrogen to form a weak, easily powdered nitride. Moreover, it cannot be obtained by electrolysis of its normal chloride, $TiCl_4$, because the chloride is a covalent, nonconducting liquid. There is no simple economical electrolytic method for obtaining titanium. The only way open is chemical reduction of the chloride with some active metal such as zinc or magnesium.

Titanium occurs in the common black mineral *ilmenite*, $FeTiO_3$. Ilmenite, a constituent of granite, is what makes many silica-sand beaches gray instead of white. Titanium is extracted from ilmenite by acids to yield the oxide TiO_2, the favorite white pigment for house paints. This oxide can be converted to the tetrachloride by reductive chlorination according to the equation[5]

$$TiO_2 + C + 2Cl_2 \xrightarrow{500°C} CO_2 + TiCl_4$$

The vapor of the tetrachloride, under pressure, is then heated with magnesium:

$$TiCl_4 + 2Mg \longrightarrow Ti + 2MgCl_2$$

The magnesium chloride can be dissolved from the mixture of products with hot water, leaving spongy gray titanium. The sponge is melted in vacuum.

This process is costly, even if the by-product $MgCl_2$ is converted to magnesium and chlorine and recycled through the process. However, the expense of titanium seems justifiable when we look at its properties—pure titanium is as strong as stainless steel 302, and it keeps its strength better when it is heated. Furthermore, like all other structural metals, titanium can be strengthened by alloying. Some titanium alloys are much stronger than stainless steel. One excellent titanium alloy contains 5 percent aluminum (to minimize effects on properties due to temperature changes), 2 percent each of iron, chromium, and molybdenum (to harden and toughen the metal), and 0.5 to 2 percent tin (to improve machinability). This alloy retains four-fifths of its strength at 400°C, and it is structurally useful even at 600°C. Such considerations explain why titanium alloys, even at $5 to $15 per pound, are the favored design materials for the engine housings and mounts for modern aircraft. The fact that titanium is only 57 percent as dense as steel is also greatly in its favor, of course.

Titanium alloys are strong and can be used at high temperatures.

---

[5] This chemical equation is oversimplified. The actual reaction also produces CO, $COCl_2$, and some CCl_4, but both $COCl_2$ and CCl_4 also convert to TiO_2 and $TiCl_4$. Eventually CO_2 and $TiCl_4$ become the chief products.

15-3 Ceramics

The ancient craft of making earthenware pottery by firing shaped and dried vessels of clay reached considerable artistic and technical heights 5000 years ago. It continues today, along with its derivative arts, in the making of porcelain and fine china. Brick and tile are also descendents of ancient pottery. In all such ceramic products, tiny particles of alumina, silica, and myriad aluminosilicates of Group I and Group II elements are cemented together by a glassy phase that is sticky and semifluid at 1000°C, but strong and vitreous at temperatures of use.

The chemical stability and inertness of ceramic ware are remarkable. Pieces of ancient pottery endure, chemically unchanged, right up to the present, long after most artifacts of those times have perished. Often such pottery sherds provide archeologists with the only permanent records of old cultures and clues to the early development of technology. The history of pottery also chronicles the beginnings of the chemical arts, providing an excellent case study of how to make do with what there is. Common clay was transformed into jars and bowls for homes and shops, and in the hands of creative artisans into objects of exceptional beauty. From the potter's craft emerged a versatile medium of artistic expression. Ceramics continue to be invaluable materials. Any modern discussion of ceramics involves far more chemistry than can be considered in this book, but a brief look at the family of materials we call **glass** is appropriate in this chapter and will provide us with one example of a ceramic material.

Glass is not a single substance, not a single kind of chemical composition. It is a state of matter. Many substances, which ordinarily are crystalline solids at low temperatures, may be kept as noncrystalline supercooled liquids if they are chilled rapidly from the melted state. As such a liquid cools, its viscosity becomes too great to allow the component molecules or ions to take up ordered positions in a regular crystal lattice; they are left in a state of frozen disorder. When the material is cooled sufficiently to be rigid, still without any crystallization, it is called a glass. There are three distinguishing characteristics of a glass: (1) it is **isotropic** (like a liquid, it has the same physical properties in all direction); (2) it breaks with an irregular **conchoidal fracture** (not with cleavage along flat crystal planes, because there are no such planes); and (3) it softens to a viscous liquid again, when heated, without undergoing chemical change. This last characteristic allows a glass to be fabricated as a plastic mass just by heating it to the softening point. For prosaic items, the shaping is done by machine, but in the hands of a skillful artist, the plastic medium yields striking results (see Figure 15-2).

Glasses may be organic (synthetics like polystyrene and polymethylacrylate, or natural resins like amber) or inorganic (such as window glass, bottle glass, and optical glass). Most inorganic glasses are silicates or contain silica, although there are phosphate, germanate, and borate glasses that contain no silicon at all. Pure crystalline B_2O_3, when melted and then cooled quickly, becomes a colorless and transparent glass. Pure SiO_2 and pure GeO_2 behave the same way. Almost all inorganic glasses in practical use are silicates, and

Glasses are especially suitable materials for recycling. Why?

so are the glassy phases that hold together earthenware and porcelain. The vitreous enamels that are used to coat steel (for protection) and gold (for decoration) are also silicate glasses.

FIGURE 15-2 Glass, the oldest of the clear plastics, lends itself to many arts and techniques. This striking figure was made at the Steuben Art Glass studios in Corning, N.Y. (Photo courtesy of Steuben Galleries, New York, N.Y.)

The properties that distinguish one glass from another, such as color, density, and hardness, are determined by chemical composition, not by structure. Hence glasses can be designed in a chemical way to perform particular tasks. We shall consider one example: the design of a glass for making beakers and flasks to be used in a chemical laboratory. We begin by looking at window glass, a glass that serves well for windows, but is not suited for beakers, for reasons that will develop.

Ordinary window glass is made from silica sand (SiO_2), limestone ($CaCO_3$), and sodium carbonate via the reactions

$$Na_2CO_3 + SiO_2 \xrightarrow{1000°C} Na_2SiO_3 + CO_2$$

$$CaCO_3 + SiO_2 \xrightarrow{1000°C} CaSiO_3 + CO_2$$

The solution of sodium and calcium silicates in each other constitutes the glass. To extend the temperature range of plasticity and workability of the glass, potassium carbonate is added to the reaction mixture to give a silicate with somewhat higher melting point than Na_2SiO_3:

$$K_2CO_3 + SiO_2 \xrightarrow{1000°C} K_2SiO_3 + CO_2$$

Extra silica also dissolves in the red-hot viscous mixture of silicates. Window glass is made continuously in enormous boxlike furnaces, as large as a classroom, heated by roaring gas flames. Fresh raw materials are introduced at the upper rear, while molten glass is withdrawn continuously at the lower front. The glass may be drawn out into sheets, or it may flow to automatic bottle-blowing machines.

This ordinary soda-lime glass is all right for windows, some bottles, and electric light bulbs, but it will not do for laboratory glassware. For one thing, it has too high a thermal coefficient of expansion; if heated at one spot by a Bunsen burner flame, it expands locally so much that the adjacent glass is strained beyond the breaking point. For another thing, ordinary window and bottle glass dissolves[6] too readily in alkaline solutions to serve satisfactorily in a chemical laboratory; the glass would contaminate many of the solutions and reagents it holds. To change the properties of glass, we might substitute other glass-forming substances for the window-glass constituents.

To reduce thermal expansion of a glass intended for laboratory ware, B_2O_3 is added to the melt in the form of sodium borate. Since B_2O_3 has a lower melting point than SiO_2, it acts as a flux and reduces the proportions of Na_2O and K_2O (or Na_2CO_3 and K_2CO_3) necessary to make the glass workable at 800°C. This is particularly advantageous for the glass designer, because these two water-soluble oxides contribute to the water solubility of glass. Addition of Al_2O_3 further increases the resistance to water and aqueous solutions, since alumina and aluminates are more resistant to water than are the alkali silicates. Finally, since a low softening temperature is considered unnecessary (and indeed even undesirable) for chemical glassware, the proportion of high-melting and chemically resistant SiO_2 is put as high as the automatic beaker-blowing equipment permits. A comparison of the final composition of laboratory glass versus window and bottle glass is given in Table 15-3.

Beakers should be made of a glass that will not crack when heated with a direct flame.

TABLE 15-3
Chemical Composition of Two Kinds of Glass

	Window and Bottle Glass	Laboratory Borosilicate Glass
SiO_2	78.4 percent	84.0 percent
B_2O_3	—	10.5 percent
Al_2O_3	0.2 percent	1.3 percent
CaO	8.4 percent	—
Na_2O	10.0 percent	3.8 percent
K_2O	3.0 percent	0.4 percent
	100.0	100.0

---

[6] It may seem strange to think of glass dissolving in water, but try an experiment. Thoroughly grind a 10 g sample of window glass in a porcelain mortar. Stir the powdered glass with 100 ml of distilled water in a Pyrex beaker, and heat the beaker and its contents overnight in an oven set at 105°C. What is the residue left on the sides of the beaker?

The only way known to increase the chemical inertness of laboratory glassware further is to avoid Na_2O, K_2O, and B_2O_3 entirely and to make the vessels of pure fused silica. Very pure silica sand is melted in an oxyhydrogen flame and is formed by machine into somewhat irregular tubing of chilled vitreous pure SiO_2. This is sometimes improperly called "quartz glass," but it is not quartz! Quartz is crystalline, and this fused silica is a glass. Fused silica melts quite suddenly at 1700°C, and there is a very narrow temperature range of plasticity, but a glassblower of extraordinary skill is able to make simple vessels and apparatus from it. Fused silica glass has an extremely low thermal coefficient of expansion, so it withstands sudden heating and cooling very well. Unfortunately, its narrow working temperature range does not adapt it for use with automatic machinery for making beakers and flasks.

If transparency is not required, porcelain is usually a better choice than glass as a chemically resistant, nonmetallic material. Porcelain is much stronger mechanically than glass, because it is not homogeneous and isotropic; small scratches do not develop into cracks. Porcelain is also very resistant chemically, because it consists of particles of alumina and aluminosilicates embedded in a matrix of high-alumina glass. And it has another advantage over glass—porcelain articles are formed at room temperature by pressing or molding a mixture of powders which later is dried and fired, whereas glass must always be fabricated hot. Hence intricate shapes can be mass-produced more readily in porcelain than in glass.

15-4 Polymers and Plastics

No area of synthetic chemistry gives more opportunity for creative design than the molecular architecture of synthetic polymers. Desired properties can be designed right into the molecular structure from the start. The only limitations are the availability of suitable raw materials, and the ingenuity to convert them to a polymer of high molecular weight with the desired molecular structure.

A mixture of polymer with coloring agents and reinforcing fibers or powders is popularly called a *plastic*. The term "plastic" actually is broader (or should be), covering everything that can be shaped by molding, and encompassing the range of materials from wet clay to glass. Within the narrower definition, the constituent polymers can be of organic, inorganic, or in-between composition. The only chemical requirement is that a stable structural unit (a monomer) be joined to other identical units by strong bonds, so that an aggregate of very high molecular weight results. This material then derives additional properties from the very fact that it is **macromolecular,** and not just a collection of separate small molecules of the same composition. For example, a polymer has greater physical strength than a collection of unbonded small monomer molecules, because covalent chemical bonds must be broken in order to tear the polymer apart. A macromolecular polymer is less soluble in any solvent than a collection of small molecules of the same composition. A polymer also has a higher melting point than its monomer.

The requirements for macromolecularity are met by some glasses that have silicate rings and chains as repetitive structural units,[7] and also by many organic and organometallic polymers. We shall consider first some representative organic polymers, and then one inbetween, organometallic polymer.

Organic polymers may be classified as addition-type or condensation-type polymers, according to the type of reaction by which they achieve their very high molecular weight. Addition reactions were discussed in Section 14-8, where it was seen that ethene can be induced to add to other molecules of ethene to form polyethylene, a popular plastic with many general uses. Various substituted ethenes, including propene, chloroethene, cyanoethene, and tetrafluorethene, can also be induced to polymerize to macromolecular form, as set forth in Table 14-2. If the addition takes place by a free-radical mechanism, the polymer has a random configuration. But if the addition takes place while the alkene is held in a constrained and activated position as part of a coordination complex,[8] then the addition takes place in a stereospecific way, resulting in a regularity of structure that encourages some localized crystallization. In general, a polymer that develops crystalline domains upon stretching or rolling is stronger than a purely amorphous polymer. A picture of crystals in polyoxymethylene, an organic polymer, is given in Figure 15-3.

Addition-type polymers usually are *thermoplastic*. That is, they achieve their maximum molecular weight during polymerization and do not undergo any further reaction during hot molding; they simply soften enough to be molded or blown, just as glass does. It follows that thermoplastic polymers can soften during use, too, if allowed to become hot. The temperature range of service is correspondingly limited. Sometimes heat softening can be an advantage, as in the heat sealing of plastic packages and the "electronic welding" of vinyl upholstery. There also is a way around the limitation—when molded articles of polyethylene are subjected to intense radiation, protons are knocked out and the polymer chains become cross-linked to one another. This cross-linking raises the melting point and increases the strength of the surface material.

The degree of regularity of structure and the extent of cross-linking affect properties significantly.

---

[7] See H. J. Emeleus and J. S. Anderson, *Modern Aspects of Inorganic Chemistry*, 3rd ed., New York: Van Nostrand, pp. 349–366, chap. IX (1960).

[8] See E. G. Rochow, *Organometallic Chemistry*, New York: Reinhold (1964). The first organometallic compound ever reported (by Zeise, in 1830) was such a coordination compound, in which the negative ion had the composition $C_2H_4PtCl_3^-$. In it the platinum orbitals are dsp^2 hybrids, with the four ligands aranged in square planar configuration

$$\begin{array}{c} \text{Cl} \\ | \\ \text{Cl}-\text{Pt}-\overset{\text{CH}_2}{\underset{\text{CH}_2}{\|}} \\ | \\ \text{Cl} \end{array}$$

The ethene contributes a pair of electrons to an empty orbital of platinum, just as a molecule of carbon monoxide would. Ethene adds similarly to solutions of copper, silver, mercury, and palladium salts, coordinating to the metal ions. In the stereospecific catalysts used for polymerization, the metal often is titanium (III), reduced from $TiCl_4$ or R_2TiCl_2 by $Al(CH_3)_3$.

FIGURE 15-3 Spiral growths of crystals in polyoxymethylene. (Photo courtesy of E. I. du Pont de Nemours & Co., Wilmington, Del.)

Condensation-type polymers develop their high molecular weight by elimination reactions between functional groups, as explained in Section 14-9, in connection with the formation of phenol-formaldehyde (Bakelite) plastics. Usually there are enough reactive groups left so that further condensation reactions can take place during molding or pressing of the article. Thus, condensation polymers are thought of as *thermosetting* instead of thermoplastic. This view is not always correct; nylon and Dacron are condensation polymers, and yet are thermoplastic and are molded or drawn into fibers in a heat-softened form. The best example of a truly thermosetting plastic is Bakelite, which hardens to an insoluble and infusible form within the mold at the time the article is fabricated. This comes about because care is taken during the earlier stages of manufacture to leave plenty of reactive groups for bonding and cross-linking. The benefit is that the finished piece will not soften or distort at elevated temperature, even at the point where thermal decomposition sets in.

The details of the molecular architecture of polymers, in which the effects of all the various types of backbone linkage and side-chain substitution are enumerated, are best left to books on polymers and plastics.[9] We are concerned here with the design and production of materials for the greatest public benefit over the longest period of time, and so consideration of resources and effects on the environment take precedence over niceties of structural detail. For our present purposes in this chapter, it may suffice to consider the general question of the availability of raw materials for organic plastics, and then to use nylon as an illustration of the entire chemical history of a polymer.

---

[9] See P. J. Flory, *Principles of Polymer Chemistry*, Ithaca, N.Y.: Cornell University Press (1953); *Modern Plastics Encyclopedia*, New York: Plastics Catalogue Corp. (1949); T. Alfrey, A. Bohrer, and H. Mark, *Copolymerization*, New York: Interscience (1952).

Carbon is a relatively rare element (see Table 15-2), constituting only 0.08 percent of the earth's crust. Moreover, most of the carbon on earth is tied up as calcium and magnesium carbonates (limestone, marble, magnesite, and dolomite—whole mountains of them in some places). The only plentiful sources of carbon in reduced form are coal, oil,[10] and natural gas, and these are irreplaceable resources that took millions of years to develop. Any large-volume utilization of organic plastics must go back to these resources (especially to coal and oil) as raw materials, and this puts plastics manufacture in competition with the use of the same resources as fuel. Instinctively it seems better to use coal or oil to make plastics and textile fibers than simply to burn them, because the plastics and textiles at least remain in existence and are not dispersed in the air as carbon dioxide. However, the mere continued existence of plastics does no good unless the waste plastic articles are collected and used as fuel, or are recycled as starting material for making something else. To clutter up the landscape with plastic jugs and polyethylene bags is only to contribute more litter and pollution; it would be better in that case to build into the polymer molecules some structures which would make them biodegradable, so that bacteria could destroy them after use in much the same way as bacteria destroy biodegradable detergents after use. Perhaps the best immediate solution is to collect all plastic wrappers and articles along with wastepaper and garbage, and to burn this refuse in specially built incinerators which will generate electricity and steam for heating buildings at the same time they dispose cleanly of the waste. Eventually, on a really long-term basis, it may be that we cannot use coal and oil either for fuel or for making plastics, because it may be needed for manufacturing basic foodstuffs.

Continued exploitation of organic plastics requires care in selection of raw materials and planning for disposal.

We turn now to nylon, which is a polyamide made by condensation of a dicarboxylic acid with a diamine. The manufacturing process may be designed so that the raw material to be used is coal, oil, or even agricultural waste. Let us first consider the steps for making nylon from coal. Ethyne (acetylene) is made by the hydrolysis of calcium carbide, which is produced by the treatment of lime with coke.

$$CaO \text{ (from limestone)} + 3C \text{ (coke)} \xrightarrow{\text{in electric furnace}} CaC_2 + CO$$

$$CaC_2 + 2H_2O \longrightarrow HC{\equiv}CH + Ca(OH)_2$$

Formaldehyde also is made from coal, by way of methanol, according to the sequence of reactions given in Section 14-9 in connection with the formation of phenol-formaldehyde polymer (Bakelite). In the present instance, the formaldehyde undergoes an addition reaction with ethyne:

$$HC{\equiv}CH + 2HCHO \longrightarrow HOCH_2{-}C{\equiv}C{-}CH_2OH$$
$$\text{2-butyne-1,4-diol}$$

---

[10] Including both liquid petroleum and oil shale.

Catalytic hydrogenation of this alcohol saturates the triple bond:

$$HOCH_2C \equiv CCH_2OH + 2H_2 \xrightarrow[200 \text{ atm}]{\text{Ni catalyst}} HOCH_2CH_2CH_2CH_2OH$$
$$\text{butane-1,4-diol}$$

The diol is dehydrated to a cyclic ether, which is then cleaved with HCl to form a dichloride via the reactions

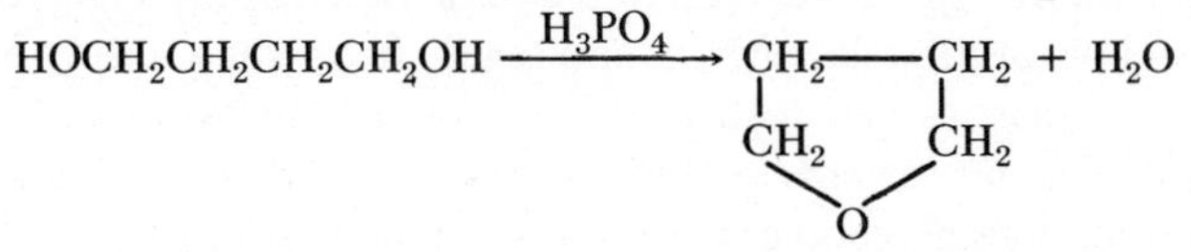

$$HOCH_2CH_2CH_2CH_2OH \xrightarrow{H_3PO_4} \begin{array}{c} CH_2-CH_2 \\ | \quad\quad | \\ CH_2 \quad CH_2 \\ \diagdown O \diagup \end{array} + H_2O$$

tetrahydrofuran, called THF

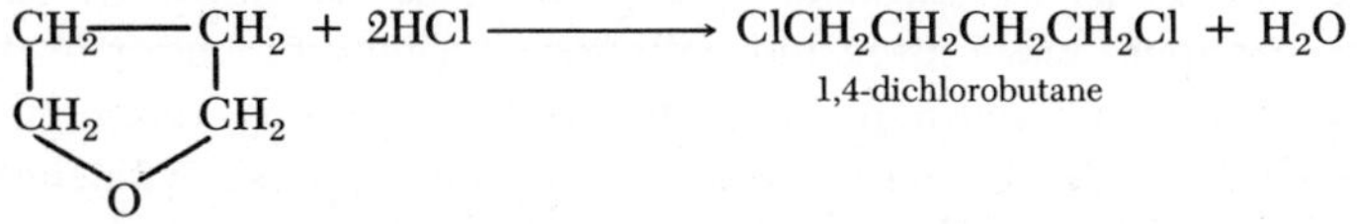

$$\begin{array}{c} CH_2-CH_2 \\ | \quad\quad | \\ CH_2 \quad CH_2 \\ \diagdown O \diagup \end{array} + 2HCl \longrightarrow ClCH_2CH_2CH_2CH_2Cl + H_2O$$

1,4-dichlorobutane

The dichloride is treated with sodium cyanide, made from sodium carbonate as described in Section 14-4, forming a dicyanide and splitting out sodium chloride:

$$Cl(CH_2)_4Cl + 2NaCN \longrightarrow N \equiv C-(CH_2)_4-C \equiv N + 2NaCl$$
1,4-dicyanobutane, called adiponitrile

Adiponitrile next undergoes *both* reactions described in Section 14-7 for the nitrile functional group. Under reductive conditions (and in the presence of an appropriate catalyst), each nitrile group is converted to an amino group via the reaction

$$N \equiv C-(CH_2)_4-C \equiv N + 4H_2 \xrightarrow{\text{Ni catalyst}} H_2N-CH_2(CH_2)_4CH_2-NH_2$$
1,6-diaminohexane

In water solution, with acid present, a dicarboxylic acid is produced from adiponitrile:

$$N \equiv C-(CH_2)_4-C \equiv N + 4H_2O \xrightarrow{H^+} HOOC-(CH_2)_4-COOH + 2NH_3$$
1,6-dihexanoic acid,
called adipic acid

We now have synthesized the two specific reagents for the final condensation reaction. As a preview of what happens, consider the general reaction of a carboxylic acid with an amine (an organic base) to form water and an amide:

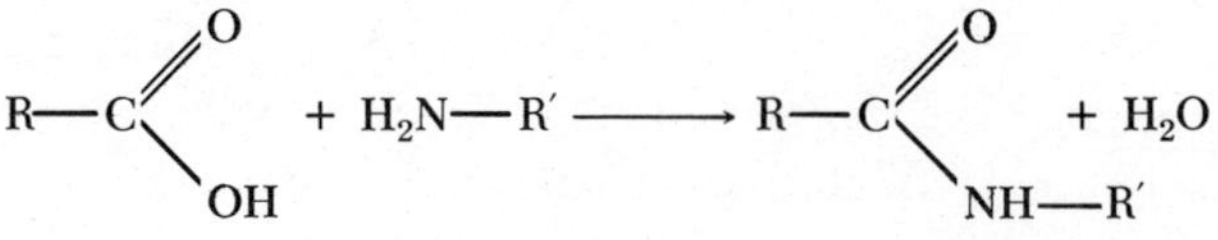

$$R-C\begin{array}{c}\diagup O \\ \diagdown OH\end{array} + H_2N-R' \longrightarrow R-C\begin{array}{c}\diagup O \\ \diagdown NH-R'\end{array} + H_2O$$

Since adipic acid is a *di*carboxylic acid, and since it reacts with a *di*amine, the product is still left with an acid group and an amine group after the condensation reaction:

$$HOOC(CH_2)_4COOH + H_2N(CH_2)_6NH_2 \longrightarrow HOOC(CH_2)_4C\begin{smallmatrix}\nearrow O \\ \searrow NH(CH_2)_6NH_2\end{smallmatrix} + H_2O$$

The remaining acid group is free to react with another molecule of the diamine, and the remaining amino group can react with another molecule of adipic acid:

$$H_2N(CH_2)_6NH_2 + HOOC(CH_2)_4\overset{O}{\underset{\|}{C}}-\overset{H}{\underset{|}{N}}-(CH_2)_6NH_2 + HOOC(CH_2)_4COOH$$

$$\longrightarrow H_2N(CH_2)_6-\overset{H}{\underset{|}{N}}-\overset{O}{\underset{\|}{C}}(CH_2)_4\overset{O}{\underset{\|}{C}}-\overset{H}{\underset{|}{N}}-(CH_2)_6\overset{H}{\underset{|}{N}}-\overset{O}{\underset{\|}{C}}(CH_2)_4COOH + 2H_2O$$

This product also has a free acid group and a free amino group, and so the condensation reaction can continue on indefinitely. The water vapor produced by the reactions is pumped out of the heated reaction vessel, and the mass slowly thickens as the molecular weight builds up to form a polymer with the repeating structure

$$\cdots -(CH_2)_6-\overset{H}{\underset{|}{N}}-\overset{O}{\underset{\|}{C}}(CH_2)_4\overset{O}{\underset{\|}{C}}-\overset{H}{\underset{|}{N}}-(CH_2)_6-\overset{H}{\underset{|}{N}}-\overset{O}{\underset{\|}{C}}(CH_2)_4\overset{O}{\underset{\|}{C}}-\overset{H}{\underset{|}{N}}-(CH_2)_6-\cdots$$

having a molecular weight of 50,000 or more. This is *nylon*, the first synthetic polyamide ever made on a large scale. Silk, wool, and other naturally occurring protein fibers are also polyamides.

Nylon can also be made from petroleum, although somewhat less conveniently. Isolation of *n*-butane from straight-run gasoline sets up the four-carbon-atom chain, which is then dehydrogenated in order to provide functional groups for the eventual attachment of nitrile groups:

$$n-C_4H_{10} \xrightarrow[\text{heat}]{\text{catalyst}} H_2C=CH-CH=CH_2 + 2H_2$$
1,3-butadiene

However, the yield of product is too poor, and the supply of starting material is too limited to make this an economical method.

A third synthetic route to nylon is attractive because it utilizes some of the agricultural waste that usually is allowed simply to rot on the ground. Given a choice, the preferred waste material to use for making nylon is oat hulls. Oat hulls can be fermented to yield *furfural*, a cyclic ether with an attached aldehyde group. The furfural is converted to furan and then to tetrahydrofuran by a sequence of reactions which is straightforward, although lengthy. First, the aldehyde is both reduced to an alcohol and oxidized to an acid in the **disproportionation** (autooxidation) process

Each monomer has two functional groups, ready for polymerization.

(autooxidation) process

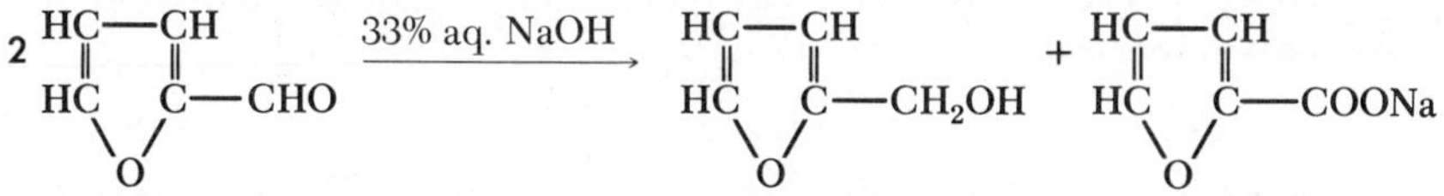

$$2 \underset{\text{furfural, from fermentation of oat hulls}}{\text{furfural}} \xrightarrow{\text{33\% aq. NaOH}} \underset{\text{furfuryl alcohol}}{\text{furfuryl-CH}_2\text{OH}} + \underset{\text{sodium salt of furoic acid}}{\text{furyl-COONa}}$$

The air oxidation of the alcohol yields the corresponding acid:

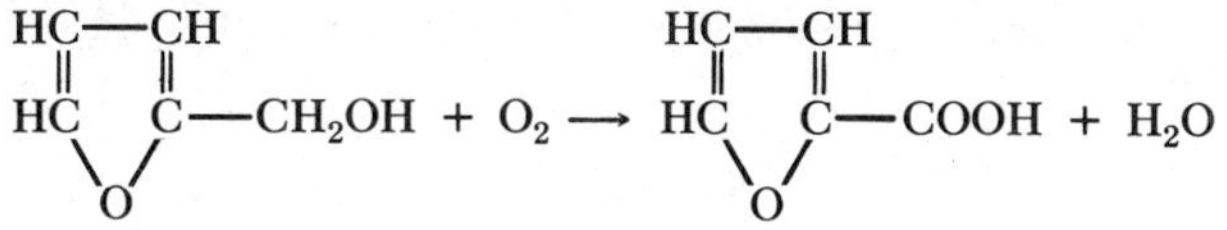

$$\underset{\text{furfuryl alcohol}}{\text{furyl-CH}_2\text{OH}} + O_2 \longrightarrow \underset{\text{furoic acid}}{\text{furyl-COOH}} + H_2O$$

Furoic acid readily decarboxylates to furan:

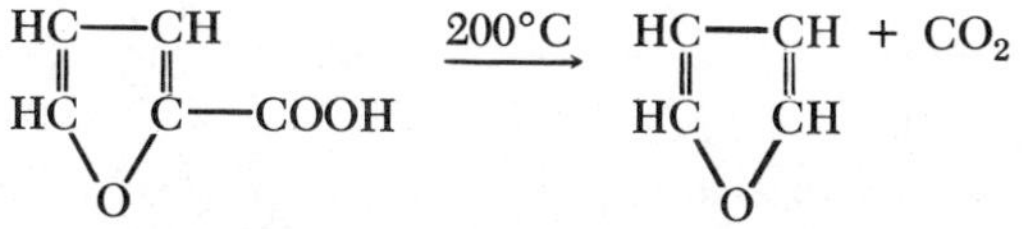

$$\underset{\text{furoic acid}}{\text{furyl-COOH}} \xrightarrow{200°C} \underset{\text{furan}}{\text{furan}} + CO_2$$

Catalytic hydrogenation of furan yields tetrahydrofuran according to the reaction

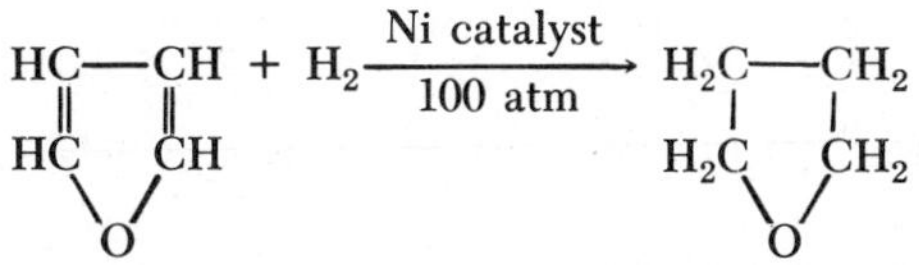

$$\underset{\text{furan}}{\text{furan}} + H_2 \xrightarrow[\text{100 atm}]{\text{Ni catalyst}} \underset{\text{tetrahydrofuran, "THF"}}{\text{THF}}$$

From here on, the THF is used just as in the steps already given for the preparation of nylon from coal.

Having seen this connection (raw material to manufactured material) between oat hulls and nylon, you may wonder whether there are other ways in which agricultural waste—such as cornstalks, corncobs, wheat straw, or bran—might be used for the production of useful chemical substances. Some material can be converted to ethanol by fermentation, and the alcohol is an acceptable motor fuel as well as being useful in chemical synthesis. It was proposed during the economic depression of the 1930s, as an aid to farmers, that production of alcohol in this way be subsidized by requiring all motor fuel to contain 10 to 20 percent alcohol. The plan fell through when it was demonstrated that only *anhydrous* ethanol is completely soluble in gasoline. As soon as some water is introduced (as for example via moisture in the air, condensing in storage tanks), the ethanol separates as a water-alcohol solution; the chemical similarity makes alcohol prefer water to the hydrocarbons, because alcohol can hydrogen-bond to water.

Think of the waste of photosynthetic effort in producing corn ears, stalks and roots, then using only the grain.

The failure of that particular scheme must not discourage us from trying to find significant chemical uses for the millions of tons of agricultural material that goes to waste each year. There must be something else besides nylon and alcohol that can and should be made from the unwanted parts of plants. If the reader will put enough ingenuity into the problem, some new approaches are likely. These new ideas will be most welcome.

As we have seen, polymers need not be based solely on frameworks of carbon-to-carbon bonds. True, there are some natural advantages to staying with carbon frameworks in designing new polymers: The raw materials are available (at least at present); the reactions for transforming them are available in the neatly organized package of science called organic chemistry; and there are some natural examples to emulate or improve upon. There are also some disadvantages: Almost all the principal avenues of synthesis have already been explored; the products have a certain sameness; and all the products suffer from the stark thermodynamic fact that carbon-carbon bonds and carbon-hydrogen bonds are unstable in the presence of oxygen. Hydrocarbon polymers have a strong chemical tendency to become water and carbon dioxide; polyesters and polyamides are not much different in this respect.[11] Think of the highly flammable acrylic sweaters and rugs! Imagine an assemblage of clothing and furniture and instruments and electrical equipment, all made of the best polymers and fibers money can buy. Put them all in one room (in the actual instance, an Apollo spaceship on the launch pad on January 27, 1967). Admit oxygen to one atmosphere pressure. Then introduce a tiny amount of activation energy (a spark from a switch, or an overheated resistor). The result is a furious, tragic, devastating fire. The laws of chemical thermodynamics will not be denied.

Even in ordinary air (pressure of oxygen is only 0.20 atm), organic polymers suffer oxidation and decomposition if they are heated too much. They also crack and break if they are cooled too much. Their temperature range of usefulness is too small. We can resort to porcelain and glass (and fancier ceramics) for service over a wider range of temperatures, but these materials are brittle. They are thermodynamically stable, being fully oxidized, but they cannot be applied to a house as paint, nor molded in the shape of a boat at room temperature. So a wide gap in properties exists between ceramic materials and organic polymers, a gap that simply pleads for the design of new materials to fill it.

Successful combination of some of the properties of both ceramics and of organic plastics has been achieved with one new kind of polymeric substance. These comprise the family of synthetic materials called **silicones**.[12] The strongly bonded and fully oxidized polymeric framework of glass, porcelain, and the natural silicates was borrowed by the chemical designers, who altered that framework by attaching sufficient organic groups to make the framework flexible. By proper choice of organic groups, it has been possible to make

The silicones combine features of both organic and inorganic polymers.

---

[11] About the only exception is Teflon (polytetrafluoroethene), and even this compound depolymerizes and oxidizes when a fry pan coated with it is heated above 300°C. (In vacuum, a temperature of 500°C is necessary to produce 1 mm pressure of monomer by depolymerization.)

[12] See E. G. Rochow, *An Introduction to the Chemistry of the Silicones*, 2nd ed., New York: Wiley (1952); Ann Arbor, Mich.: University Microfilms (1965).

the resulting material soluble, or moldable, or capable of being further polymerized at room temperature. Introduction of organic groups is, of course, a compromise, because carbon atoms bring along with them the familiar woes of burnability and limited stability. By limiting the organic content to the practical minimum, and by avoiding the carbon-to-carbon bonds entirely, enough difference in properties (compared to purely organic polymers) has been obtained to make the compromise worthwhile.

The polymer framework of the silicates is an arrangement of alternate silicon and oxygen atoms, the **siloxane** chain:

$$\begin{array}{c}\text{Si}-\text{O}-\text{Si}-\text{O}-\text{Si}-\text{O}-\text{Si}-\text{O}-\text{Si}\end{array}$$

The bond angles at the silicon atom are tetrahedral, and the

$$\text{O}\diagdown^{\text{Si}}_{\text{Si}}$$

bond angle is about 150°. The silicon orbitals are tetrahedrally oriented sp^3 hybrids. Only two silicon orbitals are required to link with oxygen atoms to form the simple framework chain. The other two orbitals may be used to link exclusively to more oxygen atoms, as in SiO_2, or to one oxygen atom and to one carbon atom of an organic group, or to carbon atoms of two organic groups. An important structural feature of the siloxane polymers is that any organic groups attached to the chain should be linked directly to silicon by C—Si bonds.[13] These are strong bonds (the bond dissociation energy is 78 kcal/mole) that are not hydrolyzable. The one organic group that meets the desiderata of minimum number of carbon and hydrogen atoms and no C—C bonds is the *methyl group*. This is the reason why methyl groups have been used, for almost all practical silicones, to modify the siloxane chains. On occasion, phenyl groups have also been used because their tightly bound rings resist oxidation.

Utilizing these structural features, we can contemplate three types of polymeric units in methyl silicones. First, there are chain-building organosiloxane units of the type

$$\begin{array}{c}\text{CH}_3\\|\\-\text{Si}-\text{O}-\\|\\\text{CH}_3\end{array}$$

Dimethylsiloxane bifunctional unit

---

[13] This requirement places silicones in the category of organometallic compounds, an interesting class of substances increasingly useful throughout all of chemistry. See E. G. Rochow, *Organometallic Chemistry*, New York: Reinhold (1964) (a paperback in the series *Selected Topics in Modern Chemistry*).

Then there are cross-linking organosiloxane units of the type

$$\begin{array}{c} \text{CH}_3 \\ | \\ -\text{Si}-\text{O}- \\ | \\ \text{O} \\ | \end{array}$$

Methylsiloxane trifunctional unit

Finally, there are chain-ending units like

$$\begin{array}{c} \text{CH}_3 \\ | \\ \text{CH}_3-\text{Si}-\text{O}- \\ | \\ \text{CH}_3 \end{array}$$

Trimethylsiloxane monofunctional unit

Imagine using essentially only the bifunctional dimethylsiloxane units, strung together in very large numbers; the result is a linear dimethylsiloxane polymer of high molecular weight. In actual practice, if the molecular weight is *extremely* high, the polymer is found to be *rubbery*. On the other hand, if some trifunctional units are incorporated here and there along the siloxane chain, the result is a cross-linked *resin*, hard and infusible. Lastly, if some monofunctional units are introduced, they terminate the polymeric chains and limit the molecular weight, making *liquids;* the viscosity of these silicone oils depends on the number of monofunctional units.

Practical silicone oils and rubbers are predominantly methylpolysiloxanes, with a small proportion of phenyl groups sometimes added to improve resistance to oxidation. Silicone resins are usually methylphenylpolysiloxanes, with a substantial proportion of phenyl groups attached to the silicon atoms.

Desired properties can be built into silicone molecules by a chemical designer.

Silicone polymers are made from plentiful raw materials. Their production does not encroach upon the supply of fuels and potential foods as much as does the manufacture of organic polymers. The starting materials are sand, coke, water, and salt. Steps in the manufacture of silicones may be summarized as follows:

1. Silicon is made by reducing sand with coke in an electric furnace:

$$\text{SiO}_2 + 2\text{C} \xrightarrow{2000°\text{C}} \text{Si} + 2\text{CO}$$

The elemental silicon so obtained is a shiny, metallic-looking, brittle solid that is used not only for making silicones, but also as an alloying agent for aluminum aircraft alloys and for steels.

2. Methanol is made from coal and water, as already described in Section 14-9, and is converted to methyl chloride by anhydrous hydrogen chloride. The HCl is obtained from a later stage of the process. The reaction that gives methyl chloride is

$$\text{CH}_3\text{OH} + \text{HCl} \xrightarrow{\text{ZnCl}_2} \text{CH}_3\text{Cl} + \text{H}_2\text{O}$$

3. The methyl chloride is heated and blown through a mixture of powdered silicon metal and 10-percent copper dust. The product is mostly dimethyldichlorosilane, but there are also some other methylchlorosilanes and a host of minor by-products. Thus, we write

$$2CH_3Cl + Si \xrightarrow[Cu]{300°C} (CH_3)_2SiCl_2$$

(with some CH_3SiCl_3, $(CH_3)_3SiCl$, CH_3SiHCl_2, and $SiHCl_3$)

All the products are separated by distillation and are stored. These are the intermediates from which the various silicone polymers are made.

4. Dimethyldichlorosilane is hydrolyzed in a limited amount of water, resulting in exchange of oxygen atoms for chlorine atoms:

$$4(CH_3)_2SiCl_2 + 4H_2O \longrightarrow [(CH_3)_2SiO]_4 + 8HCl\uparrow$$

The anhydrous hydrogen chloride that comes off is returned to the process to convert more methanol to methyl chloride. The principal hydrolysis product is the cyclic tetramer of dimethylsiloxane

$$\begin{array}{c} \text{CH}_3 \quad\quad \text{CH}_3 \\ | \quad\quad\quad | \\ \text{H}_3\text{C—Si—O—Si—CH}_3 \\ | \quad\quad\quad | \\ \text{O} \quad\quad\quad \text{O} \\ | \quad\quad\quad | \\ \text{H}_3\text{C—Si—O—Si—CH}_3 \\ | \quad\quad\quad | \\ \text{CH}_3 \quad\quad \text{CH}_3 \end{array}$$

This eight-membered ring is analogous to similar siloxane rings found in asbestos and mica. The reaction also yields other cyclic and linear dimethylsiloxanes. The linear molecules have terminal —OH groups. All the variants can be converted to the cyclic tetramer by heating with a little NaOH.

5. Silicone rubber is made from the pure cyclic tetramer by heating it with a trace (0.03 percent) of anhydrous potassium hydroxide. The KOH attacks a few Si—O bonds at a time by the reaction

$$-\overset{|}{\underset{|}{Si}}-O- + KOH \longrightarrow -\overset{|}{\underset{|}{Si}}-OK + H_2O$$

The potassium silanolate then hydrolyzes in the water produced, regenerating KOH and reforming a siloxane bond. The reaction can be summarized by

$$2\left(-\overset{|}{\underset{|}{Si}}-OK\right) + 2H_2O \longrightarrow -\overset{|}{\underset{|}{Si}}-O-\overset{|}{\underset{|}{Si}}-O- + 2KOH$$

This process is repeated millions of times, each time opening a ring or chain and allowing nearby fragments to rejoin in random fashion.[14] Eventually a

---

[14] This is a near-perfect example of a reaction driven only by entropy change, since just as many Si—O bonds are formed as are broken. The only driving force is the trend toward randomization of an initially orderly cyclic structure.

gumlike mixture of all possible molecular sizes and molecular weights is obtained. The smaller molecules are distilled as volatile material, leaving a mixture of extraordinarily large molecules—the average weight of this polymeric material is about 3,000,000. When mixed with high-surface-area SiO_2 as reinforcing agent, and "vulcanized" with benzoyl peroxide, this gum becomes strong and elastic *silicone rubber*.

6. Silicone resins are made by cohydrolyzing a mixture of chlorosilanes at low temperatures in a solvent. Consider what happens when the compounds $(CH_3)_2SiCl_2$, CH_3SiCl_3, $C_6H_5SiCl_3$, and $(C_6H_5)_2SiCl_2$ react together with water. We have the four simultaneous reactions:

$$(CH_3)_2SiCl_2 + 2H_2O \longrightarrow (CH_3)_2Si(OH)_2 + 2HCl$$

$$CH_3SiCl_3 + 3H_2O \longrightarrow CH_3Si(OH)_3 + 3HCl$$

$$C_6H_5SiCl_3 + 3H_2O \longrightarrow C_6H_5Si(OH)_3 + 3HCl$$

$$(C_6H_5)_2SiCl_2 + 2H_2O \longrightarrow (C_6H_5)_2Si(OH)_2 + 2HCl$$

Molecules of these resulting hydroxy compounds (called *silanols*) are unstable in the presence of one another. They start to *condense with one another* as soon as they are formed, splitting out water and forming siloxane chains via the reaction

$$-\overset{|}{\underset{|}{Si}}-OH + HO-\overset{|}{\underset{|}{Si}}- \longrightarrow -\overset{|}{\underset{|}{Si}}-O-\overset{|}{\underset{|}{Si}}- + H_2O$$

The condensation does not proceed to completion at room temperature, and this medium-molecular-weight, partially polymerized material stays dissolved in the solvent used for hydrolysis. After washing out all the hydrochloric acid, we have an organosiloxane solution that is stable enough to be used as a vehicle for making paints, or as an impregnant for glass fibers. When the solvent is evaporated and heat is applied, the condensation reaction continues to completion, producing an insoluble, infusible, resinous silicone polymer.

7. Silicone oils are made by cohydrolyzing $(CH_3)_2SiCl_2$ and $(CH_3)_3SiCl$ in the right proportions to give the desired average chain length, or by equilibrating a mixture of bifunctional and monofunctional methylsiloxanes. For example, cohydrolysis of 4 moles of $(CH_3)_2SiCl_2$ with 2 moles of $(CH_3)_3SiCl$ gives a compound of the structure

$$\underbrace{CH_3-\underset{\underset{CH_3}{|}}{\overset{\overset{CH_3}{|}}{Si}}-O}_{\text{(from }(CH_3)_3SiCl)}-\overbrace{\underset{\underset{CH_3}{|}}{\overset{\overset{CH_3}{|}}{Si}}-O-\underset{\underset{CH_3}{|}}{\overset{\overset{CH_3}{|}}{Si}}-O-\underset{\underset{CH_3}{|}}{\overset{\overset{CH_3}{|}}{Si}}-O-\underset{\underset{CH_3}{|}}{\overset{\overset{CH_3}{|}}{Si}}-O}^{\text{(from }(CH_3)_2SiCl_2)}-\underset{\underset{CH_3}{|}}{\overset{\overset{CH_3}{|}}{Si}}-CH_3$$

and a mixture of related compounds that average out to the same chain length. These linear "chain-stopped" siloxanes are colorless oily liquids of high stability. Their viscosity changes only slightly with temperature. A similar mixture of compounds, also designed to have a particular viscosity and a particular

Silicone products, designed and made by chemists, have properties unknown in other materials.

average molecular weight, can also be made by randomizing or equilibrating a mixture of $[(CH_3)_3SiO]_4$ and $(CH_3)_3Si-O-Si(CH_3)_3$—substances obtained by the separate hydrolysis of $(CH_3)_2SiCl_2$ and $(CH_3)_3SiCl$—with a small amount of anhydrous H_2SO_4 as a catalyst.

All the methyl silicone polymers that have been described are more stable than purely organic polymers, reflecting the strong and fully oxidized silicon-oxygen linkages which make up their framework. They also resist oxidation and are not attacked by salts or acids (except HF, which will dissolve both silicones and glass). Silicone rubber can be used for a limited time at a temperature of 400°C (750°F) and gives extended service at 315°C (600°F). Similarly, silicone resins and oil have upper service temperatures that are higher than those of all natural or synthetic organic resins and oils. What is more surprising is that the desired mechanical and electrical properties also extend to *lower* service temperatures than those of organic counterparts. For example, silicone rubber stays flexible down to −80°C, and silicone oil stays fluid down to −70°C, far below the temperature at which a hydrocarbon oil of comparable viscosity hardens to a solid. These small changes of physical properties with large changes in temperature result from weak intermolecular attractions in methyl siloxane polymers of all kinds. This is fortunate, because it provides another advantage over organic polymers and allows silicones to do jobs that the more ordinary materials could not do. For example, silicone rubber seals the windows and doors of jet aircraft even at the −60° temperature common at high altitudes; natural rubber and synthetic organic rubbers are no longer elastic at that temperature, and will not do. At the other end of the scale, electric motors and transformers insulated with silicone resins can stand considerable overheating (through temporary overload) without burning out.

FIGURE 15-4 All those sharp footprints on the moon were made by boots of silicone rubber, the only material that will work under the extreme conditions. The alternate heat and cold endured by the equipment during space flight would melt and freeze ordinary rubber, but the silicone rubber used in these boots (and many other parts of space suits) remains flexible at −160°F and yet withstands temperatures high enough to melt solder. Here the boot meets a lunar surface temperature of 250°F. (Photo courtesy of General Electric Co., Silicone Products Dept., Waterford, N.Y.)

A characteristic use for silicone rubber is shown in Figure 15-4. For this application, as for many others involving spaceships and space suits, intensive study of the properties of this particular silicone rubber made it certain that the material could do its job (and equally certain that no conventional material or natural product could succeed). The designer of a new material always hopes, in addition, that some new, useful, but totally unexpected property will pop up, just to give him an added "dividend" and to give construction engineers something new to think about. Figure 15-5 illustrates one such unexpected property.

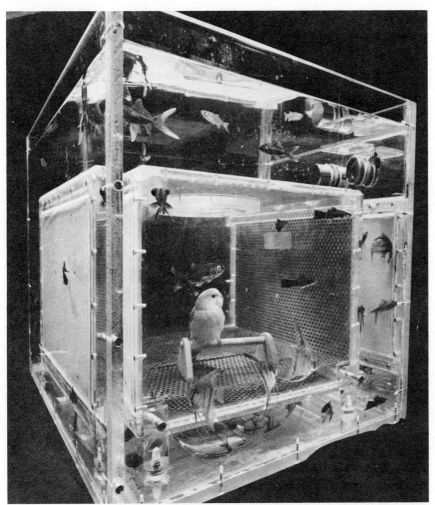

FIGURE 15-5 An unexpected property of silicone polymers comes to light. The bird's cage is watertight and has sides of thin silicone rubber sheet, which allows oxygen (from the dissolved air) to diffuse through so that the bird breathes normally, even under water. Silicone rubber is the only material so permeable to oxygen. The possibilities for human habitation under water are obvious. (Photo courtesy of General Electric Co., Silicone Products Dept., Waterford, N.Y.)

Many benefits have come out of this one excursion by chemists into the interdisciplinary region between ceramic chemistry and organic chemistry, so chemists should be encouraged to search for other extraordinary polymeric materials, perhaps polymers that are even more inorganic than the silicones.[15]

15-5 Elastomers

All polymeric materials "freeze" hard and become brittle at very low temperatures (say at the boiling point of liquid hydrogen, $-253°C$). As they warm up, each polymer has its own characteristic temperature at which it ceases to be glassy and brittle and becomes somewhat flexible. This is the *glass transition temperature;* for polystyrene it is $+100°C$, for natural rubber $-73°C$, and for silicone rubber $-109°C$. Upon further warming, most polymers also enter upon an elastic stage, where they seem *rubbery* rather than simply flexible. Usually this elastic stage occurs outside the range of indoor or climatic temperatures, and so we do not think of the substance as rubbery. The elastic range may also be very narrow. Both the range and its position on the temperature scale may be altered by copolymerization or by modification with soluble substances. For example, the familiar polyethylene is known to us as a flexible plastic at room temperature, and so is polypropylene (used especially for marine ropes and for water-skiing lines), but a **copolymer** of ethylene and propylene is a serviceable and chemically stable rubber at room temperature. As another example, polyvinyl chloride is a tough, hard, somewhat brittle solid at room temperature, with an elastic range that sets in only above $100°C$. By incorporating into it a soluble nonvolatile liquid such as tricresylphosphate (called a *plasticizer*), a satisfactory rubber can be made for use as noninflammable wire insulation.

When the elastic range of a polymer (or a copolymer or a plasticized polymer) happens to include room temperature, we speak of it as an **elastomer.** The relation between an elastomer and a serviceable rubber is the same as that between polymer and plastic—to make a rubber, one takes an elastomer and adds to it a reinforcing filler, plus various coloring materials and **vulcanizing** agents. The gum obtained from the exuded sap of the rubber tree *hevea brasiliensis* is an elastomer; by milling into it extensive amounts of carbon black, plus some sulfur and a number of vulcanizing aids, natural rubber for rubber tires is made. Other elastomers require other reinforcing agents and different vulcanizing agents, according to their molecular structures and compositions. Thus, silicone rubber is not strengthened much by carbon black, the way a carbon-chain polymer is, but requires a more sympathetic and chemically similar reinforcing agent, such as SiO_2, to strengthen its siloxane-chain polymer. Furthermore, since its methyl groups contain no double bonds, it cannot be vulcanized by sulfur; it needs oxidizing agents to break off some of its methyl-group hydrogen atoms. But if some vinyl groups are substituted for a few of its methyl groups, then it can be vulcanized with sulfur. From

---

[15] Although a great deal of money has been spent on a search for "inorganic polymers" (meaning materials that are stable, flexible, and elastic, but entirely inorganic in composition), nothing successful has resulted. This is an area where chemistry has failed.

these few facts, the reader may appreciate that the compounding and "curing" of the very many different rubbery materials is an elaborate art, where experience and technology count for more than science.

In this section, we shall concentrate only on the molecular architecture of a few elastomers, principally to understand the effect of molecular structure on their properties and "synthesize—ability." This is one area where a historical approach is the most direct and satisfactory, so we shall consider natural rubber first. We can then go on to how related elastomers came to be synthesized.

Natural *hevea* rubber[16] is a hydrocarbon. It is a polymer of *isoprene*, which is 2-methylbutadiene,

$$CH_2=C-CH=CH_2$$
$$|$$
$$CH_3$$

When natural rubber is heated (in the absence of air) hot enough to depolymerize it, isoprene distills off. For a long while, it was not known how the isoprene units in rubber were connected, but the advent of suitable X-ray diffraction methods revealed that the isoprene units polymerize head to tail, exchanging their terminal bonds for one internal double bond in the process. Further, the configuration at this bond is that of a *cis* geometric isomer, with the methyl group and the neighboring hydrogen atom both on one side of the double bond:

$$-CH_2-CH_2-\underset{\underset{H}{\overset{|}{CH_3}}}{C}=\underset{}{C}-CH_2-CH_2-\overset{\overset{CH_3}{|}}{C}=\overset{\overset{H}{|}}{C}-CH_2-CH_2-$$

$$\longleftarrow \text{identity period 9.13 Å} \longrightarrow$$

Oddly enough, another tropical tree, the Malayan *palaquim*, synthesizes the *opposite* isomer, with a *trans* configuration:

$$-CH_2-CH_2-\overset{\overset{H}{|}}{C}=\underset{\underset{CH_3}{|}}{C}-CH_2-CH_2-\overset{\overset{H}{|}}{C}=\underset{\underset{CH_3}{|}}{C}-CH_2-CH_2-$$

$$\longleftarrow \text{identity period 5.04 Å} \longrightarrow$$

This is called *gutta percha*, or *balata*. It is tough and flexible, but not elastic; it is used for the covers of golf balls.

Hevea brasiliensis grew in scattered regions of Central and South America, mostly wild and inaccessible. After Charles Goodyear discovered how to vulcanize rubber in 1844, thereby founding the rubber industry, the demand for raw rubber soon exceeded the supply from wild trees. Seeds from a specimen in Kew Gardens in England were found to grow well in Malaya, and later in the Dutch East Indies, so plantation rubber became available. The rubber

---

[16] The Central American Indians called natural rubber *caoutchouc*, and this name persists in German. The English term *rubber* came from the early use of the unvulcanized gum to rub out pencil marks.

industry flourished with the development of the automobile, but the supply of rubber from the Pacific remained variable, especially during wartime. Between 1905 and 1925, the price of raw rubber fluctuated from 18 cents to $3.50 a pound, and the material had all of the lack of uniformity that characterizes a natural product. People began to look forward to a way of synthesizing natural rubber. An English process was developed to convert turpentine to isoprene, and to polymerize the isoprene with metallic sodium, but this just exchanged dependence on one natural product for another. A German process was announced in 1912 to make isoprene from alcohol derived from potatoes, but this placed rubber in competition for the food supply. There seemed to be no easy way to make isoprene from coal or oil. Interest lapsed until 1930, when it was discovered that an analogue of isoprene could be made from acetylene in only two steps. In the first, acetylene adds to itself in a solution of CuCl dissolved in concentrated HCl:

$$HC\equiv CH + HC\equiv CH \xrightarrow{\mathrm{Cu^+/HCl}} HC\equiv C-CH=CH_2$$
"vinylacetylene"

In the second step, the vinylacetylene is treated with hydrogen chloride gas

$$HC\equiv C-CH=CH_2 + HCl\ \text{(anhydrous)} \longrightarrow H_2C=\underset{Cl}{C}-CH=CH_2$$
2-chlorobutadiene

The product, 2-chlorobutadiene, is almost an exact geometric replica of isoprene (2-methylbutadiene), because the chlorine atom has nearly the same volume as a methyl group. Polymerization of 2-chlorobutadiene proved to be easier and more dependable than that of isoprene, giving a rubber that was useful, even though it was not the equal of natural rubber for making tires. It was found to have a significant advantage over natural rubber: It does not swell or soften as much when exposed to gasoline or lubricating oil. So it was used for gasoline hoses and engine parts, and in this way the first commercially successful synthetic rubber came on the scene. Initially it was called chloroprene, and later neoprene.

The Pacific War of 1941 to 1945 cut off all supplies of natural rubber, and at the same time greatly increased the demand for rubber tires. Great efforts were made to find a suitable hydrocarbon rubber that could be manufactured from coal or oil, and that would wear as well as natural rubber. A German synthetic rubber made from acrylonitrile ($H_2C=CH-C\equiv N$, produced from C_2H_2 and HCN) was moderately successful, but the eventual choice was a copolymer of butadiene and styrene called GR-S (government rubber, styrene type). The butadiene could be made from butane by dehydrogenation, but it was easier to make it from acetylene by way of tetrahydrofuran (for the preparation of THF, see the discussion of the synthesis of nylon in Section 15-4). The reaction for dehydration of THF is

$$\underset{\text{THF}}{\begin{array}{c}H_2C-CH_2\\ |\qquad\quad |\\ H_2C\quad CH_2\\ \diagdown\ \diagup\\ O\end{array}} \xrightarrow[\text{catalyst}]{Na_2HPO_4} CH_2=CH-CH=CH_2 + H_2O$$
butadiene

The styrene was made from benzene, as previously described, and the benzene came both from soft coal and from the cyclization of hexane from petroleum. The copolymerization was best carried out while the mixed monomers were emulsified in water, giving a milky latex that could be handled like *hevea* latex:

$$\text{phenylethene (styrene)}\ \ C_6H_5\text{-}CH=CH_2 + H_2C=CH-CH=CH_2 \ \ \xrightarrow[\text{in emulsion}]{\text{copolymerization}} \ \ \text{GR-S elastomer}$$
(butadiene)

Other successful rubbers of the decade 1940 to 1950 were *polysulfide rubber* (Thiokol, made from sodium polysulfide and $ClCH_2CH_2Cl$, and consisting of chains of sulfur atoms joined by $-CH_2CH_2-$ groups) and *butyl rubber* (polymerized isobutene, $(CH_3)_2C=CH_2$). Various improvements were made in all these types, so that by the 1950s better-wearing tires could be made from a combination of synthetics than could be made from natural rubber alone.

There remained one problem: The synthetics generated more internal heat when they were flexed than did natural rubber. Natural rubber, having a more ordered internal structure than the amorphous synthetics, simply crystallized and stored the mechanical input as potential energy, like a coiled spring, for release later. So there was a continuing demand for development of a synthetic rubber of ordered internal structure. By 1960 several such rubbers had appeared. The one that eventually prevailed was a true "synthetic natural rubber," an artificial polyisoprene with a *cis* configuration, made by polymerizing synthetic isoprene with a stereospecific catalyst (a transition-metal coordination compound). The entire process for synthesizing this rubber may be summarized in this way:

$$2H_2C=CH-CH_3 \ \xrightarrow{\text{dimerize}}\ H_2C=\underset{CH_3}{\overset{|}{C}}-CH_2-CH_2-CH_3$$
(from petroleum)
2-methyl-1-pentene

$$\downarrow \text{isomerize}$$

$$CH_3-\underset{CH_3}{\overset{|}{C}}=CH-CH_2-CH_3$$
2-methyl-2-pentene

$$\downarrow \ \text{thermal cracking with HBr}$$

$$CH_2=\underset{CH_3}{\overset{|}{C}}-CH=CH_2 \ + \ CH_4$$
isoprene

$$\downarrow \ \text{polymerize with stereospecific catalyst}$$

$$(-CH_2-\underset{CH_3}{\overset{|}{C}}=\underset{H}{\overset{|}{C}}-CH_2-)$$
cis-polyisoprene ("Natsyn")

So the design of elastomers has come full circle, back to polyisoprene, but from petroleum instead of from trees in the jungle.

15-6 Dyes

It was a drab world during the Middle Ages. It was still drab in the times of Charles Dickens. The bright colors we see in clothing today simply did not exist then. Textile fabrics were dyed, if at all, with vegetable colors, and these colors faded in sun and air to a neutral gray or brown. Until the first synthetic **dyestuff** (Sir William Perkin's *mauve*, synthesized in 1856, leading to the "mauve decade"), there was nothing better.[17] Now specialized dyes are available for all fabrics and fibers, natural and synthetic, in all colors. How this abundance came about is, in itself, an exemplary case history in the design of materials.

All substances are "colored" in the broad sense that they absorb or reflect radiation in certain wavelength ranges but not in others, leading to a characteristic absorption or reflection spectrum for the substance. Even quartz, which is one of the most transparent substances known, absorbs radiation in the far ultraviolet region, and window glass absorbs in the near ultraviolet. We ordinarily use the adjective *colored* to describe a substance that absorbs or reflects light selectively within the narrow range of radiation to which our eyes are sensitive. A colored solution or a piece of stained glass absorbs part of white light and *transmits* the rest, so we see it as colored. An opaque solid (such as bright yellow cadmium sulfide or red mercury oxide, both of which are pigments) absorbs part of white light and *reflects* the rest to our eyes. We shall consider only substances that give colored solutions in water *and* impart that color permanently to paper or cloth. Such substances are called *dyes*.

What is the color of an object that absorbs all wavelengths? that transmits all wavelengths? that reflects all wavelengths?

Absorption of visible light involves electron transitions of rather modest energy, much less than that required to ionize atoms. The transition takes place by displacement of an electron to a higher *molecular* energy level, followed by either (1) dissipation of the absorbed energy as heat, or (2) reradiation in all directions, spherically (whereas the original absorption was from a directionally defined beam of light). In either case, the residual transmitted light appears colored by subtraction. Such absorption of quanta of visible light is quite common among hydrated and coordinated ions of transition metals,

---

[17] Some might insist that Royal Purple was an exception. This was a coloring substance extracted in Roman times from the glands of a snail, and since it required the color from 24,000 snails to dye one toga, it can scarcely be considered a practical dyestuff. The reader might be interested in comparing the structure of Royal Purple with those of modern dyes to follow, and in wondering how the snail synthesizes it:

but is rather rare in organic substances. In fact, only a few organic molecular configurations give rise to color, and practically all of these include double bonds. A particular structural configuration that absorbs visible light and conveys color is called a **chromophore**. The principal chromophores are

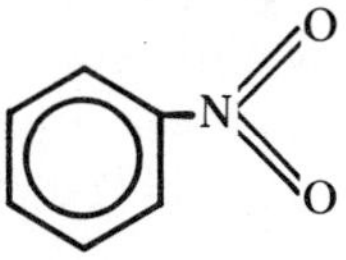

the nitro group, when attached to an aromatic ring

the acridine group

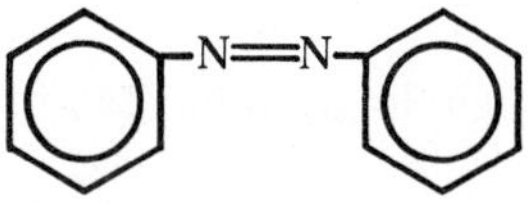

Note the presence of aromatic rings in the molecules of the nitro group.

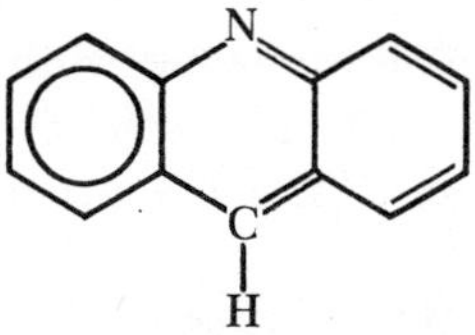

the diazo group, when attached to an aromatic ring

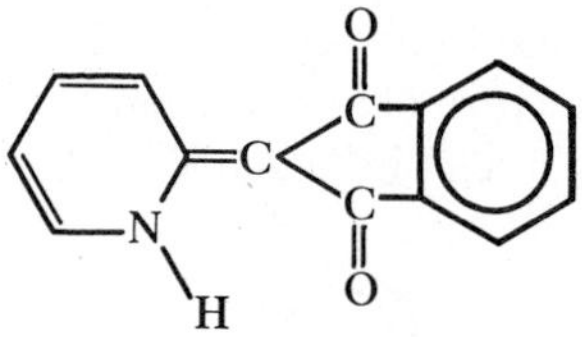

the quinoline group

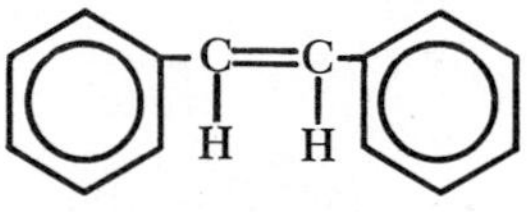

the stilbene group

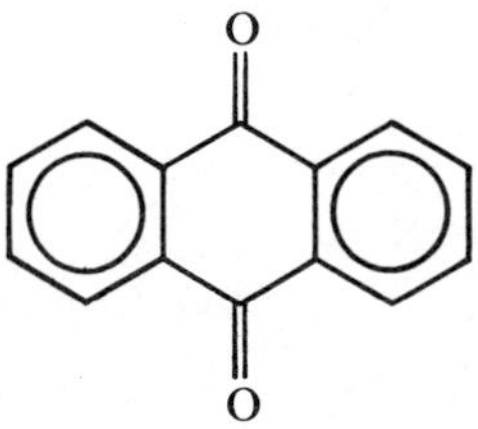

the anthraquinone group

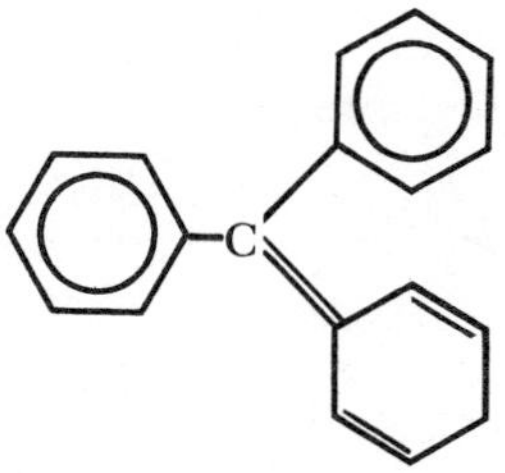

the triarylmethane group

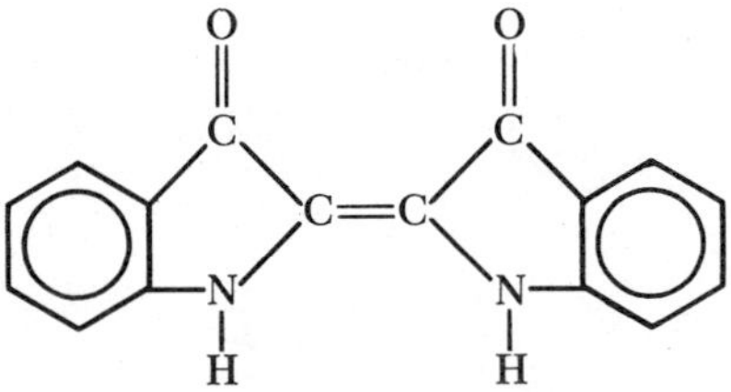

the indigoid group

Each of these structures has characteristic **absorption bands,** but to make a successful dye out of any one of them requires *two other* molecular features:

1. a chemically reactive group which will undergo reaction with functional groups on the fiber (such as the —OH groups of cotton or the amide groups of wool), thereby affixing the dye to the cloth, and
2. a group which alters the structure in a subtle way, causing the absorption bands to be shifted up or down the wavelength scale in order to achieve exactly the shade or hue which is desired. Such a group is called an **auxochrome**.

In addition to these essentials, it may also be necessary to add a hydroxyl group or a sulfonic acid group (or the like) to make the dye sufficiently water soluble to be useful.

Given all these components, the molecular architect puts them together to get the desired color, the necessary adhesion to the fiber, and the light-fastness that are required. It might take many trials, but he is guided by accurate measurements with spectrophotometers, as well as by extensive experience. When he comes up with a really successful dye, it goes into the Color Index[18] and serves as part of the huge fund of general knowledge about dyes.

The structures of most commercial dyes are very complicated, and there is no need to go into detail about them here. Our present purposes are served, instead, by considering the relationship between chromophore and auxochrome in one simple series of azo dyes that can be made quite easily, even as a lecture demonstration.

We start by making aniline, and from it phenyldiazonium chloride

$$\text{benzene} + HNO_3 \xrightarrow{H_2SO_4} \text{nitrobenzene-}NO_2 + H_2O$$

$$\text{-}NO_2 + 3H_2 \text{ (from Fe + HCl)} \longrightarrow \text{-}NH_2 \text{ (aniline)} + 2H_2O$$

$$\text{-}NH_2 + NaNO_2 + 2HCl \longrightarrow \text{-}N\equiv NCl \text{ (phenyldiazonium chloride)} + 2H_2O$$

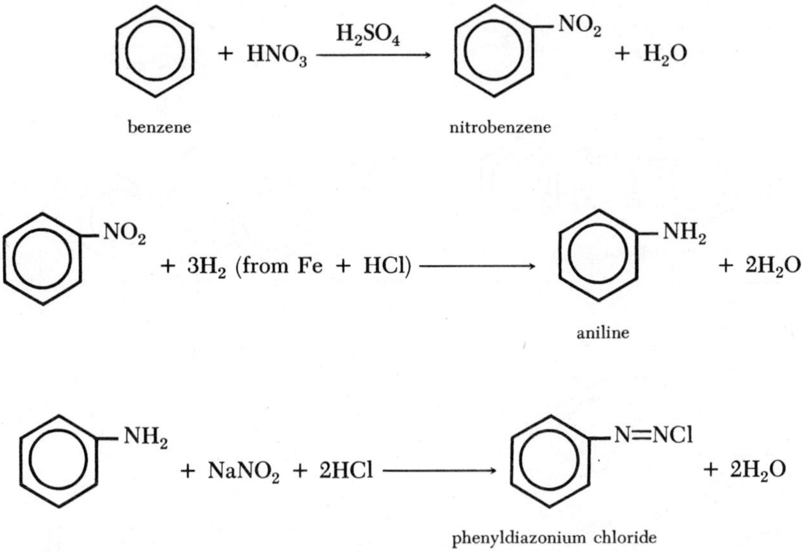

The diazo compound is then "coupled" with an amine or phenol, thereby supplying the desired auxochrome group to influence the shade:

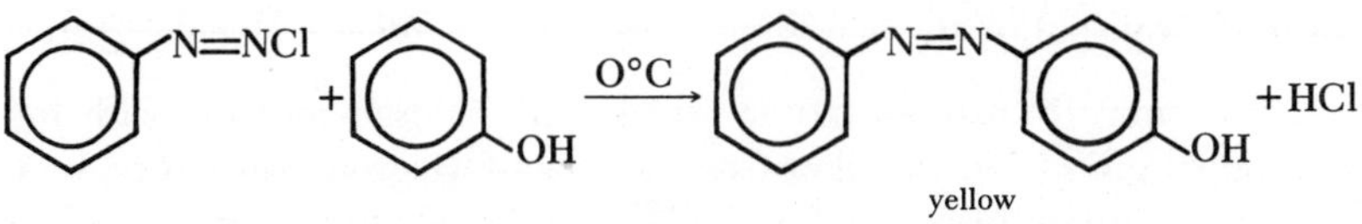

$$\text{-}N\equiv NCl + \text{-}OH \xrightarrow{0°C} \text{-}N=N\text{-}\text{-}OH \text{ (yellow)} + HCl$$

---

[18] See *McGraw-Hill Encyclopedia of Science and Technology*, New York: McGraw-Hill, "Dye," vol. 4, pp. 297–310 (1960) for explanation of the Color Index and for practical information about dyes.

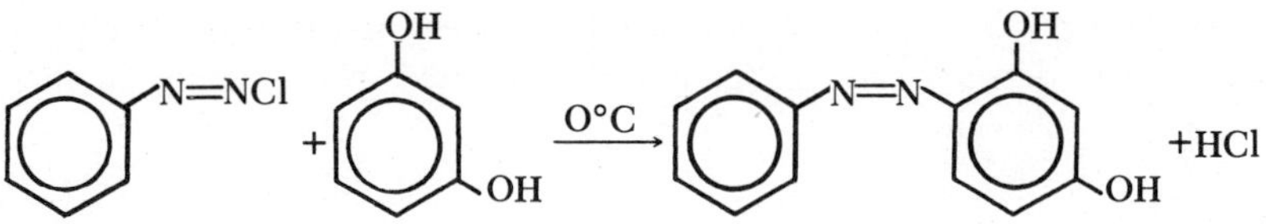

[Reaction scheme: phenyl-N=NCl + resorcinol → azo dye (bright red) + HCl at 0°C]

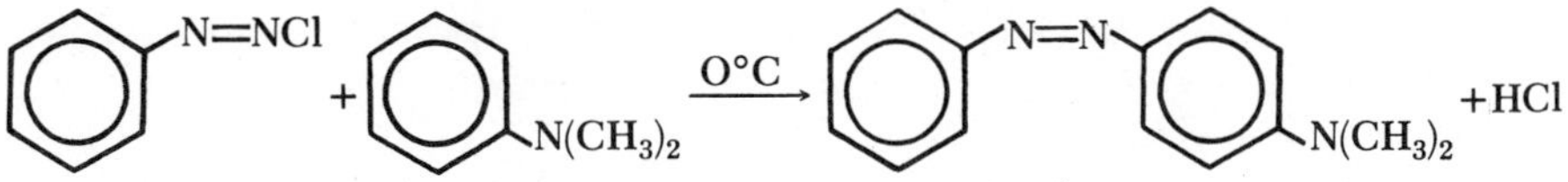

[Reaction scheme: phenyl-N=NCl + N,N-dimethylaniline → azo dye (deep orange) + HCl at 0°C]

The resulting three azo dyes show how color is achieved and varied through the incorporation of chromophore and auxochrome groups. These three were chosen for their simplicity of structure; they are not competitive commercial dyes because they have insufficient solubility in water,[19] and they lack good fixative properties. Such features could be added, of course.

15-7 Flavorings, Perfumes, and Such: Challenges for the Future

The first four topics pursued in this Chapter were all concerned with the design of materials of improved mechanical strength, or improved elasticity, by incorporating certain molecular features. The last topic was concerned with obtaining desired colors by manipulating the molecular structure of the light-absorbing system. In all cases, the reader should notice that the chemist is aided in his search for better materials by reliable measuring instruments, which enable him to evaluate the result of any change in molecular design he may incorporate. Further back than that, there is a body of reliable theory behind the design of the instruments, theory that relates the property being measured to the molecular structure of the sample. Such a situation does not exist in some other fields. For example, there is no comprehensive theory concerning the sensation of taste. There are widely different substances which taste sweet; think of sucrose (cane sugar) and saccharin:

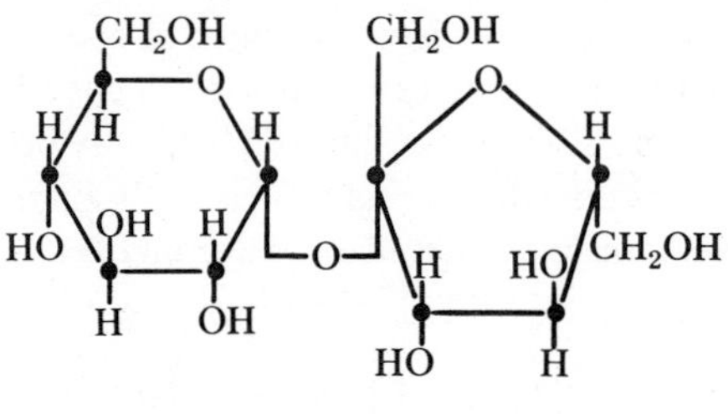

[Structures of sucrose and saccharin]

sucrose
(each ● represents a carbon atom)

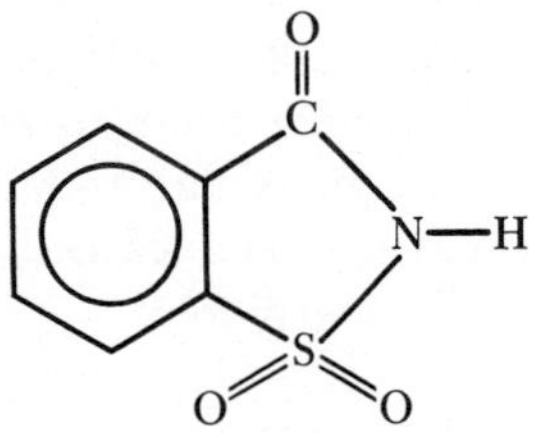

saccharin

---

[19] If these azo dyes are made as a lecture demonstration, they may be dissolved in methanol to show their true colors.

No one knows why sugar and saccharin taste sweet, nor how the sensation of sweetness is detected and registered. Or consider two isomers of aminonitropropoxybenzene:

2-amino-4-nitro-propoxybenzene (4000 times sweeter than sugar)

4-amino-2-nitro-propoxybenzene (tasteless)

Exchange the positions of the amino and nitro groups on the ring, and the sweetness is gone. Or make both substituents nitro groups, and the compound turns out to be bitter! Imagine the frustration of trying to lay out a strictly logical research program on "artificial" sweeteners without having a comprehensive theory on which to base the program. And there is no such theory at this time.

Similarly, there are widely different chemical compounds that have almost the same odor, and there are compounds with closely similar molecular structure that have widely different odors. There is as yet no general theory to explain the sense of smell,[20] and hence there is no systematic guide for the chemist who would synthesize perfume constituents. Molecular architecture has its present-day limitations. Huge areas remain to be explored, and chemistry should move forward together with physics and physiology and psychology into these areas of ignorance. Very likely the design of materials, through understanding and manipulation of molecular architecture, is just in its infancy. Much that is new and fascinating and useful awaits the synthesist.

15-8 Summary

We have seen that many types of new materials can be designed on the molecular level to have certain properties and to meet a variety of special and particular needs. In drawing up plans for making new materials, a chemical designer needs to know as much as possible about the relationships between molecular architecture and the properties of substances. Although much has been learned about these relationships, vast areas still remain virtually unexplored.

---

[20] For various hypotheses, see *McGraw-Hill Encyclopedia of Science and Technology*, New York: McGraw-Hill, "Smell," vol. 12, p. 384 (1960); and "Taste," vol. 13, p. 399 (1960). See also E. S. Hodgson, "Taste Receptor," *Scientific American*, vol. **204**, no. 5, p. 135 (May 1961); A. D. Hasler and J. A. Larsen, "The Homing Salmon," *Scientific American*, vol. **193**, no. 2, p. 72 (August 1955); and A. J. Haagen-Smit, "Smell and Taste," *Scientific American*, vol. **186**, no. 3, p. 28 (March 1952).

Many factors are involved in the design of materials in addition to the primary task of fitting the synthetic material to its intended use. Supplies of raw materials are seldom inexhaustible, and only with careful planning can certain critical resources be conserved, recycled, or diverted to the most socially advantageous usages. Manufacturing processes can introduce pollutants into the environment and can consume substantial quantities of irreplaceable fossil fuels. It is clear that design of new materials, besides taking into account the overall impact of the introduction of those materials, must be intimately involved with development of manufacturing processes. Finally, most materials are fabricated into articles which wear out, and the designer must consider the implications of various methods of disposal or reutilization of the materials.

We have examined the interplay of these various chemical factors through case studies of the design of metals, glass, plastics, rubber, and dyes. We have seen that a designer of materials must take into account engineering, economic, social, and artistic considerations as he works with these new chemicals. We have observed how the design of materials crosses the historical but outmoded lines between organic and inorganic chemistry. Finally, we have speculated that this young art and science has only just begun a long, and we expect, spectacular trip into a world of the new materials that will be the stuff for constructing the things of the future. Within certain limitations, these new materials can have properties which we specify. It is important that chemists learn to make materials according to new specifications. It is even more important that all people learn to think critically about the properties that are desirable for new materials, and about the economic, social, and environmental costs of achieving otherwise desirable properties.

GLOSSARY

absorption bands: those regions of the visible or infrared or ultraviolet spectrum that are reduced in intensity as the radiation passes through a substance. Absorption bands are recorded as peaks by a spectrophotometer (an instrument which records radiation intensity versus wavelength) or show up as gaps in the photographed spectrum.

alloy: a solid solution of one metal in another (as of copper in silver to make sterling silver), or of several metals (as in stainless steel, an alloy of Fe, Ni, and Cr), or of an intermetallic compound dissolved in an elementary metal (as in brass).

auxochrome: a grouping of atoms that intensifies the absorption of specific wavelengths by a chromophore, or that changes the electron density within the chromophore sufficiently to shift the absorption maximum and thus modify the color.

chromophore: a grouping of atoms that is specifically responsible for the absorption of some portion of visible light by a compound. Thus, all dyes must contain chromophoric groups.

conchoidal fracture: the shell-like swirls noticed on a piece of broken glass; random breakage along curved surfaces, rather than along definite planes established by crystal structure.

copolymer: a polymeric material made by the interpolymerization of two or more monomeric substances, both of which enter into the polymerization reaction under the conditions imposed on the mixture.

disproportionation: an oxidation-reduction reaction in which one molecule of a particular compound is oxidized at the expense of another molecule, which is reduced; an autooxidation (see Section 8-6).

dyestuff: any compound which will impart desired color to a textile fiber, preferably by chemical combination with active groups on the surface of the fiber.

elastomer: a polymeric substance (usually synthetic) which has (or generates) elastic properties and can be used in suitable admixture to manufacture useful rubbery articles.

glass: a general term for an isotropic, noncrystalline, vitreous solid (usually transparent and inorganic), of which window glass and bottle glass are but examples from the silicate field. Pyrex is an example of a borosilicate glass; there are phosphate, borate, and germanate glasses as well. All glasses are supercooled (or undercooled) liquids, brittle when cold but plastic when heated sufficiently.

intermetallic compound: a specific compound of two metals (seldom more), or of a metal with a metalloid, without benefit of a nonmetallic element. Examples are Mg_2Ni, Cu_5Zn_8, $LiAg$, and Cu_3Si.

isotropic: having the same physical and optical properties in all directions; invariant with respect to direction.

macromolecular: composed of very large molecules, usually of average molecular weight in the thousands.

silicone: a polymeric organosiloxane; a synthetic polymer (oil, resin, or rubber) having a framework of alternate silicon and oxygen atoms, the silicon atoms of which bear organic (usually methyl) groups.

siloxane: a compound having a sequence of alternate silicon and oxygen atoms as its principal feature. In an organosiloxane, the silicon atoms also bear organic groups, attached directly by C—Si bonds.

vulcanizing: earlier, the process (invented by Charles Goodyear) for converting sticky, flowable crude rubber into elastic, nonsticky articles by heating the milled rubber mixture with sulfur; more recently, any suitable chemical process by which the molecular chains of an elastomer become cross-linked, resulting in useful articles of the corresponding rubber.

PROBLEMS

15-1 (A reading project) Considering the world's resources, its food supply, and the environment, and considering also the molecular-structural qualifications of a cleaning agent (in terms of today's empirical principles), design a detergent that will be satisfactory 20 years from now.

15-2 (Another reading project) Paper is a fairly satisfactory material, since it can be recycled, is biodegradable, and can be fabricated in many useful forms by a well-established technology. But paper comes from wood, and wood contains a large proportion of lignin as well as the cellulose fibers which go into paper.

Look up the composition of lignin, find out what makes it so objectionable a pollutant when it is discharged into our streams, and "dream up" a list of beneficial ways in which lignin might be put to good use in the future.

15-3 Plastics are so varied in composition and properties that articles made from them cannot all be mixed together for recovery and recycling. For this reason, it is desirable for some end uses of plastics that they be biodegradable. What are the first structural requirements for biodegradability, as revealed by today's biodegradable detergents? What kind of polymer would you choose today as the best candidate for modification to achieve biodegradability?

15-4 As seen in this chapter, titanium has come from the ranks of the abundant but unfamiliar (unused) metals to a position of some usefulness. What other metallic elements are abundant enough, supposedly strong enough, and chemically stable enough to warrant study of their possible extraction and use on a large scale?

SUGGESTIONS FOR FURTHER READING

Metals. The occurrence, extraction, refining, and testing of metals are treated succinctly in the *McGraw-Hill Encyclopedia of Science and Technology*, New York: McGraw-Hill, vol. 8, pp. 266–290 (1960). Alloys are considered in vol. 1, pp. 258–265, of the same encyclopedia. See also: LEIGHOU, R. B., *The Chemistry of Engineering Materials*, 4th ed., New York: McGraw-Hill (1942).

Glasses. An authoritative account of the scientific (as distinct from the artistic) aspects of glass is given in the two-volume work by: WEYL, W. A., and MARBOE, E. C., *The Constitution of Glasses: A Dynamic Interpretation*, New York: Interscience (1962). Volume 1 covers the constituents of inorganic solids and liquids, the interaction of these constituents, and the conditions for forming glasses. Volume 2 deals with the properties of glasses as functions of composition and temperature. See also: MOREY, G. W., *Properties of Glass*, 2nd ed., New York: Reinhold (1954).

Plastics and Polymers. A good treatment at the upperclass level is given by: BILLMEYER, F. W., in *Textbook of Polymer Science*, 2nd ed., New York: Interscience (1962). All you may ever want to know about polymers is covered in the 14-volume *Encyclopedia of Polymer Science and Technology*, New York: Interscience (1964–1970).

General. An interesting treatment of natural products is given by BROWN, HARRISON, "Human Materials Production as a Process in the Biosphere," *Scientific American*, vol. 223, no. 3, p. 194 (Sept. 1970).

Almost all materials and substances (including pure water) can be harmful to the human body if handled improperly, or ingested in too large a quantity, or allowed to come in contact with sensitive parts of the body. Some substances, of course, are far more harmful or poisonous than others. It is well to know about the dangers inherent in the use of any material, or of any compound used in the manufacture of a common material. If you are not sure, LOOK IT UP! A good one-volume source is: SAX, N. I., *Dangerous Properties of Industrial Materials*, 3rd ed., New York: Van Nostrand (1968). You will be surprised at the misconceptions held by the general public.

16
Measurements on Chemical Systems: The Chemical Study of What is in Sea Water, and How Much

16-1 Abstract

Many words have been spent in a dozen earlier chapters on the fundamental principles (or "laws") that describe the behavior of matter. While helpful in understanding the universe and in meeting most of the requirements of a first course in chemistry, these principles by themselves lack direction. In this book, the study of such principles was given some direction at the outset by considering mankind's No. 1 problem (Chapter 1) and what things chemical can be done about it. All the terms and principles necessary to a full understanding of Chapter 1 have been covered and explained in detail by now, and the "what to do" about many other current problems has been woven into the subsequent chapters. For instance, the big question of what materials we can make, and at what cost to our resources and to the environment, was considered in Chapter 15; what to do about future energy requirements was taken up in Chapters 8, 9, and 10. Hundreds of other matters relating closely to everyday life are sprinkled throughout all the chapters.

But even if *all* that is actually known about how we are to feed, clothe, and house tomorrow's billions were written out (say in a hundred volumes), it would still be only half of chemistry. This is so because chemistry (like all other sciences, be they physical, social, or political) has two equally important sides: *analysis* (or taking things apart to learn what makes them tick; collecting facts in preparation for action), on the one hand, and *synthesis* (putting things together; designing and making new materials; taking constructive action), on the other. So far, this book has leaned mostly toward synthesis (note especially Chapter 15) and has introduced analysis here and there only to establish a background for some point of principle. Now at last we have a chapter devoted entirely to analysis. And following precedents established in the earlier chapters, this one does not present a list or outline of all methods of analysis, but chooses one important problem and illustrates how some selected modern methods of analysis can be brought to bear on that problem to yield dependable facts. The problem is a basic and highly relevant one: the analysis of *water*, that pervasive, essential, and abundantly scarce stuff. The actual examples are taken from the analysis of sea water, but the same methods apply to the study of "fresh" water, brackish and contaminated lake water, and, of course, to drinking water. The methods are (1) **gravimetric analysis**, which involves weighing an isolated reaction product; (2) **titration**, or **volumetric analysis**, in which the volume of reacting solution (of known concentration) is measured at some precise **end point** of the reaction; (3) **electrochemical analysis**, in which the concentration of some ion of interest is measured by ingenious ion-selective electrodes, based on the principles of Chapter 8; (4) **spectrophotometry**, in which the quantitative absorption of light (or ultraviolet or infrared radiation) at a selected frequency leads to a measurement of concentration of a particular element or compound; and **(5) chromatography,** an especially versatile method in which tiny amounts of pure substances are separated from a complex mixture and then are identified by ingenious physical methods.

These five general methods, used in concert, enable us to discover everything that may be in water: salts, acids, bases, dissolved gases, organic and inorganic pollutants, pesticides, and poisons. Variations of the same methods enable us to analyze fish, meat, fruit, grains, and other foods. The bare figures obtained from such analyses may mean little, however, unless the **reliability** of the methods is established through a careful study of the **systematic errors**, the **random errors**, and the **sampling errors**. Only after such a study can the analyst certify the degree of reliability of his measurements. The reader (as a member of the concerned public) should learn to assess the figures he hears or reads about, in terms of the percentage of error, before he forms his opinion or goes flying off into action.

16-2 The Significance of Chemical Measurements

Measurements on chemical systems are very much in the news these days. The pollution of the environment, the exhaustion of mineral resources, the estimation of world food needs, and many other current issues center around the results of chemical *analyses* of air, soil, water, ores, and foodstuffs. If we are to think clearly and act rightly on these issues, it ought to be on the basis of a clear understanding of the facts. For these, we need to know about the methods, the meanings, and the limitations of these chemical measurements. For example, it has recently become evident that many food fish in America are alarmingly contaminated by heavy metals and pesticide residues. Reasonable civic response to the situation depends on answers to questions like the following:

1. How much of what poisonous metals and pesticide residues are present in the fish intended (or available) for human consumption?
2. How accurate are the published values for these contaminant levels (that is, the values determined by public health agencies)?
3. Are the analytical methods used subject to interferences (that is, does the presence of harmless substances cause a false reading)?
4. Have the metal contaminants always been present in fish, or do recently caught fish have significantly more of these metals than older specimens taken from museums? What is the "allowable" level for each contaminant?
5. What are the specific sources of the contaminant? Where does the polluted water come from?
6. Are the fish used in the measurement "typical," or did the sampling process bias the result?
7. Are the contaminants present throughout the flesh of the fish, or are they concentrated in specific organs or tissues, such as those not ordinarily eaten?

The business of answering questions like these, and of developing new methods to meet new questions, constitutes the science of *analytical chemistry*. Clearly, the analytical chemist will play a vital role in our efforts to bring under control our technology's impact on nature. In this chapter, we shall survey briefly some old and new methods of analytical chemistry, using as a vehicle an imaginary (but realistic) chemical characterization of a sample of sea water. The list of methods we shall describe is by no means a complete survey of modern analytical chemistry, but is intended to give you some impression of the ways in which much of the chemical theory we have developed in this book contributes to the solution of a definite, practical problem.

16-3 Methods Based on Reaction Stoichiometry

Stoichiometric methods constitute the oldest analytical techniques still in use. They are based on the fact that substances react in definite and simple proportions (Chapter 3). **Gravimetric analysis** is based on the measurement of the quantity of product formed by the substance to be determined (that is, what we are interested in measuring) when it reacts with an excess of a precipitating agent. **Volumetric analysis** is based on the measurement of the volume of a standard reactant (a known reagent used in known concentration) required to react with a sample. Because stoichiometric methods are simple, reliable, and (in favorable cases) capable of high precision, they continue to be the most frequently used, in spite of their relatively ancient origins.

THE CHLORIDE ION CONTENT OF SEA WATER: A GRAVIMETRIC DETERMINATION.

In this method, a sample of sea water is treated with an excess of silver nitrate ($AgNO_3$) solution, after being made acidic to prevent the formation of silver oxide (because sea water is naturally alkaline). Silver chloride is very insoluble in water, and virtually all of the chloride ion in the sample is precipitated as AgCl. This solid is filtered out of the reaction mixture, washed, dried, and weighed. From the mass of AgCl found, the number of moles of AgCl (and thus of Cl^- originally present in the sample) may be calculated.

Pure solid AgCl with almost all the sample's chloride is separated from the sample and weighed.

If the chloride ion present in 10.00 ml of sea water produced 0.7868 g of pure dry AgCl, what was the molar concentration of Cl^- in the sample?

$$\text{Molarity} = \text{moles } Cl^-/\text{liters of sample}$$

$$= \frac{\text{moles AgCl}}{\text{sample volume}}$$

$$= \frac{\text{mass of AgCl/formula wt of AgCl}}{\text{sample volume}}$$

Consulting the table of atomic masses (Chapter 2, Table 2-1), we may insert the appropriate numbers to obtain

$$M_{Cl^-} = \frac{0.7868 \text{ g}/143.323 \text{ g-mole}^{-1}}{0.01000 \text{ liter}}$$

$$= 0.549 \text{ mole/liter chloride}[1]$$

---

[1] This corresponds to 3.2 percent of NaCl in the sea water, which is about typical for waters of the Atlantic and Pacific Oceans. Some seas, like the shallow Baltic, contain less salt; others, like the landlocked Dead Sea, are much saltier.

Other ions that form insoluble salts with Ag^+, such as OH^-, Br^-, I^-, and so forth, interfere with this method by adding to the mass of the precipitate. Interference by OH^- was suppressed by acidifying the sample, and none of the others is present in sea water to a large enough extent to invalidate the above result, which is, in fact, typical of water from the major oceans.

TOTAL ALKALINITY BY ACID-BASE TITRATION. The total alkalinity of a natural water sample is defined as the total moles of a strong Brønsted acid (Chapter 6) that are required to react completely with all of the Brønsted bases in a liter of sample. Since the greater part of the alkalinity of a natural water sample is almost always represented by carbonate (CO_3^{2-}) and bicarbonate (HCO_3^-) ions, the sample is considered to be completely neutralized when enough strong acid is added to reach the pH of a carbonic acid (H_2CO_3) solution.[2] Generally, an acid-base indicator is used to signal this condition; the indicator methyl red, for example, turns from yellow to red at the proper pH.

An acid solution (of known concentration) in a buret is added to the sample until the indicator turns red.

Suppose, for example, that a 100-ml sample of sea water required 11.3 ml of a standard 0.00978 M solution of HCl to reach a methyl red endpoint. What is the total alkalinity of the sea water?

$$\text{alkalinity} = \frac{\text{no. of moles HCl to titrate}}{\text{volume of sample}}$$

$$= \frac{\text{volume of HCl soln.} \times \text{molarity of HCl}}{\text{volume of sample}}$$

$$= \frac{(0.0113 \text{ liter})(0.00978 \ M)}{(0.100 \text{ liter})}$$

$$= 0.00111 \ M$$

MAGNESIUM AND CALCIUM IONS BY EDTA TITRATION. Coordinating compounds such as ethylenediaminetetraacetate (EDTA) were introduced in Chapter 13. They have found wide usefulness as voracious reactants with Lewis-acid metal ions. Even relatively poor Lewis acids such as Ca^{2+} and Mg^{2+} react with EDTA to form stable complexes:

$$Mg^{2+} + HEDTA^{3-} \rightleftharpoons MgEDTA^{2-} + H^+$$

(The EDTA is shown in the form of its conjugate acid, which predominates in the pH range in which these titrations are usually done.) To signal the endpoint of the titration, a *metallochromic* indicator is often used. This is a colored Lewis base, whose complex with a metal ion is of another color. The most commonly used indicator has the chromophoric structure of an azo dye (see Chapter 15)

The volume of EDTA solution needed to give a color change is measured.

---

[2] By means of algebraic methods introduced in Chapter 7, we may calculate that a solution of carbonic acid in equilibrium with the CO_2 of the air should have a pH of about 4.5 at 25°C.

eriochrome black T

and bears the name Eriochrome Black T (EBT). At the beginning of a titration of magnesium ion, for example, all of the indicator is present as the wine-red complex $MgEBT^-$. When all of the free Mg^{2+} in the sample has been titrated with EDTA, the next drop of EDTA solution added causes the equilibrium

$$HEDTA^{3-} + MgEBT^- \rightleftharpoons MgEDTA^{2-} + HEBT^{2-}$$
$$\text{(red)} \hspace{4cm} \text{(blue)}$$

to shift to the right, and the color of the solution changes from wine red to blue.

Both magnesium and calcium ions are present in appreciable concentrations in the sea, and both react readily with EDTA. Therefore, titration of a sample of sea water consumes a number of moles of EDTA equal to the *sum* of the number of moles of Mg^{2+} and Ca^{2+} present. This sum can be resolved into its components by means of a simple chemical trick: The addition of sodium oxalate ($Na_2C_2O_4$) to a fresh sample results in the precipitation of insoluble calcium oxalate, CaC_2O_4, while the magnesium ions remain in solution and can be titrated alone. The calcium content of the water is then calculated as the difference in the two titrations: the first yielding a value for the sum of Ca^{2+} and Mg^{2+}, and the second the value for Mg^{2+} alone (see Figure 16-1).

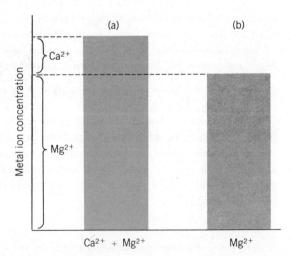

FIGURE 16-1 Stoichiometric relationships in the titrations of Ca^{2+} and Mg^{2+}. (a) Titration of a mixture of calcium and magnesium ions consumes a quantity of titrant proportional to the sum of the concentrations $[Ca^{2+}]$ and $[Mg^{2+}]$ in the sample. (b) After removal of Ca^{2+} by precipitation as CaC_2O_4, a second titration consumes titrant proportional to $[Mg^{2+}]$ alone. The value of $[Ca^{2+}]$ in the sample is calculated from the difference in these two titrations.

A 10.00 ml sample of sea water is adjusted to pH 10 with an ammonia-ammonium ion buffer (Chapter 7); a few drops of EBT indicator solution are added; and the solution is titrated with standard 0.02004 M EDTA solution (prepared by dissolving a weighed portion of the disodium salt $Na_2H_2EDTA \cdot 2H_2O$). A volume of 32.03 ml is required to reach the red-to-blue color change. A second 10.00 ml sample is treated with excess sodium oxalate; the precipitated CaC_2O_4 is removed by filtration and washed (the washings being combined with the original filtrate). The accumulated filtrate is then buffered and titrated as before. This time, with the Ca^{2+} removed, the sample requires 27.10 ml of 0.02004 M EDTA to reach the red-to-blue endpoint. What were the molar concentrations of magnesium and calcium ions in the sea water?

The first titration provides a value for the sum of the two concentrations. Since both Mg^{2+} and Ca^{2+} react in a one-to-one ratio with EDTA, the number of moles of EDTA used in the titrations is equal to the number of moles of metal ion titrated. (The concentrations, of course, are given by the number of moles of metals per liter of sample.) In the first titration,

$$\text{total } ([Ca^{2+}] + [Mg^{2+}]) = \frac{\text{no. of moles EDTA used}}{\text{volume of sample}}$$

$$= \frac{(0.03203 \text{ liter})(0.02004 \ M)}{0.01000 \text{ liter}}$$

$$= 0.06419 \ M$$

In the second titration, by the same reasoning,

$$[Mg^{2+}] = \frac{(0.02710 \text{ liter})(0.02004 \ M)}{0.01000 \text{ liter}}$$

$$= 0.05431 \ M$$

Subtracting the Mg^{2+} concentration from the sum of the two produces a value for $[Ca^{2+}]$:

$$[Ca^{2+}] = 0.06419 - 0.05431$$

$$= 0.00988 \ M$$

As in our earlier examples, these values are typical for water from the major oceans.

16-4 Ion-Selective Electrodes

In Chapter 7, we discussed the use of pH as a master variable to characterize acid-base equilibria in solutions. Our titration for total acidity in Section 16-3 tells us something about the acid-base properties of ocean water. However, such a titration says nothing about the pH of the water in its original state, and thus its effect on other acid-base equilibria (such as the respiratory chemistry that introduced Chapter 7). The pH of the ocean is very important to its capacity for sustaining life, and a pH measurement would be an important part of the chemical characterization of sea water.

A visit to a water-quality laboratory would reveal a collection of instruments labeled "pH meters." These instruments were described briefly in Chapter 8; their applicability has proved to extend beyond the measurement of pH alone. Let us review the principles of operation of a pH meter, and we shall see that those principles may be extended to the measurement of many other ionic concentrations.

The use of a pH meter is diagrammed in Figure 16-2. The pH of a solution is proportional to the electrical potential difference between two electrode systems (which in some commercial designs are incorporated into one glass sleeve). Because this assembly is, in reality, capable of estimating only the *difference* of pH of two solutions, one or (preferably) more standard solutions of known pH must be used to calibrate the electrical response of the electrodes.[3]

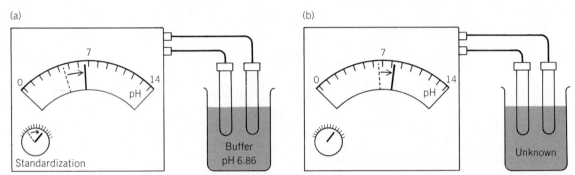

FIGURE 16-2 Use of a pH meter. (a) The measuring and reference electrodes are immersed in a standard buffer of a defined pH value near that expected for the unknown. The voltage difference between the electrodes is proportional to the pH of the solution and is displayed as pH on the meter. The meter scale is "standardized" electrically by adjusting a standardization control until the meter reads the defined pH of the buffer. (b) Without changing the standardization setting, the buffer is removed, and the electrodes are rinsed clean and immersed in the unknown solution. The meter reads the pH of the unknown solution; in this illustration, the unknown has a pH of 8.1.

A closer look at a pair of pH electrodes (Figure 16-3) reveals that they consist of two internal electrodes that are permanently surrounded by their own internal solutions. Each of these internal electrodes is constructed so as to have a constant potential relative to its surrounding internal solution. One (the *reference electrode*) is there only to complete a circuit between the meter inputs and the sample through a *liquid junction* that is designed to maintain a small, nearly constant electrical potential between its internal solution and the sample. The other (the variously named *measuring, indicator, membrane,* or *ion-selective electrode*) develops an electrical potential that is proportional to the pH difference between its internal solution and the sample. Recall (Chapter 8) that a concentration difference between two solutions of the same ion implies an electrical potential, through the Nernst equation, of $0.059/n$

---

[3] Buffers (Chapter 7) of composition and pH specified by the U.S. National Bureau of Standards are generally used. See A. Wilson, *pH Meters*, New York: Barnes & Noble, pp. 77–78 (1970).

V for every power-of-ten difference in the concentrations, where n is the charge on the ion. In the case of a pH electrode, the membrane is a thin glass bulb, and the internal solution is dilute hydrochloric acid.

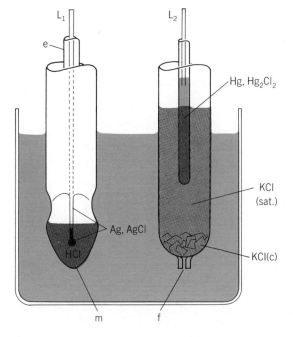

FIGURE 16-3 A cell for determining the pH of the solution in the beaker. The pH-dependent potential is established across the glass membrane m. The cell potential is measured between the leads L_1 and L_2 by means of an electronic voltmeter calibrated in pH (that is, a pH meter). Because of the high resistance of the glass membrane, L_1 is electrically shielded by the conducting sleeve e (which is isolated from L_1). The liquid junction between the KCl solution and the external solution, whose pH is being measured, is established at an asbestos fiber, f.

A glass-membrane pH electrode responds *selectively* to the difference in pH between the internal solution and the sample. Its response is directly proportional to pH, and nearly independent of other changes in the sample's composition, over a range from about pH 0 to pH 12 or 13. Useful response continues to and slightly beyond pH 14; that is, to hydrogen ion concentrations less than 10^{-14} moles/liter.[4] Such a large responsive range is made possible by the membrane's selectivity; of all the species that might be present in a sample, only hydrogen ions (and, very weakly, sodium ions) have any effect on the electrode response. The reasons for membrane selectivity are still incompletely understood; you will find a useful survey of theories and applications among the "Suggestions for Further Reading" at the end of this chapter.

At the present writing, ion-selective electrodes for over 20 species are available. Another typical and useful member of this group is the nitrate ion electrode shown in Figure 16-4. Note that the essential features seen in the

---

[4] This useful response range of over 14 powers of ten in H^+ represents a remarkable feat of measurement—roughly comparable to that of a ruler graduated in millimeters at one end and astronomical units at the other.

glass-membrane pH electrode are also present in this case, though in a somewhat altered form. The most straightforward technique for measurement with electrodes like this one is to prepare a number of standard solutions of the ion to be measured, and then to interpolate the measurement of an unknown sample. Since the electrode response is proportional to the logarithm of the measured ion's concentration (through the Nernst equation), it is useful to plot the known points on a logarithmic graph, since the result is a straight-line standard.[5]

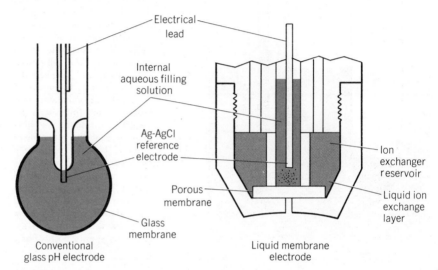

FIGURE 16-4 Comparison of conventional pH electrode and liquid-membrane "ion-selective" electrode. The theory of operation of these two electrodes is nearly the same, and their structures are analogous. The pH electrode produces an electrical signal with a voltage proportional to the difference in pH between the internal filling solution and the external solution, which are separated by the glass membrane. This ion-selective nitrate electrode produces an electrical signal with a voltage proportional to the difference in $-\log[NO_3^-]$ (pNO_3) between the internal filling solution and the external solution, which are separated by a porous membrane impregnated with a nitrate-containing liquid ion exchanger. See "Suggestions for Further Reading" for additional details. (Drawing courtesy of Orion Research, Inc.)

The results of electrode measurements on known potassium nitrate solutions in otherwise nitrate-free artificial sea water, and in a real sea water sample, are given in Table 16-1. Since most ion-selective electrodes are not as ideally selective as the glass pH electrode, all water samples are first treated with silver sulfate to precipitate (as AgCl) the chloride ion in them; otherwise, the nitrate electrode's response would have been primarily determined by its (relatively slight) sensitivity to the large concentration of chloride ion in sea water.

---

[5] Logarithms and graphs are discussed in Appendix II.

TABLE 16-1
Nitrate Ion-Selective Electrode Potentials in Artificial Sea Water Containing Known Nitrate Ion Concentrations, and in a Natural Sea Water Sample

$[NO_3^-]$	Electrode Potential[a]
0.03 M	0.000 V
0.01 M	0.028 V
3×10^{-3} M	0.059 V
1×10^{-3} M	0.087 V
3×10^{-4} M	0.118 V
1×10^{-4} M	0.147 V
3×10^{-5} M	0.173 V
1×10^{-5} M	0.191 V
3×10^{-6} M	0.206 V
1×10^{-6} M	0.210 V
Unknown (sample of natural sea water)	0.181 V

[a] Potentials measured relative to a reference electrode immersed in the same sample.

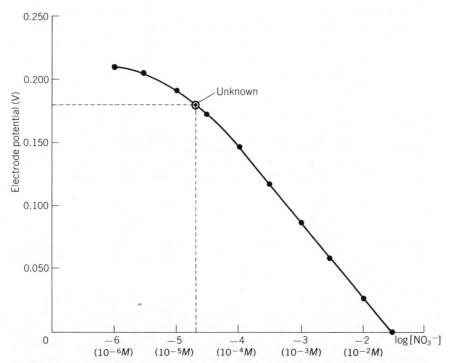

FIGURE 16-5 Calibration curve for a nitrate-selective electrode. Electrode-potential values observed in known solutions of NO_3^- are denoted by ●; the unknown is indicated by ⊙.

The data from the known solutions, and from the real sample, are plotted in Figure 16-5. Considering first the appearance of the standard curve defined by the known points, we can see that the electrode response is linearly related to the logarithm of the nitrate ion concentration over a range that extends down to about 10^{-4} M, and that the electrode response is nonlinear, though still useful, down to just above 10^{-6} M. Below 10^{-6} M, further decreases in nitrate ion content cause no further change in the electrode potential. The unknown sample, producing an electrode potential of 0.181, falls in the non-linear, but still usefully sensitive, range of the electrode. Thus, we may read from the graph that the logarithm of the value of $[NO_3^-]$ in this particular sea water sample was -4.70. Mathematical adepts will confirm that this result is equivalent to

$$[NO_3^-] = 2.0 \times 10^{-5} M$$

Such a result is typical of sea water samples taken in mid-ocean, away from large nutrient inputs such as coastal sewage dumps.

THE OVERALL COMPOSITION OF THE OCEANS. After examining all the foregoing methods for determining the concentrations of various ions in sea water and learning their average values, it is interesting to consider all the water in the oceans as a whole, and to learn what resources are stored up there. The major oceans are interconnected, so they mix together and have about the same composition, as is shown in Table 16-2. All the oceans, taken together, cover 145,000,000 square miles of the earth's surface to an average depth of 2.2 miles. Hence there are *320 million cubic miles* of sea water, and each cubic mile of this vast quantity contains useful materials to the extent shown in Table 16-3.

All the information about chemical composition of the seas comes from chemical analysis.

TABLE 16-2
Composition of the Solids Obtained by Evaporating Water From Various Oceans

	Na^+	K^+	Ca^{2+}	Mg^{2+}	Cl^-	Br^-	SO_4^{2-}
	Percent						
Atlantic	30.6	1.1	1.2	3.7	55.3	0.19	7.69
Arctic	30.9	0.9	1.2	3.8	55.3	0.14	7.78
Indian	30.9	1.1	1.2	3.7	55.4	0.13	7.79
Mediterranean	30.5	1.2	1.2	3.8	55.3	0.16	7.72

It is obvious that we are not going to run out of salt, even though people have been taking it out of the sea for thousands of years. Similarly, there is no practical limit to the amount of magnesium that could be removed from the sea for use as structural metal, although only small amounts are being taken out today. Potassium sulfate is an acceptable nutrient for plants; at present we get all we need from underground deposits left from the evaporation of ancient landlocked seas, but we may need to turn to the sea soon.

TABLE 16-3
Wealth of the Oceans: Riches in a Cubic Mile*

NaCl	128,000,000 tons
$MgCl_2$	17,900,000 tons
$MgSO_4$	7,820,000 tons
K_2SO_4	4,068,000 tons
Bromine, as Br^-	350,000 tons
Iodine, as I^-	100 tons
Silver, as Ag^+	280 lb
Gold, in unknown form	9 lb
Radium, in unknown form	5 g

* Each cubic mile of water in the major oceans (there are 320,000,000 cubic miles available) contains these materials in the quantity shown.

Bromine is recovered from the sea in several installations, using cheap chlorine to oxidize Br^- to free Br_2, and air to sweep out the volatile Br_2 from the water. Silver is not yet mined from the sea, but this may soon be necessary because its many uses (photography, electrical equipment, tableware, jewelry) have scattered the metal in permanently dispersed forms, and the ore deposits are no longer equal to the demand; even silver coins have had to be withdrawn and melted down. As for gold, the lure is always there, but the cost of recovery still runs higher than the recovered metal is worth. The gold content of sea water is highly variable, being greatest near the coasts of mountainous areas where gold deposits occur. An ingenious new approach that would trap the dissolved gold in a solid substance (rather than expending expensive solutions to treat large volumes of water) would be most welcome.

16-5 Spectrophotometry: Analysis Via the Absorption of Light

We found in our discussion of wave mechanics (Chapter 2) that the energy of electrons in atoms and molecules is *quantized;* that is, it is restricted to certain values that are characteristic of the structure of the atomic or molecular system in which the electrons are held. This relationship between structure and energy quantization gives rise to the science of *spectroscopy.* Light of the proper energy (frequency) may be absorbed or emitted by an atom or molecule, and the pattern of absorptions or emissions (absorption or emission spectra) may, as we noted earlier, give information about the structure of the atom or molecule involved. In this section we shall show how measurement of *the fraction of light absorbed* by a sample permits quantitative analysis by spectrophotometry. We shall give two illustrative examples.

THE BEER-LAMBERT LAW. It is a familiar experience that a solution of a colored (thus light-absorbing) substance looks darker (absorbs more light) the more concentrated the solution is, and the greater the thickness of the absorbing layer. These qualitative and common-sense observations are described quantitatively by the Beer-Lambert law: For a light-absorbing medium held in a beam of light of a single frequency (Figure 16-6), the relation between the incoming (incident) light intensity and the transmitted (emerging) light intensity is given by

$$\log\left[\frac{I_{\text{in.}}}{I_{\text{trans.}}}\right] = \varepsilon_\lambda \cdot b \cdot c \qquad (16\text{-}1)$$

where $I_{\text{in.}}$ and $I_{\text{trans.}}$ are the incident and transmitted light intensities; b is the breadth of the sample; c is the concentration (the molarity) of the light-absorbing component of the sample; and ε_λ is a constant characteristic of the light-absorbing component and the wavelenght (λ) of the incident light. The quantity $\log(I_{\text{in.}}/I_{\text{trans.}})$ is usually given the name *absorbance, A*.

Ways to obtain light of a single frequency are discussed in Appendix III.

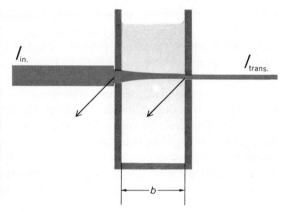

FIGURE 16-6 Absorption of light by a colored solution. A light beam of initial intensity $I_{\text{in.}}$ passes through a transparent cell containing a colored solution. Its intensity (represented in this figure by the thickness of the arrow) decreases because of reflection at the surfaces of the cell and because of absorption (according to the Beer-Lambert law) by the colored solution. The final intensity of the light is $I_{\text{trans.}}$. The light path through the solution is b cm.

When a sample (placed in a transparent cell of known breadth b) is illuminated with light of a chosen wavelength, the difference in intensities of the incoming and transmitted light can be measured (using a photoelectric cell and associated electronic circuitry) and interpreted via the Beer-Lambert law to yield a value for the concentration of the light-absorbing substance.[6]

ATOMIC ABSORPTION SPECTROPHOTOMETRY AND THE DETERMINATION OF MERCURY. At the end of Chapter 2, we mentioned in passing that the characteristic line spectra of the atoms can be used to identify them in natural samples. The Beer-Lambert law may be applied to atomic line spectra to obtain quantitative as well as qualitative data. The technique is then known as *atomic absorption spectrophotometry*. In order to observe or

---

[6] In some cases, more than one component in a solution may absorb light at a given wavelength. In this case, another wavelength may be found at which only the component of interest absorbs. Alternatively, the interfering substances may be removed chemically, or measurements at several wavelengths may be combined algebraically to solve for all unknown concentrations.

use atomic line spectra, the components of a sample must, of course, be present in atomic form; otherwise, molecular, and not atomic, spectra would be observed. Generally, this is done by injecting the sample solution into a high-temperature flame, where all chemical bonds are broken, and the compounds present are literally atomized. In the particular case of mercury, however, another strategy presents itself; elemental mercury is a liquid with a significant vapor pressure. If the sample solution is treated with a reducing agent to convert all ionic or covalently bound mercury to the elemental state, then mercury can be swept out of the solution by a stream of air. This air stream carries mercury atoms into the light path of a spectrophotometer (Figure 16-7). Since no other element or compound combines strong absorption lines in the region near those of mercury with volatility at room temperature, the only component of the air stream capable of absorbing radiation at the frequencies of mercury's line spectrum is mercury itself. And to insure that just those frequencies will be passed through the air stream, a mercury-vapor discharge lamp is used as a light source. By means of this sensitive and selective technique, mercury concentrations as low as one microgram of mercury per liter of water (one part per billion, about 10^{-8} M) may be measured.

The vapor pressure of toxic mercury makes it a hazard when spilled.

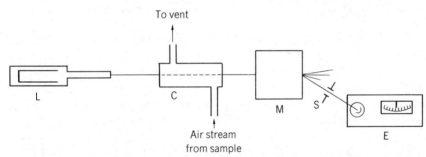

FIGURE 16-7 Atomic absorption spectrophotometric determination of mercury (schematic). The line spectrum of mercury atoms is emitted by electrically excited mercury vapor in the lamp L, and passes through cell C. Mercury vapor from the sample absorbs part of this light according to the Beer-Lambert law, and the remainder enters the monochromator M, where it is separated according to wavelength (compare Figure 2-13). A single wavelength is selected by the exit slit S, and the intensity of the transmitted light is measured by a phototube and associated electronics E.

ELECTRONIC BAND SPECTRA AND THE SPECTROPHOTOMETRIC DETERMINATION OF OXYGEN. When atoms are present in molecules, the sharp line spectra are found to be replaced by diffuse *band spectra*. Typical band spectra are shown in Figure 16-8.

This effect is due primarily to the fact that electronic energy levels in molecules (which would produce line spectra in motionless molecules) are constantly changed by the vibrational distortions of the molecule. As vibrations alter bond lengths and bond angles, electronic energy levels vary over a rather wide range. Since part of any large population of identical molecules will be found at any given time in each possible vibrational deformation, the observed energy-level transitions (and thus the emission or absorption spectrum) cover a range of wavelengths.

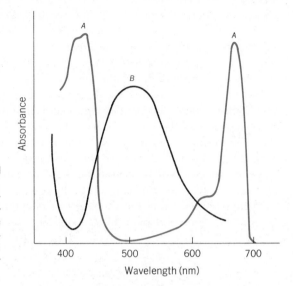

FIGURE 16-8 Molecular band spectra. (a) Chlorophyll. (b) The Mn(III)-CyDTA complex. You know the significance of chlorophyll; for the significance of Mn(III)-CyDTA, see text. Is the absorption spectrum of chlorophyll consistent with its color?

The Beer-Lambert law applies to light absorption in band spectra, and we shall now describe its use in the measurement of dissolved oxygen in sea water through the formation of the colored transition-metal complex of Mn(III) and cyclohexanediaminetetraacetate ion (CyDTA). In this technique, the following sequence of reactions is carried out by the successive addition of $MnCl_2$, NaOH, CyDTA, and sulfuric acid to a sea water sample:

$$Mn^{2+} + 2OH^- \rightarrow Mn(OH)_2(s) \tag{16-2}$$

$$4Mn(OH)_2 + O_2 + 4CyDTA^{4-} + 12H^+ \rightarrow 4MnCyDTA^- + 10H_2O \tag{16-3}$$

In Equation (16-2), a precipitate of solid manganese(II) hydroxide is prepared in the sea water sample. In reaction (16-3), this solid is simultaneously dissolved, oxidized by the oxygen in the sea water from the +2 to the +3 state, and that ordinarily strongly oxidizing state is stabilized by complexation with $CyDTA^{4-}$ ion. The Mn(III)-CyDTA complex is red-purple; its band spectrum is shown in Figure 16-8. The quantity of it that is formed is limited by the quantity of O_2 dissolved in the sample (4 moles of $MnCyDTA^-$ for every mole of O_2 originally present). The quantity of $MnCyDTA^-$ formed is determinable through the Beer-Lambert law by measuring the absorbance of the red-purple complex at a wavelength of 500 nm.

The developers of this method[7] report the equation

$$A = (0.0424) \times (ppm\ O_2) + 0.006 \tag{16-4}$$

---

[7] G. S. Sastry, R. E. Hamm, and K. H. Pool, "Spectrophotometric Determination of Dissolved Oxygen in Water," *Analytical Chemistry*, vol. 41, p. 857 (1969).

for the absorbance of MnCyDTA$^-$ [produced in a sample via reactions (16-2) and (16-3)] in cells of one-centimeter light path. In Equation (16-4), 0.0424 is the product of the factor ε_λ of Equation (16-1) and the proportionality factor relating the molar concentration of MnCyDTA$^-$ to the O_2 content of the sample in parts per million (milligrams of O_2 per kilogram of sample). The additive term (0.006) does not appear in the Beer-Lambert law. It says that, for this method and apparatus, a reading of 0.006 would result even in the absence of oxygen, and that this constant error appears at every oxygen concentration. [This could result, for example, from light absorption by the excess reagents used in reactions (16-2) and (16-3).]

Suppose that treatment of a particular sea water sample resulted in an absorbance value of 0.345 in a one-centimeter cell. Let us calculate the O_2 content to which this result corresponds. Solving Equation (16-4) for ppm O_2,

$$\text{ppm } O_2 = \frac{(A - 0.006)}{0.0424}$$

$$= \frac{0.339}{0.0424}$$

$$= 8.0 \text{ ppm}$$

This is a fairly typical value for air-saturated sea water from the surface of the ocean.

16-6 Chromatography and the Measurement of Traces of Volatile Substances

Much has been discovered and written recently about the presence of traces of organic compounds in natural water systems. Some of these substances are vitally important to life in such waters: organic nutrients, Lewis bases that complex metal ions, and those external hormones called pheremones that bring opposite sexes of the same species together for the purpose of reproduction. Other trace organics are harmful; these include naturally occurring toxic products of metabolism and, unfortunately, man-made pesticides and herbicides.

It is, of course, expected that a pesticide will not remain precisely where it is put. However, one of the unpleasant surprises of the recent past has been the realization of just how widely pesticides have spread over the earth. DDT has, for example, been found in the body fat of Arctic Indians and Antarctic penguins, thousands of miles from the point of original application. Snows deposited at the summit of Mt. Olympus after 1944 contain detectable quantities of DDT (see "Suggestions for Further Reading"). The quantities of pesticides so distributed are, of course, limited, so that, except when a pesticide is actively concentrated by a plant or animal, their concentrations in the environment are quite low. The snows of Mt. Olympus, for example, contain about 0.3 μg of DDT per kilogram of snow. Such concentrations are far, far below any that can be smelled or tasted. The detection and quantitative measurement of organic compounds like DDT at such low concentration levels is a considerable accomplishment. It is done by applying the solubility equilibria mentioned briefly in Section 7-5.

THE DETERMINATION OF CHLORINATED HYDROCARBONS IN NATURAL WATERS.

As an example of the analytical techniques that have allowed us to follow the global career of a pesticide applied locally, let us consider in detail a standard method for isolating and measuring the quantity of DDT in a sample of sea water.[8]

The molecular structure of DDT (the initials stand for *p,p'*-*dichlorodiphenyltrichloroethane*) is shown in Figure 16-9, along with that of a structural variant. The overall symmetry of this molecule and its large, nonpolar benzene rings make it readily soluble in nonpolar solvents. It is for this reason that DDT and similarly structured pesticides tend to be stored in body fat rather than to be kept in the metabolic mainstreams of the body, or to be excreted via the bloodstream and the kidneys. Advantage is taken of this property in the first step of the analysis:

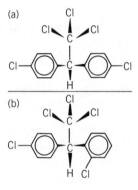

FIGURE 16-9 The structure of dichlorodiphenyltrichloroethane (DDT). (a) The *p,p'* isomer, in which the chlorine atoms on the phenyl rings are *para* (that is, at ring position 4) to the point of attachment to the rest of the molecule. (b) The *o,p'* (*ortho* on one ring, meaning at ring position 2 or 6 relative to the point of attachment to the rest of the molecule; *para* on the other ring) isomer.

1. DDT and all other nonpolar constituents are extracted from sea water by exposure to the nonpolar liquid hexane. This operation is accomplished in a separatory funnel (Figure 16-10). When the immiscible liquids are shaken together, they break apart into many globules, and the resultant large interface area allows solubility equilibrium (Chapter 7) to be established readily. A total of 75 ml of hexane is used, in three successive 25 ml extractions, on 900 ml of water. In this operation, all nonpolar solutes in the sample are separated cleanly from the water and the water-soluble ions of which the sea is so predominantly composed.

---

[8] The method is described in a paper by W. L. Lamar, D. F. Goerlitz, and L. M. Law in *Organic Pesticides in the Environment* (*Advances in Chemistry* Series No. 60), Washington, D.C.: American Chemical Society (1966).

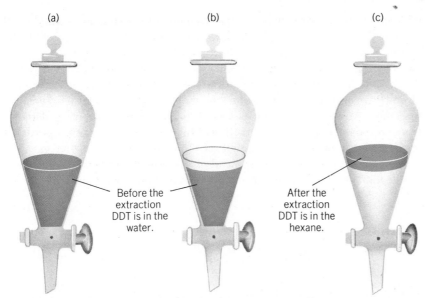

FIGURE 16-10 Separation of DDT and the other nonpolar constituents from sea water by extraction with the nonpolar liquid hydrocarbon hexane, C_6H_{14}. (a) A 900 ml sea water sample is placed in a separatory funnel. (b) A 25 ml portion of hexane is added to the separatory funnel, and then the funnel is shaken so that small droplets of hexane can mix intimately with the sea water and establish solubility equilibrium rapidly. (c) After standing, the hexane droplets rise to the top, bringing along in solution most of the DDT originally in the water. The water is drawn off, and this recovered water sample is extracted again with another fresh portion of hexane.

2. The combined hexane extracts are evaporated to a volume of 5 ml. All analytical methods are more reliable on samples of moderately high concentration than at extremely low concentration. The extraction step concentrates the DDT (with all other nonpolar components) present in 900 ml of sea water to only 5 ml of hexane solution, a 180-fold increase in concentration. During the evaporation, the temperature of the mixture does not exceed the boiling point of hexane (69°C), so the higher-boiling pesticides remain behind, while the hexane is removed.

There are two separations in this DDT analysis.

3. Substances present in the hexane extract are separated, identified, and quantitatively measured by **gas chromatography**. Gas chromatography (also called vapor-phase chromatography) is the method at the heart of this feat of analysis. So powerful, sensitive, reliable, and widely used is this technique that a description of modern chemistry (particularly that of environmental analysis) would be unrealistic without including it. Although some of the hardware used in gas chromatography looks formidable, the fundamental principle of its operation is not difficult to understand.

A schematic diagram of a gas chromotograph is shown in Figure 16-11. The heart of the instrument is a hollow column or tube of metal through which a steady stream of a *carrier gas* (in our case pure N_2) flows. A very small sample of an unknown mixture (such as a portion of our concentrated hexane extract) is injected into this stream at the upstream end of the column. Naturally, all components of the mixture move downstream through the column with the carrier gas. The separation, identification, and measurement of the components of the unknown mixture result, since each component moves downstream at a different rate because of the column filling (see below). The rates of downstream travel are reproducible and characteristic of each chemical substance. At the downstream end of the column, there awaits a detector whose response, recorded as a function of time (Figure 16-12), is proportional to the quantity of substance carried off the column by the carrier gas. The unknown mixture injected at the head of the column has been resolved into a collection of separate, identified, and measured pure substances.[9]

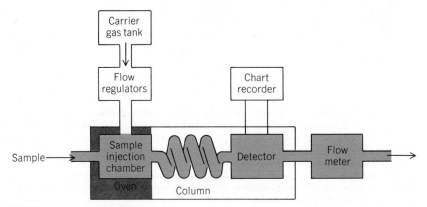

FIGURE 16-11   Block diagram of a gas chromatographic apparatus.

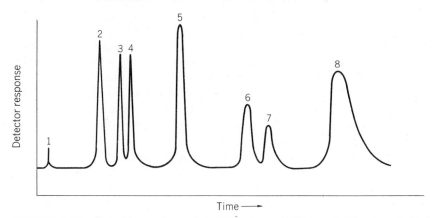

FIGURE 16-12   Chromatographic peaks (detector response) for a mixture of pesticides. Peak 1: air injected with the sample; 2: aldrin; 3: malathion; 4: parathion; 5: dieldrin; 6: o,p'-DDT; 7: DDD; 8: p,p'-DDT.

---

[9] It should be emphasized that the chromatograph does not just measure an attribute of each substance in a mixture, the way a spectrophotometer does, but rather it *actually separates* the pure components of a mixture before identifying them. Of course, the amounts separated ordinarily are very small, in the interests of saving time and money, but much larger *preparative chromatographs* are built on the same principles to do just the separating.

Having described *what* a gas chromatograph does, let us now look at how, in this particular case, the deed is done.

THE COLUMN. The components of the unknown mixture are separated by the time they reach the detector at the downstream end of the column, because they undergo different interactions with the material inside the column. This separation is based upon the same principles as that by which the nonpolar components of the sea water were separated from the polar ones in the hexane extraction step: The chromatographic column contains a stationary liquid film, and each component of the injected mixture is allowed to establish an equilibrium between the stationary liquid phase and the moving gas phase. In a commonly used form of chromatographic column, the stationary phase is in the form of a thin surface film on a finely divided, inert solid, with which the column is packed. For a given substance X, the equilibrium established at the boundary between the phases could be symbolized:

$$X(\text{solution, stationary}) \rightleftharpoons X(\text{gas, moving}) \quad (16\text{-}5)$$

Following the rules developed in Chapter 7 for writing equilibrium constants, we may write, for Equation (16-5),

$$\frac{[X]_{\text{moving gas}}}{[X]_{\text{stationary liquid}}} = K_x \quad (16\text{-}6)$$

where K_x is an equilibrium constant *that is different for every different component of the unknown mixture.*

Equation (16-6) describes the ratio in which a large collection of X molecules will partition themselves between the moving gas phase and the stationary liquid phase. However, it has another equally valid interpretation that hinges on the fact (developed in Chapters 5 and 7) that a molecular equilibrium state is characterized by equal and opposing *rates* of the reaction. A particular X molecule spends part of its time in the stationary liquid phase and the remainder in the moving gas phase, and Equation (16-6) also describes the ratio of these *times spent* by the average X molecule.

For example, if K_x happened to have the numerical value 2.00, the average time spent by X molecules in the moving gas would be twice that spent in the stationary liquid. To put it another way, molecules of X would, on the average, spend $\frac{2}{3}$ of their time in the moving gas and $\frac{1}{3}$ of their time in the stationary liquid.

But the ratio of the time spent in the moving gas to that spent in the stationary liquid directly controls the overall rate at which a substance moves through the chromatographic column. The larger the value of K_x, the larger the proportion of time spent by X molecules in moving with the carrier gas, and the earlier it arrives at the detector.

Considering the same example as before, a value K_x of 2.00 would mean that X molecules would move through the column, on the average, $\frac{2}{3}$ as fast as the carrier gas, and would therefore make the journey from injection port to detector in $\frac{3}{2}$ the time that the carrier gas requires.

The very great power of gas chromatography to resolve and identify components of unknown mixtures results from the fact that the numerical value of K_x, reflecting as it does the effect of the molecular structure of substance X on its solubility in the stationary liquid, is very sensitive to small differences in molecular structure. For example, the time required for o,p'-DDT (Figure 16-9) to traverse a common chromatographic column is only 70.8 percent of that required for p,p'-DDT; thus, these two rather similar substances are cleanly separated in the chromatograph shown in Figure 16-12.

The *identification* of each component depends upon the reproducibility and independence with which each component of an injected mixture arrives at the detector. In fact, the time between injection and detection is—for specified conditions of temperature, carrier gas flow rate, and stationary liquid—quite reproducible and can be established once and for all for a given compound by injecting the pure, known substance alone.[10]

The brains (or nose) of the instrument

THE DETECTOR. The exquisite ability of a chromatographic column to separate the components of a mixture would accomplish nothing if there were no means of detecting the separated components as they emerge from the end of the column. In gas chromatography, the need for a detector is especially acute, since the components to be detected emerge in the form of extremely dilute gaseous solutions. It is common for a component of interest to be present in less than nanogram (10^{-9} g) quantities; the detector we are about to describe is easily capable of detecting 10^{-13} g of chlorinated hydrocarbons in one milliliter of carrier gas. For a substance with a molecular weight of 200, this amounts to "only" 300 million molecules, a small number in the molecular world.

There are many types of gas chromatographic detectors, and new types appear frequently. For the present purpose, we shall describe the operating principles of an *electron capture* detector, because it is especially sensitive to chlorinated hydrocarbons and is thus frequently used in pesticide-residue analysis.

Figure 16-13 is a schematic picture of an electron-capture detector. The principle of its operation is not difficult to understand, although to build a working model would require attention to some engineering details that we suppress here. High-energy electrons from a β-particle emitter (see Chapter 10) strike N_2 molecules (remember that the carrier gas is N_2) and knock other electrons from them, losing some of their own velocity in the process. Thus, a low-density mixture of electrons and positive ions is created inside the detector. These electrons have a fairly short free lifetime, but the supply is constantly renewed, so a steady-state concentration of electrons is built up within the detector. Every 10^{-4} sec or so, this electron population is sampled by a positively charged collector; the resulting pulse of electrical current is amplified to provide a base-line signal for the recorder. When a mob of DDT

---

[10] To prevent the inevitable coincidence that two different substances might have closely similar values of K_x, it is general practice to establish these values, and to chromatograph unknown mixtures, on two or more different columns with different stationary phases.

molecules enters the detector from the end of the chromatographic column, the high electron affinity of chlorine atoms allows DDT molecules to capture electrons (hence the detector's name) out of the standing population. The resulting decrease in free electrons results in a decreased current through the positively charged collector as long as some DDT molecules are present in the detector. Thus, the passage of this chlorinated hydrocarbon off the end of the chromatography column, through the detector, and out of the instrument is signaled on the recorder by a spike-shaped decrease in collector current. These spike-shaped signals ("peaks") are clearly shown, for a series of pesticides, in Figure 16-12.

The analytical data appear as a pen line on the recorder chart paper.

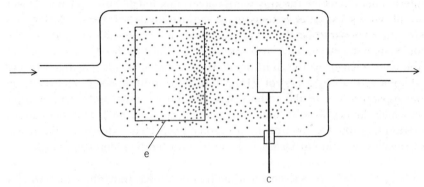

FIGURE 16-13 An electron-capture detector for gas chromatography (schematic). e: β-particle emitter. c: collector. The carrier gas enters from the chromatographic column at the left and exits at the right.

Finally, since the decrease in the free electron population in the detector is directly proportional to the number of DDT molecules passing through it at any given time, the size of the peak on the recorder chart (strictly speaking, the *area* under it) is a measure of the *quantity* of DDT injected with the sample. By comparison with known standard pesticide mixtures, both the identity and quantity of DDT and other pesticide residues in the hexane extract, and thus in the water sample, may be established. May they continue to decrease.

16-7 The Reliability of Chemical Measurements

Large flows of money, material, and human effort depend upon the results of chemical measurements (for example, when the government requests the withdrawal from sale of a batch of contaminated food, or when a chemist announces a new compound on the basis of his measurements of its qualitative and quantitative properties). Therefore, an estimate of the reliability of a chemical measurement is no less important than the measured value itself. It is unfortunately true that any single chemical measurement is only an estimate, more or less reliable, of the numerical quantity sought. We shall not burden you with a long discussion of the statistics of chemical measurements. We shall point out some of the problems to be overcome in arriving at a realistic appraisal of the reliability of a measurement, and in insuring that the reliability is as high as required as a basis for action. We also include, among the "Suggestions for Further Reading," some more-complete introductions to the subject.

SYSTEMATIC ERRORS. Many analytical techniques produce numbers that consistently over- or underestimate the quantity sought, or are subject to interferences. The additive term in Equation (16-4), had it gone unrecognized, would have produced a *systematic* positive error in O_2 concentration. The interferences to which the gravimetric determination of chloride ion is subject are likewise a source of systematic overestimation unless the interfering ions are either known to be absent, are removed, or are measured and accounted for. Systematic errors may arise through flaws in the theory of the analysis (the color change of the indicator EBT does not coincide exactly with the addition of the first excess of EDTA in the titration of some metal ions), in the apparatus (one of the authors once titrated several samples with a buret on which the engraved volume scale *omitted* the 32nd ml!), or in the pattern of sampling (the O_2 concentration in some deep lake waters goes through a seasonal cycle, and fair-weather chemists consistently underestimate it).

Facing up to uncertainty

RANDOM ERRORS. Even for a method of known reliability, the sad fact is that repeated analyses do not give the same number every time. Small, unrecognized changes in the conditions under which the sample is treated (temperature fluctuations may change rates and equilibria of reactions and the volume of apparatus, for example) result in a pattern of results scattered around some average value. The natural human tendency in the face of this situation is not to repeat a measurement, lest the error be found to be too large, but a better strategy in the long run is to accumulate several replicate measurements and to take an average. Not only is such an average the best estimate of the "true" value of the variable sought, but the scatter of the results gives some indication of the uncertainty with which the average value is to be regarded. For example, a fish may be analyzed by atomic absorption spectrophotometry for its mercury content, and a result (say 0.30 ppm) announced. If this is the result of a single trial, one has no way of telling how "good" the number is; if a second trial gave a value of 0.05 ppm, one's reaction would be different than if the second trial gave a value of 0.32 ppm. Furthermore, two results of 0.10 and 0.52 ppm would produce yet another response, even though the average value, 0.31 ppm, is the same as the average of 0.30 and 0.32 ppm.

SAMPLING ERRORS. There is an old saying that "Figures don't lie, but liars figure." One of the best ways of lying with figures (with or without meaning to) is to base an estimate of some variable on the results of analyses of a biased sample. Suppose, for example, that a large body burden of DDT were to make fish less agile, and thus less able to avoid the fisherman's net.

Then analyses for DDT in the flesh of fish caught by trawling would contain a systematic positive bias with respect to the whole population of fish. We have already mentioned existence of cyclic changes in the values of chemical parameters in lakes. Either a cyclic or a systematic drift in the system one is studying may go undetected unless the analyst makes sure to avoid patterning his samples in step with the change in the system.

In practice, it is very difficult to anticipate and take into account all of the sampling errors that may creep into a study, particularly if the system studied is large and complicated. For example, the chemical investigation of a body of water even as large as a lake may present a choice between comprehensive sampling in space and sampling over seasonal variations, and one may have to compromise between these two if, for example, the gathering of samples involves a significant input of time, effort, and money. Here it is that scientific and civic-political questions are most likely to intersect. A chemist announces that he suspects that Lake Erie (to take a famous example) is undergoing rapid accumulation of undesirable substances in the form of urban wastes and the algae that grow on them, and he presents a set of data that appear to verify his claim. Environmental activists call for an immediate ban on further pollution (in this case, perhaps, the removal of phosphates from detergents)[11] and stiff fines for identifiable polluters. Those most likely to be inconvenienced by these measures, such as detergent manufacturers, issue the familiar cry for "more research." How much research is enough? Were all of the significant variables taken in account in the sampling process? How much random error is tolerable when the average of the random fluctuations implies a crisis? Should the limited public funds available be used to buy more samples and more reliable chemical analyses, to give tax rebates to those who clean up their effluents, or to support research on sewage treatment? And how reliable are the data gathered in such research?

We think that the reader of this book, if he never goes near another chemistry course, should be ready to tackle such questions on the basis of an appreciation for the "hard data" *as well as* (not instead of) his personal feelings.

Give me a fruitful error any time, full of seeds, bursting with its own corrections. You can keep your sterile truth for yourself.
—Vilfredo Pareto, commenting on Kepler

16-8 Summary

Analytical chemistry pervades all the practical aspects of life, because there is always the necessity of finding out *how much* of *what* is present in food, water, air, clothing, building materials, ores, minerals, and biological specimens. Without reliable analyses, there can be only idle and meaningless talk about any problem of pollution or scarcity or production; given dependable analytical information, there can be constructive discussion and then remedial planning. Analytical information does not come easily; much care and labor are necessary to get reliable figures. Often a new method must be devised to suit the requirements, or an existing method must be revised drastically. The development of new methods requires a rare combination of skill, inven-

---

[11] The phosphate issue is discussed in Chapter 18.

tiveness, and background knowledge, but once the method is established and published, the actual measurements can be made anywhere in the world by adequately trained operators.

Most analyses are carried out today by means of instruments, rather than by the simple gravimetric and volumetric procedures that were used 30 years ago. Analytical instruments vary greatly in complexity, from the relatively simple pocket pH meter (costing perhaps $25) to elaborate spectrophotometers and spectrometers which may cost $100,000 or more. In this chapter, the elaboration of the simple pH meter into a variety of ion-selective direct-response meters was described, and the application of these (plus some classical gravimetric and volumetric methods) to the analysis of sea water was considered. It was found that sea water contains a large number of substances washed down from the land and stored in the water. Many of these are useful; some are intriguing; and a few are objectionable.

The absorption of light (or other radiation) by specific reaction products is the basis for a large group of analytical instruments. The Beer-Lambert law, $A \equiv \log (I_{in.}/I_{trans.}) = \varepsilon_\lambda \cdot b \cdot c$, relates measured light intensities to concentration. Armed with this fact, the quantitative determination of how much of any specified substance (amenable to this method) is in a given sample becomes a matter of measuring the ratio of the two intensities. This is the basis of optical, infrared and ultraviolet spectrophotometers, invaluable tools for modern analytical chemistry.

In addition to spectrophotometers (and other instruments which also measure a particular attribute, such as magnetic properties, of one substance in a mixture), there is a large family of analytical instruments which depend on the *actual separation* of minute quantities of each substance of interest in a mixture, and on the subsequent identification of each substance so isolated. This is done by the principle of *chromatography*, in which a stream of fluid carrying several components to be separated is allowed to interact with a solid which has differential reactivity toward the components. The simplest example of chromatography is provided by the separation of the various dyes present in a bottle of ink, as the ink seeps slowly into a piece of filter paper (see Problem 16-1). The next simplest is the separation of the pigments in a leaf, by percolating an alcohol extract of chopped-up leaf slowly through a tube packed with aluminum oxide (see Problem 16-2). The word *chromatography* comes from such separations of colors. In the present chapter, the isolation and identification of DDT in sea water (or lake water, or drinking water) by gas chromatography was taken up step by step. The separation of a series of volatile components carried in a stream of gas, by interaction with a thin coating of liquid on a porous solid within the "column" through which the gas passes, depends upon differential solubility of the components in the liquid film. The separated components may be identified by several methods (notably by different heat conductivities of the gases and vapors), but here a highly specific electron-capture detector was used to identify the chlorinated substances like DDT.

No analytical data are of any real use unless their reliability is established by independent determination of the random errors of the method. In addition, every method has to be calibrated by using known amounts of the substance of interest, in order to allow for systematic errors. And lastly, all analytical measurements on a given sample can only reflect what is in that one particular sample, so thought has to be given to possible errors introduced by the method of sampling. When all this is done, analytical measurements finally become "hard data," which can become the reliable basis for opinion or action.

GLOSSARY

chromatography: the technique of separating mixtures into pure substances by repetitive and continuous establishment of solubility equilibrium across a phase boundary between a moving phase and a stationary one.

detector (in chromatography): a device that produces a signal (usually an electrical one) when components of a mixture emerge from a chromatography column.

electrochemical analysis: a method of determining the concentration of some species by measuring the emf of an appropriate electrochemical cell.

end point: in a titration, the signal that corresponds to the addition of an equivalent amount of reagent to the substance being titrated. For an acid-base visual indicator, the end point is the color change that accompanies a rapid change in pH.

gas chromatography (vapor-phase chromatography): the chromatographic technique in which the moving phase is a stream of a carrier gas, and the solubility equilibrium is established for the components of a gaseous (vaporized) mixture.

gravimetric analysis: the analytical technique in which the component whose determination is desired is separated in the form of a pure substance, and that substance weighed to determine its amount.

random errors: numerical fluctuations in the value of a measured quantity that are caused by uncontrolled and randomly varying experimental conditions.

reliability: the degree (usually expressed as an estimated uncertainty) to which a numerical result may be taken as true.

sampling errors: errors in a numerical measurement or its interpretation caused by the fact that the sample taken is not representative of the whole system being studied.

spectrophotometry: the analytical technique that uses the Beer-Lambert law to establish the concentration of a given component in a solution.

systematic errors: errors arising from faults in the theory or execution of a measurement; as opposed to random errors, which tend to average out to zero, systematic errors generally give results consistently higher or lower than the true value.

titration: the process of stepwise addition of a reagent to a component, whose quantity is to be determined, in such a way that the first addition of the titrating reagent beyond a chemically equivalent amount will be detectable.

volumetric analysis: a titration in which stepwise addition of the titrating reagent is achieved by successive addition of small increments from a buret. The number of moles of titrating reagent added at the end point is calculated from the volume and molarity of the solution used.

PROBLEMS

16-1 (A do-it problem) Analyze any bottle or cartridge of ordinary writing ink (Skrip permanent blue-black No. 22, for example) by paper chromatography, as follows: Take one piece of filter paper, 10 or 12 cm in diameter, and put a drop of the ink in its center. Next, make two parallel scissor cuts $\frac{1}{4}$ in. apart from the edge toward the center. This will leave a $\frac{1}{4}$ in. strip of paper hanging from the center of the mutilated disk. Fill a drinking glass to within 1 in. from the top, and let the strip hang into the water so that water can ascend the strip by capilarity and spread outward from the center of the paper. Let it stand for 20 or 30 min. What separation of the components of the ink has been achieved? How many components are there?

16-2 Go into the library and look over several biochemistry textbooks and laboratory manuals until you find one that describes the chromatographic separation of the pigments in a green leaf (look under *chlorophyll* in the index). If you have the laboratory facilities available, do the experiment. If not, read a description of the results in a biochemistry textbook. From your observations (or facts gathered by reading about such chromatographic experiments), formulate a theory about why leaves turn color in the autumn.

16-3 Look up the article entitled "The Gold Content of Sea Water" in the *Journal of Chemical Education,* vol. **30,** p. 576 (Nov. 1953), and if possible, read several of the references listed under "Literature Cited" at the end of the article. You will be interested especially in Fritz Haber's attempt to pay off Germany's World War I reparations by extracting gold from sea water. After your reading, reflect on the advances in chemistry since the date of that article (or ask your instructor about such advances), and try to devise a more efficient method for the recovery of gold from the sea. Allowing 50 percent of the selling price (look up the price of gold today; it is important to you!) for labor, and 30 percent for capital equipment costs, how much can you afford to spend for the reagents necessary to extract the gold?

16-4 The sale of carbon tetrachloride as a household cleaning fluid is now prohibited, because inhalation of the vapor of CCl_4 causes damage to the liver. Suppose an old bottle of spot remover were brought to you, and suppose further that the true odor of the mixture of solvents were masked purposely by a perfumelike substance (called in the trade a "reodorizer"). How would you analyze the spot remover for carbon tetrachloride?

16-5 A rock is analyzed for sulfate ions (SO_4^{2-}) as follows: a 10.00 g sample of the rock is dissolved in nitric and hydrochloric acids, and an excess of barium chloride added. The sulfate precipitates as the insoluble $BaSO_4$, which is isolated by filtration, washing, and drying. The $BaSO_4$ weighs 0.1035 g. How many moles of SO_4^{2-} were present in the rock sample and what percentage by weight does it represent?

16-6 An analysis similar to that in Problem 16-5 is carried out to determine the iron content of an ore. In this case, however, the iron is precipitated as hydrated $Fe(OH)_3$, a gelatinous known substance that cannot be dried to a pure compound. Instead, it is "ignited" (heated at a high temperature) to drive the reaction

$$2Fe(OH)_3 \cdot xH_2O \longrightarrow Fe_2O_3 + (2x + 3)H_2O$$

to the right, and the pure Fe_2O_3 is the substance weighed. Show by means of equations how the iron content of the ore is to be calculated in this case.

16-7 A series of freshwater samples is analyzed for chloride ion by titration with a standard $0.001000\ M$ silver nitrate solution, using a chloride-ion selective electrode to determine the end point. The following data are obtained (averages of replicate samples).

Sample	Volume of Sample	Source of Sample	End-Point Volume of $AgNO_3$
1	100 ml	Lake Erie, 1906	24.54 ml
2	25 ml	Lake Erie, 1968	17.34 ml
3	100 ml	Lake Ontario, 1906	21.72 ml
4	25 ml	Lake Ontario, 1968	19.39 ml
5	10 ml	Seneca Lake, N.Y., 1968	50.92 ml

a. Calculate the chloride content of these waters, expressed as moles per liter and as milligrams Cl^- per liter.
b. Compare these results to the chloride content of sea water (Table 16-2).
c. What conclusions can you draw about the spatial and time distribution of chloride ion in these freshwater reservoirs?
d. If you were concerned about the implications of any of these results, what would your next move be?

16-8 Small concentrations of Hg(II) may be determined spectrophotometrically by measuring the concentration of the colored compound $HgBr_3^- \cdot BG^+$, where BG^+ is a positive ion of the dye Bindschedler's Green [4,4'-bis-(dimethylamino) diphenylamine]. The green ion pair is extracted from the aqueous sample in which it is formed into the water-immiscible solvent 1,2–dichloroethane. In the latter solvent, the molar absorptivity (ε) of the Hg-BG complex is reported to be 1.7×10^5 mole^{-1} -cm^{-1}-liter.

a. Why should the molar absorptivity have such odd-looking units? [Contemplate Equation (16-1).]
b. A solution of $HgBr_3^- \cdot BG^+$ in a 1.0 cm cell gave an absorbance reading of 0.340. Calculate the molar concentration of Hg(II) in that solution.
c. If the solution in the spectrophotometer cell was taken from a dichloroethane solution whose total volume was 10.0 ml, calculate the total number of grams of Hg in that solution.
d. The mercury in the solution of Part c was extracted from 1.00 g of lake trout. What was the mercury content of this trout?
e. What other information would be required in order to decide whether or not lake trout from the same source are safe to eat?

SUGGESTIONS FOR FURTHER READING

American Chemical Society, *Organic Pesticides in the Environment* (*Advances in Chemistry* Series No. 60), Washington, D.C.: American Chemical Society (1966) is a collection of loosely related papers, among which is the description of the isolation and identification of DDT discussed in this chapter. It includes scientific papers written by industrial, academic, and government scientists. You may enjoy browsing through to see the various viewpoints represented.

Durst, R., ed., *Ion-Selective Electrodes* (National Bureau of Standards Special Publication No. 314) Washington, D.C.: U.S. Government Printing Office (1969). Some of the papers in this collection (which is the standard reference work in the field) are more technical than you will want to handle at this point; others are not very difficult, however, and provide a good survey of the state of development of this field up to 1969.

Guenther, W. B., *Quantitative Chemistry: Measurements and Equilibrium*, Reading, Mass.: Addison-Wesley (1968). This book does an excellent job of relating the equilibrium concepts which we surveyed in Chapter 7 to the problems of chemical analysis. Since it is an undergraduate textbook, it is well supplied with worked-out examples and learning aids.

Pecsok, R. L., and Shields, L. D., *Modern Methods of Chemical Analysis*, New York: Wiley (1968) is an undergraduate textbook in analytical chemistry that concentrates on spectroscopy, separations, and other topics likely to be of interest to biologists and biologically inclined chemistry students.

Wilson, E. B., *An Introduction to Scientific Research*, New York: McGraw-Hill (1952). This book has a very good discussion of the statistically sound planning and interpretation of chemical measurements. Also, it is fun just to browse through to learn about the kind of things most researchers spend most of their time coping with: balky equipment, limitations of funds, time, and energy, reducing the scope of interesting but unmanageable questions to solvable dimensions.

17
Chemical Control in a Living Cell

17-1 Abstract

Living organisms are highly organized systems of molecules in which a multitude of simultaneous and competing chemical reactions are controlled in such a way that an overall steady state is achieved most of the time. The control is inherent, but not static as it is in laboratory experiments on equilibrium; the control permits the functioning of life processes and also provides for the reproductive cycle. The operation of so complicated a chemical system can best be understood in terms of control by adjustment of the *rates* of competing reactions, through limitation of concentrations, and control of mechanism by means of highly specific organic catalysts called enzymes. This area of chemistry is the domain of **biochemistry,** which usually is taught as a separate subject. This chapter presents a short treatment of it, not in the usual pedestrian terms of the structures and reactions of carbohydrates, fats, and proteins, but by considering a few specific examples of biochemical control of **metabolism** and synthesis.

17-2 General Aspects of Biochemistry

Even though human beings have been very much a part of the living world for hundreds of thousands of years, biochemistry is a new science. We all know that we can function adequately as biochemical entities without even being aware of molecules and chemical reactions. Only when confronted with illness, injury, or disease do we seem to need knowledge about our chemical constitution in order to deal with the situation. That knowledge does not come intuitively or easily; man has had to learn about his own chemistry by laboratory studies of such diverse materials as minced pigeon breast muscle, yeast cell extracts, and by-products of the slaughterhouse.

Nevertheless, biochemists have made enormous progress in elucidating the chemical processes of synthesis in living cells, in describing the molecules associated with reproduction, and in formulating models for control systems that regulate the whole dynamic collection of ongoing chemical reactions. In order to accomplish this, biochemists obviously must have had a successful strategy. This strategy might be summarized in a few operating principles, or "articles of faith":

1. Living organisms are composed of molecules, many of which can be isolated from tissues, purified, and studied in the laboratory.
2. The processes of metabolism may be described in terms of the superposition of coupled chemical reactions, many of which can be studied individually and manipulated outside the living systems by use of conventional chemical techniques.
3. The fundamental metabolic processes are similar in most living cells. Hence detailed study of a single aspect of the metabolism of a single organ of a single species may have application to a wide variety of very different living organisms.

There is a fundamental kinship in all living systems.

These are operating principles, which have been followed by two generations of biochemists[1] who have initiated and sustained a continuing revolution in man's understanding and control of living systems and life processes. We shall examine only a small sampling of the results of these investigations.

17-3 Biological Catalysts

A living cell, comprising many different kinds of molecules mixed together and reacting together, owes its stability and integrity to an elaborate internal regulatory system. A living cell contains many of the same molecules as does a pot of chicken broth simmering on the kitchen stove. Yet there is a crucial difference between chicken broth and living chicken cells. Much of this difference is due to the complex and detailed *organization* of the system of chemical molecules within each living cell, and to the intricate system of chemical *control* mechanisms which regulate the concentrations of the various chemical species in each region of a cell.

Dramatic differences in overall chemical composition can be found in nature between different living cells. Anyone who has chewed on a clove from the spice cabinet, eaten a slice of pineapple, pulled apart a cotton boll, or bounced a rubber ball knows that cloves, pineapple, cotton, and natural rubber are very different things indeed. Yet each is a plant product, and in each case there is a plant, or a part of a plant, in which the biochemical processes are directed overwhelmingly toward the production and accumulation of a single chemical substance.

The fundamental metabolic processes are directed in different directions in different organisms.

Whole cloves as sold in the grocery store are the dried unopened flower buds of the clove tree. A volatile oil comprises 15 to 20 percent of these dried buds, and 70 to 90 percent of the oil is a single compound, eugenol, which has the structure

eugenol

---

[1] Not all biochemists would subscribe to this credo. The biochemist R. J. Williams asserts that some aspects of biochemistry can be fruitfully approached by emphasizing biochemical *individuality* and *variability* within a single species. Such variability, he points out, becomes most important in complex organisms, especially in man. A lucid exposition of the significance of biochemical individualtiy is R. J. Williams, *Free and Unequal: The Biological Basis of Individual Liberty*, New York: Wiley (1964).

Most of the fruit of the pineapple plant is water, but 94 percent of the dry weight of the fruit is accounted for by sugars, of which 70 percent is the single familiar compound, sucrose, which is composed of glucose and fructose. Sucrose appears in chemically pure form on the dinner table as cane sugar. Its structure is

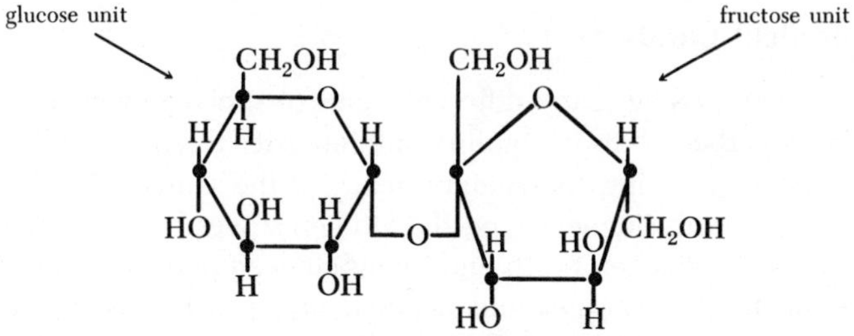

[structure of sucrose, with glucose unit and fructose unit labeled]

sucrose

(each ● represents a carbon atom)

A cotton fiber or lint is composed of essentially pure α-cellulose surrounded by a waxy outer membrane. Each molecule of α-cellulose is composed of many glucose units combined into a *polymer* having the repeating structure

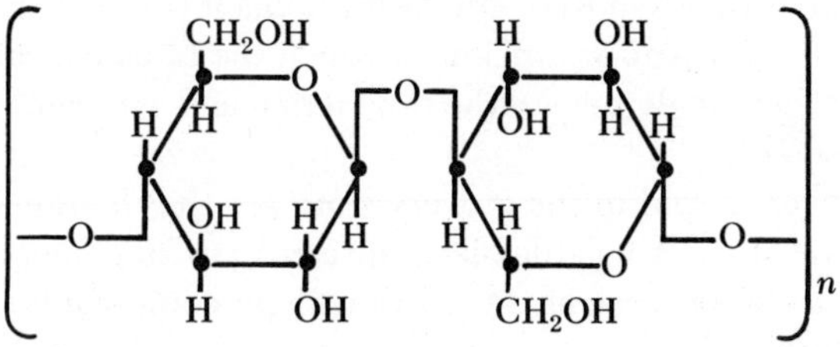

[structure of α-cellulose repeating unit, subscript n]

α-cellulose

Some 25 percent of the material in the roots of the Russian rubber-bearing plant, *Scorsonera tau-saghyz,* is a polymer of the hydrocarbon

[structure of natural rubber repeating unit, subscript n]

Natural rubber

Is this the same as the rubber described in Ch. 15?

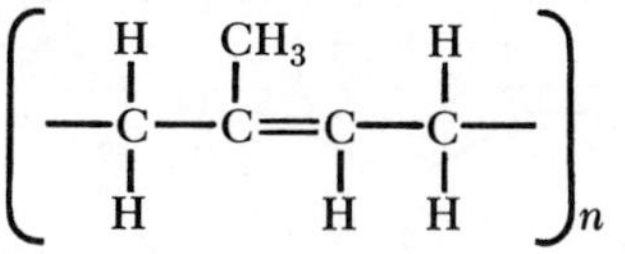

These are just a few examples of a widespread phenomenon. Even though most of the fundamental metabolic reactions occur in each living cell, somehow the regulatory mechanisms often are able to guide and direct the chemical processes overwhelmingly in certain directions in one cell, and in very different directions in other cells. One of the most effective ways available to a cell for controlling its own composition is *via the control of rates of chemical reaction.*

There is a fantastically large number of possible chemical reactions involving the molecules known to be present in a living cell. Most of these reactions are too slow at body temperature to be of any importance in changing the chemical composition of the cell. However, if selected reaction rates could be increased 100-fold, 10,000-fold, or even more, then just a few reactions and their reaction products would predominate. The concentrations of various reactants and products can be controlled by controlling the availability of reagents, as by diffusion through a membrane, or by controlling the relative rates of competing simultaneous chemical reactions. Such selective and controlled enhancement of reaction rates does occur on a significant scale in living cells, being one of the most important controlling mechanisms in the cells. The agents for this enhancement are very efficient biological catalysts, molecules called **enzymes**.

Enzymes are polymers of **amino acids,** often with molecular weights of several hundred thousand.[2] Typically, a particular enzyme catalyzes only a very few reactions, and sometimes only one specific reaction. We shall look at the particular enzyme called *fumarase* as an example.

Selective enhancement of reaction rates is accomplished by enzymes.

The compounds fumaric acid

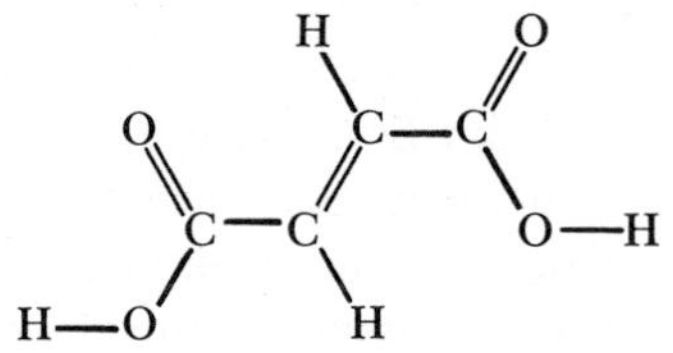

fumaric acid

and malic acid

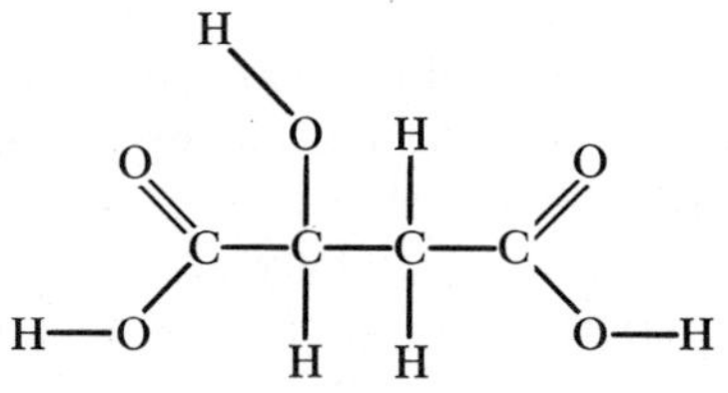

malic acid

can each be dissolved in water to give colorless solutions. The fumaric acid solution absorbs ultraviolet light at a particular wavelength, and this absorbance can be used to measure the concentration of fumaric acid in a solution. Malic acid does not absorb light of this same wavelength, because its structure is slightly different.

We can imagine a reaction for interconversion of fumaric and malic acids:

$$\text{fumaric acid} + H_2O \rightleftharpoons \text{malic acid} \tag{17-1}$$

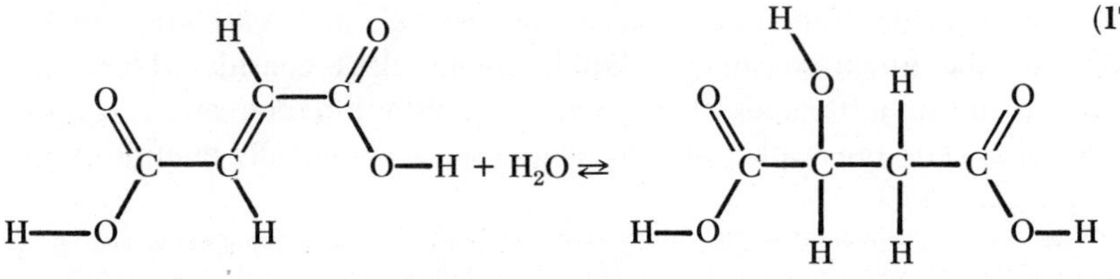

[2] Amino acid polymers are called **proteins.** Structural details of some proteins and amino acids are given later in this chapter. Not all proteins are enzymes; since our emphasis is on metabolic control, we shall concentrate on enzymes as we discuss proteins.

At room temperature (or for that matter, at body temperature), this reaction is extremely slow. However, an enzyme (which is called *fumarase*) has been found[3] which, when added to a solution of either malic acid or fumaric acid, greatly enhances the rate of reaction (17-1). Within a few seconds, detectable changes in the ultraviolet absorption occur, and these changes can be measured with a recording spectrophotometer.[4] Researchers analyzed and interpreted these changes in terms of a chemical reaction mechanism such as

$$E + F \underset{k_{-1}}{\overset{k_1}{\rightleftharpoons}} E \cdot F$$

$$E \cdot F + H_2O \underset{k_{-2}}{\overset{k_2}{\rightleftharpoons}} E \cdot M \qquad (17\text{-}2)$$

$$E \cdot M \underset{k_{-3}}{\overset{k_3}{\rightleftharpoons}} E + M$$

where the symbols E, F, and M represent enzyme, fumaric acid, and malic acid, and the symbols E · F and E · M represent complexes of the enzyme with fumaric acid and with malic acid. The rate-of-reaction data experimentally obtained at a particular temperature and in solutions having a particular pH value are consistent with mechanism (17-2). Several aspects of this mechanism are especially significant:

1. Enzyme molecules take part in the reaction both as reactants and as products.
2. Enzyme complexes E · F and E · M are much more reactive (they react much more rapidly) than the uncomplexed molecules F and M. The enzyme has *activated* F and M.
3. Only a relatively few enzyme molecules, reused over and over again, can catalyze the interconversion of many fumaric acid and malic acid molecules.
4. The rate of reaction is dependent upon the concentration of uncomplexed enzyme. Thus, control of this reaction rate could be achieved by controlling the concentration of enzyme.

Amazingly, in spite of a systematic search, no other reaction has been found for which fumarase is a catalyst. In fact, only one of the two mirror-image isomers of malic acid is produced by the enzymatic reaction, and only that isomer will react with fumarase to give fumaric acid. In common with many other enzymes, fumarase is exceedingly specific in its catalytic effect.

The fact that fumarase can be isolated from muscles is considered evidence that malic and fumaric acids, if present, are rapidly interconverted in these muscle cells. Thus, one aspect of the metabolism of chemically complex mus-

---

[3] Many of the experimental investigations have been performed using fumarase isolated from pig hearts. Pig hearts were convenient muscles because of the proximity of the research laboratory to a large slaughterhouse for pigs. Pig heart fumarase has a molecular weight of 197,000.

[4] Spectrophotometers are described in Appendix III and in Section 16-5.

cle cells—the interconversion of malic and fumaric acids—has been inferred from chemical studies of simple solutions of just the acids and the enzyme in water. It turns out, as we shall see, that this interconversion is implicated in a vital sequence of reactions that enables an animal to utilize foodstuffs to obtain energy.

Enzymes that catalyze other reactions of malic acid (such as its oxidation to oxaloacetic acid) and of fumaric acid (such as its reduction to succinic acid) have also been isolated from muscle tissue. In fact, an entire series of enzymes has been isolated, a series of individual enzymes which can be studied separately in simple test-tube-and-beaker experiments, but which together catalyze the cyclic sequence of reactions given in skeleton form in Figure 17-1. If every one of these enzymes can be isolated from the same muscle tissue,[5] and if all the various reactants are present in that tissue, then it is considered established that the particular muscle harbors and encourages this *tricarboxylic acid cycle*. The reactions and the possibilities for chemical interconversion are real; the cycle itself is a clarifying chemical invention, which aids the biochemist in focusing attention on a few aspects of a complex system.

Many of the conversions indicated in Figure 17-1 must be coupled with other reactions. Thus, the several oxidations (isocitric to oxalosuccinic acids, succinic to fumaric acids, malic to oxaloacetic acids) must each be coupled with a reduction reaction. Indeed, much of the importance of this cycle to biochemists is that it is a mechanism for using pyruvic acid[6] as a chemical energy source for other metabolic reactions by means of these coupled reductions. The carbon atoms of the oxidized pyruvic acid are released stepwise as carbon dioxide. The first carbon atom is released in the conversion of oxalosuccinic acid to α-ketoglutaric acid, and the second carbon atom is released in the conversion of α-ketoglutaric acid to succinic acid.

Actual metabolism in a cell involves the superposition of many simultaneous chemical reactions, coupled through common reactants and products. There are many ways of thinking about these reactions in terms of cycles, and one may think of metabolism as the superposition of many reaction cycles of the sort outlined in Figure 17-1 with different molecules being interconverted. It is important to remember, however, that a particular molecule in solution reacts when it encounters a favorable partner, and that a particular molecular complex dissociates when the energy requirements are just right, quite independently of the existence of a biochemical cycle written on a piece of paper. A particular molecule may not continue all the way around a reaction cycle, for there are always reverse reactions, and there are usually alternative competing reactions. Citric acid, for instance, can undergo a series of transformations which eventually lead to fats. Many of the amino acids can be synthetized from α-ketoglutaric acid or oxaloacetic acid. Oxaloacetic acid can also be converted into sugars. Succinic acid is a precursor of such molecules as chlorophyll and hemoglobin. In fact, such reactions lead to the for-

Many compounds are involved in simultaneous competitive metabolic reactions.

---

[5] In the pioneering work, these enzymes were all isolated from pigeon breast muscles.

[6] Pyruvic acid is a metabolic decomposition product of sugars, and thus in animals pyruvic acid can be obtained from foodstuffs.

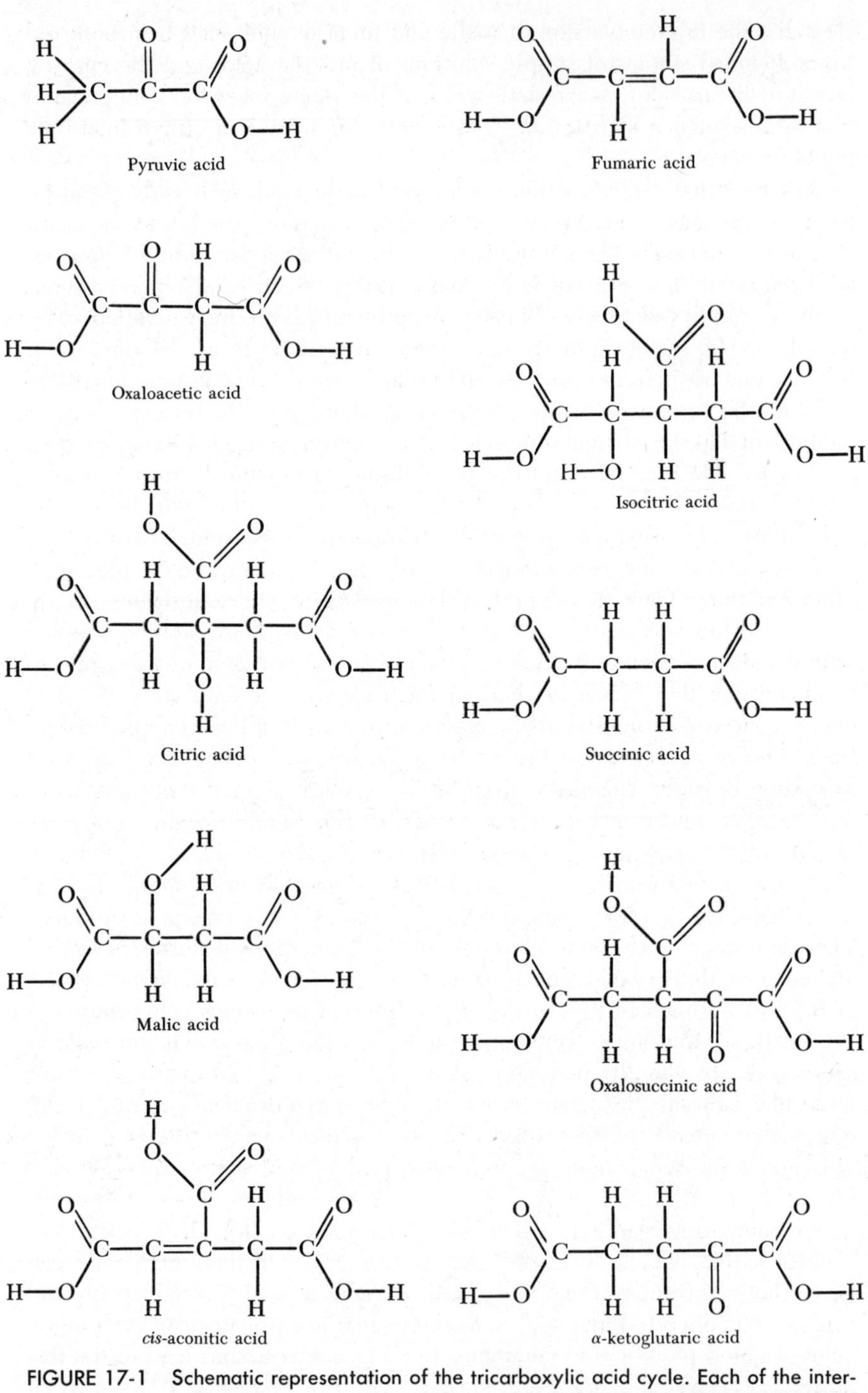

CHEMICAL CONTROL IN A LIVING CELL

FIGURE 17-1 Schematic representation of the tricarboxylic acid cycle. Each of the interconversions is catalyzed by one or more enzymes. Note that these schematic reactions are not balanced; reactants and products such as H^+, H_2O, CO_2, as well as oxidizing and reducing agents, have been omitted so that we can focus attention on the interconversions of the carboxylic acids.

mation of fats, carbohydrates, and proteins (all in the same plant) during photosynthesis.

Primary control of this network of chemical reactions involves the collection of enzymes in the cell. We shall devote much of the remainder of this chapter to an examination of enzymes as proteins and to the mechanisms of genetic control of protein synthesis in a cell. The next section is about proteins in general, and enzymes in particular.

17-4 Proteins: Molecules of Immense Variety

All enzymes are proteins.

Proteins are polymers with molecular weights ranging from a few thousand to over 40 million. The common feature of protein structure is the presence of long chains of amino acids, joined together by **peptide bonds.** To learn about proteins, we need to know about amino acids and about peptide bonds.

The 20 amino acids shown in Figure 17-2 have been isolated from naturally occurring proteins. Each amino acid contains the structural entity

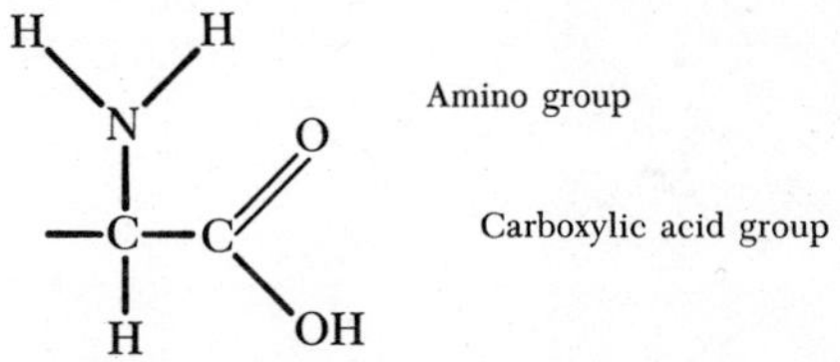

Amino group

Carboxylic acid group

except in proline where the amino nitrogen is bonded to two carbon atoms. The diversity arises from the 20 different side groups

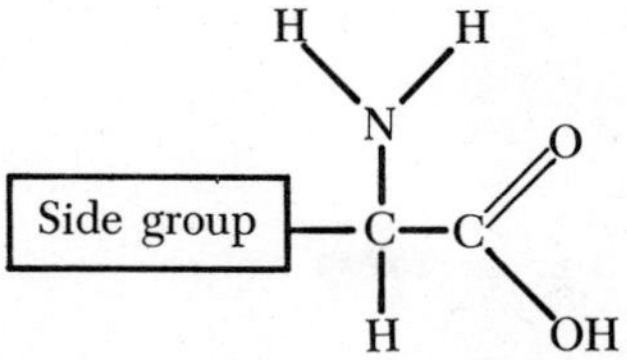

Polymerization of amino acids to form a protein molecule occurs by formation of bonds of the type

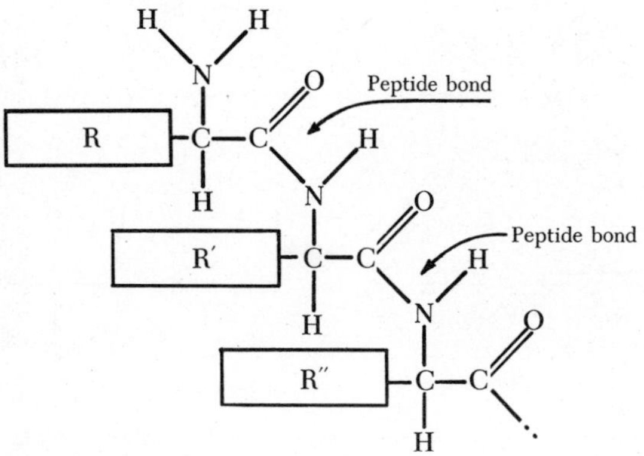

Peptide bond

Peptide bond

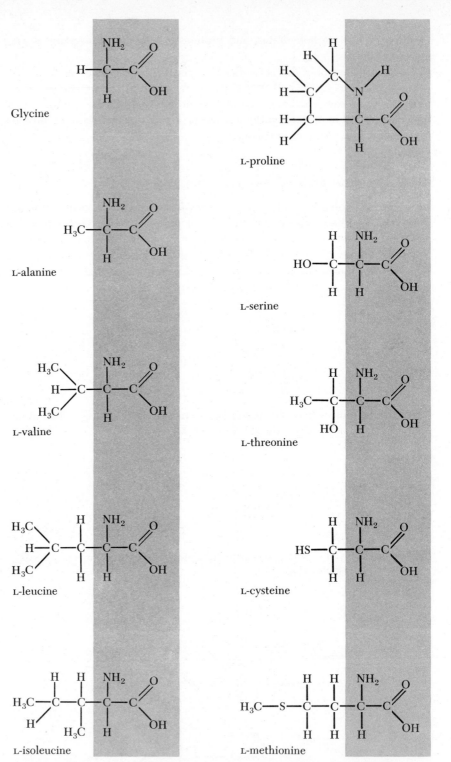

FIGURE 17-2 Structures of the amino acids found in proteins. Note the structural similarities, particularly in the shaded portions of these drawings. The 20 different side groups produce the differences among these 20 compounds.

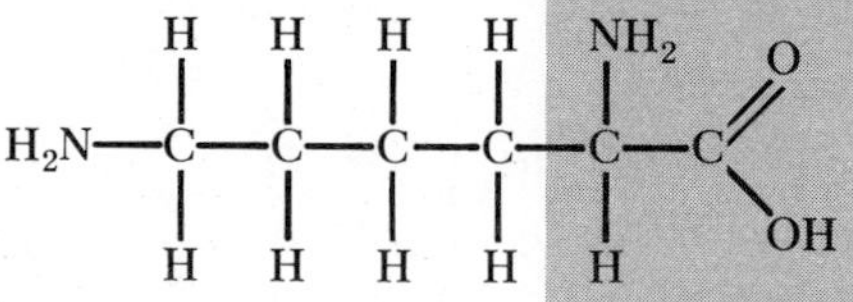

L-lysine

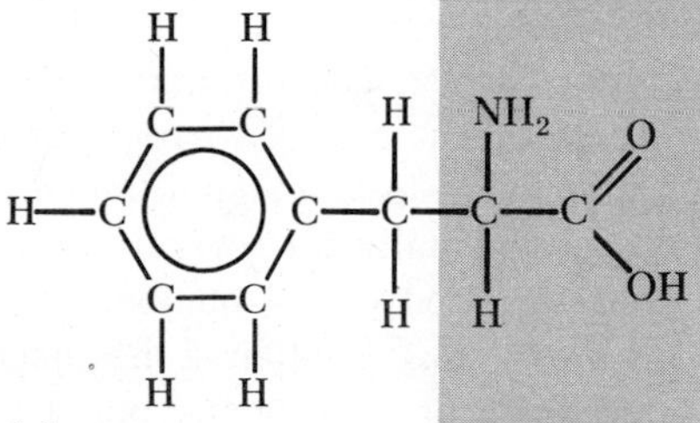

L-tryptophan

L-phenylalanine

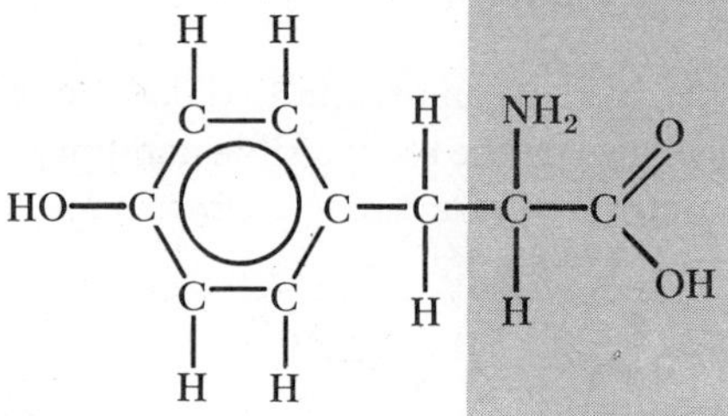

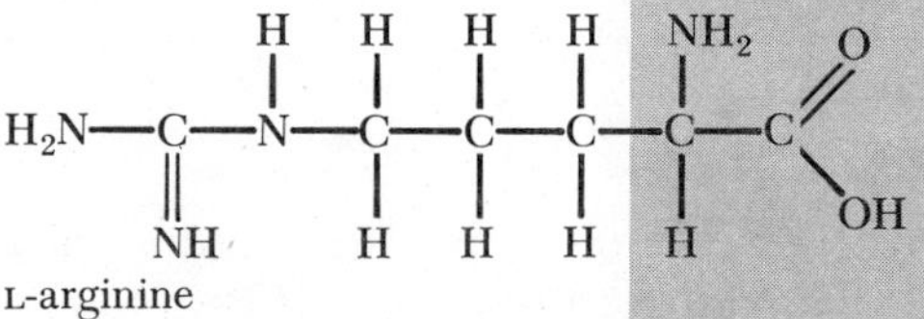

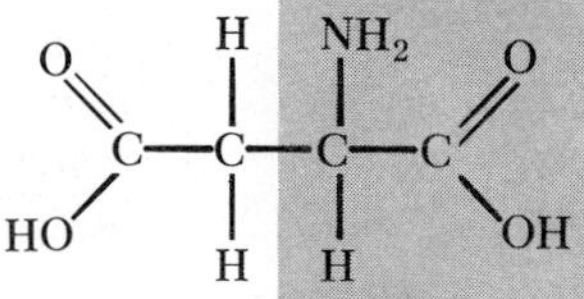

L-aspartic acid

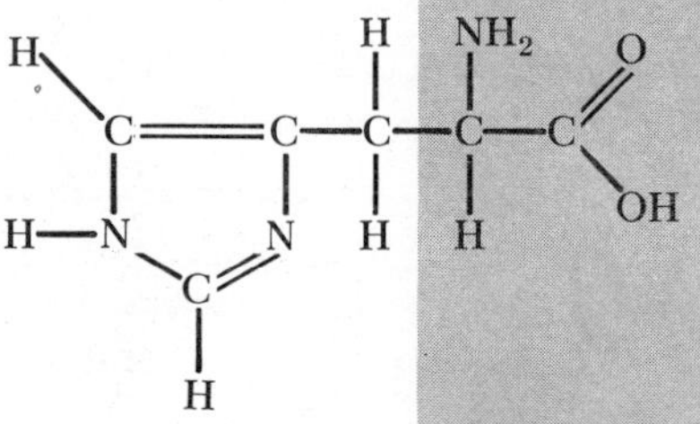

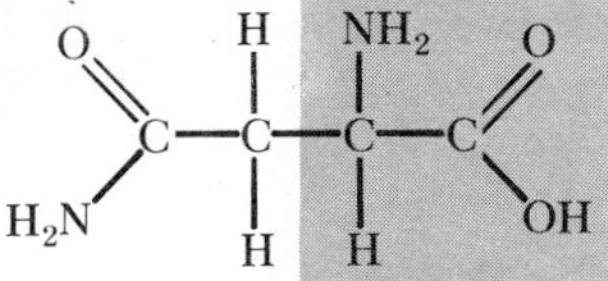

L-tyrosine

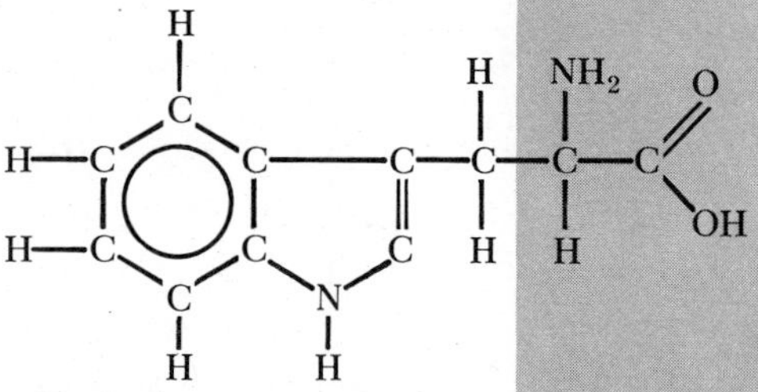

L-asparagine

L-arginine

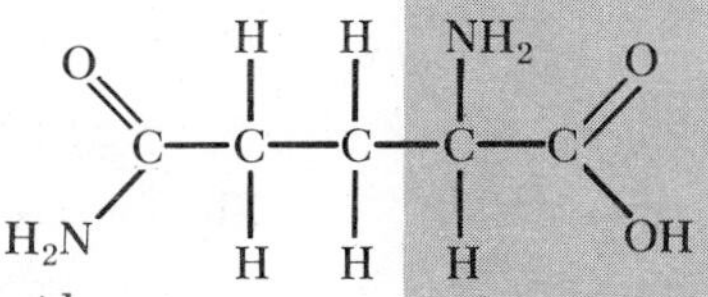

L-glutamic acid

L-histidine

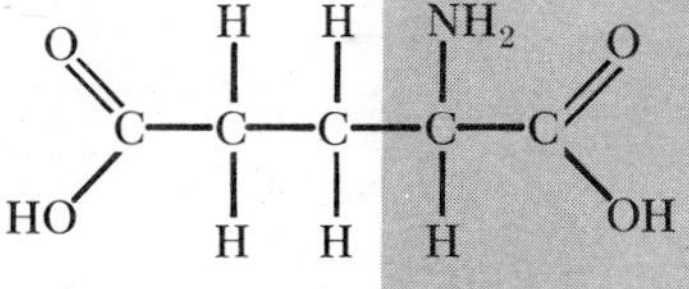

L-glutamine

by the reaction

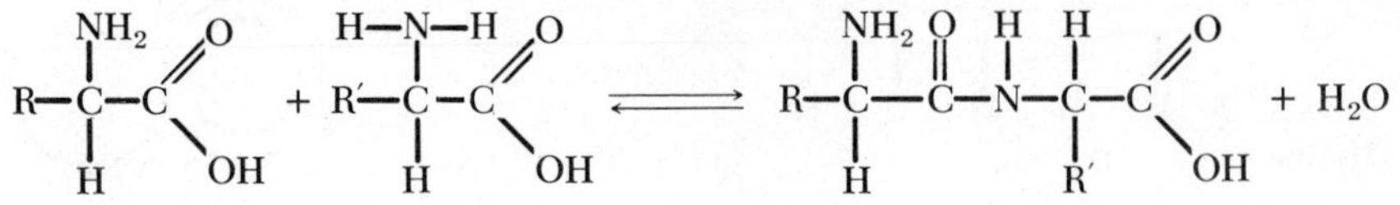

$$R-\underset{\underset{H}{|}}{\overset{\overset{NH_2}{|}}{C}}-\underset{OH}{\overset{O}{\overset{\|}{C}}} + R'-\underset{\underset{H}{|}}{\overset{\overset{H-N-H}{|}}{C}}-\underset{OH}{\overset{O}{\overset{\|}{C}}} \rightleftharpoons R-\underset{\underset{H}{|}}{\overset{\overset{NH_2}{|}}{C}}-\underset{}{\overset{\overset{O}{\|}}{C}}-\underset{}{\overset{\overset{H}{|}}{N}}-\underset{\underset{R'}{|}}{\overset{\overset{H}{|}}{C}}-\underset{OH}{\overset{O}{\overset{\|}{C}}} + H_2O$$

Here the water is the product of condensation of an amino hydrogen with the —OH of a carboxylic acid group on another molecule. This type of N—C chemical bond is called a **peptide bond.** The sequence of amino acids along this peptide chain is the primary structural feature of a particular protein molecule. The sequence of amino acids for at least a significant portion of the molecule has been determined experimentally for several hundred different proteins. It is clear from a consideration of the number of possible combinations of amino acids that only a very small fraction of the conceivable sequences actually correspond to naturally occurring proteins. But amino acid sequence is far from the whole story of protein structure; *geometry,* from the asymmetry of the constituent amino acids to the spatial configuration of the peptide chain, is of critical importance to the shape and function of the protein molecules.

Each of the amino acids (except glycine) can exist in two distinct mirror-image configurations. The difference arises in the two possible nonsuperimposable arrangements of the four groups around a single carbon nucleus (see discussion of optical isomerism in Sections 13-5 and 14-6):

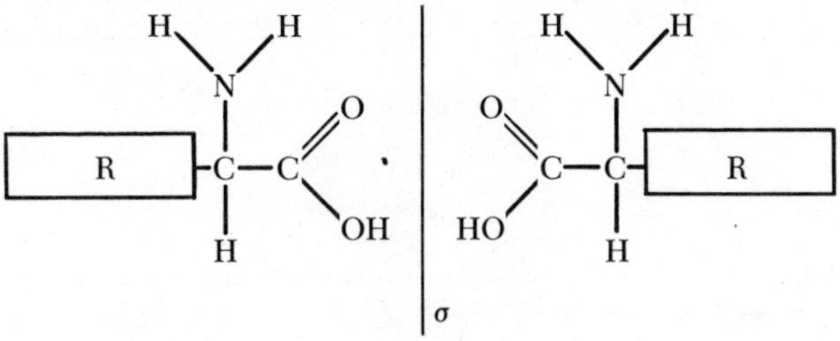

The two isomers are named differently by use of the prefixes D and L. Only L-amino acids have been isolated from proteins. *If* only D or *if* both D- and L-isomers were protein constituents, then the geometry of protein molecules would be very different from that actually found.

Sometimes sections of the peptide chain

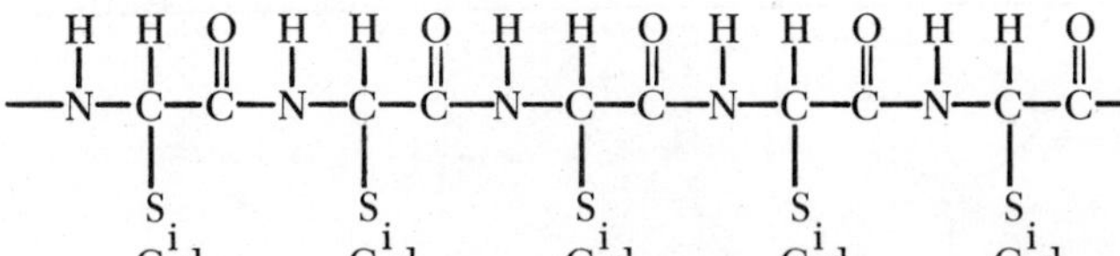

are coiled into a helix of the structure

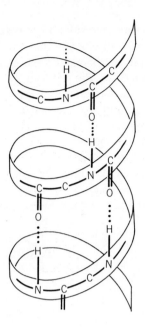

with 3.7

$$\left(\begin{array}{c} H \quad H \quad O \\ -N-C-C- \\ | \end{array} \right)$$

units for each turn of the helix. This helix is stabilized by hydrogen bonds between the

$$\rangle C=O \text{ and } H-N\langle$$

groups in adjacent turns of the helix, as shown by the dotted lines above. The various side groups are on the outside of the central rodlike helix. X-ray diffraction studies of crystalline proteins have revealed that many proteins have helical sections. But a typical protein is not just one long rodlike helix; it is a matted or interlaced network of such structures.

Proline will not fit into such a helix, and so proline in the peptide sequence breaks the rodlike helical structure. Moreover, the presence of an S—S bond between two cysteines, perhaps cysteines very far apart in the sequence, can require the peptide chain to bend. Such disulfide bridges between sections of the protein molecule are often important factors in determining the geometry of the whole molecule.

Often the resulting molecule is a compact, globular mass with various protruding side groups. How can this macromolecule catalyze a chemical reaction? The answer may involve some of these protruding side groups, which turn out to be positioned in just the right places in space.

Structural studies on the particular protein called *lysozyme* have yielded some tentative suggestions about the molecular basis for enzyme action. One of the early protein structure determinations[7] was performed on crystalline chicken-egg white lysozyme, an enzyme that hydrolyzes the sugar polymer which is a major structural component of the cell walls of bacteria. When the sugar polymer is broken down, the cell walls are "lysed" (that is, weakened and then ruptured), killing the bacteria. Lysozymes are thus part of an important defense mechanism for higher organisms against onslaught by bacteria. X-ray diffraction studies reveal that the peptide chain of 129 amino acids in a chicken-egg white lysozyme molecule is folded and twisted in such a way as to leave a long cleft in the surface of the enzyme into which a portion of sugar polymer can fit. Two glutamic acid side groups are oriented so that their free carboxyl groups can simultaneously approach the sugar and, in a cooperative, concerted attack, facilitate the hydrolysis of the

Two functional groups are positioned so that they can attack simultaneously.

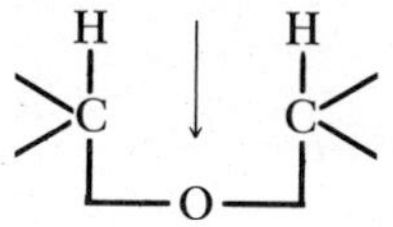

linkage in the sugar polymer.

17-5 Synthesis of Proteins

How is protein synthesized in a living cell? How is the unique sequence of amino acids, which is characteristic of a special enzyme molecule, assembled? And why does one cell manufacture lots of a particular enzyme, whereas a different kind of cell makes almost none? Nucleic acids (discovered in 1868 by Friedrich Miescher, who isolated them from the nuclei of cells) are the key substances for our inquiry. They are compounds which carry genetic information, and which are actively and intimately involved in the synthesis of protein molecules which are the expressions of that genetic information.

Nucleic acids are polymers composed of a chain of sugar-phosphate-sugar-phosphate units to which are attached side groups. A portion of such a polymer is shown in Figure 17-3. For *deoxyribose nucleic acids* (**DNA**), the usual side groups are cytosine, thymine, adenine, and guanine, and the sugar is deoxyribose

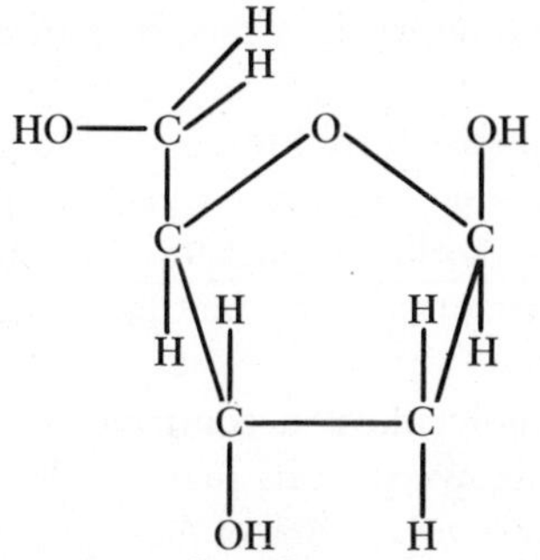

deoxyribose

---

[7] For an excellent presentation of this structure, see D. C. Phillips, "The Three-Dimensional Structure of an Enzyme Molecule," *Scientific American*, vol. **215**, no. 5, p. 78 (Nov. 1966).

Side group

Cytosine

Thymine

Adenine

Guanine

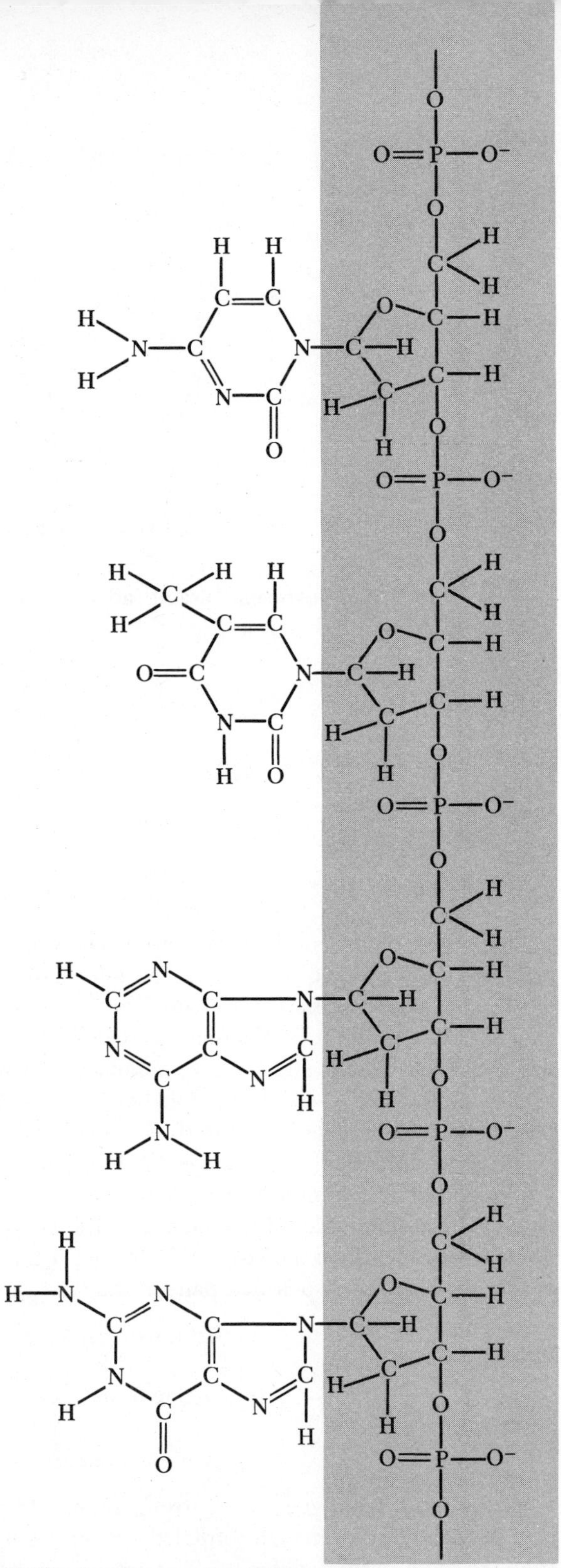

FIGURE 17-3 Constitution of DNA, showing a portion of the repetitious phosphate-sugar-phosphate-sugar backbone. Each of the four side groups found in DNA is depicted.

Another class is composed of *ribose nucleic acids* (**RNA**), polymers which have the sugar ribose

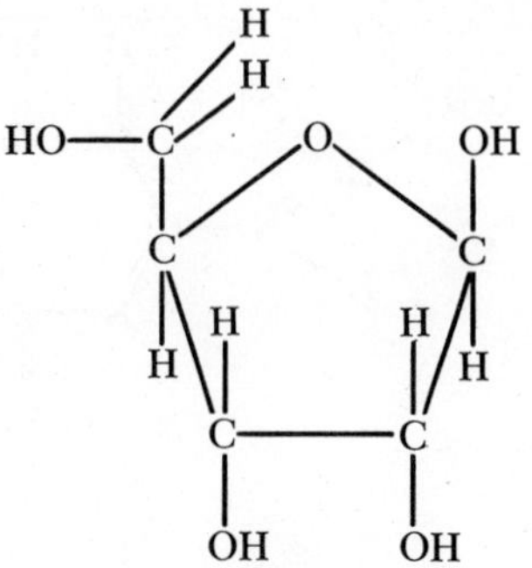

ribose

and the side groups cytosine, uracil, adenine, and guanine. A uracil side group (which is not shown in Figure 17-3) has the structure

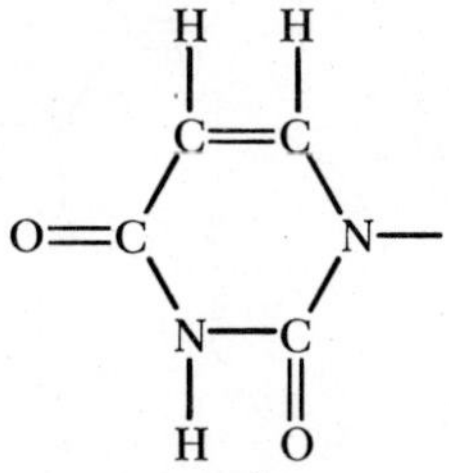

uracil

According to present-day biochemists, DNA molecules are the information carriers of the chromosomes, with the information carried in the *sequence* of side groups along the DNA chain. The DNA alphabet has four letters—A (adenine), G (guanine), T (thymine), and C (cytosine). These letters spell out amino acid sequences, and we have a molecular explanation of how information coded in this A—G—T—C alphabet becomes expressed in the form of particular protein molecules. We shall examine a portion of this explanation.

An *m*-RNA (called a *messenger* RNA because of its role in protein biosynthesis) can be assembled adjacent to a DNA molecule by sequential side-chain pairing. The order of side chains on the DNA molecule determines the order of side chains on the *m*-RNA molecule. It turns out that the geometries of the side chains are such that pairs of the type

guanine-cytosine

adenine-uracil

thymine-adenine

can form weak hydrogen bonds, strong enough to permit the proper lineup of side chains, weak enough so that the resulting *m*-RNA can be dissociated from the DNA without difficulty. The process can be thought about schematically as

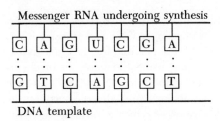

With its own transcription of the genetic information carried in its side-chain sequence, the *m*-RNA molecule moves away and eventually gets to a site of protein synthesis where the appropriate amino acids and enzymes are available. These amino acids have been bonded to molecules called *t*-RNA (*transfer RNA*) which can bond (probably pairing with three *m*-RNA side chains) to line up the amino acids in the proper sequence. Schematically, the lining up of amino acids molecules might be pictured as

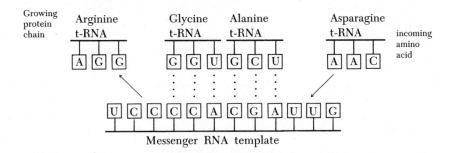

Many questions remain to be answered about this scheme, more questions now in fact than there were before the scheme was invented. But the questions are becoming increasingly chemical and molecular. The final answers about the details of molecular control of metabolic processes may never be found, but the framework in which the questions are to be asked is becoming clearer.

17-6 Summary

The way in which the enzyme *fumarase* catalyzes the conversion of fumaric acid to malic acid in muscle tissue (a conversion which would be much more difficult in glassware, in the absence of fumarase, even at much higher temperature) provides one illustration of how enzymes work. Since fumarase does not catalyze any other known reaction, we see how specific is catalysis by enzymes, and we can begin to appreciate the degree of control of reactions within living systems.

Related conversions of biologically important acids within a tricarboxylic acid cycle (involving familiar citric acid, the main constituent of lemon juice) show how structurally important intermediates for the synthesis of carbohydrates, fats, and proteins in plant tissues can all be derived from one simple progenitor. The *structures* of enzymes were then considered, leading to a survey of protein structures. Lastly, the *synthesis* of proteins within living organisms was considered in terms of the reproduction of specific structures through the agency of messenger RNA and transfer RNA.

We can thus see how it is possible for different living cells to accumulate different reaction products, using the same set of chemical reactions. With genetic information transcribed on DNA molecules in its chromosomes, an individual cell produces its own specialized assortment of enzymes. The relative abundances of various enzymes determine relative rates of competing chemical reactions, thus controlling the steady-state chemical composition of the cell and ultimately determining the biochemical character of that particular cell.

The tale told in this chapter is faithful to the research results that have been reported during the past decade. Yet this is a new story, and many biochemical researchers are active in filling in experimental details and in challenging current explanations. Ten years from now the story will be more complete, and some of the details will be changed. It is possible that there will be some dramatic and fundamental new discoveries. Biochemistry promises to continue to provide us with intellectual adventures in the coming years.

GLOSSARY

amino acids: organic (carboxylic) acids that contain an amino group ($-NH_2$) as substituent. The amino acid constituents of proteins are all α-amino acids (having the $-NH_2$ group attached to the carbon atom immediately adjacent to the carboxylic acid group), and they are all of the L-type isomers.

biochemistry: that area of chemistry which deals with reactions within living organisms and with the explanation of life processes.

DNA: deoxyribose nucleic acid, a nucleic acid (q.v.) based on the C_5 sugar called deoxyribose (see Section 17-5) and a variety of amino acids.

enzyme: an organic substance of complicated structure, produced by a living organism, and capable of inducing chemical change within the biochemical system by specific catalytic action.

metabolism: the sum of all the chemical and physical processes by which living tissue is produced and maintained from nutrients.

nucleic acid: a complex organic compound, made up of sugars, phosphate groups, and nitrogen-containing organic structures, found in the nuclei of living cells.

peptide: a combination of two or more amino acids in a particular configuration, to yield a di-, tri-, tetra-, ..., polypeptide, depending upon the number of amino acid constituents. Peptides are structural entities within proteins and are derivable from such proteins by hydrolysis.

peptide bond: the $-\overset{O}{\overset{\|}{C}}-NH-\overset{|}{\underset{|}{C}}-$ bond formed by the condensation of the carboxylic $-OH$ group of one amino acid with the amino hydrogen of another amino acid. See explanatory diagrams at the beginning of Section 17-4.

photosynthesis: the photochemical process by which plants convert carbon dioxide and water to carbohydrates, fats, proteins, and other biochemical products by absorption of sunlight or equivalent artificial light.

proteins: complex nitrogen-containing organic compounds of high molecular weight, formed by organized condensation of amino acids; the chief constituents of cell protoplasm in plants and animals.

RNA: ribose nucleic acid, a nucleic acid (q.v.) based on the C_5 sugar *ribose* (see Section 17-5) and a variety of amino acids.

***m*-RNA:** "messenger" RNA, a nucleic acid responsible for establishing a certain configuration in protein synthesis.

***t*-RNA:** "transfer" RNA, the carrier of a particular configuration of amino acids in protein synthesis.

PROBLEMS

17-1 Certain vitamins are required as *co-factors* for certain enzymes. Some vitamins have other metabolic functions. Choose a particular vitamin, and write a report describing its molecular structure, its metabolic role, and the relationship between the structure of the vitamin and its chemical function.

17-2 Phenylketonuria is a congenital human metabolic defect that can be detected soon after birth. If treatment is delayed past the first month of life, irreversible brain damage occurs. Find in the library information about the chemistry of phenylketonuria detection, its treatment, and its metabolic basis. Write a summary paper of your research.

17-3 Find out how the human converts ingested starch into some compound of the tricarboxylic acid cycle. Write equations for the individual chemical transformations, including structures for the compounds involved. Identify the reactions involving enzymes secreted into the digestive tract, and the reactions involving enzymes contained within cells.

SUGGESTIONS FOR FURTHER READING

The following articles, written for the nonspecialist, but chemically literate, reader, extend the discussion of this chapter.

Bassham, J. A., "The Path of Carbon in Photosynthesis," *Scientific American*, vol. **206**, no. 6, p. 88 (June 1962).

Horecker, B. L., "Pathways of Carbohydrate Metabolism and their Physiological Significance," *Journal of Chemical Education*, vol. **42**, p. 244 (1965).

Price, C. C., ed., "The Synthesis of Living Systems," special feature report of an American Chemical Society symposium, *Chemical and Engineering News*, vol. **45**, no. 33, p. 144 (August 7, 1967).

Polyanyi, M., "Life Transcending Physics and Chemistry," *Chemical and Engineering News*, vol. **45**, no. 35, p. 54 (August 21, 1967).

A detailed discussion of fumarase is given by Alberty, R. A., a chemist who did many of the studies of mechanism on this enzyme, in his chapter in: Poyer, P. D., Lardy, H., and Myrbäck, eds., *The Enzymes*, 2nd ed., New York: Academic, vol. **5**, part B, p. 531 (1961).

An authoritative review of investigations involving the tricarboxylic acid cycle is given by J. M. Lowenstein in his chapter in GREENBERG, D. M., ed., *Metabolic Pathways*, 3rd ed., New York: Academic, vol. 1, chap. 4 (1967).

SCHROEDER, W. A., *The Primary Structure of Proteins: Principles and Practices for the Determination of Amino Acid Sequence*, New York: Harper & Row (1968) is a discussion of experimental methods which have been used for determination of amino acid sequences in proteins. Results of protein structure determinations are presented in: DAYHOFF, M. O., ed., *Atlas of Protein Sequence and Structure*, Silver Spring, Md.: National Biomedical Research Foundation, biennial cumulative volumes. About 700 protein sequences were reported in volume **45** (1972).

Protein biosynthesis is explored in its varied aspects in a series of review articles in: ANFINSEN, C. B., ed., *Aspects of Protein Biosynthesis*, New York: Academic, parts A and B (1970).

DNA and RNA play central roles in the biosynthesis of proteins, and an excellent source of information about these molecules is the book: SPENCER, J. H., *The Physics and Chemistry of DNA and RNA*, Philadelphia: Saunders (1972). See also: WOLD, F., *Macromolecules: Structure and Function*, Englewood Cliffs, N.J.: Prentice-Hall (1971).

Representative of contemporary textbooks, surveying the broad sweep of biochemistry, are: KARLSON, P., *Introduction to Modern Biochemistry*, 3rd ed., New York: Academic (1968); LEHNINGER, A. L., *Biochemistry*, New York: Worth Publishers (1970); and MAHLER, H. R. and CORDES, E. H., *Biological Chemistry*, 2nd ed., New York: Harper & Row (1971).

If biochemical processes are regulated by enzymes, then the absence of an enzyme (ordinarily present in a particular kind of cell) can be expected to have serious consequences. Ten human diseases involving deficiencies of enzymes that catalyze the cleavage of fat-sugar complexes are discussed in: BRADY, R. O., "Hereditary Fat-Metabolism Diseases," *Scientific American*, vol. **229**, no. 2, pp. 88–97 (August 1973).

Microbiological fermentation of carbohydrates has been used for thousands of years to produce ethanol (in wines, meads, beers, ales, and distilled spirits). Fermentation biochemistry has even greater possibilities. Recently, carefully controlled microbiological processes have been developed for the production of a wide range of organic chemicals by fermentation of agricultural wastes, the sludge from municipal sewage disposal plants, animal manures, and other "useless" materials. Computer control of the process variables, utilizing detailed information about metabolic enzyme reactions (this chapter) and chemical-reaction mechanisms (Chapter 11), is essential if high yield of the desired substance is to be achieved. See "Fermentation Heads for Higher Productivity," *Chemical and Engineering News*, vol. **51**, no. 12, pp. 32–34 (March 19, 1973).

The human body cannot synthesize all of the amino acids required for its own protein synthesis. Tryptophan, leucine, isoleucine, lysine, valine, threonine, either tyrosine or phenylalanine, and either cysteine or methionine must be present in the diet in appropriate quantities. Many dietary protein sources are deficient in some of these *essential amino acids*, but prudent nutritional planning should produce menus that supply these needed nutrients. A readable source of such food information, with recipes and a good bibliography, is: LAPPÉ, F. M., *Diet for a Small Planet*, New York: Friends of the Earth (1971).

18
Case Studies in Environmental Chemistry

18-1 Abstract

Rain falls on the mountains and runs to the sea. The mountains weather; minerals dissolve in the streams and are carried to the sea. Water from the sea evaporates, and the water vapor eventually condenses into pure water in the form of clouds and rain, some of which falls again on the land. The energy to fuel this cycle pours lavishly on the earth from the sun, which also provides the energy for the growth and metabolism of plants and the animals that feed on them. The surface on which we live is in fact a dynamic system, enormously complex, vast by ordinary human standards, and now the subject of great concern, for it has proved less vast, more complex, and more fragile, in the chemical sense, than we had thought. Some of the understanding that we need to prevent further degradation of the life-supporting capability of the earth can be found in the study of chemistry. However, no one but the most fanatic reductionist would claim that the study of chemistry alone provides a *sufficient* insight into the workings of the environment. The point of this chapter is that since the biosphere is composed of ordinary matter, it is subject to the limitations of all other ordinary matter, and hence subject to all the generalizations of chemists.

18-2 Equilibrium and Cyclic Models of the Ecosphere

One of the basic generalizations of chemists is that of *chemical equilibrium.* We may at first doubt the usefulness of equilibrium models for vast, blooming, buzzing, heating, cooling, metabolizing areas of the earth as though those areas were specimens in a laboratory. Clearly, the surface of the earth is not in a state of chemical equilibrium. We may be grateful for that, for life would be impossible if all were indeed at equilibrium—all carbon would be oxidized; the atmosphere would contain much carbon dioxide but little oxygen; there would be no food available, and so on.

It is possible that processes which are rapid compared to the rate at which they are perturbed by inputs of matter and energy, or which are slow compared to the rate of a cyclically fluctuating disturbance (such as the daily and annual variations of temperature that effectively average out over a period of days or years) may indeed be close to equilibrium, or at least moving toward it. Such behavior will be discernible and comprehensible if we know what the state of equilibrium should be, and what variables affect its attainment. We can then make periodic observations and find out in what direction the system is changing.

Let us take a familiar example. We know that the atmosphere contains variable amounts of water vapor; we feel the variation when, after a humid, sticky day, a cool front brings drier air. Since 71 percent of the surface of the earth is covered with water, we might expect that a liquid-vapor equilibrium should be set up, and that the atmosphere should be saturated with water vapor at all times. A simple closed laboratory model reaches liquid-vapor equilibrium within a few hours (Figure 18-1); surely the full 4.5 billion year age of the earth would be more than a sufficient time. In the real system, however, the rotation of the earth and the influx of energy from the sun combine to produce winds that, far from promoting equilibrium by stirring, actually prevent it. A commonly observed weather pattern involves the condensation of moisture out of air flowing off the ocean as that air passes over high mountains, leaving the air drier. As a result, we have, for instance, the wet climate of the Pacific Northwest and the dry climate of the desert to the east of it. So we should not expect all air to be at equilibrium with liquid water at all times. Instead, we expect local variations due to fluctuations of temperature and wind. At any one time and place, however, we fall back on the equilibrium condition as a standard of reference, and we think of the moisture content of the air in terms of a *ratio* between the actual measured content of water vapor per unit volume, and the maximum (or equilibrium) quantity of water vapor which that volume of air could hold at the same temperature. This ratio we call the *relative humidity*.

There has been enough time to achieve equilibrium, but there are clearly contrary forces.

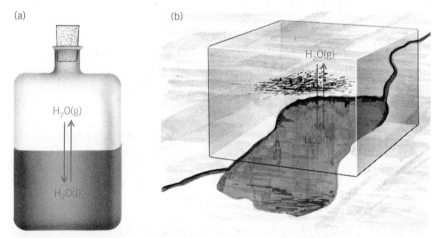

FIGURE 18-1   Liquid-vapor equilibria of water. (a) As actually realized in a glass bottle. (b) As approached over a large lake on the surface of the earth. To insure equilibrium, this macroscopic system would have to be enclosed in a large, airtight, lightproof box.

The notion of an equilibrium state is thus useful as a reference framework for the real system. It also plays a less theoretical role in nature, as we saw in Chapter 9, as a goal toward which the real system must always drift. Hence an understanding of the equilibrium state is even more necessary for a more sophisticated description of the earth as a steady state or a slowly changing system. The state of a real system results from a balance between the *rate* at which the system would approach equilibrium if it were isolated (in our simple example, the rate of evaporation of water from the ocean) and the *rate* at which the system is being perturbed (for example, by the input of heat from the sun). Since in most geochemical cases, neither of these rates is known accurately, all we can do is to deduce the equilibrium model from chemical data and compare the actual system with it. In a surprising number of cases, the agreement is good enough to be encouraging.

In the following section, we shall examine the "fit" between the real ocean and atmosphere, on the one hand, and an equilibrium model based on the solubility of the common chemical compound calcium carbonate in sea water. To the extent that we find agreement between the real system and the model, we may have learned something, but very much about even this relatively simple system remains undecided.[1] Quite possibly, few sweeping generalizations about the total earth system will ever be possible.

18-3 Limestone, Carbonic Acid, and Carbon Dioxide

One of the field tests that geologists use to identify limestone ($CaCO_3$) and other carbonate-containing rocks depends on the following scheme of linked equilibria:

$$CaCO_3(s) = Ca^{2+} + CO_3^{2-} \qquad (18\text{-}1)$$
$$+$$
$$H_2O$$
$$\|$$
$$HCO_3^- + H_2O = H_2CO_3 + OH^-$$
$$+ \qquad\qquad \|$$
$$OH^- \qquad\qquad CO_2(aq) = CO_2(g)$$
$$+$$
$$H_2O$$

The solubility of $CaCO_3$ in water is normally rather small, but when a rock is treated with hydrochloric acid, the OH^- ions are neutralized, and the carbonate-carbonic acid equilibria are shifted toward the right in Equation (18-1) (see Chapter 7). Hence the $CaCO_3$ dissolves in the acid to the accompaniment of fizzing bubbles of CO_2.

Shifting equilibria removes and forms carbonate rocks.

The equilibria in Equation (18-1) are important even in the absence of inquisitive geologists. They account, for example, for the formation of caves when rainwater and CO_2 combine to dissolve limestone:

$$H_2O + CO_2 + CaCO_3(s) = Ca^{2+} + 2HCO_3^- \qquad (18\text{-}2)$$

They also account for the formation of stalactites (and other cave oddities); when the equilibrium of Equation (18-2) shifts to the left as evaporation removes water and CO_2 from hanging drops of solution within a cave, the Ca^{2+} and CO_3^{2-} ions must precipate as new limestone.

Because all of the equilibria in Equation (18-1) are linked together, the solubility of $CaCO_3$, the distribution of carbonate among the conjugate acid-base forms, the pH of the aqueous solution, and the CO_2 pressure in the gas phase are all interdependent. If an equilibrium model applies to the real

---

[1] For an argument against the validity of this model, see K. E. Chave, "Chemical Reactions and the Composition of Sea Water," *Journal of Chemical Education*, vol. 48, pp. 148–151 (March 1971).

system, we should find that the concentrations of seven species (see Table 18-1) are consistent with the equilibrium constants governing it.

TABLE 18-1
Species Involved in Calcium Carbonate Equilibria

Solution phase	Observed value	Calculated value
Ca^{2+}	1.0×10^{-2} M	1.8×10^{-2} M
CO_3^{2-}	2.1×10^{-4} M	1.2×10^{-4} M
HCO_3^-	2.3×10^{-3} M	1.5×10^{-3} M
H_2CO_3	5.5×10^{-8} M	9.5×10^{-8} M
CO_2	2.4×10^{-5} M	1.6×10^{-5} M
H_3O^+	8×10^{-9} M	------------
Gas phase		
CO_2	3×10^{-4} atm	------------

In Table 18-1 we confront the predictions of the equilibrium model with the reality of analytical data.

Mathematically, the resolving of a system of seven unknown quantities requires seven independent equations. Five such equations are provided by the various equilibrium quotients which are known. We have not included both H_3O^+ and OH^- as unknown species, but if we had, we could include another equilibrium constant, the K_w, to relate them. The five equations are listed in Table 18-2.[2] In the complicated real system, our seven species are just part of the total picture, and charge and proton balances would bring in other ions (such as Na^+, SO_4^{2-} and Cl^-) that would only complicate things. Hence we shall take direct measurements of the concentrations of two species to provide the necessary additional information.

TABLE 18-2
Independent Conditions Governing Calcium Carbonate Equilibria

Equilibrium Quotients:

1. $CaCO_3(s) = Ca^{2+} + CO_3^{2-}$ $\qquad K_{sp} [Ca^{2+}][CO_3^{2-}]$

2. $CO_2(aq) + H_2O = H_2CO_3(aq)$ $\qquad K_{hyd} = \dfrac{[H_2CO_3]}{[CO_2(aq)]}$

3. $H_2CO_3 + H_2O = H_3O^+ + HCO_3^-$ $\qquad K_1 = \dfrac{[H_3O^+][HCO_3^-]}{[H_2CO_3]}$

4. $HCO_3^- + H_2O = H_3O^+ + CO_3^{2-}$ $\qquad K_2 = \dfrac{[H_3O^+][CO_3^{2-}]}{[HCO_3^-]}$

5. $CO_2(g) = CO_2(aq)$ $\qquad K_H = \dfrac{[CO_2(aq)]}{P_{CO_2(g)}}$ (Henry's law)

---

[2] In a truly isolated laboratory model, we could supply further independent equations such as a charge balance, and either a proton balance or some conservation of matter equations.

Because all of the species are interdependent in the equilibrium model, it does not matter in principle which two we choose to measure. In fact, the pH of the ocean and the partial pressure of CO_2 in the atmosphere are reliably known and are nearly constant over all the globe, so we shall take those two concentration terms as "given." If we take the values of $[H_3O^+]$ and p_{CO_2} to be 8×10^{-9} M and 3×10^{-4} atm, respectively, as in Table 18-1, we can then calculate expected values by means of the equations

$$[CO_2(aq)] = K_H p_{CO_2} \tag{18-3}$$

$$[H_2CO_3] = K_{hyd}[CO_2(aq)] \tag{18-4}$$

$$[HCO_3^-] = \frac{K_1[H_2CO_3]}{[H_3O^+]} \tag{18-5}$$

$$[CO_3^{2-}] = \frac{K_2[HCO_3^-]}{[H_3O^+]} \tag{18-6}$$

$$[Ca^{2+}] = \frac{K_{sp}}{[CO_3^{2-}]} \tag{18-7}$$

The values that result from such calculations are given in the third column of Table 18-1, following the observed values in the real ocean. The correspondence between the concentrations calculated from equilibrium constants and those actually observed by analysis of sea water is remarkably close, especially in view of local fluctuations of temperature and pressure. Such agreement encourages us to believe that all reliable quantitative information gained in the laboratory, for whatever purpose, can help us to understand what is going on within complicated natural systems. The agreement does not prove that the natural system is in equilibrium (it is not good enough for that), but it does tell us that our idealized equilibrium model portrays the situation effectively, and so enables us to figure out what to expect if this or that happens, or even how we may influence or control the situation. In this respect, the exercise we have gone through is like building a model of a dam and experimenting with it, in order to understand the situation and foretell what might happen before it overwhelms us.

18-4 Cycles of Nutrients

Complex as the carbonate equilibria may seem (and are), they are only a small part of carbon's chemical pattern in the environment. As we all know, carbon dioxide is present in the breath of animals as a product of metabolism, and it is both produced and consumed by photosynthetic plants. The cycle of carbon utilization is strongly dominated by living organisms, and the carbonate equilibria we have discussed must adjust themselves to the disturbing influences of the various forms of life. Indeed, there is some evidence that when such disturbing influences fluctuate in a regular way, the solubility of $CaCO_3$ follows them. Let us reconsider Equation (18-2):

$$H_2O + CO_2 + CaCO_3(s) = Ca^{2+} + 2HCO_3^-$$

In warm surface waters that are rich in photosynthetic plankton, the concentration of CO_2 in the water drops to a low level, causing $CaCO_3(s)$ to precipitate as the equilibrium of Equation (18-2) shifts to the left. At night, when the lower temperature and absence of photosynthesis combine to increase the solubility and the survival of CO_2 in the water, $CaCO_3$ becomes more soluble, as Equation (18-2) shifts to the right. In protected spots, it is possible to observe a daily cycle of $[Ca^{2+}]$: low during the day, higher at night.[3]

Plankton consumes CO_2 in the sunlight, and produces CO_2 in the dark

THE CARBON CYCLE. After hydrogen and oxygen (mostly present as water), carbon makes up by far the largest part of the mass of living organisms. As we have seen in Chapter 14, the variety of forms and reactions of carbon compounds is enormous. The most important compound of carbon in nature, the coin of the carbon-cycle realm, is carbon dioxide (Figure 18-2). All of the carbon present in organic compounds of biological origin at any given time was once carbon dioxide and was reduced to "organic" carbon through photosynthesis:

$$nCO_2 + nH_2O \xrightarrow[\text{chlorophyll}]{\text{sunlight}} (CH_2O)_n + nO_2 \tag{18-8}$$

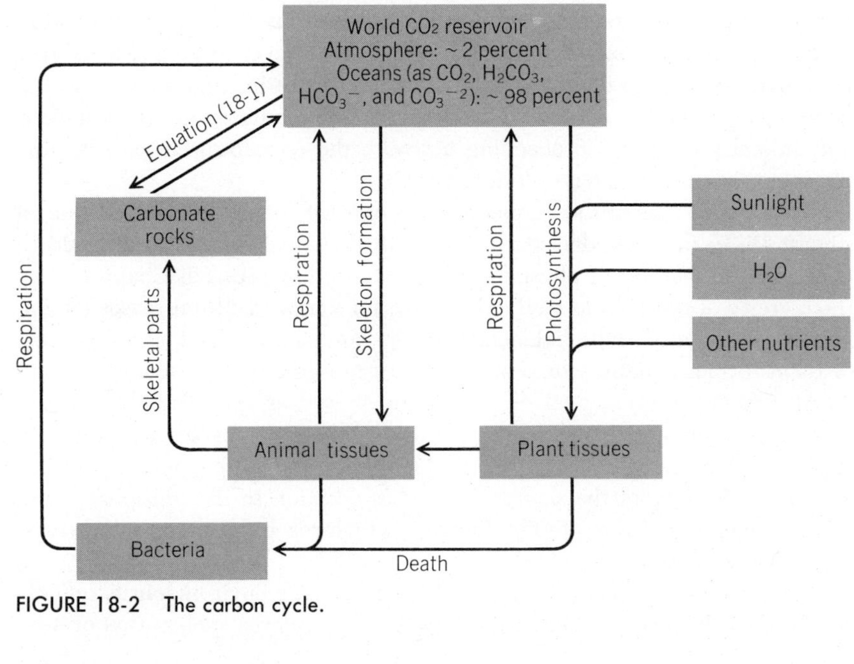

FIGURE 18-2 The carbon cycle.

---

[3] For a thorough and logical argument that the composition of the ocean is entirely dominated by biological cycles, see W. S. Broecker, "A Kinetic Model for the Chemical Composition of Sea Water," in *Quaternary Research*, vol. 1, pp. 188–207 (1970).

The conversion actually is stepwise and very complicated; our equation is a vast oversimplification which just shows the overall result as CO_2 is reduced and converted to a carbohydrate. Other equations would be needed for the formation of fats and proteins, and so on.

The annual conversion of CO_2 into organic carbon via photosynthesis (which is nearly balanced by that returned via respiration and combustion) amounts to about 0.1 percent of the CO_2 present in the air and ocean at any given time. Thus, there is ample carbon dioxide on the earth to sustain much greater conversion to organic carbon *(primary productivity)* than is actually realized. It is characteristic of living systems that they will grow and reproduce as fast as conditions permit. We might wonder, then, why more CO_2 than this trifling 0.1 percent is not used; usually, the answer lies in the limited availability of other nutrients that are not shown in Equation (18-8), chiefly nitrogen and phosporus.

THE PHOSPHORUS CYCLE. The principal locations and pathways of phosphorus in an aquatic ecology are shown in Figure 18-3. The phosphate reservoir is smaller and more complex than the carbon reservoir. Much (possibly half) of the phosphorus is in the form of organic phosphates (Chapters 14 and 17), some of which can be used by plants for the growth of new tissue. Of the inorganic phosphates, much may be present in the form of quite stable (and thus biochemically inactive) Lewis acid-base complexes (for example, of Ca^{2+} or Fe^{3+}), depending on the mineral content of the water. The net total phosphate available for plant growth from these sources varies greatly from place to place and from season to season, but it is never in great excess over the amount built into plant and animal tissues except for regions (such as the deep ocean) where productivity is very slight. Where photosynthesis and plant and animal growth are proceeding actively, the concentration of available phosphorus is nearly always close to zero.

Phosphorus availability often limits plant growth.

Thus, in an undisturbed aqueous environment (below the dotted line in Figure 18-3), the reproduction and growth of living organisms are held in check by the scarcity of phosphorus (and, as we have seen in Chapter 1, by the scarcity of nitrogen as well). However, for various cultural reasons, mankind has intervened in this balanced cycle by mining phosphate minerals (most of them former aquatic sediments) and returning them to the contemporary aquatic scene. A great deal of phosphate is used as agricultural fertilizer; whatever phosphate is not used by plants and thereby recycled runs off with rainfall. Other phosphates are used as detergents, and these reach the streams and ponds by way of the sewers. This is in addition to the phosphates and nitrogen compounds excreted by humans, which reach the streams and rivers by the same path.

Some of the phosphate thus added to the aquatic environment is quickly removed by the formation of metal-phosphate complexes, such as that of iron:

$$Fe^{3+} + HPO_4^{2-} = FeHPO_4^+ \qquad (K = 2 \times 10^8) \qquad (18\text{-}9)$$

Another fate is to become a solid sediment:

$$3PO_4^{3-} + OH^- + 5Ca^{2+} = Ca_5(PO_4)_3(OH)\ (s) \qquad (K = 4 \times 10^{55}) \quad (18\text{-}10)$$

[where $Ca_5(PO_4)_3(OH)$ (s) is hydroxyapatite, a common mineral and the chief constituent of bones and teeth] provided the proper metal ions are present. But much of it is taken up by growing plants which, as we have seen, generally have available all the carbon dioxide they need to form new tissues. During a "bloom" of algae, it is estimated that the average phosphate ion in the ecosystem has a free lifetime of 2 or 3 min before it is incorporated by the hectically growing algae. In calmer times, this may extend to weeks.

The phosphorus cycle is dynamic, involving some rapid exchange rates.

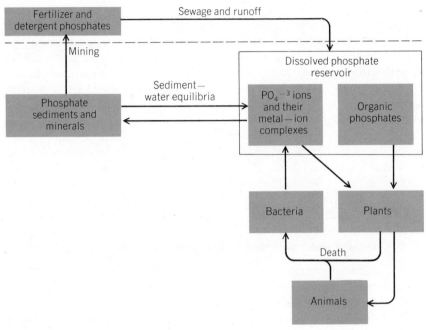

FIGURE 18-3 The phosphorus cycle.

The growth of plants and animals requires, of course, more than just carbon, nitrogen, and phosphorus in the form of appropriate compounds. Many other elements are required in greater or lesser quantity, some of them only in traces, and in a particular situation, the lack of any one of these may inhibit productivity. As in any chemical process, the amount of product cannot exceed the quantity chemically equivalent to the scarcest reagent. But in a rough way, the overall "reaction" for the formation of plant tissue may be written:[4]

$$106CO_2 + 16NO_3^- + HPO_4^{2-} + 122H_2O + 18H^+ + \text{trace nutrients}$$
$$\longrightarrow \underset{\text{"algae"}}{C_{106}H_{263}O_{110}N_{16}P} + 138O_2 \qquad (18\text{-}11)$$

---

[4] See "Suggestions for Further Reading."

This overall change results in the production of algae (of average composition) and is powered by sunlight. It is reversible; the leftward direction represents the respiratory oxidation of tissues that powers both plant and animal activity.

The "algae" product on the right-hand side of Equation (18-11) is, of course, not a single chemical compound, but thousands of different compounds, structured into living plants. Consequently, the composition of even a single species of algae is variable within limits. Nitrogen- or phosphorus-starved algae will grow, though not rapidly, to produce new tissue that is lower in N or P than Equation (18-11) indicates, and conversely, algae may hoard up "luxury" nitrogen or phosphorus if they grow in a nutrient-rich medium. Thus, Equation (18-11) describes the stoichiometry of plant growth only within relatively broad limits. The absence of absolute stoichiometry for large collections of molecules is in fact rather general; this fact does not deter us from writing simple equations as convenient summaries of chemical change.

Subject to these considerations, we define that nutrient which is in scarcest supply relative to the demand for it as a *limiting nutrient*.[5] More than one nutrient may play this role; thus, a common situation is one in which all of the substances on the left-hand side of Equation (18-11) are present in abundance except for nitrate and phosphate ions, which are present in low concentrations at approximately the 16:1 ratio required by the equation. In this case, both act jointly as limiting nutrients.

Under the condition of nitrogen and phosphorus limitation of algae growth, the "cultural" addition (through sewage and agricultural runoff) of these nutrients to a body of water may lead to exuberant growth. Geochemists classify bodies of water according to three extreme conditions of productivity:

Plant growth may be an indicator of chemical changes in a body of water.

1. Oligotrophic ecosystems support only relatively sparse growths of algae and diatoms (and, because these are the basis of the aquatic food chain, of larger organisms).
2. Eutrophic waters are rich in nutrients and support a blooming growth of life.
3. Dystrophic waters may contain ample nutrients, but, because of other chemical factors (such as natural poisons leached from the surrounding vegetation or minerals, or industrial pollution), cannot support life.

Thus, for example, the upper Great Lakes (Huron and Superior) are oligotrophic. Lake Superior, in particular, is mostly surrounded by relatively wild

---

[5] The availability of a nutrient in a dynamic system is not solely reflected in its free concentration in the environment. What is more important is the *rate* at which it can be supplied. Thus, even though there is an ample supply of CO_2 in the air, carbon may, in some cases, act as a limiting nutrient if there is no rapid mechanism for the transfer of this carbon to the metabolic sites of submerged aquatic plants. In most waters, the reservoir of dissolved carbonates acts as this intermediating mechanism. In a few carbonate-poor freshwater bodies, it appears that growth rates are limited by the (rather slow) rate of diffusion of CO_2 from the air into the water, and in this case, CO_2 is a limiting nutrient. The discovery of such cases sparked considerable controversy, partly scientific and partly economic in motivation, when it was announced at the height of the controversy over the use of phosphates in detergents. For a fuller discussion, see "Suggestions for Further Reading."

country, and its deep, cold waters contain few nutrients. Lake Erie, by contrast, is warmer, shallower (so that rooted plants can grow farther out from the shoreline), and subject to nutrient discharges accumulated from large cities (Detroit, Toledo, Cleveland, and Erie) and widespread agriculture in the surrounding countryside. Lake Erie is, therefore, eutrophic. The Cuyahoga River, which runs into Lake Erie at Cleveland, is eutrophic in its upper reaches, where it collects agricultural runoff, but dystrophic farther downstream due to industrial and urban dumping of toxic and oxygen-consuming substances.

The dystrophication of rivers and lakes through the indiscriminate dumping of sewage, industrial wastes, and pesticides is, of course, unambiguously undesirable. In contrast, the eutrophication process is less easily handled, both technologically and philosophically. All inland bodies of water have a natural life course from oligotrophic toward eutrophic, as the surrounding vegetation slowly adds nutrients through death and decay. Unless a given pond or lake becomes dystrophic, its destiny is to become more and more laden with algae and aquatic weeds and the animal food chain they support. As organic debris accumulates on the bottom, the water becomes shallower and increasingly subject to further eutrophication. A lake becomes a weedy pond, then a marsh, and finally dry land (on a time scale from hundreds to tens of thousands of years, depending on its original size and depth and the mineral content of the surrounding land). Most of the eutrophicating activities of mankind (agricultural fertilization and sewage disposal) are not of the sort that can be stopped on demand. They may be said to contribute to a process that goes on more slowly in our absence, but still goes on.

There have been some promising experiments in the extraction of nutrients from sewage and their use as fertilizer where they will do some good, but at present such recycling schemes present economic and technological problems that are still unsolved. These problems are a challenge to the new generation of scientists and are typical of the many unglamorous but highly necessary scientific tasks we now face.

18-5 Air Pollution and Photochemical Smogs

Many cases of air pollution are somewhat analogous to the industrial dystrophication of waters—a toxic or otherwise dangerous substance (for example, carbon monoxide from automobile exhausts, or sulfur dioxide from sulfur-bearing coal) is released to the atmosphere and hence dumped into the environment, and the only unsolved problem the situation presents is the economic one of stopping it. However, in at least one case, a good deal of chemistry intervenes between the initial pollution and the resulting insult to the environment. That is the case of the photochemical, or "Los Angeles-type" smog.

Most urban smogs are rich in reducing agents. They often result from the incomplete oxidation of fuels to give CO and SO_2 along with excess unburned fuel (generally hydrocarbons). The potential for further oxidation to CO_2 and SO_3 or sulfates accounts for the reducing properties of these "London-type" smogs.

However, cities which combine bright sunshine and heavy automobile traffic (like Los Angeles, Mexico City, and many others) are plagued by smogs which are characteristically *oxidizing* in character. These smogs are relatively rich in ozone (up to one part per million), in sulfuric acid aerosol, and in a large array of oxidized and irritating organic compounds, including formaldehyde, peroxides, and peroxyacyl nitrates (see below). For some time our understanding of these oxidizing mixtures was as hazy as downtown Burbank; only the basic reactions have recently been clarified, and some of the details are still obscure.

To begin with, let us remember that in a modern high-compression internal combustion engine, some nitrogen is burned along with the hydrocarbon fuel (see Chapters 1 and 3). Part of the exhaust gas from such engines, therefore, contains oxides of nitrogen (mainly NO, but with some N_2O and NO_2).[6] The NO may be oxidized fairly rapidly by oxygen and by other smog-produced oxidants to NO_2, itself highly toxic to both plants and animals. (The reddish-brown color of NO_2 is responsible for the characteristic color of many urban smogs.) But the real trouble begins with the reaction

Remember NO_2? See Ch. 9.

$$NO_2 \xrightarrow{\text{sunlight}} NO + O \quad (18\text{-}12)$$

A photochemical reaction like Equation (18-12) requires that a photon of light be absorbed by the NO_2, and that the photon carry enough energy to break a nitrogen-oxygen bond. Most of the energy of sunlight striking the earth's atmosphere is absorbed by atmospheric gases; only a few narrow ranges of wavelengths survive to reach the atmosphere near the surface. The highest-energy light that does so is the visible and near-ultraviolet radiation. Unfortunately, NO_2 is able to absorb photons in the high-energy blue and near-ultraviolet region, and it is these photons that lead to reaction (18-12) (Figure 18-4).

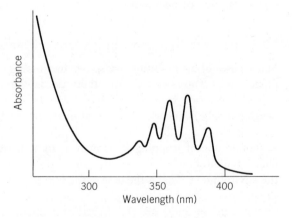

FIGURE 18-4 Absorption spectrum of NO_2, showing absorption of light in the ultraviolet region (wavelength less than 300 nm) and in the blue and near-ultraviolet region (wavelength between 340 and 380 nm).

---

[6] Because the exact mixture of nitrogen oxides depends on how the engine is operated, this mixture is often referred to in environmental journals as NO_x.

With the production of O atoms, we have a really reactive species that can, in turn, oxidize many other substances:

$$O + SO_2 \longrightarrow SO_3 \quad \text{(soluble in water; see below)} \quad (18\text{-}13)$$

$$O + O_2 \longrightarrow O_3 \quad \text{(ozone)} \quad (18\text{-}14)$$

These two products account for much physical destructiveness of an oxidizing smog. Sulfur trioxide is the anhydride of sulfuric acid; that is, the reaction

$$SO_3 + H_2O \longrightarrow H_2SO_4 \quad (18\text{-}15)$$

produces sulfuric acid which, in this case, takes the form of an aerosol of tiny drops. Sulfuric acid is a strong Brønsted acid (Chapter 6), destructive to otherwise stable materials by protonation reactions. Smog-produced sulfuric acid is credited with considerable damage to limestone structures and to marble statues and fountains in Rome and Vienna, and to the rapid deterioration of leather, nylon, and other complex organic materials in urban settings. Ozone, for its part, does not live up to its romantic image as "mountain air"; it is also highly toxic. One part per million of ozone (which is about as high as a smog-produced concentration can go) causes bronchitis and related troubles in rats, although such damage has not been demonstrated in humans (possibly because oil aerosols have been shown to protect lung tissues against ozone). Ozone also causes cracking and deterioration of rubber and similar polymers, particularly stretched rubber, which is chemically less stable than relaxed rubber. (Can you see why? See Chapter 9 for a clue.)

Ozone and atomic oxygen also play important indirect roles in photochemical smogs. They attack double bonds in unsaturated hydrocarbons (Chapter 14), which are provided by automobile exhaust and by evaporation from gasoline tanks and from storage areas. It is well established that ozone reacts with such hydrocarbons to form a variety of irritating and destructive oxygen-containing organic compounds. Atomic oxygen also is a voraciously reactive substance, which accounts for the fact that we never find it in nature except as a trace constituent of gas mixtures. It can react with both saturated and unsaturated hydrocarbons, as by the "abstraction" of a hydrogen atom from hydrocarbons

$$R\text{--}H + O \longrightarrow R\cdot + \cdot OH \quad (18\text{-}16)$$

followed by dimerization of the ·OH radicals

$$2 \cdot OH \longrightarrow H_2O_2 \quad \text{(hydrogen peroxide)} \quad (18\text{-}17)$$

The creation of free radicals (molecules with an unpaired electron) and peroxide also leads to other free-radical chain reactions, in which the unpaired electron is passed from one molecule to the next, while producing a mixture of oxygenated products. These include the common eye irritant formaldehyde (the substance familiar to biology students as a tissue preservative) and acrolein, an irritating substance of whose vapors the *Merck Index of Chemicals and Drugs* says, "Strong irritant of skin and exposed mucosae."

formaldehyde

acrolein

The combination of NO_2 and organic peroxides also leads to a new class of irritating substances called the *peroxyacetyl nitrates*, the simplest and most prevalent of which is peroxyacyl nitrate (PAN):

peroxyacyl nitrate

PAN is a powerful eye irritant and plant toxin. It has been called the leading eye irritant of city dwellers and, in addition, is estimated to have caused an annual economic loss of $100,000,000 to agriculture in California. It is definitely a product of our time. The reagents for its large-scale formation can come only from automobiles.

So it has come about that a combination of California's population, affluence, and bright sunshine has made that state's air pollution so severe that it has necessarily become a leader in the investigation and control of photochemical smogs. The existing technology for the reduction of automobile air pollution has been strongly influenced by the crisis in the late 1950s. Ways now are known by which hydrocarbons (and carbon monoxide, for that matter) can be nearly eliminated from automobile exhaust by assuring the complete oxidation of the exhaust gas. In one scheme, air is injected into the exhaust manifold, and the air-exhaust mixture is left hot to complete the combustion. In another, a considerable excess of air is drawn into the cylinder, and then the fuel is injected as a spray into a cavity in the piston head. However, such schemes have no effect on the emissions of nitrogen oxides, and thus on the initiation of photochemical air pollution [Equation (18-12)]. The hotter and "leaner" (more oxidizing) the fuel/air ratio in the engine's cylinders, the more oxides of nitrogen are produced. So we reach an impasse in the attempt to eliminate all pollutants. One suggestion has been to keep the burning cooler and the cylinder gases on the reducing side by using richer fuel-air mixtures, and then completing combustion of the exhaust gas *outside* the cylinder. Such a solution, of course, is even more wasteful of scarce hydrocarbons than a normal, air-polluting engine; once again we find a trade-off between economics and chemistry. How to minimize environmental degradation and how to balance the need for cheap automobiles against the need for desirable places to which to drive them are more of the complex and difficult problems that we offer the next generation of chemists.

18-6 Thermodynamics and Ecology

Thermodynamicists have always insisted that things are going from bad to worse. The first and second laws of thermodynamics have been paraphrased in this way: (I) You Can't Win, and (II) You Can't Even Tie. At the start of this chapter, we looked at one form of thermodynamic statement about the earth: the equilibrium model of a natural system. In this section, we shall review more of thermodynamics in the particular context of pollution.

FIGURE 18-5 A fossil-fuel power plant on Seneca Lake, New York. Note fuel and cold-water inputs, electric-power and combustion-products outputs. (Courtesy New York State Electric and Gas Corporation.)

ENTROPY POLLUTION. If it were not for the second law, there would be no pollution problems. Consider a typical pollution source: a coal-burning electric power plant, situated on the banks of a lake (Figure 18-5). Viewing it from afar, we notice a smoke plume (which contains carbon dioxide, some soot and ash, and traces of nitrogen oxides). Trainloads of coal go in at one door; truckloads of ash and cinders (metal oxides and silicates) go out another. Another activity may be more difficult to spot—the plant is pumping cold water from the deep part of the lake to cool its steam condensers and is returning warm water to the surface waters. All of these visible effects are more or less damaging to the health of the environment. The question arises—why must the power plant do these things?[7] Let us scrutinize the scene through thermodynamic glasses:

1. The gaseous pollutants coming out of the smokestacks would cause no harm if they would remain concentrated near the power plant, to be collected and processed chemically into useful by-products. But the entropy of an expanded gas is greater than that of a concentrated one, and the entropy of a mixture is higher than that of pure substances (see Chapter 9). Consequently, these gases, like other pollutants of air, water, and soil, spread themselves uniformly over the earth, where they defy entropy-decreasing attempts to gather them together again. To release a mole of SO_2 to the environment is to break the string on a necklace of 6.023×10^{23} poisonous beads.

2. Very well then, why are these products not liquids or solids, whose scattering propensities are not so troublesome as those of gases? We have seen that some of the products of the combustion of coal are solids (cinders), but there is a good reason for the prevalence of gaseous products from the heat-releasing reactions that power plants need. In the burning of coal, oil, or gas, weakly bonded molecules (hydrocarbons and organic sulfides) are converted to simple, tightly bonded ones, predominantly H_2O and CO_2. The loss of molecular potential energy appears as heat. In none of the product molecules are there strong *inter*molecular forces; they simply contain too few electrons to develop appreciable van der Waals forces, and their possibilities of extended bonding to form covalent substances like the fuel that formed them are nonexistent (see Chapter 5). But unless van der Waals forces cause considerable heat to be released to the surroundings in the process of condensation, the conversion of a gas to a liquid or solid costs the universe too much entropy (see Table 9-1, Chapter 9).

3. How about retaining the stack gases and recombining them with the cinders to reconstitute coal, or something like it? Can we not "recycle" these substances as we can glass containers? Alas, spontaneous chemical change is unidirectional, according to the second law. Coal once burned cannot be "unburned" except through intervention of elaborate and energy-wasting processes (see "Two Impossible Machines," Figure 9-5). Thus, the power plant must convert fuel to simple gaseous pollutants if it is to generate by chemical means the heat required for its steam turbines, and those pollutants will scatter widely over the earth when once released. Such are the dictates of the second law.

4. But what about the importation of cold water and the exportation of warm water; what effect does this transaction have on the lake, and why is it necessary?

The answer to this last question is moderately complex and depends somewhat on the kind of aquatic environment we are considering. If the lake is

---

[7] In one sense, it must do those things because shops and factories are turning out consumer goods on electrically driven machines; housewifes are vacuuming rugs; authors are writing on electric typewriters; and students are studying by the light of 40 W fluorescent lights and to the sound of 140 W stereo outfits. In short, it must, because we, the consumers, tell it to do so.

located in a temperate climate, it probably undergoes an annual *thermal stratification* in which a layer of wind and wave-mixed warm water floats on top of denser, cold water. This stratification is produced during warm weather in the spring and summer, and then erased in the fall when the temperature and density of the upper layer (epilimnion) become equal to those of the cold layer (hypolimnion). Once the lake has stratified in the early summer, the hypolimnion is cut off from contact with the atmosphere. Fish and other animals that prefer cold water retire to the hypolimnion to wait out the time of stratification; they must make do there with whatever oxygen and nutrients were present in the deep waters when stratification occurred (Figure 18-6). To pump cold water from the hypolimnion and return warm water to the epilimnion is both to reduce the quantity of oxygenated water that is accessible to cold-water organisms and to prolong the period during which stratification persists. Even without this perturbing effect, many deep lakes become nearly anoxic (oxygen free) in the hypolimnion by the end of summer. If the power plant is located on a river rather than a lake, the effects may be less severe; yet if we recall the effect of temperature on the rates of chemical reactions (Chapter 11), we should expect that warming the habitat of animals that do not have internal thermostats must have a considerable effect on their lives.

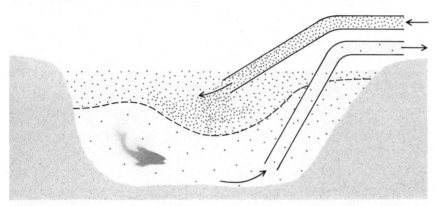

FIGURE 18-6 Use of a stratified lake as a source of cold water and a sink for warm water by a thermal polluter. Dotted line: the *thermocline,* or boundary between lighter warm water and denser cold water. Note the depression of the thermocline by the inflow of warm water (intensity of color proportional to temperature).

But *must* power plants and other heat engines export thermal energy to their environments? Again, the second law provides an unambiguous and pessimistic guide: They must. Heat engines take advantage of the difference in free energy between hot and cold matter, generally in allowing the hot matter to do work by expanding and cooling (Figure 18-7). The catch is that *some* heat must be transferred from a hot to a cold environment in the process. For example, in a steam engine, the hot steam that drives the piston must be condensed, giving up its heat of condensation to a heat sink; otherwise, there would be no way for the piston to return for a second stroke. (If you doubt the validity of this argument, imagine that the whole world were at the temperature of the steam engine's firebox.) When the second law is applied to such heat engines, it develops through a simple but lengthy argument, which we shall not reproduce here,[8] that the efficiency of conversion of heat energy to work is given by the equation

$$\text{efficiency (percent)} = \frac{T_{\text{hot}} - T_{\text{cold}}}{T_{\text{hot}}} \times 100 \qquad (18\text{-}18)$$

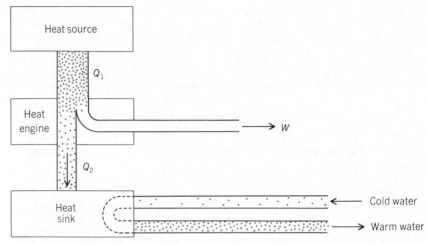

FIGURE 18-7   Energy flow in a heat engine. Energy is drawn as heat (Q_1) from the heat source, partly converted to "useful" work (W) by the heat engine. The remainder of the energy must be discarded as heat (Q_2) to the heat sink. Energy flow will continue, and the engine will produce work only as long as the heat sink is at a lower temperature than the heat source. By Equation (18-18), the fraction of Q_2 convertible to work increases as the difference in temperature between the source and sink increases. Thus, the heat sink is continuously kept cool by a flow of coolant—in most cases, cold water from a natural body of water.

In Equation (18-19), T_{hot} is the temperature of the heat reservoir (essentially, that of the steam just before it enters the piston or turbine) and T_{cold} is the temperature of the cold environment (in our example, that of the condensers which are cooled by lake water), where all the temperatures are on the Kelvin (absolute) scale. Examination of Equation (18-18) yields two considerations of great ecological (and economic) importance: (1) The efficiency of a heat engine is greater, the greater the temperature difference $T_{\text{hot}} - T_{\text{cold}}$; thus, it is economical to locate power plants near cold-water bodies so that a given

---

[8] In fact, the second law had its birth in intellectual history in a monograph on steam engines by the French engineer Sadi Carnot, *Réflexions sur la Puissance Motrice du Feu*, Paris (1824); English trans., *Reflections on the Motive Power of Fire*, New York: Dover paperback (1960). We have encountered the result of his equation (18-18) before, in the form of a graph in Chapter 8, where it was important to the discussion of fuel cells (Section 8-11).

investment in heat (that is, the burning of a given quantity of fuel) will produce as much power as possible. [Again contemplate, this time with the aid of Equation (18-18), what would happen if the whole world were at the temperature of a steam engine's firebox.] (2) Since T_{cold} cannot be at 0°K (which is unattainable), the efficiency of any heat engine will always be less than 100 percent. If not all of the heat energy can be converted to work, then by the first law, the rest must be dumped into the surroundings as heat. It is this heat that appears in Figure 18-7 as Q_2 and that we can observe in the effluent water pipe of any power plant.

NUCLEAR POWER PLANTS: A FINAL NOTE. In Chapter 10 we outlined some of the features of nuclear reactions and indicated that, in the long run, nuclear energy would have to replace the burning of fossil fuels as a power source. Besides the obvious point that fossil-fuel sources will be exhausted within the foreseeable future (about 10 yr for natural gas, about 100 yr for coal), nuclear power plants are not sources of atmospheric CO_2, CO, SO_2, and NO_x. However, as long as nuclear reactions are used only as heat sources, they will be subject to Equation (18-18), and power plants will be serious sources of "thermal pollution" (which, as we have seen, is only a special case of "entropy pollution"). Further, because of the efficiency and relative cleanliness of nuclear fuels, and because of the ever-rising demand for electric power, nuclear power plants will be large ones. At present, the only alternative to the direct pumping of heat into rivers and lakes is the use of cooling towers (Figure 18-8), which dump heat into a larger reservoir—the atmosphere.

Maybe there is a way to convert nuclear energy directly to electrical energy.

FIGURE 18-8 A fossil-fuel power plant that uses atmospheric cooling towers instead of water as a waste heat sink. Compare with Figures 18-5 and 18-7. (Courtesy New York State Electric and Gas Corporation and Pennsylvania Electric Company.)

18-7 Science, Technology, Politics, and Information

"Surely a technology that has created manifold environmental problems is also capable of solving them." So runs the plausible refrain.[9] Yet anyone who has lived with a young child can testify that the ability (and will) to clean up a mess bears very little relationship to the ability to create it. Furthermore, the problem of human ecology is more than technological; there persists abundant ignorance, both about the details and about the broad governing principles of the biosphere. Science has progressed to its present position by simplifying the systems it studies and making ever more abstract formulations of what it knows, until we can now claim to have achieved complete mastery of the innermost secrets of matter—in a few simple cases! What we do not appear to have is a complete enough understanding of the living environment we occupy and its interactions with our culture.

The nature of answers given by science depends on the nature of questions asked—by science.

So we end this text where we began it, with an appeal to the energy, sympathy, and intelligence of the reader. Control of any complex situation, such as a culture's impact on its environment, depends completely upon a feedback of information from the points of impact to the points at which decisions are made. For example, your understanding of the behavior of matter has political significance when you decide how to vote on a bond issue for the construction of a sewage plant. (In this case, you need to understand the impact of reducing agents and plant nutrients on rivers, lakes, and oceans.) If you have read through this book, you should be on your way to a better understanding of the material world, in which and of which we were created. But education does not end with a chemistry course, and some would argue that it barely begins there. Your mind and your senses are equally important channels for information about the world. Keep them open.

18-8 Summary

Throughout this book, the material problems of our planet have been discussed along with the chemical principles which enable us to understand the nature and the origin of those problems. Frequently (but not always), an understanding of what goes on in a chemical way also points out a logical solution to the problem. Economic or political objections to the solution may raise further questions, which require further chemical research for their answers, but at least a pathway is defined which holds out hope for an eventual logical solution. Blind fury, on the other hand, leads us nowhere and only uses up time which could be devoted to remedial action.

Ignorance never settles a question.
—Disraeli

This last chapter follows the same theme as it takes up a variety of interconnected problems of our environment and shows what we can and cannot do about them. The use of model systems as convenient mental frameworks for thinking about specific problems is justified by using the CO_2-carbonate system as an example, and applying the necessary equilibrium constants to

---

[9] In this case, the quotation is from the editors' introduction to the excellent and useful series, *Advances in Environmental Science,* New York: Wiley.

a calculation of concentrations to be expected at equilibrium in sea water. The carbon cycle in living organisms was considered next, especially in terms of the rate of growth of aquatic organisms and the limitations imposed by scarce nutrients. This led to consideration of the nitrogen and phosphate cycles in ecology and to some specific problems of water pollution. Air pollution was then considered in specific chemical terms: the polluting substances, their formation from all-too-readily-available materials, the intricate reactions that lead to photochemical smog, and what to do about it. Finally, the place of thermodynamics in the broad attack on pollution of the environment was brought up, and the lines of "can do" and "cannot do" were drawn very clearly. The hard facts of entropy and the second law lead inevitably to scattering of soot, cinders, and combustion gases; we can restrain the first two (by great effort), but the "recycling" of all products of combustion is impossible. Our only hope is to impose sensible restraints on automobiles, power plants, and factories (as we are beginning to do), and to pursue alternate means of transportation and production, with thorough study in advance of the chemical and ecological consequences. Nuclear power plants offer some relief in one small area, but do not carry any less (or more) of a penalty in thermal pollution.

The chemical concepts and principles required for understanding and applying Chapter 18, like those for Chapter 1 and all chapters in between, are all incorporated within this book, although not in the classical 1-2-3-4 order. The reader again is encouraged to use the glossaries, the index, the appendices, and the cross-references. The "Suggestions for Further Reading" are especially important in obtaining a full grasp of our subject. And lastly, not all teaching is done best by machines and books. Turn to your instructors and to the person in charge of this course—they are as interested as you are!

GLOSSARY

aerosol: a mist or dispersion of tiny drops of liquid in air or other gas.

algae: primitive aquatic one-celled or multicellular plants.

dystrophic: incapable of supporting life, because of poisons or adverse chemical factors.

ecology: the science of the relationships between organisms and their environments.

ecosystem: an ecological "system"; that is, the collection of reagents, products, organisms, and all other forms of matter and energy that constitute a particular region or area under ecological consideration. An ecosystem corresponds, in the wide world of reality, to the chemist's laboratory "system" which he isolates and studies intensively.

epilimnion: the warm upper layer of a still body of water, after thermal stratification sets in.

eutrophic: having abundant nutrients, and capable of supporting a blooming growth of living organisms.

hypolimnion: the lower, cooler layer in a stratified body of water.

limiting nutrient: that foodstuff or chemical substance which is in shortest supply, yet essential to the growth and activity of the organism.

oligotrophic: capable of supporting only a sparse growth of algae and diatoms (and of organisms which feed on these), because of a limiting supply of nutrients.

photochemical: produced by the agency of light or similar radiation; involving the action of photons as they supply energy to the reactants and cause a chemical change to proceed.

plankton (from Greek for "wanderer"): minute plants (phytoplankton) and animals (zooplankton) suspended in body of water and carried with it. The plant life includes bacteria, molds, and algae; the animal life includes protozoa, animalcules, and crustaceans up to 2 mm long.

smog: a smoky fog; a combination of particulate pollutants and noxious or toxic gases, usually of photochemical origin, which is irritating to throat and lungs.

PROBLEMS

18-1 Automobile manufacturers are now facing new limitations on the emissions of their cars, limitations based on *grams per vehicle mile,* rather than on percent of exhaust gas. What effect will this have on the kind of cars built in the future?

18-2 Compare current European, American, and Japanese automobiles, on the basis of manufacturer's information, in terms of potential and actual air pollution as a function of design.

18-3 Automobiles (especially old cars) emit more polluting matter when they stop and start repeatedly than in steady running. The idea behind freeways in Los Angeles was to keep the cars running, instead of stopping often for stoplights. Yet residents near freeways complain more of smog than residents near the corner stoplights formerly did. What chemical explanations can you propose for this fact?

18-4 What is there about the oxides of nitrogen that leads writers to refer to them collectively as NO_x rather than designating particular compounds?

18-5 Giving as much detail as you can about methods and procedures, outline courses of research and action toward the solution of the following environmental problems:

  a. What nutrient is limiting the growth of algae in the body of water nearest you?

  b. Does the lake nearest you undergo thermal stratification during the winter? If so, does the hypolimnion become oxygen-free by springtime?

  c. What would be the environmental impact of the creation of a residential subdivision on the outskirts of your town?

  d. Assuming that such a development should take place, who should pay for the installations (sewage plants, public maintenance and landscaping) required to minimize environmental degradation:

   i. If the housing is "luxury" housing intended for wealthy owners?

   ii. If it is public housing intended for former inner-city dwellers displaced by a freeway?

   iii. If it is low-cost private housing?

e. Which, if any, of the following reactions are at equilibrium in the lake nearest you?

i. $CO_2 + H_2O = H_2CO_3$

ii. $H_2CO_3 + H_2O = HCO_3^- + H_3O^+$

iii. $CO_3^{2-} + Ca^{2+} = CaCO_3(s)$

iv. $O_2 + 2Mn^{2+} + 2H_2O = 2MnO_2 + 4H^+$

v. $3Al_2Si_2O_5(OH)_4 + 2K^+ = 2KAl_3Si_3O_{10}(OH)_2 + 3H_3O + 2H^+$
    kaolinite   K-mica
(Kaolinite and K-mica are minerals commonly found in clay.)

18-6 H. T. Odum (*Environment, Power, and Society*; see "Suggestions for Further Reading") makes the point that we do not really know whether our civilization can run on nuclear power, since the development of nuclear power to the present time has been heavily "subsidized" by the burning of fossil fuels. In how many ways (direct or indirect) is the extraction, sale, and consumption of fossil fuels involved in the construction and operation of a nuclear power plant?

18-7 For every direct or indirect use of fossil fuels in society, try to conceive (in principle, at least) of a way in which nuclear power could be used instead. (You may have to burn some midnight oil to answer this one!)

18-8 The problem on pages 598–599 is placed in this chapter not because it refers exclusively to same, but because we have reached the end of the book. It is a crossword puzzle, the solutions to which are chemical words, ideas, or (in one case) a person. To get all the answers, you will have to have read this book, but you will also have to seek outside this book. But is that not what we told you at the beginning?

SUGGESTIONS FOR FURTHER READING:

Subcommittee on Environmental Improvement, Committee on Chemistry and Public Affairs, American Chemical Society, *Cleaning Our Environment: The Chemical Basis For Action,* Washington, D.C.: American Chemical Society (1969). The record of the American Chemical Society on environmental problems has been a somewhat ambivalent one, which may reflect mixed industrial and academic influence on its policies. Thus, its official organ *Chemical and Engineering News* subjected Rachel Carson's *Silent Spring* to a hostile treatment entitled, "Silence, Miss Carson!" *Chemical and Engineering News* (October 1, 1962). But the present volume represents a monumental effort to collect in one place what was known and conjectured about environmental pollution and its possibilities for cure, up to 1969. As a bibliographic source alone, it is invaluable.

The September, 1970, issue of *Scientific American* is a single-topic issue devoted to 11 articles under the general title, "The Biosphere." Many of the articles are devoted to cycles of matter and energy.

GOLDMAN, MARSHAL, ed., *Controlling Pollution: The Economics of a Cleaner America,*
    Englewood Cliffs, N.J.: Prentice-Hall (1967). Environmental problems are social and economic as well as chemical ones, and the flows of matter in the environment are reflected by flows of money in society. Goldman's book is both scientifically

and economically sound and represents a good place for those who would like to learn more about this aspect of the environmental crisis.

Air Conservation Commission of the American Association for the Advancement of Science, *Air Conservation,* Washington, D.C.: American Association for the Advancement of Science (1965). The AAAS has presented a more consistently proenvironmental posture than most other scientific and professional organizations. This book and the one following are collections of scholarly but readable treatments of a variety of problems connected with society and environment.

BRADY, N. C., ed., *Agriculture and the Quality of our Environmental Sciences,* Washington, D.C.: American Association for the Advancement of Science (1967).

PITTS, J. N., and METCALF, R. L., eds., *Advances in Environmental Science,* New York: Wiley. Articles generally written at a higher level than the ACS and AAAS volumes, about more scientific topics. Volume I of this continuing series was published in 1969, Volume II in 1971.

Environmental Science and Technology is the leading technical journal in the field, and its monthly issues contain both research reports on scientific investigation and broader essays, easily readable by anyone who has read this text, on larger issues. The September, 1970, issue has a survey ("The Great Phosphorus Controversy") of the role of phosphates in the eutrophication process, and a discussion of alternative theories.

RUSSELL-HUNTER, W. D., *Aquatic Productivity,* New York: Macmillan (1970). Micro- and macrolife in natural waters.

ODUM, H. T., *Environment, Power, and Society,* New York: Wiley (1971). "Systems analysis" is a quasi-mathematical method that has been developed for the purpose of developing models of large and complex systems like societies and ecosystems. Odum's book, in addition to offering an introduction to systems analysis on a level accessible to the reader of this book, represents an attempt to make chemical, biological, and social "forces" commensurate by expressing them all in terms of the flow of power (in the physical sense). Odum's viewpoint is startling and provocative.

EHRLICH, P. R., and EHRLICH, A. H., *Population/Resources/Environment,* San Francisco: Freeman (1970). The emphasis in this textbook of human ecology is on the impact of a large, affluent, and wasteful population on a precariously balanced environment. The tone is strong, urgent, or scary, depending on your own view of the immediacy of the environmental crisis, but the treatment is always factual, and the bibliographies are good.

CARSON, RACHEL, *Silent Spring,* Boston: Houghton-Mifflin (1962). If you have not read this classic, you should.

MEADOWS, DONELLA H., MEADOWS, DENNIS L., RANDERS, JORGEN, and BEHRENS, III, WILLIAM W., *The Limits To Growth,* New York: Universe Books (1972). Systems analysis can be applied to the whole world, and it can be used as a vehicle for deducing the logical consequences of hard data, guesses, and speculations about the present state and probable future of the world ecological system. Not only have these authors done the job in a comprehensive and understandable way, they have also clearly labeled what parts of their input data are hard facts and what parts are speculation. The most probable future based on current trends is remarkably insensitive to the kinds of guess made, and it is remarkably unpleasant. If you want to do something about mankind's future, you ought to read this book.

MORGAN, JAMES J., and STUMM, WERNER, *Aquatic Chemistry*, New York: Wiley (1971). This rather advanced textbook will be more understandable to you after you have studied more chemistry, but you might browse through it to see how chemical principles may be applied even to very complex natural water systems. It is the source of Equation (18-11).

BOCKRIS, J., "A Hydrogen Economy," *Science*, vol. **176**, p. 1323 (June 23, 1972). In this brief and nontechnical article, Bockris reviews the potential advantages of basing our power needs on the production and distribution of hydrogen gas at central power stations and using fuel cells or combustion engines at the point of use. Bockris' list of the advantages of this approach constitutes a review of much of the material of this textbook.

Entropy pollution as a chemical and social concern is discussed in: BENT, H.A., "Haste Makes Waste: Pollution and Entropy," *Chemistry*, vol. 44, no. 9, pp. 6–15 (October 1971).

598 CROSSWORD PUZZLE

CASE STUDIES IN ENVIRONMENTAL CHEMISTRY

ACROSS

- 2 Molecules need this type of energy before they will react.
- 7 A type of radiation emitted by radioactive elements.
- 10 Two carbon atoms.
- 12 Bonds very common in hydrocarbons.
- 13 To donate electrons in a coordinate bond.
- 15 When molecules are in this phase they always fill the containing vessel.
- 16 A large amount of this is liberated if mass is destroyed.
- 17 For an ideal gas the product of the pressure and volume is this.
- 18 Chlorine-containing insecticide.
- 19 Polymeric organosiloxanes which find use as oils, water-proofing materials and electrical insulators.
- 22 Heavy metal, liquid at room temperature.
- 23 The alkoxy group.
- 24 Coin formerly of silver, now degraded to Ni and Cu.
- 27 Oxidation state of the alkali metals in their compounds.
- 30 Unit of electrical resistance.
- 31 This element has the electronic arrangement 2.8.8.
- 33 An author's initials.
- 34 Binary compound(s) of oxygen.
- 36 Many useful materials are obtained from here.
- 38 Molecules show these in their spectrum.
- 40 If for an acid this is greater than 5 the acid is weak.
- 41 Unit of energy.
- 43 This compound evolves hydrogen on treatment with water and gives a lavender color in the Bunsen flame.
- 44 This type of bond involves the sharing of electrons between atoms.
- 48 Second element.
- 49 Hydrohalogen acid with highest heat of formation.
- 50 The halogen which is a liquid at room temperature.
- 51 Same as 41 across.
- 52 Poisonous oxide of carbon.
- 53 First two letters of 31 across.
- 54 A heavy, old-fashioned chemistry book.
- 57 Sodium salts of high-molecular-weight straight-chain fatty acids.
- 58 Most chemistry has been discovered in this part of time.
- 59 Cyanide group.
- 60 Lightest molecule.
- 62 Principal molecular species in sulfur vapor.
- 63 This term embraces about ¾ of all the elements in the periodic table.
- 65 This mineral is added to molten alumina to increase its conduc-

(ACROSS, cont'd)

tivity during the preparation of aluminum by electrolysis.
67 Negative ion of hypoiodous acid.
68 Diatomic molecule of mass 4.
69 Two-thirds of 17 across written backwards.
71 Atomic number of 14 down.
74 One of the ions formed when H₂S dissociates.
75 This linkage is found in all peroxides.
77 Common weak organic acid.
80 This salt would (a) give yellow precipitate with silver-nitrate solution which would be insoluble in dilute nitric acid (b) give a lavender color in a Bunsen flame.
81 Found in most American homes, once made of black phenolic plastic but now often of colored cellulose acetate.
83 Reverse of 85 across.
84 Oxide of nitrogen used for anesthesia.
85 This ion is found in excess when salts of strong bases and weak acids are hydrolyzed.
88 Sodium metal should *never* be put in these. Neither should filter paper or trash!
89 If this hot concentrated solution is electrolyzed and stirred, sodium chlorate and hydrogen are obtained.
91 This ion gives a white precipitate with barium chloride, insoluble in dilute acid.
92 Principal ores of this common metal are hematite and magnetite.

DOWN

1 Comparative size of I⁻ ion.
2 Initials of well-known chemical society.
3 Bonds very common in hydrocarbons.
4 A gas is this if (a) the volume occupied by the molecules themselves is negligible compared with the volume occupied by the gas (b) there is no attractive force between the molecules (2 words).
5 Often obtained when an organic substance is heated with sulfuric acid.
6 Most abundant element.
7 Linkage between atoms.
8 The rare-_____elements.
9 Trinitrotoluene.
11 Carboxylic acid group.
14 Most electronegative element.
19 Symbol for selenium.
20 In the photochemical reaction between hydrogen and chlorine this is the amount of light absorbed by the hydrogen.
21 Lecture demonstrations enable students to _____ chemical processes.
24 Type of bond found in olefins.
25 86 down reversed.
26 One-tenth of 78 down.
28 One of these could always be found at the end of the chapters in most elementary chemistry textbooks.
29 This man's counter is used in nuclear chemistry.
32 First two elements of Group 6.
34 Allotropic form of oxygen.
35 Diatomic molecule of molecular weight 3.
37 The diameter of one is about 10^{-8} cm.
39 Polymer obtained by condensing adipic acid and hexamethylenediamine. Damaged by smog compounds.
40 If this is less than 7 the solution is acid.
42 This factor is usually positive but may be negative.
43 If this substance is mixed with 49 across and electrolyzed, fluorine is obtained.
44 Examples are $Cu(NH_3)_4^{2+}$, $Ag(NH_3)_2^+$, AlO_2^- and $Zn(CN)_4^{2-}$.
45 The diseases of scurvy, beriberi and pellagra result from a deficiency of this type of chemical.
46 In an electrolytic process the oxidation step takes place here.
47 Germanium is used in this.
49 If water is this, it is valuable and used for atomic energy processes.
50 Although the chlorite ion is known, this analogous ion has never been found.
53 Symbol of element below phosphorus in periodic table.
55 In aqueous solution this plus something else always equals 14.
56 Atoms which differ only in mass are _____.
59 Ion obtained by passing chlorine into cold base.
61 Compound of hydrogen; SiH_4, B_2H_6 and PH_3 are examples.
64 Metal in Group IV.
66 This carries CO_2 from the air into the soil.
70 General formula for an alcohol.
72 All salts have crystals containing this building block.
73 If a reaction is_____thermic, heat is evolved.
78 Unit of length.
79 Type of sulfate used in photography.
82 Source of gasoline.
86 On heating, this compound reversibly dissociates into hydrogen and iodine.
87 ½ mv² equals ³/₂ times this.
89 A very hard material.
90 This gas gives a brown colored gas on mixing with air.

EPILOGUE

At the beginning of this book, you were asked to assess your own personal understanding of chemical terms and processes, gathered from your previous reading and training. Now that you have come this far with us, it is time to go back through Chapter 1 and mark off those queries that are now answered in your mind. If there still are portions of that chapter you do not understand, look up some of the key words in the index, or thumb through the pertinent chapters ("Organic Chemistry," "The Design of Materials" and so on) until you find the explanations. They are there.

Each chapter has had its glossary, but you may not remember in which chapter to look up the meaning of a particular term that still bothers you. If so, refer to the index. All words that are defined in the separate glossaries are listed in the index, which will send you back to the proper chapter, section, and glossary page.

Practical people (and that always includes prospective employers!) often worry about today's simplified or reorganized teaching of chemistry, feeling that all chemistry ought to be taught just the way it was 30 or 40 years ago. In so doing, they overlook the extent to which chemistry has changed (in content and in emphasis) since those days. However, they frequently express special concern because students are not taught per se that copper sulfate is blue, that sodium nitrate and lithium chloride melt easily, that magnesium has a density of 1.74, and that uranium hexafluoride is volatile. Well, all of these facts are in this book. They do not clutter up the text, but you *can* get to know them, if you are that much interested (and we are delighted to learn that you are!). The melting points, boiling points, densities, colors, and states of all of the compounds mentioned in this book (insofar as we could catch them all) are listed in Appendix I, alphabetically according to formula. These facts are not put there to provide sticky questions for examinations, but just to provide more information about all the substances that have been brought into the discussion for cogent reasons. In your reviewing or rereading, it would be well to look up some of the compounds you come across. We have tried to make it easy for you; if we did not include Appendix I, you would have to buy a Chemical Handbook and keep it by your side all the time.

Naturally, we hope that you have learned something about the place of chemistry in the current scene from this book. More than that, we hope that you have learned at least the essential principles of the science and are able to see how those principles apply to all material problems today. We hope you have learned the power of dispassionate investigation of every material question, and that you can now sift the facts from all the rhetoric more easily. And most of all, we hope that you will DO SOMETHING with your new knowledge and will help to make our planet a safer, more pleasant place to live.

Appendixes

APPENDIX I CHEMICAL COMPOUND HANDBOOK

Properties tabulated are for the commonly observed form of each substance. Density of solids and liquids is given in units of gm-cm^{-3}; water has a density of about 1.00, thus being a convenient reference substance for thinking about the meaning of density values. The temperature at which a liquid boils depends on the atmospheric pressure, and sometimes (always noted) the boiling point is given at a pressure other than 1.00 atm. Some substances decompose rather than melt or boil, and in these cases a decomposition temperature is listed. Other substances pass directly from solid to gas by a process called sublimation; for these substances, a sublimation temperature is given. The following abbreviations are used in the table:

mp	melting point
bp	boiling point
atm	atmospheres pressure
dec	decomposes
expl	explodes
sub	sublimes

Compounds are arranged alphabetically by formula. Subscripts are considered (C_1 compounds precede C_2 compounds). Alphabetizing generally considers a formula to be written with the most electropositive element first, and the other elements in alphabetical order; for most carbon compounds without a metal, the order is C, H, and then all others in alphabetical order. Some formulas are printed to give structural clues, but this does not affect the order of listing (ethanol is printed C_2H_5OH to show the —OH group, but is alphabetized as C_2H_6O). Hydrogen is often ignored when alphabetizing acids.

formula	name	mp (°C)	bp (°C)	density	usual color and state
Ac	actinium	1050	~3200		radioactive metal
Ag	silver	961	2212	10.49	bright gray metal
AgBr	silver bromide	432	>1300dec	6.473	yellow solid
$AgC_2H_3O_2$	silver acetate	dec		3.259	colorless crystals
AgCl	silver chloride	455	1550	5.56	photosensitive white powder
Al	aluminum	660.2	2467	2.702	silvery metal
$Al(CH_3)_3$	trimethylaluminum	0	130		self-inflammable liquid
$AlCl_3$	aluminum chloride	190	177.8sub	2.44	colorless crystals
AlN	aluminum nitride	2200(in N_2)	2000sub	3.26	white crystals
Al_2O_3	aluminum oxide	2045	2980	3.9	colorless crystals
$Al(OH)_3$	aluminum hydroxide	300dec		2.42	colorless crystals
Am	americium	995	2600	13.7	radioactive metal
Ar	argon	−189.2	−185.7	1.784 g/liter	colorless gas
As	arsenic	817	613sub	5.727	dull-gray solid
$H_3AsO_4 \cdot \frac{1}{2}H_2O$	arsenic acid	35.5	160dec	2.0–2.5	deliquescent white crystals
At	astatine				properties unknown
Au	gold	1063	2966	19.32	bright-yellow metal
B	boron	2300	2550sub	2.34	yellow crystals
BF_3	boron trifluoride	−126.7	−99.9	2.99 g/liter	colorless fuming gas
B_2H_6	diborane	−165.5	−92.5		colorless gas
B_2O_3	diboron trioxide	460	1860	2.46	colorless crystals
B_2O_3	diboron trioxide glass	450	1860	1.812	colorless glass
Ba	barium	725	1140	3.51	reactive metal
Be	beryllium	1278	2970	1.848	silvery light metal
BeO	beryllium oxide	2530	3900	3.01	white powder
Bi	bismuth	271.3	1560	9.80	reddish-gray metal
Bk	berkelium				properties unknown
Br	bromine	−7.2	58.8	3.12	dark-red liquid
BrCl	bromine chloride	~−66	~5		reddish gas
C	diamond	>3500	4827	3.51	colorless crystals
C	graphite	3652sub	4827	2.25	black crystalline solid
CBr_4	carbon tetrabromide	90.1	189.5	3.42	colorless crystal

formula	name	mp (°C)	bp (°C)	density	usual color and state
CCl_2O	carbonyl chloride	−118	8.3	1.392	poisonous gas
CCl_4	carbon tetrachloride	−22.99	76.8	1.595	colorless liquid
CF_4	carbon tetrafluoride	−184	−128	1.96(−184°)	colorless gas
CI_4	carbon tetraiodide	171dec		4.34	red solid
CO	carbon monoxide	−199	−191.5	0.793(−190°)	colorless poisonous gas
CO_2	carbon dioxide	−56.6(5 atm)	−78.5sub	1.56(−79°)	colorless gas
CS_2	carbon disulfide	−111.5	46.5	1.263	colorless inflammable liquid
$CHCl_3$	"chloroform"	−63.5	61.7	1.483	colorless liquid
CH_2Cl_2	dichloromethane	−95.1	39.8	1.326	colorless liquid
CH_2F_2	difluoromethane	−92	−51.6	0.909	colorless gas
CH_2O	formaldehyde	−92	−21	0.815	colorless gas
CH_2O_2	formic acid	8.40	100.7	1.220	colorless liquid
CH_3Br	bromomethane	−93.6	3.56	1.732(0°)	colorless gas
CH_3Cl	chloromethane	−97.7	−24.2	0.916 g/liter	colorless gas
CH_4	methane	−182.5	−164	0.555 g/liter(0°)	colorless inflammable gas
CH_3OH	methanol	−93.9	64.96	0.791	colorless liquid
$CO(NH_2)_2$	urea	135	dec	1.323	white crystalline solid
$C_2Cl_2F_4$	1,2-dichloro-1,1,2,2-tetrafluoroethane		3.8	1.531(0°)	colorless gas
C_2Cl_6	hexachloroethane	187sub		2.091	colorless crystals
C_2F_4	tetrafluoroethene	−142.5	−76.3	1.519	colorless gas
$C_2HCl_3O_2$	trichloroacetic acid	57.5	197.5	1.630	colorless crystals
C_2H_2	ethyne	−80.8	−83.6sub	0.73(−85°)	colorless inflammable gas
$C_2H_2Br_2$	cis-1,2-dibromoethene	−53	112.5	2.246	colorless liquid
$C_2H_2Br_2$	trans-1,2-dibromoethene	−6.5	108	2.231	colorless liquid
$C_2H_2Br_4$	1,1,2,2-tetrabromoethane	0	243.5	2.969	colorless liquid
$C_2H_2O_4$	oxalic acid	189.5dec	157sub	1.90	white crystals
C_2H_3Cl	"vinyl chloride"	−153.8	−13.4	0.911	colorless gas
C_2H_3ClO	acetyl chloride	−112	51	1.105	colorless fuming liquid
$C_2H_3ClO_2$	chloroacetic acid	63	189	1.58	colorless crystals
C_2H_4	ethene	−169.2	−103.7	1.269 g/liter(0°)	colorless inflammable gas
$C_2H_4Br_2$	1,2-dibromoethane	9.79	131.4	2.179	colorless liquid

formula	name	mp (°C)	bp (°C)	density	usual color and state
C$_2$H$_4$Cl$_2$	1,1-dichloroethane	−96.98	57.3	1.176	colorless liquid
C$_2$H$_4$Cl$_2$	1,2-dichloroethane	−35.4	83.5	1.235	colorless liquid
C$_2$H$_4$O	ethanal	−121	20.8	0.783	colorless fuming liquid
C$_2$H$_4$O	"ethylene oxide"	−111.3	13.5	0.882 g/liter	colorless gas
C$_2$H$_4$O$_2$	acetic acid	16.6	118	1.049	colorless liquid
C$_2$H$_4$O$_2$S	mercaptoacetic acid	−16.5	105	1.325	colorless liquid
C$_2$H$_5$Cl	chloroethane	−136.4	12.2	0.898(0°)	colorless gas
C$_2$H$_5$NO$_2$	glycine	290dec		1.161	white powder
C$_2$H$_6$	ethane	−183.5	−88.63	1.357 g/liter	colorless inflammable gas
C$_2$H$_6$O	dimethyl ether	−138.5	−23.65		colorless inflammable gas
C$_2$H$_5$OH	ethanol	−117.3	78.5	0.789	colorless liquid
C$_2$H$_6$O$_2$	"ethylene glycol"	−11.5	197.3	1.109	colorless liquid
C$_2$H$_6$O$_4$S	ethanesulfonic acid	−17	123	1.334	colorless oily liquid
C$_2$H$_7$N	ethylamine	−80.6	16.6	0.706(0°)	colorless liquid
C$_2$H$_7$N	dimethylamine	−93	6.9	0.680(0°)	colorless gas
C$_2$H$_7$NO	ethanolamine	10.5	172.2	1.018	colorless liquid
C$_2$H$_8$N$_2$	ethylenediamine	10	118	0.963	colorless liquid
C$_3$Cl$_8$	octachloropropane	160	268		white solid
C$_3$H$_3$N	acrylonitrile	−83.5	78	0.806	colorless liquid
C$_3$H$_4$O	propenal	−86.95	52.5	0.841	colorless liquid
C$_3$H$_4$O$_3$	pyruvic acid	13.6	165	1.227	colorless liquid
C$_3$H$_4$O$_4$	malonic acid	135.6	dec		white crystals
C$_3$H$_5$Cl	3-chloropropene	−134.5	45	0.938	colorless liquid
C$_3$H$_5$N$_3$O$_9$	trinitropropane	2.9	256expl	1.593	yellow liquid
C$_3$H$_6$	propene	−185.2	−47.4	0.519 g/liter	colorless inflammable gas
C$_3$H$_6$Cl$_2$O	2,3-dichloro-1-propanol		182	1.361	colorless liquid
C$_3$H$_6$O	2-propanone (acetone)	−95	56.24	0.791	colorless liquid
C$_3$H$_6$O$_2$	propanoic acid	−20.8	141.0	0.993	colorless liquid
C$_3$H$_7$Cl	2-chloropropane	−117.2	36.7	0.861	colorless liquid
C$_3$H$_7$NO$_2$	α-alanine	297dec			colorless crystals
C$_3$H$_7$NO$_2$ ·	β-alanine	196dec			colorless crystals
C$_3$H$_7$NO$_2$S	cysteine	178			white powder

formula	name	mp (°C)	bp (°C)	density	usual color and state
C$_3$H$_7$NO$_3$	serine	228dec			colorless crystals
C$_3$H$_8$	propane	−189.7	−42.07	0.501 g/liter	colorless inflammable gas
C$_3$H$_8$O	isopropanol	−88.5	82.3	0.785	colorless liquid
C$_3$H$_5$(OH)$_3$	glycerol	17.9	290dec	1.261	viscous colorless liquid
C$_3$H$_9$N	trimethylamine	−117.2	2.87	0.662(−5°)	colorless malodorous gas
C$_4$H$_4$	"vinylacetylene"	−85.7	5.1	0.710(0°)	colorless gas
C$_4$H$_4$O	furan	−85.7	32	0.951	colorless liquid
C$_4$H$_4$O$_4$	fumaric acid	287	165sub	1.635	colorless crystals
C$_4$H$_5$Cl	2-chlorobutadiene		59.4	0.958	colorless liquid
C$_4$H$_5$N$_3$O	cytosine	320dec			white powder
C$_4$H$_4$O$_5$	oxaloacetic acid	270			white powder
C$_4$H$_6$	1,3-butadiene	−109	−4.41	0.621(−6°)	colorless gas
C$_4$H$_6$O$_2$	2-butyne-1,4-diol	58	238		colorless crystals
C$_4$H$_6$O$_2$	cis-2-butenoic acid	15.5	169.3	1.027	colorless liquid
C$_4$H$_6$O$_2$	trans-2-butenoic acid	71.5	185	1.018	colorless liquid
C$_4$H$_6$O$_2$	"vinyl acetate"	−93.2	72.2	0.932	colorless liquid
C$_4$H$_6$O$_2$	3-butenoic acid	−35	163	1.009	colorless liquid
C$_4$H$_6$O$_4$	succinic acid	185	235dec	1.572	colorless crystals
C$_4$H$_6$O$_5$	malic acid	100	140dec	1.60	colorless crystals
C$_4$H$_7$NO$_4$	aspartic acid	270		1.66	white powder
C$_4$H$_8$	1-butene	−185.4	−6.3	0.595 g/liter	colorless inflammable gas
C$_4$H$_8$	cis-2-butene	−138.9	3.7	0.621 g/liter	colorless inflammable gas
C$_4$H$_8$	trans-2-butene	−105.6	0.88	0.604 g/liter	colorless inflammable gas
C$_4$H$_8$	isobutene	−140.4	−6.9	0.594 g/liter	colorless inflammable gas
C$_4$H$_8$Cl$_2$	1,4-dichlorobutane	−37.3	153.9	1.141	colorless liquid
C$_4$H$_8$Cl$_2$	2,2-dichlorobutane	−74	104	1.430	colorless liquid
C$_4$H$_8$N$_2$O$_3$	asparagine	236dec		1.543	colorless crystals
C$_4$H$_8$O	2-butanone	−86.4	79.6	0.805	colorless liquid
C$_4$H$_8$O	2-buten-1-ol	<−30	118	0.873	colorless liquid
C$_4$H$_8$O	butanal	−99.0	75.7	0.817	colorless liquid
C$_4$H$_8$O	tetrahydrofuran	−65	64.5	0.889	colorless liquid
C$_4$H$_8$O$_2$	iso-butyric acid	−47.0	154.4	0.949	colorless liquid
C$_4$H$_8$O$_2$	n-butyric acid	−7.9	163.5	0.959	colorless liquid
C$_4$H$_8$O$_2$	ethyl acetate	−83.6	77.06	0.900	colorless liquid

formula	name	mp (°C)	bp (°C)	density	usual color and state
C_4H_9Br	1-bromobutane	−112.4	101.6	1.296	colorless liquid
C_4H_9Br	2-bromobutane	−111.9	91.3	1.258	colorless liquid
$C_4H_9NO_3$	threonine	255dec			white powder
C_4H_{10}	n-butane	−135	−0.6	0.60(−1°)	colorless inflammable gas
C_4H_{10}	isobutane	−145	−10.2	0.603(−11°)	colorless inflammable gas
C_4H_9OH	n-butanol	−89.5	117.3	0.810	colorless liquid
C_4H_9OH	tert-butanol	25.5	82.2	0.789	colorless crystals
$(C_2H_5)_2O$	diethyl ether	−116.2	34.6	0.713	colorless inflammable liquid
$C_4H_{10}O_2$	butane-1,4-diol	20.1	230	1.017	colorless oily liquid
$C_4H_{11}O_2$	diethanolamine	28	268	1.097	colorless crystals
$C_4H_{12}N_2$	1,4-diaminobutane	27	158	0.877	odorous white crystals
$C_4H_{13}N_3$	2,2′-diaminodiethylamine		207.1	0.959	yellow liquid
$C_5H_4O_2$	furfural	−38.7	161.7	1.160	liquid; almond odor
$C_5H_4O_3$	furoic acid	133	230		white needles
C_5H_5N	pyridine	−42	115.5	0.982	colorless liquid
$C_5H_5N_5$	adenine	360	220sub		colorless crystals
$C_5H_5N_5O$	guanine	360dec			colorless crystals
$C_5H_6N_2O_2$	thymine	270dec			colorless crystals
$C_5H_6O_2$	furfuryl alcohol		171	1.130	yellow liquid
$C_5H_6O_5$	α-ketoglutaric acid	115			white powder
C_5H_8	2-methylbutadiene	−146	34	0.681	colorless liquid
$C_5H_8O_2$	methyl methacrylate	−48	100	0.944	colorless liquid
$C_5H_9NO_2$	proline	222dec			white powder
$C_5H_9NO_4$	glutamic acid	213dec			white powder
C_5H_{10}	1-pentene	−138	30	0.640	colorless inflammable liquid
$C_5H_{10}N_2O_3$	glutamine	185dec			white powder
$C_5H_{10}O_4$	deoxyribose	91			white powder
$C_5H_{10}O_5$	ribose	95			white powder
$C_5H_{11}NO_2$	valine	293dec		1.230	white powder
C_5H_{12}	n-pentane	−129.7	36.1	0.626	colorless liquid
C_5H_{12}	neopentane	−16.6	9.5	0.613(0°)	colorless inflammable gas
$C_6H_3N_3O_7$	"picric acid"	122	300expl	1.763	yellow crystals

formula	name	mp (°C)	bp (°C)	density	usual color and state
C_6H_5Cl	chlorobenzene	−45.6	132	1.106	colorless liquid
$C_6H_5NO_2$	nitrobenzene	5.7	210.8	1.204	colorless liquid
C_6H_6	benzene	5.51	80.09	0.879	colorless liquid
C_6H_6O	phenol	41	182	1.072	colorless crystals
$C_6H_6O_3S$	benzenesulfonic acid	50	dec		colorless crystals
$C_6H_6O_6$	cis-aconitic acid	130dec			white powder
$C_6H_6O_7$	oxalosuccinic acid				white powder
C_7H_7N	aniline	−6.3	184.1	1.022	colorless liquid
$C_6H_8N_2$	1,6-dinitrilehexane	1	295	0.968	colorless liquid
$C_6H_8N_2O_2S$	sulfanilamide	165	dec	1.08	white powder
$C_6H_8O_7$	citric acid	153	dec	1.542	colorless crystals
$C_6H_8O_7$	isocitric acid	105dec			colorless crystals
$C_6H_9NO_6$	nitrilotriacetic acid	259			crystals
$C_6H_9N_3O_2$	histidine	287dec			white powder
$C_6H_{10}O_4$	adipic acid	153	265(0.13 atm)	1.366	white solid
C_6H_{12}	cyclohexane	6.5	80.7	0.779	colorless inflammable liquid
C_6H_{12}	methyl cyclopentane	−142.4	71.8	0.749	colorless liquid
C_6H_{12}	2-methyl-1-pentene	−135.7	60.7	0.680	colorless liquid
C_6H_{12}	2-methyl-2-pentene		67.5	0.690	colorless liquid
$C_6H_{12}O_5$	fucose	145	dec	1.598	white crystals
$C_6H_{12}O_6$	glucose	146	dec	1.544	white powder
$C_6H_{13}NO_2$	isoleucine	284			white powder
C_6H_{14}	2,3-dimethylbutane	−128.5	58.1	0.662	colorless liquid
C_6H_{14}	n-hexane	−94.3	69.0	0.660	colorless liquid
$C_6H_{14}N_2O_2$	lysine	224dec			white powder
$C_6H_{14}N_4O_2$	arginine	217dec			white powder
$C_6H_{15}N_2$	1,6-hexanediamine	39	196		soft crystals
$C_7H_5NO_3S$	saccharin	224	sub		sweet-tasting white powder
$C_7H_5N_3O_6$	2,4,6-trinitrotoluene	82	240expl	1.654	colorless explosive solid
$C_7H_6O_2$	benzoic acid	122.4	249	1.266	colorless crystals
$C_7H_6O_3$	salicylic acid	159	76sub	1.443	colorless crystals
C_7H_7Cl	o-chlorotoluene	−35.1	159.2	1.083	colorless liquid

formula	name	mp (°C)	bp (°C)	density	usual color and state
C_7H_8	toluene	−95	110.6	0.866	colorless liquid
$C_7H_8O_2$	salicyl alcohol	87	sub	1.161	colorless crystals
C_7H_9N	o-toluidine	α, −23.7; β, −14.7	200.2	0.998	colorless liquid
C_7H_{16}	n-heptane	−90.6	98.4	0.684	colorless inflammable liquid
C_7H_{16}	2,2,3-trimethylbutane	−24.2	80.9	0.690	colorless inflammable liquid
$C_8H_6O_3$	piperonal (heliotropine)	37	263		aromatic white crystals
$C_8H_6O_4$	phthalic acid	208	dec	1.593	white crystals
C_8H_7Cl	p-chlorostyrene		74(0.016 atm)	1.155	colorless liquid
C_8H_8	phenylethene (styrene)	−30.6	145.2	0.906	colorless liquid
C_8H_{10}	ethylbenzene	−95.0	136.2	0.867	colorless liquid
C_8H_{18}	3,4-dimethylhexane		117.7	0.720	colorless inflammable liquid
C_8H_{18}	n-octane	−56.8	125.7	0.703	colorless liquid
C_8H_{18}	2,2,4-trimethylpentane	−107.4	99.3	0.692	colorless inflammable liquid
$C_9H_8O_4$	acetylsalicylic acid (aspirin)	135	140dec		colorless crystals
$C_9H_{11}NO_2$	phenylalanine	283dec	sub		white powder
$C_9H_{11}NO_3$	tyrosine	310dec		1.456	white powder
C_9H_{20}	n-nonane	−51	150.8	0.718	colorless liquid
$C_{10}H_8$	naphthalene	80.55	217.9	1.025	soft colorless crystals
$C_{10}H_{12}O_2$	eugenol	−7.5	253.2	1.065	colorless liquid
$C_{10}H_{16}N_2O_8$	ethylenediaminetetraacetic acid				white powder
$C_{10}H_{22}$	n-decane	−29.7	174.1	0.730	colorless liquid
$C_{11}H_{12}N_2O_2$	tryptophan	282			white powder
$C_{11}H_{24}$	n-undecane	−25.6	195.9	0.741	colorless liquid
$C_{12}H_{10}$	biphenyl	71	255.9	0.866	colorless crystals
$C_{12}H_{10}N_2$	azobenzene	68.5	293	1.203	orange crystals
$C_{12}H_{10}N_2O$	p-hydroxyazobenzene	155	220(0.026 atm)		yellow crystals

formula	name	mp (°C)	bp (°C)	density	usual color and state
$C_{12}H_{10}N_2O_2$	2,4-dihydroxyazo-benzene	170			dark-red solid
$C_{12}H_{22}O_{11}$	sucrose	186dec		1.581	colorless crystals
$C_{12}H_{26}$	n-dodecane	−9.6	216.3	0.749	colorless liquid
$C_{13}H_9N$	acridine	111	346	1.005	colorless crystals
$C_{13}H_{10}$	fluorene	116	295	1.203	colorless crystals
$C_{14}H_8O_2$	anthraquinone	288sub	379.8	1.419	yellow solid
$C_{14}H_{10}$	anthracene	216.3	340	1.283	colorless crystals
$C_{14}H_{12}$	trans-1,2-diphenyl-ethene (stilbene)	124.5	305	0.970	colorless crystals
$C_{14}H_{15}N_3$	N,N-dimethyl-p-phenylazoaniline	117	dec		orange crystals
$C_{15}H_{11}N_4NO_4$	thyroxine	235dec			white needles
$C_{16}H_{10}N_2O_2$	indigo	392	sub	1.35	purple crystals
$C_{16}H_{32}O_2$	palmitic acid	63	390	0.853(63°)	soft white solid
$C_{18}H_{14}$	1,2-diphenylbenzene (o-terphenyl)	58	332		colorless crystals
$C_{18}H_{14}$	1,3-diphenylbenzene (m-terphenyl)	89	365		colorless crystals
$C_{18}H_{36}O_2$	stearic acid	71.7	360dec	0.940	soft white solid
$C_{19}H_{16}$	triphenylmethane	94	359.2	1.014	colorless crystals
$C_{20}H_{14}O_4$	phenolphthalein	261		1.277	white powder
$(C_6H_{10}O_5)_n$	α-cellulose	dec		1.27–1.61	white amorphous powder
Ca	calcium	842	1487	1.54	gray reactive metal
$CaBr_2$	calcium bromide	730dec	806dec	3.353	deliquescent white crystals
CaC_2	calcium carbide	>1150	2300	2.22	gray powder
$CaCN_2$	calcium cyanamide		1300sub		white crystalline solid
$CaCO_3$	calcium carbonate	898dec		2.710	colorless crystals
CaC_2O_4	calcium oxalate	dec		2.2	colorless crystals
$CaCl_2$	calcium chloride	772	>1600	2.15	deliquescent white solid
CaF_2	calcium fluoride	1360	~2500	3.180	colorless crystals
CaO	calcium oxide	2580	2850	3.25–3.38	colorless crystals
$Ca(OH)_2$	calcium hydroxide	dec		2.24	white powder
$CaSO_4$	calcium sulfate	1450		2.61	colorless crystals
$CaSiO_3$	α calcium silicate	1540	dec	2.905	white powder
					colorless crystals

formula	name	mp (°C)	bp (°C)	density	usual color and state
Cd	cadmium	320.9	765	8.642	gray metal
CdS	cadmium sulfide	1750(100 atm)	sub	4.82	bright-yellow powder
Ce	cerium	795	3468	6.768	gray metal
CeO$_2$	cerium (IV) oxide	~2600		7.132	brownish-white crystals
Cf	californium				properties unknown
Cl$_2$	chlorine	−100.98	−34.6	3.214 g/liter	greenish suffocating gas
HCl	hydrogen chloride	−114.8	−84.9	1.000 g/liter	colorless gas
Cm	curium	1340			radioactive metal
Co	cobalt	1495	2900	8.85	gray metal
CoCl$_2$	cobalt(II) chloride	724	1049	3.356	blue crystals
[CoCl$_2$(NH$_3$)$_4$]Cl	cis-dichlorotetramminecobalt(III) chloride				blue crystals
[CoCl$_2$(NH$_3$)$_4$]Cl	trans-dichlorotetramminecobalt(III) chloride				green crystals
[CoCl(NH$_3$)$_5$]Cl$_2$	chloropentamminecobalt(III) chloride	dec		1.819	purple crystals
[Co(NH$_3$)$_6$]Cl$_3$	hexamminecobalt(III) chloride	215dec		1.710	yellow crystals
[CoBrNH$_3$(H$_2$N—C$_2$H$_4$—NH$_2$)$_2$]Br$_2$	bromoamminebis-(ethylenediamine)cobalt(III) bromide				red-violet crystals
[CoCl$_2$(H$_2$N—C$_2$H$_4$—NH$_2$)$_2$]Cl	cis-dichlorobis(ethylenediamine)cobalt(III) chloride				violet crystals
[CoCl$_2$(H$_2$N—C$_2$H$_4$—NH$_2$)$_2$]Cl	trans-dichlorobis(ethylenediamine)cobalt(III) chloride				dull-green powder
Cr	chromium	1890	2482	7.20	bluish hard metal
CrCl$_3$	chromium(III) chloride	1150	1300sub	2.76	infusible purple crystals
Cr$_2$O$_3$	dichromium trioxide	2435	4000	5.21	dark-green powder
Cr$_2$(SO$_4$)$_3$	chromium(III) sulfate	dec		3.012	violet powder
Cs	cesium	28.5	690	1.879	silvery reactive metal
CsBr	cesium bromide	636	1300	4.44	colorless crystals
CsCl	cesium chloride	646	1290sub	3.988	colorless deliquescent crystals

formula	name	mp (°C)	bp (°C)	density	usual color and state
CsI	cesium iodide	621	1280	4.510	colorless deliquescent crystals
CsOH	cesium hydroxide	272.3		3.675	yellow deliquescent solid
Cu	copper	1083	2582	8.96	red soft metal
CuAl$_2$	copper aluminum	595		5.8	brownish metal
CuCl	copper(I) chloride	430	1490	4.14	white crystalline solid
CuCl$_2$	copper(II) chloride	620	993dec	3.386	brown powder
CuFeS$_2$	chalcopyrite			4.2	brass-colored crystals
Cu$_5$FeS$_4$	bornite			5.07	purple iridescent solid
Cu(NO$_3$)$_2 \cdot$ 3H$_2$O	copper(II) nitrate	114.5	170dec	2.32	blue crystals
Cu$_2$(OH)$_2$CO$_3$	malachite	200dec		5.0	dark-green solid
Cu$_3$(OH)$_2$(CO$_3$)$_2$	azurite	220dec		3.88	bright-blue solid
CuS	copper(II) sulfide	103	220dec	4.6	black crystals
Cu$_2$S	copper(I) sulfide	1100		5.6	black crystals
CuSO$_4$	anhydrous copper(II) sulfate	200	650dec	3.603	white powder
CuSO$_4 \cdot$ 5H$_2$O	copper(II) sulfate pentahydrate	110dec		2.284	bright blue crystals
Cu$_5$Zn$_8$	copper zinc	748		7.56	yellow metal
Dy	dysprosium	1407	2600	8.556	gray metal
Er	erbium	1497	2900	9.164	gray metal
Es	einsteinium				properties unknown
Eu	europium	826	1439	5.245	gray metal
F$_2$	fluorine	−219.6	−188	1.69 g/liter(15°)	pale-yellow poisonous gas
HF	hydrogen fluoride	−83.1	19.54	0.991(19°)	colorless fuming liquid
Fe	iron	1535	3000	7.86	silvery metal
Fe$_3$C	iron carbide	1837		7.694	hard gray solid
Fe(CO)$_5$	pentacarbonyliron	−21	102.8	1.457	viscous yellow liquid
FeCl$_3$	iron(III) chloride	306	315dec	2.898	brown crystals
Fe(NO$_3$)$_3 \cdot$ 9H$_2$O	iron(III) nitrate	47.2	125dec	1.684	pale-violet deliquescent crystals
FeO	iron(II) oxide	1420		5.7	black powder
Fe$_3$O$_4$	iron(II,III) oxide	1538dec		5.18	black powder
Fe$_2$O$_3$	iron(III) oxide	1565		5.24	reddish-brown solid
FeS	iron(II) sulfide	1193	dec	4.74	black crystals

formula	name	mp (°C)	bp (°C)	density	usual color and state
Fe(SCN)$_3$	iron(III) thiocyanate				dark-red solid
FeTiO$_3$	iron(II) titanate			4.44	shiny black crystals
Fm	fermium				properties unknown
Fr	francium				properties unknown
Ga	gallium	29.78	2403	5.907	silvery metal, usually liquid
Gd	gadolinium	1312	~3000	7.948	gray metal
Ge	germanium	937.4	2830	5.323	yellowish brittle metalloid
GeH$_4$	germane	−165	−88.5	1.523(−142°)	colorless gas
GeO$_2$	germanium dioxide	1086		6.239	white powder
H$_2$	hydrogen	−259.2	−252.8	0.070(−253°)	colorless inflammable gas
HCN	hydrogen cyanide	−14	26	0.688	colorless poisonous liquid
He	helium	−272.2(26 atm)	−268.6	0.177 g/liter	colorless inert gas
Hf	hafnium	2150	5400	13.31	gray metal
Hg	mercury	−38.9	356.58	13.59	silvery liquid metal
Hg(CH$_3$)$_2$	dimethylmercury		96	3.069	colorless liquid
HgCl$_2$	mercury(II) chloride	276	302	5.44	white solid
Hg$_2$Cl$_2$	mercury(I) chloride	400sub		7.15	white solid
HgO	mercury(II) oxide	500dec		11.14	deep-red solid
HgSO$_4$	mercury(II) sulfate	dec		6.47	white powder
Ho	holmium	1461	2600	8.803	gray metal
I$_2$	iodine	113.5	184.3	3.94	black crystals
ICl	iodine monochloride	27.2	97.4	3.18	red solid
ICl$_3$	iodine trichloride	101(16 atm)	77dec	3.117	yellow-brown or red solid
IF$_5$	iodine pentafluoride	9.6	98	3.75	colorless liquid
IF$_7$	iodine heptafluoride	5.5	4.5sub	2.8(5°)	colorless gas
Ir	iridium	2410	4527	22.42	dense gray metal
In	indium	156.6	2000	7.31	soft silver-white metal
K	potassium	63.65	774	0.86	soft silvery metal
K$_2$CO$_3$	potassium carbonate	891	dec	2.428	white powder
KCl	potassium chloride	776	1500sub	1.984	colorless crystals
KClO$_3$	potassium chlorate	356	400dec	2.32	colorless crystals
K$_2$Cr$_2$O$_7$	potassium dichromate	398	500dec	2.676	bright-orange crystals
KF	potassium fluoride	846	1505	2.48	colorless deliquescent crystals

formula	name	mp (°C)	bp (°C)	density	usual color and state
KHF$_2$	potassium hydrogen fluoride	~225dec		2.37	colorless deliquescent crystals
KI	potassium iodide	686	1330	3.13	colorless crystals
KMnO$_4$	potassium permanganate	240dec		2.703	purple crystals
KNO$_3$	potassium nitrate	334	400dec	2.109	colorless crystals
K$_2$O	potassium monoxide	350dec		2.32	gray powder
KOH	potassium hydroxide	360.4	1320	2.044	white deliquescent solid
K$_2$SO$_4$	potassium sulfate	1069	1689	2.662	colorless crystals
K$_2$SiO$_3$	potassium silicate	976			white solid
Kr	krypton	−156.6	−152.3	3.74 g/liter	colorless gas
La	lanthanum	920	3469	6.189	gray metal
Li	lithium	179	1317	0.534	silvery light metal
LiAlH$_4$	lithium aluminum hydride	125dec		0.917	white powder
LiF	lithium fluoride	842	1676	2.635	white crystalline solid
LiH	lithium hydride	680	dec	0.82	white powder
LiNO$_3$	lithium nitrate	264	600dec	2.38	colorless crystals
Lu	lutetium	1652	3327	9.849	gray metal
Lw	lawrencium				properties unknown
Mg	magnesium	651	1107	1.74	silvery metal
MgBr$_2$	magnesium bromide	700		3.72	white crystalline solid
MgCO$_3$	magnesium carbonate	350dec		2.958	white powder
MgCl$_2$	magnesium chloride	708	1412	2.316	white crystalline solid
MgF$_2$	magnesium fluoride	1266	2239	2.9	white solid
MgI$_2$	magnesium iodide	700dec		4.43	white deliquescent crystalline solid
MgO	magnesium oxide	2800	3600	3.58	white powder
Mg(OH)$_2$	magnesium hydroxide	350dec		2.36	white powder
Mg$_2$Si	magnesium silicide	1102		1.94	blue powder
Mn	manganese	1244	2097	7.20	gray metal
MnCl$_2$	manganese(II) chloride	650	1190	2.977	pink solid
MnO$_2$	manganese dioxide	535dec		5.026	black powder
Mo	molybdenum	2610	5560	10.22	gray metal

formula	name	mp (°C)	bp (°C)	density	usual color and state
MoCl$_5$	molybdenum pentachloride	194	268	2.93	dark-gray crystals
MoO$_3$	molybdenum trioxide	795	sub	4.692	light-yellow crystals
N$_2$	nitrogen	−210	−196	0.808(−196°)	colorless gas
NCl$_3$	nitrogen trichloride	<−40	<71 (95expl)	1.653	yellow liquid
NH$_3$	ammonia	−77.7	−33.35	0.817(−79°)	colorless gas
(NH$_2$)$_2$	hydrazine	1.4	113.5	1.011	colorless liquid
N$_3$H	hydrazoic acid	−80	37		colorless liquid
NH$_4$Cl	ammonium chloride	340sub	520	1.527	colorless crystals
NH$_4$NO$_3$	ammonium nitrate	169.6	210	1.725	white crystals
NH$_2$OH	hydroxylamine	33.05	56.5	1.204	white crystals
(NH$_4$)$_2$SO$_4$	ammonium sulfate	235dec		1.769(50°)	colorless crystals
HNO$_3$	nitric acid	−42	83	1.503	colorless fuming liquid
NO	nitrogen oxide	−163.6	−151.8	1.269(−152°)	colorless gas
NO$_2$	nitrogen dioxide	−11.20	21.2	1.449(0°)	red-brown gas
N$_2$O	dinitrogen oxide	−90.8	−88.5	1.977 g/liter	colorless gas
NOCl	nitrosyl chloride	−64.5	−5.5	1.417(−12°)	yellow gas; yellow-red liquid
Na	sodium	97.81	892	0.971	soft silvery metal
Na$_3$AlF$_6$	trisodium aluminum fluoride	1000		2.90	colorless crystals
NaCN	sodium cyanide	563.7	1496		colorless deliquescent poisonous crystals
Na$_2$CO$_3$	sodium carbonate	851	dec	2.532	white powder
Na$_2$C$_2$O$_4$	sodium oxalate	250dec		2.34	colorless crystals
NaHCO$_3$	sodium hydrogen carbonate	270dec		2.159	white powder
NaCl	sodium chloride	801	1413	2.165	colorless crystals
NaClO · 5H$_2$O	sodium hypochlorite	24.5			colorless crystals
Na$_3$N	sodium nitride	300dec			dark-gray powder
NaNH$_4$HPO$_4$ · 4H$_2$O	sodium ammonium phosphate	79dec		1.554	white crystalline solid
NaNO$_3$	sodium nitrate	306.8	380dec	2.261	white crystalline solid
Na$_2$O	sodium oxide	1275sub		2.27	grayish-white deliquescent powder

formula	name	mp (°C)	bp (°C)	density	usual color and state
Na_2O_2	sodium peroxide	460dec	657dec	2.805	yellow-white powder
NaOH	sodium hydroxide	318.4	1390	2.130	white deliquescent solid
Na_2SiO_3	sodium metasilicate	1088		2.4	colorless crystals
Nb	niobium	2468	4927	8.57	gray metal
Nd	neodymium	1024	3027	7.00	gray metal
Ne	neon	−248.6	−245.9	0.900 g/liter	colorless inert gas
Ni	nickel	1453	2732	8.90	gray metal
NiO_2	nickel(IV) oxide				dark powder
$Ni(OH)_2$	nickel(II) hydroxide	230dec		4.15	green solid
Np	neptunium	640		20.4	radioactive metal
O_2	oxygen	−218.4	−183.0	1.14(−183°)	colorless gas
H_2O	water	0	100	1.000(4°)	colorless liquid
H_2O_2	hydrogen peroxide	−0.41	150.2	1.422	colorless liquid
Os	osmium	2700	4600	22.57	hard blue-gray metal
P_4	phosphorus	44.1	280	1.82	yellow waxy solid
PBr_5	phosphorus pentabromide	100dec	106dec	3.46	yellow solid
PCl_5	phosphorus pentachloride	166.8	162sub	4.65	pale-yellow solid
H_3PO_2	hypophosphorous acid	26.5	130dec	1.493	colorless crystals
H_3PO_3	orthophosphorous acid	73.6	200dec	1.651	yellow crystals
H_3PO_4	orthophosphoric acid	42.35	213dec	1.834	colorless deliquescent crystals
$H_4P_2O_7$	pyrophosphoric acid	61			colorless crystals
P_2O_5	diphosphorus pentoxide	580	300sub	2.39	white deliquescent powder
Pa	protactinium	~1230		15.37	radioactive metal
Pb	lead	327.5	1744	11.34	dull-gray metal
$Pb(CH_3CH_2)_4$	tetraethyllead		200dec	1.650	colorless liquid
$PbCl_2$	lead chloride	501	950	5.85	white crystalline solid
PbO_2	lead dioxide	290dec		9.375	brown crystals
$PbSO_4$	lead sulfate	1170dec		6.2	white powder
Pd	palladium	1552	2927	11.97	gray metal
$PdCl_2$	palladium chloride	500dec		4.0	dark-red solid

formula	name	mp (°C)	bp (°C)	density	usual color and state
Pm	promethium				properties unknown
Po	polonium	254	962	9.4	radioactive metal
Pr	praseodymium	935	3127	6.782	soft silvery metal
Pt	platinum	1769	3827	21.45	silvery metal
Pu	plutonium	639.5	3235	19.84	radioactive metal
PuO$_2$OH	plutonium(V) basic oxide				dark solid
PuO$_2$(OH)$_2$	plutonium(VI) basic oxide				dark solid
Ra	radium	700	<1737	5.0	radioactive metal
Rb	rubidium	38.89	688	1.532	reactive gray metal
Re	rhenium	3180	5627	20.53	gray metal
Rh	rhodium	1966	3727	12.44	gray metal
Rn	radon	−71	−61.8	9.73 g/liter	colorless radioactive gas
Ru	ruthenium	2250	3900	12.3	gray metal
S$_8$	α-sulfur	112.8	444.6	2.07	yellow rhombic crystals
S$_8$	β-sulfur	119.25	444.6	1.96	pale-yellow monoclinic crystals
S$_8$	γ-sulfur	~120	444.6	1.92	pale-yellow amorphous solid
SF$_6$	sulfur hexafluoride	−50.5	63.8*sub*	6.60 g/liter	colorless gas
H$_2$S	hydrogen sulfide	−85.5	−60.7	1.539 g/liter	colorless poisonous gas
SO$_2$	sulfur dioxide	−72.7	−10.0	2.927 g/liter	colorless suffocating gas
SO$_3$	sulfur trioxide	16.83	44.8	1.92	colorless fuming liquid
H$_2$SO$_3$	sulfurous acid				known only in solution
H$_2$SO$_4$	sulfuric acid	10.36	338	1.841	colorless oily liquid
H$_2$SO$_5$	peroxomonosulfuric acid	45*dec*			white crystals
H$_2$S$_2$O$_8$	peroxodisulfuric acid	65*dec*			hydroscopic crystals
Sb	antimony	630.5	1380	6.684	dull-gray metalloid
Sc	scandium	1539	2727	2.99	gray metal
Se	selenium	217	684.8	4.79	brick-red powder
H$_2$Se	hydroselenic acid	−64	−42	3.614	g/liter colorless gas
H$_2$SeO$_4$	Selenic acid	58	260*dec*	3.004	white deliquescent crystals
Si	silicon	1414	2355	2.33	brittle blue-gray solid
CH$_3$SiHCl$_2$	dichloromethylsilane	−90.6	40.4	1.135	colorless acid liquid
CH$_3$SiCl$_3$	trichloromethylsilane	−77.5	66.1	1.270	colorless acid liquid

formula	name	mp (°C)	bp (°C)	density	usual color and state
(CH$_3$)$_2$SiCl$_2$	dichlorodimethylsilane	−76.1	70.2	1.067	colorless acid liquid
(CH$_3$)$_2$Si(OH)$_2$	dimethylsilanediol	−57.7	dec	1.097	unstable colorless liquid
(CH$_3$)$_3$SiCl	chlorotrimethylsilane	−57.7	57.3	1.388	colorless acid liquid
(CH$_3$)$_3$SiOH	trimethylsilanol	−4.5	98.9	0.814	colorless liquid
SiC$_4$H$_{12}$O$_4$	tetramethoxysilane		122	1.052	colorless liquid
C$_6$H$_5$SiCl$_3$	trichlorophenylsilane		201.5	1.326	colorless liquid
(C$_6$H$_5$)$_2$SiCl$_2$	dichlorodiphenylsilane		305.2	1.220	colorless liquid
(C$_6$H$_5$)$_2$Si(OH)$_2$	diphenylsilanediol	137			white crystals
SiH$_3$Cl	chlorosilane	−118.1	−30.4	3.033 g/liter	colorless gas
SiH$_2$Cl$_2$	dichlorosilane	−122	8.3	4.599 g/liter	colorless gas
SiHCl$_3$	trichlorosilane	−126.5	31.8	1.342	colorless acid liquid
SiCl$_4$	tetrachlorosilane	−70	57.57	1.483	colorless acid liquid
SiH$_4$	silane	−185	−111.8	1.44 g/liter	colorless gas
SiO$_2$	silicon dioxide (quartz)	1610	~2500	2.65	colorless crystals
Si$_2$C$_6$H$_{18}$O	hexamethyldisiloxane	−68	100.1	0.762	colorless liquid
Si$_2$Cl$_6$	hexachlorodisilane	−1	145	1.58	colorless fuming liquid
Si$_2$H$_6$	disilane	−132.5	−14.5	0.686(−25°)	colorless gas
Si$_3$Cl$_8$	octachlorotrisilane		216		colorless crystals
Si$_3$H$_8$	trisilane	−117.4	52.9	0.725	colorless liquid
Si$_4$C$_8$H$_{24}$O$_4$	octamethylcyclotetrasiloxane	17.5	175	0.956	colorless liquid
Si$_4$H$_{10}$	tetrasilane	−108	84.3	0.79(0°)	colorless liquid
Sm	samarium	1072	1900	7.49	gray metal
Sn	α-tin	13.2(trans. to β)		5.75	gray crystalline metal
Sn	β-tin	231.89	2260	7.28	white metal
SnCl$_2$	tin(II) chloride	246	652	3.95	white solid
SnCl$_4$	tin(IV) chloride	−33	114.1	2.226	colorless liquid
Sr	strontium	769	1384	2.60	gray metal
Ta	tantalum	2996	5425	16.6	gray metal
Tb	terbium	1356	2800	8.272	gray metal
Tc	technetium	2200		11.46	gray powder
Te	tellurium	452	1390	6.24	gray powder
TeH$_2$	tellurium hydride	−48.9	−2.2	4.49 g/liter	colorless poisonous gas
Th	thorium	1700	4000	11.7	gray radioactive metal

formula	name	mp (°C)	bp (°C)	density	usual color and state
Ti	titanium	1675	3260	4.51	blue-gray metal
TiC	titanium carbide	3140	4820	4.93	metallic crystals
Ti(C$_5$H$_5$)$_2$Cl$_2$	dicyclopentadienyl-titanium chloride	290	dec		orange crystals
TiC$_{12}$H$_{18}$O$_4$	tetrapropoxytitanium	20	58(0.0017 atm)	0.955	colorless liquid
TiCl$_4$	titanium tetrachloride	−25	136.4	1.726	colorless fuming liquid
TiN	titanium(III) nitride	2930		5.22	bronze-red crystals
TiO$_2$	titanium dioxide	1825		4.17	white powder
Tl	thallium	303.5	1457	11.85	bluish metal
Tm	thulium	1545	1727	9.332	gray metal
U	uranium	1132	3818	19.07	gray radioactive metal
UF$_6$	uranium hexafluoride	69.2(2 atm)	56.2sub	4.68	colorless volatile crystals
V	vanadium	1890	~3000	5.96	gray metal
W	tungsten	3410	5927	19.35	bluish-gray metal
WC	tungsten carbide	~2870	~6000	15.63	black solid
WCl$_6$	tungsten hexachloride	275	346.7	3.52	black cubic crystals
WO$_3$	tungsten trioxide	1473		7.16	yellow crystals
Xe	xenon	−111.9	−107.1	5.887 g/liter	colorless gas
XeF$_4$	xenon tetrafluoride				white crystals
Y	yttrium	1495	2927	4.34	gray metal
Yb	ytterbium	824	1427	6.959	gray metal
Zn	zinc	419.4	907	7.14	blue-gray metal
ZnCl$_2$	zinc chloride	283	732	2.91	white crystals
ZnI$_2$	zinc iodide	446	624dec	4.736	colorless crystals
ZnO	zinc oxide	1975		5.606	white powder
Zn(OH)$_2$	zinc hydroxide	125dec		3.053	colorless crystals
ZnS	zinc sulfide (wurtzite)	1185sub		3.98	colorless crystals
ZnS	zinc sulfide (zinc blende)	1020dec		4.102	colorless crystals
ZnSO$_4$	zinc sulfate	740dec		3.54	colorless crystals
Zr	zirconium	1852	3578	6.49	gray metal
ZrO$_2$	zirconium dioxide	2700	5000	5.89	white powder

APPENDIX II MATHEMATICS IN CHEMISTRY

"The question is," said Humpty Dumpty, "which is to be master—that's all."

LEWIS CARROLL
From *Alice in Wonderland*

Mathematics is a shorthand method of thinking logically (that is, consistently) about relationships between quantities. It is also, for many people, a forbidding hurdle to the study of science. For that reason, we have generally stressed nonmathematical approaches in this text. However, the difficulties that some students have with mathematical discussions may arise not so much from an inability to think logically, as from unfamiliarity with the abstract and compact notation of mathematics. We cannot provide the necessary familiarity in a short appendix and have no intention to try. But we may provide a useful review of some of the mathematical language that the student should already have seen, and try to place mathematics in its proper context in chemistry.

To take up the second point first, it is certainly not true, as some assert, that "numbers rule the universe." Elementary particles have no understanding of mathematics. Mathematics provides a purely metaphorical language for the construction of more or less abstract *models* of physical systems, and (though some would dispute this) for the conversion of these abstract physical models into intuitive, tangible ones that satisfy our need for a "feel" for the physical system. Particles and waves, to take two fundamental examples, are tangible metaphors for particular mathematical models. So is the concept of energy, in the special sense of something that is conserved, as described by the first law of thermodynamics (Chapter 9). To the extent that these mathematical constructs do not contradict the results of observation and experiment, they may be more or less useful ways of describing physical systems. They have no more fundamental status than that, any more than an artist's portrait fully describes its subject, much less "rules" it. However, just as a skillful artist may show us more about his subject than we had casually noticed, so the unique language of mathematics can be useful and revealing in modeling nature if we take time to master some of its vocabulary and methods. Humpty Dumpty's famous dictum about words, which begins this appendix, applies just as well to the relationship between people and mathematics—from the arithmetic of balancing a checkbook (a model of your actual financial holdings) to the most abstruse algebra, calculus, group theory, or what have you, that is used to model matter.

Functionality

The most valuable concept in mathematics as it is used in chemistry is that of a *functional* relationship between variables. In exploring the world, we generally want to know what will *happen* (that is, how some variable will change) when we *do something* (that is, alter the values of other variables). For example, the ideal gas law (Chapter 5) provides a mathematical model of what happens to the volume of a gas if its temperature is changed. In relating these two variables, the ideal gas law is said to express the volume V of a gas as a *function* of its temperature T, or vice versa. Symbolically, we write

$$V = (\text{constant}) \times T \quad (\text{at constant pressure}) \quad \text{(II-1)}$$

Given the numerical value of the constant[1] in Equation (II-1), we may calculate a unique value of V for any given T. Thus, a function is an algebraic rule for relating values of one variable (in our example, volume) to those of another (such as temperature). Give me a value of the temperature, and I can use the function in Equation (II-1) to give you the corresponding value of the volume.

INVERSE FUNCTIONS. You may be motivated to reply, with perfect justification, "Give me a value of the volume, and I will use Equation (II-1) backward to give you the corresponding temperature."[2] And in the case of Equation (II-1), you would be correct. The function

$$T = \frac{V}{\text{constant}} \quad (\text{at constant pressure})$$

is the inverse of the function in Equation (II-1). It is always possible to define an inverse to any function, though it is not always true, as it is for this one, that the inverse function provides an unambiguous result. For example, let us suppose that your height increases normally as you are growing up, but that after middle age, you, as most people do, "grow" a little shorter because of changes in your spinal column. In principle, a function could be found to relate your height to your age, such that one could say, "Give me a value of age, and I will give you your height at that age." However, the inverse information may be ambiguous. There may be two answers to the question: "At what age is your height exactly 5'10"?"—one age on the way up, and another on the way down. Thus, any function relating age to height would be, for a certain range of heights, a two-to-one relationship. A hypothetical function relating age to weight would probably be a many-to-one relationship. None of these considerations alters the usefulness of the functionality idea. Let us now consider some functions that arise in chemistry, along with their inverse functions.

LINEAR FUNCTIONS. Equation (II-1) is known (for reasons that will be evident later) as a *linear* function. All variables in it appear "straight," without being squared (or raised to any power but 1) or operated on in any way

---

[1] If you have read Chapter 5, you will recall that it is equal to nR/P; the "constant" is itself a function of n and P.

[2] Probably one of those replies that never got replied.

except for being multiplied by a constant. Another linear function is that appearing in Section 5-10:

$$PV = Rt + b \tag{II-2}$$

provided that we treat the product PV as a single variable, even though it is made of two other variables. The addition of the constant term b does not alter the fact that each variable appears to the first power only.

The inverse of a linear function is also a linear function. Thus, the temperature of a gas varies linearly with its volume at constant pressure.[3]

In the equation relating kinetic energy $E_{kinetic}$ to the mass and velocity of a moving body,

$$E_{kinetic} = \tfrac{1}{2}mv^2 \tag{II-3}$$

$E_{kinetic}$ is a linear function of the body's mass (m appears to the first power) and a *quadratic* function of its velocity v.

Exponential Functions and Logarithms

Consider the functional relationship

$$y = b^x \tag{II-4}$$

involving the variables y and x, and a constant b. Since x appears as an exponent, Equation (II-4) is known as an *exponential* relationship. One of the characteristics of such a function is that, for positive values of x, y depends very sensitively on x. For example, for $b = 10$, $y\ (= 10^x)$ increases tenfold for every unit change in x.

The inverse of an exponential function is a *logarithmic* one. That is, if Equation (II-4) is true for two variables x and y, we may say that x is the *exponent to which b must be raised to equal y*. This awkward phrase (in italics) is replaced by a shorter one: x is the *logarithm of y to the base b*. Or,

$$x = \log_b y$$

All three phrases mean the same thing; a logarithm is an exponent. If x is the logarithm of y, then y is known as the *antilogarithm* of x.

Although any number may serve as the base for a logarithmic relationship, two are in general use: 10, because it is the basis of the decimal number system;

---

[3] We often find it useful to distinguish between *dependent* and *independent* variables. It is experimentally inconvenient to control the temperature of a gas by altering its volume at constant pressure; rather, the volume is governed by an experimentally fixed temperature. In that case, the volume is thought of as depending on the temperature, and not the other way around; volume is thought of, that is, as a dependent variable, temperature as an independent variable. There is no fundamental mathematical distinction between the two categories of variables even when, as in the case of the variation of height with age, the experimental difference is crucial; however nice it might be to do so, one cannot control one's age by stretching or shrinking oneself!

and a constant known as e that has the approximate value

$$e = 2.71828\ldots^4$$

Exponential functions using e as a base arise "naturally" in the course of certain operations in calculus, and logarithms to the base e are thus known as natural logarithms. The student is referred to any calculus text for further word on e. For purposes of both mental and pencil-and-paper manipulation, logarithms to the base 10 (known as "common" logarithms) are easier to use, since the approximate value of the logarithm gives an immediate indication whether the corresponding antilogarithm is a number in the 10's (10^1 to 10^2), hundreds (10^2 to 10^3), and so forth.

Let us review some of the properties of exponents (and thus of logarithms) that are useful in chemical mathematics. In this review, we shall use the common base, base 10; analogous statements can be written regardless of what base is used.

1. *When two exponential numbers having the same base are multiplied, the exponents are added.* For example, the equation

$$10^1 \times 10^1 = 10^2$$

expresses the well-known fact that $10 \times 10 = 100$.

2. *Division of an exponential by another having the same base results in subtraction of the exponent of the divisor from that of the dividend.* Thus,

$$\frac{10^3}{10^1} = 10^2$$

That is,

$$\frac{1000}{10} = 100$$

3. *Any number raised to the zero power equals one.* Thus,

$$\frac{10^x}{10^x} = 10^{(x-x)} = 10^0$$

is another way of writing

$$\frac{10^x}{10^x} = 1$$

4. *When an exponential number is raised to a power, the exponent is multiplied by the power to which the number is to be raised.* This is a direct consequence of Rule 1. For example, the equation

$$100^3 = 1,000,000$$

---

[4] The string of dots indicates that e is a continuing decimal which never terminates or repeats.

can as well be written

$$(10^2)^3 = 10^2 \times 10^2 \times 10^2$$
$$= 10^{(2+2+2)}$$
$$= 10^{(2\times 3)}$$
$$= 10^6$$

5. *Conversely, taking the nth root of an exponential number is accomplished by dividing the exponent by n.* Thus,

$$\sqrt[3]{10^6} = 10^2$$

6. *The exponent need not be an integer to be meaningful.* For example, consider the problem

$$\sqrt{1000} = ?$$

Writing 1000 as 10^3, and using Rule 5, we find

$$\sqrt{10^3} = 10^{3/2}$$
$$= 10^{1.5}$$

Since, after all, there is such a number as $\sqrt{1000}$, it must be that $10^{1.5}$ refers to that perfectly definite number. Extracting the square root by hand, or with a calculator, we find that

$$\sqrt{1000} = 10^{1.5} = 31.6$$

to an accuracy sufficient for our purposes. We may also, using the definition of a logarithm, say that[5]

$$\log_{10} 31.6 = 1.5$$

In short, every positive number can be expressed as 10 raised to some power, and that power is the number's logarithm to the base 10.

We may use Rule 1 to simplify the handling of nonintegral logarithms. For example, $10^{1.5}$ is the same as $10^1 \times 10^{0.5}$. Similarly, $10^{2.5}$ is the same as $10^2 \times 10^{0.5}$, and so forth. So whenever a nonintegral exponent (logarithm) arises as the result of a calculation, it may be broken into an integer and a number between zero and one. The antilogarithm of a nonintegral exponent, then, may be found with reasonable ease. Suppose

$$x = 10^{1.51}$$

We can find x, the antilogarithm of 1.51, by the following process:

$$10^{1.51} = 10^1 \times 10^{0.51}$$
$$= 10 \times 10^{0.51}$$

---

[5] Note that, since 31.6 is a number between 10 and 100, it is reasonable that its logarithm is between 1 and 2.

In order to evaluate $10^{0.51}$ (that is, to find the antilogarithm of 0.51), we must resort to some device: a logarithm table, a calculator, or a slide rule. (The latter two are particularly handy if you have them, and you should press your instructor for help in learning how to use them.) Since a calculator may not be available, or a slide rule may not be accurate enough for your needs, a

TABLE II-1
Logarithms to the Base 10

n	0	1	2	3	4	5	6	7	8	9
1.0	.0000	.0043	.0086	.0128	.0170	.0212	.0253	.0294	.0334	.0374
1.1	.0414	.0453	.0492	.0531	.0569	.0607	.0645	.0682	.0719	.0755
1.2	.0792	.0828	.0864	.0899	.0934	.0969	.1004	.1038	.1072	.1106
1.3	.1139	.1173	.1206	.1239	.1271	.1303	.1335	.1367	.1399	.1430
1.4	.1461	.1492	.1523	.1553	.1584	.1614	.1644	.1673	.1703	.1732
1.5	.1761	.1790	.1818	.1847	.1875	.1903	.1931	.1959	.1987	.2014
1.6	.2041	.2068	.2095	.2122	.2148	.2175	.2201	.2227	.2253	.2279
1.7	.2304	.2330	.2355	.2380	.2405	.2430	.2455	.2480	.2504	.2529
1.8	.2553	.2577	.2601	.2625	.2648	.2672	.2695	.2718	.2742	.2765
1.9	.2788	.2810	.2833	.2856	.2878	.2900	.2923	.2945	.2967	.2989
2.0	.3010	.3032	.3054	.3075	.3096	.3118	.3139	.3160	.3181	.3201
2.1	.3222	.3243	.3263	.3284	.3304	.3324	.3345	.3365	.3385	.3404
2.2	.3424	.3444	.3464	.3483	.3502	.3522	.3541	.3560	.3579	.3598
2.3	.3617	.3636	.3655	.3674	.3692	.3711	.3729	.3747	.3766	.3784
2.4	.3802	.3820	.3838	.3856	.3874	.3892	.3909	.3927	.3945	.3962
2.5	.3979	.3997	.4014	.4031	.4048	.4065	.4082	.4099	.4116	.4133
2.6	.4150	.4166	.4183	.4200	.4216	.4232	.4249	.4265	.4281	.4298
2.7	.4314	.4330	.4346	.4362	.4378	.4393	.4409	.4425	.4440	.4456
2.8	.4472	.4487	.4502	.4518	.4533	.4548	.4564	.4579	.4594	.4609
2.9	.4624	.4639	.4654	.4669	.4683	.4698	.4713	.4728	.4742	.4757
3.0	.4771	.4786	.4800	.4814	.4829	.4843	.4857	.4871	.4886	.4900
3.1	.4914	.4928	.4942	.4955	.4969	.4983	.4997	.5011	.5024	.5038
3.2	.5051	.5065	.5079	.5092	.5105	.5119	.5132	.5145	.5159	.5172
3.3	.5185	.5198	.5211	.5224	.5237	.5250	.5263	.5276	.5289	.5302
3.4	.5315	.5328	.5340	.5353	.5366	.5378	.5391	.5403	.5416	.5428
3.5	.5441	.5453	.5465	.5478	.5490	.5502	.5514	.5527	.5539	.5551
3.6	.5563	.5575	.5587	.5599	.5611	.5623	.5635	.5647	.5658	.5670
3.7	.5682	.5694	.5705	.5717	.5729	.5740	.5752	.5763	.5775	.5786
3.8	.5798	.5809	.5821	.5832	.5843	.5855	.5866	.5877	.5888	.5899
3.9	.5911	.5922	.5933	.5944	.5955	.5966	.5977	.5988	.5999	.6010
4.0	.6021	.6031	.6042	.6053	.6064	.6075	.6085	.6096	.6107	.6117
4.1	.6128	.6138	.6149	.6160	.6170	.6180	.6191	.6201	.6212	.6222
4.2	.6232	.6243	.6253	.6263	.6274	.6284	.6294	.6304	.6314	.6325
4.3	.6335	.6345	.6355	.6365	.6375	.6385	.6395	.6405	.6415	.6425
4.4	.6435	.6444	.6454	.6464	.6474	.6484	.6493	.6503	.6513	.6522
4.5	.6532	.6542	.6551	.6561	.6571	.6580	.6590	.6599	.6609	.6618
4.6	.6628	.6637	.6646	.6656	.6665	.6675	.6684	.6693	.6702	.6712
4.7	.6721	.6730	.6739	.6749	.6758	.6767	.6776	.6785	.6794	.6803
4.8	.6812	.6821	.6830	.6839	.6848	.6857	.6866	.6875	.6884	.6893
4.9	.6902	.6911	.6920	.6928	.6937	.6946	.6955	.6964	.6972	.6981
5.0	.6990	.6998	.7007	.7016	.7024	.7033	.7042	.7050	.7059	.7067
5.1	.7076	.7084	.7093	.7101	.7110	.7118	.7126	.7135	.7143	.7152
5.2	.7160	.7168	.7177	.7185	.7193	.7202	.7210	.7218	.7226	.7235
5.3	.7243	.7251	.7259	.7267	.7275	.7284	.7292	.7300	.7308	.7316
5.4	.7324	.7332	.7340	.7348	.7356	.7364	.7372	.7380	.7388	.7396

logarithm table is useful. A small one is provided in Table II-1, and by consulting it we find that, to three figures, 0.51 is the logarithm of 3.24. So

$$10^{1.51} = 10 \times 3.24 = 32.4$$

TABLE II-1 (continued)
Logarithms to the Base 10

n	0	1	2	3	4	5	6	7	8	9
5.5	.7404	.7412	.7419	.7427	.7435	.7443	.7451	.7459	.7466	.7474
5.6	.7482	.7490	.7497	.7505	.7513	.7520	.7528	.7536	.7543	.7551
5.7	.7559	.7566	.7574	.7582	.7589	.7597	.7604	.7612	.7619	.7627
5.8	.7634	.7642	.7649	.7657	.7664	.7672	.7679	.7686	.7694	.7701
5.9	.7709	.7716	.7723	.7731	.7738	.7745	.7752	.7760	.7767	.7774
6.0	.7782	.7789	.7796	.7803	.7810	.7818	.7825	.7832	.7839	.7846
6.1	.7853	.7860	.7868	.7875	.7882	.7889	.7896	.7903	.7910	.7917
6.2	.7924	.7931	.7938	.7945	.7952	.7959	.7966	.7973	.7980	.7987
6.3	.7993	.8000	.8007	.8014	.8021	.8028	.8035	.8041	.8048	.8055
6.4	.8062	.8069	.8075	.8082	.8089	.8096	.8102	.8109	.8116	.8122
6.5	.8129	.8136	.8142	.8149	.8156	.8162	.8169	.8176	.8182	.8189
6.6	.8195	.8202	.8209	.8215	.8222	.8228	.8235	.8241	.8248	.8254
6.7	.8261	.8267	.8274	.8280	.8287	.8293	.8299	.8306	.8312	.8319
6.8	.8325	.8331	.8338	.8344	.8351	.8357	.8363	.8370	.8376	.8382
6.9	.8388	.8395	.8401	.8407	.8414	.8420	.8426	.8432	.8439	.8445
7.0	.8451	.8457	.8463	.8470	.8476	.8482	.8488	.8494	.8500	.8506
7.1	.8513	.8519	.8525	.8531	.8537	.8543	.8549	.8555	.8561	.8567
7.2	.8573	.8579	.8585	.8591	.8597	.8603	.8609	.8615	.8621	.8627
7.3	.8633	.8639	.8645	.8651	.8657	.8663	.8669	.8675	.8681	.8686
7.4	.8692	.8698	.8704	.8710	.8716	.8722	.8727	.8733	.8739	.8745
7.5	.8751	.8756	.8762	.8768	.8774	.8779	.8785	.8791	.8797	.8802
7.6	.8808	.8814	.8820	.8825	.8831	.8837	.8842	.8848	.8854	.8859
7.7	.8865	.8871	.8876	.8882	.8887	.8893	.8899	.8904	.8910	.8915
7.8	.8921	.8927	.8932	.8938	.8943	.8949	.8954	.8960	.8965	.8971
7.9	.8976	.8982	.8987	.8993	.8998	.9004	.9009	.9015	.9020	.9025
8.0	.9031	.9036	.9042	.9047	.9053	.9058	.9063	.9069	.9074	.9079
8.1	.9085	.9090	.9096	.9101	.9106	.9112	.9117	.9122	.9128	.9133
8.2	.9138	.9143	.9149	.9154	.9159	.9165	.9170	.9175	.9180	.9186
8.3	.9191	.9196	.9201	.9206	.9212	.9217	.9222	.9227	.9232	.9238
8.4	.9243	.9248	.9253	.9258	.9263	.9269	.9274	.9279	.9284	.9289
8.5	.9294	.9299	.9304	.9309	.9315	.9320	.9325	.9330	.9335	.9340
8.6	.9345	.9350	.9355	.9360	.9365	.9370	.9375	.9380	.9385	.9390
8.7	.9395	.9400	.9405	.9410	.9415	.9420	.9425	.9430	.9435	.9440
8.8	.9445	.9450	.9455	.9460	.9465	.9469	.9474	.9479	.9484	.9489
8.9	.9494	.9499	.9504	.9509	.9513	.9518	.9523	.9528	.9533	.9538
9.0	.9542	.9547	.9552	.9557	.9562	.9566	.9571	.9576	.9581	.9586
9.1	.9590	.9595	.9600	.9605	.9609	.9614	.9619	.9624	.9628	.9633
9.2	.9638	.9643	.9647	.9652	.9657	.9661	.9666	.9671	.9675	.9680
9.3	.9685	.9689	.9694	.9699	.9703	.9708	.9713	.9717	.9722	.9727
9.4	.9731	.9736	.9741	.9745	.9750	.9754	.9759	.9763	.9768	.9773
9.5	.9777	.9782	.9786	.9791	.9795	.9800	.9805	.9809	.9814	.9818
9.6	.9823	.9827	.9832	.9836	.9841	.9845	.9850	.9854	.9859	.9863
9.7	.9868	.9872	.9877	.9881	.9886	.9890	.9894	.9899	.9903	.9908
9.8	.9912	.9917	.9921	.9926	.9930	.9934	.9939	.9943	.9948	.9952
9.9	.9956	.9961	.9965	.9969	.9974	.9978	.9983	.9987	.9991	.9996

SOURCE: *Rinehart Mathematical Tables, Formulas and Curves* by H. D. Larsen. Copyright 1953 by H. D. Larsen. Reprinted by permission of Holt, Rinehart and Winston, Inc.

The reverse calculation is no more difficult. Suppose we need to know (Who cares why?) the logarithm of 730. We write the number 730 as 7.30×10^2. The logarithm of 7.30 may be found in the logarithm table to be 0.8633; and of course the logarithm of 10^2 is exactly 2, so that

$$\log_{10} 730 = \log_{10}(7.30 \times 10^2)$$
$$= 0.8633 + 2.0000$$
$$= 2.8633$$

You should practice by converting some of your favorite numbers to their logarithms, and vice versa.

p NOTATION. The useful quantity pH was introduced in Chapter 7. Since this p notation is just a form of logarithm (pH = $-\log_{10}[H^+]$), we should have no difficulty with it. There is a slight wrinkle, however. Since most of the quantities to which the p operator is applied are very small (the numbers are less than one), their logarithms are less than zero—their logarithms are negative. For example, we remarked in Chapter 7 that the hydrogen-ion concentration of healthy blood is always close to 4×10^{-8} M. Using Rule 1, we may easily convert this concentration to the pH:

$$pH = -\log_{10}(4 \times 10^{-8})$$
$$= -(\log_{10} 4 + \log_{10} 10^{-8})$$

From Table II-1, we find that[6]

$$\log_{10} 4 = 0.6$$

And it is inescapable that $\log_{10} 10^{-8} = -8$. So

$$pH = -(0.6 - 8.0)$$
$$= -(-7.4) \qquad (11\text{-}5)$$
$$= 7.4$$

Notice that the two parts of the logarithm in Equation (11-5) have opposite signs. We shall have to bear that in mind when making the reverse calculation. For example, the pH of the ocean is almost always close to 8.1 (Chapter 18). Let us calculate the value of $[H^+]$ to which pH 8.1 corresponds. Remember our experience with Equation (11-5) and resist the temptation to rush to the log table and look up the antilogarithm of 0.1 (or, for that matter, of 8.1). Remember that a pH value is the *negative* of the logarithm of $[H^+]$, and notice that the log table contains only positive decimals. We shall have to adjust the logarithm (-8.1) so that the decimal part of it is positive. This is not hard, since

$$-8.1 = 0.9 - 9.0$$

---

[6] The logarithm of *exactly* 4 is given in Table II-1 as 0.6021. However, the 4 in this example, being an experimentally determined number, is not exact. Since this concentration is given to only one-figure accuracy, there is no point in misleading people as to the accuracy of the data by writing down extra figures in the pH value (the logarithm) that have no significance.

Now we may use Rule 1, the log table, and common sense (not necessarily in that order):

$$\log_{10} [H^+] = 0.9 - 9.0$$
$$[H^+] = 10^{+0.9} \times 10^{-9}\ M$$
$$= 8 \times 10^{-9}\ M$$

(Again notice that the use of a logarithm accurate only to the first decimal place allows calculation of the antilogarithm only to one digit, even though more accuracy is available in the logarithm table for use when the number of significant digits in the input data warrant.)

Graphs

Very often it is easier to "see" a functional relationship between variables when one can literally see it. That is the purpose of graphs. A graph is simply an arrangement for translating numerical values of a variable into geometrical quantities—distances or angles—known collectively as *coordinates*. You are probably already familiar with the Cartesian coordinate system, in which the numerical values of the variables are represented by distances from perpendicular axes. Let us look at some common functional relationships in two-dimensional Cartesian coordinates.

LINEAR FUNCTIONS.   Figure II-1 is a graph of Equation (II-1),

$$V = (\text{constant}) \times T$$

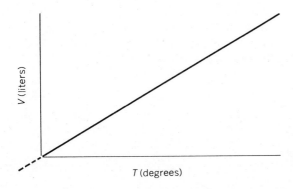

FIGURE II-1   A graph of the equation $V = (\text{constant}) \times T$. [Equation (II-1)]. The straightness of the line is the reason the term *linear* is used to describe equations like (II-1).

Now it becomes evident why such a relationship is called *linear*—its graph in rectilinear Cartesian coordinates is a straight line. Note that this particular line passes through the *origin*, because according to Equation (II-1), $V = 0$ when $T = 0$. The angle that this line makes with the coordinate axes depends on the value of (constant); Equation (II-1) says that when T increases by one degree, V increases by (constant) liters. From Chapter 5, we know that (constant) = nR/P, so that its value depends both on the quantity of gas present and on the gas's pressure. Equation (II-1) is plotted for two values of n and for two values of P in Figure II-2. Notice that the slope of the lines changes, but their straightness, being a picture of a linear function, is unchanged when (constant) changes. The numerical value of (constant) is, in fact, called the *slope* of the line. As we have seen, the slope gives the rate at which one variable (V) changes as the other (T) changes, and this is the general definition of the slope of a line on a graph.

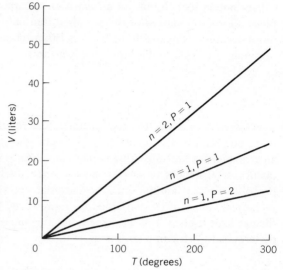

FIGURE II-2 Equation (II-1) plotted for various values of n (the number of moles of gas) and P (pressure). Changing in this way the value of (constant) in Equation (II-1) changes the slope of the line, but not its straightness. The graph retains the quality of linearity.

Now let us turn to the graph of Equation (II-2),

$$PV = Rt + b$$

This relationship, treating the product PV as a single variable, is graphed in Figure II-3. The line is straight (the function is still a linear one) and has a slope of R and an *intercept* (the value of PV when $T = 0$) of b.

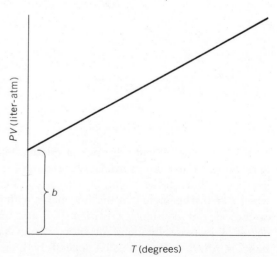

FIGURE II-3 A graph of Equation (II-2). The additive constant b has the effect of moving every point on the line upward by b units, leading to a nonzero intercept.

EXPONENTIAL FUNCTIONS. Figure II-4 is a graph of the function

$$y = a \cdot b^{cx}$$

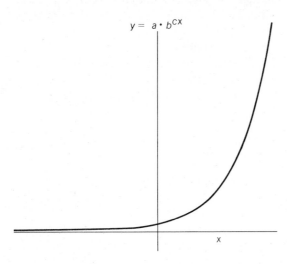

FIGURE II-4 The exponential function $y = a \cdot b^{cx}$.

Functions of this form having the same general appearance whatever the values of the constants a and c or of the base b) occur frequently in nature, as we have seen. The graph reveals two interesting features. First of all, note that as x increases in the positive (rightward) direction, y increases at first slowly, and then within a relatively short space appears to "take off." Such behavior is exhibited by populations of most living things, including people, growing under conditions of unlimited nourishment and freedom from predators (Chapter 11). As such, it plays an important role in the social concerns woven throughout this book and, we hope, your life.

If we look at Figure II-4 from the opposite direction—that is, examine increasing negative values of x—we find another important phenomenon in nature: exponential decay. A function that decays exponentially with increasing values of the exponent is exemplified by the radial part of the $1s$ wave function (see Chapter 2). Here it is important to see that although a function of the form $a \cdot b^{-cx}$ becomes very small as x increases, it never reaches zero. This means that the wave function for an electron "in" an atom, and thus the probability of finding the electron, never becomes zero, no matter how far from the nucleus (the origin of the function) one looks. Wave functions for electrons in this page (and, for that matter, in each of the authors) have a small positive value on the tip of your nose. The value is indeed very small—"vanishingly small," some would say.

POLAR COORDINATES. The Cartesian system of coordinates is a very useful one for displaying abstract mathematical functions that are models of physical systems. However, it is not the only way of translating functional into geometrical relationships. When the physical system being modeled has an important central point from which distances can be measured in various directions, it often makes the mathematical model simpler to use the *polar* coordinate system. In this system, we retain the Cartesian coordinate axes (or at least some of them), but use them as reference lines for the measurement of angles. For example, in two-dimensional polar coordinates (Figure II-5), any point in the plane is characterized, not by its x and y coordinates, but by two other geometric measurements: its distance r from the origin of coordinates, and the angle θ between a line connecting the point to the origin and a reference line (for example, the positive x axis). It should be clear that any point in the plane may be described as well by the pair of numbers $\{r, \theta\}$ as by the pair $\{x, y\}$.

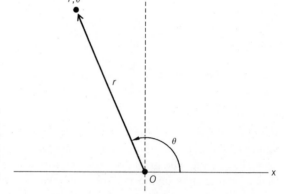

FIGURE II-5 Two-dimensional polar coordinates. The Cartesian x axis is used as a reference line for measurement of the angle θ. The Cartesian y axis, which is not needed here, is shown dashed.

The simplification of modeling functions achieved by polar coordinates is exemplified by (but by no means limited to) the equation describing a circle. In Cartesian coordinates, a circle is a line, all of whose points represent number pairs $\{x, y\}$ obeying the equation

$$\sqrt{x^2 + y^2} = a \qquad \text{(II-6)}$$

where a is a constant. The very same circle is modeled in polar coordinates by the much simpler equation

$$r = a \qquad \text{(II-7)}$$

The simplicity in Equation (II-7) compared to Equation (II-6) reflects the fact that a circle's central symmetry is ideally suited to modeling functions plotted in polar coordinates.

Since atoms have an important central point (the nucleus), polar coordinates are generally used for atomic wave functions. And since an atom is a three-dimensional object, we use three-dimensional polar coordinates (Figure II-6), in which the third geometric quantity is another angle, ϕ, measured from a second reference axis (usually designated as the z axis) at right angles to the first. Although θ varies from 0° to 360° (0 to 2π radians), ϕ need only run from 0° to 180° (0 to π radians) to reach every point in the three-dimensional space around the origin.

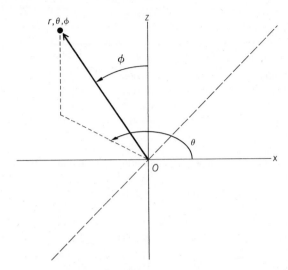

FIGURE II-6 Three-dimensional polar coordinates. As in two-dimensional polar coordinates, a radial line is drawn from the origin to the point in space. The angle ϕ is measured from the Cartesian z axis to that radial line; θ is the angle between the Cartesian x axis and the projection (dashed) of the radial line on the xy plane.

SINE AND COSINE FUNCTIONS IN CARTESIAN AND POLAR COORDINATES.

The subject of waves is central to quantum physics, and the mathematical model of a wave is always made up from the sine and/or cosine functions of trigonometry. Figure II-7 shows why. The purpose of this section is to show you how waves (modeled by sine and cosine functions) look when plotted in polar coordinates. We begin by listing (in Table II-2) values of sin θ and cos θ, and also, since these appear in wave functions whose *squares* are the physically significant quantities, their squares.

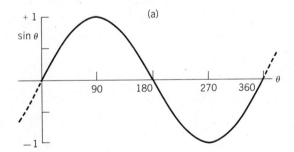

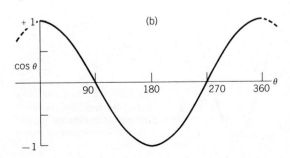

FIGURE II-7 Sine and cosine functions plotted using Cartesian coordinates. (a) sin θ and (b) cos θ plotted versus the variable θ in degrees. Notice that each function is positive during half of a cycle, and negative during the other half. Since 0° and 360° are the same point, the waves continue indefinitely in both directions, indicated by the dashed extensions.

TABLE II-2
Sine, Cosine, Sine² and Cosine² of Angles

Angle (deg.)	Sin θ	Cos θ	Sin² θ	Cos² θ
0	0.000	1.000	0.000	1.000
10	0.174	0.985	0.030	0.970
20	0.342	0.940	0.117	0.883
40	0.643	0.766	0.413	0.586
60	0.866	0.500	0.750	0.250
80	0.985	0.174	0.970	0.030
90	1.000	0.000	1.000	0.000
100	0.985	−0.174	0.970	0.030
120	0.866	−0.500	0.750	0.250
140	0.643	−0.766	0.413	0.586
160	0.342	−0.940	0.117	0.883
180	0.000	−1.000	0.000	1.000
200	−0.342	−0.940	0.117	0.883
220	−0.643	−0.766	0.413	0.586
240	−0.866	−0.500	0.750	0.250
260	−0.985	−0.174	0.970	0.030
270	−1.000	0.000	1.000	0.000
280	−0.985	0.174	0.970	0.030
300	−0.866	0.500	0.750	0.250
320	−0.643	0.766	0.413	0.586
340	−0.342	0.940	0.117	0.883
360 (= 0°)	0.000	1.000	0.000	1.000

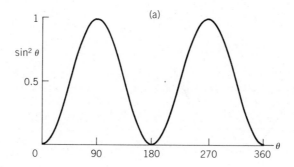

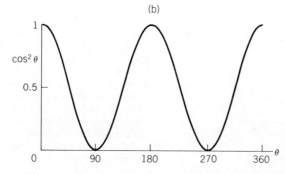

FIGURE II-8 Squared sine and cosine functions plotted in Cartesian coordinates. (a) $\sin^2 \theta$ and (b) $\cos^2 \theta$ plotted versus θ in degrees.

The Cartesian plot of the sine and cosine functions is shown in Figure II-7. The plot of the squares is Figure II-8; notice that, since the square of a negative number is positive, the squared sine and cosine functions are always positive, though they touch zero where the sine and cosine functions pass through it. These zero points, in waves, are called *nodes*.

Now let us plot these same pairs of numerical values in polar coordinates. The result (Figure II-9) is both surprising and pleasing, at least to the authors. The fact that the $x = \sin y$ function becomes zero at 0° and 180° (Figure II-8) means, in polar coordinates, that $r = \sin^2 \theta$ does so when r makes angles of 0° and 180° with the reference axis. In the two-dimensional polar coordinates shown here, this creates a *nodal line*, corresponding to the nodal points of the Cartesian plot.

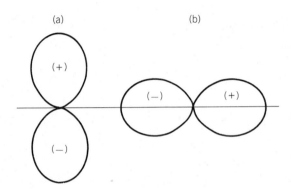

FIGURE II-9 Squared sine and cosine functions plotted in polar coordinates. The signs in parentheses are those of the original sine and cosine functions before squaring. (a) $\sin^2 \theta$ and (b) $\cos^2 \theta$. Compare Figures II-7 and II-8. The horizontal line is the Cartesian x axis. For the relationship of this graph to atomic-orbital functions (and thus to chemistry), see Figures 2-6 (Chapter 2) and 12-5 (Chapter 12).

Let us play one last trick. By plotting $r = \sin^2 \theta$ and $r = \cos^2 \theta$ in Figure II-9, we have generated graphs that bear a considerable resemblance to the p orbital functions shown in Chapters 2 and 4. This resemblance is no accident, since the p orbital wave functions do contain, and take their general shape from, squared sine and cosine functions. That is, when a wave (an electron) is placed in a centrally symmetric environment (under the attraction of the electric charge of the nucleus), it turns out to look like a sine or a cosine function plotted in centrally symmetrical polar coordinates.

EXERCISE: Make a table of values of r, θ, and $\sin^2 (2\theta)$, and compare the appearance of the function

$$r = \sin^2 (2\theta)$$

plotted in polar coordinates to another atomic-orbital function.

APPENDIX III
SOME
PHYSICS BACKGROUND
FOR CHEMISTRY

The kingdom of heaven is like a grain of mustard seed.

MATTHEW 13:31,
Revised Standard Version of the Bible

The kingdom of heaven is an abstract and complicated concept, not easy to talk about. A mustard seed can be seen and can be held in the hand. Thus, it is useful to use a mustard seed as a model for the kingdom of heaven. The model is not an exact replica, but some features are similar. So it is with many aspects of nature. Our world, even in its simpler aspects, is too complicated to analyze with exactness. We constantly use models as analogues for systems, and this book is filled with such analogies.

Physics uses some of the simplest models, analyzing them in great detail to serve as prototypes for understanding the essential features of more complicated systems. An early and dramatic success with such model building came when men were willing to say that the solar system is like a collection of moving point masses, each attracting all others; this simple model predicts the observed *motions* of the actual planets. In this book, the vocabulary and many of the ideas about chemical bonding were generated by saying that every atom is like a hydrogen atom (except for the nuclear charge and the number of electrons), and that a hydrogen atom is like two point masses, each attracting the other. The essence of a complicated system was abstracted, and then dealt with in terms of a simpler model. By confining exact analysis to simple systems, there is of course the hazard that the simple model may not be faithful to the complicated system in some important respects. This is the hazard of using analogies that people have been living with, even before the kingdom of heaven was likened to the mustard seed.

Some Fundamental Properties of Things

A set of scales in a butcher shop, a yardstick in a fabric shop, a time clock in a factory—these familiar devices for measuring mass, length, and time are essential for our way of life. The units of measurement are also familiar to us all: a pound of hamburger, a yard of muslin, an hour of work. In a world in which people talk about distances as different as an interatomic spacing and the space between galaxies, units such as angstroms and light-years take their place with the yard as practical units of measurement.

Many of the units in this book are based on the *gram* as the standard of mass, the *meter* as the standard of length (you may find the comparisons in Figure III-1 informative), and the *second* as the standard of time. Then, to

638 SOME PHYSICS BACKGROUND FOR CHEMISTRY

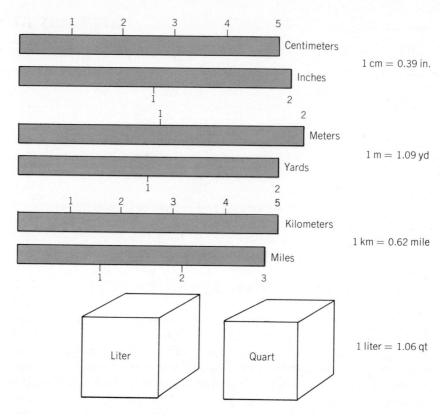

FIGURE III-1 Comparisons of relative sizes of common metric and English units of length and volume.

TABLE III-1
Metric Units of Length and Mass

Commonly Used Units of Length		
*kilo*meter	km	10^3 m
meter	m	
*deci*meter	dm	10^{-1} m
*centi*meter	cm	10^{-2} m
*milli*meter	mm	10^{-3} m
micron	μ	10^{-6} m
*nano*meter	nm	10^{-9} m
angstrom	Å	10^{-10} m
Commonly Used Units of Mass		
metric ton	ton	10^6 g
*kilo*gram	kg	10^3 g
gram	g	
*milli*gram	mg	10^{-3} g

have units of a convenient size for a particular measuring task, these basic units are made larger or smaller by factors of ten, and given new names. Table III-1 lists some common divisions and multiples of the meter and gram that we have used in this book.

It might be more convenient for a consumer of hamburger to have that pound of meat sold as 1400 kcal of beef, or as 72 g of protein. This could be done, and perhaps should be done. Caloric content or protein content might be more meaningful to an enlightened customer, although the measuring techniques are not as familiar as measurement of mass. One goal of physical science is the formulation of descriptions of nature in appropriate units, units that reveal the essence of the object or phenomenon being described, units that are closely related to the measuring instruments being used, and units that are ultimately based on international standards[1] of mass, length, and time (and a few other fundamental quantities, such as the size of the mole and the properties of the temperature scale). Since so much of this book is concerned with energy, we shall devote a large portion of this appendix to considerations of energy, and the appropriate numbers to be used in talking about energy. Note that we are able to talk about energy in its various forms in terms of mass, length, and time.

Energy: Its Conversion and Transformation

Let us examine the dynamics of a simple two-particle system in which the governing force is gravitation. The earth is one of the particles. The other particle has a mass of one gram. We focus on just three properties of the one-gram dot: its position, its velocity, and its mass; we shall call this dot a *point mass*. By changing position, and by changing velocity, this point mass gets involved in the transformations of potential energy into kinetic energy. We shall use this system to calibrate our intuition about such physical concepts as energy, force, and work.

Let us raise this lightweight particle to a height of 12,250 cm above the surface of the earth. This unlikely number is the distance that an object will fall in 5 sec; it is a little longer than a 100-yd dash, only straight up. Actually, it is the distance that the freely falling object would fall if it encountered no air molecules; a point mass is not big enough to encounter anything!

Now release the particle, and let it fall freely back to earth. This fall is described in several ways in Table III-2. The downward speed—the particle's velocity v—is zero at the instant of release (time $t = 0$), but v increases with time and has reached the value of 4900 cm/sec at the moment of impact with the earth. The velocity increases during the fall because there is a constant gravitational pull—a force F—between the earth and the one-gram par-

---

[1] An excellent discussion of the fundamental standards and the ways supplementary units are related to the base units is P. Vigoureux, *Units and Standards of Electromagnetism*, London: Wykeham Publications (1971) (paperback).

ticle. *Changes in velocity always require a force.* The constant force attributed to gravity has the value given by the equation

$$F = m \cdot (980 \text{ cm/sec}^2) \qquad \text{(III-1)}$$

where m is the mass of the particle. The *increase* in velocity—the acceleration a—is related to the force by the equation

$$F = ma \qquad \text{(III-2)}$$

Equations (III-1) and (III-2) together require that

$$a = 980 \text{ cm/sec}^2$$

Table III-2 has constant entries for a and F throughout the fall. But changes are indeed taking place in values of position x and velocity, and also in the distribution of energy between two categories named kinetic and potential.

TABLE III-2
The Fall of a One-gram Point Mass

t (sec)	F (g-cm/sec^2)	a (cm/sec^2)	v (cm/sec)	x (cm)	Energy (g-cm^2/sec^2 ≡ erg)		
					$E_{kinetic}$	$+\ E_{potential}$	$=\ E_{total}$
0	980	980	0	12,250	0	$+\ 1.20 \times 10^7$	$=\ 1.20 \times 10^7$
1	980	980	980	11,760	4.8×10^5	$+\ 1.15 \times 10^7$	$=\ 1.20 \times 10^7$
2	980	980	1960	10,290	1.92×10^6	$+\ 1.01 \times 10^7$	$=\ 1.20 \times 10^7$
3	980	980	2940	7840	4.32×10^6	$+\ 7.69 \times 10^6$	$=\ 1.20 \times 10^7$
4	980	980	3920	4410	7.69×10^6	$+\ 4.33 \times 10^6$	$=\ 1.20 \times 10^7$
5	980	980	4900	0	1.20×10^7	$+\ 0$	$=\ 1.20 \times 10^7$

The energy of motion of the particle—its kinetic energy $E_{kinetic}$—is given by the equation

$$E_{kinetic} = \tfrac{1}{2}mv^2 \qquad \text{(III-3)}$$

The faster the particle moves, the greater its energy. Where does the accelerating particle get this new kinetic energy? The traditional answer is that the particle already had a store of potential energy $E_{potential}$, and that potential energy gets converted into kinetic energy during the fall. At the last instant of our tale, just before impact, all the potential energy is gone, all transformed into kinetic energy. Total energy—the sum of $E_{kinetic}$ and $E_{potential}$—remains constant, and we say that *energy has been conserved* in this event.

The potential energy was originally given to the particle when *work* W was done to raise it to the height of 12,250 cm. We can find the amount of work done by multiplying the force acting on the particle by the distance moved (in a direction opposite to the direction the force acts). That is, if the force F acts along the x axis, then

$$W = F \triangle x$$

where $\triangle x$ is the change in distance. For our particle, the work done is $(980 \text{ g–cm/sec}^2) \cdot (12{,}250 \text{ cm}) = (1.20 \times 10^7 \text{ g–cm}^2/\text{sec}^2)$. When this amount of work has been done on a particle, the particle is said to have acquired that amount of potential energy. Since the units of force, g–cm/sec², make up too big a cluster to say easily, they have been given the name *dyne*. The energy units, g–cm²/sec², are called an *erg*. And since the erg is so tiny for talking about one-gram particles and larger objects, we call 10^7 ergs a *joule*. Table III-3 gives some of these relationships between units.

TABLE III-3
Units for Force and Energy

cgs system of units	mks system of units
distance in *centimeters*	distance in *meters*
mass in *grams*	mass in *kilograms*
time in *seconds*	time in *seconds*
force (g-cm/sec²)	**force** (kg-m/sec²)
A *dyne* is the force needed to accelerate a mass of one gram at the rate of one centimeter per second per second.	a *newton* is the force needed to accelerate a mass of one kilogram at the rate of one meter per second per second.
1 dyne = 10^{-5} newtons (N)	1 newton (N) = 10^5 dynes
energy (g-cm²/sec²)	**energy** (kg-m²/sec²)
An *erg* is the work done by a force of one dyne acting over a distance of one centimeter.	A *joule* is the work done by a force of one newton acting over a distance of one meter.
1 erg = 10^{-7} joules (J)	1 joule (J) = 10^7 ergs

There are many ways that a more complicated system can store energy, and in each case that energy is called potential energy if it can be readily converted to kinetic energy, and if it is readily available for doing work. In purely mechanical systems without friction, energy can be freely converted among these various forms. When friction is present (as it always is!), then some energy is dissipated as heat; energy is conserved, because we say that heat is another form of energy, but the transformation of mechanical energy into heat means a loss of available, freely converted energy (recall Chapter 8).

We have used some familiar words in special ways:

force: a push or a pull.

work: the arithmetic product of force times distance, with distance moved measured along the direction in which the force acts.

energy: the ability to do work.

WHEN LIKES REPEL, AND OPPOSITES ATTRACT. Another two-particle system that illustrates interrelationships among force, work, potential energy, and kinetic energy involves two electrically charged particles. Charged particles behave as if there were two kinds of electricity; Benjamin Franklin named them positive and negative electricity, a convenient naming scheme for numbers that go into algebraic equations. Charles de Coulomb found experimentally that the force F of attraction between two charged particles is given by the equation

$$F = - \frac{\kappa q_1 q_2}{r^2} \tag{III-4}$$

where q_1 and q_2 are numbers proportional to the size of the electrical charge on particles 1 and 2 (with a negative sign for negative electricity, a positive sign for positive electricity); r is the distance between the two particles; and κ is a proportionality constant whose value depends on the units in which the q's, r, and F are expressed, and also on the properties of the medium between the particles. If the two charges have different signs (one q is negative, the other positive), then the attractive force is positive, and the particles are pulled together by this coulombic force. If the charges have the same sign, the force is negative, and the particles are pushed apart.

When two particles of opposite charge are pulled apart, work is done. The work done is equal to the force multiplied by the distance moved; the calculation is harder than with a particle experiencing a gravitational force near the surface of the earth, since the coulombic force varies as the interparticle distance changes. Having been pulled apart, the two-particle system has additional energy (equal to the work done on the system); this energy is potential energy. If the particles are allowed to move toward each other, potential energy will be transformed into kinetic energy. This story sounds the same as the gravitational story, except that the force equation is different. The units for force and energy are the same as given in Table III-3.

Another tale can be told. Imagine a two-particle system composed of two magnets. Each magnet can be described as possessing a north pole and a south pole. When the two magnets are oriented so that opposite poles (a north and a south; positive and negative would be easier for an algebraicist, but more confusing for a navigator) are adjacent, the energy of the system is lower than if like poles are adjacent. Work is required to change from opposites adjacent to likes adjacent. There is a magnetic force between the two magnets; when work is done against the force, potential energy is stored. When the magnets are allowed to move freely, potential energy gets converted into kinetic energy of motion.

EXERCISE: Consider a cylinder, closed at one end, and fitted with a freely-moving piston at the other end. Imagine that a chemical reaction inside the cylinder converts liquid or solid reactant into gaseous product. The piston moves to accommodate the gases, being constrained only by the atmospheric pressure P that opposes the pressure of the gases within the cylinder. Show that the mechanical work done by the chemical reaction, against the constant force of atmospheric pressure P, is given by the equation

$$W = P\Delta V$$

where ΔV is the change of volume of the gas within the cylinder.

HEAT AND TEMPERATURE. James Joule, throughout the middle third of the nineteenth century, conducted a series of careful experiments to find how much mechanical work must be done on a system to produce the same effect as produced when a given quantity of heat is added to that system. The presently accepted value of Joule's *mechanical equivalent of heat* is

$$4.1840 \text{ J} = 1 \text{ cal}$$

where a calorie[2] is the amount of heat required to raise the temperature of one gram of water one degree from 14.5°C to 15.5°C.

When heat is added to any equilibrium system, the temperature of the system tends to increase. Consider the familiar example of ordinary water. Addition of a calorie of heat to a gram of liquid water raises the temperature about a degree; the exact temperature increase varies slightly with the temperature of the water, but it is always about one degree. We say that the *heat capacity* C of liquid water is 1 cal/g–°C. Addition of heat to a piece of solid ice raises its temperature, as does addition of heat to a quantity of gaseous steam. But add heat to a mixture of ice and water, and the temperature stays at 0°C until all the ice is melted. Or add heat to boiling water, and the temperature stays at 100°C, until all the water is vaporized. When addition of heat does not raise the temperature, we anticipate that a *phase change* (such as melting or boiling) is occurring. This thermal behavior of matter makes possible two very important experimental procedures: the determination of fixed points on a temperature scale and the calorimetric evaluation of the energy output or uptake of a chemical reaction.

The two familiar phase changes of water are used to define the size of the centigrade degree and to establish a reproducible reference point for both the Kelvin and the Celsius thermometer scales.[3] These centigrade scales are defined in terms of the boiling point of water at one atmosphere pressure (defined as 100°C) and the freezing point of water in equilibrium with water vapor (called the triple point of water, the only temperature at which gas, liquid, and solid water coexist, defined as 0.01°C). In addition to these definitions, establishment of a standard temperature scale requires a standard thermometer, because it turns out that intermediate points between 0 and 100°C (as well as temperatures outside this rather limited range) are different, for

---

[2] The "calorie" of dietetics is a thousand times larger, called a kilocalorie by chemists.

[3] A very readable introduction to practical considerations involved in the measurement of temperature and the establishment of temperature scales is J. A. Hall, *The Measurement of Temperature*, London: Chapman and Hall (1966) (paperback).

instance, on linear scales engraved on a glass thermometer when different liquids are used. The reference thermometer used for standardizing the temperature scale is the *ideal gas* thermometer. Temperature is then defined by the equation

$$T = \frac{PV}{nR} \tag{III-5}$$

where P and V are the pressure and volume of n moles of a gas, and R is the gas constant, equal to 0.082 liter-atm/deg-mole. Temperature measured according to Equation (III-5) is called Kelvin temperature and is related to Celsius temperature by the equation

$$T_{Kelvin} = T_{Celsius} + 273.15° \tag{III-6}$$

The unattainable lower limit of temperature is 0°K (−273.15°C).

When a definite amount of heat is released into a known weight of water,[4] a definite and predictable temperature rise occurs. Thus, a temperature rise can be used to determine the amount of heat produced by a chemical reaction, if all the heat can be transferred into a quantity of water. Figure III-2 shows a simple experimental arrangement for determining the heat produced during the chemical reaction of the base sodium hydroxide with hydrochloric acid in water solution. For other reactions, the reactants may be mixed inside a

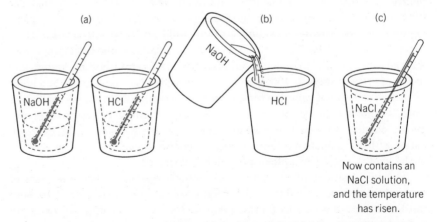

FIGURE III-2 Calorimetric determination of the heat evolved in the neutralization of an acid and a base, using styrofoam coffee cups. (a) Initially, separate solutions of sodium hydroxide and hydrochloric acid are brought to the same temperature. (b) The two solutions are mixed, initiating the neutralization reaction. (c) The resulting sodium chloride solution is hotter, because of the heat released during the chemical reaction. The temperature has risen by an amount determined by the heat released and by the heat capacity of the sodium chloride solution, the thermometer, and the coffee cup that serves as the calorimeter. The coffee cup serves as an excellent calorimeter, because it is an excellent insulator, and because it has a very low heat capacity.

---

[4] Or into a known weight of NaCl solution; the heat capacity is slightly different, but the principles are the same.

container that is immersed in the water; then the heat capacity of the water and the container must be considered in the calculations that relate temperature rise to energy of reaction. In order to determine the caloric content of a food, a sample of the food is completely burned inside a closed container (a calorimeter bomb), and the heat produced is measured by observing a temperature rise.

Energy stored in chemical compounds might be called potential energy. Energy stored in gasoline can be transformed, by controlled explosions in an internal combustion engine, into kinetic energy and heat energy. With appropriate mechanical linkages, the kinetic energy can be used to do work, allowing some energy to be stored again as another form of potential energy (with some loss, because of friction, into heat). The vocabulary—the words and the units—are largely the same for talking about a gasoline engine as for talking about a falling-to-earth mass point.

Waves, Photons, and Spectrophotometry

Visible light is a form of radiation, having properties in common with high-energy, short-wavelength, high-frequency X-rays, and with low-energy, long-wavelength, low-frequency radio waves. Chemists talk a lot about radiation. Many of the most powerful methods for learning about molecules involve interactions of radiation with matter. Such experimental techniques in which light is absorbed or emitted are collectively called *spectroscopic* methods. Light is described in terms of properties such as wavelength, frequency, and energy. Wavelength and frequency are customary terms for describing a *wave*. Light energy, however, is the energy of a *particle* of radiation called a photon. We shall talk about light as if it were *both* wavelike *and* particlelike. No totally wave model has ever been found that describes all the properties of light, and no totally particle model has been able to describe those properties. Two complementary models are required; light may be said to behave as if it were the superposition of the two models.

As a photon of light moves along in a straight line, there moves along that straight line an oscillating electric force and a perpendicular oscillating magnetic force. Figure III-3 is a representation of these oscillations—an electromagnetic wave. This wave can be characterized by specifying the wavelength λ—the distance along the path of the photon between two identical states of the electric and magnetic forces. Another closely related quantity is the frequency ν—the number of times per second that identical states (a crest of the wave, for instance) pass a point on the light path. Note that the number of passes per second, multiplied by the distance between passes, gives the velocity of propagation of the wave—the speed of light, c. That is,

$$\nu(\text{sec}^{-1}) \cdot \lambda(\text{cm}) = c(\text{cm/sec}) \tag{III-7}$$

The energy of the photon is related to the frequency by the equation

$$E = h\nu \tag{III-8}$$

where E and ν are the energy and frequency of the photon, and the proportionality constant (called Planck's constant, after Max Planck, the father of

FIGURE III-3 An electromagnetic wave, showing the changing direction and size of the electric and magnetic forces as the wave moves along the x axis. The wavelength is a distance equal to the repeat distance for the wave pattern. At every instant, the direction of the electric force is in a plane perpendicular to the plane in which the magnetic force is directed.

FIGURE III-4 The electromagnetic spectrum, showing the relationship of visible light to the radiation employed in chemical spectroscopy, and to the whole range of electromagnetic radiation.

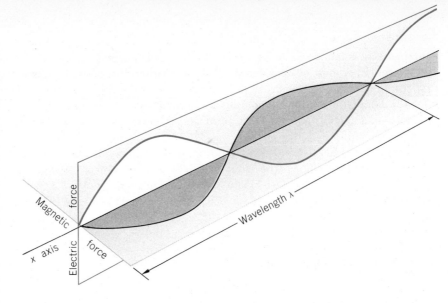

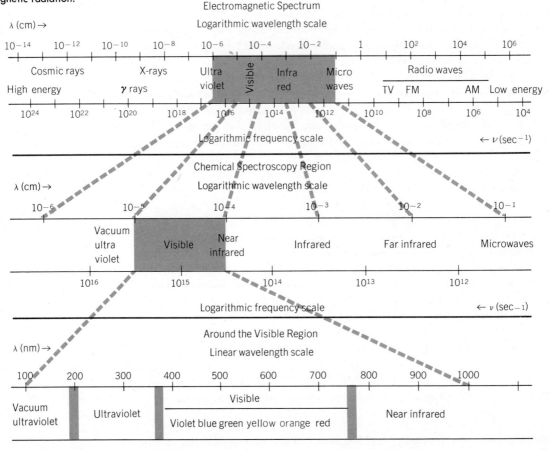

quantum theory) has a value that depends on the units in which energy is expressed. Some oft-used values for h are

$h = 6.62 \times 10^{-27}$ erg-sec $h = 1.58 \times 10^{-34}$ cal-sec

$h = 6.62 \times 10^{-34}$ J-sec $h = 4.14 \times 10^{-15}$ eV-sec

When light is absorbed by a molecule, the molecule moves to a higher energy state. Energy can be absorbed only when there is an exact match between the energy of the photon and the energy difference between two molecular states. Thus, information about the energy states of a molecule can be obtained by finding out what energies (that is, what frequencies, or what wavelengths) of light are absorbed by that compound. Such measurements are routinely made with a *spectrophotometer*. (Figure III-4 shows the electromagnetic spectrum, focusing on the portion that is exploited by chemical absorption spectroscopy.) A spectrophotometer that measures the absorption of light in the visible region of the spectrum is often called a *color*imeter, since its measurements provide a way to express the color of a sample in terms of numbers.

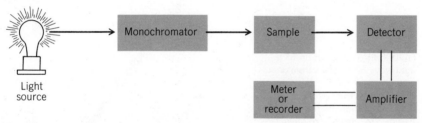

FIGURE III-5 The components of an absorption spectrophotometer for use in the visible region of the spectrum. Light from a tungsten-filament source passes through a monochromator, which selects a narrow range of wavelengths to pass through a sample and then to a detector, where the intensity of the light is converted into an electrical signal. The electrical signal is amplified, and then displayed as absorbance on a meter or a recorder. For use in the ultraviolet region, the tungsten source would be replaced by an electric arc, typically contained in a glass bulb with hydrogen or deuterium gas. For measurements in the infrared region, a heated piece of rare earth oxides, or silicon carbide (heated by passing electric current through the rod of oxide or carbide), can serve as the source, and a heat detector is used to measure the radiation. In infrared instruments, the monochromator is typically placed between the sample and the detector.

The necessary components of a typical spectrophotometer are represented schematically in Figure III-5. For visible light, a tungsten-filament light bulb (such as used in an ordinary incandescent lighting fixture) serves as a stable source of white light. The monochromator (a refracting prism or a diffraction grating) rejects almost all of the white light, allowing a beam of nearly monochromatic (one color, therefore one selected wavelength, one frequency, and one energy) light to shine through a sample (perhaps a solution in a transparent glass cell) and into the detector. A typical detector is a phototube (such as used to turn on street lights when the sun sets), or a more sensitive photomultiplier tube. The detector produces an electrical signal that is proportional to the intensity of the light beam, and this signal is amplified electronically before being displayed on a meter or on a recorder. We shall examine two types of monochromators. Ways to obtain practical chemical information from spectrophotometer readings are discussed in Section 16-5.[5]

REFRACTION.   One way to prepare a beam of monochromatic light from the mixture of colors that constitutes white light is to pass white light through a glass prism. When passing from air into glass, light slows down,[6] and when a beam enters glass at an acute angle, the beam is bent. Figure III-6 indicates why the slowing of a light beam bends the light beam. The photon moves as a wave with a planar wave front. When this plane impinges on an air-glass interface at any acute angle, part of the wave front enters the glass before the rest, thereafter moving slower through glass, while some of the wave is still moving faster through air. When the totally-in-glass plane wave front

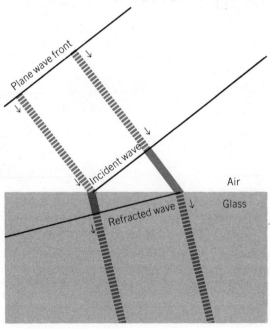

FIGURE III-6   Refraction of a light wave as it passes from air into glass. When the incident plane wave front strikes the glass, a portion of the wave begins to move relatively slowly in glass, while the rest of the wave is still moving fast in air. During the same interval of time, the wave moves the distance a in glass, and the distance b in air. In order to maintain the planar wave front, the wave in glass has changed direction. We say that the ray of light has been bent.

---

[5] An excellent source of information about spectrophotometry is M. G. Mellon, *Analytical Absorption Spectroscopy*, New York: Wiley (1950), although continuing rapid advances in instrumentation make it necessary to supplement such a book with the latest information from instrument manufacturers. A valuable guide to spectrophotometric measurements, dealing mainly with how to obtain dependable absorption data within the wavelength range 200 to 800 nm, is J. R. Edisbury, *Practical Hints on Absorption Spectrometry (Ultraviolet and Visible)*, New York: Plenum (1967). An introduction to both theory and practice of spectrophotometry in the infrared region is C. E. Meloan, *Elementary Infrared Spectroscopy*, New York: Macmillan (1963).

[6] The speed of light in a vacuum or in air, c_{vac}, is about 3×10^{10} cm/sec. In a typical glass, the speed of light, c_{glass}, is about 2×10^{10} cm/sec. The ratio $c_{vac}/c_{glass} = n_{glass}$ is called the *refractive index* of glass. The refractive-index value depends on wavelength, thus making the bending angle dependent on wavelength, and making possible the separation of white light into a rainbow of colors by a prism.

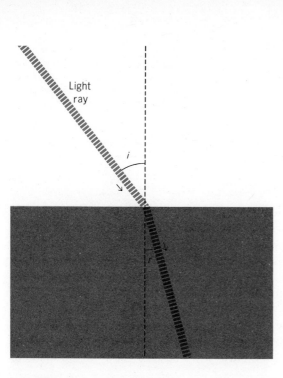

FIGURE III-7 Refraction of a light wave incident upon an air-glass interface. The angle of incidence i and the angle of refraction r are measured with respect to a line drawn perpendicular to the glass surface. Since the speed of light is slower in glass, the ray is bent *toward* the perpendicular reference line.

is finally organized, part of the wave has traveled farther than other parts. The result—occurring because the path a in Figure III-6 is shorter than the path b—is that the new direction of the wave-front motion is altered from the old direction. That is, the beam has been bent.

The incident angle (angle i in Figure III-7) and the refracted angle (angle r in Figure III-7) are related to the speeds of light in the two media by the equation

$$\frac{\sin i}{\sin r} = \frac{c_i}{c_r} \tag{III-9}$$

where c_i and c_r are the speeds of light in the first and second media.

Exercise: Assuming an index of refraction of 1.5 for glass, construct the path of a ray of light entering a glass prism according to the diagram given in Figure III-8. Given the trends of index of refraction with wavelength presented in Table III-4, predict the differences in behavior of red and violet light in passing through the prism.

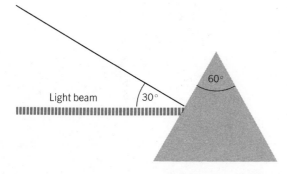

FIGURE III-8 A light beam is entering a 60° glass prism, with an angle of incidence of 30°. What happens to the light if the beam contains only monochromatic light? What happens if the beam is white light? Draw the appropriate rays for light entering, passing through, and leaving the prism.

TABLE III-4
Wavelength Dependence of the Index of Refraction of Glass

	768 nm	656 nm	589 nm	486 nm	405 nm
Borosilicate crown glass	1.519	1.522	1.524	1.530	1.538
Flint glass	1.644	1.650	1.656	1.669	1.690

DIFFRACTION. Another way to prepare a beam of monochromatic light is to reflect white light from a diffraction grating. One type of grating is made by making a series of fine parallel scratches (rulings) on a mirror, perhaps as many as 6000 lines per centimeter. An enlarged section of such a grating is depicted in Figure III-9, where incident light is reflected (as from any mirror) with the angle of incidence equal to the angle of reflection. If the plane wave front is to be reconstituted after reflection, light from the individual mirrored ridges must be "in phase," and this is the requirement that the distance [Figure III-9(c)] labeled $n\lambda$ be some multiple of the wavelength. We shall extend this argument later, but first we need to clarify the meaning of the phrase "in phase."

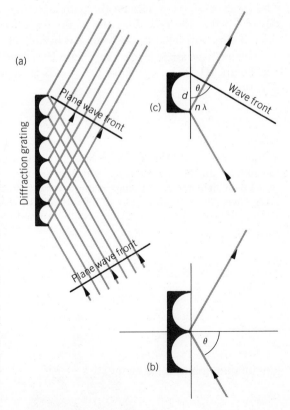

FIGURE III-9 The fate of a plane wave front incident upon a reflection grating. The grating is made from a mirror by cutting parallel grooves into the mirror surface. (a) A plane wave front impinges upon the grating, and a plane wave front is reflected, with the angle of incidence equal to the angle of reflection. (b) The angle θ is measured with respect to a normal line (a perpendicular line). (c) If constructive superposition of waves is to occur, the distance labeled $n\lambda$ must be equal to an integral multiple of the wavelength.

Let us return to the wave sketched in Figure III-3, considering only the oscillating electric force (so that we can draw diagrams without problems of perspective). A light wave f of wavelength λ and amplitude a being propagated along the x axis can be sketched as shown in Figure III-10. Note that the amplitude a is the maximum peak height. Now let us suppose that there is a second wave g with the same amplitude a and the same wavelength λ, but this time with a head start of half a wavelength. Wave f, exactly out of step with wave g, is depicted in Figure III-11. If wave f and wave g are superimposed (added point by point), it is apparent (see Figure III-12) that the sum of the two waves is zero at all points along the x axis. This is an example of *destructive interference*; no light remains. The two out-of-step waves are exactly *out of phase*.

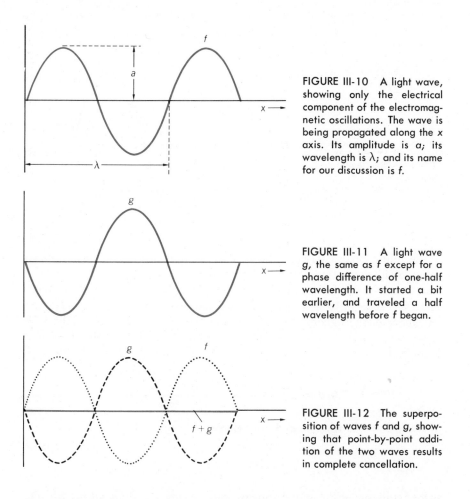

FIGURE III-10 A light wave, showing only the electrical component of the electromagnetic oscillations. The wave is being propagated along the x axis. Its amplitude is a; its wavelength is λ; and its name for our discussion is f.

FIGURE III-11 A light wave g, the same as f except for a phase difference of one-half wavelength. It started a bit earlier, and traveled a half wavelength before f began.

FIGURE III-12 The superposition of waves f and g, showing that point-by-point addition of the two waves results in complete cancellation.

Now suppose that there is still another wave h with the same amplitude a and the same wavelength λ, this time with a head start of a whole wavelength. Wave h is shown in Figure III-13. When waves f and h are superimposed, it is apparent that the result is a wave with amplitude $2a$ and wavelength λ. This is an example of *constructive interference;* the intensities add and the light becomes twice as bright. This time the two superimposed waves are exactly in phase (see Figure III-14).

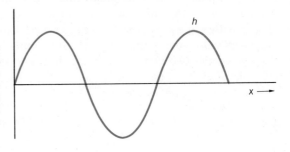

FIGURE III-13 A light wave h, out of phase with f by a full wavelength. Clearly, this phase difference makes h indistinguishable from f. Thus, to be "out of phase by a full wavelength" is to be exactly in phase.

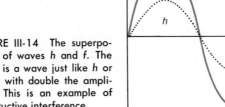

FIGURE III-14 The superposition of waves h and f. The result is a wave just like h or f, but with double the amplitude. This is an example of constructive interference.

Recalling Figure III-9(c), we see why there is a distance labeled $n\lambda$ (n being an integer); this is the requirement for reinforcement of a wave of wavelength λ. Reinforcement of that wave will occur whenever the incident angle θ is such as to make this reflected pathlength equal to $n\lambda$. [Can you remember enough geometry to show that the angle θ in Figure III-9(b) is equal to the angle θ in Figure III-9(c)?] We shall use the fact that in a right triangle,

$$\sin \theta = \frac{\text{length of the side opposite angle } \theta}{\text{length of the hypotenuse}}$$

to give us, for the triangle in Figure III-9(c),

$$\sin \theta = \frac{n\lambda}{d} \tag{III-10}$$

as the requirement for reinforcement. *Since the angle is different for different wavelengths, a diffraction grating can be used to separate white light into its constituent colors.* Gratings have several advantages for use in a monochromator. A grating is lighter than a prism, requiring a less massive mounting.

In the infrared region, inexpensive glass prisms do not transmit light (glass is not transparent to infrared radiation), and so a prism monochromator must use optics made of a more expensive and often less durable material. Since a reflection-diffraction grating does not have to be transparent, it can be made of glass or metal and serve quite well.

The Intertwining of Electricity and Magnetism

We have noted that radiation of light is simultaneously an electrical wave and a magnetic wave (Figure III-3). This is just one example of the many fundamental and practical interconnections between electrical and magnetic phenomena.

You may remember making a magnet in junior high school general science class by passing an electric current through a coil of insulated copper wire; the helical coil of wire conducted electrons around circular paths, and a magnetic field was produced. An electric charge moving in a circle *always* creates a magnetic field perpendicular to the plane of that circle, and conversely an electric charge moving in a magnetic field is directed (by that field) into a circular path. This principle is the basis for practical electromagnets, electric motors, and electric generators.

This principle also allows control of the path of charged particles moving through space, making possible cathode ray tubes (electrons being the charged particles) for television sets and oscilloscopes, and mass spectrographs (ions being the charged particles) such as diagrammed in Figure 2-2. The radius of curvature of an ion beam in a magnetic field depends on the velocity of the moving ions, the number of charges on the ions, and the strength of the magnetic field.

Index

Absolute zero, 154
Absorbance, 536
Abundance of elements
   cosmic, 326
   terrestrial, 325
Acetic acid, 99, 201, 202
   production of, 193
Acid, 188
   Brønsted, 191, 197, 214, 217
   industrial disposal of, 86–87
Acid anhydride, 202, 227
Acid-base reactions, 188–196
Acrolein, 585
Acrylonitrile, 475, 512
Actinides, 43
Activation energy, 384
Addition reactions, 468
Aerosols, 585, 593
Agent Blue, 190
Aging of living systems, 315
Agricultural wastes, 13
Aircraft materials, 490
Air pollution, 474, 583ff
Alcohol, 99
Algae, 581
Alkali metals, 62, 196
Alkaline-earth metals, 196
Alkalinity, 527
α fractional concentration, 227
Alpha particles, 28
   detectors for, 350
   effects of, 351
   in neutron production, 339
   from plutonium, 344
   in structural theory, 337
Alternating axis, 109
Aluminosilicates, 152
Aluminum, 489
   preparation of, 83–85
   production of, 278
Aluminum hydroxide
   amphoterism of, 194
Aluminum oxide, 83–85
Amalgams, 269
Amide ion, 192
Amino acids, 12, 557
Ammonia, 78
   molecular structure of, 116
   synthesis of, 7, 208
Ampere, 244
Amphoteric, 194, 197
Analytical chemistry, 81

Angstrom, 25, 51
Anhydride, 195, 197
Anisotopic element, 30
Antibonding orbitals, 409
Antifluorite structure, 149–150
Antifreeze, 158
Antilogarithm, 623
Antimony poisoning, 190
Antisymmetric, 121, 124
   orbitals, 412
Arc spectra, 48–49
Arsenic poisoning, 190
Asbestos, 152
Asymmetric stretching mode, 121, 124
Atmosphere, as pressure unit, 88
Atomic absorption spectrophotometry, 536
Atomic kernel, 407, 413
Atomic number, 26, 51, 19t–21t
Atomic orbitals, 37–41
Atomic sizes, 177–183
   periodicity of, 179–183
Atomic spectra, 536
Atomic weight, 18, 19t–21t, 51
Atoms, wave-mechanical model of, 34–41
Aufbau, 41, 51
Automobile exhaust, 77, 90
Autooxidation, 241, 281
Autoprotolysis, 208
Auxochrome, 515
Avogadro, Amadeo, 24
Avogadro's number, 24
Axis of symmetry, 106, 109, 124
Azurite, 95

Bakelite, 476, 498
Balanced equation, 85
Balancing oxidation-reduction equations, 239
Balata, 511
Balmer series, 48
Banana bonds, 76
Band spectra, 537
Barium poisoning, 190
Barn (unit of cross section), 328, 355
Barometer, 88
Base, 188
   Brønsted, 191, 197, 214, 217
Battery, 262, 267–272
Beer-Lambert Law, 535
Bending mode, 121–122, 124

Bent bonds, 76
Benzene, molecular structure of, 126, 134
Beta particles, 327, 335, 544
   detectors for, 350
   effects of, 351
Bilateral symmetry, 101
Bimolecular process, 373
Binary compounds, 63
Binding energy, 30
Binding energy of nuclei, 331
   in Bohr theory, 340
   calculation of, 332
   curve of, 333
   released in fission, 344
   released in fusion, 348
Biochemistry, operating principles of, 554
Biological catalysts, 555
Bismuth poisoning, 190
Blast furnaces, 490
Blasting powder, 95
Blue sapphire, 82
Bohr, Niels, 25, 55, 56
Bohr theory of atom, 31–33, 53
Boiling point, 131, 160, 162, 165
Bond energies, $445t$
Bonding, 151, 407, 413
Bonding orbitals, 407
Bonds
   conventional representation of, 100
   donor-acceptor, 78, 93
   double, 72, 178
   formation of, in water molecule, 66–69
   strengths of, 122
Bornite, 95
Boron hydride, molecular structure of, 191
Boron trifluoride, molecular structure of, 117, 189
Boundary conditions on a wave, 34–36
Boyle, Robert, 153
Breeding blanket, 345
Breeding reactions, 344, 355
Brick walls, 143–144, 166
British thermal units (BTU), 244
Bromides, 63
Bromine, 70
Brønsted acids and bases, 191, 197, 214, 217
   table of pK's, $218t$
BTU. *See* British thermal units
Bubbling, 160
Buffers (buffering), 202, 221, 531

C symmetry element, 109
Cacodylic acid, 190
Calcium
   atom, electronic configuration of, 46
   ion, 527
   as plant nutrient, 11
Calcium carbide, 6
Calcium carbonate, 577
Calcium cyanamid, 6
Calorie, 62, 93, 244, 643
Calorimeter, 262, 644
Candida yeast, 13
Cane sugar (sucrose), 99, 517, 556
Carbanion, 192
Carbohydrates, 13, 202
Carbon, 499
   compounds, 417, $442ff$
   cycle, 579
   radioactive, 366
Carbon dioxide, 202, 577
   as plant nutrient, 11
   molecular structure of, 134
Carbon disulfide, 297
Carbon electrodes, 86
Carbon-hydrogen bond, polarity of, 192
Carbon monoxide poisoning, 77, 190, 434
Carbon-12, as atomic weight standard, 21
Carbon-14, dating, 366
Carbonate alkalinity, 527
Carbonate mineral equilibria, 577
Carbonic acid, 202
Carnot cycle, 273
Catalysts, 555
Catenation, 442
Cells, electrochemical, 247–263, 282
   concentration, 256
   electrolytic, 275–280
   fuel, 272
   half, 248
   practical, 267
   storage, 269
   zinc-iodine, 262
Cellulose, 13, 556

Celsius temperature, 643
Center of symmetry, 108–109, 124
Ceramics, 493ff
Chalcopyrite, 95
Chaos, 153, 299
Charge, size of unit, 26
Chelate compounds, 432
Chemical bonds, 66, 100, 122, 393, 407, 413, 445
Chemical equation, 58, 80, 93
Chemical equilibrium, 202ff, 574
  and entropy, 306
Chemical kinetics, 371–385
Chlorides, 63
  in lake water, 96
Chlorinated hydrocarbons, 541
Chlorine, 70
  electrolytic preparation of, 277
Chlorine atom, 46
Chlorophyll, 538, 559
Chromatography, 222, 539ff
Chromophore, 515
Cigarette smoke, 77
Cis-dichloromethane, molecular structure of, 113
Cis, trans, 419
Citric acid, 559
Close-packing, 144–151
  in nucleus, 336
Clothing, 14
Coal, 12
  combustion of, 95
Cobalt complex ions, 418ff
Color, 514, 646
Column, chromatographic, 542, 543
Complementarity, 25, 55, 645
Compound, 58, 93
Compound nucleus, 340
Computer simulation of model, 375
Concentration, 86
Concentration cell, 256, 282
Condensation reactions, 469
Conformation, 119
Conjugate acid, 192
Conjugate base, 192
Conservation of energy, 290
Constructive interference, 652
Contour line, 413
Contour maps, 396
Coordinate covalent bond, 78, 93
Coordination compounds, 65, 78, 93
  in electrochemistry, 253
Coordination sphere, 426
Copper
  metallurgy and refining, 279
  ore, 95
Core, atomic, 407, 413
Cotton, 556
Coulomb, 243
Coulomb, Charles, 642
Coulomb's Law, 642
Covalent bond, 66, 93
Covalent radii, 177–178, 197
  periodicity of, 181–182
Covalent substance, 138, 165
Critical point, 159
Criticality in fission reactors, 347
Crookes, Sir William, 3
Cross section of nuclei, 328, 329t, 355
Cryolite, 83
Crystal, 61, 93, 143, 165
Cubic close packing, 145–147
Curie (unit of radioactivity), 351, 355
Cuyahoga River, Ohio, 583
Cyanamid process, 6
Cyanide poisoning, 434
CyDTA, 538

d orbitals, 38–41
Dacron, 498
Dalton, John, 103
Dark-line spectra, 54
Dating, carbon-14, 366
Davisson and Germer experiments, 33
DDT, 539
de Broglie, Louis, 32
de Broglie wavelength, 32, 51
Degree of disorder (entropy), 264
Deoxyribose nucleic acids (DNA), 566
Destructive interference, 651
Detector
  chromatographic, 542, 544
  photometric, 647
Diamond, 445
  model of, 134, 151
Diborane, molecular structure of, 128
Dichlorosilane, molecular structure of, 128
Diffraction, 650
  of neutrons, 33
  of X-rays, 24

Difluoromethane, molecular structure of, 126
Dimerization, 288
Dimethyl ether, 99
Dinitrogen tetroxide, structure and formation of, 289
Dipole, 68, 70, 93
Directed bonding, 151
Dispersion forces, 136–138
Dissociation constant, 207
  of acids (K_a), 207, 218
  of water (K_w), 208
Distribution coefficient, 227
DNA, 566
Donor-acceptor bond, 78, 93
Double bond, 72, 178
Doubling time, 364, 385
Dry cell (Le Clanche cell), 268
Dyestuffs, 514ff
Dynamite, 95
Dyne, 641
Dystrophic waters, 582

e (base of natural logarithms), 362
Earthenware, 493
Earths, 195–196
Eclipsed conformation, 119
Ecology, and thermodynamics, 587
Ecosphere, 574
Ecosystems, 574ff, 593
EDTA, 430, 527
Efficiency, of heat engines, 590
Einstein, 290
Einstein energy equation, 30
Elastomers, 510ff
Electrical conductance, 133, 139, 141
Electrochemistry, 235, 242
Electrode, 84, 245, 282
  ion-selective, 530
Electrode potentials
  in aqueous solution, 246–261
  change with concentration, 255
  effect of coordination on, 253
  in liquid ammonia, 247
  in molten silicates, 247
  measurement of, 249
  standard, 248–253
Electrolysis, 83–85, 93, 275, 282
  oxidation, examples of, 276
  reduction, examples of, 278
Electrolyte, 275, 282

Electromagnetic spectrum, 50, 646
Electron affinity, 176, 197
Electron configurations of atoms, 40–41
Electron-deficient compound, 191
Electron microscope, 17
Electron pair bonds, 423
Electron-transfer reactions, 234, 245
Electronegative, 61, 93
  elements, 45
Electronegativity, 184–185, 186t, 187–188, 197
  periodicity of, 187–188
Electron-pair repulsion theory, 114–118
Electrons, 25, 52
  in atoms, evidence for, 25–26
Electropositive, 61, 93
  elements, 45
Element, 18, 19t–21t, 52
  abundance of, 325, 326, 488
Elementary chemical process, 372, 385
Emission spectra, 47–48
Empirical formula, 82, 93
Endoergic, 59, 93, 235
Endothermic, 59, 93
Energy
  barrier, 119
  chemical, 242
  conservation, 640
  electrical, 244
  of formation, 291
  free, 242
  heat, 243
  kinetic, 640
  from oxidation-reduction reactions, 242
  potential, 640
  production, and environment, 587
  sources, 323
Energy-level diagrams, 47
Energy levels
  of electrons in atoms, 47
  of hydrogen atom, 48
  for vibration, 122
Enthalpy, 292
  definition of, 294
  examples of, 295–297
  of fusion, 163
  of vaporization, 163
Entropy, 262, 282, 303, 305t

Entropy, (cont.)
  definition of, 316
  measurement of, 264
  of mixing, 305t
  and pollution, 587
Enzymes, 557, 558
Epilimnion, 589
Equilibrium, 158–159, 165, 202, 227
  in aqueous solutions, 209
  in electrochemical cells, 246
  in nonpolar liquids, 222
  within solids, 225
  solubility considerations, 223
Equilibrium constant, 205
  K_a (acid dissociation constant), 207, 215
  K_p (gas reactions), 209
  K_w (water dissociation), 208, 213
  K_x (mole fractions), 209
  from standard electrode potentials, 257
Erg, definition of, 641
Eriochrome Black T, 528
Escaping tendency, 157
Ethane
  internal rotation in, 118–120
  symmetry of, 110–111
Ethanol, 99
  air oxidation of, 193
Ethene, models of, 104–106
Ethyl alcohol. *See* Ethanol
Ethylenediaminetetraacetate (EDTA), 430, 527
Ethyne, 499
  molecular structure of, 126
Eutrophication, 582
Evolution, thermodynamics of, 313
Exclusion principle, 41, 112–118
Exoergic, 59, 94, 235
Exothemic, 59, 94
Explosives, 95
Exponential decay, 370
Exponential functions, 360, 385, 623, 631
Extraction, liquid-liquid, 222

f orbitals, 38–41
Face-centered cube, 146
Faraday, 243
Fats, 12
Feasibility of reaction, 258

Fertilizer, 5, 7, 580
Fission, nuclear, 341–348
  breeding in, 344
  energy balance, 345
  fissile species in, 341
  mass number *vs*., 342
  material balance, 344
  reactions and products of, 343
  reactors, 345
  spontaneous, 342
  yield curve, 343
Fixed nitrogen, 5
Fluorides, 63
Fluorine, 61
  preparation of, 276
Fluorine atom, 61
Fluorine molecule, bonding in, 69
Fluorite structure, 149–150
Fluorocarbons, 463
Food, 12
Force, 639, 641
Formaldehyde, 585
  vibration of, 128
Formula weight, 79
Formulas, 80–83
Fossil fuels, 323, 591
Free energy, 282
Free rotation, 120
Freeze drying, 159
Freezing point, 160, 162
  lowering, 162
Fuel
  fossil, 323
  nuclear, 345
  rods, 346
Fuel cells, 272
Fumarase, 557
Functional groups, 455*ff*
Functions (mathematical), 621
  cosine, 633
  exponential, 623, 631
  inverse, 622
  quadratic, 623
  sine, 633
Fusion, heat of, 163
Fusion reactions, 348
  advantages and disadvantages, 349

Gamma rays, 327
  detectors for, 350
  effects of, 351

Gas, 130–132, 308
Gas chromatography, 541
Gasoline, 471
   combustion of, 89–90
Genetic information, 566
Geode, 142
Germer and Davisson experiments, 33
Glass, 493*ff*
   index of refraction of, 649–653
Glass transition temperature, 510
Glide symmetry, 101
Glucose, 99, 464
Glycerol, 12
Gold
   from copper refining, 280
   in sea water, 534
Gram, 637, 638
Gram-atomic weight, 22–24, 52
Gram-equivalent, 243, 282
Gram-molecular weight, 80, 94
Graphite, 445
Graphs, 629
Gravimetric analysis, 526
Ground state, 47
   of H_2O molecule, 410
Group, 42, 52
Guitar string, wave motion of, 36
Gutta-percha, 511

Haber, Fritz, 7
Haber process, 7, 9
Half-cell, 245, 282
Half-life, 334, 339, 368
   relation to rate constant, 369
Half-reactions, 237, 282
Halides, polarizability of, 183–184
Hall, Charles, 83
Halogens, 62
Hard acid, 189, 197
Hard base, 189, 197
Hardness, 141
Heat, 643
   capacity, 643
   of formation, 294, 316
   and free energy, 263
   of fusion, 163, 165
   of reaction, 294–297, 316
   of vaporization, 163, 165
Heat engines, efficiency of, 590
Heavy metal poisoning, 190

Heisenberg, Werner, 55, 123
Heisenberg uncertainty principle, 34
Helium atom, electronic configuration of, 61
Helix, of proteins, 565
Hemoglobin, 77, 559
Hess's Law, 294
Hexagonal close packing, 147–148
Hofmann, A. W., 103–104
Holes, 144–151
Homilies, 592
Horsepower, 244
Humidity, 575
Hybridization, 73, 94
Hydrides, Brønsted acid strength of, 191–193
Hydrocarbons, 13, 449*ff*
Hydrogen, 70
   atomic orbitals, 395
   preparation of, 89
Hydrogen atom
   energy levels of, 48
   wave mechanics of, 34–41
Hydrogen bonding, 136
Hydrogen electrode, 248–254
Hydrogen fluoride
   bonding in, 70
   hydrogen bonding in, 136
Hydrogen ions, 190
Hydrogen peroxide
   electrolytic preparation, 277
   molecular structure of, 128
Hydrogen selenide poisoning, 190
Hydrogen sulfide poisoning, 190
Hydronium ion, 191
Hydroxyapatite, 581
"Hypo" (photographic fixer), 210
Hypochlorous acid, 193
Hypolimnion, 589
Hypophosphorous acid, 261

I symmetry element, 109
i symmetry element, 109
Ice
   crystalline forms of, 163
   molecular structure of, 152
Ice-water equilibrium, 160
Ideal gas, 155–157, 165
   temperature scale, 643
Ideal gas law, 155

Identity, 106, 110, 124
Improper rotation, 106, 108–109, 111, 124
Index of refraction, 649
Indicator, 227
  acid-base, 214, 217t
Indium, 17
Induced dipoles, 137–138
"Inert-gas" structures, 175
Infrared spectrophotometer, 122
Inorganic compounds, nomenclature of, 64
Inorganic houses, 14
Interference, 651
Interhalogens, internuclear distances in, 177–178
Intermetallic compounds, 487
Internal genetics, 315
Internal molecular motions, 118–123
Interstellar molecules, 122
Inverse functions, 622
Inversion, 106, 108, 111, 125
Iodides, 63
Iodine, 70
Iodine heptafluoride, molecular structure of, 184
Iodine pentabromide, molecular structure of, 117
Iodine trichloride, molecular structure of, 117
Ionic band, 63, 94
Ionic character, 71
Ionic compounds, 63–64
Ionic radii, 178–179, 197
  periodicity of, 180–183
Ionic substance, 133, 165
Ionization energy, 170, 197
Ionization potential, 170, 197
  periodicity of, 170–176
Ions, 45, 52
  aquated, 210
  association of, 210
  equilibria in water, 210
Iron, as plant nutrient, 11, 432
Iron (III) ion, reaction with halide ions, 189
Isomerism, 419, 451
  geometrical, 453
  opitical, 421, 454
  structural, 452
Isomers, 99, 125

Isoprene, 511
Isotopes, 30, 52, 334t, 355t

Joule, 244, 641

K (equilibrium constant), 206
Kekulé, August, 416
Kelvin temperature scale, 88, 153–155, 643
Kernel, 100
  atomic, 407, 413
Kinetic energy, 640, 645

λ (wavelength), 645
Lanthanide, 43, 52, 180–183
Lanthanide contraction, 182, 197
Lattice, 61, 94
Lead, in drinking water, 96
Lead poisoning, 190
le Chatelier's principle, 164, 203, 207, 233
Length, units of, 638
Lewis, Gilbert Newton, 69, 188, 416, 423–425
Lewis acid, 188, 197, 209
Lewis base, 188, 197, 209
Lewis dot diagram, 69
Life expectancy of a nucleus, 369
Light
  absorption of, 140, 535
  diffraction of, 650
  refraction of, 648
  speed of, 645
Limiting nutrient, 582, 594
Line spectra, 47, 536
Linear combinations, 74, 402–410, 413
  of atomic orbitals, 427
Linear functions, 622, 629
Liquid-gas equilibria, 159
Liquid-liquid extraction, 222
Lithium
  burning of, 61
  ionization of, 173–174
Lithium atom, electronic configuration of, 46, 50, 173–174
Lithium fluoride, formation of, 62
Lithium fluoride crystal lattice, 62
Living systems, and thermodynamics, 312
Logarithms, 623, 626t

London, Fritz, 137
London dispersion forces, 136–138, 183–184
Lone-pair orbitals, 116–117
Looking-glass milk, 99
Lyman series, 48
Lysozyme, 566

Macroscopic rate constant, 381, 386
Magic numbers, 333, 338, 355
Magnesium, 59, 489
 ion, 527
 as plant nutrient, 11
 production, 279
Magnesium oxide
 enthalpy of formation, 296
 formation of, 59
 reaction with water, 195–196
Magnetic quantum number, 37
Magnetism, 427, 642, 653
Malachite, 95
Malthus, Thomas, 8
Manganese-CyDTA complex, 538
Markownikoff's rule, 469
Mass defect, 30
Mass number, 31
Mass spectrograph, 21–22, 653
Material balance in fission, 344
Mauve, 514
Mayer theory of nuclear structure, 337
Measurement, units of, 637
Mechanism of reactions, 360–385, 466
Melting point, 131, 165
Mendeleef, Dimitri, 16, 41–45
Mercaptans, 464
Mercury, 138
 analysis, 536
 in natural waters, 536
Mercury (II) ion, reaction with halide ions, 189
Mercury poisoning, 190
Mercury vapor, toxicity of, 537
Messenger RNA, 568
Metabolic cycles, 559
Metabolism, 554
 thermodynamics of, 313
Metallic bonding, 138–139
Metalloid, 45, 52
Metals, 44, 138–141, 486
Meter, 637, 638

Methane
 bonding in, 73–75
 molecular structure of, 112–115
Methanol, 82–83
Methyl alcohol. *See* Methanol
Methyl orange
 as indicator, 216
 structure of, 215
Metric ton, 86
Mica, 152
Microcurie, 351
Microscopic rate constant, 381, 386
Microwave spectra, 122
Millicurie, 351
Mirror plane, 72, 73, 107
Mirror symmetry, 421$f\!f$
Mixtures of gases, 156
Models
 mathematical, 360
 of molecules, 103–106
Moderator, 345, 355
Molar entropies, 305t
Molar volume, 89
Molarity, 86
Mole, 58, 94
Mole fraction, 157, 204, 228
Molecular disorder, 303
Molecular models, 103–106
Molecular motions, 158
Molecular orbitals, 38, 165, 392–412
 antibonding, 409
 bonding, 407
 guidelines in constructing, 402
 of H_2O molecule, 404
 by linear combination, 402
 nonbonding, 409
 symmetry properties of, 405
Molecular spectra, 537
Molecular substance, 134, 165
Molecular weight, 58, 94
Molecule, 52, 60, 94
Monochromator, 647
Monoclinic sulfur, 163
Moon boots, 508
Moon rocks, 17
Moseley, Henry, 26–27
Mueller, Erwin, 17
Multiple bonding, 71–77
Multiple bonds, 455
 effect on covalent radius, 178

Muscle, enzymes in, 558

Naming. *See* Nomenclature
Negative feedback, 164, 367, 378, 386
Neon atom, electronic configuration of, 61
Neoprene rubber, 512
Nernst Equation, 532
Neutron diffraction, 33
Neutrons, 29, 52, 328
  effects of, 351
  instability of, 335
  thermal, 329
Newman projection, 119
Nitrate ion, 534
Nitrates, 63
Nitrile, 462
Nitrobenzene, charge distribution in, 135
Nitrogen, 59, 70
Nitrogen atom, electronic configuration of, 46
Nitrogen cycle, 4
Nitrogen dioxide, 584
  structure of, 289
Nitrogen fixation, 5, 8, 9, 473
Nitrogen oxide, formation of, 59
Nitrogen shortage, 3–5
Nodes, 36
Nomenclature
  inorganic, 64
  organic, 448ff
Nonbonding orbitals, 410, 413
Nonmetals, 44
ν (frequency), 645
Nuclear energy
  calculated from mass, 339
  fission, 344
  fusion, 348
  reactor design, 345
Nuclear power, 591
Nuclear reactions, 339
  Bohr theory of, 340
  energy of, 330, 340
  equations for, 339
  fission, 341
Nucleic acids, 566ff
Nucleon, 332
Nucleus, atomic, 27, 52
  binding energy of, 331, 332, 333, 340, 344, 348
  instability of, 327
  reactions, 322
  stability of, 324
  structure of, 336, 337
Nucleus, life expectancy of, 369
Nutrients, 3, 11, 578
Nylon, 498

Oceans, 534
Octahedral holes, 145
Octahedron, 118
Octane rating system, 472
Ohm, 244
Oligotrophic waters, 582
Optical rotation, 421, 441
Orbital quantum number, 37
Orbitals, 37–41, 52, 392, 413
  atomic, 37–41
  molecular, 138, 165
  order of filling, 39–41
  overlap of, 71
Organic chemistry, 440ff
Overlap of orbitals, 71
Oxidation, 236, 282
Oxidation number, 237, 282
Oxidation potentials, 251
Oxidation-reduction equations
  examples of, 283
  method of balancing, 239
Oxidation-reduction reactions, 234
  organic, 470
Oxidation state, 65, 94
Oxides, 63, 195
Oxygen, 59, 70
  measurement of, 538
  in natural waters, 589
  transport, 77
Oxygen acids, 193–195
Oxygen atom, electronic configuration of, 67
Ozone, 584, 585

p orbitals, 38–41, 635
Paired electrons, repulsion of, 175
Pairing (protons and neutrons), 331
Palladium dichloride, crystal structure of, 152
Partial pressure, 156
  in ammonia synthesis, 208

Partial pressure, (cont.)
  effect on equilibrium, 206
Partition coefficient, 227
Pasteur, Louis, 421
Pauli, Wolfgang, 41
Pauli exclusion principle, 41, 112–118
Pauling, Linus, 71, 128, 185
Penetration of s orbitals, 175
Pentacarbonyliron (0), 78
Peptide bond, 561, 564
Perchlorate ion, structure of, 194
Perchloric acid, 195
Period, 42, 52
Periodic System, 41–45
Periodic Table, 42, 52, 172
Periodicity
  of atomic sizes, 179–183
  of Brønsted acid strength, 191
  of covalent radii, 181–182
  of electronegativity, 187–188
  of ionic radii, 180–183
  of ionization potential, 170–176
  and reactivity, 188–196
Peroxyacetyl nitrate (PAN), 586
Perpetual motion, 298
Pesticide residues, 540
Petroleum, 12
pH, 228, 628
pH meter, 256, 529
pH scale, 214
  range of indicators, 217
Phase, 157, 165
  of waves, 651
Phase boundary, 159
Pheremones, 539
Phosphate, 63, 580
  as plant nutrient, 11
Phosphate ion, structure of, 194
Phosphoric acid, 195
Phosphorus cycle, 580
Phosphorus pentabromide, molecular structure of, 117
Phosphorus pentachloride, molecular structure of, 127
Photochemical smogs, 583, 594
Photodissociation, 413
  of H_2O, 410
Photography
  "fixing," 223
  processing, 210
Photosynthesis, 11

pi bond, 72
pK scale and values, 218
Planck, Max, 32, 645
Planck's constant, 32, 645
Plane of inquiry, 400, 413
Plane of symmetry, 109
Plankton, 582, 594
Plasticizer, 510
Plastics, 474, 496ff
Plutonium synthesis, 344
Plywood adhesive, 95
Point symmetry, 101
Poisoning. See under names of specific agents, e.g., Arsenic; Mercury
Polar coordinates, 631
Polar molecule, 68, 94
Polarizability, 183–184
Pollution, 499
Polyamides, 501
Polyelectronic atoms, 395
Polyethylene, 497, 510
Polymers, 444, 496ff, 521
Polypropylene, 510
Polyprotic acids or bases, 228
Population growth, 11, 360–366
Porcelain, 496
Potassium, as plant nutrient, 11
Potential
  chemical, 243, 245
  electrical, 243
Potential energy, 640, 645
Pottery, 493
Power plants, 587, 591
ppm as concentration unit, 96
Precipitation hardening, 487
Pressure, 153–157
Pressure cookers, 158
Principal quantum number, 37
Probability, 300
Product, 83, 94
Productivity (photosynthetic), 580
Proteins, 13, 557, 561
  geometry of, 564
  synthesis of, 566ff
Proton, 29, 52
Proton-transfer reactions, 213
ψ (function in wave equation), 393
Pyruvic acid, 559

Qualitative analysis, 81
Quanta, 32

Quantitative analysis, 81
Quantum numbers, 35–38, 52

Rad (unit of radiation), 351
Radial symmetry, 101
Radiation
   allowable limits of, 352
   effects of, 349
   lifetime accumulation of, 352
   natural background, 351
   units, 351
Radioactivity, 355, 372
   emissions, 347
   shielding against, 345–352
   units, 351
Radiochemical dating, 367
Ramsey, William, 175
Rare-earth metals, 42, 180–183
Rate constant, 381–382, 386
   definition of, 386
   vs. half-life, 369
Rate of reaction, 206
Reactant, 83, 94
Reaction, 83, 94
Reaction mechanism, 371–385
   computer simulation of, 375
   testing of, 382
Reactivity, 206
   and periodicity, 188–196
Reactors, nuclear fission, 345
   advantages and disadvantages of, 348
   power of, 347
   research on, 346
Rearrangement reactions, 471
Recycling, 499
Reducing agent, 236
Reduction, 236, 282
Reduction potentials, 251
Refining (petroleum), 471*ff*
Reflection, 106–109, 111, 125
Reflection, 106–109, 111, 125
Refraction, 648
Relaxation phenomena, 378, 386
Reliability of measurements, 545*ff*
Rem (unit of radiation), 351, 355
Representative elements, 42
Resonance, nuclear, 330, 355
Respiration, 203
Restricted rotation, 124
Reversible reactions, 203

Rhombic sulfur, 163
Ribose nucleic acids, (RNA), 568
RNA, 568
Road salt, 158
Rocksalt structure, 62, 149–150
Roentgen, 351, 356
Rotation, 106–107, 109, 110, 125
   free, 120
Rubber, 510, 556
   entropy, of stretching, 302
Ruby, 82
Rutherford, Ernest, 27

s orbital, 37–41
S symmetry element, 109
Saccharin, 517
Salt bridge, 246, 248
Sandwich, symmetry of, 108–109
Sapphire, 81–82
Scattering of alpha particles, 28
Schrödinger, Erwin, 35, 55
Schrödinger equation, 35, 55, 393, 395
Screening, 172
   constant, 28
Sea urchin skeleton, 143
Sea water, 526*ff*
Second Law of Thermodynamics, 587
Seneca Lake, 96
Separatory funnel, 541
Shell structure of nuclei, 337
Shielding against radioactivity, 345, 347, 352
Sigma bond, 72
Sigma (σ) symmetry element, 109
Silica, 493*ff*
   fused, 496
Silicic acid, 194–195
Silicon carbide, 79
Silicones, 82, 503*ff*
   oils, 507
   resin, 507
   rubber, 505, 507
Silk, 501
Siloxane group, 504
Silver
   from copper refining, 280
   ions, coordination of, 210
Silver chloride, 526
Silverware tarnish, 140
Sine function, 101, 633
Skeleton, sea urchin, 143

Skewed conformations, 119
Slow neutrons, in fission, 342
Smog, 583
Snowflakes, 142
Sodium, emission spectrum of, 54
Sodium aluminum fluoride, 83–85
Sodium chloride
 effect on escaping tendency of water, 160–162
 structure of, 149–150
Soft acid, 189, 197
Soft base, 189, 197
Solar energy, 323
Solid, 130–132
Solubility, 221
 definition of, 228
 ionic, 64, 222
 of polar and nonpolar substances, 222
 product, 223, 228
Solubility equilibrium, 203, 540
Solubility product K_{sp}, 223, 228
Solute, 86, 94
 effect on escaping tendency of solvent, 160–162
Solution, 86, 94
Solvation energy, 195, 197
Solvay process, 447
Solvent, 86, 94
sp hybridization, 75
sp^2 hybridization, 75
sp^3 hybridization, 74
Spectrograph, 49
Spectrophotometry, 535ff, 647
Spectroscopy, 645
Spectrum, electromagnetic, 646
Spin quantum number, 37
Spontaneous fission, and mass number, 342
Square planar bonding, 152
Stable vs. unstable nuclei, 324
Staggered conformation, 119
Standard electric potential, 250, 252t
 definition of, 282
Standard temperature and pressure, 88–90
State function, 393, 413
Stationary states, 34
Steady-state condition, 367, 386
Steel, 488, 490
Stoichiometry, 85, 94, 526
Straight-line graph, 153, 629

Stratification, 589
Strength of acids and bases, 218
Strong acid, 191, 198
 definition of, 228
Strong bases
 definition of, 228
Structural formula, 82, 94
Structure, nuclear, 336
Substitution reactions, 466
Sucrose (cane sugar), 99, 517, 556
Sugar, 99, 556
Sulfate ion
 as plant nutrient, 11
 structure of, 194
Sulfides, 63
Sulfur
 crystalline forms of, 163
 in petroleum, 90
Sulfur dioxide, 464
 removal from stack gases, 95
Sulfur trioxide, reaction with water, 195
Sulfuric acid, 195
Supercooling, 160
Sweetness, 517
Symbols of the elements, 19–21
Symmetric, 121, 125
Symmetric stretching mode, 121, 125
Symmetry, 101–102, 106–124
 center of, 108–109, 124
 element, 106–112, 125
 molecular, 393
 operation, 106–112, 125, 404
 orbital, 393, 404
Symmetry-correct mode of vibration, 121, 125
Symmetry equivalent, 112, 125
System, chemical, 306

Tarnish, silverware, 140
Temperature, 643
 effect on equilibrium, 207
 scales, 643
Tendency to react, 242
Tensile strengths of wires, 487
Termolecular process, 373
Ternary compounds, 63
Tetrahedral geometry, 75, 112–115, 417
 of carbon compounds, 112–115
 deviations from, 115–117
Tetrahedral holes, 145

Tetrahedron, 118
   construction of, 74
Tetrahydrofuran (THF), 502, 512
Textiles, 14
Thermal neutrons, 356
Thermal pollution, 347, 589
Thermal stratification, 589
Thermochemistry, 316
Thermodynamics, 316, 587
Thermodynamics, laws of
   first, 290
   second, 303
   second, living systems and, 312
THF. See Tetrahydrofuran
Thiosulfate ion, 210
   silver complexes of, 212
Thorium conversion by breeding, 344
Three-body dilemma, 394
Three-phase equilibria, 159–162
Titanium, 492
Titanium (II) ion, electronic configuration of, 46
Toxicities of Lewis acids and bases, 190
Trace nutrients, 11
Trans-dichloromethane, molecular structure of, 113
Transfer RNA, 569
Transition metal, 43, 52
Transitivity of equilibrium, 159
Translational symmetry, 101, 143
Triangle, symmetry of, 107
Tricarboxylic acid cycle, 559
Trigonal bipyramid, 118
Trigonal holes, 144
Trimethylamine, molecular structure of, 189
Triple bond, 72, 178
Tungsten, 138

Ultimate source of energy, 324
Uncertainty principle, 34, 52
Unimolecular mechanism, 373, 374, 386
Unit cell, 143, 165
Units, 638
Uranium fission, 343
Uranium hexafluoride, molecular structure of, 184
Uranium isotopes, 342–346
   separation of, 184
Uranium metals, 42

Urban life, chemical problems of, 91
Urea, 440*ff*

Valence, 424
Valence shell, 171, 198
   orbitals, 171
van der Waals forces, 134, 165
van Tamelen, E. E., 10
Vapor phase chromatography (VPC), 541
Vapor pressure of water, 160–161
Vaporization, heat of, 163
Vibration, 120, 125
Vinegar, 99
Volt, 244
Volumetric analysis, 526
VPC. See Vapor phase chromatography
Vulcanization, 510

Water
   hydrogen bonding in, 136
   molecular structure of, 99–100, 111–112, 115–117
   phase equilibria of, 157–164
   vapor in atmosphere, 574
Water molecule
   electron distribution, 403
   vibration of, 121–122
Watt, 244
Wave, 25, 33, 645
Wave equation, 34
Wave motion, of guitar string, 36
Wave-particle duality, 25, 33, 645
Wavelength of radiation, 26, 645
Weak acid, 191, 198
   definition of, 228
Weak base, definition of, 228
Werner, Alfred, 416*ff*
Wheat, 3–5
White sapphire, 81
Wire, strength of, 487
Wool, 501
Work, 639, 640, 641
Wurtzite structure, 150–151

X-ray diffraction, 24, 178

Yield, 85

Zero of temperature scales, 154
Zinc blende structure, 150–151
Zinc sulfide structure, 150–151
Zwitterion, 228